Addison-Wesley

Pre-Alge

Addison-Wesley Pre-Algebra is a completely new program for middle school or junior high school students, designed to ease the transition from arithmetic to algebra. Building on arithmetic concepts and skills, **Addison-Wesley Pre-Algebra** introduces the idea of variable and extends it to working with algebraic expressions and equations.

For selected examples of the features of **Addison-Wesley Pre-Algebra**, please turn to the following pages.

- **Teacher support materials**
  - Comprehensive Teacher's Edition
  - Tests and quizzes
  - Enrichment activities
  - Solutions manual

    and

  - for the student . . . Skills Practice Workbook

Turn to your Reviewer's Guide on page T6.

Addison-Wesley

Pre-Algebra

Teacher's Edition

Phares G. O'Daffer

Stanley R. Clemens

Randall I. Charles

Addison-Wesley Publishing Company

Menlo Park, California   Reading, Massachusetts   Don Mills, Ontario

Wokingham, England   Amsterdam   Sydney   Singapore

Tokyo   Madrid   San Juan

Teacher's Edition Contents

Grateful acknowledgement is made to the many Addison-Wesley staff members who contributed to this teacher's edition and corresponding text, with special thanks to the following:

Editorial staff: Sandra Ward, Connie Thorpe, Robin Blakely

Design staff: Barbara Robinson, Linda Stinchfield

Photo Editor: Margee Huntzicker

This expanded version is fully substitutable with *Addison-Wesley Pre-Algebra* © 1987. The expanded material appears at the end of the text, beginning on page 509.

ISBN 0-201-29736-1

ABCDEFGHIJKL-KR-892109

Authors

Phares G. O'Daffer is Professor of Mathematics at Illinois State University, where he teaches elementary and junior high school mathematics teachers. Formerly a junior high and high school mathematics teacher, he has coordinated a comprehensive mathematics laboratory for teachers and students at Illinois State University. Dr. O'Daffer holds a Ph.D. in mathematics education from the University of Illinois and is the author or coauthor of numerous articles and books. His publications include the following books for junior high school students: *School Mathematics I and II, Investigating School Mathematics, Success With Mathematics, Mathematics in Our World,* and the recently published *Addison-Wesley Mathematics* grades K–8.

Stanley R. Clemens is Professor of Mathematics at Bluffton College, Ohio, where his teaching includes mathematics courses for elementary and junior high school teachers. He was involved in work with mathematics majors and teachers at Illinois State University for 16 years, and has been active as a conference speaker, NSF grant reviewer, and member of the editorial panel of *The Mathematics Teacher*. Dr. Clemens holds a Ph.D. in mathematics from the University of North Carolina. His writings include journal articles as well as *Laboratory Investigations in Geometry, Geometry: An Investigative Approach,* and *Geometry with Applications and Problem Solving,* all published by Addison-Wesley.

Randall I. Charles is Associate Professor of Mathematics at Illinois State University, where he works with elementary and junior high school teachers. He has taught at all levels and has been an elementary and secondary school mathematics supervisor. He has recently been involved in the development and evaluation of a nationally recognized problem-solving program for elementary and junior high school students. Dr. Charles holds a Ph.D. in Mathematics Education from Indiana University, and is coauthor of several books, including *Addison-Wesley Mathematics* grades K–8 and *Problem-Solving Experiences in Mathematics* grades K–8.

A Note to the Teacher

Ideally, authors planning to write school texts would meet with every teacher who would eventually use their books and discuss their concerns. Of course it was impossible for us to meet all of you, but we did talk with many pre-algebra teachers, visited their classrooms, met with them in focus groups, and received advice from them as consultants throughout the writing of this book.

Our teaching philosophy has developed as a result of our own teaching experience and advice from pre-algebra teachers. We feel it is important to share some of it with you, the teacher of our text.

First, we believe that the idea of variable should be carefully introduced and developed. In this text, students will progress from working with numerical expressions to manipulating algebraic expressions. Experiences with evaluating expressions, first with whole numbers, then with integers and rational numbers, reinforce the idea of variables. Special "Numbers to Algebra" activities help students make the transition from arithmetic to algebra by using numerical examples to develop key algebraic ideas.

Second, we feel that procedures for solving equations should be introduced early in the course and then reinforced and gradually extended. We use whole numbers first to introduce equations involving addition and subtraction, and then those involving multiplication and division. We present integers in Chapter 3 and rational numbers in Chapter 6, so that solving equations with positive and negative numbers can be used throughout the course.

Third, we have found that students need specific, carefully developed instruction in problem solving. Each chapter of this book contains three types of problem-solving lessons: applications, strategies for attacking nonroutine problems, and writing equations to solve problems. The sequence of the lessons is important for extending students' abilities to solve problems comfortably and their willingness to approach unfamiliar problems.

Finally, we recognize that students must develop the understanding and the thinking skills they will need for future work in mathematics. We present concepts and processes through numerous worked-out examples with explanations that encourage learning with understanding. Special mental math, estimation, calculator, and computer activities also develop thinking skills. "Extending Thinking Skills" exercises help students to reason logically, perceive spatially, discover patterns, generalize, test conjectures, organize and analyze data, and make important decisions in a variety of situations.

We have tried to create a program that is easy to teach and manage. At the same time, it should give students the security and daily success that result from a careful sequence, a reasonable pace, clear presentation, and ample practice of basic concepts and skills. A daily mixed review keeps skills fresh throughout the course, and interesting problems and enrichment activities offer variety, challenge, and stimulation.

We hope that you can use this book to help your students stretch their minds, exercise their creativity, find enjoyment, and experience the rewards of success in mathematics.

Phares G. O'Daffer
Stanley R. Clemens
Randall I. Charles

Consultants

Karen E. Bright, Math Department Chair, Fairview Intermediate Magnet School, Dayton, Ohio

Marita H. Eng, Math Department Chair, Sandalwood Junior-Senior High School, Jacksonville, Florida

James M. Foley, Teacher, Fred Moore Junior High School, Anoka, Minnesota

Bill Kennedy, Teacher, A. E. Kent Middle School, Kentfield, California

Steven Leinwand, Mathematics Consultant, Connecticut State Department of Education

Andy Reeves, Mathematics Consultant, Florida Department of Education

Freddie Lee Renfro, Coordinator of Mathematics, LaPorte Independent School District, LaPorte, Texas

Phyllis Thom, Teacher, Malaga Cove Intermediate School, Palos Verdes, California

Mary Kay Tornrose, Coordinator of Mathematics, Newton Public Schools, Newton, Massachusetts

Sandra Young, Secondary Math Supervisor, El Paso Schools, El Paso, Texas

Reviewer's Guide to Addison-Wesley Pre-Algebra

Addison-Wesley Pre-Algebra prepares eighth grade, accelerated seventh grade, and high school students for success in algebra. Written to ease the transition from arithmetic to algebra, it offers the complete program content and teacher support you need for active learning with understanding. Concepts, computational skills, thinking skills, and problem-solving skills are balanced in a carefully planned lesson sequence to build proficiency and mastery.

Complete, Balanced Program Content

- Complete coverage of topics to help students master the transition from arithmetic to algebra (see Table of Contents)
- Easy-to-teach Example–Solution–Practice format for learning with understanding (see pages T8–T9)
- Abundant exercises at three levels of difficulty to increase proficiency and reinforce concepts (see pages T10–T11)
- Skill maintenance exercises in every problem set for maximum retention (see page T11)
- Strong problem-solving strand to develop confidence and effective decision-making (see pages T12–T13)
- Interesting and relevant activities to encourage involvement (see pages T14–T15)

Supports Effective Teaching

- Management guide for minimum, regular, maximum courses
- Situational problems for every chapter, teaching suggestions, and tips for teaching problem-solving
- Teacher's Edition containing answers to all exercises
- Objectives for each lesson and suggested assignments overprinted on student page
- Complete, separate solutions manual for all problems in the text
- Tests and quizzes in blackline master form
- Extra practice supplement for skill reinforcement
- Enrichment activities and computer programs for every chapter

(See pages T7, T18–T19)

Program Components

Student Edition

Full coverage of topics that consolidate arithmetic skills while helping the student make an easy transition to first-year algebra.

Teacher's Edition

Ample teacher support in front of student edition. Annotations and answers to exercises overprinted in blue on student pages.

Assessment Supplement

Complete testing program: two quizzes per chapter; chapter tests, mid-year test, and end-of-year test. Chapter tests are provided in both free-response and multiple choice forms. 96 blackline masters.

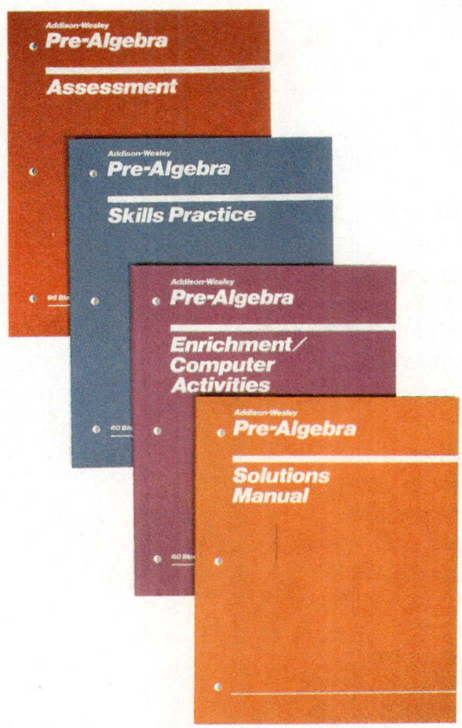

Skills Practice Supplement

Four pages of practice exercises for the major skills in each chapter. Available as blackline masters or as consumable student workbooks with a teacher's edition.

Enrichment/Computer Activities

A computer worksheet for each chapter; two activities per chapter using physical objects and constructions to teach mathematical concepts intuitively. Blackline masters.

Solutions Manual

Complete worked-out solutions to all exercises and problems in the student text, including strategy problems.

Builds Learning With Understanding

Lessons are carefully developed with concise explanations, clear examples, and opportunities to check understanding. Example–Solution–Practice organization makes lessons easy to teach and easy to learn.

Highlighted Information

Important information is highlighted in tinted boxes for easy reference.

Motivational Situations

Sections often begin with a motivational situation to connect the mathematics to the real world.

Examples

Fully worked examples show the solution steps. Side comments provide a complete, detailed explanation of the solution.
Every objective is matched to examples and practice exercises.

3-4 Solving Integer Equations: Addition and Subtraction

To solve integer equations such as $t + 34 = 28$ or $x - 18 = -35$, you need to get the variable by itself on one side of the equation. The following steps show you how to use the ideas of inverse operations or the additive inverse property and the properties of equality to do this.

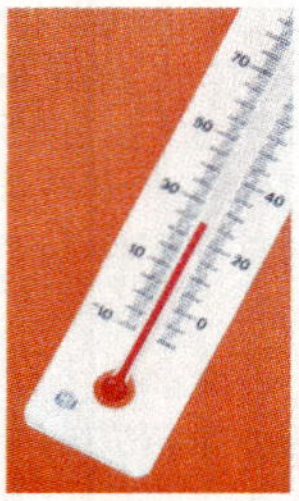

If the temperature increased 34°C to reach a high of 28°C, you could solve the equation $t + 34 = 28$ to find the original temperature.

Solving Equations Using Addition and Subtraction

1. Decide which operation (addition or subtraction) has been applied to the variable.
2. Use the inverse operation or use the additive inverse property, adding or subtracting the same number on both sides of the equation.

Examples 1 and 2 show how to use inverse operations and the same steps you used with whole number equations to solve integer equations. Example 3 shows how to use the additive inverse property to solve integer equations.

Example 1

Solve and check. $t + 34 = 28$

Solution $t + 34 = 28$ — You need to get the variable by itself on one side.

$t + 34 \mathbf{- 34} = 28 \mathbf{- 34}$ — To undo adding 34, subtract 34 from both sides of the equation so they remain equal. (Subtraction property of equality.)

$t = -6$

Check $-6 + 34 \stackrel{?}{=} 28$ — Replace t with -6 in $t + 34 = 28$.

$28 = 28$ ✓ — The solution is -6.

Practice Solve and check.

a. $x + 19 = -47$ **b.** $-42 = y + 76$ **c.** $z + (-35) = (-164)$

3-4 Solving Integer Equations: Addition and Subtraction **79**

Example 2

Solve and check. $x - 18 = -35$

Solution $x - 18 = -35$

$x - 18 + \mathbf{18} = -35 + \mathbf{18}$

$x = -17$

To undo subtracting 18, add 18 to both sides so they remain equal. (Addition property of equality.)

Check $-17 - 18 \stackrel{?}{=} -35$

$-35 = -35 \surd$

Replace x with -17 in $x - 18 = -35$.

The solution is -17.

Practice Solve and check.

a. $b - (-15) = 43$ **b.** $n - 29 = -45$ **c.** $y - (-37) = -64$

Since subtracting an integer is the same as adding the opposite integer, an equation that can be solved by subtracting can also be solved by using the additive inverse property.

Example 3

Solve and check. $n + (-25) = 46$

Solution $n + (-25) = 46$

$n + (-25) + \mathbf{25} = 46 + \mathbf{25}$

$n = 71$

To undo adding -25, use the additive inverse property and add the opposite of -25, or 25. $-25 + 25 = 0$.

Check $71 + (-25) \stackrel{?}{=} 46$

$46 = 46 \surd$

Replace n with 71 in $n + (-25) = 46$.

The solution is 71.

Practice Solve and check. **a.** $a + (-17) = -75$ **b.** $43 = p + (-26)$

Oral Exercises

To solve, what integer would you add to or subtract from each side?

1. $x + 19 = -48$ **2.** $n - 28 = -56$ **3.** $-379 = x + 85$ **4.** $78 = b - (-45)$
5. $c - 56 = -198$ **6.** $b + (-37) = 84$ **7.** $z + 53 = -53$ **8.** $0 = n - 67$

Exercises

A Solve and check.

1. $n + 27 = -84$ **2.** $x + 97 = 42$ **3.** $a + 157 = -96$
4. $y + 79 = -32$ **5.** $c + 76 = 154$ **6.** $n + 67 = -282$
7. $x + 279 = 194$ **8.** $b + 68 = -253$ **9.** $y + (-36) = 273$
10. $a + (-285) = -643$ **11.** $c + 947 = -765$ **12.** $x + (-258) = 1752$

Practice Exercises

Practice exercises immediately follow each example to reinforce the concept or skill just taught and to offer an opportunity for diagnosis and remediation of student difficulties.

Oral Exercises

Oral exercises provide whole-class readiness activities and a check for understanding before written exercises are assigned.

Increases Proficiency and Retention of Skills

Exercise sets are separated into three levels of difficulty for flexibility in meeting individual needs. A mixed review of previously learned skills is included in every lesson.

A Exercises

A Exercises correlate directly to the examples and practice exercises. All the course objectives are covered in the A set of exercises.

B Exercises

B Exercises expand upon the A Exercises and are slightly more difficult. Variables and multi-step problems are emphasized where possible.

Exercises

A Multiply. Reduce to lowest terms.

1. $\frac{3}{5}\left(\frac{3}{4}\right)$	**2.** $\frac{3}{5}\left(\frac{5}{7}\right)$	**3.** $\frac{1}{2}\left(\frac{5}{16}\right)$
4. $-\frac{8}{16}\left(\frac{1}{2}\right)$	**5.** $-\frac{4}{3}\left(\frac{5}{8}\right)$	**6.** $-\frac{6}{4}\left(-\frac{3}{5}\right)$
7. $\frac{14}{6}\left(-\frac{3}{10}\right)$	**8.** $\frac{5}{3}\left(\frac{15}{8}\right)$	**9.** $\frac{7}{12}\left(\frac{4}{3}\right)$
10. $-\frac{3}{8}\left(-\frac{2}{3}\right)$	**11.** $-\frac{5}{2}\left(-\frac{5}{10}\right)$	**12.** $\frac{12}{25}\left(\frac{2}{10}\right)$
13. $\frac{4}{12}\left(\frac{15}{24}\right)$	**14.** $\frac{24}{26}\left(-\frac{3}{5}\right)$	**15.** $-\frac{6}{7}\left(-\frac{14}{10}\right)$
16. $-3\frac{1}{10}\left(1\frac{1}{3}\right)$	**17.** $6\frac{1}{2}\left(2\frac{2}{8}\right)$	**18.** $4\frac{1}{4}\left(1\frac{2}{5}\right)$
19. $\frac{6}{2}\left(2\frac{1}{4}\right)$	**20.** $-8\frac{1}{8}\left(-\frac{3}{4}\right)$	**21.** $-4\frac{5}{6}\left(2\frac{2}{3}\right)$
22. $8\frac{7}{9}\left(-5\frac{2}{5}\right)$	**23.** $2\frac{3}{4}\left(8\frac{4}{7}\right)$	**24.** $-1\frac{1}{10}\left(3\frac{2}{10}\right)$
25. $1\frac{3}{4}\left(-\frac{4}{5}\right)$	**26.** $-2\frac{3}{4}\left(-2\frac{1}{4}\right)$	**27.** $5\left(4\frac{2}{5}\right)$

B Find the value of each expression. Reduce to lowest terms.

28. $\frac{2}{3}b$ for $b = 1\frac{1}{4}$	**29.** $-\frac{4}{3}a$ for $a = 12$
30. $1\frac{1}{5}y - \frac{3}{10}$ for $y = -1\frac{3}{5}$	**31.** $\frac{1}{8}x + 2$ for $x = 3\frac{1}{2}$
32. $1\frac{1}{2}h$ for $h = -\frac{5}{6}$	**33.** $\frac{2}{3}x + \frac{1}{4}$ for $x = -\frac{5}{6}$

Find the value of x. Reduce to lowest terms.

34. $x = \frac{2}{3}\left(\frac{3}{5} + \frac{6}{15}\right)$	**35.** $x = \left(\frac{1}{2} + \frac{3}{4}\right)\frac{2}{5}$
36. $x = -\frac{18}{24}(9 - 15)$	**37.** $x = -\frac{4}{5}(-5 + 8)$
38. $x = \frac{4}{5}\left(\frac{4}{5} + \frac{3}{4}\right)$	**39.** $x = \frac{3}{8}(6 - 7)$

Solve.

40. A cookbook recommends roasting a turkey at a low temperature $\frac{3}{4}$ hour for each pound. How long should you cook a $10\frac{1}{2}$ pound turkey?

41. Each section of a fence is $6\frac{1}{2}$ ft long. How long is this fence if it has 8 sections?

42. A certain steel bar weighs $2\frac{1}{2}$ pounds per foot. What would be the weight of a piece $3\frac{3}{4}$ ft long?

206 *Chapter 7 Rational Numbers: Multiplication and Division*

B The scale for this drawing of a patio is 1 cm : 1.25 m. Measure the dimensions of the patio to the nearest millimeter. Then use a proportion to calculate the actual dimensions.

14. dimension a

15. dimension c

16. dimension b

17. dimension d

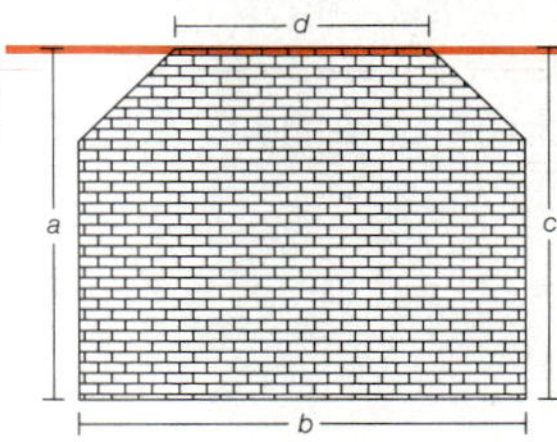

18. The distance from A to B on the map is 4.2 cm. The distance from B to C is 1.4 cm. B is on a straight road from A to C. The scale on the map is 1 cm : 125 km. What is the actual distance from A to C?

19. On a map with scale 2 cm : 25 km, what would be the dimensions of a 145 km by 80 km rectangle?

281

B Exercises

B Exercises often include problem-solving applications and calculator exercises.

C Extending Thinking Skills

32. Solve for y. $(7 \times 10^2)y = 6.3 \times 10^6$

33. Use the digits 1, 2, 3, and 4 and one negative sign to write a number in scientific notation that is as close to 0.01 as possible. $\blacksquare.\blacksquare\blacksquare \times 10^{\blacksquare}$

34. Chemical X is made in units weighing 2.5×10^{-2} g. To produce one unit of this chemical, 4×10^{-3} g of chemical A and 5.0×10^{-2} g of chemical B are needed. In the production process, 2.5×10^{-3} g of chemical A and 4.2×10^{-2} g of chemical B are burned off. What amount of each unit of chemical X comes from chemicals other than A and B?

Mixed Review

Reduce each fraction to lowest terms. **35.** $\frac{27}{135}$ **36.** $\frac{48}{192}$ **37.** $\frac{12}{140}$

Solve and check. **38.** $a - \frac{3}{4} = -\frac{5}{8}$ **39.** $h + \frac{13}{15} = 1\frac{1}{3}$

Simplify. **40.** $z \cdot z \cdot 10 \cdot 4 \cdot z$ **41.** $t + 2t + 3t + 4t$

226

C Exercises

C Exercises are challenging, non-routine problems that extend thinking skills.

Mixed Review

Mixed Review maintains previously learned skills.

More Practice

Solve and check.

1. $-2m = 144$	**2.** $26 + c = 11$	**3.** $t - 16 = 4$	**4.** $\frac{r}{4} = -28$
5. $\frac{x}{-5} = -19$	**6.** $m - 28 = 42$	**7.** $41 = a + (-19)$	**8.** $15n = -540$
9. $k - (-6) = 11$	**10.** $\frac{t}{9} = -16$	**11.** $-12a = 636$	**12.** $r + 6 = 0$
13. $97 + m = -6$	**14.** $350 = -14c$	**15.** $\frac{h}{-3} = -27$	**16.** $x - 6 = -8$
17. $12t = 144$	**18.** $16 = n - 3$	**19.** $w + 30 = 4$	**20.** $\frac{a}{16} = -23$
21. $\frac{m}{-7} = 46$	**22.** $t + 16 = 100$	**23.** $r - 5 = 0$	**24.** $-6k = -594$
25. $c - 6 = -4$	**26.** $\frac{z}{9} = -42$	**27.** $19y = -304$	**28.** $17 = n + 12$

92

More Practice

More Practice pages mix important skills from several sections for effective reinforcement. A page of Extra Practice for each chapter appears in the back of the book.

Develops Confidence and Effective Problem-Solving Skills

A three-part program, including *Applications, Writing Equations to Solve Problems,* and *Using Strategies to Solve Non-Routine Problems,* develops problem-solving skills in a clear, organized sequence.

Problem-Solving: Applications

Problems are drawn from career or consumer situations in the real world. Lessons encourage a choice of techniques, including pencil and paper, mental math, estimation, and use of a calculator.

PROBLEM SOLVING: APPLICATIONS

9-7 Determining the Best Buy.

Grocery stores often show unit prices.

The cost of one unit of an item is called the **unit price**. You can find and compare unit prices to determine the better buy. The item with the lower unit price is usually considered the better buy.

A & B Discount	
100 vitamins	2/$1.99
video cassettes	3/$17.50
earrings	2 pair for $5

Bob's Super Saver	
100 vitamins	4/$3.75
video cassettes	2/$11
earrings	4 pair for $11

282

Data Search

Students go outside the classroom to collect information.

Muffins	6 for $0.75	$0.15 each
Biscuits	6 for $0.55	$0.12 each

14. Data Search Choose two grocery stores. Find common brands for: $\frac{1}{2}$ gallon whole milk; 20 lbs dog food; 3 cans frozen orange juice; 2 loaves whole wheat bread; 1 whole chicken. Which store has the better buy for each?

What's Your Decision?

Students must use judgment in solving real-life problems.

What's Your Decision?

You have $10 to spend on items in the advertisements on page 282. What would you buy and from which store?

9-7 Determining the Best Buy **283**

PROBLEM SOLVING: WRITING EQUATIONS

4-8 Writing Equations to Solve Problems

You can use the Problem-Solving Checklist as a guide for solving problems when your plan involves the strategy **Write an Equation**.

Problem-Solving Checklist: Writing Equations

Understand the Question/Find the Needed Data
What do you need to find? Can you show the data in the problem?

Plan What to Do
Can you use a variable to represent an unknown number?
Can you represent other conditions in terms of the variable?
What is equal in the problem? Can you write and solve an equation?

Find the Answer/Check Back
Does the solution of the equation check?
What is the answer to the question in the problem?
Does the answer seem reasonable?

121

Problem-Solving: Writing Equations

The skill of translating words and word problems into mathematical expressions and equations is developed in lessons throughout the text.

PROBLEM SOLVING: STRATEGIES

3-10 Make a Table, Look for a Pattern

Problem-solving strategies called **Make a Table** and **Look for a Pattern** are helpful when solving problems involving numerical relationships. Consider the following problem.

Problem A secret agent was hired for a special assignment that would take exactly 14 days. He could choose to be paid one of the following two ways. Under payment plan A, he would receive $6000 for the job. Under payment plan B, his employer would put $1 into a safe for the first day and increase the amount in the safe to $2 the second day, $4 the third day, $8 the fourth day, and so on, doubling the amount in the safe each day. At the end of the assignment, the agent could claim the contents of the safe. Which payment plan should the agent choose?

To find the amount for payment plan B, you can use the data in the problem to begin a **table**. The red numbers show the data given in the problem. Look for a **pattern** to help extend the table and provide new information.

Notice that the numbers in the second column are found by multiplying 1 less factor of two than the number for the day. On day **4** there were **3** factors of 2, or 2(2)(2), dollars in the safe. So on day **14** there would be **13** factors of 2, or 8192 dollars in the safe.

Day	Number of dollars in the safe
1	**1**
2	**2** ← 2
3	**4** ← 2(2)
4	**8** ← 2(2)(2)
5	16
6	32
⋮	⋮
14	?

95

Problem-Solving: Strategies

Lessons in each chapter provide instruction and practice in using strategies such as Guess-Check-Revise, Draw a Picture, Make a Table, Look for a Pattern, Work Backward, and many more to solve non-routine problems. Strategy lessons develop thinking skills.

Encourages Involvement Through Interesting and Relevant Activities

Developing "Number Sense"

Estimation activities in every chapter improve students' "number sense." Students develop confidence as they learn to recognize accurate, reasonable answers.

Thinking skills: 31.

Mixed Review

Find all the factors of each number. **33.** 25 **34.** 21

Give the absolute value and opposite of each. **35.** −7 **36.** 19

Estimate, then find the value of the variable. **37.** $M = 6.5(68 \div 17)$

Solve and check. **38.** $19.8 = -4c$ **39.** $-11.62 + n = -49.33$

ESTIMATION

You can substitute compatible numbers to estimate the sum or difference of rational numbers by changing them to approximate equivalent fractions. For example, to estimate the sum $\frac{1}{4} + \frac{7}{8}$, think: $\frac{7}{8}$ is about $\frac{3}{4}$, so $\frac{1}{4} + \frac{7}{8}$ is about $\frac{1}{4} + \frac{3}{4}$, or 1.

Estimate each sum or difference.

1. $\frac{7}{8} + \frac{12}{13}$ **2.** $\frac{3}{7} + \frac{5}{8}$ **3.** $\frac{11}{12} - \frac{3}{4}$ **4.** $\frac{14}{15} + \frac{1}{2}$

6-6 Adding and Subtracting with Unlike Denominators **185**

Improving Mental Computation

Mental Math activities give students practice in and new techniques for quick mental calculation.

MENTAL MATH

Breaking apart numbers is a mental math technique based on the associative and commutative properties. Study the examples.

- To find **76** + **9**, think "70 plus **6** + **9** is 70 plus **15**, or 85."
- To find **57** + **38**, think "50 + 30 is 80. **7** + **8** is **15**. 80 + **15** is 95."
- To find **347** + **526**, think "300 + 500 plus **40** + **20** plus **7** + **6** is 800 + **60** + **13**, or 873."

Find each sum mentally by breaking apart the numbers.

1. 67 + 6 **2.** 96 + 9 **3.** 37 + 65

4. 78 + 94 **5.** 437 + 328 **6.** 256 + 493

1-5 Solving Equations: Using Mental Math **17**

Understanding Variables

Numbers to Algebra activities on key topics help students make the transition from arithmetic to algebra with confidence.

NUMBERS TO ALGEBRA

When an expression has the same number as a factor in each addend, a reversed form of the distributive property, $ab + ac = a(b + c)$ may be used to write an equivalent expression that is a product.

Check the number examples to see that they are true.

This property is used to **factor** algebraic expressions

Numbers	Algebra (factored form)
$4 \cdot 3 + 4 \cdot 5 = 4(3 + 5)$	$4b + 4c = 4(b + c)$
$6 \cdot 5 + 6 \cdot 7 = 6(5 + 7)$	$6x + 6y = 6(x + y)$
$5 \cdot 8 + 5 \cdot 4 = 5(8 + 4)$	$5m + 5n = 5(m + n)$

Use the distributive property to factor each expression.

1. $2a + 2b$ **2.** $5y + 5z$ **3.** $6n + 6(7)$

4. $8s + 16$ **5.** $3p + np$ **6.** $ax + az$

2-3 Basic Properties of Multiplication **45**

C Extending Thinking Skills

51. In a magic square, the sum of the numbers in each row, column, and diagonal is the same. Find the missing numbers in this magic square, assuming the magic sum is -2.

4	−6		
−7		2	
	−3	−8	
			−5

52. An elevator went up 6 floors, down 9 floors, down 12 more floors, up 8 floors, and down 4 floors. It stopped on the 43rd floor. On what floor did it start? **The 54th floor**

Mixed Review

Solve and check. **53.** $8y = 216$ **54.** $t + 163 = 204$

Find the sum. **55.** $-5 + 3$ **56.** $5 + 3$ **57.** $5 + (-3)$

58. $-5 + (-3)$

CALCULATOR ACTIVITY

The change-sign key [+/−] allows you to show a negative integer by changing the displayed number to its opposite.

To add $86 + (-123)$: **86** [+] **123** [+/−] [=]

To subtract $-67 - (-239)$: **67** [+/−] [−] **239** [+/−] [=]

Add or subtract, using the change-sign key.

1. $258 + (-689)$ **2.** $-9674 + 6739$ **3.** $-783 - (-975)$

4. $-7386 + 2978$ **5.** $4547 - (-8964)$ **6.** $-435 + 848$

Using Calculators

Calculator activities help students learn efficient key sequences for a variety of problem types. Use of special keys is introduced.

C Extending Thinking Skills

41. Use inverse operations to find the missing number n in this flow chart. **1776**

Start with n → subtract 1492 → multiply by 6 → add 282 → end with 1986

42. If a number is multiplied by 3 and added to 9, the result is 54. What is the number?

Mixed Review

Write as an expression. **43.** 13 added to the product of 6 and 9

Solve and check. **44.** $n - 36 = 17$ **45.** $41 + x = 63$ **46.** $9 = x - 334$

COMPUTER ACTIVITY

This computer program can be used to solve equations of the form $\frac{x}{A} = B$. **Remind students to separate their number inputs in line 30 with a comma.**

```
10 REM: SOLVE AN EQUATION X/A=B
20 PRINT "CHOOSE WHOLE NUMBERS A AND B, A<>0"
30 INPUT A,B
40 PRINT "THE EQUATION IS X/"A;"=";B
50 PRINT "THE SOLUTION IS X=";A*B
60 END
```

1. Run the program or show what the run would look like.
2. Change the program to solve equations of the form $Ax = B$.

Using Computers

Computer activities involve keying in, running, and rewriting programs that demonstrate the power of the computer as a tool in mathematics. No special technological expertise is required to teach the lessons.

Offers a Complete Range of Enrichment and Assessment Options for Each Chapter

Enrichment

Enrichment activities relate to the mathematics in the chapter. Some give a look at interesting applications, and others preview higher levels of math.

Enrichment

A Golden Rectangle

Early Greek architects, painters, and sculptors identified what they believed to be the rectangle most pleasing to the human eye. They called it the **Golden Rectangle**. The ratio of the length l to the width w of this rectangle, $\frac{l}{w}$, was called the **Golden Ratio**. The Greeks discovered that the following proportion holds only for a Golden Rectangle:

$$\frac{l}{w} = \frac{l + w}{l}$$

1. Each of the rectangles below is a Golden Rectangle. Measure the length and width of each, to the nearest millimeter. Find the ratio $\frac{l}{w}$ for each. Express each ratio as a decimal, and use them to estimate the Golden Ratio as a decimal.

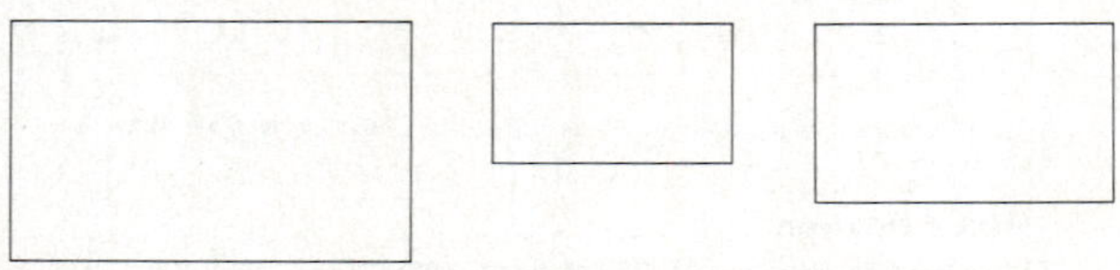

2. Measure the length and width of each rectangle below to the nearest millimeter and compute the ratio $\frac{l}{w}$ for each. For which rectangle is this ratio closest to the Golden Ratio? Is this rectangle most pleasing to your eye?

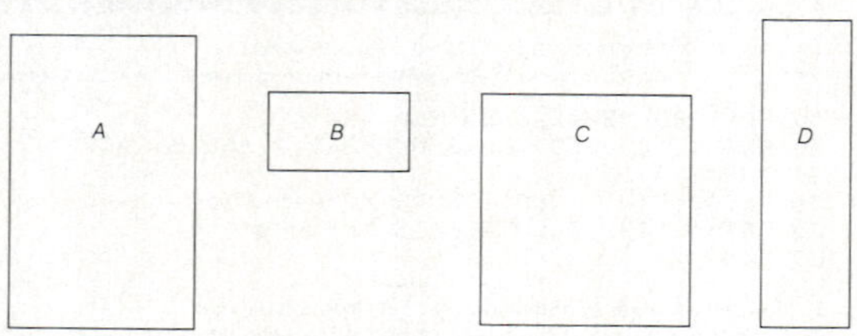

Chapter 4 Review

4-1 Write as a decimal.

1. $6(10) + 3(1) + 8\left(\frac{1}{10}\right)$ **2.** $2\left(\frac{1}{1000}\right) + 5(1) + 7\left(\frac{1}{100}\right)$

Write >, <, or = for each □.

3. −4.23 □ −4.28 **4.** 1.01 □ 1.010 **5.** 0.51 □ 0.514

4-2 Estimate the value of x by rounding.

6. $x = 94.27 - 13.66$ **7.** $x = 801.7 + 200.1$

8. $x = 47.27 + 15.25$ **9.** $x = 62.85 - 10.13$

4-3 Add or subtract.

10. 7.13 + 1.3 **11.** 67.19 + 7.17 + 11.11

12. −3.87 + 9.75 **13.** 61.3 + (−17.4) + 2.93

Evaluate.

14. $u - 16.45$ for $u = -12.3$ **15.** $93.5 - t$ for $t = 6.8$

130

Chapter Reviews

Chapter Reviews can be used as self-tests at the end of each chapter or as classwork to assess progress. Problems are keyed to the sections where the relevant skills were taught.

Chapter 4 Test

Write as a decimal.

1. $1(10) + 5(1) + 3\left(\frac{1}{10}\right)$ **2.** $4(100) + 6\left(\frac{1}{100}\right) + 8\left(\frac{1}{10}\right)$

Write >, <, or = for each □.

3. 290 □ 7.29 **4.** 6.01 □ 6.1 **5.** −3.25 □ −3.45

Estimate the value of x by rounding.

6. $33.89 - 19.25 = x$ **7.** $x = 415.2 + 75.5$

8. $x = 391.37 - 106.93$ **9.** $59.82 + 16.05 = x$

Add or subtract.

10. 32.83 + 12.93 **11.** 3.14 + 23.36

12. 45.2 + (−2.2) **13.** −0.33 + 24.81 + 2.66

14. 88.2 − 35.5 **15.** −2.76 − 67.21

Solve and check.

131

Chapter Tests

Chapter Tests correlate to problems in the lessons and to the Chapter Review.

Cumulative Review

Write as a numerical expression.

1. 7 less than 18 **2.** 15 more than 32 **3.** 12 increased by 8

Write as an algebraic expression.

4. m less than 7 **5.** c decreased by 9 **6.** a number added to t

Solve the equation for the replacement set given.

7. $c + 4 = 12$ {6, 7, 8, 9} **8.** $z - 8 = 8$ {13, 14, 15, 16}

9. $10 - n = 3$ {6, 7, 8, 9} **10.** $4 + g = 14$ {7, 8, 9, 10}

Evaluate each algebraic expression.

11. $d + 1$ for $d = 6$ **12.** $h - i$ for $h = 7, i = 0$

13. $g + (-10)$ for $g = 3$ **14.** $35 + t$ for $t = -10$

15. $-4 - c$ for $c = 3$ **16.** $-5 - z$ for $z = 2$

Find the sum or difference.

132

Cumulative Reviews

Cumulative Reviews follow every chapter to reinforce previously taught skills and point out areas for review and remediation.

Supports Effective Teaching

Two-part Teacher's Edition contains 77 pages at the front of the text with:

Teaching Suggestions

Teaching Suggestions

Chapter 1

1-1 *pages 2–5* Introduce evaluating a numerical expression by giving several examples involving parentheses. (Note that order of operations without parentheses is introduced in Chapter 2, with all four operations.) Tell students to think of a numerical expression as a name for a number. Discuss Example 1 and its Practice. In Example 2, emphasize that the variable

Ex. 41 Suggest that students guess and check. **Ex. 42** This is the Fibonacci sequence. **Mental Math** Start by showing how to think of 52 − 29 as 52 − 30 + 1.

1-3 *pages 9–12* Introduce the Commutative Property by asking in what situations order of operations does or does not make a difference. ("Put on shoes, then put on socks" or "put on socks, then put on shoes"; "Put 5 marbles in a bag, then put in

T38

Management Guide

Management Guide

The Management Chart lists assignments for minimum, regular, and maximum courses. The total number of days is based on 174, with 87 assignment days for each semester.

First Semester

Chapter	1	2	3	4	5	6	7	8
Days	10	11	11	11	10	12	10	12

T72

Tips for Teaching Problem Solving

Tips for Teaching Problem Solving

Developing a Positive Classroom Atmosphere

There are four essential ingredients to a successful problem-solving program: 1) the content of the program must be appropriate for students; 2) the problem-solving program must be part of a sound instructional program in all other basic skill areas of mathematics

can do to promote desirable problem-solving behaviors.

- Recognize and reinforce willingness and perseverance.
- Reward risk takers.
- Encourage students to play hunches.
- Accept unusual solutions.

T22

Situational Problem Solving

Problem Situations

Problem Situation 1

for use with Chapter 1

Your dog is lost and you want to run an ad in the classified section of your local newspaper. You have $25 to spend on the ad and a reward. You want to find out how much you can offer for the reward.

Solution

Cost of running the ad:

$1.66 × 3 × 3 = $14.94.

Amount for the reward:

$25 − 15 = $10

You can offer about $10 for the reward.

T28

5-1 Multiples and Factors

Objective: To write multiples and factors of a number.
MIN: 1–35; REG: 1–35 even, 36–41; MAX: 1–35 even, 36–43; ALL: Mixed Review

Section Objectives

Daily Assignment Guide

The product of two whole numbers is a **multiple** of each of the whole numbers. The multiples of a number can be found by multiplying the number by 0, 1, 2, 3, 4, and so on.

$$0 \times 2 = 0,\ 1 \times 2 = 2,\ 2 \times 2 = 4,\ 3 \times 2 = 6, \ldots$$

0, 2, 4, 6, . . . are multiples of **2**. All whole numbers that are multiples of 2 are called **even numbers**. All whole numbers that are not multiples of 2 are called **odd numbers**.

Example 1

Encourage students to practice skip-counting by 2s, 3s, 4s . . . 8s, and 9s to produce multiples of these numbers.

Class Discussion Suggestions

Write the first five nonzero multiples of 6.

Solution 6, 12, 18, 24, 30 Multiply 6 by 1, 2, 3, 4, and 5.

Practice

a. Write the first three nonzero multiples of 7. 7, 14, 21

Answers

b. Write the first five nonzero multiples of 9. 9, 18, 27, 36, 45

On some calculators, you can enter a number and press [+] [=] [=] [=] . . . *to display its multiples.*

When two or more whole numbers are multiplied to form a product, each is called a **factor** of the product.

factors of 20 — product of 4 times 5

$$4 \times 5 = 20$$

When a whole number is divided by one of its nonzero factors, the quotient is a whole number and the remainder is zero. Thus, you can find out if one number is a factor of another by dividing. If the quotient is a whole number with a remainder of 0, then the divisor and the quotient are each factors of the dividend.

Example 2

Is 9 a factor of 153?

Solution $153 \div 9 = 17$
9 is a factor of 153.

Since $153 \div 9 = 17$, with a remainder of 0, then $17 \times 9 = 153$, and 9 is a factor of 153.
17 is also a factor of 153.

134 Chapter 4 Number Theory

Annotated Teacher's Edition shows annotations in blue for ready identification.

Additional Resources from Addison-Wesley

Tailor your classwork to your students and your individual teaching style.

The Middle Grades Mathematics Project

Five teacher source books with unique teaching guides and blackline masters develop problem-solving and thinking skills. Each book includes lessons and activities to build understanding of mathematical concepts. Mouse and Elephant: Measuring Growth, Factors and Multiples, Similarity and Equivalent Fractions, Spatial Visualization, and Probability extend and enrich Pre-Algebra classes. Developed with aid from the National Science Foundation.

Sci-Math, Module One

The Sci-math Program develops the mathematics skills needed for science courses and for problem-solving in everyday, real-world situations. A self-contained worktext, Sci-Math provides problems, exercises, hands-on activities, and experiments that teach the arithmetic and logic of proportions. A comprehensive teacher's guide includes blackline masters and simple materials suggestions.

Math Motivators! Investigations in Pre-Algebra

Thirty-two self-contained enrichment investigations extend and reinforce the concepts and skills of the basic course, develop problem-solving skills, and open the door to exciting mathematical ideas. Each blackline master is accompanied by full teaching notes and extension activities.

Problem-Solving Experiences in Mathematics *Level 8*

A thoroughly tested instructional program including blackline masters for 150 problem-solving experiences and a teaching strategy for problem solving. Appealing problems encourage students to try strategies and approach problems systematically.

Addison-Wesley Mathematics Skills, Software *Level 8*

Two disks, one covering basic operations with whole numbers, decimals, fractions, and integers, and the other teaching manipulation of mathematical expressions and equations. For IBM PCjr and Apple computers.

Addison-Wesley Mathematics Problem-Solving Software *Level 8*

Classroom software with an open-ended format for problem-solving exploration. Using a checklist, students can experiment with word problems and evaluate their choices. For Apple computers.

Addison-Wesley Pre-Algebra

Builds learning with understanding

Increases proficiency and retention of skills

Develops confidence and effective problem-solving skills

Encourages involvement through interesting and relevant activities

Offers a complete range of enrichment and assessment options

Supports effective teaching

The Ideal Transition to Algebra

Tips for Teaching Problem Solving

Developing a Positive Classroom Atmosphere

There are four essential ingredients to a successful problem-solving program: 1) the content of the program must be appropriate for students; 2) the problem-solving program must be part of a sound instructional program in all other basic skill areas of mathematics (computational skills, estimation, measurement, and so on); 3) students must have ample opportunity to participate in problem-solving experiences; and 4) the teacher's actions must promote a positive classroom atmosphere related to problem solving.

A successful problem-solving program must have all of the characteristics above. Assistance with the first three is provided by the text. The importance of the teacher in developing a classroom atmosphere that is conducive to problem solving cannot be overemphasized.

There are two ways in which your actions affect the classroom atmosphere. First, your attitude about problem solving will influence your students' attitudes about problem solving. If you demonstrate that problem solving is important, exciting, and fun, most students will develop a similar attitude. Here are some things you can do to promote positive problem-solving attitudes among your students.

- Be enthusiastic about problem solving.
- Encourage students to contribute problems from their personal experiences.
- Personalize problems whenever possible; use students' names in problems, for example.
- Provide the appropriate amount and type of assistance to avoid excessive frustration.

Second, your comments about problem solving communicate to students the types of behaviors you consider desirable related to problem solving. Here are some things you can do to promote desirable problem-solving behaviors.

- Recognize and reinforce willingness and perseverance.
- Reward risk takers.
- Encourage students to play hunches.
- Accept unusual solutions.
- Praise students for getting correct solutions but emphasize the selection and use of problem-solving strategies.
- Emphasize persistence rather than speed.

Working with Small Groups

Many teachers find that using small groups can be a valuable way to organize the classroom for teaching problem solving. One of the most important things a teacher of problem solving can do is to move around the room, observing and questioning students while they solve problems. If there are 25 students in a class, it is difficult to give much attention to any one student. When small groups are formed, most teachers find it easier to monitor and assess their students' problem-solving performance.

In addition to helping in the management of instruction, there are other benefits of using small groups. Students who have not had much experience with problem solving may have considerable problem-solving anxiety. The use of small groups is one way to reduce the pressure on the individual student. In problem-solving groups, progress and success on a problem are the responsibility of the group, not the individual. Another benefit of small-group instruction is in eliciting behaviors that promote the improvement of problem-solving performance. For example, students are often required to justify their ideas, evaluate the ideas of others, and deal with contradictions. Here are other guidelines for using small groups.

- Limit the group size to three or four students.
- Accept a higher noise level in the classroom.
- Do not interrupt a group that is working well. If a group appears to be floundering, however, ask a student to tell what the group is discussing or which part of the problem is giving them difficulty.
- Ask discussion questions, rather than telling a group what to do.
- Try different grouping patterns—homogeneous, heterogeneous, teacher-selected groups, student-selected groups, for example—to find what works best for your students.

When small groups are used, it is important to keep every student involved in the problem-solving task. Here are some ideas.

- Identify a group captain for the day. This person is responsible for explaining the group's work for that day.
- Identify a recorder to write all of the group's work.
- Require that students ask for your help only when everyone in the group has the same question.
- Require that everyone in the group agree on one answer.
- Question students who appear not to be involved in the group's work. Try to determine whether the students do not understand the problem or whether they are not participating.

Evaluating Problem-Solving Performance

As problem-solving experiences play a greater role in your mathematics program, the evaluation of these experiences becomes more important. The best way to evaluate problem-solving performance is through a one-to-one interview either while the student is solving a problem or immediately after the problem is solved. Unfortunately, this type of interview takes more time than is available to most teachers.

There are, however, two evaluation techniques that do not require much time: 1) analyzing a student's written work, and 2) observing while a student solves problems in a whole-class setting. When used jointly, those techniques can provide valuable information about each student's problem-solving performance. A scheme for analyzing a student's written work and a sample problem are given below.

A Scoring Scheme for Written Work

Understanding the problem

0—Completely misinterprets the problem.
1—Misinterprets part of the problem.
2—Completely understands the problem.

Choosing and implementing a solution strategy

0—Makes no attempt or uses a totally inappropriate strategy.
1—Chooses a partly correct strategy based on interpreting part of the problem correctly.
2—Chooses a strategy that could lead to a correct solution if used without error.

Getting the answer

0—Gets no answer or a wrong answer based on an inappropriate solution strategy.
1—Makes copying error or a computational error, gets partial answer for a problem with multiple answers, or labels answer incorrectly.
2—Gets correct solution.

Sample Problem

A well is 10 meters deep. A frog climbs up 5 meters during the day but slips back 4 meters during the night. If the frog starts at the bottom of the well, how many days does it take the frog to reach the top?

According to the scoring scheme, the following examples of solutions to the problem might be scored as shown below:

Student A

Solution: 5 meters
−4 meters
1 meter gained each day

It takes 10 days for the frog to reach the top.

Score: 1, 1, 0

Student B

Solution: 5 meters each day
10 meters in all

It takes 2 days for the frog to reach the top.

Score: 1, 0, 0

Student C

Solution: 5 meters
−4 meters
1 meter gained each day

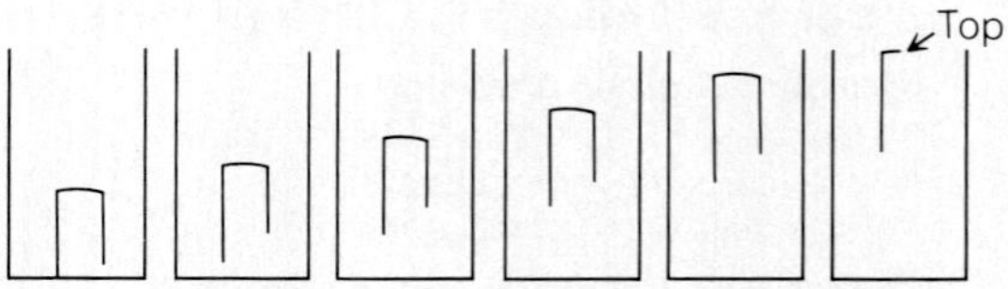

It takes 6 days for the frog to reach the top.

Score: 2, 2, 2

A complete and accurate evaluation of students' problem-solving performances cannot be achieved if evaluation techniques are limited to the point system described above. Many of the goals in teaching problem solving involve students' attitudes toward problem solving, their methods of understanding problems, choosing and implementing solution strategies, and checking solutions. Goals like these cannot be accurately assessed only by examining students' written work for problems. While your students are involved in solving problems, observe and question them about their work. Informal evaluative comments can be made directly to them at that time, or observations can be recorded and shared later, at a parent or student conference, for example. A checklist that can be used to summarize and report your observations is shown on the following page.

Helping Students Read Mathematics Problems

A student's reading ability has some influence on whether that student becomes a successful problem solver. The ideas listed below for improving students' reading skills will also increase skills in understanding problems.

- Make a bulletin board display of words that have special meanings in mathematics.
  Examples:

expression	inequality	perimeter
simplify	base	area
pattern	meter	face

- Have students write sentences illustrating the different meanings of a word.
  Examples:
  Johnny's *face* is red.
  The *face* of a cube is a square.
- Have students rewrite number words as numerals. Example: Have student write two hundred twenty-five as 225.
- Give students opportunities to write their own word problems. See the discussion below on teaching students to formulate problems.
- Have students first read problems silently, then reread the problems aloud and slowly.
- Have students substitute nouns for the pronouns if the action in the problem is confusing.
- Have a mathematics dictionary available to students.

Teaching Students to Formulate Problems

The most important element in improving students' problem-solving performance is solving and discussing many problems. Nevertheless, improving performance in problem solving involves more than simply doing a task over and over. Formulating problems focuses the student's attention on particular parts of the problem-solving process in isolation from the solution itself. Some of the skills needed in formulating problems are:

- asking a question that makes sense
- incorporating all relevant data in a story
- incorporating action in a story appropriate for the operation(s) needed to find a solution.

Problem-Solving Observation Checklist

Student ______________________________

Date ______________________________

	Frequently	Sometimes	Never
1. Selects appropriate solution strategies.	______	______	______
2. Accurately implements solution strategies.	______	______	______
3. Tries different solution strategies when stuck (without aid from the teacher).	______	______	______
4. Approaches problems in a systematic manner (clarifies the question, identifies needed data, selects and implements a solution strategy, checks solution).	______	______	______
5. Perseveres in solving problems.	______	______	______
6. Checks work (without being told to by the teacher).	______	______	______
7. Shows a willingness to try problems.	______	______	______
8. Demonstrates self-confidence.	______	______	______

9. Additional comments: ______________________________

____

____

____

____

Using Evaluation Results to Improve Instruction

Evaluation and grading are not synonymous. Regardless of whether or not grades are assigned to indicate progress in problem solving, all teachers should have a plan for evaluating progress. The goal of the plan is to obtain information that suggests how instruction could be modified to meet individual needs. Following are some ideas for the types of instructional actions that could be added or emphasized in your instructional program, depending upon the findings from your evaluation of students.

Problem Area: Understanding the Problem

1. Have discussions that focus on understanding before students start work on a problem.
   - **a.** Ask questions that focus on what it is they are asked to find (i.e., the question), the conditions and variables in the problem, and the data (needed and unneeded).
   - **b.** Have students explain problems in their own words.
   - **c.** Remind students of similar problems.
   - **d.** Have students use colored markers to highlight important phrases and data.
   - **e.** Have students list the data given.

2. Use skill activities that focus on the QUESTION and DATA phases of problem solving. Examples:
   - **a.** Given a story, write a question that can be answered using data in the story.
   - **b.** Given a problem with unneeded data, identify the data needed to find a solution.
   - **c.** Given a problem with missing data, make up the appropriate data.

Problem Area: Developing a Plan

1. Suggest a solution strategy with the problem statement.

2. Discuss possible solution strategies before students start solving a problem.
   - **a.** Have students suggest reasons why they believe particular strategies might work, but be careful not to evaluate their ideas.
   - **b.** For one-step and multi-step problems, have students tell what action is taking place that suggests a particular operation.

3. Remind students of similar problems.

4. Use skill activities that focus on the PLAN phase of problem solving. Examples:
   - **a.** Given a one-step or multi-step problem, tell the operation or operations needed to find a solution [i.e., practice choosing the operation(s)].
   - **b.** Given a one-step or multi-step problem without numbers, tell the operation or operations needed to find a solution.
   - **c.** Given a completed solution (e.g., a number sentence, an organized list, a picture), tell a story problem that could be solved with the given solution.

5. Discuss solution strategies used in solving a problem after students have completed work on the problem.
   - **a.** Have students tell why they selected particular strategies. You may wish to compare this with the discussion that took place before students started solving the problem.
   - **b.** Show different solutions (strategies) if possible.
   - **c.** Evaluate the usefulness of different solution strategies after students have completed work on the problem.
   - **d.** Show and discuss incorrect solution attempts (inappropriate strategies) used by students. Discuss which strategies were used and why those strategies were not appropriate.

Problem Area: Implementing a Plan

1. Have students evaluate the implementation of a solution strategy to determine whether it was done accurately.

2. Give students the start of a solution (strategy) and have them complete the solution to find the answer.

3. Give a hint with the problem statement telling how to start a solution (strategy).

4. Give direct instruction and practice with particular solution strategies.

5. Show solution strategies that if properly implemented would have led to the correct solution to the problem. Show where the error occurred in implementing the strategy.

Problem Area: Answering the Problem and Checking the Answer

1. Have students check to make sure that they used all important information in the problem.
2. Have students check any arithmetic they might have used in finding the answer.
3. Have students give answers to problems in complete sentences.
4. Use skill activities that focus on the ANSWER and CHECK phases of problem solving. Examples:
   - **a.** Use estimation to find answers.
   - **b.** Use estimation to check answers.
   - **c.** Given the numerical part of an answer, tell the answer in a complete sentence.
   - **d.** Given a problem and an answer, decide whether the answer is reasonable.

Problem Situations

Problem Situation 1
for use with Chapter 1

Your dog is lost and you want to run an ad in the classified section of your local newspaper. You have $25 to spend on the ad and a reward. You want to find out how much you can offer for the reward.

Directions Have students work in groups of 3 or 4 using the group guidelines given on page T22 to work together toward a solution for the problem situation above. Ask them to follow these *Guidelines for Planning* as they plan their solution to the problem.

> **Guidelines for Planning**
>
> **Formulate Problems** you will need to solve.
>
> Discuss **Assumptions** you will make and **Data** you will need.

Possible Student Plans

- Some problems students might formulate are:

  How much will the ad cost?

  How much will I be able to offer for a reward?

- Some assumptions students might make are:

  I will run the ad for 3 days.

  The ad will be 3 lines long.

  The reward will be more than $5.

- Some data students might decide to collect are:

  What is the cost for the ad per line per day?
  What is the cost for 3 days?

Sample Solution

Data
The ad will cost $1.66 per line per day.

Solution
Cost of running the ad:

$1.66 × 3 × 3 = $14.94.

Amount for the reward:

$25 − 15 = $10

You can offer about $10 for the reward.

Problem Situation 2
for use with Chapter 2

Your class is to make punch for a school party. The following punch recipe makes about 3 quarts: 1 46-oz can of fruit punch, 1 package pre-sweetened cherry soft drink mix, 1 qt ginger ale, 2 cups of water. You need to find the cost to serve about 120 people.

Directions Have students work in groups of 3 or 4 using the group guidelines given on page T22 to work together toward a solution for the problem situation above. Ask them to follow these *Guidelines for Planning* as they plan their solution to the problem.

> **Guidelines for Planning**
>
> **Formulate Problems** you will need to solve.
>
> Discuss **Assumptions** you will make and **Data** you will need.

Possible Student Plans

- Some problems students might formulate are:

  How much punch do we need for 120 people?

  How many times do we need to multiply the recipe?

  How much of each ingredient do we need?

  How much will all of the ingredients cost?

- Some assumptions students might make are:

  We will allow 8 oz of punch for each person.

- Some data students might decide to collect are:

  The cost of each ingredient.

Sample Solution

Data

Cost of the fruit punch: \$1.09
Cost of the soft drink mix: \$0.89
Cost of the ginger ale: \$0.99

Solution

Amount of punch needed:

$120 \times 8 = 960 \text{ oz}, \frac{960}{32} = 30 \text{ qt}$

Punch recipe makes about 3 qt.

$\frac{30}{3} = 10$ times the original recipe.

Cost for one recipe:

$\$1.09 + \$0.89 + \$0.99 = \$2.87$

Cost for ten recipes:

$\$2.87 \times 10 = \$28.70$

It will cost \$28.70 to make the punch for 120 people.

Problem Situation 3

for use with Chapter 3

For your birthday you want to take three of your friends bowling and then out for pizza afterwards. You need to know how much it will cost.

Directions Have students work in groups of 3 or 4 using the group guidelines given on page T22 to work together toward a solution for the problem situation above. Ask them to follow these *Guidelines for Planning* as they plan their solution to the problem.

Guidelines for Planning

Formulate Problems you will need to solve.

Discuss **Assumptions** you will make and **Data** you will need.

Possible Student Plans

- Some problems students might formulate are:

  How much will the bowling cost?

  How much will refreshments cost?

  How much will the birthday treat cost?

- Some assumptions students might make are:

  We will bowl 2 games.

  We will need 2 drinks per person at the bowling lane.

  We will order a large pizza and a pitcher of soft drinks at the restaurant.

  Everybody needs to rent shoes.

- Some data students might decide to collect are:

  How much will it cost for shoe rental?

  How much does it cost to bowl each game?

  How much will soft drinks cost at the bowling lanes? at the restaurant?

Sample Solution

Data

Shoe rental: \$1.00 a pair
Cost to bowl 1 game: \$1.60
Drinks at bowling lane: \$0.50
Large pizza and pitcher of soft drinks: \$11.99

Solution

Shoe rental for 4 people: \$4.00
Four people bowling 2 games each:

$4 \times 2 \times \$1.60 = \$12.80$

Large pizza and pitcher of soft drinks: \$11.99
Two soft drinks for each person:

$2 \times 4 \times \$0.50 = \$4.00$

Total cost for the bowling party:

$\$4.00 + \$12.80 + \$11.99 + \$4.00 = \$32.79$

Problem Situation 4

for use with Chapter 4

Your younger sister is on a Little League soccer team. The coach asked you to take the team's picture and make a 5 × 7 print for each of the 15 members on the team. The coach is willing to pay you \$4 an hour for your time and whatever expenses are involved. You want to find out how much this will cost the coach.

Directions Have students work in groups of 3 or 4 using the group guidelines given on page

T22 to work together toward a solution for the problem situation above. Ask them to follow these *Guidelines for Planning* as they plan their solution to the problem.

> ### Guidelines for Planning
> **Formulate Problems** you will need to solve.
>
> Discuss **Assumptions** you will make and **Data** you will need.

Possible Student Plans

- Some problems students might formulate are:

  How much will it cost to provide the pictures?

  How much will it cost for my time?

  How much will it cost the coach?

- Some assumptions students might make are:

  I will need to buy one roll of 24-exposure film.

  It will involve 2 hours of my time.

- Some data students might decide to collect are:

  How much does one roll of 24-exposure film cost?

  How much does it cost to develop the roll of film?

  How much does it cost to make each 5×7 print?

Sample Solution

Data

One roll of film: \$2.97
Processing fee: \$5.99
5×7 print: \$1.59

Solution

Cost of film and processing:

$\$2.97 + \$5.99 = \$8.96$

Cost for 5×7 prints:

$\$1.59 \times 15 = \$23.85$

$\$4 \times 2 = \$8$ for your time

Total cost:

$\$8.96 + \$23.85 + \$8 = \$40.81$

It will cost the coach \$40.81 for pictures for his soccer team.

Problem Situation 5

for use with Chapter 5

You are a business person who travels from Chicago to New York City twice a week. As a frequent flyer you receive 2 free round-trip tickets to anywhere in the continental USA when you have flown 30,000 miles. You want to find out how many round trips to New York you need to make to fly 30,000 miles and how much you will have paid in air fare.

Directions Have students work in groups of 3 or 4 using the group guidelines given on page T22 to work together toward a solution for the problem situation above. Ask them to follow these *Guidelines for Planning* as they plan their solution to the problem.

> ### Guidelines for Planning
> **Formulate Problems** you will need to solve.
>
> Discuss **Assumptions** you will make and **Data** you will need.

Possible Student Plans

- Some problems students might formulate are:

  How far is a round trip from Chicago to New York?

  How many round trips does it take to fly 30,000 miles?

  How much will you have paid in air fare to receive the free tickets?

- Some assumptions students might make are:

  I am traveling coach fare.

  The air fare will not change.

- Some data students might decide to collect are:

  The distance from Chicago to New York City.

  The cost of 1 round trip ticket from Chicago to New York City.

Sample Solution

Data

Round-trip coach fare ticket: \$295
Distance from Chicago to NYC: 843 miles

Solution

Round-trip mileage: 843 × 2 = 1686 miles

30,000 miles ÷ 1686 = 17.79

I will need to make 18 round trips.

$295 × 18 trips = $5310

I will have spent $5310 to receive the free trip.

Problem Situation 6

for use with Chapter 6

At the school carnival your Math Club is going to sponsor a booth where people pay $0.50 to guess the number of jelly beans in a jar. The person who guesses closest to the actual number of jelly beans wins a calculator. The second place winner receives the jelly beans. You want to find out how many people must pay to guess the number of jelly beans so that the Math Club will break even on the booth and how many people are needed to make a $25 profit.

Directions Have students work in groups of 3 or 4 using the group guidelines given on page T22 to work together toward a solution for the problem situation above. Ask them to follow these *Guidelines for Planning* as they plan their solution to the problem.

> **Guidelines for Planning**
>
> **Formulate Problems** you will need to solve.
>
> Discuss **Assumptions** you will make and **Data** you will need.

Possible Student Plans

- Some problems students might formulate are:

  How much money will be needed to buy the calculator and the jelly beans?

  How many customers will be needed to pay for the calculator and jelly beans?

  How many customers will be needed to make a $25 profit?

- Some assumptions students might make are:

  We will buy a solar-powered calculator for the first prize.

  We will buy 3 lb of jelly beans.

  First prize must be worth more than second prize.

  If there is a tie, we will toss a coin to determine the winner.

- Some data students might decide to collect are:

  The cost for the calculator.

  The price per pound for the jelly beans.

Sample Solution

Data

Calculator costs $12.97
Jelly beans cost $2.49/lb

Solution

Cost for jelly beans: $2.49 × 3 = $7.47
Cost for jelly beans & calculator:

$12.97 + $7.47 = $20.44

Guesses to break even: $\frac{\$20.44}{\$0.50} = 40.88$

Need to sell 41 guesses to break even.

$\frac{\$25}{\$0.50} = 50$ guesses

41 + 50 = 91 guesses to make a $25 profit.
We need to have 41 customers to break even and 91 to make a $25 profit.

Problem Situation 7

for use with Chapter 7

You are in charge of purchasing hamburgers, hot dogs, and buns for your class picnic. Each class member has invited a guest from another class. You need to know the total cost.

Directions Have students work in groups of 3 or 4 using the group guidelines given on page T22 to work together toward a solution for the problem situation above. Ask them to follow these *Guidelines for Planning* as they plan their solution to the problem.

> **Guidelines for Planning**
>
> **Formulate Problems** you will need to solve.
>
> Discuss **Assumptions** you will make and **Data** you will need.

Possible Student Plans

- Some problems students might formulate are:

  About how many pounds of hamburger should we buy?

  About how many packages of hot dogs should we buy?

  About how much will the hamburger cost? the hot dogs? the buns?

  What is the total estimated cost for the hamburger, hot dogs, and buns?

- Some assumptions students might make are:

  We will make quarter-pound hamburgers.

  We will provide 1 hamburger and 1 hot dog for each person.

  We will need a bun for each hamburger and hot dog.

- Some data students might decide to collect are:

  The cost per pound of hamburger.

  The cost per package of hot dogs.

  The number of hot dogs per package.

  The cost of hamburger buns and hot dog buns.

  The number of people who will be attending.

Sample Solution

Data

Cost of hamburger: \$1.80 per lb
Cost of hot dogs: pkg. of 10 for \$1.69
Cost of buns: pkg. of 8 for \$1.29
Number attending: 50

Solution

Hamburger needed:

$$50 \times \frac{1}{4} \text{ lb} = 12\frac{1}{2} \text{ lbs}$$

Cost for $12\frac{1}{2}$ lb of hamburger:

$$\$1.80 \times 12\frac{1}{2} = \$22.50$$

Packages of hot dogs needed:

$$\frac{50}{10} = 5$$

Cost for hot dogs:

$$5 \times \$1.69 = \$8.45$$

$$\frac{50}{8} = 6.25$$

7 packages of hamburger buns and 7 packages of hot dog buns will be needed.
Cost for the buns:

$$14 \times \$1.29 = \$18.06$$

Amount needed to buy the hamburgers, hot dogs, and buns:

$$\$22.50 + \$8.45 + \$18.06 = \$49.01$$

Problem Situation 8

for use with Chapter 8

Your video club is selling tickets to a film festival to help pay for a new television set. You plan to rent videos to show from 9:00 a.m. until 6:00 p.m. with a 10-minute intermission between films. You want to provide free popcorn for those who attend, You want to find out how much you should charge for tickets.

Directions Have students work in groups of 3 or 4 using the group guidelines given on page T22 to work together toward a solution for the problem situation above. Ask them to follow these *Guidelines for Planning* as they plan their solution to the problem.

Guidelines for Planning

Formulate Problems you will need to solve.

Discuss **Assumptions** you will make and **Data** you will need.

Possible Student Plans

- Some problems students might formulate are:

  How many minutes will the film festival be operating?

  How many video tapes will we need?

  How much will it cost to rent the videos and equipment?

  How much will we spend for popcorn?

  How much should we charge for each ticket?

- Some assumptions students might make are:

  We can sell 100 tickets.

  The average video runs about 100 minutes.

  We will use two rooms to show the videos.

- Some data students might decide to collect are:

  The cost of the television set.

  The cost of renting each video.

  The cost of renting a video player.

  The cost of popcorn.

  The length of time needed to show each video tape.

Sample Solution

Data

Television set will cost \$350.
Each video can be rented for \$2.00.
Each video player can be rented for \$8.00.
The popcorn will cost \$4.00.

Solution

From 9:00 to 6:00 is 9 hours or 540 min.
100 min + 10 min intermission = 110 min

$\frac{540}{110}$ min = 4; remainder, 100 min

We can show 5 video tapes with a 10-minute intermission after the first 4.

$5 \times \$2.00 = \$10.00$ (5 video tapes)
$2 \times \$8.00 = \$16.00$ (2 video players)

Cost of video tapes and players:

$\$10.00 + \$16.00 = \$26.00$

Total expenses:

$\$26.00 + \$4.00 + \$350.00 = \$380$

$\frac{\$380}{100} = \$3.80$

We should charge at least \$3.80 for each ticket.

Problem Situation 9

for use with Chapter 9

Your class is having a car wash to raise money for a camping trip. Groups of 8 students each will be washing cars from 8:30 a.m. until 3:30 p.m. You have estimated that a group of 8 students can wash 3 cars every 20 minutes. You will need to buy car wash soap and pay a \$5 fee for using the water. You want to find out what profit you might expect.

Directions Have students work in groups of 3 or 4 using the group guidelines given on page T22 to work together toward a solution for the problem situation above. Ask them to follow these *Guidelines for Planning* as they plan their solution to the problem.

> ***Guidelines for Planning***
>
> **Formulate Problems** you will need to solve.
>
> Discuss **Assumptions** you will make and **Data** you will need.

Possible Student Plans

- Some problems students might formulate are:

  How long will we be washing cars?

  How many cars can we wash?

  How much will we need to pay for the soap?

  How much profit can we expect to make?

- Some assumptions students might make are:

  There will be only 1 group of 8 working at a time.

  We will charge \$2.50 per car.

  There will be no other expenses.

- Some data students might decide to collect are:

  How much does 1 container of soap cost?

  How much soap do we need to wash each car?

Sample Solution

Data

Cost for one container of car wash soap:

\$2.49 for 532 ml

Each car will require about 35 ml of soap.

Solution

From 8:30 to 3:30 is 7 hours (420 min)

$\frac{3 \text{ cars}}{20 \text{ min}} = \frac{x}{420 \text{ min}}$

63 cars can be washed.
Earnings from cars washed:

$63 \times \$2.50 = \$157.50$

$\frac{532}{35}$ ml $= 15.2$

About 15 cars can be washed with each container of soap.

$\frac{63}{15} = 4.2$

Need 5 containers of soap.
Cost of the soap: $\$2.49 \times 5 = \$12.45$
Water fee: $5.00
Expected profit:

$\$157.50 - (\$5 + \$12.45) = \$140.05$

If we wash 63 cars, we can expect to make about $140 on the car wash.

Problem Situation 10

for use with Chapter 10

You are selling greeting cards for $3.50 per box on 8% commission. In addition, you receive one box of cards free for each 50 you order from the company. You want to know how many boxes of greeting cards you will need to sell in order to buy a new 12-speed bicycle.

Directions Have students work in groups of 3 or 4 using the group guidelines given on page T22 to work together toward a solution for the problem situation above. Ask them to follow these *Guidelines for Planning* as they plan their solution to the problem.

Guidelines for Planning

Formulate Problems you will need to solve.

Discuss **Assumptions** you will make and **Data** you will need.

Possible Student Plans

- Some problems students might formulate are:

  How much will I make on each box of cards I sell?

  How many boxes will I need to sell?

  How many boxes will I need to order?

- Some assumptions students might make are:

  The greeting card company will pay for postage and handling.

- Some data students might decide to collect are:

  How much do I need for the new bicycle?

Sample Solution

Data

The cost of the bicycle is $157.

Solution

$\$3.50 \times 0.08 = 0.28$ profit for each box sold

$\frac{\$157}{0.28} = 560.71$

I would need to sell 561 boxes.
I could order 550 boxes and get 11 boxes free.
Note: Students should also discuss whether it is realistic to assume actual sales of 561 boxes of greeting cards.

Problem Situation 11

for use with Chapter 11

You want to buy a used car and need to know how many weeks you must work at a fast-food restaurant in order to have enough money to buy a car.

Directions Have students work in groups of 3 or 4 using the group guidelines given on page T22 to work together toward a solution for the problem situation above. Ask them to follow these *Guidelines for Planning* as they plan their solution to the problem.

Guidelines for Planning

Formulate Problems you will need to solve.

Discuss **Assumptions** you will make and **Data** you will need.

Possible Student Plans

- Some problems students might formulate are:

  How much can I make in one day at the restaurant? in one week?

  How much can I save in one week?

How many weeks would it take to have enough money to buy the car?

- Some assumptions students might make are:

  I will save $\frac{3}{5}$ of my earnings each week.

  I will earn \$4.75 an hour.

  I work 5 hours per day, 5 days per week.

- Some data students might decide to collect are:

  How much does the car cost, including tax and license fees?

Sample Solution

Data

Cost of the car: \$1000

Solution

Earnings per day: $5 \times \$4.75 = \$23.75$
Earnings per week: $\$23.75 \times 5 = \$118.75$
Savings per week: $\$118.75 \times \frac{3}{5} = \$71.25$
Number of weeks needed to save enough to buy the car: $\frac{\$1000}{71.25} = 14.04$, or 14

Problem Situation 12

for use with Chapter 12

You are refinishing an old trunk that is 15 inches high, 16 inches wide, and 30 inches long. To line the trunk you have decided to glue fabric to the inside surfaces. You want to find the cost of the project.

Directions Have students work in groups of 3 or 4 using the group guidelines given on page T22 to work together toward a solution for the problem situation above. Ask them to follow these *Guidelines for Planning* as they plan their solution to the problem.

> **Guidelines for Planning**
>
> **Formulate Problems** you will need to solve.
>
> Discuss **Assumptions** you will make and **Data** you will need.

Possible Student Plans

- Some problems students might formulate are:

  How much area do I need to cover?

  How much fabric do I need to buy?

- Some assumptions students might make are:

  I'll need one 8-oz bottle of glue.

  The fabric does not need to be matched.

  I can buy a fraction of a yard.

- Some data students might decide to collect are:

  The cost of the glue.

  The cost per yard of fabric.

Sample Solution

Data

Fabric that is 45″ wide costs \$3.99 a yard. The cost for one 8-oz container of glue is \$1.59.

Solution

Front of the trunk:	30″ × 15″ = 450 sq in.
Back of the trunk:	30″ × 15″ = 450 sq in.
Bottom of the trunk:	30″ × 16″ = 480 sq in.
Top of the trunk:	30″ × 16″ = 480 sq in.
End of the trunk:	15″ × 16″ = 240 sq in.
End of the trunk:	15″ × 16″ = 240 sq in.
	2340 sq in.

2340 sq in. needed to be covered.
$45'' \times 36'' = 1620$ sq in. of fabric in 1 yard that is 45″ wide.

$$\frac{2340}{1620} = 1\frac{4}{9}$$

I will need to buy about $1\frac{1}{2}$ yards of fabric.

$$1\frac{1}{2} \times 3.99 = \$5.99$$

The fabric will cost \$5.99.
$\$5.99 + \$1.59 = \$7.58$ will be the total cost of the project.

Problem Situation 13

for use with Chapter 13

For your birthday you received an aquarium that is 60 cm by 40 cm by 30 cm with all the accessories (rock, light, filter, pump), but you need to buy fish and water conditioner. You need to find out how many fish you can put in your aquarium, how much water conditioner needs to be put in, and the cost to do this.

Directions Have students work in groups of 3 or 4 using the group guidelines given on page T22 to work together toward a solution for the problem situation above. Ask them to follow these *Guidelines for Planning* as they plan their solution to the problem.

Guidelines for Planning

Formulate Problems you will need to solve.

Discuss **Assumptions** you will make and **Data** you will need.

Possible Student Plans

- Some problems students might formulate are:

  How much water does the aquarium hold?

  How much conditioner do I need to put in?

  How many fish can I safely put in the aquarium?

  How much will the fish cost?

  What will be the total cost to get the aquarium going?

- Some assumptions students might make are:

  I want to stock the aquarium to its maximum safe capacity.

  I want a variety of fish.

- Some data students might decide to collect are:

  The cost of each fish.

  The amount of water needed for each fish.

  The cost of water conditioner.

  The amount of conditioner needed per liter of water.

Sample Solution

Data

The fish cost $1.09 each.
Each fish needs 3 L of water.
The water conditioner costs $2.99 for 100 mL.
5 mL is needed for 36 L of water.

Solution

$60 \text{ cm} \times 40 \text{ cm} \times 30 \text{ cm} = 72{,}000 \text{ cm}^3$

$72{,}000 \text{ cm}^3 = 72 \text{ L}$

$\frac{5 \text{ ml}}{36 \text{ L}} = \frac{n \text{ ml}}{72 \text{ L}}$, so 10 ml of water conditioner is needed.

$\frac{1 \text{ fish}}{3 \text{ L}} = \frac{n \text{ ml}}{72 \text{ L}}$, so 24 fish is maximum allowed.

Cost of fish: $24 \times \$1.09 = \$26.16$
Cost of fish and conditioner:

$\$26.16 + \$2.99 = \$29.15$

I need to put in 10 ml of water conditioner. The fish and water conditioner will cost me about $29.15.

Problem Situation 14

for use with Chapter 14

The faucet in the restroom at school has been dripping for 3 months. You want to find out how much money has been wasted during that time because of the dripping faucet.

Directions Have students work in groups of 3 or 4 using the group guidelines given on page T22 to work together toward a solution for the problem situation above. Ask them to follow these *Guidelines for Planning* as they plan their solution to the problem.

Guidelines for Planning

Formulate Problems you will need to solve.

Discuss **Assumptions** you will make and **Data** you will need.

Possible Student Plans

- Some problems students might formulate are:

  How much water is wasted in one hour?

  How many hours are there in 3 months?

  How much does one gallon of water cost?

- Some assumptions students might make are:

  The water drip has been constant during 3 months.

  There are 30 days in each month.

- Some data students might decide to collect are:

  The amount of water that drips in one hour.

The cost of water from public utilities.

The number of quarts in 1 gallon.

The number of gallons in a cubic yard.

The number of cubic feet in a cubic yard.

Sample Solution

Data

In 1 hour 1 quart of water is wasted.
For the first 2300 cubic feet the cost is

$\frac{\$1.12}{100}$ cubic feet.

There are 7.5 gallons in one cubic yard.
There are 27 cubic feet in a cubic yard.

Solution

1 hour—1 qt wasted
1 day (24 hours)—24 qt wasted
90 days (3 mo)—24 × 90 = 2160 qt wasted

$$\frac{2160}{4} = 540 \text{ gallons}$$

$$\frac{540}{7.5} = 72 \text{ cubic yards of water}$$

$$72 \times 27 = 1944 \text{ cubic feet}$$

$$\frac{1944}{100} = 19.44$$

$$19.44 \times \$1.12 = \$21.77$$

$21.77 has been wasted due to the dripping faucet.

Problem Situation 15

for use with Chapter 15

Your older brother has agreed to drive you and 2 of your friends to summer camp. The camp is 175 miles from your home. Your parents have agreed to pay your brother $0.15 per mile round trip. You want to find out how much he will have left after he pays for the gasoline and any oil that the car needs during the trip.

Directions Have students work in groups of 3 or 4 using the group guidelines given on page T22 to work together toward a solution for the problem situation above. Ask them to follow these *Guidelines for Planning* as they plan their solution to the problem.

> ***Guidelines for Planning***
>
> **Formulate Problems** you will need to solve.
>
> Discuss **Assumptions** you will make and **Data** you will need.

Possible Student Plans

- Some problems students might formulate are:

  How far will he drive?

  How much will he pay for gas?

  How much will he have left after he pays for the gas?

- Some assumptions students might make are:

  His car gets 24 miles per gallon on highway driving.

  Most of the trip will be on the highway.

  He will need to add 1 quart of oil before he starts his return trip.

  He will start the trip with a full tank of gasoline and end the trip with a full tank.

- Some data students might decide to collect are:

  The cost of 1 gallon of gasoline.

  The cost of 1 quart of oil.

Sample Solution

Data

Cost of 1 gallon of gas: $1.03
Cost of 1 quart of oil: $1.49

Solution

Round trip mileage: 175 × 2 = 350 miles
Pay for trip: 350 × $0.15 = $52.50
Gasoline needed: $\frac{350}{24}$ = 14.58 gal
Cost of gas: $1.03 × 15 gal = $15.45
Cost of gas and oil: $15.45 + $1.49 = $16.94
Amount left after expenses:

$52.50 − $16.94 = $35.56

Teaching Suggestions

Chapter 1

1-1 *pages 2–5* Introduce evaluating a numerical expression by giving several examples involving parentheses. (Note that order of operations without parentheses is introduced in Chapter 2, with all four operations.) Tell students to think of a numerical expression as a name for a number. Discuss Example 1 and its Practice. In Example 2, emphasize that the variable reserves a place for a number, and we replace the variable with a number to evaluate. The expressions in Example 3 involve two variables. A variable must be replaced by the same number each time it occurs in the expression. Make sure students correctly replace the variables and use parentheses in the Practice. **Ex. 57** Write the definition of subtraction "$c - b$ is the number a such that $a + b = c$" on the chalkboard, and discuss in terms of addends and sums. **Ex. 59** Help students find x and y by guessing and checking their guesses. Do not teach formal procedures for solving such equations at this time. **Calculator Activity** Ask students for another way to use memory keys to calculate 936 − (342 − 178). Example: enter 936 in memory, find 342 − 178, and use the M- key.

1-2 *pages 6–8* Discuss the first paragraph. Ask for words or phrases suggesting addition or subtraction. Give several phrases using these words, such as "nine increased by five" or "the difference between 24 and 8" for students to translate to numerical expressions. Complete Example 1 and its Practice. Then focus on the table. To translate a phrase such as "a number decreased by 4," students may find it useful to replace "a number" with a specific number and write a numerical expression, then replace the number with a variable to arrive at the algebraic expression. Complete Examples 2 and 3 and their Practices. The sequence in Example 3 helps students think clearly about variables. **Ex. 41** Suggest that students guess and check. **Ex. 42** This is the Fibonacci sequence. **Mental Math** Start by showing how to think of 52 − 29 as 52 − 30 + 1.

1-3 *pages 9–12* Introduce the Commutative Property by asking in what situations order of operations does or does not make a difference. ("Put on shoes, then put on socks" or "put on socks, then put on shoes"; "Put 5 marbles in a bag, then put in 3" or "put in 3 and then put in 5.") Illustrate with whole number examples. When discussing Example 1, write $x + 8$ and $8 + x$ on the chalkboard and ask, "If you evaluate these expressions for $x = 24$, will they have the same value? Why?" Illustrate the Associative Property with whole number examples. For Example 2, evaluate $(n + 6) + 12$ and $n + (6 + 12)$ for several replacements for n, observing that the expressions remain equal. Point out that since $(\boldsymbol{a} + \boldsymbol{b}) + c = a + (\boldsymbol{b} + \boldsymbol{c}) = a + (c + b) = (\boldsymbol{a} + \boldsymbol{c}) + b$, we can choose to add any 2 of 3 addends first. Discuss Example 3; illustrate with whole numbers. **Ex. 46** Starting at 3 on the clock and moving 2 spaces around the clock takes you to 1. Starting at 2 and moving 3 spaces also takes you to 1. Looking at all possible cases convinces us the operation is commutative. **Numbers to Algebra** Point out that a mathematical generalization is a statement that is true for all of a designated type of number. The generalization formed is used when finding 24 + 63 mentally. (Find 20 + 60 and add to 4 + 3.)

1-4 *pages 13–14* Discuss the meaning of "equation." Write $n + 9 = 17$ on the chalkboard. Emphasize that we *solve* (or find the *solution set* of) an equation by finding the set of numbers that replace the variable to make the equation true. Cover the variable with your hand and ask, "What number adds to 9 to give 17?" Do this with several equations. Discuss Examples 1 and 2 by writing the equation on the chalkboard and having students replace the variable with each number in the replacement set. **Ex. 15–18** Students can replace the variable, each time it occurs, with the numbers 0, 1, 2, 3, and 4, then use what they find to draw a conclusion about the replacement set of all whole numbers. **Ex. 19** Focus is on a replacement set other than all whole numbers.

1-5 *pages 15–17* Note that in Example 1 the replacement set is the set of all whole numbers. Tell students that from this point on, unless indicated otherwise, the replacement set for equations will be the set of whole numbers. Write the equation on the chalkboard and cover the variable with your hand. Ask students to use mental math to find the number that replaces the variable to make the equation true. Emphasize the role of mental math in solving simple equations. Discuss Example 2, emphasizing the fact that addition and subtraction are inverse operations. **Ex. 45** Do not instruct students on how to solve the equations formally; suggest that they guess and check. **Ex. 46** Students should look at the sequence of solutions to see the pattern.

1-6 *pages 18–20* Prepare students for solving equations by emphasizing the important ideas introduced in the text. To emphasize that *addition and subtraction are inverse operations*, give a mental math exercise that includes "24 + 6 − 6" and "31 − 9 + 9." Explain that to undo addition we subtract, and to undo subtraction we add. To emphasize the *properties of equality*, write "12 − 7" on one card and "3 + 2" on another. Get the students to agree that the symbols on these two cards represent the same number. Then have a student add 4 to the expression on each card. Ask, "Do the symbols on the cards represent the same number now?" Point out that a systematic procedure is needed for solving equations that can't easily be solved mentally. Discuss the steps for solving equations, emphasizing the importance of getting the variable by itself on one side of the equation. As you discuss Examples 1 and 2 and their Practices, continue to emphasize these steps.

Ex. 31–34 Point out that these equations must be simplified by finding a sum or difference before the steps for solving equations are followed. **Ex. 37** Discuss the properties of this simple magic square. **Computer Activity** Emphasize the power of the computer to solve any equation of a given type.

1-7 *pages 22–23* Discuss the Problem-Solving Checklist. Emphasize that these are not rules for solving problems, but guidelines, to remind us to consider the phases of problem solving. Read the sample problem. Discuss the terms *base list price, dealer's cost*, and *dealer's profit*, and relate them to the data in the table. Work through the sample problem, following each phase of the checklist. Focus on the words *Question, Data, Plan, Answer*, and *Check*. Stress the need to reread the original problem carefully to see whether an answer makes sense. Emphasize the role of *estimation* in the *Check* phase. Remind students to write the answer in a complete sentence. The **Data Search** requires students to collect data from outside sources. Encourage them to use the telephone, auto magazines, or newspaper advertisements to find the information. In **What's Your Decision**, students choose options from the table on p. 22. There is no "right answer." Discuss student choices and decisions.

1-8 *pages 24–26* Direct attention to the situation in the photo caption. Ask students how they would estimate the number of miles the camper had been driven. Briefly review place value by writing a number and having students give the value represented by each digit in the number. Use Example 1 to discuss the steps for estimating using rounding. Review rounding whole numbers. Discuss the procedure when the digit to the right of the place to which you want to round is 5. Front-end estimation, shown in Example 2, may be new to many students. Be sure they see that this procedure must be revised when the numbers have a different number of digits. To estimate a sum such as 4,587 + 926, for example, they could add 45 and 9 and then adjust. As students complete the Practices, discuss techniques for adjusting the estimate.

Ex. 26–27 Encourage students to guess numbers and revise their guesses as necessary. **Ex. 30** Give examples of consecutive whole numbers.

1-9 *pages 27–28* Review the Problem-Solving Checklist on p. 22, stressing the importance of the *Plan* phase. Discuss the first sample Problem, showing use of the strategy *Choose the Operations*. Then use the second sample Problem to introduce the *Guess, Check, Revise* strategy. Point out why the guess was revised up after the first guess and down after the second. Possible strategies are suggested here for each problem. **Ex. 1** Students need to find two numbers that differ by 6, and whose sum is 30. Encourage starting with any guess. Note that a student might also *Choose the Operations*, thinking, "I'll take half of 30, or 15, and go 3 below it and 3 above it to make a difference of 6. Twelve and 18 are the numbers." **Ex. 2** *Choose the Operations*. Students can subtract to find the number sold in August, subtract again to find the number sold in July, then add to find the total. **Ex. 3** *Choose the Operations*. If Eric is 9, in 4 years he will be 13. His father will be 13 more than twice 13, or 39. **Ex. 4** Students must focus on the conditions of the problem carefully. The scores are 2 multiples of 9 that differ by only 9. A look at the multiples of 9 will show that 45 and 54 are the numbers. *Choose the Operations* is used. **Ex. 5** *Guess, Check, Revise*. **Ex. 6** *Guess, Check, Revise*.

Enrichment *page 29* You may wish to encourage more able students to read the page and try the exercises independently. If you wish to introduce the ideas to the class, you might draw a circle on the chalkboard and label it "All major league baseball players." Give students names of famous people and ask whether each name should be written inside or outside the circle. Then draw a pair of intersecting circles, as shown in the text. Label Set A "Sudents in our class wearing glasses." Label Set B "Students in our class with brown eyes." Have the class decide which students' names should be written in each circle. Then draw one circle contained in another, as shown in the text. Label Set C "Students in our class" and Set D "Students in our class with black hair." Give the name of a class member and have the class decide where the name should be written. Have the students use the words *and, or, not*, and *if . . . then* to describe the parts of the Venn Diagram shown.

Chapter 2

2-1 *pages 34–37* Write the expressions $(6 + 8) \div 2$, $6 + [8 \div 2]$, and $6 + 8 \div 2$ on the chalkboard, and ask students to simplify them. Discuss the role of the grouping symbols. Direct attention to the third expression, pointing out that if we add first, the answer is 7. If we divide first, the answer is 10. Emphasize the need for an agreement on which operations to do first when no grouping symbols are given. Discuss the Examples, stressing the role of the fraction bar as a grouping symbol and pointing out that braces, {}, are also used as grouping symbols. You might give students the mnemonic "My Dear Aunt Sally" for remembering the rule for order of operations, or encourage them to make up their own mnemonics. **Ex. 40–41** Suggest use of the calculator's memory. **Ex. 42** Encourage students to write an expression equal to zero, such as $(8 \times 4) \times [12 - (6 \times 2)]$.
Calculator Activity Students using calculators not programmed for the rules on order of operations will need to follow the rules when entering the problems, and use the calculator's memory.

2-2 *pages 38–40* Ask students to think of words or phrases that suggest multiplication or division. (Multiplication: multiply, times, the product of; division: divide, quotient of, shared among.) Give several phrases using these words, such as "the product of 8 and 9" or "the quotient of 24 and 8," and ask students to translate them to verbal expressions. After completing Example 1 and its Practice, focus on the table. When translating a phrase such as "9 times a number," students may find it useful to replace "a number" with a specific number and write a numerical expression, then replace the number with a variable to arrive at the algebraic expression. Complete Examples 2 and 3 and their Practices. Note that Example 3 involves two operations. **Ex. 23–30** Have students make up other real-world situations. **Ex. 31–34** Students should be aware that subtracting or adding 1 to an odd number produces an even number, and doing the same to an even number produces an odd number. **Ex. 35** Students might use a *Guess, Check, Revise* strategy.

2-3 *pages 41–45* Briefly review the properties of addition. To illustrate the Commutative and Associative Properties of Multiplication, use the numerical examples in the text. Discuss Examples 1 and 2, emphasizing that when we use the properties to rewrite an expression, it has the same value as the original expression no matter what number is written in place of the variable. To help students feel comfortable with this idea, evaluate several of the expressions, using different numbers for the variable. The word "simplify" will be introduced in the next section. Explain the difference between "a number divided by zero" and "zero divided by a number" and the fact that we never divide by zero. The Distributive Property is emphasized again in Section 2-4 and in Chapter 8. You may want to discuss the following different forms of the property statement: $a(b + c) = ab + ac$, $(b + c)a = ba + ca$, $ab + ac = a(b + c)$, $ba + ca = (b + c)a$. **Ex. 31–32** Note that the Commutative and Associative Properties are not true for division of whole numbers. The Distributive Property of Multiplication over Subtraction is true in all cases in which subtraction of whole numbers is possible. Note that the property will be true for all *integers a, b,* and *c*. **Numbers to Algebra** To help students see this, you can write the general statement $ab + ac = a(b + c)$ on the chalkboard. Then erase "a" everywhere it occurs and write in "4." Do the same as you write "x" for "b" and "y" for "c."

2-4 *pages 46–48* Write $4(2n)$ on the chalkboard and ask students to write as many other equivalent expressions as they can. Do this for $(n + 5) + 3$ and $3n + 4n$. Check by replacing the variable with a number to see whether the values are the same. Discuss the terminology in the text, and complete the Examples. You may need to give further examples to help students understand *like terms*. Point out that the following are *not* like terms: $3a$ and $4ab$; $5x$ and $5y$; $2n$ and $2nn$. Discuss the tables, which show that two equivalent expressions always have the same value when the variable is replaced. Replace the variable with a number to demonstrate this, but note that the basic properties ensure that the expressions will be equivalent. **Ex. 30–33** Note that the extended Distributive Property works for any number of like terms. **Ex. 39** Students might write n; $n + 7$; $n + 7 - 2$, or $n + 5$; $2(n + 5)$, or $2n + 10$; $2n + 10 - 10$, or $2n$; $2n \div 2$ or n. They will end with the starting number.

2-5 *pages 49–51* Conduct an oral exercise in which the students compute mentally. Give two factors, such as 12 and 4, and ask for the product; or give the product and one factor, e.g., 45 and 3, and ask for the other factor. Direct attention to the photo and caption, and work through Example 1 to solve the equation. Discuss the relationship between multiplication and division. Work through Example 2 and its Practice. **Ex. 25–33** Students should use a calculator and the idea of inverse operations to solve these. **Ex. 35** Students can use a *Guess, Check, Revise* strategy.

2-6 *pages 52–54* Discuss the importance of getting the variable by itself on one side of an equation. To help emphasize the idea that *multiplication and division are inverse operations*, give students mental math exercises such as "$24 \times 6 \div 6$" and "$32 \div 4 \times 4$." Help them see that to undo multiplication we divide, and vice versa. To emphasize the Properties of Equality, write "$12 \div 4$" on one card and "3×1" on another card. Get the students to agree that the symbols on the cards represent the same number. Then have a student multiply the expressions on each card by 4. Ask, "Do the symbols on the cards represent the same numbers now?" Point out that we need a systematic procedure for solving equations that can't easily be solved mentally. As you discuss Examples 1 and 2 and their Practices, emphasize the steps for solving equations. **Ex. 39–40** These reinforce the meaning of and the relationships between factors and products, and also help students translate verbal statements to algebraic expressions. **Ex. 41** Students can start at the ending number and reverse the flow chart, using inverse operations. **Ex. 42** At this time, do not teach the formal procedures for solving equations such as $3x + 9 = 54$. Encourage the use of inverse operations or a *Guess, Check, Revise* strategy.

2-7 *page 56* Review the **Problem-Solving Checklist** given on p. 22. Emphasize again that these are not "rules," but guidelines. Discuss the data on calories on the page, explaining that a calorie is a measure of the energy-producing value of food. When you eat food, you "take in" calories. When you exercise, you "use up" calories. (1 Calorie = the amount of heat required to raise the temperature of 1 kg of water 1 degree Celsius). Discuss the methods students might use for finding answers to problems: pencil and paper, mental math, estimation, and use of a calculator. Emphasize the fact that they should make choices according to the type of problem they are solving. Work through the first problem, following the checklist phases *Question, Data, Plan, Answer,* and *Check.* Stress the need to reread the original problem carefully to see whether an answer makes sense. Emphasize the role of *estimation* in the *Check* phase. Note that **Data Search** requires students to weigh themselves. In **What's Your Decision**, students must use data from the tables. Note that there is no "right answer." Discuss student choices and decisions.

2-8 *pages 57–58* Discuss the first paragraph. If possible, have students measure their own heights and evaluate the formula to find predictions for their weights. Then have them weigh themselves, and compare the data. Discuss similarities and differences between evaluating an expression and evaluating a formula. A formula expresses a relationship between at least two variables, which usually represent *real-world quantities*. Discuss Example 2. You might wish to arrange in advance for students to do dips and pullups and find their own arm strength ratings. **Ex. 1–6** These provide practice in finding the value for a variable when the value of a related variable is given. Note that the value of one variable is given and the value of a related variable is to be found. **Ex. 8–15** These involve formulas in real-world situations. Extend by having students make up formulas for common situations such as Profit = Income − Expenses, Age in months = age in years × 12, etc. **Ex. 17** Rather than giving a mechanical procedure, allow students to reason logically about the situation and come up with their own formulas.

2-9 *pages 59–60* For Example 1, point out that the dot will be used to show multiplication. Give special emphasis to estimating using **compatible numbers**. You may want to precede the discussion with some oral exercises in which students replace a quotient such as $37 \div 9$ with the "closest" basic fact. A common procedure is to round the numbers first and then, if the quotient can't be found mentally, replace the dividend with a number compatible with the divisor so the estimate can be made using a basic fact. **Ex. 34** Note that the last 3-digit number subtracted cannot be greater than 900 − 640, or 350, so it must be 192.

2-10 *pages 61–62* You may want to review the Problem-Solving Checklist on p. 22, emphasizing that it gives guidelines, not rules. Present the following problem: "Nell wanted to make a rectangular pen 45 m long and 25 m wide for her pony. To make the pen, she wanted to place posts 5 m apart. How many posts will she need?" Work with students to find a solution, focusing on the value of the strategy **Draw a Picture**. Then discuss the sample Problem. Note that data is given for the lengths of parts of the fish. Start with a segment representing the body, then complete the segments for the head and the tail based on the body length. Once the picture is complete, the problem is simple.

Possible strategies are given here for each problem. **Ex. 1** *Draw a Picture*. **Ex. 2** *Draw a Picture*. **Ex. 3** *Guess, Check, Revise*. By guessing that Al answered 13 three-point questions and 9 five-point questions, you find that his score is $13(3) + 9(5) = 39 + 45 = 84$. **Ex. 4** *Draw a Picture*. **Ex. 5** *Choose the Operations*. She missed 16 problems. She completed 2(16), or 32, correctly. There were 32 + 16, or 48, problems on the test. Her score was 32(3) − 16, or 80. **Ex. 6** *Guess, Check, Revise*. **Ex. 7** *Choose the Operations*. **Ex. 8** *Draw a Picture*. **Ex. 9** *Draw a Picture* and *Guess, Check, Revise*.

Enrichment *Page 63* You may want to have students simulate the operation of a function machine: one presents an input number, another applies the rule, and another prints the results on an "input, output" card. As you discuss the ideas, emphasize the notion of a function as a *pairing* of each first element with exactly one second element. Give a real-world example such as {(Al, 2), (Bob, 3), (Carl, 4), (Denny, 5)} to emphasize this idea. Ask students to suggest other real-world examples. Give particular attention to the interpretation of $f(3)$, when $f(n) = 5n - 2$. This notation can be thought of as "replace the variable n in the function rule with the number 3 and evaluate."

Chapter 3

3-1 *pages 68–71* Discuss the photo and caption, and ask students to suggest other situations that are the opposites of each other. Point out that to describe such situations, we use numbers that are opposite the whole numbers on the number line. Direct attention to the integers on the number line in the text. Discuss the opposite of an integer. To introduce the concept, a negative integer is represented in the text with a raised minus sign, e.g. $^{-}4$. A regular minus sign is used for *the opposite of an integer*: $-(^{-}4)$ represents "the opposite of negative 4." You may want to ask such questions as, "What integer is the opposite of the opposite of −6?" In Example 1, focus on the distance an integer is from zero. Instruct students to count spaces (unit segments) rather than points to determine these distances. Use the idea of distance developed in Example 1 to introduce absolute value in Example 2. Review the symbols $<$ and $>$, and use the number line for comparing and ordering. **Basic Property Update** The properties given here for integers are identical to the properties for whole numbers. They give students some idea about how the operations +, −, ×, and ÷ work with integers, and should be reviewed carefully.

3-2 *pages 72–74* Draw a number line on the chalkboard. Point out that addition of a positive integer is shown by an arrow to the right, and addition of a negative integer is shown by an arrow to the left. Have students use the number line to find sums such as $8 + {}^{-}5$, ${}^{-}5 + 2$, and $-4 + (-3)$. Direct attention to the addition of ${}^{-}4 + 4$ shown on the number line in the text. Use the number line with other examples to introduce the Inverse Property of Addition. The rules for adding integers should follow naturally from work with the number line. You might have students close their books and encourage them to state their own rules for adding integers. Discuss and refine their rules and compare them with those given. Note that the basic properties can be used to prove that the rules are correct. For example, from the Inverse Property of Addition, we know that −6 adds to 6 to give 0. Since $({}^{-}2 + {}^{-}4) + 6 = ({}^{-}2 + {}^{-}4) + (4 + 2) = ({}^{-}2 + 2) + ({}^{-}4 + 4) = 0$, we see that ${}^{-}2 + {}^{-}4$ adds to 6 to give 0, and so ${}^{-}2 + {}^{-}4 = {}^{-}6$. We use this fact to prove that ${}^{-}6 + 4 = {}^{-}2$.

$${}^{-}6 + 4 = ({}^{-}2 + {}^{-}4) + 4$$
(Substitute ${}^{-}2 + {}^{-}4$ for ${}^{-}6$)
$= {}^{-}2 + ({}^{-}4 + 4)$ (Associative Property)
$= {}^{-}2 + 0$ (Additive Inverse Property)
$= {}^{-}2$ (Zero Property)

Ex. 27–28 These problems can be solved without the use of integers, but you can now encourage students to use integers to represent elements in the problems. **Ex. 29** Students can find the sum mentally by looking for pairs of numbers that are opposites or near-opposites. Assume numbers such as 58 and −57 are opposites and compensate by adding on the extra one later. **Ex. 30** Discuss the proof, emphasizing the importance of the basic properties. **Numbers to Algebra** A common student error is to think an expression with a variable preceded by a minus sign always gives a negative number when evaluated. The statements here help students see that such expressions can be evaluated to produce either a positive or negative number, depending upon what integer replaces the variable.

3-3 *pages 75–77* Use the number line to help give meaning to the procedure of subtracting an integer by adding its opposite. Explain that on number lines A and B, an arrow to the left shows "adding $^{-}2$" and an arrow to the right, the opposite direction, shows "subtracting $^{-}2$." Students should see that this arrow to the right also shows "adding 2," which suggests the rule for subtracting integers. Illustrate with several examples, and use the number line to verify as needed. Point out that $^{-}5$ and -5 represent the same integer, and that from this point on the regular minus sign will be used. Complete the Examples and have students check the subtraction by adding. **Ex. 47–48** These illustrate that neither the Commutative nor the Associative Property holds true for subtraction of integers. **Ex. 51–52** A *work backward* strategy might be used. Extend **Ex. 51** by having students make a 3-by-3 magic square using integers. **Ex. 52** Encourage students to use integers to represent the trips. **Calculator Activity** Ask students to use the change-sign key to verify the rule for subtracting integers.

3-4 *pages 79–81* Discuss the situation and equation described in the caption. Explain that the steps for solving equations with integers are the same as those for whole numbers. Point out that the Additive Inverse Property can be used, if desired. For Example 3, show that using the Additive Inverse Property provides a more efficient way of solving this type of equation. Some teachers feel that students make fewer errors if they use the additive inverse for *all* integer equations. If you use this approach, note that an equation such as $x - 18 = -35$ will have to be rewritten as $x + (-18) = -35$, and then solved. Another approach is to suggest using inverse operations *or* the additive inverse, depending upon which is most convenient for a particular problem. **Ex. 37** Two-step equations will be used to solve problems in Chapter 8. At this stage, use a *Guess, Check, Revise* strategy. **Mental Math** This technique is based on the formula for finding the sum of the first n consecutive whole numbers, $S = n\frac{n+1}{2}$. You might tell students the story of how Karl Frederich Gauss, the famous mathematician, was as a young child asked by his teacher to find the sum of the first 100 whole numbers. The teacher thought it would keep him busy for quite some time, but Gauss found the sum in a few seconds by thinking: "1 + 2 + 3 + . . . + 98 + 99 + 100 is the sum I want to find. When I add the sum to itself

$$\begin{array}{rrrcrrr} 1\,+ & 2\,+ & 3\,+ & \ldots\,+ & 98\,+ & 99\,+ & 100 \\ +\ 100\,+ & 99\,+ & 98\,+ & \ldots\,+ & 3\,+ & 2\,+ & 1 \\ \hline 101\,+ & 101\,+ & 101\,+ & \ldots\,+ & 101\,+ & 101\,+ & 101 \end{array}$$

I get 100 101's, or 10,100. Since I added the sum twice, I'll divide by 2 to get the sum, which is 5050."

3-5 *pages 82–83* You might wish to review the Problem-Solving Checklist on page 22. Again, emphasize that these are not rules for solving problems, but guidelines. Discuss checking accounts and the terms *credit, debit, balance,* and *overdrawn.* Show students an actual checkbook and record sheet, if available. Work through the first problem with the class, explaining how a checkbook record is kept. Stress the need to reread the original problem carefully and estimate to see whether an answer makes sense. **Data Search** This requires students to invent data, then solve a problem. You may wish to have students work in small groups. A local bank might supply sample record forms for group use. **What's Your Decision** Note that there is no "right answer." Answers should vary according to the amount of money one expects to have. Discuss student choices and decisions. Extend by having students call or visit local banks and savings and loans for data on checking accounts and decide which type of account they would like to open.

3-6 *pages 84–86* Discuss multiplication as repeated addition. Ask students how they could use this idea to interpret $4(-5)$. Use several examples. Lead students to generalize that the product of a positive times a negative integer is a negative integer. Point out that the Commutative Property allows us to conclude also that $-5(4)$ is -20. To show that the product of two negative integers is a positive integer, write $4(-5) = -20$ on the chalkboard. Then, as you erase the 4 and write -4 in its place, ask, "Since -4 is the opposite of 4, how do you think the product will change when -4 is written in place of 4?" (The product should be the opposite of the original product.) You may also wish to show the following proof that the product for $-4(-5)$ is the opposite of the product for $4(-5)$, or 20.

$$-4(-5) + 4(-5) = (-4 + 4)(-5) \quad \text{(Dist. Prop.)}$$
$$= 0(-5) \quad \text{(Additive Inverse Property)}$$
$$= 0 \quad \text{(Zero Property)}$$

Since their sum is 0, $-4(-5)$ is the opposite of $4(-5)$. So $-4(-5)$ must be 20. Discuss the rule for multiplying integers and complete Examples 1 and 2. **Ex. 29–31** Help students generalize: when the number of negative factors is odd, the product is negative, and when the number is even, the product is positive. **Ex. 49** Encourage students to develop the habit of looking for patterns in numerical situations. This is a useful higher-level thinking skill. **Ex. 50** A *Guess, Check, Revise* strategy may be used. If necessary, give students a hint to try negative numbers.

3-7 *pages 87–88* Conduct an exercise in mental math, giving students 2 factors and having them give the product. Then give them a product and one factor and have them give the other factor. Emphasize the relationship between multiplication and division, and point out that they can think of a quotient such as $24 \div (-6)$ as asking for the factor that multiplies by -6 to give 24. Discuss the method for dividing integers, and complete Examples 1 and 2. Have the students check the division by multiplying. **Ex. 33** This illustrates that there is no Distributive Property for division over addition. Extend the problem by asking students to look for values of a, b, and c that make the following generalization true: $(a + b) \div c = a \div c + b \div c$. **Ex. 38** Encourage students to use the calculator's memory keys. **Ex. 40** Do not teach formal procedures for solving a pair of equations at this time. Encourage the use of a *Guess, Check, Revise* strategy.

3-8 *pages 89–91* Discuss the situation and equation in the photo caption. Use integer examples to help students become familiar with the property of -1. Note in Example 3 that each side of the equation can be multiplied or divided by -1; the result is the same. You may want to point out that we can "take the opposite" of both sides to solve an equation such as $-n = 35$. Some students may think, "If the opposite of n is 35, then n must be -35," to solve readily. **Computer Activity** These equations have more than one solution. Encourage students to make the best guesses they can. Have them check the computer solutions.

3-9 *pages 93–94* Discuss the first paragraph. Then write a statement such as "The difference between a number and 6 is 28" on the chalkboard, and work with students to translate it into an equation. Emphasize the importance of studying each word and phrase in the statement and thinking about what it means. Write a statement such as "The glove cost \$23, which was \$8 more than the cost of the bat" on the chalkboard for students to translate into an equation. Emphasize the phrases that suggest the different operations. As you complete the Examples and their Practices, ask students to explain the rationale for their translations. While it is often not necessary to write an equation in simple problems, the exercises here are useful practice in the *process* of translating to an equation, a process often used later for more difficult problems. **Ex. 22–25** You may wish to extend by suggesting the writing of real-world problems that would be translated into the equation. **Ex. 26** Note that the equation contains two variables. Suggest a *Guess, Check, Revise* strategy. Encourage students to find more than one pair of values. **Ex. 27** Formal procedures for solving an equation such as $x + (x + 1) + (x + 2) = 828$ will be developed in Chapter 8. At this stage, encourage a *Guess, Check, Revise* strategy.

3-10 *pages 95–96* Have students read the sample problem, then close their books as you use the data in the problem to build a table for days 1, 2, 3, and 4. Have students look for a pattern and extend the table. Note that it is not necessary to extend the table to day 14. Another pattern (that of the number of factors of two in the number of dollars in the safe) allows us, as shown in the text, to figure out the number of dollars there will be on day 14. After the solution has been found, review the solution process. Possible strategies for each problem are given here. **Ex. 1** *Make a Table, Look for a Pattern* If students have difficulty finding a pattern, suggest looking at the differences between successive terms. 1 is the first whole number, 3 is the sum of the first 2 whole numbers, 6 is the sum of the first 3 whole numbers, etc. Relate to the formula given in the mental math exercise on p. 81. **Ex. 2** *Make a Table, Look for a Pattern.* If hints are needed, suggest looking at the numbers in the column for total dollars given away, 1, 4, 9, 16, 25, etc. After students find a solution, suggest using the table to find a formula for the sum of the first n odd numbers (sum of first n odd numbers $= n \cdot n$). **Ex. 3** *Look for a Pattern* If a hint is needed, suggest adding pairs of terms in the sequence. Note this is the Fibonacci Sequence: the sum of two consecutive terms is the next term. **Ex. 4** If students have difficulty, suggest a *Guess, Check, Revise* strategy. **Ex. 5** *Choose the Operations.* **Ex. 6** *Draw a Picture, Look for a Pattern.* **Ex. 7** *Make a Table, Look for a Pattern.*

Enrichment *page 97* Make sure students understand that the rule for the nth term of a sequence is an algebraic expression which, when evaluated by replacing the variable with the number of a term, produces that term. If necessary, use the sequence 1, 3, 5, 7, 9, to help students better understand these ideas. Ask them to give a specified term. Then ask them to try to find the rule for the nth term ($2n - 1$). Have them use this rule to produce the 10th term, 50th term, etc. **Ex. 3** Students may write the rule as $n \cdot n$. Note that exponents will be developed in Chapter 5. **Ex. 4** If students have trouble with this, suggest making a table:

Number of the term.	1	2	3	4	5	6
Term	−3	−7	−11	−15	−19	−23

Mention that the term is about −4 times the number of the term, and encourage students to use this idea to find the rule, $1 - 4n$. **Ex. 7** Note that whether the number of the term chosen is odd or even, the product of adjacent terms is one more than the term squared.

Chapter 4

4-1 *pages 102–104* Begin by reviewing place value for whole numbers. Discuss the meaning of the number 312 as an abbreviation for 3(100) + 1(10) + 2(1). Then review the meaning of the fractions 1/10, 1/100, and 1/1000. Show how place value notation for decimal numbers is a natural extension of whole number notation. You may want to expand the number line at the bottom of page 101 by copying it on the chalkboard and dividing the part from 0.6 to 0.7 into ten parts to represent the decimals 0.61 to 0.69. Build upon the previous chapter on integers by using negative decimals. Ordering decimals as in Example 3 is an important skill. Make sure students are thinking clearly about comparisons of negative decimals. Visualizing positions on a number line may help. Note that negative decimals have the opposite order of their absolute values. Provide practice until students become proficient at ordering. **Ex. 19–24** Have students compare two numbers at a time, i.e., compare the first two and then compare the third to each of them.

4-2 *pages 105–107* The strand on estimation developed in Sections 1.8 and 2.9 is continued in this lesson. Students become proficient estimators through practice. You should bring estimation into classroom work or discussion daily. Begin by citing examples of rounding in everyday life. You might keep a list of examples of estimation that occur in your own life. The technique for estimation introduced in this lesson is the "5 or greater, round up," "less than 5, round down" rule that we usually teach. This is not the only rounding rule used. Some calculators and computers simply drop all digits beyond a certain number of digits; this is also a form of rounding. Use a number-line model for rounding as a supplement to the rule stated in the text. It may be particularly useful to teach the rounding of negative decimals with a number line. **Ex. 34–35** Ask students to describe the estimation methods that they used. Emphasize that there is more than one way to estimate a solution.

4-3 *pages 108–110* Discuss the photo and the situation discussed in the caption. The problem solved in Example 1 gives the time in seconds for the relay team. Evaluating expressions as in Example 3 can be a vehicle for further review of the addition and subtraction algorithms for decimals. Most problems in this lesson are written horizontally, as a transition to algebra. You will need to gauge how much to focus on writing problems vertically; this may depend on how much review your class needs. **Ex. 23** This provides an opportunity to emphasize that parentheses are needed whenever an operation is not associative. Review the Associative Property. **Numbers to Algebra** This illustrates that in algebra we follow the same principles and rules for simplifying that we use in arithmetic.

4-4 *pages 111–112* The equation-solving techniques used in this chapter are identical to those used in previous chapters. The instruction in the Examples has students begin to solve an equation such as $x + 4.3 = -1.5$ by subtracting 4.3 from both sides of the equation. The right side of the equation then becomes $-1.5 - 4.3$. An alternate technique is to use the Additive Inverse Property and add -4.3 to both sides of the equation. With this approach, the right side of the equation becomes $-1.5 + (^{-}4.3)$. Some students are less prone to error when following this technique, so you should consider teaching it if you feel they will respond well to it. The equations in the A exercises all have the unknown as the first term on the left side of the equation. In the B exercises, the position of the unknown is varied. **Mental Math** Students may find it helpful to think in terms of "dollars and cents" when estimating with decimals. You may want to spend a few minutes each day reviewing and practicing the various mental math techniques introduced in text.

4-5 *pages 113–115* Estimation is an important vehicle for developing "number sense" in students. Discuss the situation described at the top of p. 113. You may want to review the method for rounding decimals on p. 105. Example 1 involves the numbers from the climber's situation; the solution can be used as an estimate of the total price of the rope. As you discuss the steps for multiplying decimals, you might like to provide a more detailed explanation of why they work: $4.12 \times 2.1 = 412/100 \times 21/10 = 412 \times 21/1000 = 8652/1000 = 8.652$. **Ex. 22–24** You might review the role of parentheses when both addition and multiplication are used in an expression. **Mental Math** This activity is an application of the Distributive Property.

4-6 *pages 116–117* Begin by reviewing the metric prefixes, and their meaning as they are applied to the basic unit *meter*. On a meter stick or meter tape, show that a decimeter is 1/10 of a meter and a centimeter is 1/100 of a meter. Note that in the metric unit chart on page 116, the unit is smaller by a factor of 0.1 as you move to the right, and larger by a factor of 10 as you move to the left. Try to have each student do some measuring to provide "hands on" experience with the metric units of length. Students should be able to show with their hands a length that is approximately one meter, one decimeter, or one centimeter. Emphasize the close relationships between metric units of capacity and mass. As shown on p. 117, a container with a volume of one cubic decimeter has a capacity of one liter. The mass of one liter of water is one kilogram. **Ex. 1–6** Students are asked to change from one metric unit to another, again giving them some "benchmarks" to help them remember the approximate sizes of the units of length. **Ex. 9** This provides a point of reference which may develop some number sense for the number one million. **Ex. 10–11** These deal with the relationships between capacity and mass. You might note how much more difficult these problems would be using standard units since they provide no easy conversions between capacity and mass. **What's Your Decision** also focuses on this idea.

4-7 *pages 118–120* You may want to review the fact that division can tell how many units are in each subset when a collection is divided into a given number of subsets. Show how this model applies to the situation described at the beginning of the lesson. Review the estimation technique of "using a compatible number" to estimate a whole-number division problem; the answer to Example 1 shows that this technique also applies to decimals. Examples 2, 3, and 4 review the division algorithm for decimals. Be sure to review the rule for determining the sign of the quotient.

4-8 *pages 121–123* Introduce the Checklist for solving word problems by writing and solving an equation. Note that this is not a list of rules, but a series of questions to ask oneself while solving a problem. Discuss the Example and Practice, focusing on the phases *Question, Data, Plan, Answer, Check*. In the *Plan* stage, focus on assigning a variable and finding what is equal in the problem. After writing and solving an equation, have students check the solution to the equation, then reread the original problem to see whether the equation is a correct translation and the answer seems reasonable. Giving the answer in a complete sentence helps students consider this.

4-9 *pages 125–126* This section puts multiplication and division of decimals in the context of equation-solving. The solution techniques presented are identical to those discussed in Sections 2.8 and 3.9. **Ex. 16–24** Some of these equations have the variable on the right side. You may want to instruct your students to first rewrite the equations into the form of those in the A exercises.
Emphasize that operations inside parentheses must be performed first.

4-10 *pages 127–128* Discuss the sample Problem. Possible strategies for each problem are given here. **Ex. 1** *Make a Table, Look for a Pattern, Simplify the Problem.* **Ex. 2** *Draw a Picture, Make a Table, Look for a Pattern, Simplify the Problem.* **Ex. 3** *Draw a Picture.* **Ex. 4** *Guess, Check, Revise.* **Ex. 5** *Draw a Picture.* **Ex. 6** *Make a table.*

Enrichment *page 129* The binary system is probably the most often used non-decimal system. You may want to encourage students to write the numbers 1–64 sequentially in binary form. You might then explain base 3 or base 4.

Chapter 5

5-1 *pages 134–136* Explain that when you "count by 5's" you are giving multiples of 5. Ask students to "count by" 2's, 3's, 4's, 6's, 7's, 8's, and 9's. Discuss the meaning of multiple. Remind students that the numbers they multiply to find a product are called *factors*. Ask them to find factors of numbers such as 18, 24, or 27 mentally. As you complete the Examples and their Practices, emphasize the concepts of multiple and factor. You may wish to discuss or have students formulate a formal definition: A number x is a factor of a number y if there is a whole number k such that $xk = y$. The number y is a multiple of the numbers x and k. **Ex. 43** The next perfect number after 28 is 496. A challenge could be to show that this is true. Extend the exercise by defining *deficient* numbers, numbers in which the sum of factors other than the number is less than the number, and *abundant* numbers, as those in which this sum is more than the number. Students could make a table classifying the numbers 1–50 as deficient, perfect, or abundant.

5-2 *pages 137–139* Briefly discuss the divisibility rules for 2 and 5. Give as oral exercises several numbers to test for divisibility. Then present the rule for divisibility by 3, and check it by testing several numbers using the rule and then dividing by 3. Note that divisibility rules for 4, 6, and 9 are given in the B exercises. You may wish to present the following rules for 7 and 8: To check for divisibility by 7, double the last digit and subtract the result from the other digits. Continue until you can tell whether the difference is divisible by 7. If it is, the original number is divisible by 7. A number is divisible by 8 if it can be halved 3 times with no remainders, or if the number formed by its last 3 digits is divisible by 8. **Ex. 47** Suggest using *Logical Reasoning*, considering each divisor in turn. **Ex. 48** An example that shows a generalization is false is called a *counterexample*. **Mental Math** This is based on the idea that if a and b are divisible by c, then $a + b$ is divisible by c. For example, in the number 645, the sum of the digits is $6 + (4 + 5)$. Since 6 is divisible by 3, the sum will be divisible by 3 provided $4 + 5$ is divisible by 3. Because of this, we can "cast out" multiples of 3 when considering divisibility.

5-3 *pages 140–141* Review the Problem-Solving Checklist for solving word problems by writing and solving an equation. Discuss the Example and Practice, focusing on the checklist phases, *Question, Data, Plan, Answer, Check*. In the *Plan* stage, focus on assigning a variable and finding what is equal in the problem. After writing and solving an equation, have students check the solution to the equation, then reread the original problem and see whether the equation is a correct translation and the answer seems reasonable. Giving the answer in a complete sentence helps students consider this. **Ex. 7** This contains extra information, the mention of packages of 4. Students must select the useful data. **Ex. 8** This requires data not given, the fact that there are 12 eggs in a dozen. **Ex. 9–10** You might want to suggest themes to get students started.

5-4 *pages 142–143* Show students a 2-by-3 rectangle drawn on graph paper and tell them that since the area is 6, you will call it a "rectangle for 6." Ask them to try to find rectangles with sides greater than 1 for each of the numbers 1–15 and show them on graph paper. The numbers for which no such rectangle can be drawn are the prime numbers. Discuss the definition of prime number given in the text and relate it to this activity. Note that since 1 has only one factor, and since every number is a factor of $0 (0 = 1 \cdot 0, 0 = 2 \cdot 0$, etc.), 1 and 0 are neither prime nor composite. Discuss the table in the text, asking students to give the number of factors for each number. You may wish to introduce the ***Sieve of Eratosthenes:*** make a table of numbers 1–100 to separate out the prime numbers. Find the first prime, 2, and mark out all multiples of 2, except 2. Find the next prime, 3, and mark out all multiples of 3, except 3. Continue this for primes less than 10. The numbers not marked out are prime numbers. A table of the first 100 primes appears on p. 161. **Ex. 24** This formula produces a prime number for every replacement for the variable less than 11. Eleven produces a prime number because in its case the formula produces a number divisible by 11. Another such formula is $P = n \cdot n - n + 19$, which produces primes for values of n less than 19.

5-5 *pages 144–147* Write 3^4 on the chalkboard and point out the base and the exponent, explaining that the exponent tells how many times the base is used as a factor. Give several examples for students to interpret, and write the products on the chalkboard. A common student error is to think that 3^4 means $3 \cdot 4$. Complete several calculations to show that this is incorrect. To prepare students for multiplying with exponents, work out several examples like the one at the top of p. 145, emphasizing that when we add the exponents, we are adding the number of factors in each power being multiplied. Discuss the method for multiplying, stressing that *the bases must be the same* and that *the rule is only for multiplying powers*. Use Example 4 to help prevent the common student error of applying the method to adding powers. **Ex. 6–8** Help students form a generalization about raising negative integers to odd and even powers.

5-6 *pages 148–150* Draw 3 factor trees starting with 24 on the chalkboard. Ask students to start each tree with 2 different factors, (e.g., $24 = 6 \cdot 4$, $24 = 8 \cdot 3$, and $24 = 12 \cdot 2$). Have a student complete each tree and observe that the last row of each tree contains the same prime factors. Discuss prime factorization and the Unique Factorization Theorem. The Examples show three different methods of finding the prime factorization of a number. Note that the method in Example 2 is similar to the factor tree, but the answers are expressed differently. In Example 3, students must see that for each new division, they must start with the smallest prime and test each subsequent prime until they find a prime that is a factor of the number. **Ex. 42** The first number is the first prime, the second number is the product of the first 2 primes, the third number is the product of the first 3 primes, etc.

5-7 *pages 151–153* Direct attention to the Venn Diagram and verify with students that the left circle (set) contains the factors of 24 and the right circle (set) contains the factors of 30. Point out that the overlap of the two circles (the intersection of the two sets) contains factors common to both circles (sets). Identify the common factors and the GCF. Use other examples. Note that in Example 1, after listing the factors of 18, we first consider the *largest* factor of 18 and progress downward until we find a factor of 18 that is also a factor of 24. In Example 2, since the factor 2, for example, occurs twice in both factors, we include it twice in the GCF. The factor 3 occurs twice in one factor and once in the other. Since the most it occurs in *both factors* is once, we include it once in the GCF. **Ex. 39** Two numbers that are relatively prime have no common prime factors. **Ex. 40** This can be solved by finding the GCF of 210 and 180. **Numbers to Algebra** The algebraic procedure is identical to that used with numbers. Use additional numerical examples as needed.

5-8 *pages 154–155* Discuss the graph and the information given below it. Work through the first problem with the class, explaining how to use the graph to find needed data. Note that this **Data Search** requires students to collect data in their neighborhoods. You might have students make a chart or poster to display their data and calculated information. The data needed for **What's Your Decision** can be found in the graph on p. 154. Note that there is no "right answer." Discuss student choices and decisions. Extend by having students look at a driver's manual or observe whether drivers seem to use these recommendations.

5-9 *pages 156–158* Direct attention to the Venn Diagram and verify with students that the left circle (set) contains the multiples of 6 and the right circle (set) contains the multiples of 8. Point out that the overlap of the two circles (ie., the intersection of the two sets) contains multiples that are common to both circles (sets). Identify the common multiples and the LCM. Use other examples. In Example 1, after listing the multiples of the larger number, 12, we first consider the *smallest* multiple of 12 and progress upward until finding a multiple of 12 that is also a multiple of 9. In Example 2, we begin by listing the prime factorization of the smaller number, then include factors as needed so the prime factorization of the larger number will also appear in the LCM. **Ex. 41–42** Solve by finding the LCM of the two numbers. **Ex. 46** There are 366 days in a leap year. **Ex. 47** Encourage students to test other cases.

5-10 *pages 159–160* Present the following problem, for which the strategy *Make an Organized List* is useful. Jeff took on a trip 3 pairs of pants (brown, grey, and blue), 4 shirts (yellow, blue, striped, and plaid), and 2 sweaters (white and maroon). How many different outfits could he have? Suggest starting a list with the brown pants and listing all the shirts and sweaters that could be worn with them. Then do the same for the other pairs of pants. After the solution, 24, has been found, review the solution process and the role of making an organized list. Discuss the sample problem in the text. Possible strategies for each problem are given here. **Ex. 1** *Make an Organized List.* Help students begin by giving them MAT, MAH, MTA, MTH, MHA, and MHT, then asking them to give the arrangements beginning with A, T, and H. **Ex. 2** *Make an Organized List.* Suggest starting with A and writing all variations of the other letters, then starting with B, C, and D. **Ex. 3** *Choose the Operations.* **Ex. 4** *Guess, Check, Revise.* **Ex. 5** *Make an Organized List.* (The list might start: AB, AC, AD, AE, AF, AG, AH, AI; BC, BD, etc.) or *Simplify the Problem, Make a Table, Look for a Pattern, Draw a Picture.* **Ex. 6** *Make a Table, Simplify the Problem, Look for a Pattern.*

Enrichment *Page 161* Tell students that in 1979, the largest known prime was $2^{44497} - 1$, a number with 13,395 digits. Finding that prime took a computer run of 2 months. In 1985, the prime number $2^{216091} - 1$ was discovered, almost accidentally, by scientists at Chevron Geosciences who were running a program in a new supercomputer to test for "glitches." Make sure students understand how to read the table of the first 99 prime numbers, and assign the exercises as independent work. **Ex. 3** Number theorists have proved that the distribution of primes gets thinner the farther we go in the sequence of whole numbers, and it is possible to exhibit as long a sequence of whole numbers as desired that contains no primes!

Chapter 6

6-1 *pages 166–169* Discuss the meaning of "fraction." Use the region shown to explain the terms *numerator* and *denominator*. Emphasize that the region is divided into parts equal in size. Draw a region on the chalkboard and divide it into fourths and then into eighths. Draw other regions as needed to explain fractions and equivalent fractions. Introduce the property and discuss Example 1. On page 167, show why the cross products are equal by multiplying 5/12 by 36/36 and 15/36 by 12/12, and pointing out that the numerators show that the cross products are equal and the denominators are the same. In Example 2 where the cross products are not equal, explain this by changing the fractions to equivalent fractions with like denominators. Define lowest terms and discuss Example 3. You may need to review GCF and prime factors. **Ex. 62–63** Remind students that each digit can be used only once in each statement. **Numbers to Algebra** Give more number examples as needed to explain the algebraic procedure.

6-2 *pages 170–172* Review how to name a fraction, given a region. Use regions like the one on page 170 to explain improper fractions and mixed numbers, emphasizing that the region is divided into equal parts. Discuss the $a \div b$ interpretation of a fraction, and work through Example 1. To develop understanding, have students rename the improper fractions in different ways. For example, students could write 19/7 = 1 11/7 = 2 5/7. **Ex. 39** Some students may believe the whole number part of the mixed number can only be the 1. Other numbers are possible. **Ex. 40** This generalization may have been discovered in completing Example 2, but students should give it here with variables rather than numbers.

6-3 *pages 173–175* The key idea here is that although a rational number can be named in many different ways, each rational number is associated with a unique point on the number line. Emphasize that whole numbers, decimals, and integers are all rational numbers because they can be given in the form a/b. In Example 1, note that the most common form is with the negative sign preceding the entire rational number. Point out that $-2/-3$ is also a rational number but can be viewed as a negative divided by a negative, which gives a positive quotient. Discuss Example 2. **Ex. 27–34** Point out that you can tell the opposite of an expression, but you cannot tell whether the value of the expression is positive or negative unless you know a value to substitute for the variable. For example, the opposite of $3/x$, $-3/x$, is negative if $x = 2$, but the opposite is positive if $x = -2$. **Ex. 39–40** Students should look for patterns by inspecting, not by computing.

6-4 *pages 176–179* Explain why we can decide which rational number is greater by its relative position on the number line. Example 1 shows the most common method for comparing fractions. The method in Example 2 is particularly useful when working with a calculator. Introduce the method for comparing, and work through Example 3. Show how this process is a variation of the method shown in Example 1 by changing to fractions with like denominators. **Ex. 47** Most students can tell that $1/2 > 1/3$, but cannot tell that $5/2 > 5/3$. This exercise helps students see that two fractions with the same numerator compare as fractions with numerators of 1. Encourage students to find a way to order these numbers without changing to mixed numbers.

6-5 *pages 180–182* Use the development that begins the lesson to explain, mathematically, the process for adding and subtracting rational numbers with like denominators. Understanding of the methods given for adding and subtracting positive rational numbers can be developed by using the number line or regions (e.g., a circle divided into fifths, to add 1/5 by shading 1 section and 3/5 by shading 3 more sections). Use the number line to illustrate addition and subtraction with positive and negative numbers. Example 2 shows how to subtract a negative rational number. Remind students that subtracting $-7/25$ is the same as adding the opposite of $-7/25$, or $7/25$. If necessary, review this idea with integers. **Ex. 49** Students should recognize that since the denominators are the same, they can write the sum of the numerator over the common denominator, $24/t$. Any factor of 24 will make this a whole number.

6-6 *pages 183–185* Discuss the situation presented in the photo caption. Note that the numbers involved are used in Example 1. You can use fraction strips (strips divided into halves, thirds, fourths, etc.) to illustrate adding fractions with unlike denominators. In Example 2, give particular attention to the case of subtracting a negative number. **Ex. 32** This shows mathematically the reasons underlying the steps for adding rational numbers with unlike denominators.

6-7 *pages 186–188* This lesson presents two methods for adding and subtracting mixed numbers: the vertical form and the horizontal form. Since one form can be more easily used than the other in a given situation, it is helpful if students are familiar with both methods. However, you may choose to introduce only one. All exercises can be completed in either format. Begin by discussing the situation described in the photo and caption. Note that the numbers in the caption are used for Example 1. Example 2 shows how the mixed number has to be renamed before subtracting. You may want to review this skill before completing the example. Students should be encouraged to use the format that is easiest for the given problem.

6-8 *pages 190–191* Begin by reviewing how to add and subtract mixed numbers. The size of the numbers in many of these exercises suggests that a vertical format may be easier than a horizontal. Explain how to read the stock sheet; note that fractions are written with a slash mark. Before assigning the problems, use the example and the key under the chart to help students interpret the data. **Data Search** The format of stock reports can vary considerably. You might want to give the class a report for the data search so they can all deal with data in the same format. **What's Your Decision** This could be extended over time: students select stocks from a list, pretend they invest so many shares in each stock, and follow the progress of the stocks over a period of time.

6-9 *pages 192–194* Begin with a review of adding and subtracting rational numbers. Discuss Examples 1 and 2. Emphasize that there are two ideas to support steps for solving the equation. The first is the Addition

Property of Equality, which states that the same number can be added to both sides of the equation and they will remain equal. The second idea is adding the additive inverse of the number that has been added to the variable. This puts the variable alone on one side, since the sum of a number and its additive inverse is 0.

6-10 *pages 195–196* You may want to review the Checklist for solving word problems by writing and solving an equation (p. 121). Discuss the Example and Practice. In the *Plan* stage, focus on assigning a variable and finding what is equal. After writing and solving the equation, have students check the solution to the equation and the reasonableness of the answer to the problem.

6-11 *pages 197–198* The name of this strategy, **Logical Reasoning**, might seem odd since one must reason logically to solve any problem. The name is used here to refer to a type of problem in which students are given some information and must use reasoning to obtain additional information to solve the problem. This strategy can be called **deductive reasoning**. Possible strategies are given here for each problem. **Ex. 1** *Use Logical Reasoning*. **Ex. 2** *Use Logical Reasoning*. **Ex. 3** *Guess, Check, Revise, Make an Organized List, Look for a Pattern, Draw a Picture*. **Ex. 4** *Simplify the Problem, Draw a Picture, Make a Table, Look for a Pattern*. **Ex. 5** *Make a Table, Choose the Operations*. **Ex. 6** *Choose the Operations*.

Enrichment *page 199* To provide further practice in using the calculator to add or subtract fractions, you may want to assign exercises from previous lessons.

Chapter 7

7-1 *pages 204–207* Introduce multiplication of fractions using the seating chart model shown at the top of the lesson. You may want to review this model as an interpretation for multiplication of whole numbers (e.g., 2 rows with 3 in each row or 2×3) before using it as a model for fractions. Introduce and discuss the method for multiplying rational numbers. You might introduce $5 \times (-1/3)$ as $(-1/3) + (-1/3) + (-1/3) + (-1/3) + (-1/3) = -5/3$, then rewrite the original problem as $5/1 \times (-1/3)$ to show that the rule for multiplying rational numbers applies to negative rational numbers. You may wish to review the rules for determining the sign of an integer product. Discuss Examples 1 and 2, emphasizing the step of changing mixed numbers to improper fractions. In Example 1, you may want to allow students to divide the 5 in the numerator and denominator before multiplying. **Ex. 34–39** Encourage use of the Distributive Property. **Mental Math** This procedure is helpful when the denominator in the mixed number is a factor of the whole number.

7-2 *pages 208–210* Discuss the Multiplicative Inverse Property, explaining that one can use the term "reciprocol" for multiplicative inverse. Use the problem on p. 208 to introduce the method for dividing rational numbers. If you want to show a physical interpretation of division of fractions, ask questions such as, "How many one-thirds are in 2?" How many one-halfs are in 3?" "How many one-fourths are in one-half?" Draw pictures to show these situations and to illustrate the method for dividing. Discuss Examples 1 and 2. **Ex. 34** Have students first list all the factors of 9 and all the factors of 15. **Ex. 35** Since the product is negative, one factor is positive and the other is negative. **Estimation** Students can round mixed numbers to whole numbers, but must still decide which inequality is appropriate.

7-3 *pages 211–213* You may want to begin by showing how to solve the equation $\frac{3}{4}x = \frac{5}{8}$ by first dividing both sides of the equation by 3/4, and discussing the fact that this is cumbersome. Review the multiplicative inverse and discuss Example 1. Then discuss Example 2, emphasizing the reason for each step in solving the equation. **Ex. 34** *Guess, Check, Revise* is a useful strategy. **Ex. 35** Have students think "1/2 of what number equals 8?" Since 1/2 of 16 equals 8, y must be 4 or −4. The same type of reasoning gives $x = 1/2$ or $-1/2$ for Exercise 36. **Computer Activity** You may want to work through the steps of solving the sample equation to show why the computer form provides the correct solution.

7-4 *pages 215–216* You may want to review the phases of the Checklist for writing an equation to solve a problem (p. 121). Discuss Example 1, emphasizing the selection of the variable and the setting up of the equation. **Ex. 10** You can refer students back to Ex. 8, since the structure is similar.

7-5 *pages 217–218* Begin by discussing the type of work a mason does. You may want to review the Problem-Solving Checklist (p. 22) while discussing the first few problems. **Ex. 2–4** Be sure students understand the meaning of "rise" and "run" in the picture given. **What's Your Decision** You might extend this to a class project by having students find an actual location for a patio and find the total cost of installing it.

7-6 *pages 219–220* Begin by reviewing the meaning of a rational number: a number which can be written in the form a/b. Review the fact that whole numbers, integers, and decimals are rational numbers since they all can be written in the form a/b. Introduce the fact that a rational number can be written as a decimal. Give several examples to clarify the difference between terminating and repeating. Give examples of decimals such as 4.23233233323333 that neither terminate nor repeat to introduce the idea of irrational numbers. Discuss Example 1. Emphasize the placement of the bar over the decimal part that repeats. You may choose to have students use a calculator for all the exercises. **Ex. 29–30** These exercises prepare students for the Enrichment at the end of the chapter. **Ex. 34** The fact that 0.43333 repeats may lead some students to think that it is greater than 0.44, which terminates.

7-7 *pages 221–223* Begin by reviewing the meaning of an exponent. Give several numbers written in exponential form (a^n) and have the students express them as the product of the same factor (base). Review the rule for multiplying two numbers in exponential form with the same base. Emphasize that this works only when the base is the same. Use the development in the text to explain the rule for negative exponents. Introduce the methods for dividing two numbers involving exponents. Discuss Examples 1, 2, and 3. **Ex. 29–40** Have students use positive or negative exponents, not fractional expressions like $4/2^3$. **Ex. 48** Suggest setting up a table with 3 rows: day number, number of people interviewed each day, and the total number interviewed. **Numbers to Algebra** In the algebra example, point out that the Commutative Property of multiplication allows us to rearrange the x and y in the denominator. You could have students factor whole numbers into prime factors to help simplify.

7-8 *pages 224–226* Discuss very large and very small numbers that students have encountered. If possible, show examples from newspapers, magazines, and other sources. Discuss the introductory paragraph, noting the examples and discussing why each is or is not in scientific notation. In Examples 1 and 2, emphasize the sign of the exponent. If necessary, write the small number as a fraction to show why the exponent is negative. **Ex. 32** Students should realize they can divide the decimal parts in the usual manner and divide the powers of 10 using the rules for dividing expressions in exponential form. **Ex. 33** Since 0.01 is less than 1, the negative sign has to go with the exponent. **Ex. 34** Students could draw a picture representing each chemical and showing the intersections of the chemicals. **Calculator Activity** If students do not have a calculator that works with exponents, discuss how they can use the calculator and mental math to find the product of numbers expressed in scientific notation.

7-9 *pages 227–228* Discuss what type of problem can be solved by working backward. Read the sample problem and work through the solution. Possible strategies for each problem are given here. **Ex. 1** *Work Backward, Choose the Operations*. You may want to encourage students to start with

Brenda's number and work backward. The inverse action of "twice as many" would be to divide by 2 or multiply by 1/2. **Ex. 2** *Work Backward, Choose the Operations, Draw a Picture*. Students should start with 2 jars, each showing 64 ml. Encourage students to draw pictures to show each step of working backward. **Ex. 3** *Guess, Check, Revise, Make an Organized List*. **Ex. 4** *Guess, Check, Revise, Work Backward*. **Ex. 5** *Draw a Picture, Look for a Pattern*. Find the total number of paths to each of the intersections near A1. Then look for a pattern to determine how many paths there are to each intersection. **Ex. 6** *Draw a Picture, Make a Table, Look for a Pattern*.

Enrichment *page 229* You may want to work through the examples before assigning the exercises.

Chapter 8

8-1 *pages 234–236* Write on the chalkboard the equation $4n - 3 = 13$. Ask students to help you complete a flow chart of the order in which operations have been applied to the variable: Start with $n \longrightarrow$ Multiply by 4 $\longrightarrow$ Subtract 3 $\longrightarrow$ End with 13

Then show how to reverse the flow chart and use inverse operations to undo the operations and find the value of the variable: Start with 13 $\longrightarrow$ Add 3 $\longrightarrow$ Divide by 4 $\longrightarrow$ End with n

Discuss the equations in the chart on p. 234, emphasizing that the additive or multiplicative inverse property could also be used to undo the operations (e.g., in the first equation, by adding -5 to each side and then multiplying each side by 1/3). Discuss the steps for solving equations, and use them to complete Examples 1 and 2. In Example 2, make sure students understand that they should divide first and then subtract. Another approach would be to first use the Distributive Property to write the equation as $5x + 35 = 105$. Do not teach or emphasize this procedure at this time. Encourage students to use a *Guess, Check, Revise* strategy.

8-2 *pages 237–238* Discuss the first paragraph. Write a statement such as "The difference between twice a number and 8 is 24" on the chalkboard, and work with students to translate it into an equation. Then write a statement such as "An $86 coat was only $5 more than three times the cost of a hat," and have students translate it into an equation. As you complete these translations, emphasize the phrases that suggest the different operations. The second statement could be translated as $86 = 3h + 5$, or $3h + 5 = 86$. As you complete the Example and its practice, ask students to explain their translations. It is sometimes not necessary to write an equation to find the missing number, but practice with the *process* of translating to an equation will be useful later. **Ex. 1–8** To extend these, ask students to *solve* their equations. **Ex. 12–13** Suggest a *Guess, Check, Revise* or *Logical Reasoning* strategy rather than writing equations.

8-3 *pages 239–240* You may want to review the phases of the Checklist for writing an equation to solve a problem (p. 121). As you discuss the Example, focus on assigning a variable to the number you are trying to find, on representing other conditions in terms of the variable, and on finding what is equal in the problem. As you write and solve an equation, be sure to have students first check the solution to the equation, then reread the original problem and see whether the equation is a correct translation, and whether the answer seems reasonable. Requiring that students give the answer using a complete sentence helps them consider this. **Ex. 6–8** Note that numerical calculations are required to simplify these equations before solving. **Ex. 9–10** You may wish to suggest a subject.

8-4 *pages 241–243* Use numerical examples to review the Distributive Property. Review simplifying an expression such as $3x + 4x$. Extend the Distributive Property to include subtraction examples such as $7x - 5x$ by writing the general form $ba - ca = b(c - a)$ on the chalkboard and, as you erase each variable in turn, writing 7 in place of b whenever it occurs, writing 5 in place of c, and writing x in place of a. After students see that $7x - 5x = (7 - 5)x = 2x$, replace the variable with numbers to show that $7x - 5x$ has the same value as $2x$ no matter what number is used in place of x. **Ex. 25–30** These extend understanding of the Distributive Property. **Ex. 35** Extend by asking students to check that this generalization is true for $(a, b, c) = (2, 3, 4)$, $(3, 5, 6)$, $(4, 7, 8)$, $(5, 9, 10)$, $(6, 11, 12)$, $(7, 13, 14)$, and write a generalization about which triples make the statement true.

8-5 *pages 244–246* Review meaning of like terms (introduced on p. 47) and extend to involve integers, fractions, and exponents. In Example 1, emphasize the use of the Distributive Property. Since $a + (\mathbf{b} + \mathbf{c}) = (\mathbf{a} + \mathbf{b}) + c$ (Associative Property), $a + (b + c) = a + (c + b)$ (Commutative Property) and $a + (c + b) = (\mathbf{a} + \mathbf{c}) + b$ (Associative Property), any pair from three numbers a, b, and c can be added first, and the sum is the same. This is sometimes called the "Rearrangement Property," and is used in Example 2. **Ex. 31–32** An equation solution is used to complete the solution to each problem. **Numbers to Algebra** You may wish to use the following language when discussing the generalization: the opposite of a difference is the reverse of the difference. The generalization can be proved as follows:

$$\begin{aligned} &-(a - b) & \\ &= -1(a - b) && \text{(Property of } -1) \\ &= -1(a + -b) && (x - y = x + (-y)) \\ &= -1a + -1(-b) && \text{(Distributive Property)} \\ &= -a + b && \text{(Property of } -1) \\ &= b - a && (x + -y = x - y) \end{aligned}$$

8-6 *pages 247–249* Discuss the photo situation described in the caption to show the need to solve equations with variables on both sides. Work through Example 1, noting that the solution to the equation $5h = 12 + 3h$ gives the number of hours an item would have to be rented in order for the two charges to be the same. Ask which plan would be best for a smaller number of hours and for a greater number of hours. For Example 1, Practice b, note that isolating the variable on the right side of the equation rather than on the left is acceptable and more appropriate in some situations. **Ex. 34–36** Equations for solving these were developed in Sections 8.1–8.3. **Ex. 37–38** These are more complex; encourage a *Guess, Check, Revise,* or *Logical Reasoning* strategy. Note in **Ex. 38** that if one imagines each large block replaced with 3 small blocks, there will be 14 small blocks, weighing a total of 112 kg, or 8 kg per block. **Mental Math** This is based on the idea that $(a - b)(a + b)$ is $a^2 - b^2$. Note that in the first example, 29(31) is $(30 - 1)(30 + 1)$.

8-7 *pages 250–251* You may want to review the guidelines in the Problem-Solving Checklist on page 22. Discuss the table on the cost of operating an automobile. Define depreciation as the loss in value of a car as it is used. Work through the first problem, explaining how to use the graph to find needed data. Emphasize the importance of writing the answer in a complete sentence. **Data Search** This requires students to collect data by phoning car dealers or reading automobile magazines. **What's Your Decision** Note that there is no "right answer." Discuss student choices and decisions. Extend by having students consult local sources to verify or change the data in the chart, then reconsider their decisions on the basis of new information. Another extension is to keep a record of the fuel expenses on a car and calculate miles per gallon and total cost.

8-8 *pages 253–255* Begin with an inequality statement such as $x < 4$, and ask students to give replacements for x that make the statement true (including negative numbers) and then give others that make the statement false. Discuss the *solution set* of an inequality, the set of all numbers that can replace the variable to make a true statement. In preparation for solving inequalities, give numerical inequalities such as $-1 < 3$ and ask students to add, subtract, multiply, and divide both positive and negative numbers on each side of the inequality. Each time, ask, "Is the number on the left side still less than the number on the right?" Give several such numerical examples to lead students to discover that the inequality sign is reversed only when multiplying or dividing each side by a negative number. Discuss the Examples and their Practices. Encourage students to check the solution of an inequality by checking both the computation and the inequality sign. Note that the check provides assurance that the solution process is correct and the direction of the inequality sign is correct. It would not be possible to check *every* solution to this type of inequality. **Ex. 23–25** The process for solving a problem by writing an inequality is similar to that for solving an equation. Work through the example with the students.

8-9 *pages 256–258* Review the steps for solving equations involving more than one operation. Explain that the inequalities in this section are solved by a similar procedure, except that the inequality sign is reversed whenever we multiply or divide each side by a negative number. In Example 1, multiplying each side by -1 gets x by itself. This reverses the inequality. In Example 2, emphasize the role of the Distributive Property. In Example 3, where the variable is on both sides of the inequality, an alternative is to subtract $5n$ from each side and solve the inequality $-2n > -6$. **Estimation** To give students a feel for determining whether a rational number is close to 0, close to 1/2, or close to 1, ask students to give numbers a bit more and a bit less than each of these benchmarks. Note that Exercise 6 could be interpreted in terms of parts of a year (365 days).

8-10 *pages 259–260* You may want to review the Problem-Solving Checklist on page 22. Work through the Sample Problem in the text. If an overhead projector is available, use to show and help students count the squares. After the table is constructed, help them see that for side length n, the number of squares is the sum of the square numbers up to n^2. Possible strategies for each problem are given here. **Ex. 1** *Make a Table, Simplify the Problem, Look for a Pattern.* This is an extension of the sample problem. A checkerboard is 8 by 8, the pattern discovered on page 259 could be used to find a total of $8^2 + 7^2 + 6^2 + 5^2 + 4^2 + 3^2 + 2^2 + 1^2$ squares. **Ex. 2** *Make an Organized List.* Start with 3 quarters, and then include as many extra coins as possible. **Ex. 3** *Draw a Picture, Make a Table.* **Ex. 4** *Draw a Picture, Guess, Check, Revise, Logical Reasoning.* You might suggest that each of the 3 links in one of the sections be cut. **Ex. 5** *Draw a Picture, Guess, Check, Revise, Use Logical Reasoning; or, Simplify the Problem, Make a Table, Look for a Pattern, Use Logical Reasoning.* You might suggest beginning by placing 3 pieces on each pan of the scale. **Ex. 6** *Make a Table, Look for a Pattern, Simplify the Problem.*

Enrichment *page 261* Discuss the meaning of the terms given at the top of the page. Review multiplying a monomial by a monomial such as $4n(5n)$ and a monomial by a binomial such as $3(x + 6)$. Write the general form of the Distributive Property on the chalkboard and, as you erase each variable, write the expressions that replace it as shown in the example. Continue the process, emphasizing the role of the Distributive Property. Work through a second example, such as $(n + 4)(n + 5)$. Discuss the shortcut and illustrate it with the example shown. Help students see how to multiply each term of the first binomial by every term of the second binomial.

Extend the exercises by giving some like the following:

1. $(x + -3)\ (x + 5)$
2. $(x + -6)\ (x + -7)$
3. $(r - 3)\ (r + 6)$
4. $(y - 4)\ (y - 7)$
5. $(8 + t)\ (t + 3)$

Chapter 9

9-1 *pages 266–267* Begin by discussing the data on instruments in an orchestra, noting that changing the number of one type of instrument would change the sound of the orchestra. Introduce the concept of ratio and show ways to write a ratio. Discuss a ratio as a type of a fraction. Discuss the Example. **Ex. 33** Point out that the units in a ratio need to be consistent. **Ex. 34** This illustrates how a ratio is different from the common interpretation of a fraction (a comparison of a part to the whole). **Ex. 37** This shows that a ratio can be used to compare several quantities.

9-2 *pages 268–271* Discuss the two ratios given and show that they both are written as 5/9 in simplest form. Introduce the meaning of proportion and the Property of Proportions. Explain the meaning of "if and only if," using the proportion given above the property. Discuss Example 1. Relate it to the use of cross products to compare fractions. With Example 2, emphasize the need to check. With Example 3, emphasize that each quantity needs to have a consistent place in the ratio.

9-3 *pages 272–273* Begin by discussing the meaning of speed limit signs. Explain that they give ratios in which the bottom number is a 1 and is not written (e.g., 80km/h = 80km/1h). Discuss Example 1, pointing out that the units are consistent in the top of each ratio and the same in the bottom. Discuss Example 2, showing how to set up the proportion. **Ex. 22** Solving involves multiple steps. Students can use ratios or make complete tables.

9-4 *pages 274–275* Begin by giving a word problem that can be solved by writing an equation. "Jackie was given 25 tickets to sell for a school raffle. So far she has sold 8. How many more does she need to sell?" ($8 + x = 25$) Point out that problems like this involve one unknown (x = the number she has to sell). Discuss Example 1, emphasizing the two unknowns and two expressions. Discuss Example 2. Do not solve the equations.

9-5 *pages 276–277* You might begin by reviewing the phases of the Problem-Solving Checklist. Review the material in 9.4, in which students gave two expressions for a problem but did not continue to solve the equations. Discuss the example, emphasizing the selection of the two expressions and the steps involved in solving the equation. **Ex. 10** This involves three expressions.

9-6 *pages 279–281* Begin with a review of proportions (9.2), and rate problems in which the units of the top numbers are consistent and the units of the bottom numbers are consistent. Discuss the scale drawing on the page. Complete Example 1, pointing out that cm are on the top and m on the bottom of each ratio. Discuss Example 2. Note that in both examples, the Property of Proportions was used in solving for the unknown. **Ex. 14–17** Answers may vary slightly depending on the accuracy of students' measurements.

9-7 *pages 282–283* Read and discuss the opening paragraph. You may want to have students examine advertisements in newspapers. Discuss the meaning of "unit price." Point out the similarity to rate, where an amount of one unit was compared to an amount of another unit. Work through the Example. **Data Search** You may want to give students newspaper advertisements and choose different items to compare. To extend, have each student budget an amount of money to buy personal and household supplies for a week and visit stores to compare prices.

9-8 *pages 284–286* Begin by discussing situations in which students have heard the words *percent* and *percentage* used. Note that the word *percent* is used in the text and in direction lines to mean both "percent" and "percentage." In word problems, the noun "percentage" is used. Introduce the meaning of percent as "per one hundred," and work through Example 1. In Example 2, students must complete several steps to arrive at a solution. Be sure they understand each step; the next few lessons will involve these skills. Discuss Example 3. Point out that a ratio with a bottom number of 100 was used in all three examples. **Ex. 44** You may need to review the meanings of "prime" and "composite."

9-9 *pages 287–290* Begin by discussing situations in which percents might be greater than 100 or less than 1. Draw attention to the situation presented in the photo and caption. Discuss Example 1, noting that it can be used to answer the question, "What percentage of his goal did he meet?" Emphasize that each fraction needs to be changed to an equivalent fraction with a denominator of 100. Explain that since 6/5 is greater than 1 (a whole), 6/5 should be greater than 100%. Thinking in these terms and using estimation helps students evaluate answers when working with percent. In Example 4, emphasize that dividing by 100 is the same as multiplying by the reciprocal of 100 (1/100). For each example, have students evaluate the answer and consider whether it is reasonable. In Example 3, for instance, note that 150% is greater than 100%, and 1 1/2 is greater than 1. **Ex. 47** You might allow students to use blocks and develop a model to help solve this problem.

9-10 *pages 291–293* Begin by reviewing the process of changing a fraction to a decimal by dividing (1/2 = 0.5; 8/10 = 0.8). Discuss Example 1. Point out that dividing by 100 moves the decimal point two places to the left. Discuss Example 2, pointing out that the decimal point is moved two places to the right. Discuss the four comparisons of the percent form to the decimal form. Use them to explain the shortcuts for changing from a percent to a decimal or a decimal to a percent.

9-11 *pages 295–296* You may want to review the Problem-Solving Checklist (p. 22). Discuss the fact that people use different strategies to come up with the same solution. Read the sample Problem, discussing the two solution methods shown. Possible strategies for each problem are given here. **Ex. 1** *Work Backward* or *Guess, Check, Revise*. **Ex. 2** *Guess, Check, Revise* or *Use Logical Reasoning*. **Ex. 3** *Guess, Check, Revise* or *Write an Equation*. **Ex. 4** *Make a Table*. **Ex. 5** *Make an Organized List*. **Ex. 6** *Make a Table* or *Choose the Operations*.

Enrichment *page 297* You may want to mention the use of the Golden Rectangle in ancient Greek architecture. The Parthenon is a good example.

Chapter 10

10-1 *pages 302–304* Begin by reviewing ideas and skills from the previous chapter that are used in this lesson: the meaning of "percent," solving proportions, and changing a percent to a decimal. Discuss the photo and caption. Note that Example 1 can be used to answer the question, "What was the amount of the tip?" Point out that use of a proportion is appropriate when there are no fractional percents; otherwise, the computations are difficult. In working through Example 2, make sure students understand how to change the percent to a decimal. **Ex. 1–27** If you prefer to introduce only one of the two methods shown in the examples, change the directions as you wish. **Ex. 41–43** These can be estimated by thinking of the fraction of each region that is shaded.

10-2 *pages 305–307* Begin by discussing the data in the chart, noting that it is difficult to compare the totals of records and tapes sold because the total of each sold is not the same. Discuss Example 1. Since 120 is the total number of tapes sold, this number is in the denominator. Note that Practice a is a comparison of the jazz records and tapes sold. Work through Example 2. Point out that the word "of" *usually* translates to multiplication. You may want to set up an equation to solve Example 1 and a proportion to solve Example 2, to show that the answers will be the same.

10-3 *pages 308–310* Begin by discussing the advertisement for computer disks, making sure students understand that the original price of the disks is not given. Discuss Examples 1 and 2, pointing out that the proportion method is usually less difficult than the equation method since the latter involves dividing by a decimal. **Ex. 40** Allow students to use blocks to make a model.

10-4 *pages 311–312* You may want to review the Checklist for solving a problem by writing an equation (p. 121). Discuss the Example. Like the problems in Chapter 9, this has two expressions that need to be defined and used in the equation. The exercises include the type introduced in the example here, along with some involving equations introduced earlier, such as $2x + 5 = 20$.

10-5 *pages 313–315* Begin by discussing situations in which the cost of something has either increased or decreased. Discuss the situation in the photo and caption, and introduce the procedures for finding the percent of increase or decrease. Example 1 can be used to answer the question, "What was the percent of increase of Ms. Hernandez's salary?" Discuss Examples 1 and 2. Point out that students could either write an equation to find the percent or use the proportion method as shown. **Ex. 23–24** These involve percent of increase greater than 100.

10-6 *pages 316–318* Introduce the ideas of discount, sale price and commission by discussing situations that students have encountered. Help them think of situations in which salespeople receive commissions. You might have them find discounts in advertisements. Discuss the introductory material and Examples 1 and 2. **Ex. 1–16** You might choose to allow students to use calculators.

10-7 *pages 320–321* Begin by discussing city and state taxes and rates. Call attention to the data in the tax tables, pointing out that depending on where you live, you may pay a state tax, both state and city tax, or neither. The sample Problem and **Ex. 1–8** involve figuring only city *or* state tax amounts; note that this might not be the full tax paid on the item (Ex. 1, for example, asks for state tax but no city tax). **Ex. 9–13** Involve figuring the *full* tax paid on items. Suggest adding the state and city percentages together, then computing the tax amount. **What's Your Decision?** To expand this to a class project, have students select two cities in which they might like to live, encouraging them to use large cities for which cost-of-living data are available. Have them identify the financial information that would help them make a decision. They can collect data at the library or by writing to Chambers of Commerce.

10-8 *pages 322–325* Discuss the meaning of "interest." Be sure students understand that interest can be *earned* (e.g., through a savings account) or *paid* (e.g., on a loan). Discuss the formula for simple interest. You may choose to mention the difference between simple and compound interest. Compound interest is the focus of the Enrichment on page 331. Discuss Example 1, noting that the units of time need to be consistent in using the formula for simple interest. Discuss Example 2. **Ex. 1–25** You might choose to allow students to use calculators. **Ex. 26–32** You may need to discuss the concept of finance charge and point out that it is a type of interest.

10-9 *pages 326–328* If possible, begin by having students examine circle graphs in newspapers and magazines. Introduce the idea that circle graphs are most appropriate for showing relationships of data to some meaningful whole. Introduce angle measurement. Explain that the sum of the measures of the angles in the circle is 360°. You might have students practice using a protractor to construct angles of various measures. Introduce the steps for constructing circle graphs, and work through the Example. You might have students use a calculator to change each amount to a decimal and then to a percent. Use the Oral Exercises to point out how parts of a graph are labeled.

10-10 *pages 329–330* Have the students come up with questions for which there is no answer (for example, how many students in the class are less than 6 years old?). Discuss the sample Problem, showing how the list is organized, starting with 3 quarters. Possible strategies for each problem are given here. **Ex. 1** Students might use the strategy *Guess, Check, Revise,* or *Make an Organized List* to find that no more people can sign up. **Ex. 2** Students can *Choose the Operations* to solve this problem. **Ex. 3** Students can *Make an Organized List* to find that there is no solution. **Ex. 4, 5** *Make an Organized List*. **Ex. 6** *Guess, Check, Revise*.

Enrichment *page 331* Students need to understand that they must divide by 2 for a rate compounded semiannually, by 4 for a rate compounded quarterly, and by 12 for a rate compounded monthly.

Chapter 11

11-1 *pages 336–337* Introduce point, line, and plane by referring to physical objects that suggest the three ideas. At the same time, emphasize that the physical objects are models, whereas point, line, and plane are ideas, or concepts. Any attempt to draw a line or plane is imperfect, because it is not possible to draw the "infinite extent" that both concepts involve. Arrows are used on pictures of lines and dots are used on pictures of line segments. A ray is drawn with a dot on one end and an arrow on the other. **Ex. 9–13** These provide an opportunity for students to practice spatial visualization. A cube or box would be a useful model for discussion. **Ex. 14** Note that the third figure is not complete. The first six numbers in this pattern are 3, 6, 10, 15, 21, and 28.

11-2 *pages 338–341* Begin by reviewing the metric units meter, decimeter, centimeter, and millimeter. As an aid in estimation, suggest points of reference for one decimeter (width of a hand), centimeter (width of a fingernail), and millimeter (thickness of a dime). In Examples 2 and 4, equations are used to find an unknown length. Writing an equation is a systematic way of completing these problems that will be helpful with more difficult problems. The equations need to be simplified before they are solved. **Ex. 21–23** These require students to interpret a figure. Although some may find it difficult, this is a skill students should learn. Note that no specific units are used. In discussing the problems, either speak of "units" or choose specific units.

11-3 *pages 342–345* Define an angle, and describe the notation used to name angles. Show a few examples of figures that are not angles. Note that an angle is a set of points, and its measure is a number. Two angles are equal only if they consist of the same two rays. However, different angles may have equal measures. Have students practice using the protractor. Have them measure some angles in which neither ray is horizontal or vertical. Have them measure some angles of a triangle of which the sides must be extended for the protractor to be read.

11-4 *pages 346–348* Emphasize that parallel lines are in the same plane. Two lines in space that have no points in common may not be parallel. Note that two lines in a plane are either parallel or intersecting, and that this is not true of lines in space. In this section, students again write and solve equations to find unknown angle measures. Another solution method for Example 1 is to begin with the equation $m\angle 1 + 61° = 90°$. For Example 2, instead of using the fact that $\angle 4$ and $\angle 2$ are corresponding angles, one could use the fact that $\angle 3$ and $\angle 4$ are supplementary angles. You may want to point out these alternatives for finding the solutions. There is also more than one way to solve many of the exercises. Ask students to describe their solution methods. **Ex. 9–12** These can be completed by combining vertical angles, supplementary angles, and corresponding angles. They do not require knowledge of the concept of alternate interior angles, which has not been introduced.

11-5 *pages 349–351* You might introduce the concept of triangle by moving from the physical to the abstract. First show students a triangular cardboard cutout. Then show them three straws that are joined by pins or other fasteners. Then use a straightedge to draw a triangular figure on the chalkboard. Finally, state the definition of triangle, and ask which of the three figures best fits the definition. You might use straw models to illustrate the classifications by triangles by angle size and by side length. You may want to have students draw one large triangle on an 8 1/2 × 11 sheet of paper and measure the three angles with a protractor to verify that their measures are equal to 180°. In Example 1, do not allow students to skip the equation-writing step. **Ex. 11–14** Write an equation to solve for x by using the fact that the sum of the angles of a triangle is 180°. Note that solving the equations for x in **Ex. 14** does not find the measures of all the angles. Students must substitute the value of x into the expressions to find the measure of each angle.

11-6 *pages 352–355* To present the polygon concept, you could use an approach similar to that for triangles in 11-5: first show cardboard cutouts, then straws fastened together, then drawings on the chalkboard.

Emphasize that for a polygon to be a regular polygon, *two* conditions must be satisfied: its sides must be equal in length, and its angles must be equal in measure. To illustrate that the sum of the angle measures of a quadrilateral equals 360°, cut off the vertices of a cardboard cutout of a quadrilateral and rearrange them as shown in the figure on p. 353. You might have students do this themselves with paper cutouts. You could also have them measure the four angles of a quadrilateral with a protractor and add the angle measures. Another way to illustrate the concept is to show that one diagonal of a quadrilateral divided it into two triangles, each having angle measures that add to 180°. **Ex. 9–11** After writing an equation and solving for x, students must substitute the value of x into the expression to find each angle measure. **Ex. 14** Encourage students to find an expression (in terms of n) for the number of diagonals from one vertex and the number of triangles formed in an n-gon. These patterns are used in **Numbers to Algebra** to discover the sum of the angle measures in a polygon.

11-7 *pages 356–357* Begin by providing some instruction in map-reading. Make sure that students can recognize an interstate highway on a map, and that they can identify the carot marks and the small red and black numbers along a road. Then review the basic equation *Distance* = *rate* × *time* and its two other forms, *rate* = *distance/time* and *time* = *distance/rate*. **Ex. 1–3, 5** These use the map on p. 356. **Ex. 10–12** These use the mileage chart shown. Students may need instruction in reading the chart. **What's Your Decision?** This requires the use of road maps or a road atlas.

11-8 *pages 358–360* Note that *radius* and *diameter* are defined as segments. The same terms are commonly used to mean the *length* of these segments. We use the terms to mean either the segment *or* its length. The ratio C/d refers to the diameter as a number. **Ex. 1–18** You may choose to have students express their answers using the symbol π. Students should realize that they are expressing an exact answer *only* when they use the symbol π. When they substitute 3.14 or the decimal carried out to more places, they are finding an approximation. **Ex. 19–20** These require that an approximation be made. **Ex. 21–23** All arcs shown are either 1/4 of a circle, a semicircle, or 3/4 of a circle. **Ex. 25** If you estimate that 6 meters is approximately 2π meters, then splicing in the 6 meters yields a circle that has a radius approximately 1 meter greater than the radius of the earth. This results in a space under which a cat could walk.

11-9 *pages 361–364* You might begin by using a large bolt and a box full of nuts to demonstrate that all nuts fit onto the one bolt because all the nuts are congruent to one another, that is, they are all the same size and shape. This lesson focuses on the simpler figures of segments, angles, triangles and quadrilaterals. Students need to see that when they write $\triangle ABC = \triangle XYZ$, the order in which they write the vertices has significance. Also, such a statement is really a statement about six congruences: 3 congruent angles and 3 congruent segments. Verifying three of these six congruences with the SAS, SSS, and ASA congruence conditions is sufficient to show that two triangles are congruent. You might teach students how to copy a triangle with compass and straightedge and point out they they are using the SSS property. You might also teach them how to copy an angle and show that, given two angles and the length of a side, one can construct a unique triangle using the SAS or with the ASA property. For the Oral Exercises, you may want to have students trace and cut out the polygons to use for checking congruence. **Ex. 19–20** Students must use the facts that in a regular hexagon, all the angles are congruent to each other and all the sides are congruent to each other. **Ex. 22** You can either treat this as a visual exercise, or have students verify congruences.

11-10 *page 365* You may want to review the Problem-Solving Checklist for using an equation to solve a problem (p. 121), noting that it provides guidelines, not rules.

11-11 *pages 366–368* This lesson requires the use of a compass, and shows five constructions. Students should duplicate each of the five constructions before starting the exercise.

11-12 *pages 369–370* Discuss the sample Problem, noting its many solutions. Possible strategies are given here for each problem. **Ex. 1** *Make an Organized List, Guess, Check, Revise*. **Ex. 2** *Draw a Picture, Simplify the Problem, Make a Table*. **Ex. 3** *Make an Organized List*. **Ex. 4** *Guess, Check, Revise*. **Ex. 5** *Make an Organized List*. **Ex. 6** *Use Logical Reasoning*.

Enrichment *page 371* This lends itself to "hands on" exploration. Students can use mirrors to check for lines of symmetry, and tracing paper to check for rotational symmetry. You may want to have students create symmetry patterns that could be used for a bulletin board display. Paperfolding activities would be a natural extension.

Chapter 12

12-1 *pages 376–379* You could begin by having students draw figures on graph paper and find the areas by counting the squares inside each figure. You might also use an overhead projector: draw a figure on a transparency, superimpose on it a grid of squares, and count the number of squares needed to cover the figure. The text focuses on metric units of area. Point out that each unit of length, including the inch, foot, and yard, is the basis for a unit of area. Make sure students realize that when applying the formula for area of a rectangle, they may consider any one of the four sides as the length. That is, the length is not necessarily the longest side. Use cardboard cutouts to demonstrate that a parallelogram can be transformed into a rectangle by cutting and moving one piece. Emphasize that the height of a parallelogram is the *perpendicular* distance between two parallel sides, and it is not necessarily a vertical distance. Note that in Oral Exercise 3, distance 8 is a height even though it is not vertical. In Oral Exercises 4 and 5, 17 is neither a height nor a base.

Ex. 26 Students should discover that the rectangle of greatest area with perimeter 100 is a square. **Calculator Activity** Explain that whenever a number is rounded before a calculation, the error will be greater than if the rounding is done after the calculation.

12-2 *pages 380–383* Draw one height on the front of a cardboard triangle and another height on the back to illustrate that each side can be a base that is associated with a height. Use this cardboard triangle as a template to draw a triangle on the chalkboard, then rotate the template and draw a second triangle to complete a parallelogram. This drawing leads to the formula for the area of a triangle. You could repeat these with a cardboard trapezoid. **Ex. 1–22** Encourage students to write the formula before beginning each problem. Discourage shortcuts that bypass writing the formula. **Ex. 23–25** You may want students to complete **Numbers to Algebra** before doing these exercises, since it is related. **Ex. 27** This shows that area formulas, such as $A = x^2$ for the area of a square, are based on a square shape for the "basic unit" of area. If an equilateral triangle were the "basic unit" of area, then the formula $A = x^2$ would be used for finding the area of larger equilateral triangles. **Numbers to Algebra** This shows that you can substitute values into a formula and then solve the unknown, or you can rewrite the formula solving for the unknown, then substitute into the rearranged formula.

12-3 *pages 384–386* Make a cardboard model of a circle cut into sectors and rearrange the pieces to form a parallelogram-like figure (as shown in the text) to show the derivation of the formula for the area of a circle. The example uses the approximation 3.14 for π, noting that this gives an approximate answer. You may want to have students express answers with the symbol π. **Ex. 14** Note that when comparing discs of various sizes, it is more efficient to express areas in terms of π than to introduce an approximation. **Ex. 24** You may want to ask students whether they prefer thinking of the small circles as having radii of 1/4 mile or 1320 feet.

12-4 *pages 387–388* Recognizing how to divide up the region and find missing dimensions is the new skill in this lesson. Begin by using the chalkboard or an overhead projector and transparencies to draw several irregularly-shaped regions. Discuss the variety of ways to divide each region into familiar shapes. Show that there is usually more than one way to divide the region and note that it is good practice to consider several of the possibilities before beginning calculations, since one method may result in easier calculations than another. Some divisions may not work (when you are unable to determine a dimension needed). **Ex. 7** The figure could be divided into several triangles and a rectangle, but some needed dimensions could not then be found. Students should divide it by one horizontal line into a trapezoid and a parallelogram.

12-5 *page 389* You may want to review the Checklist for using an equation to solve a problem (p. 121), noting that it provides guidelines, not rules.

12-6 *pages 390–393* Discuss the attributes of a prism and a cylinder. Use physical models as examples of what is and is not a prism or a cylinder. You should plan to purchase or make several models each year until you have a good collection. You can use food containers for many shapes. Review the concept of volume. You can use a shoe box, with building blocks as the unit of volume. Show the volume of the box in block units by counting the number of blocks needed to fill it. Discuss the value of using a standard unit of volume such as the cm^3, and the advantage of using a formula to calculate volume. Complete several calculations with the volume formulas. Point out the similarity in the formulas for prisms and cylinders: both cases involve calculating the area of the base times the height. Show that the formula for volume of a prism becomes length times width times height when it is applied to a rectangular prism. **Ex. 8–19** Units in some of these are mixed. Remind students that they will need to express each dimension in the same unit. **Ex. 23–24** Students should consider the small cube that seems to be sitting on top of the figure as the unit of volume, then count the number of such cubes that are glued together to complete the solids shown.

12-7 *pages 394–395* Read through the list of materials and prices. Give the students a visual sense of a cubic yard and square yard. Mention that 3 bundles of shingles are needed for a 10 ft × 10 ft square, which roofers sometimes call a "square." Complete the Example to make sure students understand that the units in all three dimensions must be compatible. Challenge Problems: For **Ex. 5** have students assume that 10% extra plywood must be bought to allow for scrap. For **Ex. 6** figure the cost of wallpaper for a room 14 × 12 feet with 8-foot ceilings. **What's Your Decision?** The pre-cut piece is 2 ft^2 too small. Discussion might involve the question of whether they could "piece" that carpet to fit and hide the hole. Also, note that the salesman wants them to buy a 12 × 19 ft carpet, wasting the 10 × 4 ft area in the corner. They might be able to piece a carpet measuring less than 30 square yards.

12-8 *pages 396–398* Discuss the attributes of a pyramid and a cone. Emphasize that a pyramid's base is *not* necessarily a triangle or square. It can be any polygon. To give a dramatic illustration of the volume formulas, use hollow models of a pyramid and a prism with the same base and height. Fill the pyramid with sand or rice three times and pour it into the prism to demonstrate that the volume of the prism is 3 times the volume of the pyramid. This explains the need for the fraction 1/3 in the formula for the volume of a pyramid. For Exercises asking for the volume of a cone, you must decide whether to have students use an approximation for π or give their answers in terms of π. The approximation 3.14 was used for the answers provided here. **Ex. 32–35** These involve using the formula $V = \frac{4}{3}\pi r^3$ to find the volume of a sphere. **Ex. 36–37** You may want to use physical models of spheres to help students visualize the stacking patterns.

12-9 *pages 399–402* Begin this lesson by taking construction-paper models of a cylinder and a prism, cutting them apart and laying them flat, as shown in the text. Use the models to explain the formulas for surface area of a cylinder and a prism. In the case of a rectangular prism, as in Example 2, each pair of parallel faces can be viewed as bases. A small box is a good model to help students visualize the base and height relationship. **Ex. 19, 20** Note that the surface area of a sphere formula is used.

12-10 *pages 403–404* Discuss the sample Problem. Note that psychologists call the moment of special insight the "aha!" experience. All the problems here require special insight; you may want to provide hints if students seem to need them.

Enrichment *page 405* This gives students practice in calculating areas of irregularly-shaped regions by dividing them into triangular and rectangular subregions, as illustrated in the Example. After using this procedure for finding the areas of the regions in Ex. 1–3, have students complete the table and discover Pick's formula. They can check the formula on several figures that you draw, then use the formula to find the area of the regons in Ex. 4, 5, and 6.

Chapter 13

13-1 *pages 410–412* Use the election ballot on page 410 to introduce the term *outcome* by discussing the sample outcome given. Have students suggest other outcomes. Explain that a *tree diagram* is one method for listing all possible outcomes of an event. Note the connection to the problem-solving strategy *Make an Organized List*. Explain how to follow a branch to show an outcome. Have students read all the possible outcomes on the tree diagram from top to bottom. Introduce the counting principle, and discuss Example 1. You might have students draw a tree diagram to show its relation to the counting principle. **Ex. 15** Students will need to list flights to check the conditions for time between flights.

13-2 *pages 413–415* Begin with an activity in which the class finds the number of arrangements of 3 students in a row. You may want to have a student record the different arrangements on the chalkboard as the three rearrange themselves. Introduce the term *permutation*, and relate this activity to the books pictured in the text. Use the picture of the three books to explain how the counting principle can give the total number of outcomes or permutations. Introduce the term *factorial* and the steps for finding the total number of permutations of n objects. Make sure students understand the meaning of the three dots in the formula. Discuss Example 1. You may want to have the students try to find all 24 arrangements. Introduce the ordering of r objects out of a set of n objects by finding the ways to arrange 2 out of 3 students in front of the room. Introduce the steps for finding the permutations of n objects taken r at a time. Work through Example 2 and its Practice. **Ex. 26–27** See Solutions Manual for detailed discussions.

13-3 *pages 416–418* Begin with an activity in which 2 out of 4 students are to be selected to be class representatives. Have 4 students stand in front of the room, and record on the chalkboard the names of possible pairs of representatives. Emphasize that the order does not matter when we are concerned with a *selection*. Compare this with the permutation activity, in which order was important. Discuss the two steps shown on page 416: first listing the number of permutations, then, since order is not relevant, eliminating 10 of the outcomes. Introduce the term *combination* and stress the difference between the 20 permutations and the 10 combinations. Introduce the steps for finding the number of combinations, relating them to the steps for finding permutations (page 413). Point out that dividing by $r!$ removes the outcomes that have duplicate elements in different orders. Discuss Example 1 and its Practice. **Ex. 27** Students could *Guess, Check, Revise* to find the number of combinations of 8 people taken 4 at a time.

13-4 *pages 419–421* Begin by having students examine a die or other cube with faces numbered 1 through 6. Introduce the term "equally likely." Discuss situations in which outcomes are and are not equally likely. Introduce the terms *probability* and *event*. Examples 1a and 1b illustrate how an event can be made up of several outcomes. Example 1c illustrates that the *probability of an impossible event is 0*. That the *probability of a certain event is 1* is illustrated in Practice Exercise b. Introduce the term *mutually exclusive*. Discuss Example 2 and its Practice. Before assigning the exercises, check to see that students are familiar with the elements of a standard deck of 52 cards. **Ex. 31** Students could *Make an Organized List* of all possible outcomes.

13-5 *pages 422–424* Write the numbers 1, 2, 3 and the letters A, B, E on the chalkboard. Ask how many outcomes there would be for selecting 1 number and 1 letter. (By the counting principle, there are 9.) Ask how many of these outcomes have an odd number *or* a vowel (4). Discuss the probability of getting an odd number *and* a vowel, 4/9. Direct attention to the text's discussion of the two spinners, noting that each outcome on each spinner is equally likely. Introduce the term *independent* by discussing the two spinners. Introduce the formula for independent events. Complete Example 1 and its Practice. Explain the term *dependent*. You may want to illustrate the text's discussion of marbles in a bag with a demonstration, using a bag of colored marbles. Emphasize that whether the first marble is *replaced* determines whether the events are independent or dependent. Discuss Example 2 and its Practice. **Ex. 25–28** These require the student to extend the formula to multiple selections without replacement.

13-6 *page 425* You may want to begin by reviewing the Checklist for using equations to solve problems (p. 121). Point out that students may use strategies other than *Write an Equation*, such as drawing pictures or using simpler numbers, as an aid to solving.

13-7 *pages 426–428* Students should be familiar with the word *data*. Discuss the *frequency table*. Have students count the tally marks to verify the frequencies given in the table. Explain how the *range* and *mode* are calculated, and point out that both are ways to describe how data are distributed. Discuss the Example and its Practice. The data in Exercises 1–6 are arranged in order from smallest to largest, making it easier for the students to construct frequency tables. In Exercises 7–13, the data are not ordered, so students need to be careful to record the tally marks accurately. **Ex. 13** Students should use the largest score in the top interval and the lowest score in the bottom interval to calculate the range. The mode will be an interval. **Estimation** Note that reading a graph often involves estimating the position of each bar or line.

13-8 *pages 429–431* Begin by reviewing range and mode. Discuss with students uses and meanings of the word "average." Discuss the data on amounts spent by restaurant chains on advertising. Introduce the terms *mean* and *median*, and work through the Example. Note that the data do not have to be listed in order to find the mean, but they do to find the median. The Example has an even number of numbers, so there are two middle numbers. The Practice shows how to find the median of an odd number of numbers (middle number). **Ex. 24** Note that "batting average" involves a different use of the term "average" (i.e., this average is not found like a mean). **Ex. 25** Students should look for pairs of numbers that can be added without paper and pencil (e.g., 195 + 305 = 500). **Ex. 26** Students might use the strategy *Guess, Check, Revise,* or *Logical Reasoning*.

13-9 *pages 432–433* Discuss the computer store setting and have students suggest ways in which a manager might use mathematics. Discuss reasons a computer store manager might want to show data in the form of a graph, or might have to get data from graphs to answer questions. Complete the Example with the class, following the Problem-Solving Checklist if you wish. **Data Search** Allow students to find prices of fewer than 5 computers if necessary. You might suggest that they look for advertisements if there are no stores nearby. **What's Your Decision** Students should consider factors other than cost (e.g., service and length of warranty) that might influence their decision.

13-10 *pages 434–437* Show sample bar graphs and line graphs from newspapers and magazines, if possible, and review the reading of the graphs. Emphasize the key components of the graph: title, labels for all parts, and scale. Discuss the steps for constructing a bar graph. Point out that determining the scale is usually the most difficult step. Discuss Example 1. Ask why $5 was selected for the scale, and discuss how other scales might affect the graph. Discuss Example 2, again asking why a scale of 1 unit = 1 person was selected and what other scales could have been used. **Ex. 5** Discuss the influence a graph's scale can have on the perception of data.

13-11 *pages 438–440* Show pictographs in newspapers and magazines, if possible, and review the reading of pictographs. Introduce the steps for making a pictograph. As with bar and line graphs, selecting the appropriate scale is usually the most difficult step. Discuss the Example, emphasizing that the actual numbers are "rounded" to the scale selected for the graph. Note that the numbers here are rounded to the nearest multiple of 5; make sure students understand the technique. **Ex. 8** This illustrates the role and importance of rounding. Many correct sets of data could be given.

13-12 *pages 441–442* You may want to begin by reviewing the Problem-Solving Checklist on p. 22. Emphasize that more than one strategy can be used to solve a given problem. Possible strategies for each problem are given here. **Ex. 1** *Draw a Picture, Make a Table, Simplify the Pattern, Look for a Pattern*. **Ex. 2** *Use Logical Reasoning, Draw a Picture*. **Ex. 3** *Make an Organized List*. **Ex. 4** *Work Backward*. **Ex. 5** *Guess, Check, Revise*. **Ex. 6** *Guess, Check, Revise*. **Ex. 7** *Make an Organized List* **Ex. 8** *Guess, Check, Revise*.

Enrichment *page 443* To conduct this experiment, each student will need a sheet of paper and a toothpick cut to a 2-cm length.

Chapter 14

14-1 *pages 448–450* Introduce the terms "square" and "square root" and the use of the radical symbol. Give several examples. Note that though the radical symbol refers to the positive square root, a positive number has *two* square roots. A negative sign must be used when referring to a negative square root. Note that the square root of a number that is not a perfect square is an irrational number; square root is an increasing function. That is, as a variable gets larger, its square also gets larger: $16 < 18$, and $\sqrt{16} < \sqrt{18}$. Point out that when evaluating expressions as in Example 3, it is important to notice whether or not operations are under the radical sign. **Ex. 29, 32, 34**, etc. Point out that the square root of a sum is not the sum of the square roots. **Ex. 42–44** focus on multiplying a square root by a whole number.

14-2 *pages 451–453* Emphasize that since square roots of numbers that are not perfect squares are irrational numbers, the decimals shown are only approximations. You may wish to begin the lesson by verifying a sequence of successive approximations for $\sqrt{2}$, such as: $1.4 < \sqrt{2} < 1.5$; $1.41 < \sqrt{2} < 1.42$; $1.414 < \sqrt{2} < 1.415$, etc. Discuss the roles of tables and calculators in finding square roots. **Ex. 1–27** These provide practice in using a table and calculator. Note that **Ex. 38–39** do not require finding a square root of a non-perfect square. The calculator activity provides a method of approximating a square root when using a calculator without a square root key.

14-3 *pages 454–455* Begin by listing examples of the types of equations the students will be solving in this lesson. Include cases with perfect squares and non-perfect squares. Introduce the Square Root Property, the key to solving these equations. Show that $\sqrt{x^2} = x$ is true only for positive values of x. For example, $\sqrt{(-3)^2} \neq -3$. In the Oral Exercise, students must recognize perfect squares. **Ex. 1–24** Note that when an answer is not a perfect square, the square root form (e.g., $\sqrt{47}$) is used instead of an approximation. **Ex. 25–28** You may want to give the hint to focus on small positive integers in the *Guess and Check* process. **Estimation** Students must know the perfect squares up to 100.

14-4 *pages 456–459* Introduce the terms "hypotenuse" and "leg." A physical model that illustrates the Pythagorean Theorem would be helpful. If you do not already have a model, you could make a tracing of the 3-4-5 triangle figure that appears on the student text, and make separate cutouts of the 25 unit squares. Put both on an overhead projector, and arrange 9 unit square cutouts on the square on one leg and 16 on the other leg. Then arrange the same 25 squares on the square on the hypotenuse. To make sure your students do not just memorize the equation $a^2 + b^2 = c^2$, you might draw a right triangle with hypotenuse length a and leg lengths b and c, with the hypotenuse either a or b. Then state the Pythagorean Theorem as $a^2 = b^2 + c^2$. **Ex. 1–24** Remind students again that the radical form is an exact answer, not an approximation.

14-5 *pages 460–461* Discuss the data on official dimensions of playing fields. **Ex. 1–5, 7** require the use of the Pythagorean Theorem. Students must decide when the data table is needed. **Ex. 8** requires the use of ratios. **What's Your Decision** Discuss students' decisions.

14-6 *pages 462–464* You might begin by discussing equipment such as microscopes that enlarges images without changing their shapes. You could draw a triangle on a transparency and project it onto a screen with an overhead projector, then have students measure the sides and angles of the two triangles to discover their equivalence. Review the terminology of corresponding angles and sides and the notation used for similarity. Explain the use of the arc marks to show angles that are congruent. Show how knowing corresponding angles determines how to find the corresponding sides. Use the example to show how to write and solve a proportion. **Ex. 7–8** Point out that the triangles are not similarly positioned. **Ex. 9–12** Encourage students to write down the similarity relationship before trying to complete the problem. **Ex. 12** Note that the ratio of the two altitudes is equal to the ratio of two corresponding sides.

14-7 *pages 465–467* Discuss the photo caption, showing that the 45-45 right triangle is half of a square. Use the Pythagorean Theorem to show that the hypotenuse of the 45-45 right triangle is $\sqrt{2}$ times the length of the leg. Then show that the 30-60 right triangle is half of an equilateral triangle, that its hypotenuse is twice the length of the short leg, and that the long leg is $\sqrt{3}$ times the length of the short leg. Emphasize the development of these relationships to help students remember them. **Oral Exercises** The figures are drawn to look like 45-45 or 30-60 right triangles, but the lengths of the sides are not all labeled correctly. **Ex. 1–8** Encourage students to use the two theorems as illustrated in Examples 1 and 2. Some may find it easier to rederive the relationship by using the Pythagorean Theorem each time. **Ex. 9–10** Students need to recognize that a 30-60 right triangle is part of the problem. **Ex. 11–12** Encourage students to draw figures. **Ex. 14–15** The secret is to recognize that ABC is an equilateral triangle. Once the side lengths are found, the altitude can be found and the area calculated.

14-8 *pages 468–471* Discuss the photo and the fact that we cannot always measure directly. Then define the ratios of sine, cosine, and tangent as ratios of lengths of sides of a right triangle. As a memory device, point out that the sine and cosine involve the use of the hypotenuse, whose length is always in the denominator, and that the tangent is a ratio between the lengths of the two legs. The skill of finding the ratios is shown in Example 1. Return to the problem in the photo. Point out that the values for all the trigonometric ratios for all angles can be found in tables or by using a scientific calculator. You may want to illustrate how these values can be found. Solve the problem in the photo again using different numbers for length AC and angle BAC. Before assigning **Ex. 11–18**, review the properties of the 45-45 triangle and the 30-60 triangle. You may want to ask students to complete a table of values for the sine, cosine, and tangent for 30°, 45°, and 60°. Express these values using radical symbols rather than decimal approximations. Students can refer to this table when completing **Ex. 22–31**. Point out to your students that the values on the right side of the equation in **Ex. 22–31** are values that appear in the table of trigonometric values that they made.

14-9 *page 472* You may want to review the guidelines on solving problems by writing equations (p. 121).

14-10 *pages 473–474* Review the Problem-Solving Checklist (p. 22). Possible strategies for each problem are given here. **Ex. 1** *Choose the Operations.* **Ex. 2** *Write an Equation.* **Ex. 3** *Make a Table, Look for a Pattern.* **Ex. 4** *Logical Reasoning.* **Ex. 5** *Make a Table.* **Ex. 6** *Use Logical Reasoning, Draw a Picture.* **Ex. 7** *Simplify the Problem, Look for a Pattern.* **Ex. 8** *Work Backward.* **Ex. 9** *Make an Organized List.* **Ex. 10** *Make a Table, Use Logical Reasoning.*

Enrichment *page 475* The main focus of this enrichment is that irrational numbers such as $\sqrt{2}$ or $\sqrt{3}$ cannot be expressed exactly as decimals. However, these irrational numbers can be approximated with rational numbers. Students should use a calculator to square each term in these sequences to see that the square is approximately 2 or 3. This means the fraction is approximately $\sqrt{2}$ or $\sqrt{3}$. Successive terms of a sequence are better and better approximations.

Chapter 15

15-1 *pages 480–481* Use a number line to introduce the concept of *real numbers*. Review the techniques learned in Chapter 8 for solving an inequality such as $2x < 6$. In Example 1, point out the preliminary step of adding 5 to both sides of the inequality. Emphasize the similarities between solving an inequality and solving an equation, but note that when you multiply or divide both sides of an inequality by a negative number, you reverse the sign of inequality. Make sure students understand that they must first solve an inequality, then graph it. Point out that the arrow on the graph shows that the solution set is infinite in one direction. **Ex. 21–28** These involve double inequalities. Point out that an inequality such as $5 < x < 3$ does not have a solution, since a number cannot at the same time be greater than 5 and less than 3. Explain that we write a double inequality only when both inequalities are in the same direction. We do not write $3 > x > 5$. **Ex. 29** Suggest drawing a diagram:

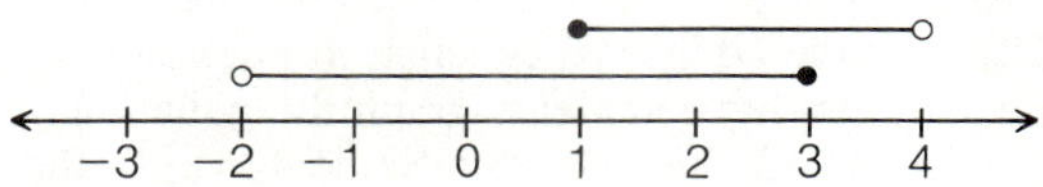

This will lead to the answer:

Ex. 30 In discussion, provide the geometric interpretation of the solution: all numbers of distance 3 from the number 5.

15-2 *pages 482–485* Discuss the road map on page 482. Ask students to think of other situations in which a pair of numbers or a number and a letter record a position (e.g., row and seat in an auditorium; aisle and side in a grocery store). Introduce the coordinate axes and the rectangular coordinate system. Include examples of points whose coordinates are not both integers. Emphasize the fact that each pair of real numbers determines a unique point, and each point has a unique set of coordinates.

15-3 *pages 486–488* Begin by giving examples of linear equations and of equations that are not linear. Note that linear equations are straight lines. The form $y = ax + b$ rather than the form $ax + by + c = 0$ is used here to define linear equation because it is the easiest form for finding a table of solutions. Point out that in a linear equation, both variables have an exponent of 1. Any equation with variables raised to a power greater than 1 or with xy terms cannot be linear. Have students practice determining whether or not a given point satisfies a given equation before trying to complete tables of solutions. In Example 2, stress that we could be choosing *any* values for x, not just the numbers -2, -1, 0, 1, 2, 3, and 4. Note that a line contains an infinite number of points, and all are solutions to a line's equation. After drawing the graph of a linear equation in Example 3, choose a noninteger value for x, find the corresponding value of y, and graph the point to show that it is also on the line. **Ex. 1–12** These lead up to **13–20**, which represent the main focus of this lesson. To provide additional practice in graphing equations, have students graph the equations in Ex. 1–12. **Ex. 21–28** Students verify that the given equations are linear equations. **Ex. 30** To extend this exercise, have students consider the general equation $x/a + y/b = 1$.

15-4 *page 489* You may want to begin by reviewing the Checklist of guidelines for using equations to solve problems (p. 121). Point out that students may use other strategies as well.

15-5 *pages 490–493* Draw a rectangular coordinate system. Label the horizontal axis "hours," and the vertical axis "miles." Using the example of a car traveling at 40 mi/h, graph the points (1, 40), (2, 80), (3, 120), etc. Show that as the distance increases at a constant rate, the distance traveled increases at a constant rate. Show that the points, which are on a line through the origin, satisfy the equation $y = 40x$. These properties characterize "y varies directly with x." On another rectangular coordinate system, label the horizontal axis "hours," and the vertical axis "km/h." Using the example of a car that needs to travel 40 km, graph the points (2, 20), (1, 40), (1/2, 80), (1/3, 120), etc. Draw the line through these points and show that as the time increases, the speed decreases, and that these points satisfy the equation $xy = 40$. These properties characterize "y varies inversely with x." Students should have a geometric feeling for direct and inverse variation as well as an algebraic feeling. **Ex. 11–13** Students may find it helpful to graph the points, as suggested above, to determine which equation to check for direct or inverse variation.

15-6 *pages 494–497* Use accurate graphs—either a transparency grid on an overhead projector, or a grid on the chalkboard. Draw a line on the grid. Choose two points A and B on the line, and count the number of horizontal unit steps from A to B, then the number of vertical unit steps from A to B. Draw the "rise to run" triangle, as shown in the text, and calculate the slope of the line. Introduce the definition of *slope*. Complete Examples 1 and 2 and their Practices, emphasizing that to find the x-intercept you let $y = 0$, and to find the y-intercept you let $x = 0$. **Estimation** You get equal temperatures if for Celsius you count by 5's beginning with zero, and for Fahrenheit you count by 9's beginning with 32.

15-7 *pages 498–499* Bring in some pages of newspaper ads, having students identify types of ads (some sold by the line, some blocked out and sold by the inch, some one column wide, some several columns wide). Discuss the fact that most newspaper revenue comes from advertising sales. Make sure students understand the information on column width and column length and the advertising costs listed in the table.

15-8 *pages 500–501* Complete the Example, explaining each step carefully. This lesson focuses on the concepts of a system of equations and a solution of a system of equations. Students find the solution by graphing each equation, visually determining the point of intersection, and then verifying algebraically that the visual determination is correct. If students are not able to complete the **Oral Exercises** mentally, allow them to use paper and pencil. **Ex. 10–12** Students should determine solutions by graphing the equations. **Ex. 15** Students should first write an equation for the cost of each of the rental cars, then graph the equations and find the solution.

15-9 *pages 502–503* Begin by graphing the equation $y = x + 2$ and verifying that (1, 3) is a point on the graph. Then verify that (1, 4), (1, 5), (1, 6), etc., all satisfy $y > x + 2$, and that (1, 2), (1, 1), (1, 0), etc. all satisfy $y < x + 2$. Repeat this process, beginning with the point $(-3, -1)$ on the line and checking points $(-3, 0)$, $(-3, 1)$, $(-3, 2)$, etc. and $(-3, -2)$, $(-3, -3)$, $(-3, -4)$, etc. Show that points below the line $y = x + 2$ satisfy the inequality $y < x + 2$, and that points above the line satisfy the inequality $y > x + 2$. **Ex. 1–9** Students should draw the boundary line and then shade in the appropriate region for the graph. **Ex. 10–15** Have students first rewrite the inequality into the $y = ax + b$ form. **Ex. 18–20** Ask students to graph both inequalities on the same graph and shade each with a different color of pencil. The region that is shaded with both colors is the graph of the system of equations.

15-10 *page 504* Possible strategies for each problem are given here. **Ex. 1** *Choose the Operations*. **Ex. 2** *Write an Equation, Choose the Operations*. **Ex. 3** *Simplify the Problem, Make a Table, Look for a Pattern, Draw a Picture*. **Ex. 4** *Use Logical Reasoning*. **Ex. 5** *Use Logical Reasoning, Guess, Check, Revise*. **Ex. 6** *Make a Table, Look for a Pattern*.

Enrichment *page 505* The graph of the "open" box shows that the "endpoint" does not belong to the graph. This Enrichment illustrates that if you try to sketch a graph after plotting only a few points, you might sketch it incorrectly. In addition to plotting points, it is important to study the "rules" behind the graph. In this case, you observe that the charge for long-distance service is constant throughout a 1-minute period, and that the graph therefore shows a straight-line segment for each minute.

Management Guide

The Management Chart lists assignments for minimum, regular, and maximum courses. The total number of days is based on 174, with 87 assignment days for each semester.

First Semester

Chapter	1	2	3	4	5	6	7	8
Days	10	11	11	11	11	12	10	11

Second Semester

Chapter	9	10	11	12	13	14	15
Days	12	12	13	11	13	13	13

Supplementary materials are referenced in the right-hand column. A key to the abbreviations used in the Management Chart is given below.

MR Mixed Review
PS Practice Supplement
AS Assessment Supplement

Note: Numbers to Algebra, Mental Math, Estimation, Calculator Activities and Computer Activities are suitable for all levels and may be assigned at the teacher's option.

Section	Minimum	Regular	Maximum	Supplements
1-1	1–44, 59, MR	1–44 even, 45–50, MR	1–44 even, 45–59, MR	
1-2	1–36, MR	1–36 even, 37–40, MR	1–36 even, 37–42, MR	PS 1
1-3	1–29, MR	1–29 even, 30–42, MR	1–29, 30–46, MR	
1-4	1–14, MR	1–14 even, 15–18, MR	1–14 even, 15–19, MR	PS 2, AS Quiz p. 1
1-5	1–30, 37, MR	1–30 even, 31–44, MR	1–30 even, 31–46, MR	
1-6	1–30, MR	1–30 even, 31–36, MR	1–30 even, 31–39, MR	PS 3
1-7	1–11	1–11	1–11	AS Quiz p. 2
1-8	1–22, 29–30, MR	1–22 even, 23–30, MR	1–22 even, 23–30, MR	PS 4
1-9	1–2	1–4	1–6	
Review/ Testing	Chapter 1 Review, Chapter 1 Test, Cumulative Review			AS pp. 31–32, 61–62

Section	Minimum	Regular	Maximum	Supplements
2-1	1–33, 42–43, MR	1–33 even, 34–43, MR	1–33 even, 34–43, MR	
2-2	1–30, MR	1–30 even, 31–34, MR	1–30 even, 31–36, MR	PS 5
2-3	1–24, MR	1–24 even, 25–33, MR	1–24 even, 25–35, MR	
2-4	1–29, MR	1–29 even, 30–37, MR	1–29 even, 30–39, MR	PS 6
2-5	1–24, 34–35, MR	1–24 even, 25–35, MR	1–24 even, 25–35, MR	AS Quiz p. 3
2-6	1–32, 41–42, MR	1–32 even, 33–42, MR	1–32 even, 33–42, MR	PS 7
2-7	1–9	1–9	1–9	
2-8	1–9, 16, MR	1–16, MR	1–17, MR	AS Quiz p. 4
2-9	1–27, MR	1–27 even, 28–33, MR	1–27 even, 28–35, MR	PS 8
2-10	1–2	1–5	1–9	
Review/ Testing	Chapter 2 Review, Chapter 2 Test, Cumulative Review			AS pp. 33–34, 63–64

Section	Minimum	Regular	Maximum	Supplements
3-1	1–25, 38–40, MR	1–25 even, 26–41, MR	1–25 even, 26–41, MR	
3-2	1–24, MR	1–24 even, 25–28, MR	1–24 even, 25–30, MR	PS 9
3-3	1–39, 51–52, MR	1–39 even, 40–52, MR	1–39 even, 40–52, MR	
3-4	1–30, 37–38, MR	1–30 even, 31–37, MR	1–30 even, 31–38, MR	AS Quiz p. 5
3-5	1–9	1–9	1–9	PS 10
3-6	1–28, 49, MR	1–49 even, MR	1–50 even, MR	
3-7	1–24, 39–40, MR	1–42 even, 43–44, MR	1–42 even, 43–45, MR	PS 11
3-8	1–28, MR	1–28 even, 29–38, MR	1–28 even, 29–38, MR	AS Quiz p. 6
3-9	1–21, MR	1–21 even, 22–25, MR	1–21 even, 22–27, MR	PS 12
3-10	1–2	1–4	1–7	
Review/ Testing	Chapter 3 Review, Chapter 3 Test, Cumulative Review			AS pp. 35–36, 65–66

Section	Minimum	Regular	Maximum	Supplements
4-1	1–18, 31–32, MR	1–18 even, 19–32, MR	1–18 even, 19–32, MR	
4-2	1–30, 36–37, MR	1–30 even, 31–37, MR	1–30 even, 31–39, MR	PS 13
4-3	1–22, 32, MR	1–22 even, 23–32, MR	1–22 even, 23–34, MR	
4-4	1–18, MR	1–18 even, 19–25, MR	1–18 even, 19–27, MR	PS 14
4-5	1–18, MR	1–18 even, 19–32, MR	1–18 even, 19–34, MR	AS Quiz p. 7
4-6	1–11	1–11	1–11	
4-7	1–21, MR	1–21 even, 22–28, MR	1–21 even, 22–30, MR	PS 15
4-8	1–6, MR	1–6 even, 7–9, MR	1–6 even, 7–11, MR	AS Quiz p. 8
4-9	1–15, MR	1–15 even, 16–26, MR	1–15 even, 16–28, MR	PS 16
4-10	1–2	1–4	1–6	
Review/ Testing	Chapter 4 Review, Chapter 4 Test, Cumulative Review			AS pp. 37–38, 67–68

Section	Minimum	Regular	Maximum	Supplements
5-1	1–35, MR	1–35 even, 36–41, MR	1–35 even, 36–43, MR	
5-2	1–32, MR	1–32 even, 33–46, MR	1–32 even, 33–49, MR	PS 17
5-3	1–5, 9–10, MR	1–5 even, 6–10, MR	1–5 even, 6–10, MR	
5-4	1–20, 27, MR	1–20 even, 21–24, 27, MR	1–20 even, 21–27, MR	PS 18
5-5	1–40, MR	1–50 even, 51–53, MR	1–50 even, 51–55, MR	AS Quiz p. 9
5-6	1–29, MR	1–29 even, 30–39, MR	1–29 even, 30–42, MR	PS 19
5-7	1–32, MR	1–32 even, 33–40, MR	1–32 even, 33–42, MR	
5-8	1–12	1–12	1–12	AS Quiz p. 10
5-9	1–28, MR	1–28 even, 29–45, MR	1–28 even, 29–47, MR	PS 20
5-10	1–2	1–4	1–6	
Review/ Testing	Chapter 5 Review, Chapter 5 Test, Cumulative Review			AS pp. 39–40, 69–70

Section	Minimum	Regular	Maximum	Supplements
6-1	1–44, 62, MR	1–44 even, 45–60, 62, MR	1–60 even, 61–63, MR	
6-2	1–32, MR	1–32 even, 33–39, MR	1–32 even, 33–40, MR	PS 21
6-3	1–26, 39, MR	1–26 even, 27–39, MR	1–26 even, 27–40, MR	
6-4	1–32, MR	1–32 even, 33–46, MR	1–32 even, 33–48, MR	PS 22
6-5	1–27, MR	1–27 even, 28–48, MR	1–27 even, 28–51, MR	AS Quiz p. 11
6-6	1–24, 31, MR	1–24, 25–31, MR	1–24 even, 25–32, MR	
6-7	1–18, MR	1–18 even, 19–29, MR	1–18 even, 19–30, MR	PS 23
6-8	1–10	1–10	1–10	AS Quiz p. 12
6-9	1–27, MR	1–27 even, 28–36, MR	1–27 even, 28–38, MR	
6-10	1–5, MR	1–5 even, 6–8, MR	1–5 even, 6–10, MR	PS 24
6-11	1–2	1–4	1–6	
Review/ Testing	Chapter 6 Review, Chapter 6 Test, Cumulative Review			AS pp. 41–42, 71–72

Section	Minimum	Regular	Maximum	Supplements
7-1	1–27, MR	1–27 even, 28–50, MR	1–50 even, 51–53, MR	
7-2	1–28, MR	1–28 even, 29–33, MR	1–28 even, 29–35, MR	PS 25
7-3	1–24, MR	1–24 even, 25–33, MR	1–24 even, 25–36, MR	
7-4	1–5, MR	1–8, MR	1–10, MR	AS Quiz p. 13
7-5	1–10	1–10	1–10	PS 26
7-6	1–16, 34–35, MR	1–16 even, 17–35, MR	1–16 even, 17–36, MR	PS 27
7-7	1–28, 45, MR	1–28 even, 29–45, MR	1–28 even, 29–48, MR	AS Quiz p. 14
7-8	1–24, MR	1–24 even, 25–31, MR	1–24 even, 25–34, MR	PS 28
7-9	1–2	1–4	1–6	
Review/ Testing	Chapter 7 Review, Chapter 7 Test, Cumulative Review			AS pp. 43–44, 73–74

Section	Minimum	Regular	Maximum	Supplements
8-1	1–24, MR	1–24 even, 25–31, MR	1–24 even, 25–33, MR	
8-2	1–8, MR	1–11, MR	1–8 even, 9–13, MR	PS 29
8-3	1–5, 9–10, MR	1–5 even, 6–10, MR	1–5 even, 6–10, MR	
8-4	1–24, MR	1–24 even, 25–34, MR	1–24 even, 25–36, MR	PS 30, AS Quiz p. 15
8-5	1–24, 33–34, MR	1–24 even, 25–34, MR	1–24 even, 25–34, MR	
8-6	1–24, MR	1–24 even, 25–36, MR	1–24 even, 25–38, MR	
8-7	1–11	1–11	1–11	PS 31
8-8	1–18, MR	1–18 even, 19–25, MR	1–18 even, 19–27, MR	AS Quiz p. 16
8-9	1–27, MR	1–27 even, 28–37, MR	1–27 even, 28–39, MR	PS 32
8-10	1–2	1–4	1–6	
Review/ Testing	Chapter 8 Review, Chapter 8 Test, Cumulative Review			AS pp. 45–46, 75–76, 91–93

Section	Minimum	Regular	Maximum	Supplements
9-1	1–20, MR	1–20 even, 21–35, MR	1–20 even, 21–37, MR	
9-2	1–32, MR	1–32 even, 33–47, MR	1–32 even, 33–49, MR	PS 33
9-3	1–17, MR	1–17 even, 18–20, MR	1–17 even, 18–22, MR	
9-4	1–9, MR	1–10, MR	1–12, MR	AS Quiz p. 17
9-5	1–5, MR	1–5 even, 6–8, MR	1–5 even, 6–10, MR	PS 34
9-6	1–13, MR	1–13 even, 14–19, MR	1–13 even, 14–22, MR	
9-7	1–14	1–14	1–14	
9-8	1–40, MR	1–40 even, 41–44, MR	1–40 even, 41–47, MR	PS 35, AS Quiz p. 18
9-9	1–40, MR	1–40 even, 41–45, MR	1–40 even, 41–47, MR	
9-10	1–48, MR	1–48 even, 49–54, MR	1–48 even, 49–58, MR	PS 36
9-11	1–2	1–4	1–6	
Review/ Testing	Chapter 9 Review, Chapter 9 Test, Cumulative Review			AS pp. 47–48, 77–78

Section	Minimum	Regular	Maximum	Supplements
10-1	1–27, MR	1–27 even, 28–40, MR	1–27 even, 28–44, MR	
10-2	1–28, MR	1–28 even, 29–37, MR	1–28 even, 29–40, MR	PS 37
10-3	1–26, MR	1–26 even, 27–37, MR	1–26 even, 27–40, MR	
10-4	1–5, MR	1–8, MR	1–10, MR	PS 38
10-5	1–16, MR	1–16 even, 17–24, MR	1–16 even, 17–26, MR	AS Quiz p. 19
10-6	1–16, MR	1–16 even, 17–21, MR	1–16 even, 17–23, MR	
10-7	1–13	1–13	1–13	PS 39
10-8	1–25, MR	1–25 even, 26–35, MR	1–25 even, 26–37, MR	AS Quiz p. 20
10-9	1–7, MR	1–8, MR	1–9, MR	PS 40
10-10	1–3	1–5	1–7	
Review/ Testing	Chapter 10 Review, Chapter 10 Test, Cumulative Review			AS pp. 49–50, 79–80

Section	Minimum	Regular	Maximum	Supplements
11-1	1–8, MR	1–8 even, 9–13, MR	1–8 even, 9–14, MR	
11-2	1–16, MR	1–16 even, 17–20, MR	1–16 even, 17–23, MR	
11-3	1–16, MR	1–16 even, 17–22, MR	1–16 even, 17–24, MR	PS 41
11-4	1–8, MR	1–8 even, 9–11, MR	1–8 even, 9–13, MR	AS Quiz p. 21
11-5	1–10, MR	1–10 even, 11–14, MR	1–10 even, 11–15, MR	
11-6	1–8, MR	1–8 even, 9–13, MR	1–8 even, 9–14, MR	PS 42
11-7	1–13	1–13	1–13	AS Quiz p. 22
11-8	1–18, MR	1–18 even, 19–23, MR	1–18 even, 19–25, MR	
11-9	1–18, MR	1–18 even, 19–20, MR	1–18 even, 19–22, MR	PS 43
11-10	1–4, MR	1–4 even, 5–8, MR	1–4 even, 5–10, MR	
11-11	1–9, MR	1–8 even, 10–18, MR	1–8 even, 10–20, MR	PS 44
11-12	1–6	1–6	1–6	
Review/ Testing	Chapter 11 Review, Chapter 11 Test, Cumulative Review			AS pp. 51–52, 81–82

Section	Minimum	Regular	Maximum	Supplements
12-1	1–11, MR	1–10 even, 12–24, MR	1–10 even, 12–26, MR	
12-2	1–22, MR	1–22 even, 23–26, MR	1–22 even, 23–27, MR	PS 45
12-3	1–12, MR	1–12 even, 13–23, MR	1–12 even, 13–24, MR	AS Quiz p. 23
12-4	1–6, MR	1–6 even, 7–9, MR	1–6 even, 7–11, MR	PS 46
12-5	1–5, MR	1–7, MR	1–5 even, 6–9, MR	
12-6	1–7, MR	1–6 even, 8–22, MR	1–6 even, 8–25, MR	PS 47
12-7	1–8	1–8	1–8	AS Quiz p. 24
12-8	1–22, MR	1–22 even, 23–35, MR	1–22 even, 23–37, MR	
12-9	1–15, MR	1–14 even, 16–20, MR	1–14 even, 16–22, MR	PS 48
12-10	1–2	1–4	1–6	
Review/ Testing	Chapter 12 Review, Chapter 12 Test, Cumulative Review			AS pp. 53–54, 83–84

Section	Minimum	Regular	Maximum	Supplements
13-1	1–10, MR	1–10 even, 11–14, MR	1–10 even, 11–15, MR	
13-2	1–18, MR	1–18 even, 19–25, MR	1–18 even, 19–27, MR	
13-3	1–20, MR	1–20 even, 21–24, MR	1–20 even, 21–27, MR	PS 49
13-4	1–25, MR	1–25 even, 26–30, MR	1–25 even, 26–31, MR	
13-5	1–24, MR	1–24 even, 25–32, MR	1–24 even, 25–36, MR	AS Quiz p. 25
13-6	1–5, MR	1–5 even, 6–8, MR	1–5 even, 6–10, MR	PS 50
13-7	1–12, MR	1–12 even, 13, MR	1–12 even, 13–16, MR	
13-8	1–22, 25–26, MR	1–22 even, 23–26, MR	1–22 even, 23–26, MR	
13-9	1–15	1–15	1–15	PS 51, AS Quiz p. 26
13-10	1–4, 6, MR	1–6, MR	1–6, MR	
13-11	1–5, MR	1–5 even, 6–7, MR	1–5 even, 6–8, MR	PS 52
13-12	1–2	1–4	1–8	
Review/ Testing	Chapter 13 Review, Chapter 13 Test, Cumulative Review			AS pp. 55–56, 85–86

Section	Minimum	Regular	Maximum	Supplements
14-1	1–37, MR	1–37 even, 38–45, MR	1–37 even, 38–47, MR	
14-2	1–27, MR	1–27 even, 28–40, MR	1–27 even, 28–42, MR	PS 53
14-3	1–16, MR	1–16 even, 17–24, MR	1–16 even, 17–28, MR	
14-4	1–24, MR	1–24 even, 25–36, MR	1–24 even, 25–40, MR	PS 54
14-5	1–9	1–9	1–9	AS Quiz p. 27
14-6	1–4, 5–11 even, MR	1–11, MR	1–13, MR	
14-7	1–8, MR	1–8 even, 9–13, MR	1–8 even, 9–15, MR	PS 55
14-8	1–10, MR	1–10 even, 11–27, MR	1–10 even, 11–33, MR	AS Quiz p. 28
14-9	1–6, MR	1–6 even, 7–8, MR	1–6 even, 7–10, MR	PS 56
14-10	1–4	1–7	1–10	
Review/ Testing	Chapter 14 Review, Chapter 14 Test, Cumulative Review			AS pp. 57–58, 87–88

Section	Minimum	Regular	Maximum	Supplements
15-1	1–20, MR	1–20 even, 21–28, MR	1–20 even, 21–30, MR	
15-2	1–22, 32, MR	1–22 even, 23–37, MR	1–22 even, 23–39, MR	PS 57
15-3	1–20, MR	1–20 even, 21–28, MR	1–20 even, 21–30, MR	
15-4	1–5, MR	1–5 even, 6–8, MR	1–5 even, 6–10, MR	PS 58, AS Quiz p. 29
15-5	1–10, MR	1–10 even, 11–13, MR	1–10 even, 11–16, MR	
15-6	1–28, MR	1–28 even, 29–31, MR	1–28 even, 29–34, MR	
15-7	1–12	1–12	1–12	PS 59
15-8	1–9, MR	1–9 even, 10–14, MR	1–9 even, 10–16, MR	AS Quiz p. 30
15-9	1–9, MR	1–9 even, 10–17, MR	1–9 even, 10–20, MR	PS 60
15-10	1–2	1–4	1–6	
Review/ Testing	Chapter 15 Review, Chapter 15 Test, Cumulative Review			AS pp. 59–60, 89–90, 94–96

Additional Answers

Page 20

Computer Activity Change corresponding lines as follows:

```
10 REM: SOLVE AN EQUATION X−A=B
20 PRINT "CHOOSE WHOLE NUMBERS A AND B"
50 PRINT "EQUATION; X − "; A; " = ";B
60 PRINT "SOLUTION: X = ";B+A
```

Page 40

Exercises **15.** $35 + 2n$ **16.** $(n + 3) - 6$ **17.** $2n - 12$ **18.** $\frac{x}{1,000} - 100$ **19.** $3r - 24$ **20.** $8(18 + x)$ **21.** $5n + 4$ **22.** $\frac{n}{6} + 5$

Page 45

Exercises **34.** Answers will vary. Example:

3 @ 4 = $2 \cdot 3 + 4 = 10$

4 @ 3 = $2 \cdot 4 + 3 = 11$

$10 \neq 11$, so the commutative property does not hold for this symbol.

(4 @ 3) @ 2 = $(2 \cdot 4 + 3)$ @ 2 = $(8 + 3)$ @ 2 =
11 @ 2 = $11 \cdot 2 + 2 = 22 + 2 = 24$

4 @ (3 @ 2) = 4 @ $(3 \cdot 2 + 2)$ = 4 @ $(6 + 2)$ =
4 @ 8 = $4 \cdot 2 + 8 = 8 + 8 = 16$

$24 \neq 16$, so the associative property does not hold for this symbol.

Page 54

Computer Activity **2.**

```
10 REM: SOLVE AN EQUATION AX=B
40 PRINT "THE EQUATION IS "A;"X = ";B
50 PRINT "THE SOLUTION IS X=";B/A
```

With versions of BASIC that do not allow use of the commands shown, replace line 30 with the following two lines:

```
30 PRINT "ENTER A";:INPUT A
35 PRINT "ENTER B";:INPUT B
```

Page 63

Exercises **5.** Rule: $f(b) = (b + 3)(b + 3)$

Input	1	2	3	4	5	6	7	8
Output	16	25	36	49	64	81	100	121

6. Domain: $\{2, 4, 6, 8, 10\}$

Range: $\{2, 3, 4, 5, 6\}$

Function rule: $f(x) = \frac{x}{2} + 1$

Page 86

Exercises **42.** $2b + 3b = 5b$; $2 \cdot 4 + 3 \cdot 4 = 5 \cdot 4$;
$8 + 12 = 20$; $20 = 20$
$2(-8) + 3(-8) = 5(-8)$; $-16 + -24 = -40$;
$-40 = -40$
$2(-34) + 3(-34) = 5(-34)$; $-68 + -102 = -170$;
$-170 = -170$ distributive property

43. $-9(a + 25) = -9a + (-9)(25)$;
$-9(-9 + 25) = -9(-9) + (-9)(25)$;
$-9(16) = 81 - 225 = -144$
$-9(12 + 25) = -9(12) + (-9)(25)$;
$-9(37) = -108 - 225$;
$-333 = -333$; $-9(-24 + 25) = -9(-24) + (-9)(25)$;
$-9(1) = 216 - 225$; $-9 = -9$

Page 94

Exercises **22.** 9 more than a number is 23.
23. A number divided by −4 is 9.
24. A number multiplied by −8 gives 72.
25. 16 decreased by a number is 7.

Page 104

Exercises **19.** 3.069, 3.07, 3.7 **20.** 0.004, 0.0041, 0.039
21. 6.010, 6.0101, 6.101 **22.** 4, 4.001, 4.01
23. −5.404, −5.044, −5.040
24. −0.101, −0.011, −0.01, −0.001

Page 107

Computer Activity **2.** Change to:

```
20 PRINT "TYPE A DECIMAL THAT HAS TWO OR
   MORE DECIMAL PLACES."
40 PRINT N; "ROUNDED TO TENTHS IS";
50 PRINT INT (N*10+0.5)/10
```

3. Change to:

```
20 PRINT "TYPE A DECIMAL THAT HAS FOUR OR
   MORE DECIMAL PLACES."
40 PRINT N; "ROUNDED TO THOUSANDTHS IS";
50 PRINT INT (N*1000+0.5)/1000
```

Page 135

Exercises **1.** 3, 6, 9, 12 **2.** 5, 10, 15, 20 **3.** 7, 14, 21, 28
4. 2, 4, 6, 8 **5.** 6, 12, 18, 24 **6.** 8, 16, 24, 32
7. 9, 18, 27, 36 **8.** 10, 20, 30, 40 **9.** 4, 8, 12, 16
10. 11, 22, 33, 44 **11.** 20, 40, 60, 80
12. 100, 200, 300, 400
28. 1, 2, 4, 8, 16 **29.** 1, 2, 4, 5, 10, 20 **30.** 1, 2, 4, 7, 14, 28
31. 1, 2, 3, 4, 6, 7, 12, 14, 21, 28, 42, 84
32. 1, 2, 4, 5, 8, 10, 20, 40
33. 1, 2, 3, 4, 5, 6, 10, 12, 15, 20, 30, 60 **34.** 1, 2, 17, 34
35. 1, 3, 5, 7, 15, 21, 35, 105

Page 143

Exercises **27.** Answers will vary. Examples: $2 + 2 = 4$;
$3 + 3 = 6$; $3 + 5 = 8$; $3 + 7 = 10$; $5 + 7 = 12$;
$3 + 11 = 14$; $5 + 11 = 16$; $7 + 11 = 18$;
$7 + 13 = 20$; $11 + 11 = 22$; $11 + 13 = 24$;
$13 + 13 = 26$; $11 + 17 = 28$; $7 + 23 = 30$

Page 146

Exercises **1.** $10 \cdot 10 \cdot 10 \cdot 10 \cdot 10 = 100,000$
2. $5 \cdot 5 \cdot 5 \cdot 5 = 625$ **3.** $7 \cdot 7 \cdot 7 = 343$

4. $12 \cdot 12 \cdot 12 \cdot 12 \cdot 12 = 248{,}832$ **5.** $9 \cdot 9 \cdot 9 = 729$
6. $(-10)(-10)(-10)(-10)(-10)(-10) = 1{,}000{,}000$
7. $8 \cdot 8 = 64$ **8.** $(-6)(-6)(-6) = -216$ **9.** $n \cdot n \cdot n \cdot n$
10. $b \cdot b \cdot b \cdot b \cdot b$

Page 150
Exercises

4. $2^5 = 32$

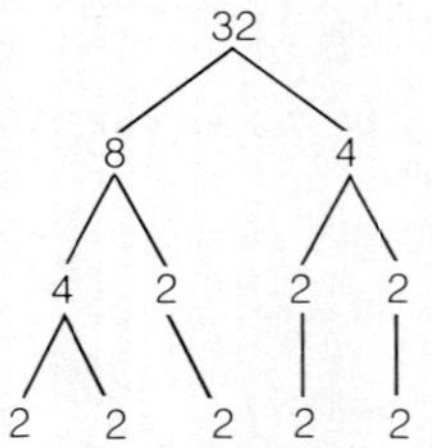

5. $2^4 \cdot 3 = 48$

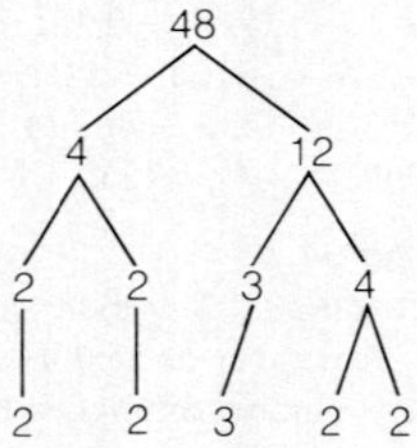

6. $3^2 \cdot 7 = 63$

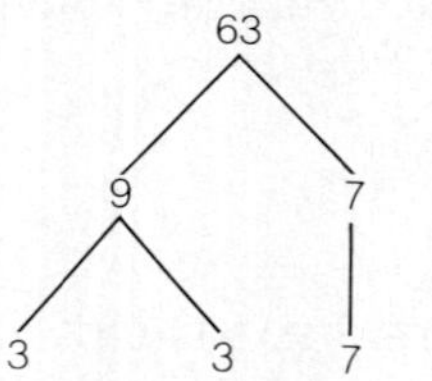

7. $2^2 \cdot 5 \cdot 7 = 140$

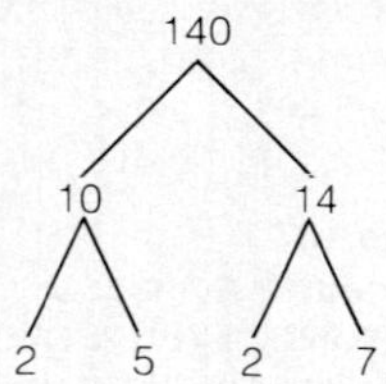

8. $2 \cdot 3 \cdot 5^2 = 150$

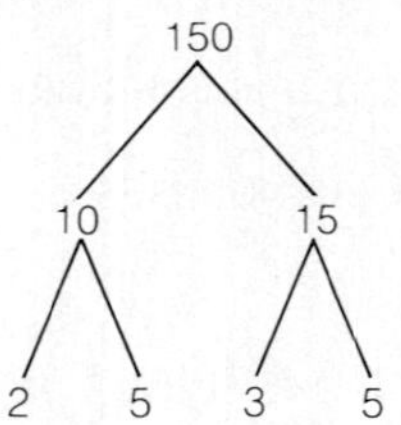

Page 169
Exercises **63.** $\frac{2}{4} = \frac{3}{6}; \frac{2}{4} = \frac{4}{8}; \frac{2}{4} = \frac{5}{10}; \frac{3}{6} = \frac{2}{4}; \frac{4}{8} = \frac{2}{4};$
$\frac{5}{10} = \frac{2}{4}; \frac{4}{2} = \frac{6}{3}; \frac{4}{2} = \frac{8}{4}; \frac{4}{2} = \frac{10}{5}; \frac{6}{3} = \frac{4}{2}; \frac{8}{4} = \frac{4}{2}; \frac{10}{5} = \frac{4}{2};$
$\frac{3}{6} = \frac{4}{8}; \frac{3}{6} = \frac{5}{10}; \frac{4}{8} = \frac{3}{6}; \frac{5}{10} = \frac{3}{6}; \frac{6}{3} = \frac{8}{4}; \frac{6}{3} = \frac{10}{5}; \frac{8}{4} = \frac{6}{3};$
$\frac{10}{5} = \frac{6}{3}; \frac{4}{8} = \frac{5}{10}; \frac{5}{10} = \frac{4}{8}; \frac{8}{4} = \frac{10}{5}; \frac{10}{5} = \frac{8}{4}; \frac{2}{3} = \frac{4}{6};$
$\frac{2}{3} = \frac{6}{9}; \frac{4}{6} = \frac{2}{3}; \frac{6}{9} = \frac{2}{3}; \frac{3}{2} = \frac{6}{4}; \frac{3}{2} = \frac{9}{6}; \frac{6}{4} = \frac{3}{2}; \frac{9}{6} = \frac{3}{2};$
$\frac{4}{6} = \frac{6}{9}; \frac{6}{9} = \frac{4}{6}; \frac{6}{4} = \frac{9}{6}; \frac{9}{6} = \frac{6}{4}; \frac{3}{4} = \frac{6}{8}; \frac{6}{8} = \frac{3}{4}; \frac{4}{3} = \frac{8}{6};$
$\frac{8}{6} = \frac{4}{3}; \frac{2}{5} = \frac{4}{10}; \frac{4}{10} = \frac{2}{5}; \frac{5}{2} = \frac{10}{4}; \frac{10}{4} = \frac{5}{2}; \frac{3}{5} = \frac{6}{10}; \frac{6}{10} = \frac{3}{5};$
$\frac{5}{3} = \frac{10}{6}; \frac{10}{6} = \frac{5}{3}; \frac{2}{6} = \frac{3}{9}; \frac{3}{9} = \frac{2}{6}; \frac{6}{2} = \frac{9}{3}; \frac{9}{3} = \frac{6}{2}; \frac{4}{5} = \frac{8}{10};$
$\frac{8}{10} = \frac{4}{5}; \frac{5}{4} = \frac{10}{8}; \frac{10}{8} = \frac{5}{4}$

Page 175
Exercises

17.
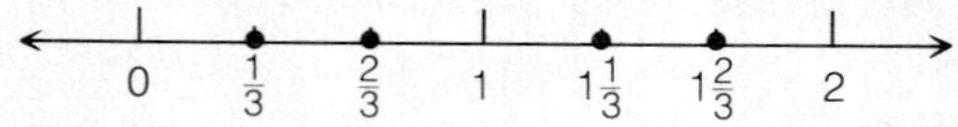

18.
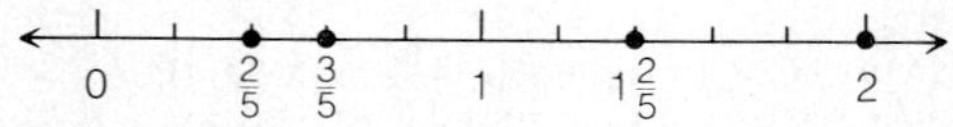

19.
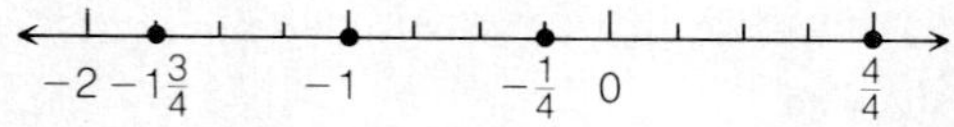

20.
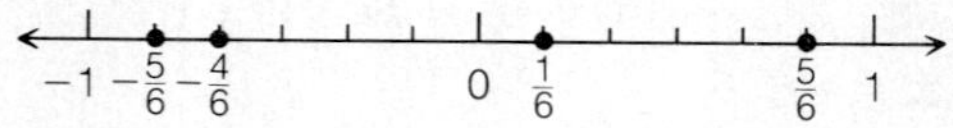

21.
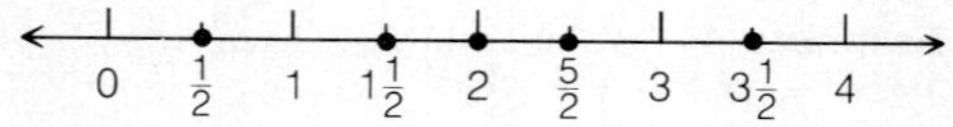

22.
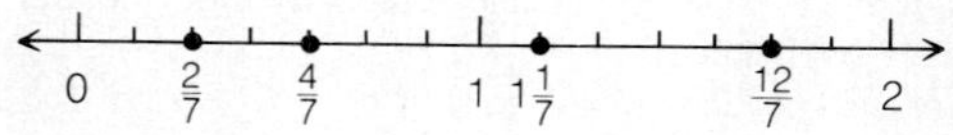

23.
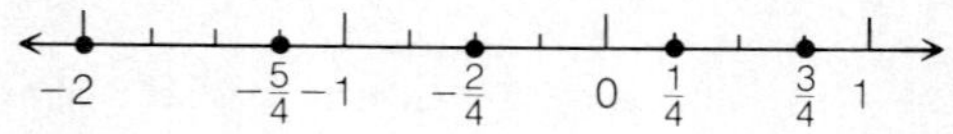

24.
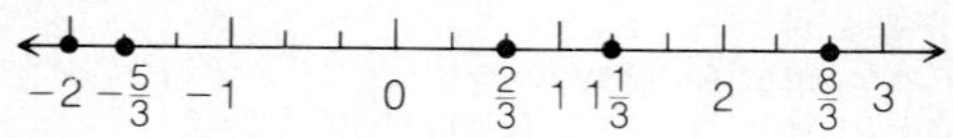

25.
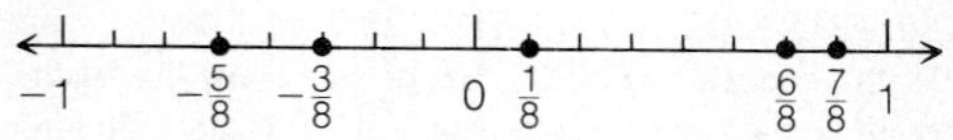

26.
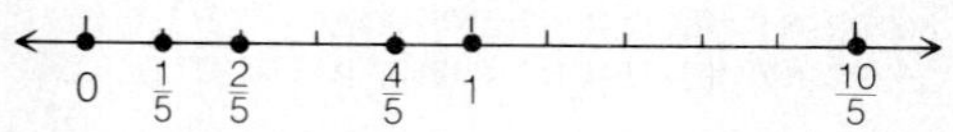

Page 178
Exercises **48.** $\frac{2}{3}, \frac{2}{4}, \frac{2}{5}, \frac{2}{6}, \frac{2}{7}, \frac{2}{8}, \frac{2}{9}, \frac{2}{10}, \frac{3}{4}, \frac{3}{5}, \frac{3}{6}, \frac{3}{7}, \frac{3}{8},$
$\frac{3}{9}, \frac{3}{10}, \frac{4}{5}, \frac{4}{6}, \frac{4}{7}, \frac{4}{8}, \frac{4}{9}, \frac{4}{10}, \frac{5}{6}, \frac{5}{7}, \frac{5}{8}, \frac{5}{9}, \frac{5}{10}, \frac{6}{7}, \frac{6}{8}, \frac{6}{9}, \frac{6}{10}, \frac{7}{8},$
$\frac{7}{9}, \frac{7}{10}, \frac{8}{9}, \frac{8}{10}, \frac{9}{10}$

Page 213
Computer Activity **5.**

```
10 PRINT "SOLVING EQUATIONS OF THE FORM
   A * N = B/C"
30 IF A=0 OR C=0 THEN GOTO 70
40 N=B/(A*C)
70 PRINT "NEITHER A NOR C CAN EQUAL ZERO."
```

6.

```
10 PRINT "SOLVING EQUATIONS OF THE FORM
   A/B * N = C/D"
20 INPUT " ENTER A, B, C, AND D.";A,B,C,D
30 IF A=0 OR D=0 THEN GOTO 70
40 N = C*B/(D*A)
70 PRINT "NEITHER A NOR D CAN EQUAL ZERO."
```

Page 223

Exercises **43.** $4^{-3} \cdot 4^5 = 4^{-3+5} = 4^2 = 16$
$4^5 \cdot 4^{-3} = 4^{5+-3} = 4^2 = 16$
Yes, $a^n \cdot a^m = a^m \cdot a^n$ is true for all values of a, n, and m because of the commutative property.

Page 225

Exercises **1.** 35,000 **2.** 0.062 **3.** 405,000 **4.** 0.00007 **5.** 1.35×10^2 **6.** 2.3×10^4 **7.** 3.45×10^5 **8.** 8.4×10^0 **9.** 1.24×10^3 **10.** 6.5×10^5 **11.** 4.55×10^6 **12.** 6×10^2 **13.** 9.9×10^4 **14.** 1×10^6 **15.** 7.8×10^{-2} **16.** 4×10^{-1} **17.** 6.77×10^{-4} **18.** 5.5×10^{-3} **19.** 1×10^{-6} **20.** 5×10^{-2} **21.** 4.05×10^{-5} **22.** 7×10^{-6} **23.** 1.01×10^{-1} **24.** 3×10^{-8}

Page 228

Exercises **3.** Jana could have bought 5 granola bars and 6 carob chews, or 6 granola bars, 2 honey drops, and 3 carob chews, or 7 granola bars and 4 honey drops.
4. Answers may vary. Example: Fill the 9-qt pail; pour into 4-qt pail, leaving 5 quarts. Empty 4-qt pail and fill from the larger pail, leaving 1 qt in the larger pail. Empty 4-qt pail and transfer 1 qt to the smaller pail. Fill the 9-qt pail and pour into the smaller pail. It will take 3 quarts to fill the smaller pail, leaving 6 quarts in the larger pail.

Page 250

Exercises
7.

	New	After 1 yr	After 2 yr	After 3 yr
price of car	$9,450	$7,560	$6,142.50	$4,725

Page 255

Exercises **26.** Answers may vary. Example: Fill up the 5-liter container and pour it into the 8-liter container. Fill up the 5-liter container again and start pouring it into the 8-liter container. When the 8-liter container is filled, empty it, and pour the remaining 2 liters from the 5-liter container into it. Fill up the 5-liter container and pour it into the 8-liter container. It now contains exactly 7 liters.

Page 260

Exercises **4.** Cut each of the 3 links of *one* section. Use each of the 3 cut links to connect the other sections. Weld each of the 3 pieces together again. 3 cuts = 3 × 10 = 30; 3 welds = 3 × 20 = 60; 30 + 60 = 90.
5. Dividing the gold into groups of 4 requires 3 weighings, so the prospector divided the gold pieces into 3 groups. Group A has 3 pieces of gold, group B has 3 pieces, and group C has 2 pieces. He places group A on one side of the balance and group B on the other side. If they balance, he removes the six pieces and places one piece of gold on each side of the balance from group C. He will then be able to recognize the heavier piece. If group A and group B do not balance, he removes the lighter group. Then he selects 2 pieces from the heavier group (A or B) and places one on each balance. If these 2 pieces balance then the 3rd piece is the heavier piece. If they don't balance, the prospector will recognize the heavier piece of gold.

Page 273

Exercises **21.** $\frac{a}{b} = \frac{c}{d}$; $\frac{a}{b} + 1 = \frac{c}{d} + 1$;
$\frac{a}{b} + \frac{b}{b} = \frac{c}{d} + \frac{d}{d}$; $\frac{a+b}{b} = \frac{c+d}{d}$

Page 328

Exercises **1.** adventure 180°; science fiction 72°; children's 54°; comedy 36°; other 18°

Page 340

Exercises **7.** Let $x = AB$; $x + 5.8 = 12.2$; $x = 6.4$; $AB = 6.4$ mm
8. Let $y = EF$; $y + 5.1 + 3.4 = 10.9$; $y = 2.4$; $EF = 2.4$ cm
9. Let $z = YZ$; $z + 9.6 + 4.7 = 18.3$; $z = 4$; $YZ = 4$ m
10. Let $b = ST$; let $2b = RS$; $2b + b + 11.7 = 32.4$; $3b = 20.7$; $b = 6.9$; $ST = 6.9$ cm; $RS + 13.8$ cm
11. $P = 2(12.8) + 2(5.1)$; $P = 25.6 + 10.2$; $P = 35.8$ cm
12. $P = 5 + 6 + 4$; $P = 15$ cm
13. $P = 13 + 7 + 11 + 4$; $P = 35$ cm
14. $30 + 5 + x + 8 + 8 + 13 = 86$; $x = 22$ cm
15. $31.4 + 11.8 + y + 17.1 = 76.6$; $y = 16.3$ m
16. $z + z + 21.3 = 68.6$; $z = 13$ m

Page 341

Exercises **21.** $y = 14 - 3 = 11$; $x = 22 - (18 + 2) = 2$; $P = 22 + 14 + 2 + 11 + 18 + 11 + 2 + 14$;
The perimeter is 94.
22. $x + 16 = 13 + 5$; $x = 2$; $y + 11 = 9 + 8$; $y = 6$; $P = 9 + 5 + 8 + 13 + 6 + 2 + 11 + 16$;
The perimeter is 70.
23. $y + 14 + 3 = 21$; $y = 4$; $x + 4 = 14$; $x = 10$; $P = 21 + 15 + 3 + 6 + 14 + 4 + 10 + 2 + 14 + 15$;
The perimeter is 104.

Page 344

Exercises **12.** $x + 23 = 90$; 67° **13.** $x + 53 = 90$; 37° **14.** $x + 123 = 180$; 57° **15.** $x + 153 = 180$; 27° **16.** $x + 23 = 180$; 157° **17.** $x + 17 = 90$; 73°

Page 351

Exercises **5.** $40 + 55 + a = 180$ $m\angle A = 85°$
6. $61 + 58 + a = 180$ $m\angle A = 61°$
7. $15 + 70 + a = 180$ $m\angle A = 95°$
8. $40 + 70 + a = 180$ $m\angle A = 70°$

Page 355

Exercises **9.** $x + x + 2x + 2x = 360$; $6x = 360$; $x = 60$; The angles are 60°, 60°, 120°, 120°.
10. $x + 3x + x + 3x = 360$; $8x = 360$; $x = 45$; The angles are 45°, 45°, 135°, 135°.
11. $(4x - 10) + (x + 30) + (2x + 25) + (3x + 15) = 360$; $10x + 60 = 360$; $10x = 300$; $x = 30$; $4(30) - 10 = 110$; $30 + 30 = 60$; $2(30) + 25 = 85$; $3(30) + 15 = 105$; The angles are 110°, 60°, 85°, and 105°.

Page 361

Practice $A \cong G, B \cong H, C \cong I, D \cong J, E \cong K, F \cong L$; $\overline{AB} \cong \overline{GH}$, $\overline{BC} \cong \overline{HI}$, $\overline{CD} \cong \overline{IJ}$, $\overline{DE} \cong \overline{JK}$, $\overline{EF} \cong \overline{KL}$, $\overline{FA} \cong \overline{LG}$

Page 392

Exercises **12.** 3,617.28 cm^3 **13.** 7,218.86 m^3 **14.** 37.68 m^3 **15.** 602.88 dm^3 **16.** 56.06784 cm^3 **17.** 1,969,408 cm^3 **18.** 80,384,000 cm^3 **19.** 565.2 cm^3

Page 401

Exercises **5.** 1004.8 cm^2 **6.** 150.72 cm^2 **7.** 678.24 m^2 **8.** 137.1552 m^2 **9.** 439.6 dm^2 **10.** 2486.88 cm^2 **11.** 3165.12 m^2 **12.** 602.88 cm^2

Page 402

Computer Activity

1a. Volume = 1.45870002 E + 10
Surface Area = 28866095.4

b. Volume = 3.9217527 E + 10
Surface Area = 55812419.8

c. Volume = 2.60579713 E + 11
Surface Area = 197259435

Page 404

Exercises **5.** Fill the 3 oz can and pour it into the 8 oz can. Fill the 3 oz can again and pour it into the 8 oz can. Fill the 3 oz can and pour 2 oz into the 8 oz can (to the top of the can). Empty the 8 oz can. Pour the remaining 1 oz in the 3 oz can into the 8 oz can. Now fill the 3 oz can and pour it into the 8 oz can. There are now 4 oz in the 8 oz can.

Page 463

Exercises

2. $\angle M$ and $\angle Y$
$\angle N$ and $\angle Z$
$\angle P$ and $\angle X$
$\overline{MP}$ and $\overline{YX}$
$\overline{PN}$ and $\overline{XZ}$
$\overline{NM}$ and $\overline{ZY}$

3. $\angle P$ and $\angle S$
$\angle T$ and $\angle R$
$\angle PQT$ and $\angle RQS$
$\overline{TP}$ and $\overline{RS}$
$\overline{PQ}$ and $\overline{SQ}$
$\overline{QT}$ and $\overline{QR}$

Page 469

Exercises

1. $\sin A = \frac{6}{17}$, $\cos A = \frac{15}{17}$, $\tan A = \frac{2}{5}$

2. $\sin A = \frac{20}{29}$, $\cos A = \frac{21}{29}$, $\tan A = \frac{20}{21}$

3. $\sin A = \frac{1}{4}$, $\cos A = \frac{29}{36}$, $\tan A = \frac{9}{29}$

4. $\sin A = \frac{5}{13}$, $\cos A = \frac{12}{13}$, $\tan A = \frac{5}{12}$

5. $\sin A = \frac{1}{3}$, $\cos A = \frac{5}{7}$, $\tan A = \frac{7}{15}$

6. $\sin A = \frac{4}{5}$, $\cos A = \frac{9}{10}$, $\tan A = \frac{8}{9}$

Page 470

Exercises

19. $\sin A = \frac{7}{25}$, $\cos A = \frac{24}{25}$, $\tan A = \frac{7}{24}$

20. $\sin A = \frac{9}{17}$, $\cos A = \frac{14.42}{17}$, $\tan A = \frac{9}{14.42}$

21. $\sin A = \frac{4}{7}$, $\cos A = \frac{5.74}{7}$, $\tan A = \frac{4}{5.74}$

Page 473

Exercises Company A receives C's stationery and B's envelopes. Company B receives A's stationery and C's envelopes. Company C receives B's stationery and A's envelopes.

Page 481

Exercises

1.
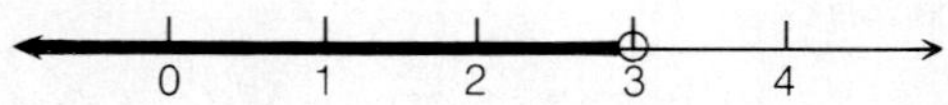

2.
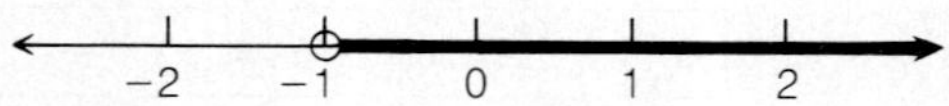

3.
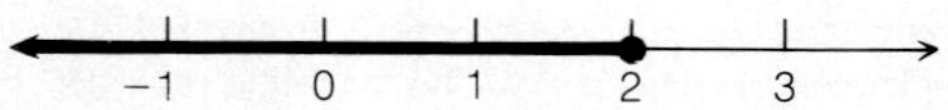

4.
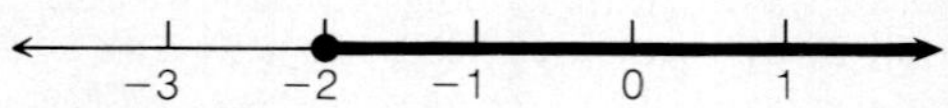

5.
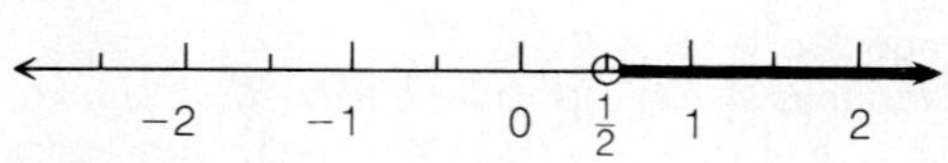

6.
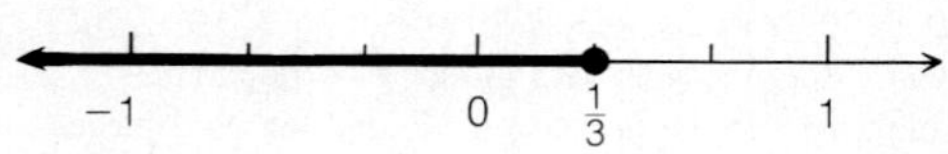

7. 

8. 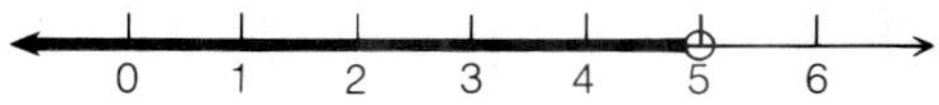

9. 

10. 

11. 

12. 

13. 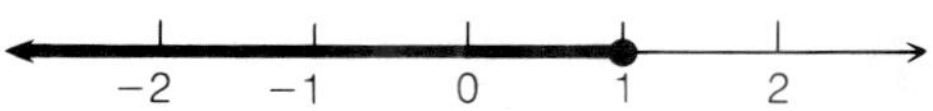

14. 

15. 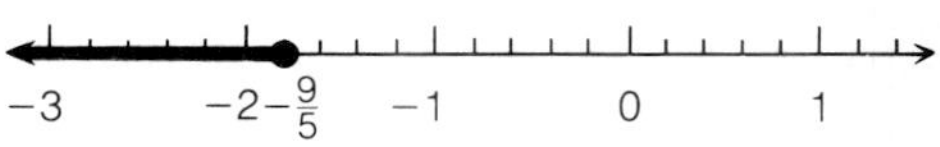

16. 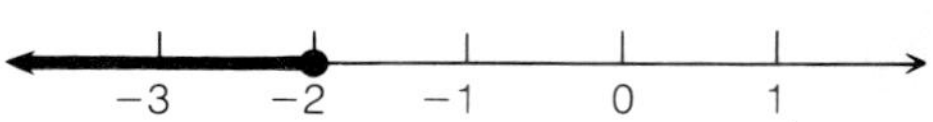

17. 

18. 

19. 

20. 

21. $-1 < x \leq 3$

22. $1 \leq x \leq 3$

23. $-2 < x < 3$

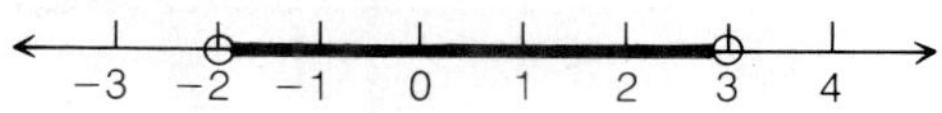

24. $-3 \leq x < -1$

25. $0 \leq x < 2$

26. $-3 < x \leq 4$

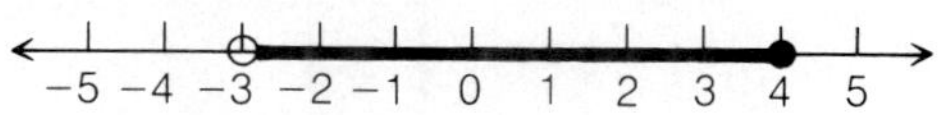

27. $-1 \leq x \leq 2$

28. $-3 < x \leq 0$

29. $1 \leq x \leq 3$

30. $2 < x < 8$

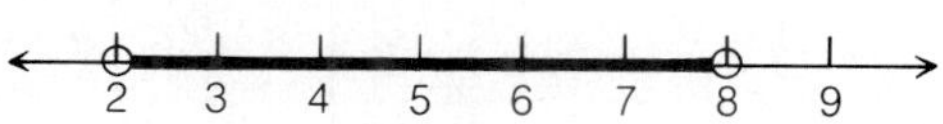

Since it is an absolute value, the expression could be positive or negative.

Case 1. It is positive: $x - 5 < 3$; $x < 8$

Case 2. It is negative: $-(x - 5) < 3$; $-x + 5 < 3$; $-x < -2$; $x > 2$

Combine the two inequalities: $2 < x < 8$

Page 483
Practice
2.

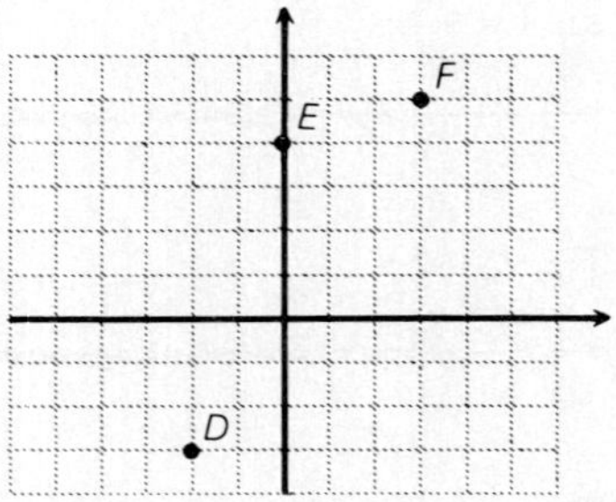

Page 487
Practice
$y = 2x - 1$

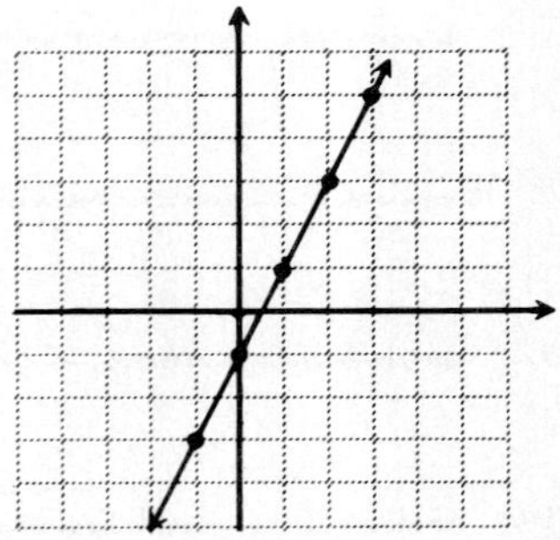

Page 484–485
Exercises
13–24.

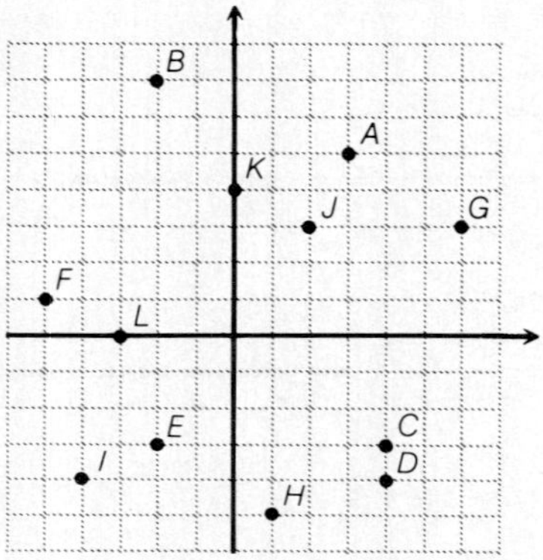

29–37.

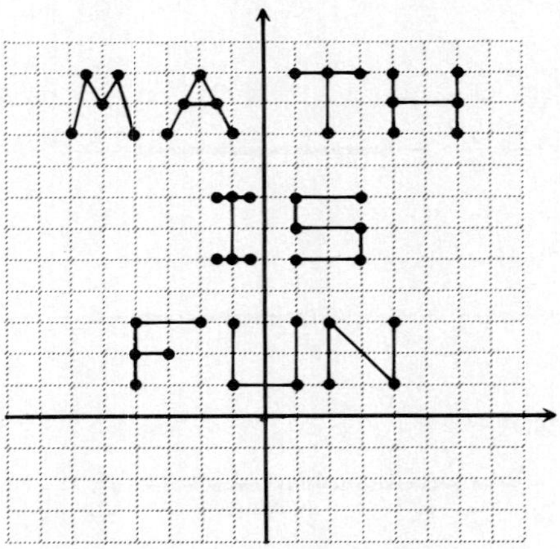

38. $y = x$

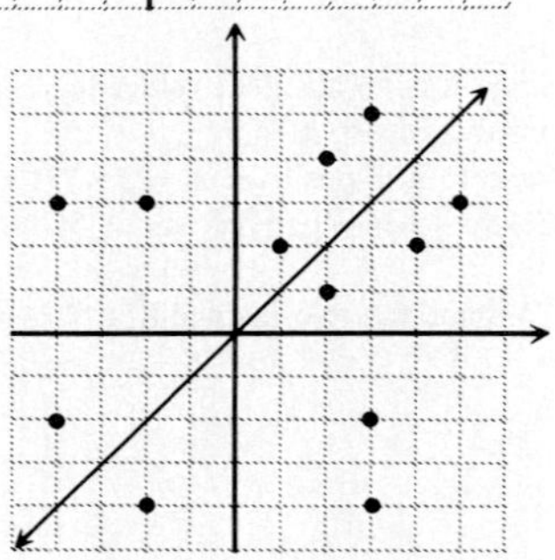

Page 488
Exercises

5. $y = 2x - 1$

x	y
−2	−5
−1	−3
0	−1
1	1
2	3
3	5
4	7

6. $y = 3x + 2$

x	y
−2	−4
−1	−1
0	2
1	5
2	8
3	11
4	14

7. $-2x + 5 = y$

x	y
−2	9
−1	7
0	5
1	3
2	1
3	−1
4	−3

8. $y = -x + 3$

x	y
−2	5
−1	4
0	3
1	2
2	1
3	0
4	−1

9. $y = 4x - 7$

x	y
−2	−15
−1	−11
0	−7
1	−3
2	1
3	5
4	9

10. $y = -2x + 3$

x	y
−2	7
−1	5
0	3
1	1
2	−1
3	−3
4	−5

11. $y = x - 12$

x	y
−2	−14
−1	−13
0	−12
1	−11
2	−10
3	−9
4	−8

12. $y = 2x - 5$

x	y
−2	−9
−1	−7
0	−5
1	−3
2	−1
3	1
4	3

13. $y = x + 1$

x	y
−2	−1
−1	0
0	1
1	2
2	3
3	4
4	5

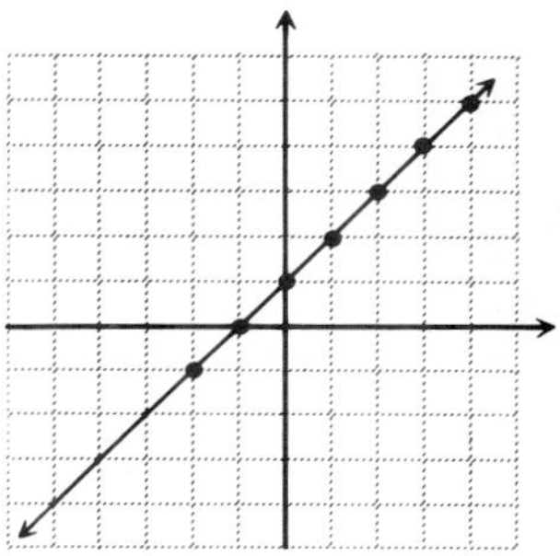

14. $y = 2x - 3$

x	y
−2	−7
−1	−5
0	−3
1	−1
2	1
3	3
4	5

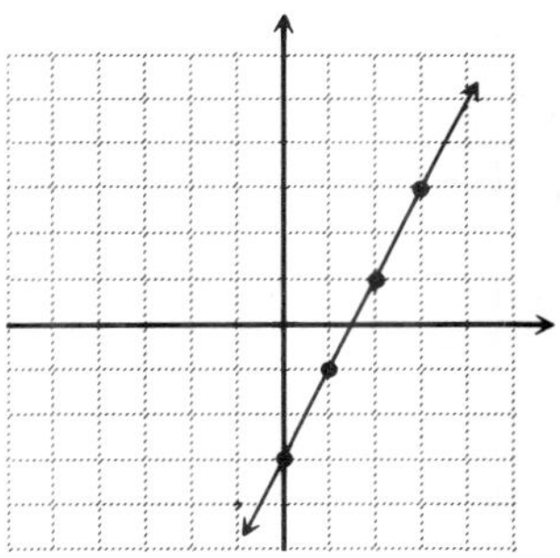

15. $y = -x + 2$

x	y
−2	4
−1	3
0	2
1	1
2	0
3	−1
4	−2

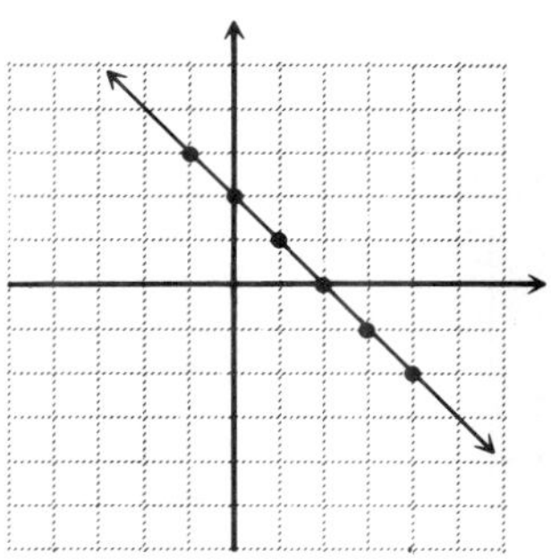

16. $y = 3x - 4$

x	y
−2	−10
−1	−7
0	−4
1	−1
2	2
3	5
4	8

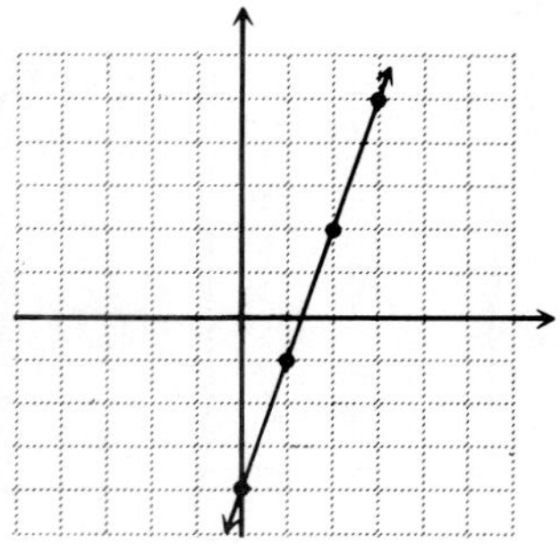

17. $y = x + 3$

x	y
−2	1
−1	2
0	3
1	4
2	5
3	6
4	7

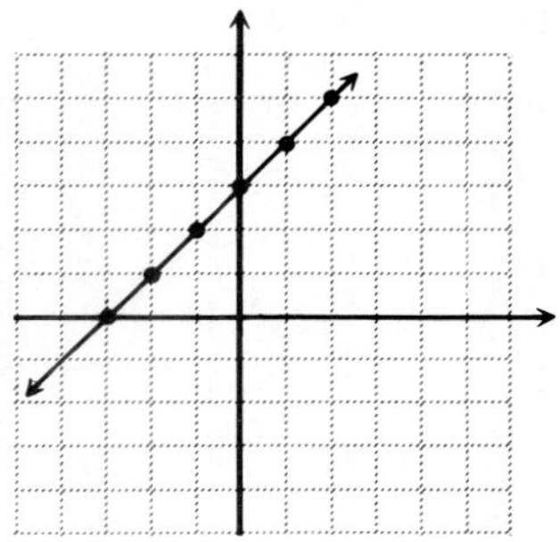

18. $y = -x - 2$

x	y
−2	0
−1	−1
0	−2
1	−3
2	−4
3	−5
4	−6

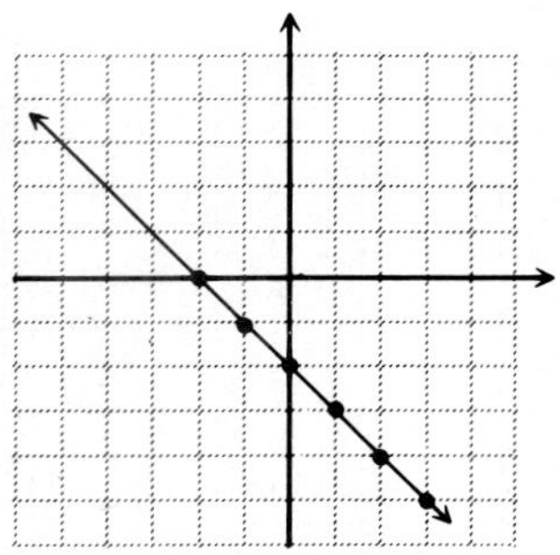

19. $y = -2x + 1$

x	y
−2	5
−1	3
0	1
1	−1
2	−3
3	−5
4	−7

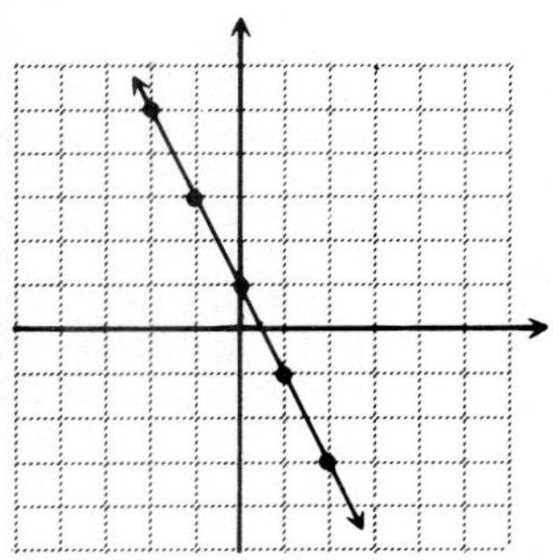

20. $y = -x + 3$

x	y
-2	5
-1	4
0	3
1	2
2	1
3	0
4	-1

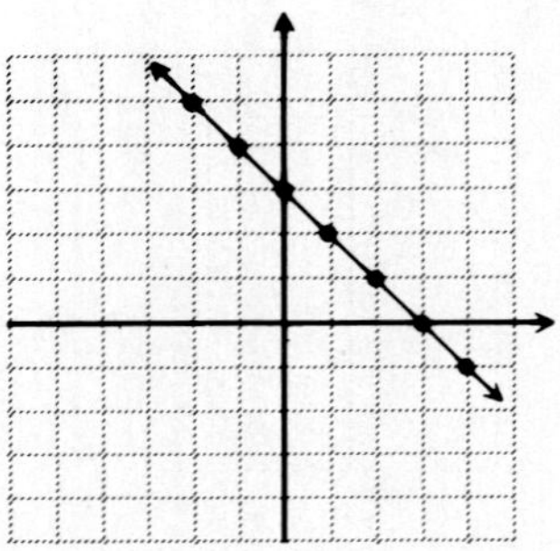

21. $4x + y = 4;$
$y = -4x + 4$

x	y
-2	12
-1	8
0	4
1	0
2	-4
3	-8
4	-12

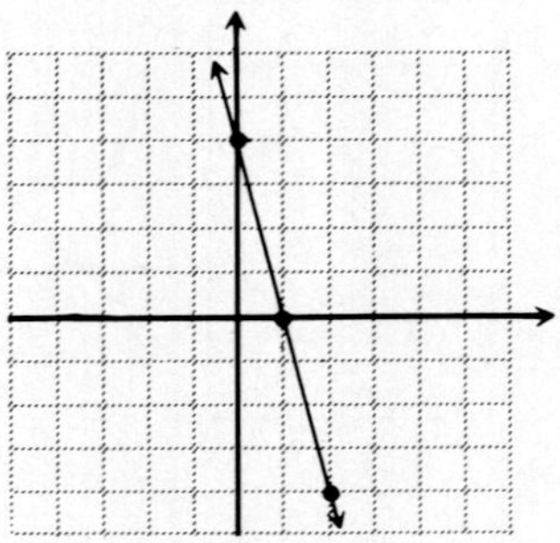

22. $-2x + y = 6;$
$y = 2x + 6$

x	y
-2	2
-1	4
0	6
1	8
2	10
3	12
4	14

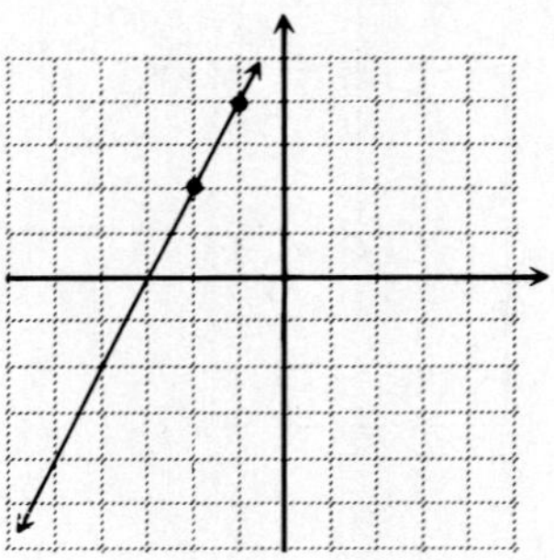

23. $2x + y = 8;$
$y = -2x + 8$

x	y
-2	12
-1	10
0	8
1	6
2	4
3	2
4	0

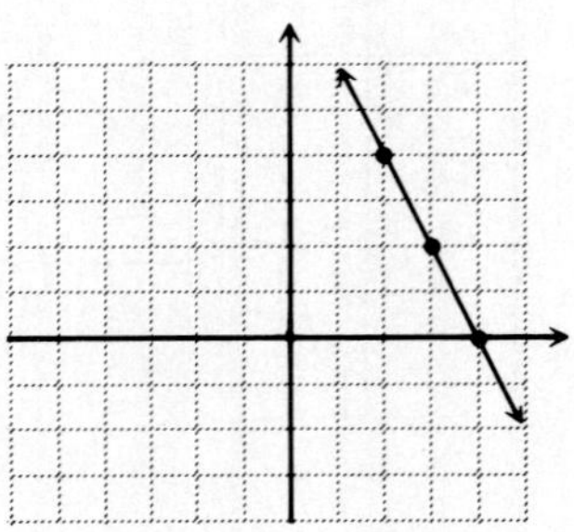

24. $y - 2x = 1;$
$y = 2x + 1$

x	y
-2	-3
-1	-1
0	1
1	3
2	5
3	7
4	9

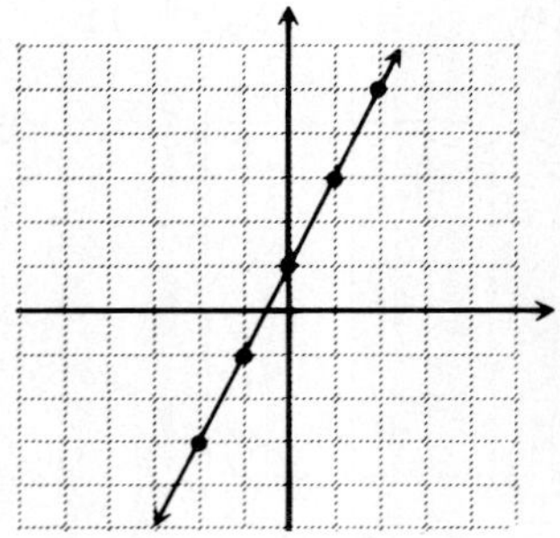

25. $3x = 2y + 4;\ 2y = 3x - 4;\ y = \frac{3}{2}x - 2$

x	y
-2	-5
-1	$-3\frac{1}{2}$
0	-2
1	$-\frac{1}{2}$
2	1
3	$2\frac{1}{2}$
4	4

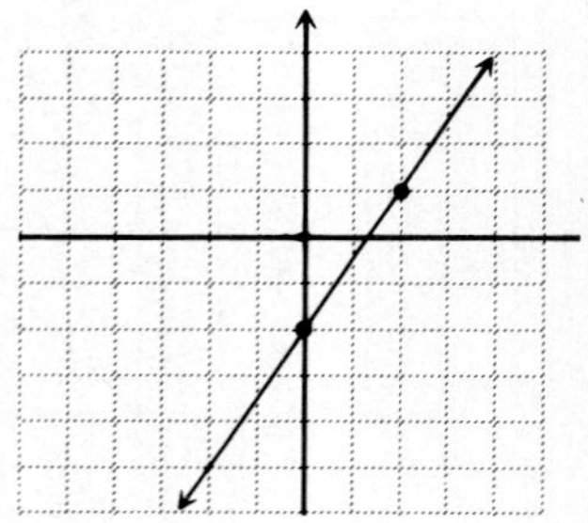

26. $3y + 6x = 6;\ 3y = -6x + 6;\ y = -2x + 2$

x	y
-2	6
-1	4
0	2
1	0
2	-2
3	-4
4	-6

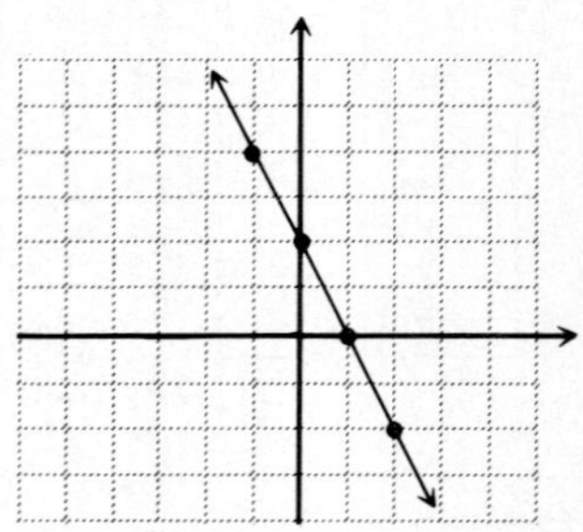

27. $2x = 4y;\ 4y = 2x;\ y = \frac{1}{2}x$

x	y
-2	-1
-1	$-\frac{1}{2}$
0	0
1	$\frac{1}{2}$
2	1
3	$1\frac{1}{2}$
4	2

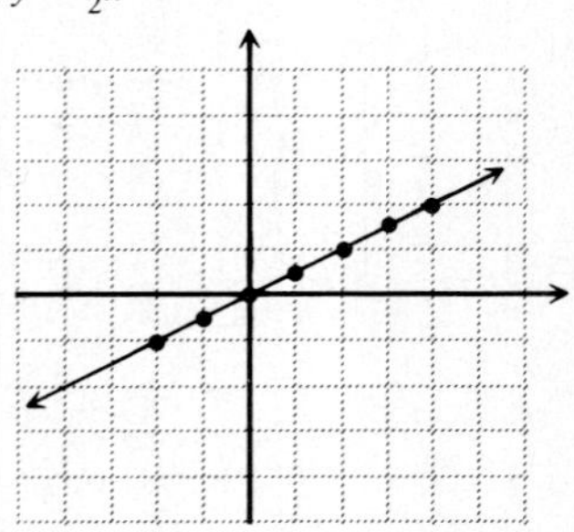

28. $3x - 4y = 8$; $-4y = -3x + 8$; $y = \frac{3}{4}x - 2$

x	y
-2	$-3\frac{1}{2}$
-1	$-2\frac{3}{4}$
0	-2
1	$-1\frac{1}{4}$
2	$-\frac{1}{2}$
3	$\frac{1}{4}$
4	1

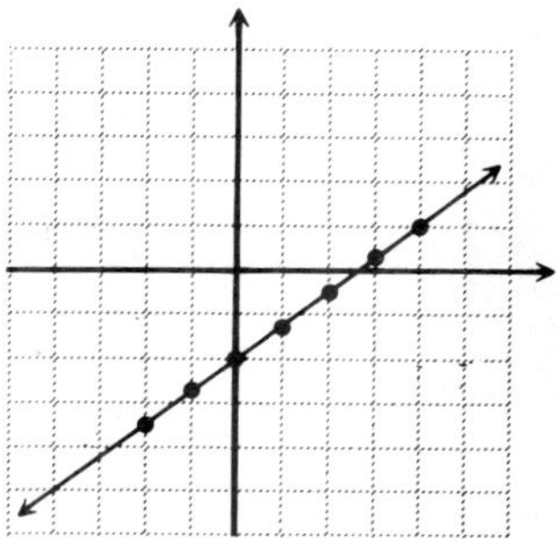

Page 493
Exercises

14. $k = 20$, $y = 20x$

x	y
1	20
2	40
3	60
4	80

15. $k = 320$, $y = \frac{320}{x}$

x	y
1	320
2	160
3	$106\frac{2}{3}$
4	80

16. $k = 5$, $y = 5x^2$

x	y
1	5
2	20
3	45
4	80

Page 493
Calculator Activity

2. $y = 123x$

x	y
35	4,305
70	8,610
105	12,915
140	17,220
175	21,525
210	25,830
245	30,135
280	34,440

3. 123 **M+** **MR** $\div x = y$

x	y
1	123
3	41
6	20.5
10	12.3
12	10.25
15	8.2
20	6.15
123	1

Page 495
Practice
a.

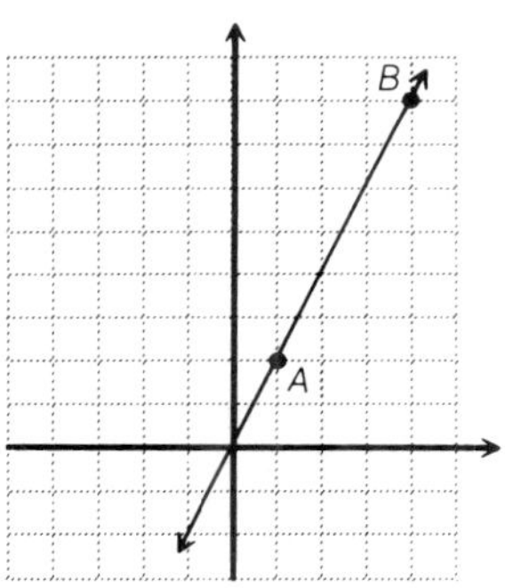

b.

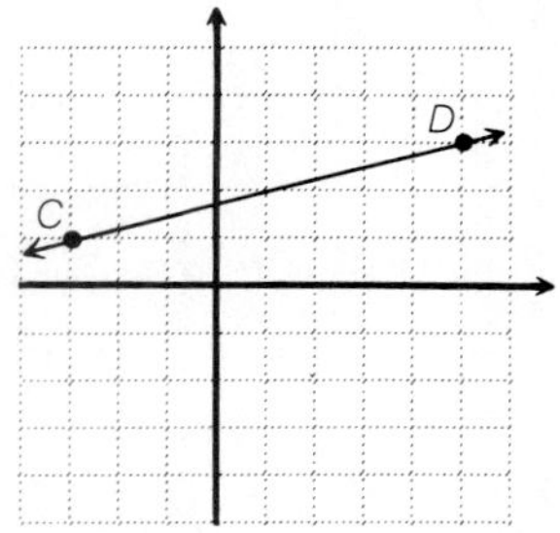

Page 496
Exercises
1. slope $= \frac{2}{3}$

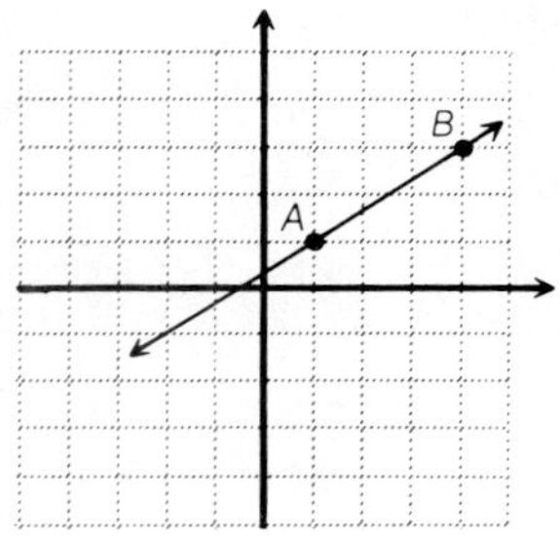

2. slope $= \frac{5}{4}$

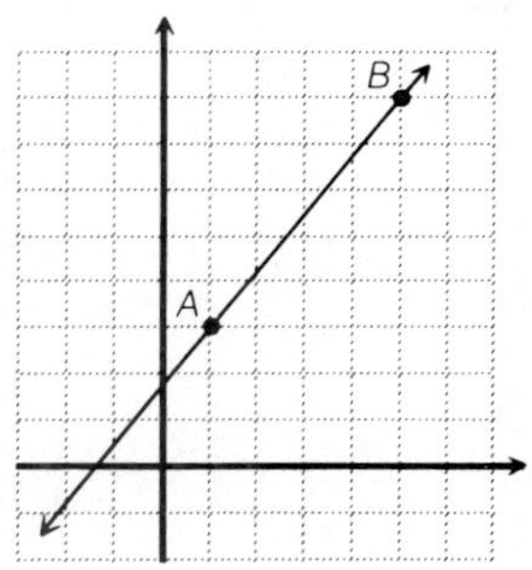

3. slope $= \frac{3}{2}$

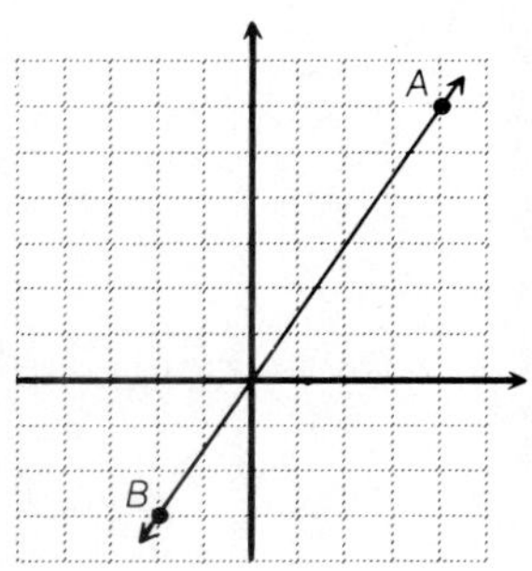

4. slope = $\frac{1}{3}$

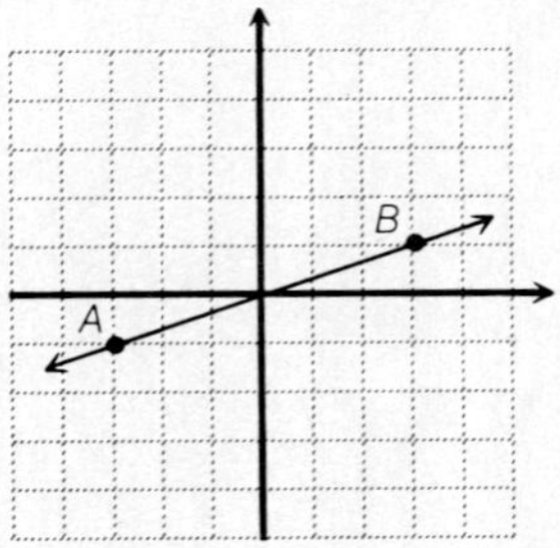

5. slope = -1

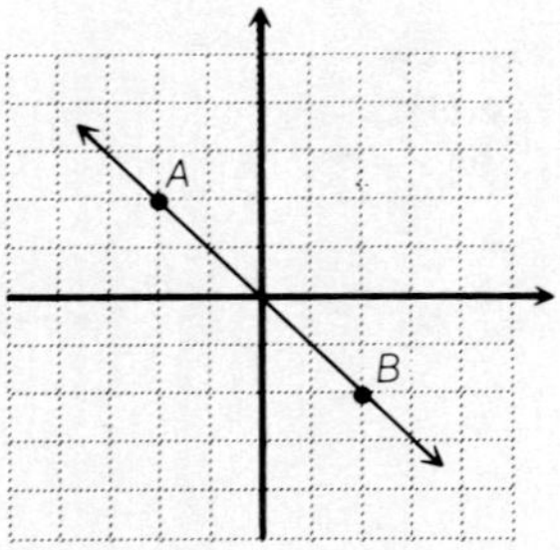

6. slope = -3

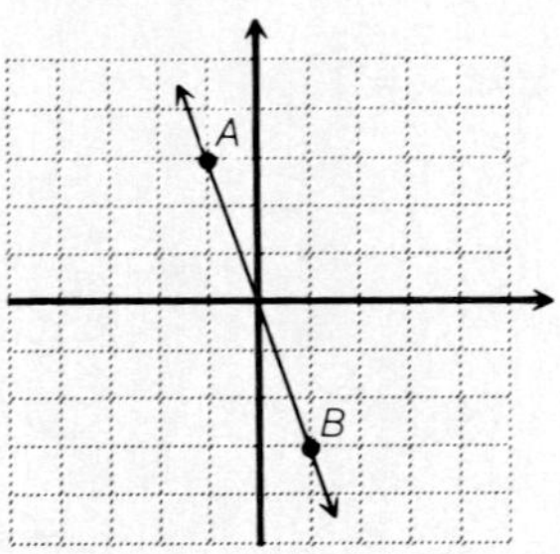

7. slope = -2

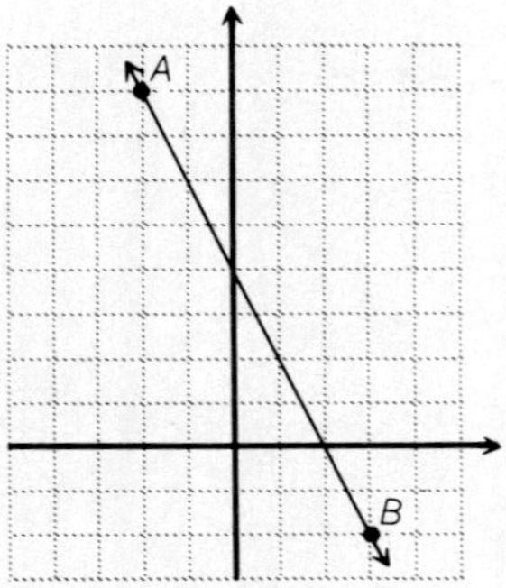

8. slope = $-\frac{3}{2}$

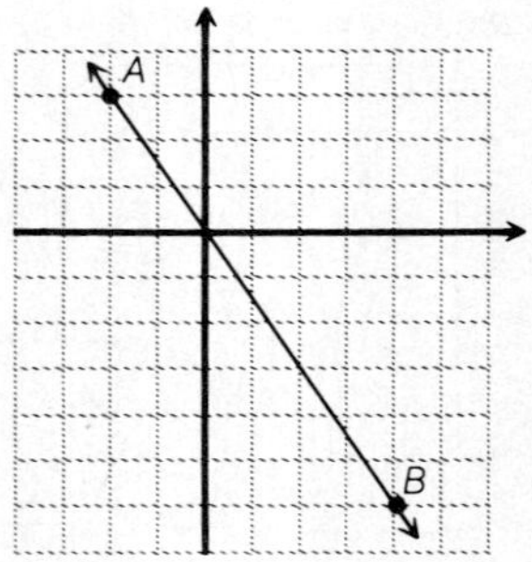

9. slope = $\frac{1}{7}$

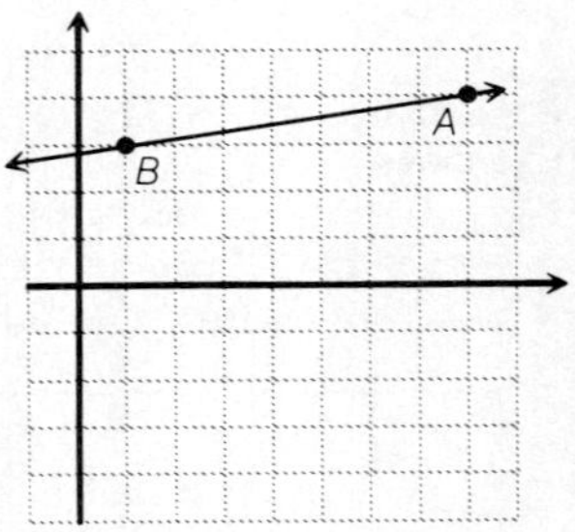

10. slope = $\frac{8}{5}$

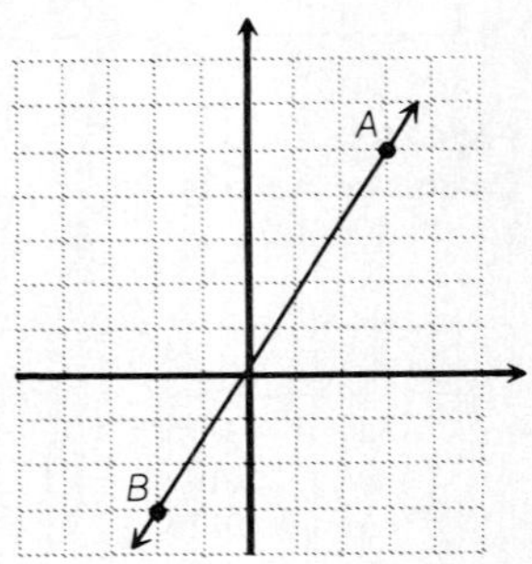

11. slope = 3

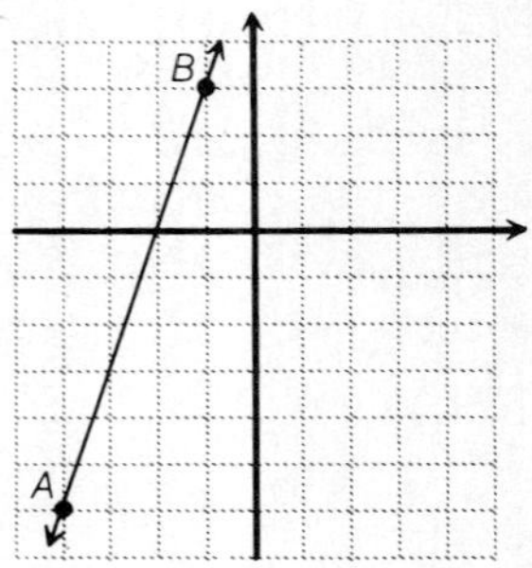

12. slope $= -\frac{13}{6}$

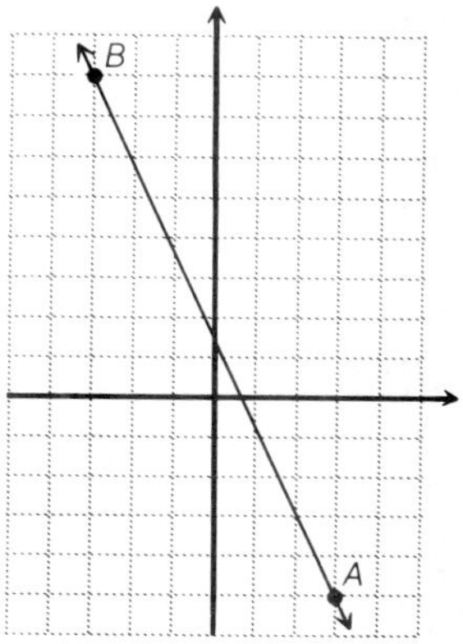

Page 497

Exercises **21.** x-int. 5, y-int. 5 **22.** x-int. 2, y-int. 4 **23.** x-int. 6, y-int. -2 **24.** x-int. 12, y-int. -8 **25.** x-int. 5, y-int. 4 **26.** x-int. 4, y-int. -3 **27.** x-int. 2.5, y-int. -5 **28.** x-int. 3, y-int. 5

29. $y = 2x - 3$
slope $= 2$

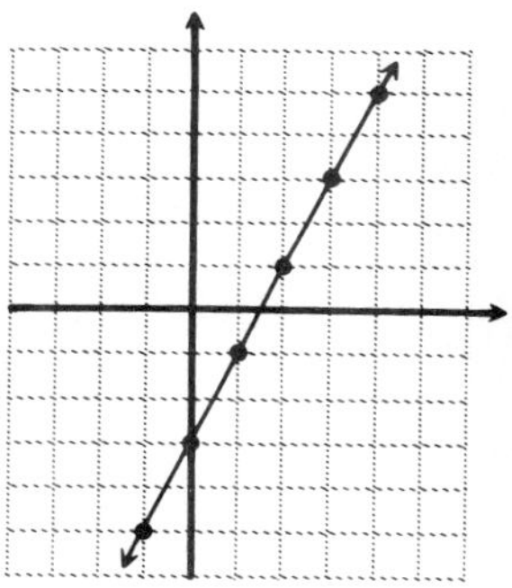

30. $y = -3x + 5$
slope $= -3$

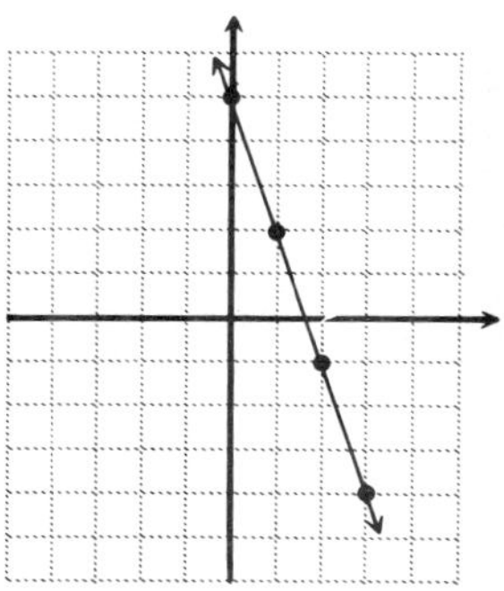

31. $y = \frac{1}{2}x + 1$
slope $= \frac{1}{2}$

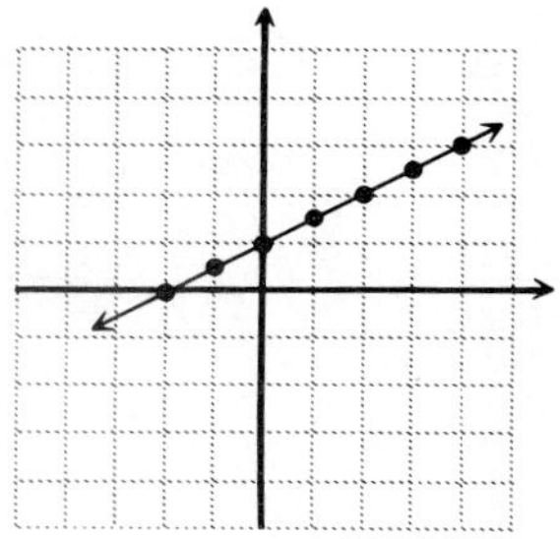

From Exercises 35–37, students should conclude that the slope of the line is equal to the x-coefficient. The teacher may now wish to point out that this is true only when the equation is in the form $y = mx + b$.

Page 500
Practice

The lines intersect at (2, 1).

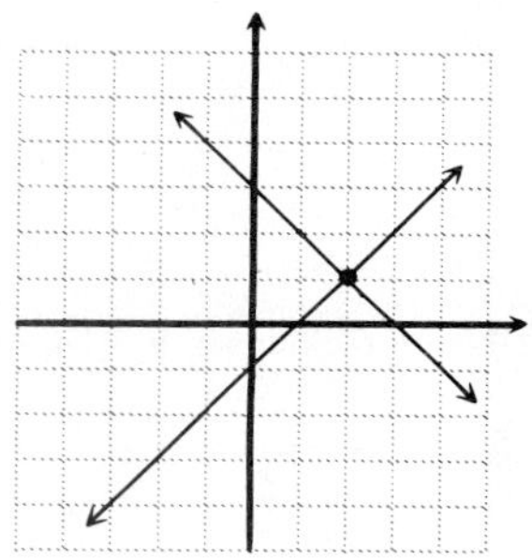

Page 501
Exercises

1. $x - y = 1$;
$x + 2y = 1$ (1, 0)

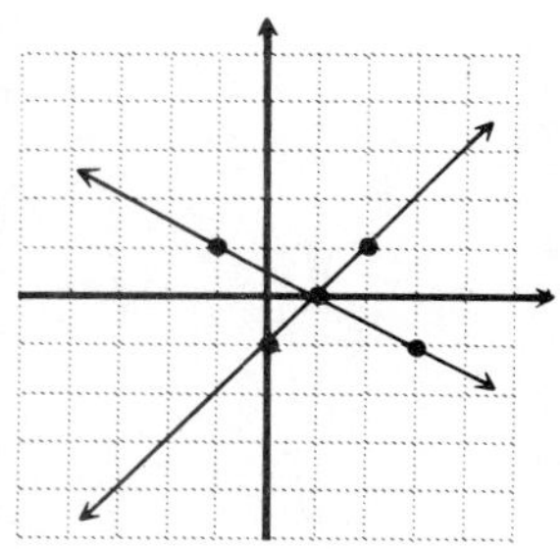

2. $x + y = 2$;
$2x - y = 1$; (1, 1)

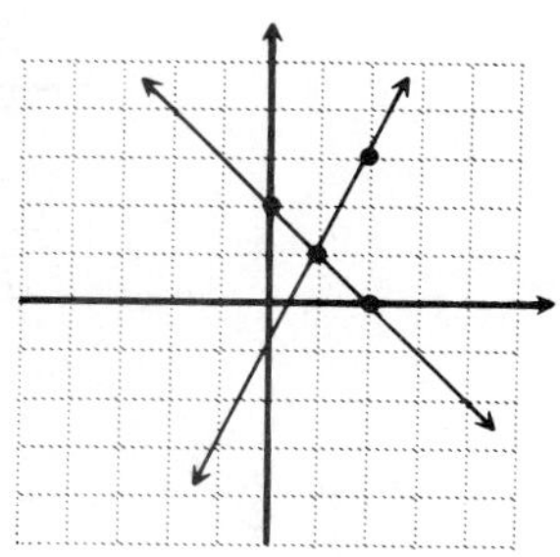

3. $x - y = 0$;
$2x + y = 6$; (2, 2)

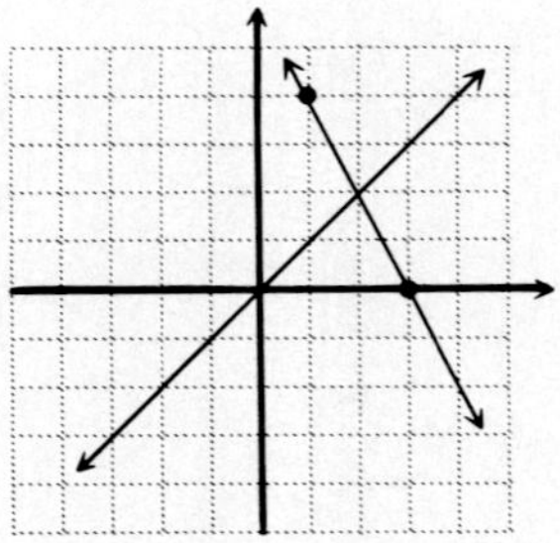

4. $x + y = 0$;
$x - y = -2$;
$(-1, 1)$

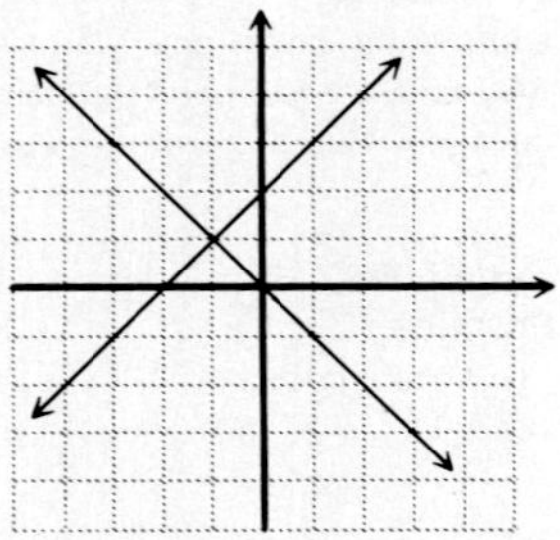

5. $2x - y = -1$;
$x + y = -2$;
$(-1, -1)$

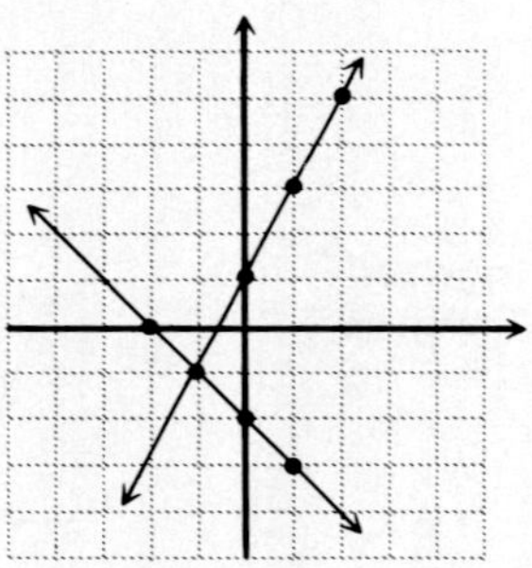

6. $2x + y = 5$;
$-2x + y = 1$;
(1, 3)

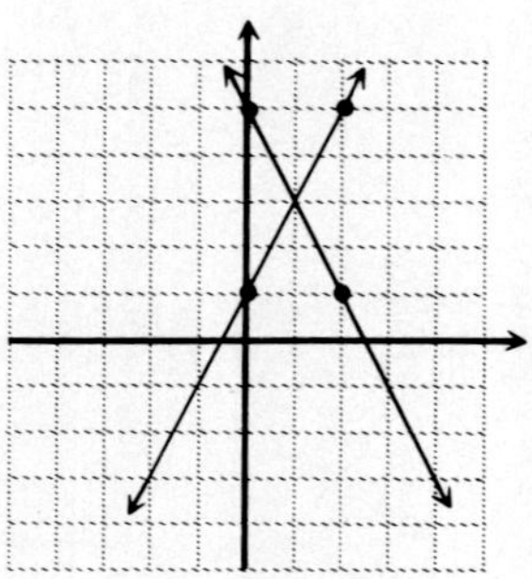

7. $x + y = 3$;
$x - y = 1$; (2, 1)

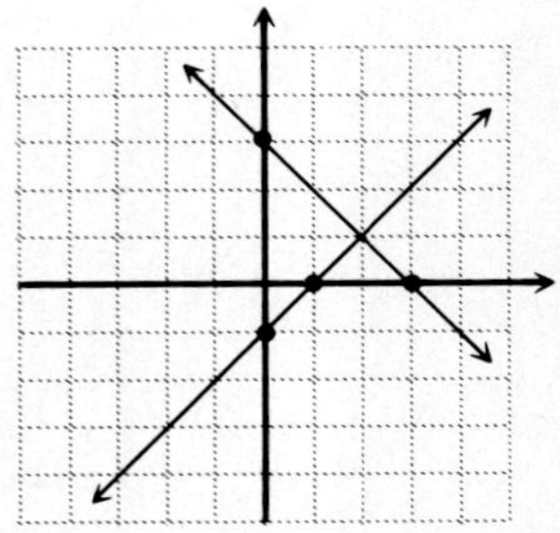

8. $-2x + y = -4$;
$2x + y = 0$;
$(1, -2)$

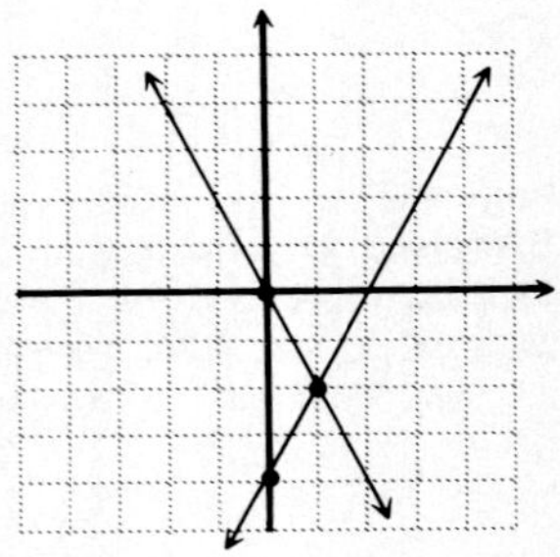

9. $x - 2y = 1$;
$2x - y = 5$;
(3, 1)

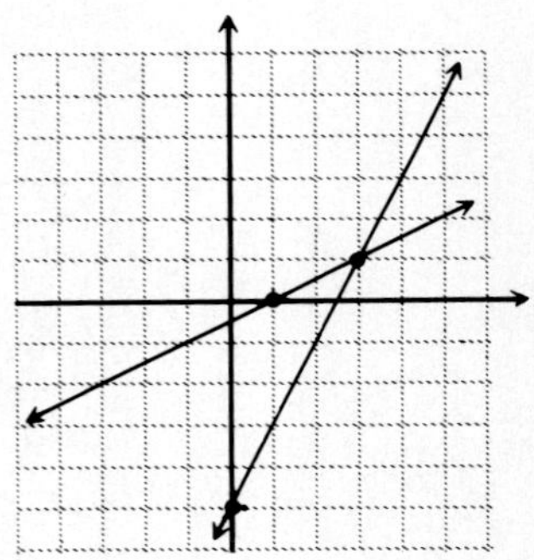

Page 502
Practice

a. $y > 2x - 1$

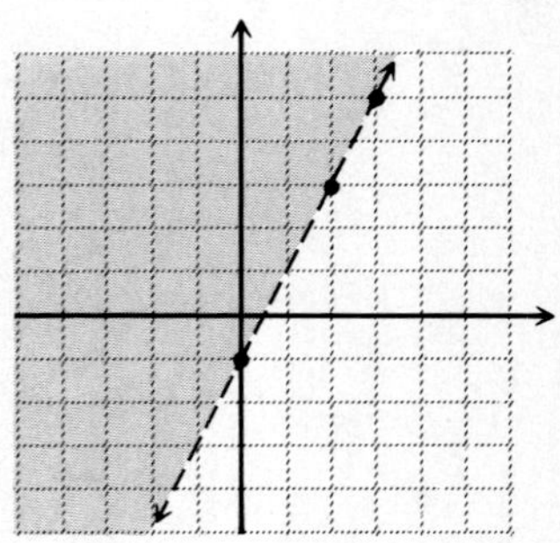

b. $y \le \frac{1}{2}x + 1$

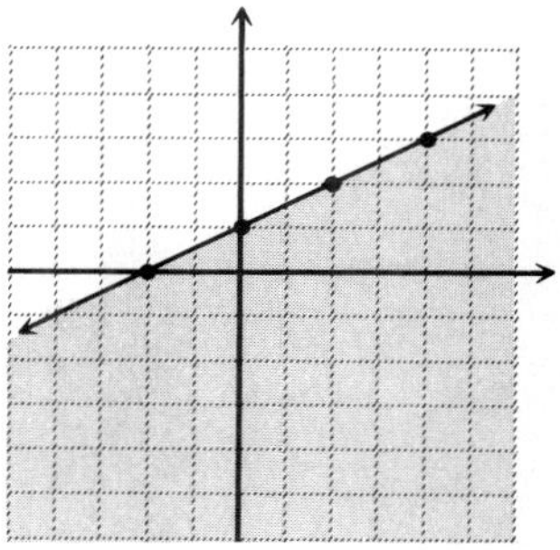

4. $x \ge 1$

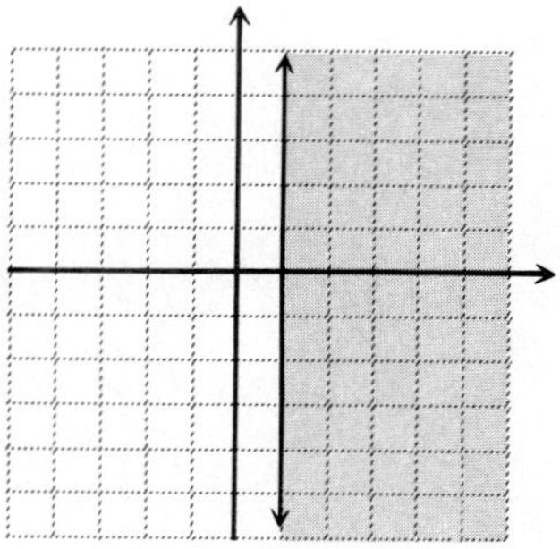

Page 503
Exercises

1. $y < 3x - 2$

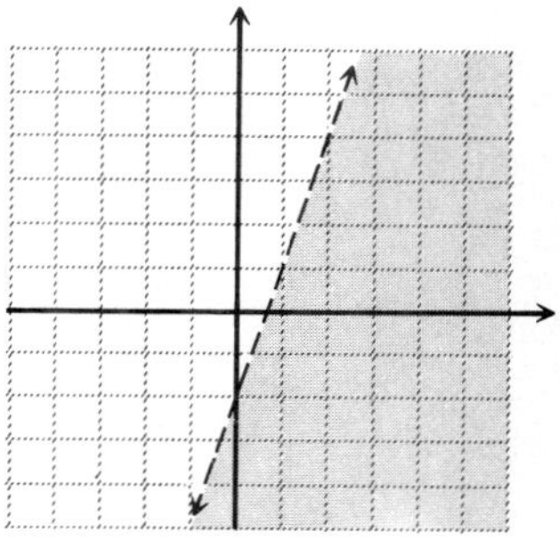

5. $y < 5$

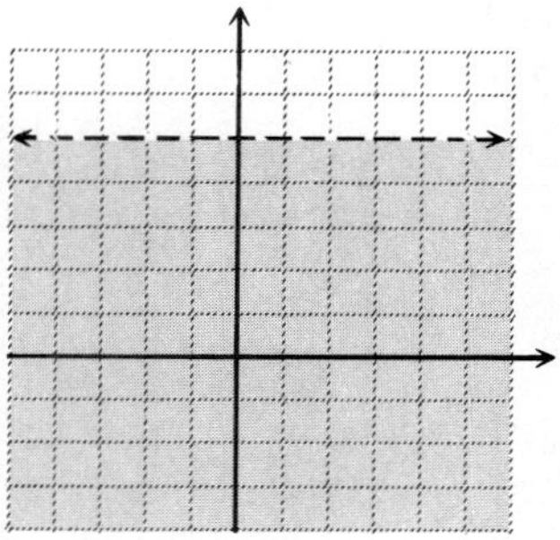

2. $y \ge -x + 2$

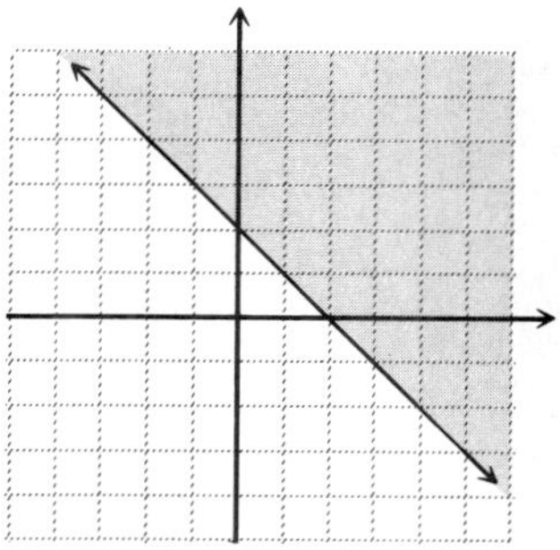

6. $y < \frac{1}{2}x + 3$

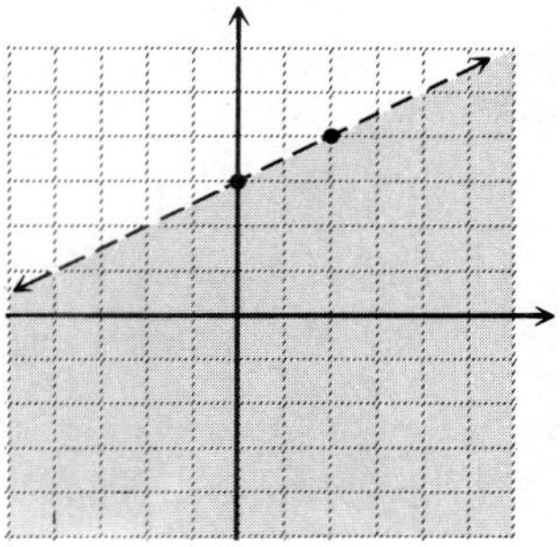

3. $y > -2x + 1$

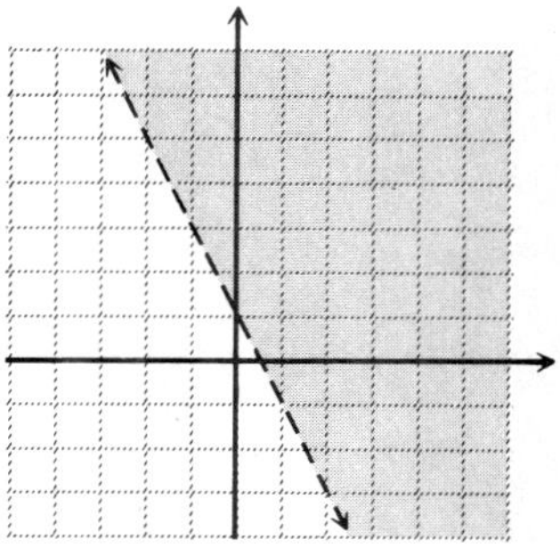

7. $y < 2x - 5$

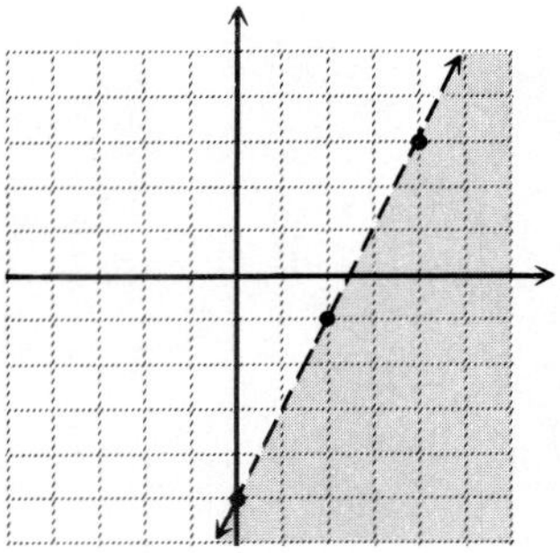

8. $x + 4 \geq y$

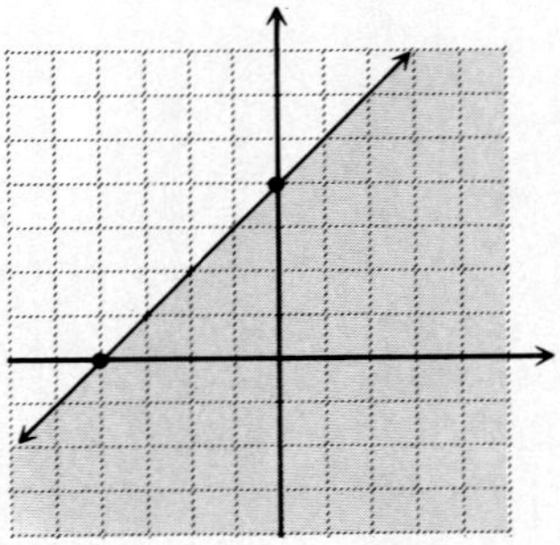

9. $y < 3 - x$

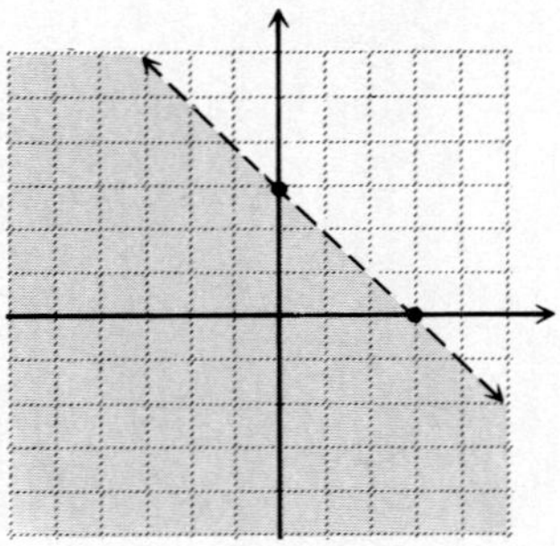

10. $2x + 3y \geq 6;$
$3y \geq 6 - 2x;$
$y \geq 2 - \frac{2}{3}x$

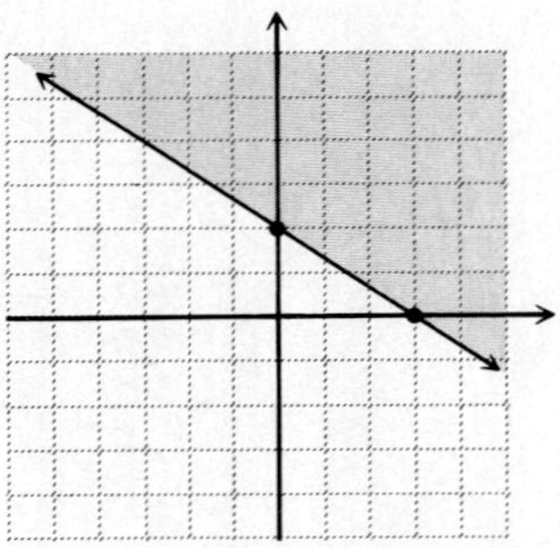

11. $3(x + 2y) \geq 6;$
$(x + 2y) \geq 2;$
$2y \geq -x + 2;$
$y \geq -\frac{1}{2}x + 1$

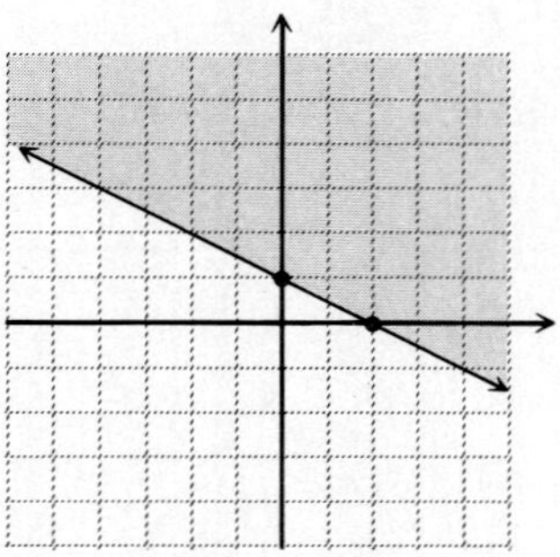

12. $x < 2y + 1;$
$x - 1 < 2y;$
$\frac{1}{2}x - \frac{1}{2} < y;$
$y > \frac{1}{2}x - \frac{1}{2}$

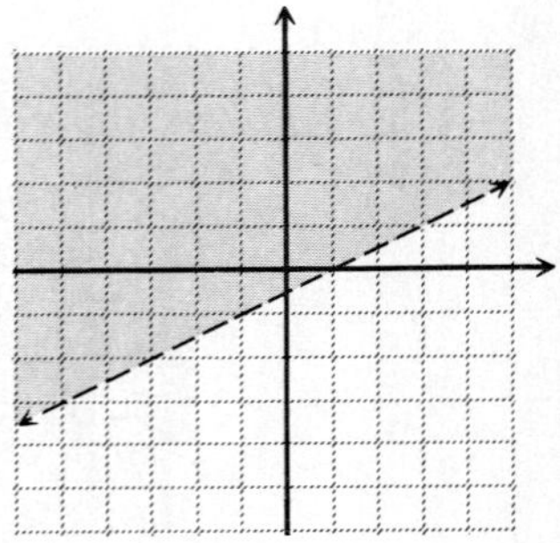

13. $3 \leq \frac{6}{x};\ x \leq 2$

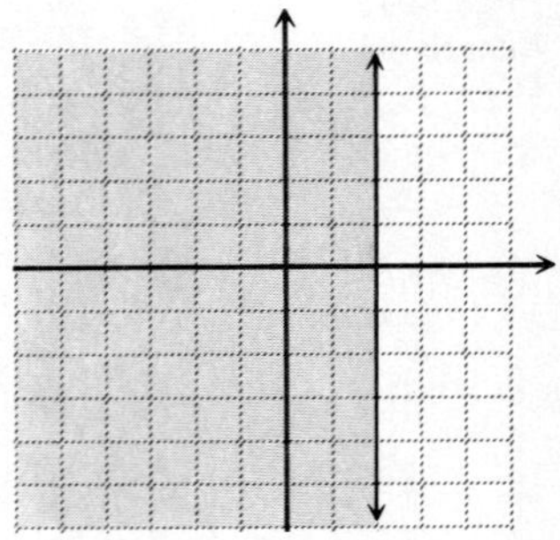

14. $2x + 7 > y - 3x;$
$5x + 7 > y$

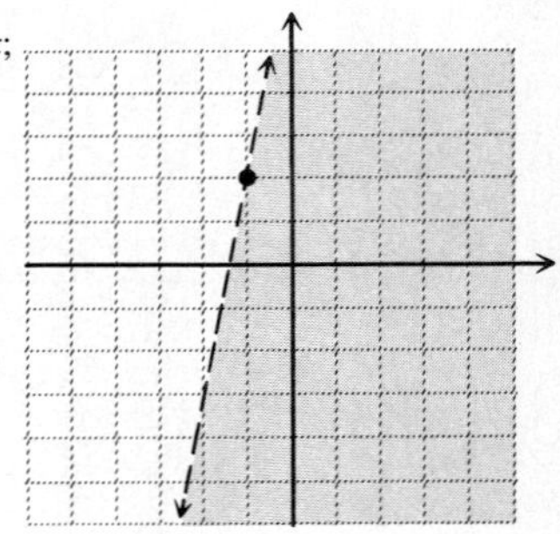

15. $4y - 3x < 12;$
$4y < 3x + 12;$
$y < \frac{3}{4}x + 3$

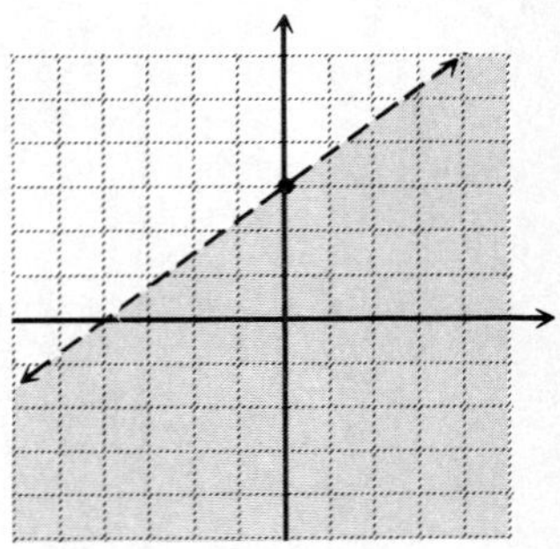

18. $x \leq 3$; $y \geq -1$

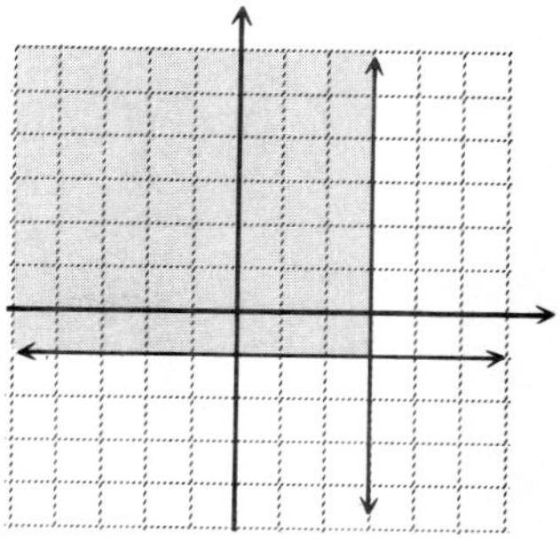

19. $y \leq x + 2$;
$y \leq -x + 1$

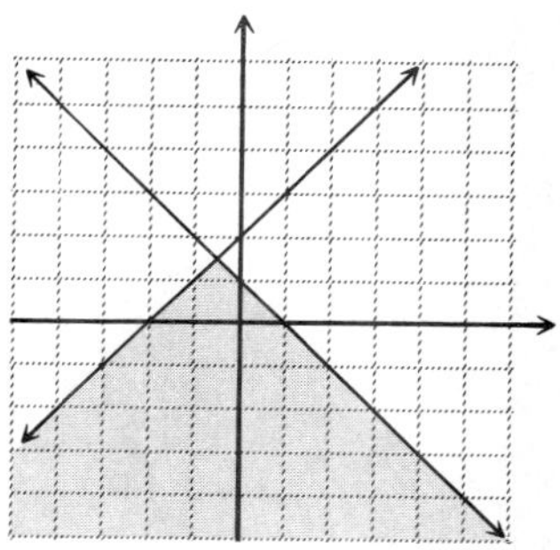

20. $y \geq \frac{1}{2}x - 2$;
$y \leq -2x + 6$

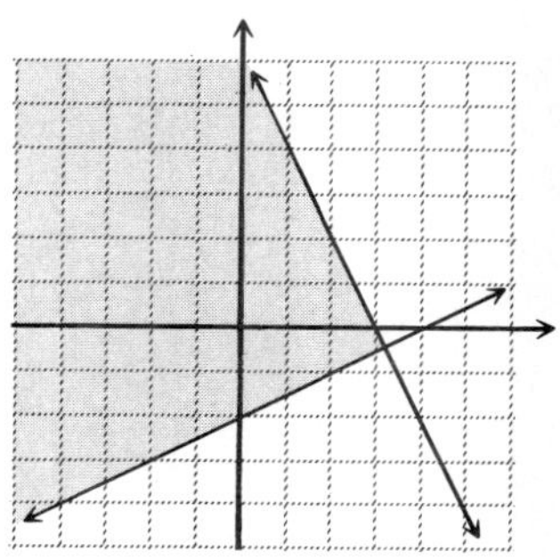

Page 505
Exercises

1.

time	cost
0.5	0.34
1	0.34
1.5	0.58
2	0.58
2.5	0.82
3	0.82
3.5	1.06
4	1.06
4.5	1.30
5	1.30

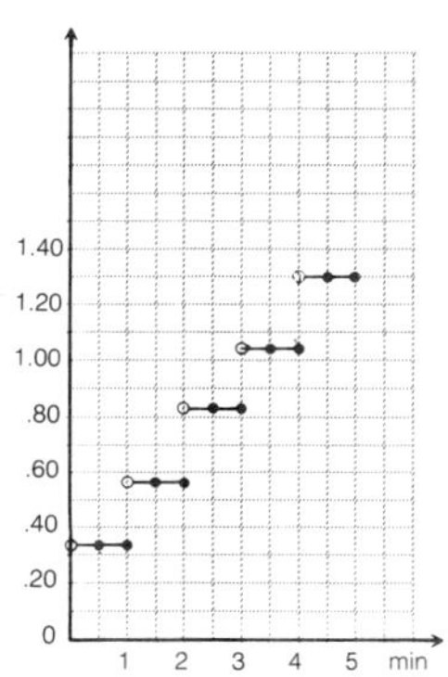

2.

time	cost
0.5	0.23
1	0.23
1.5	0.39
2	0.39
2.5	0.55
3	0.55
3.5	0.71
4	0.71
4.5	0.87
5	0.87

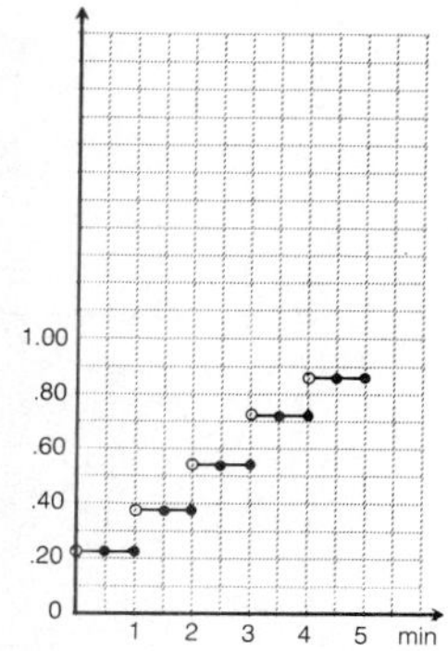

3.

time	cost
0.5	1.78
1	1.78
1.5	1.94
2	1.94
2.5	2.10
3	2.10
3.5	2.26
4	2.26
4.5	2.42
5	2.42

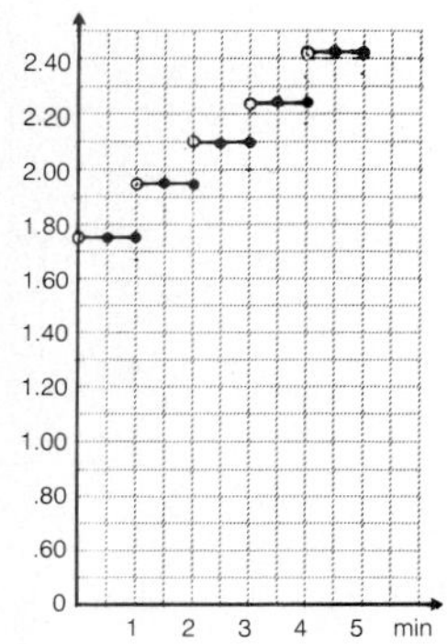

Page 506
Exercises

1. $x < 0$

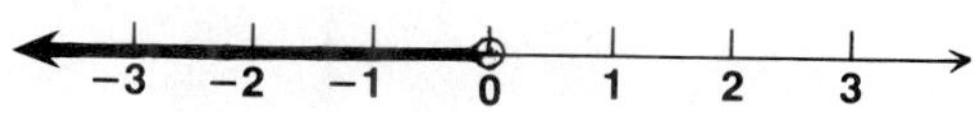

2. $x = 1 \geq 0$; $x \geq -1$

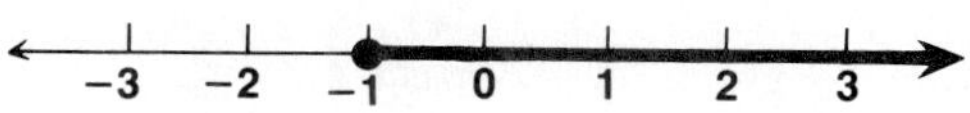

5–6.

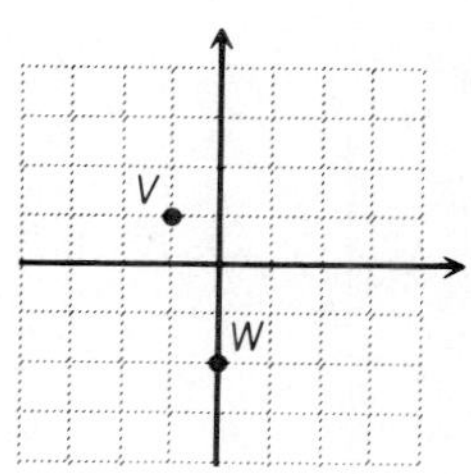

8. $y = -x - 1$

x	y
−2	1
−1	0
0	−1
1	−2
2	−3
3	−4

19.

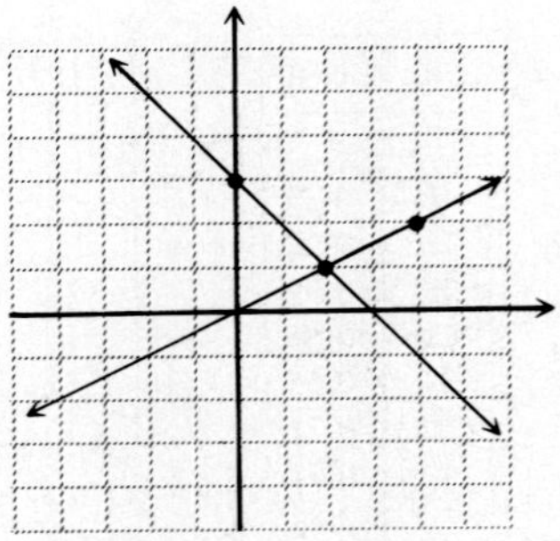

20. $y \geq 2x - 1$

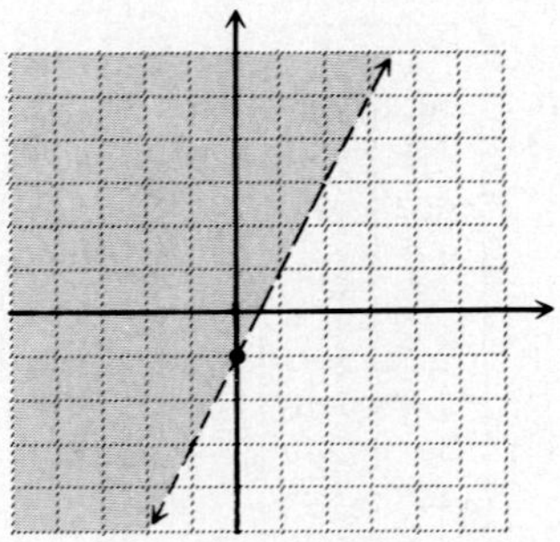

Page 507 Exercises

1. $x > -2$

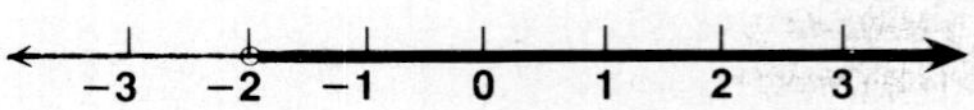

2. $2x - 1 > -1$; $2x > 0$; $x > 0$

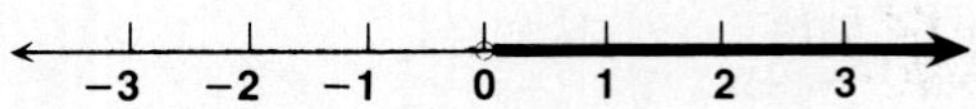

5–6.

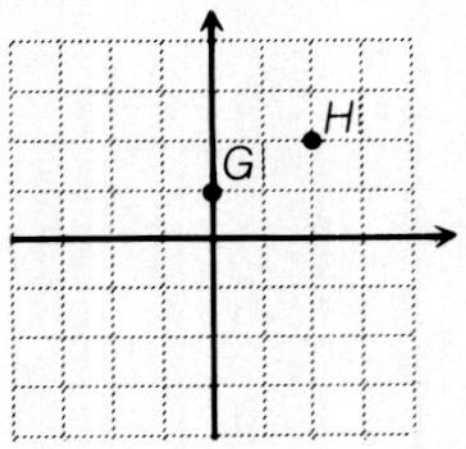

19. $x + 2y = 1$
$x - y = -2$

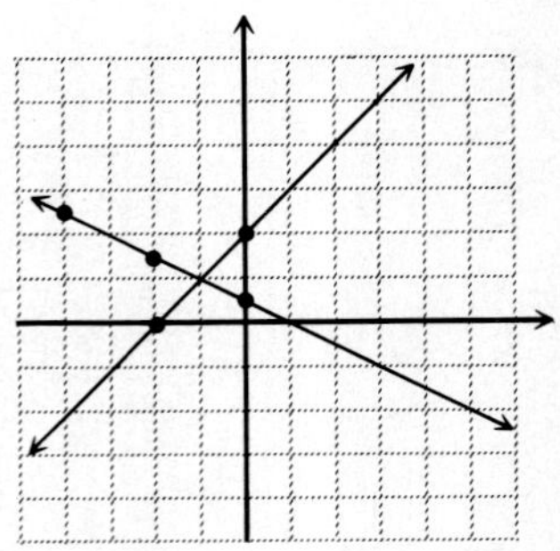

20. $y < (2 - x)$

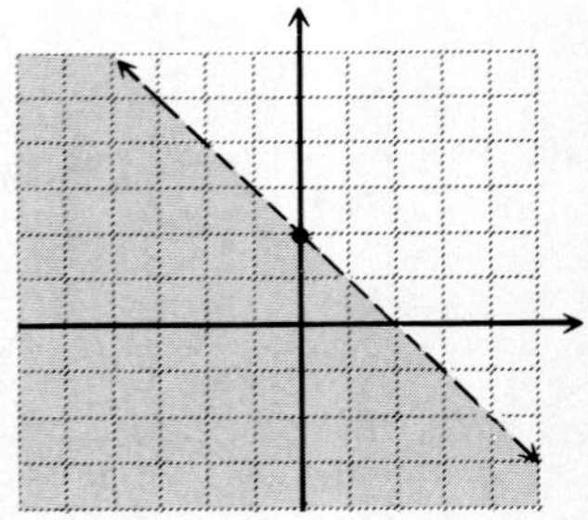

Page 525 Exercises

5–6. $C(2, -3)$, $D(-1, -1)$

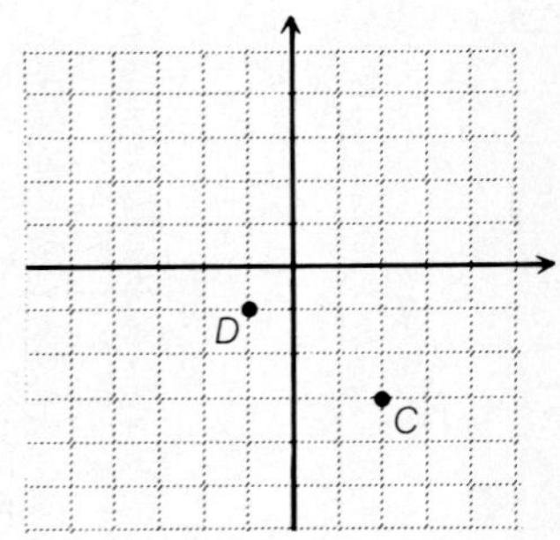

8. $y = -2x + 1$

x	y
−2	5
−1	4
0	3
1	2
2	1
3	0

Addison-Wesley

Pre-Algebra

Phares G. O'Daffer

Stanley R. Clemens

Randall I. Charles

Addison-Wesley Publishing Company

Menlo Park, California   Reading, Massachusetts   Don Mills, Ontario

Wokingham, England   Amsterdam   Sydney   Singapore

Tokyo   Madrid   San Juan

Photograph Acknowledgments

Cover Craig Aurness/West Light

CR Studio: 1. Tom Stack/Tom Stack & Associates: 2. Ben Rose/The Image Bank West: 15. Wayland Lee*/Addison-Wesley Publishing Company: 18. Bill Ross/West Light: 24. George B. Fry III*: 28. Mitchell Funk: 33. David Madison/Bruce Coleman Inc.: 40. Doug Lee/Tom Stack & Associates: 49. Mark Perlstein/Black Star: 57. Brett Froomer/The Image Bank West: 59. NASA: 62. Arthur Meyerson/The Image Bank West: 67. W. H. Hodge/Peter Arnold Inc.: 68L. Phelps/West Light: 68R. Wayland Lee*/Addison-Wesley Publishing Company: 79. Wayland Lee*/Addison-Wesley Publishing Company: 89. Keith Murakami/Tom Stack & Associates: 96. Jay Freis/The Image Bank West: 101. Wayland Lee*/Addison-Wesley Publishing Company: 105. Horst Schafer/Peter Arnold Inc.: 108. P. Kelly/The Image Bank West: 113. Jeanne Carley*: 117. Erika Stone/Peter Arnold Inc.: 118. Peter Menzel/Stock, Boston: 129. J. Carmichael/The Image Bank West: 133. Wayland Lee*/Addison-Wesley Publishing Company: 134. Stephen Frisch*: 142. Rick McIntyre/Tom Stack & Associates: 147. Jeanne Carley*: 160. Flip Chalfant/The Image Bank West: 165. David Madison/Bruce Coleman Inc.: 170. David Madison/Bruce Coleman Inc.: 183. Stacy Pick/Stock, Boston: 186. Andy Levin/Black Star: 191. David Madison/Bruce Coleman Inc.: 198. Lou Jones/The Image Bank West: 203. Michael Salas/The Image Bank West: 215. Bill Gallery/Stock, Boston: 217. Tom Stack & Associates: 224. Ken Cooper/The Image Bank West: 233. Wayland Lee*/Addison-Wesley Publishing Company: 234. Al Saiterwhite/The Image Bank West: 239. Joan Saxe/Bruce Coleman Inc.: 247. Richard Choy/Peter Arnold Inc.: 251. NASA: 253. Brian Parker/Tom Stack & Associates: 260. Michael Melford/The Image Bank West: 265. Cecile Brunswick/Peter Arnold Inc.: 266. John Coletti/Stock, Boston: 272. Charles Gupton/Stock, Boston: 282. Jay Freis/The Image Bank West: 287. National Baseball Hall of Fame: 293. Wayland Lee*/Addison-Wesley Publishing Company: 296. Craig Aurness/West Light: 301. Cary Wolinsky/Stock, Boston: 302. Wayland Lee*/Addison-Wesley Publishing Company: 308. Mike Mazzaso/Stock, Boston: 312. D. Brewster/Bruce Coleman Inc.: 313. Gary Milburn/Tom Stack & Associates: 319. Suzi Barnes/Tom Stack & Associates: 330. Ted Speigel/Black Star: 335. Steve Dunwell/The Image Bank West: 342T. Wayland Lee*/Addison-Wesley Publishing Company: 342B. NASA: 360. Wayland Lee*/Addison-Wesley Publishing Company: 368. Marc Solomon/The Image Bank West: 375. Eric Simmons/Stock, Boston: 384. Alvis Upitis/The Image Bank West: 395. T. Schneps/The Image Bank West: 399. Nicholar Foster/The Image Bank West: 409. Focus On Sports: 418. Marc Romanelli/The Image Bank West: 425. © Elliot Varner Smith 1975: 428. Focus On Sports: 431. Rich McIntyre/Tom Stack & Associate: 439. Ron Sherman/Bruce Coleman Inc.: 442. David Hiser/The Image Bank West: 447. Stephen Frisch*: 451. Focus On Sports: 461. Gus Schonefeld/Berg & Associates: 465. Wayland Lee*/Addison-Wesley Publishing Company: 468. Stephen Frisch*: 474. George B. Fry III*: 475. © Greg Pease 1986: 479. Gary Withey/Bruce Coleman Inc.: 490. Miguel/The Image Bank West: 494.

*Photographs provided expressly for the publisher

ISBN 0-201-29727-2

ABCDEFGHIJKL-KR–892109

Contents

Chapter 1
Expressions and Equations: Addition and Subtraction

Chapter 2
Expressions and Equations: Multiplication and Division

Chapter 3
Integers

Chapter 4
Decimals

Chapter 5
Number Theory

Chapter 6
Rational Numbers: Addition and Subtraction

Chapter 7
Rational Numbers: Multiplication and Division

Chapter 8
Equations and Inequalities

Chapter 9
Ratio, Proportion, and Percent

Chapter 10
Using Percent

Chapter 11
Equations in Geometry

Chapter 12
Area and Volume Formulas

Chapter 13
Probability, Statistics, and Graphs

Chapter 14
Square Roots and Special Triangles

Chapter 15
Graphs of Equations and Inequalities

Chapter One

Expressions and Equations: Addition and Subtraction

1-1 Variables and Expressions

Objective: To evaluate numerical and algebric expressions.
MIN: 1–44, 59; REG: 1–44 even, 45–50, 59; MAX: 1–44 even, 45–59; ALL: Mixed Review
An exploratory lesson opener can be found on page 545.

The burner in a hot-air balloon heats air inside the balloon. The hot air makes it rise. When the burner is turned off, the air in the balloon cools and the balloon drops.

With its burner on, a balloon rose to an altitude of 300 meters above sea level. Then the pilot turned the burner off and the balloon dropped 50 meters. If the pilot turned the burner on again and the balloon rose 100 meters, what altitude would the balloon reach?

The solution to this problem can be found by using the **numerical expression** $(300 - 50) + 100$. A numerical expression is a name for a number. To **evaluate** a numerical expression you find the number it represents. When an expression includes parentheses ( ), do the operation inside the parentheses first.

Example 1

Evaluate. $(300 - 50) + 100$

Solution $(300 - 50) + 100$

$= 250 + 100$ Do the operation inside the parentheses first. $300 - 50 = 250$.

$= 350$ Complete the calculation. $250 + 100 = 350$. **The balloon would reach an altitude of 350 meters.**

Practice Evaluate.

a. $16 - 7$ **9** **b.** $8 + (17 - 9)$ **16** **c.** $(27 + 6) - 19$ **14**

A **variable** is a letter, such as n, that reserves a place for a number. An expression such as $\boldsymbol{n + 7}$ that contains at least one variable is called an **algebraic expression**. To evaluate an algebraic expression, you replace each variable with a number, or value, and evaluate the numerical expression that results.

Example 2

Evaluate $n + 7$ for $n = 9$.

Solution $n + 7$

$9 + 7$ Replace the variable n with the number 9.

$= 16$

Practice Evaluate.

a. $32 - x$ for $x = 8$ 24 **b.** $a + (13 - 8)$ for $a = 6$ 11

An algebraic expression may contain more than one variable.

Example 3

Evaluate $(a + b) - 10$ for $a = 8$ and $b = 9$.

Solution $(a + b) - 10$

$(8 + 9) - 10$ Replace a with 8 and b with 9.

$= 17 - 10$

$= 7$

Practice Evaluate. Tell students that when a variable appears more than once, they should replace it with the same number each time.

a. $(x - 6) + z$ for $x = 11$ and $z = 9$ 14 **b.** $c - (c - d)$ for $c = 17$ and $d = 4$. 4

Oral Exercises

What letter is used as a variable? What number replaces it?

1. $x + 5$ x
$3 + 5$ 3

2. $5 + a$ a
$5 + 7$ 7

3. $12 - b$ b
$12 - 8$ 8

4. $r - 7$ r
$9 - 7$ 9

5. $9 + s$ s
$9 + 2$ 2

6. $13 - c$ c
$13 - 4$ 4

7. $y - 8$ y
$10 - 8$ 10

8. $t + 8$ t
$6 + 8$ 6

Exercises

A Evaluate each numerical expression.

1. $8 + 7$ 15

2. $17 - 9$ 8

3. $13 - 13$ 0

4. $49 + 9$ 58

5. $45 - 10$ 35

6. $(36 - 8) + 5$ 33

7. $43 - (6 + 7)$ 30

8. $(19 + 9) - 10$ 18

9. $(35 + 4) + 7$ 46

Evaluate each numerical expression.

10. $26 + (12 - 4)$ 34
11. $(16 - 5) + 4$ 15
12. $(20 - 15) + 6$ 11
13. $49 - (8 + 2)$ 39
14. $23 + (8 + 10)$ 41
15. $(19 - 8) - 6$ 5
16. $(11 + 12) - 13$ 10
17. $33 - (22 + 10)$ 1
18. $28 + (19 - 4)$ 43

Evaluate each algebraic expression.

19. $x + 4$ for $x = 6$ 10
20. $n - 6$ for $n = 9$ 3
21. $n + 5$ for $n = 7$ 12
22. $b - 13$ for $b = 17$ 4
23. $4 + c$ for $c = 29$ 33
24. $16 - a$ for $a = 7$ 9
25. $(x - 9) + 8$ for $x = 75$
26. $8 + (6 + c)$ for $c = 53$
27. $(e - 5) - 9$ for $e = 64$
28. $n + 7$ for $n = 5$, for $n = 7$, and for $n = 15$ 12, 14, 22

25. 74 26. 67 27. 50

29. $x - 8$ for $x = 12$, for $x = 14$, and for $x = 76$ 4, 6, 68

Evaluate each expression for $a = 5$, $b = 8$, and $c = 9$.

30. $a + b$ 13
31. $c - a$ 4
32. $c - (13 - b)$ 4
33. $(c - b) + 26$ 27
34. $(b + b) - 8$ 8
35. $(29 + 57) + (c - a)$ 90
36. $(c + a) - (12 - 4)$ 6
37. $(a + b) - (17 - 8)$ 4
38. $(13 - c) + (14 - b)$ 10

Evaluate.

39. $(x + 8) + (x - 7)$ for $x = 15$ 31
40. $(p + 4) + (p - 7)$ for $p = 12$ 21
41. $(9 - b) - (2 + b)$ for $b = 3$ 1
42. $(12 - q) + (8 - q)$ for $q = 5$ 10
43. $(z + 7) - (z + 5)$ for $z = 8$ 2
44. $(8 + d) + (c - d)$ for $d = 5, c = 12$ 20

B Evaluate each expression for $w = 47$, $x = 56$, $y = 94$, and $z = 123$.

45. $(w + x) + (z - y)$ 132
46. $(z - x) - (y - w)$ 20
47. $(z - y) + (x - w)$ 38

Copy and complete each table, evaluating the algebraic expression for the numbers given.

	n	$n + n - 1$
	5	$5 + 5 - 1 = 9$
48.	10	$10 + 10 - 1 = 19$
49.	9	$9 + 9 - 1 = 17$
50.	20	$20 + 20 - 1 = 39$

	x	y	$x + x - y$
	8	7	$8 + 8 - 7 = 9$
51.	7	6	$7 + 7 - 6 = 8$
52.	9	8	$9 + 9 - 8 = 10$
53.	6	9	$6 + 6 - 9 = 3$

	a	b	$a + b + b$
	1	7	$1 + 7 + 7 = 15$
54.	3	10	$3 + 10 + 10 = 23$
55.	6	8	$6 + 8 + 8 = 22$
56.	15	20	$15 + 20 + 20 = 55$

C Extending Thinking Skills

57. Choose numbers for a, b, and c so that $c - b = a$. Do this three times with different numbers. Does $a + b = c$ each time? Yes.

58. What is the value of the next expression in this pattern if $n = 9$?
$n + 1, n + 3, n + 6, n + 10, n + 15,$? 30

59. Find number replacements for x and y so that $x + y = 16$ and $x - y = 2$. $x = 9, y = 7$ Encourage students to guess and check.

Thinking skills: 57. Generalizing 58. Discovering a pattern 59. Testing a conjecture

Mixed Review

Add or subtract.

Examples:

```
   1 11        5 9 9 13
  3,346         6,003
+   894       -   978
  -----         -----
  4,240         5,025
```

60. 104 + 86 **190**	**61.** 171 + 74 **245**	**62.** 16 + 191 **207**
63. 54 + 47 **101**	**64.** 936 + 243 **1,179**	**65.** 511 + 106 **617**
66. 518 + 166 **684**	**67.** 110 + 749 **859**	**68.** 219 + 371 **590**
69. 183 + 128 **311**	**70.** 308 − 225 **83**	**71.** 436 − 88 **348**
72. 221 − 96 **125**	**73.** 178 − 65 **113**	**74.** 2,104 − 196 **1,908**
75. 3,096 − 854 **2,242**	**76.** 2,004 − 857 **1,147**	**77.** 40,000 − 32,651 **7,349**
78. 7,841 − 538 **7,303**	**79.** 1,240 − 141 **1,099**	**80.** 14,326 − 2,134 **12,192**
81. 30,001 − 5,642 **24,359**	**82.** 50,129 + 11,297 **61,426**	**83.** 3,590 + 7,246 **10,836**
84. 65,667 − 25,340 **40,327**	**85.** 42,208 − 3,175 **39,033**	**86.** 90,128 + 3,808 **93,936**
87. 12,645 − 3,810 **8,835**	**88.** 7,775 + 3,816 **11,591**	**89.** 20,251 − 18,532 **1,719**
90. 31,190 + 4,325 **35,515**	**91.** 4,261 − 1,705 **2,556**	**92.** 30,672 + 3,488 **34,160**
93. 50,119 + 5,679 **55,798**	**94.** 24,307 − 2,867 **21,440**	**95.** 10,146 − 5,982 **4,164**
96. 10,085 + 4,709 **14,794**	**97.** 3,290 − 1,701 **1,589**	**98.** 128,104 + 9,685 **137,789**

CALCULATOR ACTIVITY

You can use the memory keys on a calculator to evaluate expressions.

Use the [M+] key to add the displayed number to the total in the memory.

Use the [M−] key to subtract the displayed number from the total in the memory.

Use the [MR] key to display the total in the memory.

Evaluate $956 - (a - b)$ for $a = 342$ and $b = 178$.

342 [−] **178** [=] [M+] Find the difference in parentheses first and store it in memory.

936 [−] [MR] [=] **772** Then subtract the difference in the memory from 936.

Use the calculator memory keys to evaluate each expression.

1. $5{,}392 - (x + y)$ for $x = 849$, $y = 768$ **3,775**
2. $4{,}238 + (c - d)$ for $c = 6{,}982$, $d = 748$ **10,472**
3. $3{,}479 - (a - b)$ for $a = 1{,}603$, $b = 479$ **2,355**
4. $965 + (p - q + 467)$ for $p = 640$, $q = 398$ **1,674**

Certain calculators have [STO] and [RCL] keys in place of [M+] and [MR]. The sequence in the example would be as follows.

342 [−] 178 [=] [STO] 936 [−] [RCL] [=]

PROBLEM SOLVING: WRITING EXPRESSIONS

1-2 Translating Phrases to Algebraic Expressions

Objective: To translate a verbal phrase to a numerical or algebraic expression.
MIN: 1–36; REG: 1–36 even, 37–40; MAX: 1–36 even, 37–42; ALL: Mixed Review

Verbal phrases that suggest addition or subtraction can be translated into numerical or algebraic expressions.

Example 1

Write as a numerical expression. "5 increased by 3"

Solution $5 + 3$ The phrase "increased by 3" suggests adding 3.

Practice Write as a numerical expression.

a. 6 less than 10 $10 - 6$

b. 9 decreased by 4 $9 - 4$

c. the sum of 8 and 7 $8 + 7$

d. 7 more than 8 $8 + 7$

The phrase "5 increased by 3" is translated to the numerical expression $5 + 3$. The phrase "a number increased by 3" becomes the algebraic expression $n + 3$.

number	number increased by 3	
5	$5 + 3$	"5 increased by 3"
10	$10 + 3$	"10 increased by 3"
n	$n + 3$	"a **number** increased by 3"

Example 2

Write as an algebraic expression. "a number decreased by 4"

Solution $n - 4$ Think of a specific number, say 6. "6 decreased by 4" is $6 - 4$, so "a *number* decreased by 4" is $n - 4$.

Practice Write as an algebraic expression.

a. a number plus 5 $n + 5$

b. a number decreased by 8 $n - 8$

c. 7 more than a number $n + 7$

d. 5 less than a number $n - 5$

e. the difference of a number and 3 $n - 3$

Example 3

Write as an expression. Jeff has \$68. How much will he have if:

a. he earns 43 dollars? **b.** he earns x dollars?
c. he spends 29 dollars? **d.** he spends y dollars?

Solutions

a. he earns 43 dollars? $68 + 43$
b. he earns x dollars? $68 + x$ Think about a number, write a variable.
c. he spends 29 dollars? $68 - 29$
d. he spends y dollars? $68 - y$ Think about a number, write a variable.

Practice Write as an expression. Tina's house is 23 years old. How old:

a. will it be in 9 years? **b.** will it be in y years?
c. was it 14 years ago? **d.** was it t years ago?

a. $23 + 9$
b. $23 + y$
c. $23 - 14$
d. $23 - t$

Oral Exercises

Tell whether the phrase suggests addition or subtraction.

1. the sum of + **2.** decreased by − **3.** minus −
4. increased by + **5.** make it greater + **6.** added to +
7. the difference of − **8.** make it less − **9.** plus +
10. subtracted from − **11.** make it more + **12.** a total of +

Exercises

A Write as a numerical expression.

1. the sum of 8 and 6 $8 + 6$
2. the difference of 12 and 4 $12 - 4$
3. 7 more than 9 $9 + 7$
4. 3 less than 10 $10 - 3$
5. 6 increased by 7 $6 + 7$
6. 14 decreased by 5 $14 - 5$
7. 8 added to 12 $12 + 8$
8. 12 subtracted from 15 $15 - 12$

Write as an algebraic expression.

9. the sum of x and 6 $x + 6$
10. 4 taken away from t $t - 4$
11. 7 more than n $n + 7$
12. 3 less than s $s - 3$
13. 8 added to r $r + 8$
14. 12 subtracted from z $z - 12$
15. y decreased by 5 $y - 5$
16. x more than 16 $16 + x$
17. the difference of a number and 9 $n - 9$
18. the sum of a number and 8 $n + 8$
19. 6 more than a number $n + 6$
20. a number increased by 11 $n + 11$
21. 16 added to a number $n + 16$
22. a number decreased by 56 $n - 56$
23. 126 less than a number $n - 126$
24. 9 increased by a number $9 + n$

Write an expression for each question.
Mr. Henderson is 64 years old. How old:

25. will he be in 9 years? $64 + 9$

26. will he be in y years? $64 + y$

27. was he 16 years ago? $64 - 16$

28. was he x years ago? $64 - x$

Beth weighs 45 kg. How much will she weigh:

29. after she gains 7 kg? $45 + 7$

30. after she gains n kg? $45 + n$

31. after she loses 4 kg? $45 - 4$

32. after she loses y kg? $45 - y$

Todd earned \$75 by washing cars. How much money will he have:

33. after he earns 45 dollars more? $75 + 45$

34. after he earns x dollars more? $75 + x$

35. after he buys some \$37 shoes? $75 - 37$

36. after he buys an item costing d dollars? $75 - d$

B The letter n represents an even number 0, 2, 4, 6, 8, ...
Write an algebraic expression for:

37. the whole number just after n. $n + 1$

38. the whole number just before n. $n - 1$

39. the even number just after n. $n + 2$

40. the even number just before n. $n - 2$

C *Extending Thinking Skills*

41. Encourage students to guess and check.

41. Choose two variables and write an algebraic expression for their sum. Write an expression for the difference of the same two variables. Find values for the variables that have a sum of 31 and difference of 3. $x + y = 31, x - y = 3, x = 17, y = 14$

42. Look for a pattern in this sequence: 1, 1, 2, 3, 5, 8, 13, 21, 34, 55, . . . Let a represent a number in the sequence and b represent the next number after a. Write an algebraic expression for the next number in the sequence after b. $a + b$

Thinking skills: 41. Testing a conjecture 42. Discovering a pattern

Mixed Review

Evaluate. **43.** $850 + 143$ 993 **44.** $248 - (20 + 28)$ 200 **45.** $466 - (96 - 66)$ 436

Evaluate for $s = 136$, $s = 238$, and $s = 312$. **46.** $70 + (s - 86)$ 120, 222, 296 **47.** $1000 - (s + s)$ 728, 524, 376

MENTAL MATH

You can use **compensation** to find certain sums and differences mentally.

- To find 752 − **398**, think "752 − **400** is 352. I subtracted **2** too much, so I'll add it back. The difference is 352 + **2**, or 354."
- To find 645 + **296**, think "645 + **300** is 945. I added **4** too much, so I'll subtract it. The sum is 945 − **4**, or 941."

Use compensation to find each sum or difference mentally.

1. 642 − 299 343

2. 8,546 − 3,998 4,548

3. 874 − 549 325

4. 387 + 497 884

5. 4,638 + 2,995 7,633

6. 529 + 348 877

1-3 Properties of Addition

Objective: To use the basic properties of addition to write an algebraic expression equivalent to a given algebraic expression.
MIN: 1–29; REG: 1–29 even, 30–42; MAX: 1–29 even, 30–46; ALL: Mixed Review

The **whole numbers** are the numbers 0, 1, 2, 3, 4, 5, . . . The basic properties for addition of whole numbers can make computation with numerical expressions easier. You can also use the properties to write **equivalent** expressions. Two algebraic expressions are equivalent if (and only if) they have the same value for any number that replaces the variable.

You can change the order in which two numbers are added and the sum will stay the same. For example, 578 + 964 = 964 + 578.

$$\begin{array}{r} 578 \\ +\ 964 \\ \hline 1{,}542 \end{array} \qquad \begin{array}{r} 964 \\ +\ 578 \\ \hline 1{,}542 \end{array}$$

Variables can be used to state this property for all whole numbers.

Commutative Property of Addition

A change in the order in which two whole numbers are added does not change their sum.

For all whole numbers ***a*** and ***b***, $a + b = b + a$

Example 1

Use the commutative property to write an equivalent expression. $n + 5$

Solution $5 + n$ $5 + n$ has the same value as $n + 5$ for any number that replaces n.

Practice Use the commutative property to write an equivalent expression.

a. $45 + x$ $x + 45$ **b.** $a + 5{,}138$ $5{,}138 + a$ **c.** $479 + t$ $t + 479$

You can change the way whole numbers are grouped for addition and the sum will stay the same.
For example, (87 + 649) + 151 = 87 + (649 + 151).

$$\begin{array}{r} 87 \\ +\ 649 \\ \hline 736 \end{array} \quad \begin{array}{r} 736 \\ +\ 151 \\ \hline 887 \end{array} \quad \text{and} \quad \begin{array}{r} 649 \\ +\ 151 \\ \hline 800 \end{array} \quad \begin{array}{r} 87 \\ +\ 800 \\ \hline 887 \end{array}$$

Note that changing the grouping may allow you to find the sum mentally.

Variables can be used to state this property for all whole numbers.

Associative Property of Addition

A change in the grouping of whole numbers for addition does not change the sum.

For all whole numbers a, b, and c, $a + (b + c) = (a + b) + c$

Example 2

Use the associative property to write an equivalent expression.
$(x + 57) + 43$

Solution $x + (57 + 43)$ $x + (57 + 43)$ has the same value as $(x + 57) + 43$ for any number that replaces x.

Practice Use the associative property to write an equivalent expression.

a. $(n + 145) + 68$ a. $n + (145 + 68)$

b. $125 + (75 + a)$ b. $(125 + 75) + a$

c. $m + (123 + 77)$ c. $(m + 123) + 77$

d. $(118 + 12) + c$ d. $118 + (12 + c)$

Zero is called the **additive identity** because when it is added to a number, the result is that same number. For example, $689 + 0 = 689$.

Identity Property of Addition

The sum of an addend and zero is the addend.

For every whole number a, $a + 0 = a$ $0 + a = a$

When zero is subtracted from a number, the result is that same number: $a - 0 = a$. When a number is subtracted from itself, the result is zero: $a - a = 0$.

Example 3

Use the identity property to write an equivalent expression. $x + 0$

Solution x x has the same value as $x + 0$ for any number that replaces x.

Practice Use the identity property to write an equivalent expression.

a. $0 + t$ t **b.** $a + 0$ a **c.** $z + 0$ z **d.** $0 + n$ n

Oral Exercises

Name the property shown by each equation.

1. $9 + 0 = 9$ ident.
2. $7 + 8 = 8 + 7$ commut.
3. $0 + 6 = 6$ ident.
4. $6 + (3 + 5) = (6 + 3) + 5$ assoc.
5. $8 + 23 = 23 + 8$ commut.
6. $18 + (2 + 9) = (18 + 2) + 9$ assoc.
7. $47 = 0 + 47$ ident.
8. $(14 + 6) + 15 = 14 + (6 + 15)$ assoc.

Exercises

A Use the commutative property to write an equivalent expression.

1. $y + 4$ $4 + y$
2. $9 + x$ $x + 9$
3. $12 + n$ $n + 12$
4. $x + 28$ $28 + x$
5. $137 + b$ $b + 137$
6. $c + 236$ $236 + c$
7. $507 + y$ $y + 507$
8. $p + 1578$ $1578 + p$

Use the associative property to write an equivalent expression.

9. $(n + 2) + 8$
10. $(y + 8) + 12$
11. $(t + 9) + 7$
12. $(a + 4) + 16$
13. $(z + 16) + 34$
14. $(n + 78) + 22$
15. $17 + (8 + b)$
16. $26 + (4 + c)$
17. $38 + (12 + r)$

9. $n + (2 + 8)$
10. $y + (8 + 12)$
11. $t + (9 + 7)$
12. $a + (4 + 16)$
13. $z + (16 + 34)$
14. $n + (78 + 22)$
15. $(17 + 8) + b$
16. $(26 + 4) + c$
17. $(38 + 12) + r$

Use the identity property to write an equivalent expression.

18. $x + 0$ x
19. $0 + y$ y
20. $a - a$ 0

Write an equivalent expression and name the property you used.

21. $n + 7$
22. $y + 0$
23. $(n + 7) + 3$
24. $57 + x$
25. $p + 23$
26. $83 + (17 + x)$
27. $8 + s$
28. $0 + n$
29. $0 + (x + y)$

21. $7 + n$, commut.
22. y, ident.
23. $n + (7 + 3)$, assoc.
24. $x + 57$, commut.
25. $23 + p$, commut.
26. $(83 + 17) + x$, assoc.
27. $s + 8$, commut.
28. n, ident.
29. $(0 + x) + y$, assoc. or $x + y$, ident.

B Use the basic properties to find the value for each variable.

30. $7 + n = 9 + 7$ 9
31. $6 + x = 6$ 0
32. $b + 0 = 12$ 12
33. $34 - z = 0$ 34
34. $(c + 9) + 12 = 7 + (9 + 12)$ 7

35. Evaluate $a - b$ for $a = 15$, $b = 8$. If the answer must be a whole number, can $b - a$ be evaluated for the same numbers? Is the commutative property true for subtraction of whole numbers? The commutative property is not true for subtraction of whole numbers.

36. Evaluate $(p - q) - r$ for $p = 12$, $q = 8$, $r = 4$. Evaluate $p - (q - r)$ for the same values. Is the associative property true for subtraction of whole numbers? The associative property is not true for subtraction of whole numbers.

Use the commutative and associative properties to change the pairings of the numbers. Compute mentally. Look for **compatible numbers** such as $75 + 25$ that make easy sums.

37. $(98 + 75) + 25$ 198
38. $55 + (45 + 87)$ 187
39. $(975 + 492) + 25$ 1492
40. $750 + (958 + 250)$ 1958
41. $64 + 9 + 36 + 11 + 3 + 79 + 97$ 299
42. $125 + 555 + 375 + 445 + 237$ 1737

C Extending Thinking Skills

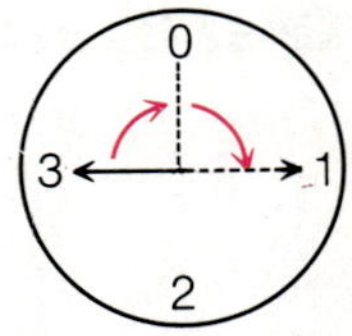

This "clock" can be used to find "clock sums." For example, $3 \oplus 2 = 1$, because two hours after pointing to 3 o'clock, the hand would point to 1 o'clock.

Find each clock sum. Do you think the operation $\oplus$ is commutative? yes

43. $2 \oplus 1$ 3 **44.** $2 \oplus 2$ 0 **45.** $2 \oplus 3$ 1 **46.** $3 \oplus 2$ 1

Thinking skill: 43–46. Generalizing

Mixed Review

Add or subtract. **47.** 954 − 765 189 **48.** 340 + 565 905 **49.** 2,506 − 987 1,519

Evaluate. **50.** (45 − 18) − 16 11 **51.** 23 + 49 + 85 157 **52.** 65 + 25 90

Evaluate for $a = 10$. **53.** $a + 16$ 26 **54.** $a + 8$ 18 **55.** $25 - a$ 15

Evaluate for $b = 12$, $c = 13$. **56.** $b + c$ 25 **57.** $c - b$ 1

Write the word name. **58.** 2,017 **59.** 452 **60.** 6,906

58. two thousand seventeen 59. four hundred fifty-two 60. six thousand nine hundred six

NUMBERS TO ALGEBRA

A generalization that can be made about numbers can be expressed with variables. Check to see if the number examples are true. Then see how the variables are used to express the generalization.

Numbers	Algebra
(35 + 49) + 15 = (35 + 15) + 49	For whole numbers ***a*, *b*,** and ***c*,**
(75 + 89) + 25 = (75 + 25) + 89	$(a + b) + c = (a + c) + b$
(88 + 76) + 12 = (88 + 12) + 76	
(148 + 97) + 52 = (148 + 52) + 97	

Give two more number examples, then copy and complete the generalization.

Number Examples

(50 + 6) + (30 + 7) = (50 + 30) + (6 + 7)

(40 + 9) + (60 + 8) = (40 + 60) + (9 + 8)

(200 + 52) + (600 + 48) = (200 + 600) + (52 + 48)

Generalization

For whole numbers *a*, *b*, *c*, and *d*,

$(a + b) + (c + d) =$ ___?___

$(a + c) + (b + d)$

Number examples answers will vary. Example: (16 + 4) + (7 + 3) = (16 + 7) + (4 + 3)
(5 + 32) + (9 + 11) = (5 + 9) + (32 + 11)

1-4 Equations

Objective: To solve an equation for a replacement set that is given.
MIN: 1–14; REG: 1–14 even, 15–18; MAX: 1–14 even, 15–19; ALL: Mixed Review
An exploratory lesson opener can be found on page 545.

An **equation** is a mathematical sentence that uses an equal sign to state that two expressions represent the same number or are equivalent.

The expression on the left side	names the same number as	the expression on the right side
$24 + 6$	$=$	30

An equation that contains only numbers can be either true or false. For example, $24 + 6 = 30$ is true, but $24 + 6 = 31$ is false. An equation that contains at least one variable is an **open sentence** and is neither true nor false. For example, $x + 6 = 30$ is neither true nor false because x has not been replaced with a number.

The set of numbers from which you can select replacements for the variable is called the **replacement set**. A replacement for a variable that makes an equation true is called a **solution** of the equation. You can **solve** an equation by finding all of its solutions. The collection of all the solutions is called the **solution set** of the equation.

When the replacement set contains only a few numbers, one way to solve the equation is to try all the numbers.

Example 1

Solve $x + 38 = 42$ for the replacement set $\{1, 2, 3, 4, 5\}$.

Solution

$x + 38 = 42$

$1 + 38 = 42$ (false) Replace the variable x with each number in the replacement set.

$2 + 38 = 42$ (false)

$3 + 38 = 42$ (false)

$4 + 38 = 42$ (true) ✓ The number that makes the equation true is the solution. $x = 4$.

$5 + 38 = 42$ (false)

Practice Solve the equation for the replacement set $\{0, 2, 4, 6\}$.

a. $51 - n = 45$ 6 **b.** $b + 79 = 81$ 2 **c.** $31 = 27 + y$ 4

Example 2

Solve the equation $x - 8 = 16$ for the replacement set {23, 25, 27}.

Solution

$x - 8 = 16$

$23 - 8 = 16$ (false) Replace the variable x with each number in the replacement set.

$25 - 8 = 16$ (false)

$27 - 8 = 16$ (false) Since no number makes the equation true, it has no solution for the replacement set given.

There is no solution.

Practice Solve the equation for the replacement set given.

a. $n + 15 = 45$ {31, 32, 33} **b.** $x + 5 = 2$ {0, 1, 2, 3} a. no solution b. no solution

Oral Exercises

Is the value given for the variable a solution to the equation?

1. $n + 5 = 12, n = 8$? no **2.** $9 + p = 13, p = 4$? yes **3.** $e - 5 = 9, e = 15$? no

4. $9 = a - 6, a = 15$? yes **5.** $15 - y = 7, y = 8$? yes **6.** $13 = 6 + c, c = 8$? no

Exercises

A Solve the equation for the replacement set given.

1. $n + 3 = 12$ {8, 9} 9

2. $q - 7 = 9$ {14, 15, 16, 17} 16

3. $9 + n = 13$ {0, 2, 4, 6} 4

4. $12 - r = 7$ {1, 3, 5, 7, 9} 5

5. $z + 8 = 17$ {7, 8, 9, 10} 9

6. $11 - s = 7$ {1, 2, 3, 4, 5} 4

7. $31 = 28 + x$ {0, 2, 4} no solution

8. $12 + n = 20$ {2, 4, 6, 8, 10} 8

9. $17 = n - 3$ {20, 21, 22} 20

10. $24 = 5 + y$ {15, 16, 17, 18} no solution

11. $p + 0 = 8$ {0, 4, 8, 12, 14} 8

12. $28 = t - 10$ {17, 18, 19, 20} no solution

13. $z + 34 = 187$ {150, 151, 152} no solution

14. $y - 37 = 241$ {277, 278} 278

B Describe the solution set. The replacement set is the set of all whole numbers.

15. $n + 4 = 0$ no solution

16. $x + 5 = 5 + x$ all whole numbers

17. $y - y = 0$ all whole numbers

18. $a + a = a$ 0

C Extending Thinking Skills

19. Write an equation with no solution. Replacement set: all odd whole numbers.
Answers will vary. Example: $m + 1 = 15$ Thinking skill: Reasoning

Mixed Review

20. 7 + 12 21. 7 − 6 22. (25 + 75) + 42

Write a numerical expression for each. **20.** 12 added to 7 **21.** 7 decreased by 6

Use the associative property to write an equivalent expression. **22.** $25 + (75 + 42)$

Evaluate for $a = 20$, $a = 30$, and $a = 40$. **23.** $(165 + a) + 35$ 220, 230, 240 **24.** $a + 80$ 100, 110, 120

1-5 Solving Equations: Using Mental Math

Objective: To solve simple equations mentally.
MIN: 1–30, 37; REG: 1–30 even, 31–44; MAX: 1–30 even, 31–46; ALL: Mixed Review

When the replacement set for an equation is the set of all whole numbers, it is impossible to try every number to find the solution. If an equation involves small numbers and a single operation, it often can be solved mentally.

To find the number of pins knocked down when 6 are left standing, you can solve the equation $p + 6 = 10$ mentally.

Example 1

Solve and check mentally. $p + 6 = 10$

Solution $p + 6 = 10$ Think: what number added to 6 gives 10?

Check $4 + 6 \stackrel{?}{=} 10$ Mentally substitute 4 for p in $p + 6 = 10$.

$10 = 10 \checkmark$ The equation is true, so 4 is the solution.

Practice Solve and check mentally.

a. $8 - a = 5$ 3 **b.** $7 + n = 15$ 8 **c.** $y - 6 = 2$ 8

These equations illustrate the relationship between addition and subtraction.

addend		addend		sum		sum		addend		addend
65	+	35	=	100	$\longrightarrow$	100	−	35	=	65

Since addition can undo subtraction and subtraction can undo addition, addition and subtraction are called **inverse operations**. You can use this idea to solve equations mentally.

Example 2

Use the inverse operation to solve mentally. $x + 7 = 29$

Solution $x + 7 = 29$ Think "7 has been added to a number to produce 29,

$x = 29 - 7$ so 29 minus 7 should equal the number."

$x = 22$

Check $22 + 7 \stackrel{?}{=} 29$ Mentally substitute 22 for x in the equation.

$29 = 29 \checkmark$ The equation is true, so 22 is the solution.

Practice Use the inverse operation to solve mentally.

a. $n - 5 = 31$ 36 **b.** $24 = y + 8$ 16 **c.** $d + 7 = 21$ 14

Oral Exercises

1. the number that adds to 5 to give 13 (etc.)

The solution of the equation $6 + x = 10$ is "the number that adds to 6 to give 10." Describe the solution of each equation.

1. $n + 5 = 13$ **2.** $4 + c = 20$ **3.** $x - 5 = 12$

4. $x - 9 = 7$ **5.** $24 = a + 12$ **6.** $13 = c - 7$

To solve the equation $x + 7 = 15$, you can use the inverse operation and find $15 - 7$. Tell how you can solve each equation.

7. $a + 4 = 15$ find 15 − 4 **8.** $y - 7 = 8$ find 8 + 7 **9.** $23 = x + 19$ find 23 − 19

10. $8 = n - 9$ find 8 + 9 **11.** $6 + r = 19$ find 19 − 6 **12.** $p - 16 = 2$ find 2 + 16

Exercises

A Solve and check mentally. The replacement set is the set of all whole numbers.

1. $n + 2 = 8$ 6 **2.** $a + 4 = 9$ 5 **3.** $7 + x = 9$ 2

4. $8 + b = 9$ 1 **5.** $c + 3 = 12$ 9 **6.** $7 + z = 15$ 8

7. $6 + p = 11$ 5 **8.** $a + 7 = 12$ 5 **9.** $t - 4 = 9$ 13

10. $x - 8 = 9$ 17 **11.** $c - 5 = 6$ 11 **12.** $b - 10 = 8$ 18

13. $13 - n = 5$ 8 **14.** $17 - d = 8$ 9 **15.** $a - 6 = 0$ 6

16. $x - 7 = 9$ 16 **17.** $50 = n + 20$ 30 **18.** $40 + x = 90$ 50

19. $b - 60 = 10$ 70 **20.** $20 = 80 - n$ 60 **21.** $31 = 28 + x$ 3

22. $17 = n - 3$ 20 **23.** $24 = 5 + y$ 19 **24.** $28 = 10 + x$ 18

25. $a - 19 = 2$ 21 **26.** $120 = p + 70$ 50 **27.** $0 = n - 13$ 13

28. $125 + z = 200$ 75 **29.** $n + 25 = 100$ 75 **30.** $t - 300 = 700$ 1000

B Use a calculator and inverse operations to solve the equations in Exercises 31–41. For example, to solve $x + 346 = 923$, think: In the equation, 346 has been *added* to a number to produce the sum 923. Start with 923 on your calculator and *subtract* 346 to find the number.

				Display	
923	[−]	346	[=]	577	$x = 577$

31. $n + 479 = 1{,}235$ **32.** $b - 567 = 494$ **33.** $c + 5{,}926 = 13{,}415$

34. $y - 8{,}219 = 6{,}437$ **35.** $r + 3{,}643 = 9{,}008$ **36.** $t - 54{,}876 = 67{,}954$

31. 756 32. 1,061 33. 7,489 34. 14,656 35. 5,365 36. 122,830

37. After giving away 2,575 campaign buttons, a politician had 4,925 left. Solve the equation $b - 2{,}575 = 4{,}925$ to find how many she had at the beginning of her campaign. 7,500

38. The population of a city increased by 5,689 during a certain year, making a total of 157,743. Solve the equation $p + 5{,}689 = 157{,}743$ to find the population before the increase. 152,054

39. At a time when few people watch, a television commercial cost \$235,650. This was \$195,750 less than its cost during prime time. Solve the equation $p - 195{,}750 = 235{,}650$ to find the cost of the commercial during prime time. **\$431,400**

40. After a salesman had flown enough miles to receive two free tickets, he flew an extra 5,478 miles. His total mileage was 81,453. Solve the equation $m + 5{,}478 = 81{,}453$ to find the number of miles he had to fly to win two free tickets. **75,975**

41. The attendance at a major league baseball game on Tuesday was 5,679 less than the attendance on Sunday. Tuesday's attendance was 31,685. Solve the equation $s - 5{,}679 = 31{,}685$ to find Sunday's attendance. **37,364**

For all whole numbers, a, b, and c, if $a - b = c$, then $a = c + b$. Use this relationship between addition and subtraction to rewrite each equation with the variable by itself on one side. Then solve for the variable.

42. $x - 38 = 72$ **$x = 72 + 38$ $x = 110$**

43. $y - 35 = 96$ **$y = 96 + 35$ $y = 131$**

44. $60 = z - 23$ **$z = 60 + 23$ $z = 83$**

C Extending Thinking Skills

45. Find numbers for x and y that make both equations true. **Encourage students to guess and check.**

$x + y = 24 \qquad x - y = 2$ **$x = 13, y = 11$**

46. These 3 equations have the same solution. Find the solution. Look for a pattern and write two more equations that have the same solution.

$x + 2 = 5$ **$x = 3$**
$x + 5 = 8$ **$x + 11 = 14$**
$x + 8 = 11$ **$x + 14 = 17$**

Thinking skills: 45. Testing a conjecture 46. Finding a pattern

Mixed Review

Evaluate. **47.** $b + 45$ for $b = 55$ **100** **48.** $35 - (7 + x)$ for $x = 8$ **20**

Use the commutative property to write an equivalent expression. **49.** $28 + y$ **$y + 28$**

Solve the equation for the replacement set {10, 15, 20}. **50.** $y - 5 = 15$ **20**

MENTAL MATH

Breaking apart numbers is a mental math technique based on the associative and commutative properties. Study the examples.

- To find 76 + 9, think "70 plus 6 + 9 is 70 plus 15, or 85."
- To find 57 + 38, think "50 + 30 is 80. 7 + 8 is 15. 80 + 15 is 95."
- To find 347 + 526, think "300 + 500 plus 40 + 20 plus 7 + 6 is 800 + 60 + 13, or 873."

Find each sum mentally by breaking apart the numbers.

1. 67 + 6 **73** **2.** 96 + 9 **105** **3.** 37 + 65 **102**

4. 78 + 94 **172** **5.** 437 + 328 **765** **6.** 256 + 493 **749**

1-6 Solving Equations: Using Addition and Subtraction

Objective: To use addition or subtraction to solve equations involving whole numbers.
MIN: 1–30; REG: 1–30 even, 31–36; MAX: 1–30 even, 31–39; ALL: Mixed Review

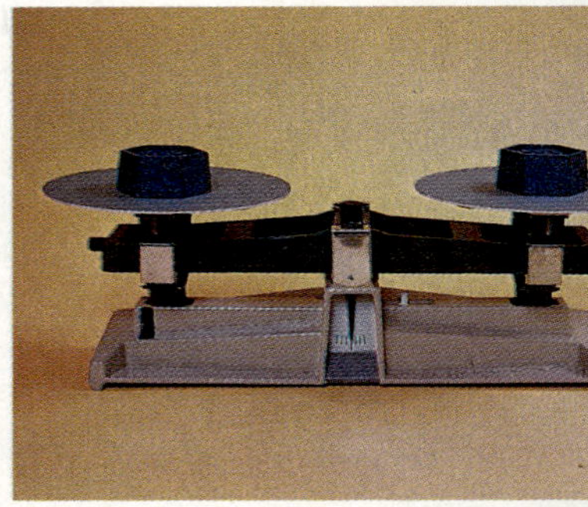

Although some simple equations can be solved mentally, other methods are useful with equations that have larger numbers or more operations. Two ideas are used in solving equations involving addition or subtraction. The first is that addition and subtraction are inverse operations and each can undo the other.

Subtraction can undo addition.	Addition can undo subtraction.
$10 + 9 - 9 = 10$	$10 - 9 + 9 = 10$
$n + 9 - 9 = n$	$n - 9 + 9 = n$

The second idea is that an equation is like a balanced scale. To keep the two sides of a scale balanced, whatever is done to one side must also be done to the other side. The properties of equality state this idea.

Addition and Subtraction Properties of Equality

You can add or subtract the same number on both sides of an equation and the two sides will remain equal.

For all numbers ***a*, *b*,** and ***c*,** if ***a* = *b*,** then ***a* + *c* = *b* + *c***
and if ***a* = *b*,** then ***a* − *c* = *b* − *c***

The following steps for solving equations show how to use the idea of inverse operations and the properties of equality to get the variable by itself on one side of the equation.

Solving Equations Using Addition and Subtraction

1. Decide which operation has been applied to the variable.
2. Use the inverse of this operation, adding or subtracting the same number on both sides of the equation.

The following examples show how to use these steps to solve equations.

Example 1

Solve and check. $n + 89 = 134$

Solution $n + 89 = 134$ You need to get the variable by itself.

$n + 89 - 89 = 134 - 89$ To undo adding 89, subtract 89 from both sides.

$n = 45$

Check $45 + 89 \stackrel{?}{=} 134$ Replace n with 45.

$134 = 134$ ✓ The solution is 45.

Practice Solve and check. **a.** $x + 77 = 394$ 317 **b.** $123 = x + 87$ 36

Example 2

Solve and check. $x - 89 = 176$

Solution $x - 89 = 176$ You need to get the variable by itself.

$x - 89 + 89 = 176 + 89$ To undo subtracting 89, add 89 to both sides.

$x = 265$

Check $265 - 89 \stackrel{?}{=} 176$ Replace x with 265.

$176 = 176$ ✓ The solution is 265.

Practice Solve. **a.** $m - 76 = 158$ 234 **b.** $146 = y - 89$ 235

Oral Exercises

Give the inverse of each operation.

1. Adding 58 sub. 58
2. Subtracting 97 add. 97
3. Subtracting 123 add. 123
4. Subtracting 29 add. 29
5. Adding 456 sub. 456
6. Subtracting 43 add. 43

Name the operation that would be used to solve the equation.

7. $x + 19 = 56$ sub.
8. $n - 28 = 56$ add.
9. $467 = x + 85$ sub.
10. $98 = b - 35$ add.
11. $c - 56 = 178$ add.
12. $b + 278 = 562$ sub.

Exercises

A Solve and check.

1. $n + 38 = 84$ 46
2. $x + 56 = 92$ 36
3. $a + 47 = 85$ 38
4. $y + 29 = 92$ 63
5. $c + 76 = 154$ 78
6. $n + 67 = 282$ 215
7. $x + 79 = 194$ 115
8. $b + 68 = 290$ 222
9. $x - 87 = 63$ 150
10. $y - 69 = 145$ 214
11. $x - 58 = 139$ 197
12. $y - 77 = 229$ 306
13. $c - 167 = 85$ 252
14. $n - 258 = 197$ 455
15. $c - 376 = 488$ 864

16. $b - 96 = 685$ 781
17. $178 = n + 89$ 89
18. $164 = n - 85$ 249
19. $254 = a + 67$ 187
20. $182 = y - 347$ 529
21. $x + 87 = 264$ 177
22. $183 = y - 259$ 442
23. $n - 575 = 391$ 966
24. $743 = c + 88$ 655
25. $a + 86 = 842$ 756
26. $b - 248 = 560$ 808
27. $5{,}684 = n - 975$ 6659
28. $x + 437 = 1{,}065$ 628
29. $a + 455 = 738$ 283
30. $2{,}256 - x = 394$ 1862

B Solve and check.

31. $x + (157 + 29) = 342$ 156
32. $a - (104 - 78) = 28$ 54
33. $(358 + 76) + y = 500$ 66
34. $(56 + 87) - b = 17$ 126

35. Write and solve an equation in which the sum is 2,475 and the addend 1,697 is added to a number represented by the variable n. $2475 = 1697 + n$, $n = 778$

36. Write and solve an equation in which 3,478 is subtracted from a number represented by the variable y to produce the difference 5,607. $y - 3478 = 5607$, $y = 9085$

C Extending Thinking Skills

37. $y = 6$, $n = 4$, $x = 11$

In a magic square, the sum of the numbers in each row, column, and diagonal is the same.

x	2	14
12	9	y
n	16	7

↓ 27 (Sum)

37. Write and solve equations as needed to find x, y, and n.

38. Andy had more nickels than dimes, and more dimes than pennies. Can you tell the value of his 6 coins?
No; you would have to know he has at least one penny.

39. Use the variables a and b to complete this generalization about solving equations of the type $x - a = b$ for x: $x =$ ___?___ $b + a$

Thinking skills: 37, 38. Using logic 39. Generalizing

Mixed Review

Write as an algebraic expression. **40.** 125 added to a number $n + 125$

Evaluate. **41.** $927 + 386$ 1313 **42.** $400 - 157$ 243 **43.** $98 + 837 + 369$ 1304

Solve mentally and check. **44.** $n + 5 = 11$ 6 **45.** $y - 6 = 10$ 16

COMPUTER ACTIVITY

This computer program can be used to solve equations of the form $x + A = B$. Remind students to separate their number inputs with a comma.

```
10 REM: SOLVE AN EQUATION X + A = B
20 PRINT "CHOOSE WHOLE NUMBERS A AND B WITH A < B"
30 INPUT A,B
40 PRINT "EQUATION: X + ";A;" = ";B
50 PRINT "SOLUTION: X = ";B-A
60 END
```

1. Run the program or show what the run would look like.
2. Change the program to solve equations of the form $x - A = B$. See additional answers.

More Practice

Solve and check.

1. $x - 31 = 19$ 50
2. $b + 83 = 109$ 26
3. $39 = y - 75$ 114
4. $m - 104 = 127$ 231
5. $z - 64 = 241$ 305
6. $r + 27 = 46$ 19
7. $b + 60 = 82$ 22
8. $c + 535 = 1{,}024$ 489
9. $m - 54 = 147$ 201
10. $111 = w + 37$ 74
11. $r - 97 = 406$ 503
12. $t + 42 = 150$ 108
13. $r - 104 = 90$ 194
14. $a - 276 = 519$ 795
15. $92 = k + 67$ 25
16. $a + 94 = 163$ 69
17. $h + 139 = 297$ 158
18. $x - 192 = 12$ 204
19. $w + 484 = 620$ 136
20. $5{,}104 = r - 628$ 5732
21. $n - 4 = 97$ 101
22. $516 = b + 377$ 139
23. $267 = m - 734$ 1001
24. $c + 139 = 297$ 158
25. $y + 84 = 110$ 26
26. $p - 181 = 190$ 371
27. $k - 4 = 97$ 101
28. $c - 26 = 87$ 113
29. $x + 19 = 27$ 8
30. $62 = z - 18$ 80
31. $t + 140 = 816$ 676
32. $r - 152 = 28$ 180
33. $a + 493 = 602$ 109
34. $n - 218 = 672$ 890
35. $120 = b + 49$ 71
36. $f - 19 = 91$ 110
37. $657 = k + 298$ 359
38. $h - 46 = 47$ 93
39. $w - 26 = 590$ 616
40. $z + 97 = 406$ 309
41. $n - 27 = 261$ 288
42. $t + 184 = 901$ 717
43. $r + 77 = 311$ 234
44. $p - 19 = 74$ 93
45. $z - 116 = 493$ 609
46. $204 = m - 79$ 283
47. $k + 183 = 291$ 108
48. $n + 882 = 960$ 78
49. $h + 68 = 390$ 322
50. $406 = r + 297$ 109
51. $b - 332 = 49$ 381
52. $a + 76 = 375$ 299
53. $54 = x - 127$ 181
54. $117 = m - 83$ 200
55. $86 = n + 53$ 33
56. $m + 30 = 214$ 184
57. $a + 599 = 604$ 5
58. $w + 38 = 52$ 14
59. $162 = t + 38$ 124
60. $c - 745 = 267$ 1012
61. $z - 56 = 39$ 95
62. $k + 137 = 514$ 377
63. $r - 11 = 86$ 97
64. $b - 129 = 92$ 221
65. $y - 356 = 25$ 381
66. $33 = p + 25$ 8
67. $k + 16 = 108$ 92
68. $r + 177 = 219$ 42
69. $a - 36 = 25$ 61
70. $93 = m - 15$ 108
71. $x + 692 = 781$ 89
72. $z + 63 = 87$ 24
73. $r - 273 = 26$ 299
74. $211 = x + 16$ 195
75. $m - 47 = 44$ 91
76. $c + 67 = 207$ 140
77. $a - 669 = 241$ 910
78. $k + 48 = 60$ 12
79. $129 = t - 65$ 194
80. $r + 114 = 177$ 63
81. $b - 56 = 12$ 68
82. $60 = r + 48$ 12
83. $b + 18 = 29$ 11
84. $w + 18 = 92$ 74

PROBLEM SOLVING: APPLICATIONS

1-7 Buying a Car

Objective: To use the Problem-Solving Checklist to solve applied problems.
All: 1–11

When solving a word problem, you can use the five guidelines on this Problem-Solving Checklist.

Problem-Solving Checklist

- Understand the **QUESTION**
- Find the Needed **DATA**
- **PLAN** What to Do
- Find the **ANSWER**
- **CHECK** Back

Problem The dealer adds $59 to the base list price of a Cyclone 500 for preparation and handling. If the car cost the dealer $7,189, what is the dealer's profit?

Price List Cyclone 500	
Base List Price:	$8,012
Extra Options, List Price:	
Air Conditioner	$575
Cruise Control	$139
Rear Defrost	$126
Tinted Glass	$ 98
PinStripe	$ 70
AM/FM Stereo Radio	$184
Power Steering	$209
Automatic Transmission	$439
Selling Price:	$9,852

Note that the price list shows the customer's cost. The customer's "base list price" equals the dealer's cost plus the dealer's profit.

Understand the QUESTION.

Read the problem carefully and decide what question is asked.

You are asked to find the amount of profit the dealer has included in the base list price.

Find the Needed DATA.

Decide what information is needed to solve the problem.

You can look at the data chart to find that the base list price is $8,012.

PLAN What to Do.

Choose a strategy that will help you solve the problem.

To *compare* the base price with the dealer's cost, first choose the operation *subtraction*.
To *include* the extra profit from the handling charge, choose the operation *addition*.

Find the ANSWER.

Complete the reasoning or computation needed to find the answer.

(8,012 + 59) − 7,189 = 882.

Add the preparation cost of 59 to the base list price of 8,012, then subtract the dealer's cost, 7,189.

The total dealer's profit is $882.

CHECK Back.

Reread the problem. Does your answer seem reasonable?

The dealer's cost was about $7,200. If a profit of about $800 was added, the base list price would be about $8,000. The answer is reasonable.

Problems

These problems are well-suited to small-group work.

Solve. Use the data on p. 22.

1. What is the total cost of the extra options for the Cyclone 500? **$1840.**

2. The extra options actually cost the dealer $1,146. How much profit would the dealer make on the extra options? **$694.**

3. The dealer's profit is the difference between the total dealer's cost and the total listed selling price. What is the dealer's profit for the Cyclone 500 with all the extra options? **$1576.**

4. Suppose a customer convinces the dealer to reduce the selling price by $500. How much profit would the dealer make? **$1076.**

5. What would the listed selling price be if a customer chose all the extra options except air conditioning? **$9277.**

6. How much lower would the selling price be without air conditioning, cruise control, power steering, and automatic transmission? **$1362.**

7. What would the total cost be if a customer chose all the options and paid the full selling price plus a sales tax of $493? **$10,345 *or* $10,404. See below.**

8. Suppose the dealer reduced the selling price by $375 and then the buyer decided to have the car rustproofed at a cost of $195. What would the total cost be before the addition of sales tax? **$9672 *or* $9731. See below.**

9. Which two extra options could a customer omit to save about $400? **AM/FM Stereo Radio and Power Steering**

10. The dealer wants to sell the car and is willing to take only $275 profit. What will the selling price be for a Cyclone 500 with all the extra options? **$8669.**

11. ***Data Search*** Find the listed selling prices of two of your favorite cars. By how much do the prices of these cars differ? **Answers will vary.**

What's Your Decision?

You need a new car and have chosen the Cyclone 500. You can afford to spend only $8,700 plus tax. The dealer will not reduce the listed selling price. If you buy the car, which extra options will you choose? **Answers will vary.**

Check addition to be sure sum of cost of extra options does not exceed $629.

7,8,9. Answers depend upon whether the dealer or the customer absorbs the preparation and handling charge. Discuss "hidden costs" in real life.

1-8 Estimating

Objective: To estimate whole number sums and differences by using rounding or front-end estimation.
MIN: 1–22, 29–30; REG, MAX: 1–22 even, 23–30; ALL: Mixed Review

Three years ago, the Garcias bought a camper with an odometer reading of 16,329. Now the odometer reads 38,647. You could estimate to figure out about how many miles they had driven the camper.

It is often useful to **estimate** to find an approximate answer to a problem or to see if an answer found with a calculator makes sense. There are several ways to estimate sums and differences of whole numbers. One way is called **estimating using rounding**.

Estimating Using Rounding

- Round the numbers to the desired place.
- Compute with the rounded numbers to find the estimate.

A procedure for rounding is described on page 105.

Example 1

Round to the nearest thousand and estimate the value of the variable.
$n = 38{,}647 - 16{,}329$

Solution

$$\begin{array}{r} 39{,}000 \\ -\ 16{,}000 \\ \hline 23{,}000 \end{array}$$

Round 38,647 to 39,000.
Round 16,329 to 16,000.

The estimated value for n is 23,000.

Practice Round to the nearest thousand and estimate the value of the variable.

a. $a = 9{,}586 + 4{,}299$ 14,000 **b.** $76{,}953 - 5{,}466 = y$ 72,000

Front-end estimation is another useful technique for estimating sums and differences. It is most often used when the numbers have the same number of digits.

Front-end Estimation

- Add (or subtract) the first digits to get a rough estimate.
- Adjust your estimate by using the remaining digits and looking for numbers that are compatible.

Example 2

Use front-end estimation to estimate the value of the variable.
$3{,}527 + 7{,}969 + 5{,}493 = n$

Solution Rough Estimate: 15,000 — Add the "front-end" digits. **3**,527 + **7**,969 + **5**,493 is about **15** thousand.

Adjusted Estimate: 17,000 — Look at the other digits, 3,**527** + 7,**969** + 5,**493**, for compatible numbers. **527** + **493** is about 1,000. **969** is about 1,000. Increase the estimate by 2,000.

Practice Use front-end estimation to estimate.

a. $347 + 598 + 754$ 1700 **b.** $8{,}802 - 2{,}319$ 6,500

Oral Exercises

Round to the place indicated.

1. 658 to nearest ten 660
2. 864 to nearest hundred 900
3. 6,523 to nearest thousand 7000
4. 74,214 to nearest ten thousand 70,000
5. 91,096 to nearest hundred 91,100
6. 6,325 to nearest ten 6,330
7. 4,352 to nearest thousand 4000
8. 54,986 to nearest hundred 55,000
9. 65,847 to nearest thousand 66,000
10. 724,063 to nearest ten thousand 720,000

Exercises

A Round to the nearest ten and estimate the value of the variable.

1. $59 + 87 = x$ 150
2. $92 - 58 = b$ 30
3. $94 + 78 = y$ 170
4. $z = 149 - 62$ 90
5. $s = 263 + 695$ 960
6. $e = 865 - 243$ 630

Round to the nearest hundred and estimate the value of the variable.

7. $n = 725 + 487$ 1200
8. $1{,}248 - 519 = x$ 700
9. $z = 938 + 694$ 1600

Round to the nearest thousand and estimate the value of the variable.

10. $y = 8{,}399 + 6{,}508$ 15,000
11. $12{,}778 - 7{,}499 = c$ 6,000
12. $56{,}468 + 79{,}122 = x$ 135,000

Use front-end estimation to estimate the value of the variable. **Possible answers are given.**

13. $346 + 657 + 498 = n$ **1500**

14. $873 + 449 + 628 + 858 = y$ **2,800**

15. $2{,}543 + 7{,}986 + 4{,}478 = x$ **15,000**

16. $9{,}245 + 6{,}508 + 7{,}796 = n$ **23,500**

17. $36{,}247 + 43{,}572 = a$ **80,000**

18. $875{,}652 + 925{,}468 = y$ **1,800,000**

19. $9{,}704 - 6{,}218 = x$ **3,500**

20. $8{,}703 - 4{,}687 = a$ **4,000**

21. $74{,}907 - 35{,}186 = c$ **40,000**

22. $973{,}477 + 23{,}885 = c$ **997,000**

B Solve.

23. The first owner of a car drove it for 56,478 miles. The second owner put 37,517 miles on it. Estimate the total mileage, to the nearest thousand miles. **94,000 mi**

24. The total cost of a sports car was $17,485. An economy car cost $4,965. Estimate the difference in cost, to the nearest thousand dollars. **$12,000**

25. A bike trip around the mainland of the United States covered 12,092 miles. A trip directly from the west coast to the east coast covered 2,964 miles. About how much longer was the trip around the mainland? **9000 mi**

26. Write two numbers, each of which is 24,000 when rounded to the nearest thousand. **Answers will vary.**

27. A number, rounded to the nearest thousand, is 6,000. What letters on the number line below could show the position of this number? ***F, G, H, I, J, K, L, M, N, O***

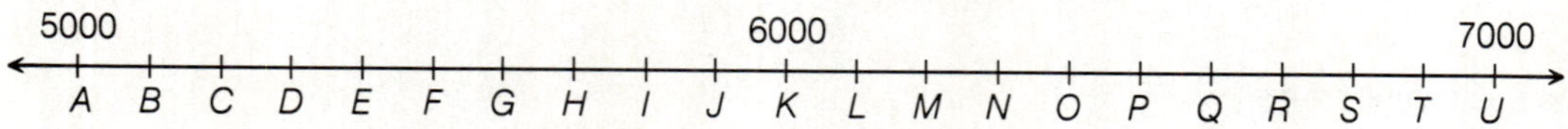

28. A used-car salesperson says that a car has been driven about 70,000 miles. If this number has been rounded to the nearest ten thousand, what is the greatest number of miles the car could have been driven? What is the smallest? **74,999 and 65,000.**

C Extending Thinking Skills **29,30. Encourage students to guess and check.**

29. Estimate the sum of the whole numbers from one to 25. Use a calculator to check your estimate. **325**

30. Whole numbers in order, such as 8, 9, and 10, are called **consecutive** whole numbers. Find 3 consecutive whole numbers whose sum is 111. **36, 37, 38**

Thinking skills: 29. Estimating 30. Testing a conjecture

Mixed Review

Use the associative property to write an equivalent expression. **31.** $(78 + 65) + 135$ **78 + (65 + 135)**

Write as an algebraic expression. **32.** a number decreased by 650 ***n* − 650**

Solve and check. **33.** $65 - x = 38$ **27** **34.** $48 + y = 93$ **45** **35.** $n + 37 = 129$ **92**

Add or subtract. **36.** $1{,}295 + 1{,}128$ **2,423** **37.** $5{,}007 - 3{,}985$ **1,022**

38. $22{,}423 + 10{,}525$ **32,948** **39.** $13{,}254 + 8{,}098$ **21,352** **40.** $500{,}821 - 39{,}827$ **460,994**

PROBLEM SOLVING: STRATEGIES

1-9 Guess, Check, Revise

Objective: To solve nonroutine problems, using the strategies Choose the Operations or Guess, Check, Revise.
MIN: 1–2, REG: 1–4; MAX: 1–6

In Section 1-7 you had to decide which operations to use to solve a problem. This strategy is called **Choose the Operations**. Think about what operations are needed to solve the following problem.

Problem A television set costs \$469. You could pay that amount in cash, or you could choose an installment plan and first pay a \$95 down payment and then pay 24 installment payments totaling \$456. How much more would you pay if you chose the installment plan?

To solve the problem, you might plan to use the operation of addition first, then subtraction.

$95 + 456 = 551$ The television costs \$551 on the installment plan.

$551 - 469 = 82$ You would pay \$82 more on the installment plan.

You need to use other strategies for word problems that cannot be solved by simply choosing the operations. You can solve some by guessing a solution, checking the guess, and using what you learned from the check to revise the guess. This strategy is called **Guess, Check, Revise.** You can use it to solve the following problem.

Problem A newspaper ad offered a soccer ball and soccer shoes for a total price of \$59. If the shoes cost \$15 more than the ball, what did each cost?

The chart below shows how you might use the Guess, Check, Revise strategy to find the solution to this problem.

Guess	Check	Revise
First Guess **Ball: \$20**	Shoes: \$20 + \$15 = \$35 Total Cost: \$20 + \$35 = \$55	55 is less than 59. Revise guess up. **Revised guess, \$25**
Second Guess **Ball: \$25**	Shoes: \$25 + \$15 = \$40 Total Cost: \$25 + \$40 = \$65	65 is more than 59. Revise guess down. **Revised guess, \$22**
Third Guess **Ball: \$22**	Shoes: \$22 + \$15 = \$37 Total Cost: \$22 + \$37 = \$59	This is correct!

The ball cost \$22 and the shoes cost \$37.

Problem-Solving Strategies

Choose the Operations

Guess, Check, Revise

Additional strategies will be introduced in later chapters.

Problems

Solve.

1. Denise scored 6 more points in this week's basketball game than she scored in last week's game. She scored a total of 30 points in the two games. How many points did she score in each game? 12, 18

2. Mrs. Gentry sold 36 more computers in August than she did in July. She sold one more computer in September than she sold in August. In September, she sold 55 computers. How many did she sell in the three months? 127

3. Eric is 9 years old. In 4 years his father will be 13 years older than twice what Eric's age will be then. How old is Eric's father now? 35

4. While playing darts, Raul noticed that if he reversed the two digits of his score, it would produce Linda's score. Linda's score was a multiple of 9 and only 9 more than Raul's score. What were their scores? (Note: multiples of nine are 0, 9, 18, 27, 36 . . .) Linda = 54, Raul = 45

5. The difference between Richard's and Harriet's weekly earnings is $80. The sum of their earnings is $560. Harriet earns the greater amount. How much does each earn? Harriet = $320, Richard = $240

6. Will is one year older than Phil and Phil is one year older than Gil. The sum of their ages is 75. How old is each? Gil = 24, Phil = 25, Will = 26

Enrichment

Logical Reasoning and Venn Diagrams

The words ***and, or, not,*** and ***if-then*** are used to express ideas in **logical reasoning**. A **Venn Diagram** is a drawing used to show logical relationships among members of sets. The set members can be numbers, people, cards, or any other elements. The Venn Diagrams below show some relationships.

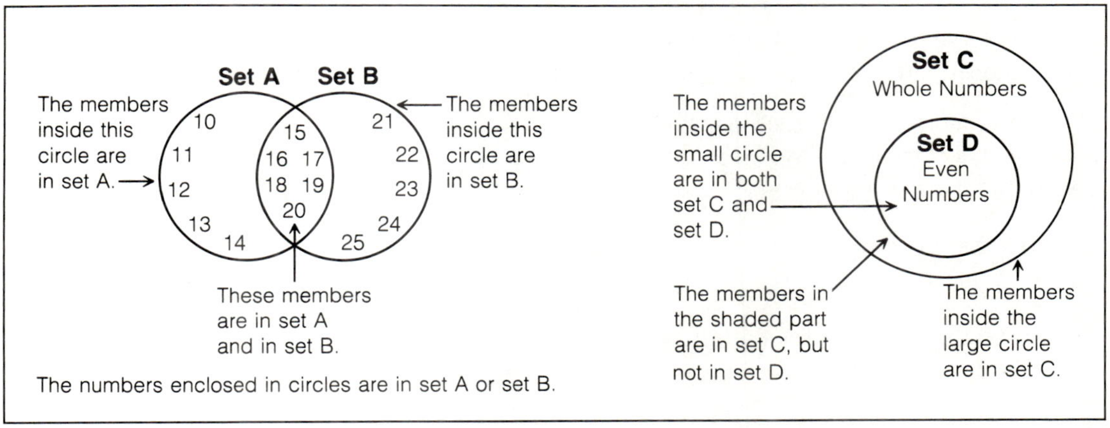

Refer to the Venn Diagrams above for Exercises 1–7. List the numbers in:

1. both set A ***and*** set B **15–20**

2. set B, but ***not*** in set A. **21–25**

3. set A, but ***not*** in set B. **10–14**

4. set A ***or*** set B. **10–25**

5. List the smallest five numbers that are in set C, but ***not*** in set D. **1, 3, 5, 7, 9**

6. Are there any numbers that are in set D, but ***not*** in set C? **No**

7. ***If*** a number is in set D, ***then*** what can you conclude? **It is also in set C.**

Make a Venn Diagram to show: Set A = brown-haired people. Set B = green-eyed people. Describe the people in:

8. both set A ***and*** set B.

9. set A, but ***not*** set B.

10. set B, but ***not*** set A.

11. ***neither*** set A ***nor*** set B.

Make a Venn Diagram to show: Set C = people from the United States. Set D = people from Texas.

12. A person is in set C, but ***not*** in set D. What can you conclude?

13. A person is ***not*** in set D. What can you conclude?

14. A person is ***not*** in set C. What can you conclude?

8. brown-haired, green-eyed people
9. brown-haired people *without* green eyes
10. green-eyed people *without* brown hair
11. people without brown hair *or* green eyes.
12. that person is from the U.S., but not from Texas
13. that person is not from Texas, and may or may not be from the U.S.
14. that person is not from the U.S., and not from Texas.

Chapter 1 Review

1-1 Evaluate each numerical expression.

1. $24 + (15 - 6)$ 33 **2.** $(16 + 4) - 8$ 12 **3.** $44 - (12 + 6)$ 26

4. $38 - (8 - 5)$ 35 **5.** $27 + (6 - 4)$ 29 **6.** $(13 + 3) - 6$ 10

Evaluate each algebraic expression.

7. $r + 4$ for $r = 4$ 8 **8.** $12 - s$ for $s = 8$ 4 **9.** $(5 + a) + 1$ for $a = 3$ 9

10. $18 - (3 + b)$ for $b = 4$ 11 **11.** $x - y$ for $x = 7, y = 2$ 5 **12.** $(a + 10) - 9$ for $a = 7$ 8

1-2 Write as an algebraic expression.

13. 1 more than z $z + 1$ **14.** 9 added to p $p + 9$ **15.** v decreased by 2 $v - 2$

Write an expression for each question.
Mark walks 9 miles a day. How far will he walk if

16. he walks m miles farther? $9 + m$ **17.** he walks 2 miles less? $9 - 2$

1-3 Use the basic properties. Write an equivalent expression.

18. $(c + 1) + 3$ $c + (1 + 3)$ **19.** $k + 5$ $5 + k$ **20.** $0 + w$ w or $w + 0$

1-4 Solve the equation for the replacement set given.

21. $f - 4 = 6$ {9, 10, 11, 12} 10 **22.** $b + 3 = 11$ {8, 9, 10, 11} 8

1-5 Solve.

23. $15 - m = 6$ 9 **24.** $4 + x = 17$ 13 **25.** $b + 8 = 17$ 9

1-6 Solve and check.

26. $t + 28 = 39$ 11 **27.** $q - 45 = 22$ 67 **28.** $691 + p = 777$ 86

29. $121 = 193 - z$ 72 **30.** $238 + a = 238$ 0 **31.** $144 - m = 32$ 112

1-7 Solve.

32. Teresa rode her bicycle 5 kilometers farther today than she did yesterday. She rode 17 kilometers yesterday. How far did she ride today? 22 km

33. Josh had 22 tapes. He gave all but 5 away. How many did he give away? 17

1-8 Round to the nearest thousand and estimate the value of the variable.

34. $9{,}862 + 6{,}490 = p$ 16,000 **35.** $n = 2{,}714 - 1{,}152$ 2,000 **36.** $3{,}561 + 5{,}126 = k$ 9,000

Round to the nearest hundred and estimate the value of the variable.

37. $r = 927 + 610$ 1500 **38.** $683 - 274 = y$ 400 **39.** $3{,}596 - 834 = c$ 2800

Use front-end estimation to estimate the value of the variable. Possible answers are given.

40. $m = 798 + 413 + 112$ 1300 **41.** $(3{,}418 + 1{,}113) + 2{,}976 = f$ 7500

Chapter 1 Test

Evaluate each numerical expression.

1. $17 - (3 + 6)$ 8
2. $(25 - 5) + 4$ 24
3. $12 + (8 + 6)$ 26
4. $(7 - 4) + 8$ 11
5. $18 - (7 - 5)$ 16
6. $(16 - 7) + 5$ 14

Evaluate each algebraic expression.

7. $y + 7$ for $y = 6$ 13
8. $z - 3$ for $z = 9$ 6
9. $(3 + 5) - m$ for $m = 4$ 4
10. $13 - (x + 7)$ for $x = 2$ 4
11. $r - k$ for $r = 7, k = 2$ 5
12. $(a + b) - 3$ for $a = 8, b = 6$ 11

Write as an algebraic expression.

13. 5 more than p $p + 5$
14. v decreased by 4 $v - 4$
15. 7 added to a number $n + 7$

Write an expression to answer each question.
Hitoshi has \$45. How much will she have if:

16. she earns \$18 more? 45 + 18
17. she buys earrings for \$36? 45 – 36

Use the basic properties. Write an equivalent expression.

18. $t + 0$ $0 + t$ or t
19. $(n + 5) + 4$ $n + (5 + 4)$ or $n + 9$
20. $6 + y$ $y + 6$

Solve the equation for the replacement set given.

21. $x + 9 = 12$ {1, 2, 3, 4} 3
22. $n - 7 = 7$ {13, 14, 15, 16} 14

Solve and check.

23. $16 - n = 8$ 8
24. $1 + y = 8$ 7
25. $z + 7 = 16$ 9
26. $g + 14 = 67$ 53
27. $s - 74 = 12$ 86
28. $255 + w = 500$ 245
29. $166 = 212 - h$ 46
30. $413 = 62 + r$ 351
31. $113 - f = 113$ 0

Solve.

32. Wayne practiced the piano 7 hours more this week than he did last week. He practiced 9 hours last week. How long did he practice in the two weeks? 25 hours

33. Margaret gave 19 of her magazines to a friend. She has 24 magazines left. How many magazines did she have to start with? 43

Round to the nearest thousand and estimate the value of the variable.

34. $4{,}679 + 3{,}125 = a$ 8,000
35. $s = 11{,}591 - 2{,}139$ 10,000
36. $7{,}825 + 3{,}368 = p$ 11,000

Round to the nearest hundred and estimate the value of the variable.

37. $j = 853 + 234$ 1100
38. $891 - 255 = w$ 600
39. $1{,}316 - 402 = b$ 900

Use front-end estimation to estimate the value of the variable. Possible answers are given.

40. $n = 143 + 978 + 245$ 1,400
41. $1{,}708 + 1{,}125 + 6{,}855 = a$ 9,700

Cumulative Review

Evaluate each numerical expression.

1. $13 - (7 + 4)$ 2
2. $(12 - 9) + 8$ 11
3. $23 + (5 + 6)$ 34

Use front-end estimation to estimate the value of the variable. **Possible answers are given.**

4. $b = 212 + 583 + 785$ 1,600
5. $5{,}258 + 9{,}295 + 4{,}502 = p$ 19,000

Write as an algebraic expression.

6. 2 less than n $n - 2$
7. k increased by 3 $k + 3$
8. 7 added to a $a + 7$

Write an expression to answer each question.
Susan has \$15. How much will she have if:

9. she earns \$22 more? 15 + 22
10. she buys a record for \$9? 15 − 9

Use the commutative, associative, or zero properties. Write an equivalent expression.

11. $r + 5$ $5 + r$
12. $(y + 4) + 6$ $y + (4 + 6)$ *or* $y + 10$
13. $6 + 0$ 6 *or* $0 + 6$

Solve the equation for the replacement set given.

14. $z + 4 = 12$ {7, 8, 9, 10} 8
15. $a - 5 = 7$ {12, 13, 14, 15} 12
16. $20 - m = 14$ {6, 7, 8, 9} 6
17. $4 + g = 14$ {7, 8, 9, 10} 10

Solve.

18. $7 + x = 14$ 7
19. $a - 8 = 15$ 23
20. $13 - b = 6$ 7

Solve and check.

21. $x + 16 = 35$ 19
22. $k - 45 = 17$ 62
23. $743 + n = 850$ 107
24. $102 = 314 - f$ 212
25. $555 = 770 - r$ 215
26. $225 - p = 115$ 110

Solve each problem.

27. Mickey played baseball 10 hours more this week than he did last week. He played baseball 12 hours last week. How long did he play this week? 22 hours
28. Barbara gave 11 of her books to a friend. She had 31 left. How many did she have to start with? 42

Round to the nearest ten and estimate the value of the variable.

29. $55 + 23 = v$ 80
30. $z = 693 - 58$ 630
31. $47 + 21 = d$ 70

Round to the nearest hundred and estimate the value of the variable.

32. $c = 697 + 162$ 900
33. $836 - 642 = h$ 200
34. $147 + 121 = d$ 200

Evaluate each algebraic expression.

35. $t + 9$ for $t = 6$ 15
36. $(4 + 7) - c$ for $c = 5$ 6
37. $s + y - 2$ for $y = 6$, $s = 3$ 7

Chapter 2

Expressions and Equations: Multiplication and Division

2-1 Evaluating Expressions: Order of Operations

Objective: To evaluate an expression involving all four operations, with or without grouping symbols.
MIN: 1–33, 42–43; REG, MAX: 1–33 even, 34–43; ALL: Mixed Review

An exploratory lesson opener can be found on page 545.

Numerical expressions involving multiplication and division can be written in several ways.

9×8 can be written $9 \cdot 8$ or $9(8)$

$72 \div 8$ can be written $\frac{72}{8}$

Parentheses ( ) and brackets [ ] indicate the order in which operations should be done, and are called **grouping symbols**. The fraction bar, which serves as a division symbol, is also a grouping symbol.

Expression	Meaning of grouping symbols	Results
$(9 + 6) \div 3$	*Parentheses* indicate that addition is to be done first.	$15 \div 3 = 5$
$9 + (6 \div 3)$	*Parentheses* indicate that division is to be done first.	$9 + 2 = 11$
$7 - [(6 + 9) \div 5]$	*Inner parentheses* indicate that addition is to be done first. The division inside the *brackets* is to be done next.	$7 - [15 \div 5]$ $7 - 3 = 4$
$\frac{9 + 3}{3 \cdot 2}$	The *fraction bar* indicates that the addition above and the multiplication below are to be done before dividing.	$\frac{12}{6} = 12 \div 6 = 2$

When no grouping symbols are given, as in the expression $5 + 4 \cdot 7$, we need a rule for the order of operations. If addition is done first, $5 + 4 \cdot 7$ is 63. If multiplication is done first, $5 + 4 \cdot 7$ is 33. The following rule ensures that everyone gets the same answer.

Order of Operations

First do all multiplying and dividing in order from left to right.
Then do all adding and subtracting in order from left to right.

Example 1

Evaluate. $[18 + (4 \cdot 6)] \div 7$

Solution

$[18 + (4 \cdot 6)] \div 7$	Look for the inside grouping symbols. Compute $4 \cdot 6$ first.
$= [18 + 24] \div 7$	Next do the operation in the outer pair of grouping symbols. Compute $18 + 24$.
$= 42 \div 7$	
$= 6$	

Practice Evaluate.

a. $27 \div (5 + 4)$ **3**

b. $23 - [(56 + 7) \div 9]$ **16**

Example 2

Evaluate. $8 \cdot 2 + 45 \div 9$

Solution

$8 \cdot 2 + 45 \div 9$	Do multiplication and division from left to right. First compute 8×2.
$= 16 + 45 \div 9$	Then compute $45 \div 9$.
$= 16 + 5$	Next do addition and subtraction from left to right. Compute $16 + 5$.
$= 21$	

Practice Evaluate.

a. $5 + 4 \cdot 7$ **33** **b.** $24 + 40 \div 8 - 3$ **26** **c.** $16 + 30 \div 3 \cdot 2$ **36**

Like numerical expressions, algebraic expressions involving multiplication and division can be written in several ways. For example, $\mathbf{9 \times s}$ can be written $\mathbf{9 \cdot s}$, $\mathbf{9(s)}$, or $\mathbf{9s}$. An expression such as $\mathbf{7 \times a \times b}$ is most often written $\mathbf{7ab}$.

The expression $\mathbf{y \div 3}$ can be written $\mathbf{\frac{y}{3}}$.

Example 3

Evaluate $x(7 + y)$ for $x = 9$ and $y = 3$

Solution $x(7 + y)$

$9(7 + 3)$	Replace x with 9 and y with 3. Add within the parentheses first.
$= 9(10)$	Remember that 9(10) means 9×10.
$= 90$	

Practice Evaluate.

a. $3(a + 5)b$ for $a = 7$ and $b = 9$ **324**

b. $8 + \left(\frac{yz}{40}\right)$ for $y = 10$ and $z = 4$ **9**

Example 4

Evaluate. $\frac{x + 7}{6}$ for $x = 47$

Solution $\frac{x + 7}{6}$

$\frac{47 + 7}{6}$ Replace x with 47. Do the operations above the fraction bar first. Find $47 + 7$.

$= \frac{54}{6}$

$= 9$

Practice Evaluate.

a. $\frac{9 + 6}{a}$ for $a = 3$ 5 **b.** $\frac{7 + 9}{z + 6}$ for $z = 10$ 1 **c.** $\frac{3y + 9}{2z}$ for $y = 9$ and $z = 9$ 2

Oral Exercises

Which operation would you do first?

1. $5 + 3 \cdot 2$ **multiply**
2. $(7 + 4) \cdot 6$ **add**
3. $(36 - 32) \div 4$ **subtract**
4. $48 - 24 \div 6 \cdot 2$ **divide**
5. $15 - 8 \div 2$ **divide**
6. $7 + 6 \cdot 3 - 3$ **multiply**

Exercises

A Evaluate each numerical expression.

1. $13 + (54 \div 9)$ 19
2. $63 \div (21 \div 3)$ 9
3. $(7 \cdot 8) - 50$ 6
4. $[9 + (9 \cdot 7)] \div 8$ 9
5. $[15 - (4 \cdot 2)] + 9$ 16
6. $[48 \div (32 - 26)] \cdot 12$ 96
7. $16 - 7 \cdot 2$ 2
8. $9 \cdot 4 \div 2$ 18
9. $543 - 196 \div 28$ 536
10. $12 + 9 \cdot 3 - 28$ 11
11. $45 - 20 \div 4 + 9$ 49
12. $96 \div 4 - 3 \cdot 5$ 9
13. $54 + 24 \div 3 - 30$ 32
14. $36 \div 3 + 4 \cdot 5 - 3$ 29
15. $5 \cdot 6 - 56 \div 7 + 7 \cdot 4$ 50

Evaluate each algebraic expression.

16. $5x$ for $x = 9$ 45
17. $4bc$ for $b = 16$, $c = 17$ 1088
18. $(18 + 7)ab$ for $a = 2$, $b = 5$ 250
19. $\frac{x}{8}$ for $x = 32$ 4
20. $\frac{a}{b}$ for $a = 945$, $b = 9$ 105
21. $\frac{3cd}{6}$ for $c = 18$, $d = 25$ 225
22. $\frac{54}{b}$ for $b = 6$ 9
23. $12(52 - p)$ for $p = 47$ 60
24. $24 + bc$ for $b = 39$, $c = 16$ 648
25. $\frac{s + 8}{16}$ for $s = 24$ 2
26. $\frac{5x}{13}$ for $x = 26$ 10
27. $\frac{43 - t}{9}$ for $t = 7$ 4
28. $6y - 2y$ for $y = 24$ 96
29. $\frac{5a}{b + 4}$ for $a = 9$, $b = 1$ 9
30. $\frac{y + 7}{x - 5}$ for $x = 14$, $y = 20$ 3
31. $n(n) + 2$ for $n = 8$ 66
32. $\frac{56}{n} + 24$ for $n = 8$ 31
33. $\frac{2a + 4}{b - 7}$ for $a = 20$, $b = 18$ 4

B Copy and complete each table.

	a	b	$6a - b$
	3	3	$6(3) - 4 = 14$
34.	5	30	$6(5) - 30 = 0$
35.	9	4	$6(9) - 4 = 50$
36.	7	5	$6(7) - 5 = 37$

	x	y	$x \div 3 + y \div 3$
	6	9	$6 \div 3 + 9 \div 3 = 5$
37.	12	18	$12 \div 3 + 18 \div 3 = 10$
38.	24	45	$24 \div 3 + 45 \div 3 = 23$
39.	63	99	$63 \div 3 + 99 \div 3 = 54$

Find the final answer without writing down intermediate answers.

40. Evaluate. $959 \div 7 - 16 \cdot 8$ 9 **41.** Simplify. $24 \cdot 57 + 1{,}424 \div 89$ 1384

C Extending Thinking skills

Thinking skills: 42, 43. Testing a conjecture

42. Use each of the numbers 2, 4, 6, 8, and 12 exactly once, with any operation signs and grouping symbols you wish, to write an expression for the smallest possible whole number. Expressions for 0 will vary. Example: $(12 + 4) - (8 + 6 + 2)$

43. Use the digits for this year with any grouping symbols and operations to write expressions for as many of the whole numbers from 0 to 10 as you can. Answers will vary.

Mixed Review

44. 15 45. 25 46. 100

Solve mentally. **44.** $120 + n = 135$ **45.** $80 = 105 - m$ **46.** $m - 45 = 55$

Write as an expression. **47.** x less than 27 $27 - x$ **48.** x more than 12 $12 + x$

Round to the nearest hundred and estimate the sum. **49.** 3,647 + 2,891 6500

CALCULATOR ACTIVITY

Some calculators are programmed to follow the order of operation rules. With others, you must think about the rules and enter the computations into the calculator in the order in which they should be performed. Test your calculator by entering $4 \cdot 3 + 8 \div 2$ as follows:

4 [×] 3 [+] 8 [÷] 2 [=]

What answer will the calculator display if it uses the order of operation rules? 16
What answer will it display if it simply performs the operations from left to right? 10
Which answer is correct? 16

Use a calculator to evaluate each expression.

1. $56 + 9 \cdot 57$ 569 **2.** $84 - 240 \div 16$ 69

3. $832 \div 8 + 4 \cdot 26$ 208 **4.** $254 - 19 \cdot 13 + 68$ 75

5. $16 + 4 \cdot 7 - 36 \div 9$ 40 **6.** $536 + 18 \cdot 53 - 9$ 1481

PROBLEM SOLVING: WRITING EXPRESSIONS

2-2 Translating Phrases to Algebraic Expressions

Objective: To translate a verbal phrase to a numerical or algebraic expression.
MIN: 1–30; REG: 1–30 even, 31–34; MAX: 1–30 even, 31–36; ALL: Mixed Review

Verbal phrases that suggest multiplication or division can be translated to numerical or algebraic expressions.

Example 1

Write as a numerical expression. "the product of 9 and 8"

Solution 9×8 The phrase "the product of" suggests multiplying the two numbers.

Practice Write as a numerical expression.

a. the quotient of 24 and 6 **24 ÷ 6**

b. twice 8 **2 · 8**

c. 12 multiplied by 7 **12 · 7**

The phrase "9 times 8" translates to the numerical expression 9×8. The phrase "9 times a number" translates to the algebraic expression $9n$.

number	9 times the number	
8	9×8	"9 times 8"
10	9×10	"9 times 10"
n	$9n$	"9 times a **number**"

Example 2

Write an algebraic expression for "a number divided by 6."

Solution $\frac{n}{6}$ Think of a specific number, say 24. "24 divided by 6" is $\frac{24}{6}$, so "a *number* divided by 6" is $\frac{n}{6}$.

Practice Write as an algebraic expression.

a. the product of 6 and a number **6*n***

b. double a number **2*n***

c. the quotient of a number and 5 **$n \div 5$**

A phrase may suggest a combination of operations, as in the following example.

Example 3

Write an algebraic expression for "6 less than twice a number."

Solution $2n - 6$ "6 less than" suggests subtracting 6.

Practice Write as an algebraic expression.

a. 5 increased by twice a number $5 + 2n$

b. 7 times the sum of a number and 4 $7(n + 4)$

Example 4

Write as an expression.
Rita earned $24. How much will she have if:

a. she earns 7 times as much? **b.** she earns n times as much?

c. she divides it among 3 people? **d.** she divides it among n people?

Solutions

a. she earns 7 times as much? 24×7

b. she earns n times as much? $24n$ Think about a number, write a variable.

c. she divides it among 3 people? $24 \div 3$

d. she divides it among n people? $24 \div n$ Think about a number, write a variable.

Practice Write as an expression.

A machine produces 645 items in an hour. How many does it produce in: **a.** 8 hours? $645(8)$ **b.** h hours? $645h$

How many items are in each box if 645 items are divided equally among: **c.** 5 boxes? $645 \div 5$ **d.** b boxes? $645 \div b$

Oral Exercises

Tell whether the phrase suggests multiplication or division.

1. the product of mult
2. divided by div
3. the quotient of div
4. doubled mult
5. multiplied by mult
6. times mult
7. one-third of div
8. halved div
9. tripled mult

Exercises

A Write as a numerical expression.

1. the product of 8 and 7 $8 \cdot 7$
2. 7 times 9 $7 \cdot 9$
3. twice 46 $2 \cdot 46$
4. 6 multiplied by 7 $6 \cdot 7$
5. 48 shared among 4 $48 \div 4$
6. 45 divided by 15 $45 \div 15$

Write as an algebraic expression.

7. the product of a number and 9 $9n$
8. the quotient of a number and 8 $n \div 8$
9. a number times 15 15x
10. a number multiplied by 11 $11n$

11. 6 times a number 6w

12. half of a number $t \div 2$

13. a number divided by 5 $n \div 5$

14. double a number 2y

15. 35 increased by twice a number

16. 6 less than the sum of a number and 3

17. 12 less than twice a number

18. 100 less than a number divided by 1,000

19. 24 less than 3 times a number

20. 8 times the sum of 18 and a number

21. 5 times a number, plus 4

22. 5 more than the quotient of a number and 6

15–22 See additional answers.

Write an expression for each question. 25. $36 \div 4$ 26. $36 \div n$

There are 36 cars in the parking lot. How many will there be when there are:

23. 9 times as many? 36(9)

24. y times as many? 36y

25. the number there now divided by 4?

26. the number there now divided by n?

Kristy sold 48 calculators in one week. How many will she sell in a week when:

27. her sales are multiplied by 4? 48(4)

28. her sales are multiplied by x? 48x

29. her sales are divided by 12? $48 \div 12$

30. her sales are divided by z? $48 \div z$

B The expression $2n + 1$ represents an odd number such as 1, 3, 5, 7, 9, . . .
Write an algebraic expression for:

31. the even number that comes just after $2n + 1$ $2n + 2$

32. the even number that comes just before $2n + 1$ $2n$

33. the odd number that comes just after $2n + 1$ $2n + 3$

34. the odd number that comes just before $2n + 1$ $2n - 1$

C *Extending Thinking Skills*

35. In a championship football game, the Rams scored 25 points more than the Dolphins. If the Dolphins had scored 3 times as many points, they would have scored 1 point more than the Rams. How many points did each team score? Dolphins 13, Rams 38

Thinking skills: 35. Testing a conjecture 36. Finding a pattern, generalizing

36. Look for a pattern in the equations below.

Sum of the first **1** odd number	$1 = 1$
Sum of the first **2** odd numbers	$1 + 3 = 4$
Sum of the first **3** odd numbers	$1 + 3 + 5 = 9$
Sum of the first **4** odd numbers	$1 + 3 + 5 + 7 = 16$

Write an algebraic expression for the sum of the first n odd numbers. Use it to find the sum of the first 100 odd numbers.

The sum of the first n odd numbers is n^2 or $n \cdot n$. The sum of the first 100 odd numbers is 10,000.

Mixed Review

Solve and check. **37.** $n - 163 = 291$ 454 **38.** $x + 62 = 157$ 95 **39.** $26 = r - 45$ 71

Solve for the replacement set given. **40.** $c + 16 = 42$ {24, 25, 26, 27} 26

Use the associative property to write an equivalent expression. **41.** $(x + 7) + 3$ $x + (7 + 3)$

2-3 Basic Properties of Multiplication

Objective: To use the basic properties of multiplication to write an algebraic expression that is equivalent to a given algebraic expression.
MIN: 1–24; REG: 1–24 even, 25–33; MAX: 1–24 even, 25–35; ALL: Mixed Review

The basic properties of multiplication and division can be used to write equivalent expressions.

Remind students that in Chapter 1 they learned how to use the commutative, associative, and identity properties of addition to write equivalent expressions.

You can change the order in which two numbers are multiplied and their product will stay the same.
For example, $32 \times 15 = 15 \times 32$.

$$\begin{array}{r} 32 \\ \times 15 \\ \hline 160 \\ 32 \\ \hline 480 \end{array} \qquad \begin{array}{r} 15 \\ \times 32 \\ \hline 30 \\ 45 \\ \hline 480 \end{array}$$

With variables, this property can be stated for all whole numbers.

Commutative Property of Multiplication

A change in the order in which two numbers are multiplied does not change their product.

For all whole numbers ***a*** and ***b***, ***ab* = *ba***

Example 1

Use the commutative property to write an expression equivalent to $5y$

Solution $y(5)$ $y(5)$ has the same value as $5y$ for any number that replaces y.

Practice Use the commutative property to write an equivalent expression. **a.** xy **yx** **b.** $z(6)$ **6z** **c.** $24p$ **p(24)**

You can change the way three numbers are grouped for multiplication and the product stays the same.
For example, $(17 \times 25) \times 4 = 17 \times (25 \times 4)$.

$$(17 \times 25) \times 4 = 425 \times 4 = 1700$$
$$17 \times (25 \times 4) = 17 \times 100 = 1700$$

Note that changing the grouping makes this problem easy to solve.

With variables, this property can be stated for all whole numbers.

Associative Property of Multiplication

A change in the grouping of three whole numbers for multiplication does not change the product.

For all whole numbers **a, b,** and **c**, $\quad a(bc) = (ab)c$

Example 2

Use the associative property to write an expression equivalent to $6(4n)$.

Solution $\quad (6 \cdot 4)n \quad$ $(6 \cdot 4)n$ has the same value as $6(4n)$ for any number that replaces n.

Practice Use the associative property to write an equivalent expression.

a. $(n \cdot 12)7$ **b.** $25(4a)$ **c.** $(6a)b$

a. $n(12 \cdot 7)$
b. $(25 \cdot 4)a$
c. $6(ab)$

The number **1** is called the **multiplicative identity** because when it is multiplied by a number, the result is that same number. For example, $\mathbf{49 \times 1 = 49}$.

Identity Property of Multiplication

The product of a factor and one is the factor.

For all whole numbers **a**, $\quad a(1) = a \quad 1(a) = a$

Any number divided by 1 equals that same number: $a \div 1 = a$.
Any number divided by itself equals one: $a \div a = 1$.

The number 0 also has special properties. When any number is multiplied by 0, the product is always 0: $\mathbf{6 \times 0 = 0}$.

Zero Property of Multiplication

The product of any number and zero is zero.

For all whole numbers **a**, $\quad a(0) = 0 \quad 0(a) = 0$

When 0 is divided by another number, the quotient is always 0. When you consider dividing by 0, you will see that every number could be a solution to the equation $0 \div 0 = n$. An equation such as $24 \div 0 = n$ would have no solution. Division by 0 is undefined for these reasons.

Example 3

Use the zero or identity property to write an equivalent expression.

a. $0 \cdot x$ **b.** $1 \cdot x$

Solutions **a.** 0 $0 \cdot x = 0$ for any number that replaces x.

b. x $1 \cdot x = x$ for any number replacing x.

Practice Use the zero or identity property to write an equivalent expression.

a. $0 \cdot t$ 0 **b.** $1n$ n **c.** $p(0)$ 0

The numbers below show that $9 \cdot (10 + 6) = 9 \cdot 10 + 9 \cdot 6$. The result is the same whether the two numbers are added first and their sum is multiplied by 9, or the products are found first and then added.

$9 \cdot (10 + 6)$	$9 \cdot 10 + 9 \cdot 6$
$9 \cdot (16)$	$90 + 54$
144	**144**

This important property ties multiplication and addition together.

Distributive Property of Multiplication Over Addition

When two numbers have been added and then multiplied by a factor, the result will be the same when each number is multiplied by the factor and the products are then added.

For whole numbers **a, b, and c,** $a(b + c) = ab + ac$

$(b + c)a = ba + ca$

Example 4

Use the distributive property to write an expression equivalent to $5(x + y)$

Solution $5x + 5y$ $5(x + y)$ has the same value as $5x + 5y$ for any numbers that replace the variables.

Practice Use the distributive property to write an equivalent expression.

a. $6(r + 4)$ **b.** $(5 + 3)n$ **a.** $6r + 24$ **b.** $5n + 3n$

Oral Exercises

Name the property shown by each equation. Do not compute.

1. $12 \times 0 = 0$ zero
2. $17 \times 8 = 8 \times 17$ commut. ×
3. $16 \times (3 \times 5) = (16 \times 3) \times 5$ assoc. ×
4. $1 \times 26 = 26$ identity
5. $83 \times 14 = 14 \times 83$ commut. ×
6. $6 \times (12 + 7) = 6 \times 12 + 6 \times 7$ distrib.
7. $8 + 23 = 23 + 8$ commut. +
8. $9(14) + 9(6) = 9(14 + 6)$ distrib.

Exercises

A Use the commutative property to write an equivalent expression.

1. $y(4)$ 4y
2. $9x$ x(9)
3. cd dc
4. $x(28)$ 28x
5. $68b$ b(68)
6. $c(36)$ 36c
7. $15r$ r(15)
8. $9(43)$ 43(9)

Use the associative property to write an equivalent expression.

9. $(n \cdot 3)8$ $n(3 \cdot 8)$
10. $(y \cdot 7)13$ $y(7 \cdot 13)$
11. $8(5 \cdot a)$ $(8 \cdot 5)a$
12. $50(4n)$ $(50 \cdot 4)n$

Use the zero or identity property to write an equivalent expression.

13. $1y$ y
14. $0 \cdot s$ 0
15. $y \div y$ 1
16. $0 \div m$ 0

Use the distributive property to write an equivalent expression.

17. $4(x + 3)$ 4x + 12
18. $8(s + 7)$ 8s + 56
19. $y(6 + 9)$ 6y + 9y
20. $(n + 4)3$ 3n + 12
21. $(6 + s)5$ 30 + 5s
22. $(8 + 7)y$ 8y + 7y
23. $5(z + 3)$ 5z + 15
24. $y(2 + 8)$ 2y + 8y

B

25. Evaluate $(12a)b$ and $12(ab)$ for $a = 7$ and $b = 13$. Are the values equal? yes
26. Evaluate $a(7 + b)$ and $7a + ab$ for $a = 9$ and $b = 6$. Are the values equal? yes

Use basic properties to solve the equations.

27. $8n = 13(8)$ 13
28. $8(16 + 24) = 8(16) + 8m$ 24
29. $(54 \cdot 36)23 = 54(b \cdot 23)$ 36
30. $8(4x) = (8 \cdot 4)100$ 100
31. Evaluate $a \div b$ for $a = 10$ and $b = 5$. 2
If only whole number answers are allowed, why can't you evaluate $b \div a$ using these same numbers? Is the commutative property true for division of whole numbers? No For $b \div a$, the answer is a fraction.
32. Evaluate $a(b - c)$ for $a = 7$, $b = 14$, and $c = 8$. 42
Evaluate $ab - ac$ for the same numbers. 42
Try some other numbers for a, b, and c. When would a distributive property involving multiplication with subtraction be true? True when $b > c$.
33. Choose values for a, b, and c to show that a distributive property for addition over multiplication, $a + (b \cdot c) = (a + b) \cdot (a + c)$, does not hold true for all whole numbers. Answers will vary.

C Extending Thinking Skills 35. a. $26 \cdot 10 = 260$ b. $38 \cdot 100 = 3800$ c. $56 \cdot 100 = 5600$

34. Suppose the symbol @ represents the operation "doubleadd" so that $3 @ 4 = 2(3) + 4 = 10$. That is, double the first number and add the second. Give examples to show that neither the commutative nor the associative properties hold true for this operation. See additional answers.

35. How could you calculate answers to these problems quickly? Use the distributive property.

a. $26(6) + 26(4)$ 260 **b.** $38(57) + 38(43)$ 3,800 **c.** $56(64) + 56(36)$ 5,600

Thinking skills: 34. Seeking a counterexample 35. Computing mentally/Generalizing

Mixed Review

Find the product or quotient.

Examples:

$$\begin{array}{r} 376 \\ \times\ 23 \\ \hline 1128 \\ 752 \\ \hline 8648 \end{array} \qquad \begin{array}{r} 306 \\ 45\overline{)13,770} \\ 13\ 5 \\ \hline 270 \\ 270 \\ \hline 0 \end{array}$$

36. $236 \cdot 43$ 10,148 **37.** $95 \cdot 307$ 29,165 **38.** $119 \cdot 45$ 5355

39. $472 \cdot 11$ 5192 **40.** $8586 \div 53$ 162 **41.** $513 \div 27$ 19

42. $1848 \div 56$ 33 **43.** $1292 \div 76$ 17 **44.** $67 \cdot 29$ 1943

45. $1311 \div 57$ 23 **46.** $94 \cdot 280$ 26,320 **47.** $8100 \div 36$ 225

48. $732 \cdot 58$ 42,456 **49.** $35322 \div 42$ 841 **50.** $3000 \div 25$ 120

NUMBERS TO ALGEBRA

When an expression has the same number as a factor in each addend, a reversed form of the distributive property, $\boldsymbol{ab + ac = a(b + c)}$ may be used to write an equivalent expression that is a product.

Check the number examples to see that they are true.

This property is used to **factor** algebraic expressions

Numbers	Algebra (factored form)
$\mathbf{4} \cdot 3 + \mathbf{4} \cdot 5 = \mathbf{4}(3 + 5)$	$4b + 4c = 4(b + c)$
$\mathbf{6} \cdot 5 + \mathbf{6} \cdot 7 = \mathbf{6}(5 + 7)$	$6x + 6y = 6(x + y)$
$\mathbf{5} \cdot 8 + \mathbf{5} \cdot 4 = \mathbf{5}(8 + 4)$	$5m + 5n = 5(m + n)$

Use the distributive property to factor each expression.

1. $2a + 2b$ $2(a + b)$ **2.** $5y + 5z$ $5(y + z)$ **3.** $6n + 6(7)$ $6(n + 7)$

4. $8s + 16$ $8(s + 2)$ **5.** $3p + np$ $p(3 + n)$ **6.** $ax + az$ $a(x + z)$

2-4 Simplifying Algebraic Expressions

Objective: To use the associative and distributive properties to simplify algebraic expressions.
MIN: 1–29; REG: 1–29 even, 30–37; MAX: 1–29 even, 30–39; ALL: Mixed Review

To **simplify an algebraic expression**, you replace it with a simpler equivalent expression. An expression in its simplest form does not contain parentheses. The table below suggests that $6n$ has the same value as $3(2n)$ for every number that replaces the variable n. $3(2n)$ can be simplified to $6n$ because $6n$ is a simpler expression that is equivalent to $3(2n)$.

n	$3(2n)$	$6n$
1	6	6
2	12	12
3	18	18
⋮	⋮	⋮

You can use the basic properties to simplify algebraic expressions.

Example 1

Use the associative property of multiplication to simplify. $3(2n)$

Solution $3(2n) = (3 \cdot 2)n$ $(3 \cdot 2)n$ has the same value as $3(2n)$ for any value of n.

$= 6n$

Practice Simplify. **a.** $5(6x)$ **30x** **b.** $9(7b)$ **63b** **c.** $12(8y)$ **96y**

Example 2

Use the associative property of addition to simplify. $(x + 7) + 6$

Solution $(x + 7) + 6 = x + (7 + 6)$ $x + (7 + 6)$ has the same value as $(x + 7) + 6$ for any value of n.

$= x + 13$

Practice Simplify. **a.** $(n + 24) + 8$ **n + 32** **b.** $(z + 13) + 7$ **z + 20** **c.** $(r + 25) + 7$ **r + 32**

The parts of an algebraic expression that are separated by an addition or subtraction sign are called **terms**. The expression $4x + 2y - 3$ has three terms.

$4x + 2y - 3$

terms

Terms with the same variable factors are called **like terms**. $2n$ and $3n$ are like terms, but $4x$ and $2y$ are unlike terms because their variable factors, x and y, are different.

To simplify expressions with addends that are like terms, you can **combine like terms** by using the distributive property: $\boldsymbol{ba + ca = (b + c)a}$. The number replacements for n in the table below suggest that $2n + 3n$ can be simplified to $5n$.

n	$2n + 3n$	$5n$
1	5	5
2	10	10
3	15	15
⋮	⋮	⋮

Example 3

Use the distributive property to simplify by combining like terms. $5x + 8x$

Solution $5x + 8x = (5 + 8)x$ $(5 + 8)x = 5x + 8x$ for any number replacing x.

$= 13x$

Practice Combine like terms. **a.** $6n + 5n$ $11n$ **b.** $25b + 15b$ $40b$ **c.** $37z + 4z$ $41z$ **d.** $x + 5x$ $6x$

When combining like terms such as x + 5x, it may help students to think of x as 1x.

Oral Exercises

Name the property that could be used to simplify the expression.

1. $5(4x)$ assoc. **2.** $(n + 9) + 6$ assoc. **3.** $7(5 \cdot b)$ assoc.

4. $6y + 9y$ distrib. **5.** $(c \cdot 5)7$ assoc. **6.** $12c + 17c$ distrib.

7. $8 + (9 + r)$ assoc. **8.** $23t + 1t$ distrib. **9.** $6(2n)$ assoc.

Exercises

A Use the associative property of multiplication to simplify.

1. $13(16c)$ $208c$ **2.** $5(3 \cdot s)$ $15s$ **3.** $8(6p)$ $48p$ **4.** $6(10 \cdot n)$ $60n$

5. $7(8x)$ $56x$ **6.** $7(4 \cdot y)$ $28y$ **7.** $27(3y)$ $81y$ **8.** $10(100n)$ $1000n$

9. $46(3b)$ $138b$ **10.** $4(9 \cdot z)$ $36z$ **11.** $34(46z)$ $1564z$ **12.** $56(89x)$ $4984x$

Use the associative property of addition to simplify.

13. $(x + 7) + 9$ $x + 16$ **14.** $(p + 13) + 8$ $p + 21$ **15.** $(b + 27) + 5$ $b + 32$

16. $(z + 45) + 25$ $z + 70$ **17.** $(n + 18) + 5$ $n + 23$ **18.** $(c + 24) + 9$ $c + 33$

19. $(b + 36) + 47$ $b + 83$ **20.** $(q + 67) + 8925$ $q + 8992$ **21.** $(y + 135) + 68$ $y + 203$

Use the distributive property to simplify by combining like terms.

22. $4x + 6x$ $10x$ **23.** $7a + 5a$ $12a$ **24.** $9p + 7p$ $16p$ **25.** $4z + 12z$ $16z$

26. $8s + 9s$ $17s$ **27.** $15y + 8y$ $23y$ **28.** $5a + 36a$ $41a$ **29.** $86c + 37c$ $123c$

B The distributive property can be extended. For whole numbers $a, b, c, d, \ldots, a(b + c + d + \ldots) = ab + ac + ad + \ldots$

Use the **extended distributive property** to simplify.

30. $5x + 4x + 3x$ $12x$ **31.** $6a + 8a + 9a + 7a$ $30a$

32. $23y + 6y + 5y$ $34y$ **33.** $27c + 23c + 6c$ $56c$

Use the basic properties as needed to simplify each expression.

34. $(5 + y) + 9$ $y + 14$ **35.** $5(3x + 4x)$ $35x$ **36.** $7(x + 8)$ $7x + 56$ **37.** $\frac{6a + 4a}{7a + 3a}$ 1

C Extending Thinking Skills

38. Nancy sold tickets to the school play. Compared to the number she sold on Monday, Tuesday's number was double, Wednesday's was triple, and Thursday's was quadruple. If Nancy sold a total of 80 tickets, how many did she sell each day? Mon. = 8 Tues. = 16 Wed. = 24 Thurs. = 32

39. Try this "number trick": choose a number; add 7; subtract 2; multiply by 2; subtract 10; divide by 2. What is the result? Choose a variable to express the number and write an algebraic expression to help explain why this works.

Thinking skills: 38. Testing a conjecture 39. Generalizing, Formulating a proof.

Mixed Review

Write as an algebraic expression. **40.** 9 less than 6 times a number $6x - 9$

Solve and check. **41.** $x - 19 = 6$ 25 **42.** $304 = x - 24$ 328 **43.** $167 + x = 540$ 373

39. $\frac{[(x + 7 - 2)2] - 10}{2}$, or $\frac{(2x + 10) - 10}{2}$, or $\frac{2x}{2}$, or x

MENTAL MATH

Because of the distributive property, you can find products such as 3(24) by "breaking apart" 24 and thinking "3(20) plus 3(4), or 60 + 12. The product is 72." Break apart numbers to find each product mentally.

1. 2(34) 60 + 8 = 68 **2.** 3(26) 60 + 18 = 78 **3.** 4(23) 80 + 12 = 92

4. 5(33) 150 + 15 = 165 **5.** 6(43) 258 **6.** 7(24) 140 + 28 = 168

2-5 Solving Equations: Using Mental Math

Objective: To solve simple equations mentally.
MIN: 1–24, 34–35; REG, MAX: 1–24 even, 25–35; ALL: Mixed Review

An exploratory lesson opener can be found on page 546.

To find the average speed of a raft on a 5-hour trip with a distance of 35 km you can solve the equation 5n = 35.

An equation that involves small numbers and a single operation, either multiplication or division, can often be solved mentally.

Example 1

Solve and check mentally. $5n = 35$

Solution $5n = 35$ Think: what number multiplied by 5 gives 35?
$n = 7$

Check $5(7) \stackrel{?}{=} 35$ Mentally substitute 7 for n in the equation.
$35 = 35 \checkmark$ The equation is true, so 7 is a solution.

The average speed of the raft was 7 km/h.

Practice Solve and check mentally.

a. $\frac{x}{4} = 9$ **36** **b.** $56 = 7r$ **8** **c.** $9 = \frac{c}{3}$ **27**

The equations below show the relationship between multiplication and division.

Product		Factor		Factor		Factor		Factor		Product
240	÷	**40**	=	**6**	⟷	**6**	·	**40**	=	**240**

Since multiplication can undo division and division can undo multiplication, multiplication and division are inverse operations. You can use this idea to solve simple equations mentally.

Example 2

Use the inverse operation to solve mentally. $\frac{x}{6} = 5$

Solution $\frac{x}{6} = 5$ Think "A number has been divided by 6 to produce 5, so 5 times 6 should equal the number."

$x = 5 \cdot 6$

$x = 30$

Check $\frac{30}{6} \stackrel{?}{=} 5$ Mentally substitute 30 for the x in the equation.

$5 = 5 \checkmark$ The equation is true, so 30 is the solution.

Practice Use the inverse operation to solve mentally.

a. $r \cdot 8 = 40$ 5 **b.** $\frac{a}{6} = 9$ 54 **c.** $6y = 48$ 8

Oral Exercises

The solution of the equation $6x = 42$ is "the number that multiplied by 6 gives 42." Describe the solution of each of these equations.

1. $\frac{n}{5} = 9$ **2.** $4c = 20$ **3.** $\frac{x}{7} = 4$

4. $x \cdot 9 = 7$ **5.** $4 = \frac{a}{8}$ **6.** $63 = 7c$

1. The number that divided by 5 gives 9, etc.

To solve the equation $\frac{x}{21} = 7$, you can use the inverse operation and find $21 \cdot 7$. How can you solve each of these equations?

7. $\frac{a}{4} = 9$ find 9(4) **8.** $7y = 28$ $28 \div 7$ **9.** $7 = \frac{x}{9}$ 7(9)

10. $72 = 9n$ $72 \div 9$ **11.** $6r = 42$ $42 \div 6$ **12.** $\frac{p}{6} = 6$ 6(6)

Exercises

A Solve and check mentally.

1. $3n = 27$ 9 **2.** $4a = 28$ 7 **3.** $7x = 49$ 7 **4.** $8b = 72$ 9

5. $\frac{t}{4} = 9$ 36 **6.** $\frac{x}{8} = 9$ 72 **7.** $\frac{c}{5} = 6$ 30 **8.** $\frac{b}{10} = 8$ 80

9. $30p = 150$ 5 **10.** $40x = 80$ 2 **11.** $\frac{b}{60} = 4$ 240 **12.** $20 = \frac{n}{4}$ 80

13. $240 = 8x$ 30 **14.** $10 = \frac{n}{3}$ 30 **15.** $100 = 5y$ 20 **16.** $250 = 10x$ 25

17. $\frac{a}{50} = 250$ **18.** $3 = \frac{p}{70}$ 210 **19.** $100 = \frac{n}{5}$ 500 **20.** $80 = \frac{t}{4}$ 320

21. $25n = 100$ 4 **22.** $350t = 700$ 2 **23.** $125z = 250$ 2 **24.** $\frac{r}{1000} = 3$

17. 12500 24. 3000

B Use a calculator and inverse operations to solve the equations in Exercises 25–32. For example, to solve $\frac{x}{38} = 26$, think: a number has been *divided* by 38 to produce the factor 26. Start with 26 and *multiply* by 38.

Display

26 [×] 38 [=] 988 $x = 988$

25. $\frac{n}{49} = 135$ 6615 **26.** $57c = 684$ 12 **27.** $\frac{c}{78} = 41$ 3198 **28.** $\frac{y}{82} = 137$ 11,234

29. An amusement park director divided the total attendance for a two-week period by 14 to find that the average daily attendance was 8,654. Solve the equation $\frac{t}{14} = 8{,}654$ to find the total two-week attendance. 121,156

30. A detective divided her earnings for an 8-week case by 8 to find that her average weekly salary had been $4,535. Solve the equation $\frac{a}{8} = 4{,}535$ to find the total amount she earned. $36,280

31. If the monthly sales dollars made by a salesman were multiplied by 9, he would reach his yearly sales goal of $508,275. Solve the equation $9s = 508{,}275$ to find the amount of his monthly sales. $56,475

32. A scientist knew that the time a satellite had traveled, 3,699 seconds, times its speed in miles per second would equal the 25,893 miles it had traveled. Solve the equation $3{,}699s = 25{,}893$ to find the speed of the satellite. 7 miles per second

33. For all whole numbers a, b, and c, if $\frac{a}{b} = c$, then $a = c \cdot b$. Use this relationship to rewrite each equation with the variable by itself on one side. Then solve for the variable. **a.** $\frac{x}{28} = 42$ **b.** $\frac{n}{24} = 53$ **c.** $\frac{z}{125} = 17$
a. 1176 b. 1272 c. 2125

C *Extending Thinking Skills*

34. Find numbers for x and y that make both $x \cdot y = 144$ and $x \div y = 4$ true. $x = 24, y = 6$

35. The product of the ages of a teenager and her grandmother was 1024. The sum of their ages was 80. What was the quotient of their ages? 4 (Grandmother is 64, teenager is 16)

Thinking skills: 34–35 Testing a conjecture.

Mixed Review

Evaluate each expression. **36.** $(9 - 6) \cdot 3$ 9 **37.** $21 \div 7 - 6 \div 3$ 1 **38.** $(6 \cdot 2) - 9$ 3

Solve and check. **39.** $x + 71 = 90$ 19 **40.** $27 = x - 14$ 41 **41.** $x + 26 = 53$ 27

33. a. $x = 42(28)$, $x = 1176$ b. $n = 53(24)$, $n = 1272$ c. $z = 17(125)$, $z = 2125$

ESTIMATION

Since the addends in the sum 598 + 613 + 587 cluster around 600 you can use **clustering** to estimate this sum as 3 × 600, or 1800. Use clustering to estimate the following sums.

1. 275 + 314 + 326 3 × 300 = 900

2. 923 + 879 + 894 3 × 900 = 2700

3. 1232 + 1187 + 1199 3 × 1200 = 3600

4. 708 + 674 + 719 + 697 4 × 700 = 2800

5. 196 + 226 + 175 + 213 4 × 200 = 800

6. 2417 + 2373 + 2386 + 2394 4 × 2400 = 9600

2-6 Solving Equations: Using Multiplication and Division

Objective: To use multiplication and divison to solve equations involving whole numbers.
MIN: 1–32, 41–42; REG, MAX: 1–32 even, 33–42; ALL: Mixed Review

Although some simple equations can be solved mentally, other methods are useful with equations that have larger numbers or more operations. Two ideas are used to solve equations involving multiplication and division. The first is that multiplication and division are inverse operations, and each can be used to undo the other.

Division can undo multiplication.	Multiplication can undo division.
$\frac{4 \cdot 8}{4} = 8$	$\frac{8}{4} \cdot 4 = 8$
$\frac{4 \cdot n}{4} = n$	$\frac{n}{4} \cdot 4 = n$

The second idea is that an equation is like a balanced scale. Whatever is done to one side must also be done to the other side. The properties of equality state this idea.

Multiplication and Division Properties of Equality

You can multiply or divide by the same number on both sides of the equation and the two sides will remain equal.
For all numbers **a, b, and c,** if **a = b,** then **ac = bc**

and, when **c ≠ 0,** if **a = b,** then $\frac{a}{c} = \frac{b}{c}$

To solve an equation, you need to change it into an equation with the variable by itself on one side. The following steps use the idea of inverse operations and the properties of equality to help you do this.

Solving Equations Using Multiplication and Division

1. Decide which operation has been applied to the variable.
2. Use the inverse of this operation, multiplying or dividing by the same number on both sides of the equation.

Example 1

Solve and check. $4n = 72$

Solution $4n = 72$

$\frac{4n}{4} = \frac{72}{4}$ To undo multiplying by 4, divide by 4.

$n = 18$ Divide both sides by 4 so they remain equal.

Check $4 \cdot 18 \stackrel{?}{=} 72$ Replace n with 18 in $4n = 72$.

$72 = 72 \checkmark$ The solution is 18.

Practice Solve and check. **a.** $6x = 78$ 13 **b.** $192 = 8n$ 24

Example 2

Solve and check. $\frac{n}{4} = 26$

Solution $\frac{n}{4} = 26$

$\frac{n}{4} \cdot 4 = 26 \cdot 4$ To undo dividing by 4, multiply by 4.

$n = 104$ Multiply both sides by 4 so they remain equal.

Check $\frac{104}{4} \stackrel{?}{=} 26$ Replace n with 104 in $\frac{n}{4} = 26$.

$26 = 26 \checkmark$ The solution is 104.

Practice Solve and check. **a.** $\frac{n}{6} = 12$ 72 **b.** $4 = \frac{x}{8}$ 32

Oral Exercises

Name the operation that would be used to solve the equation.

1. $27c = 513$ div. **2.** $17r = 765$ div. **3.** $504 = 56x$ div.

4. $\frac{b}{36} = 24$ mult. **5.** $125 = z \div 94$ mult. **6.** $35 = \frac{y}{16}$ mult.

Exercises

A Solve and check.

1. $6y = 48$ 8 **2.** $2c = 98$ 49 **3.** $9n = 99$ 11 **4.** $4c = 68$ 17

5. $5x = 115$ 23 **6.** $9c = 234$ 26 **7.** $12a = 144$ 12 **8.** $21x = 168$ 8

9. $32b = 192$ 6 **10.** $43n = 172$ 4 **11.** $54y = 162$ 3 **12.** $72n = 432$ 6

13. $\frac{n}{24} = 8$ 192 **14.** $\frac{x}{32} = 9$ 288 **15.** $\frac{c}{41} = 5$ 205 **16.** $\frac{b}{26} = 5$ 130

17. $\frac{y}{16} = 12$ 192 **18.** $\frac{a}{23} = 11$ 253 **19.** $\frac{x}{32} = 16$ 512 **20.** $\frac{n}{44} = 15$ 660

21. $\frac{a}{36} = 54$ 1944 **22.** $\frac{b}{75} = 28$ 2100 **23.** $\frac{c}{38} = 69$ 2622 **24.** $\frac{y}{54} = 183$ 9882

25. $432 = 36b$ 12 **26.** $38 = \frac{y}{78}$ 2964 **27.** $527 = 31n$ 17 **28.** $23 = \frac{c}{35}$ 805

29. $16n = 5216$ 326 **30.** $1924 = 52y$ 37 **31.** $\frac{y}{361} = 27$ 9747 **32.** $94 = \frac{x}{57}$ 5358

B Solve and check.

33. $(32 + 45)n = 231$ 3 **34.** $(54 - 9) = 15y$ 3 **35.** $\frac{a}{27(3)} = 24 + 36$ 4860

36. $\frac{x}{3 \cdot 18} = 4$ 216 **37.** $(126 \div 9) = \frac{z}{5}$ 70 **38.** $266 = (46 - 8)p$ 7

39. Write and solve an equation in which the product is 612, one factor is 3, and the other factor is represented by the variable n. $3n = 612$, $n = 204$

40. Write and solve an equation in which m is divided by 9 and the quotient is 5.
$m \div 9 = 5$, $m = 45$

C Extending Thinking Skills

41. Use inverse operations to find the missing number n in this flow chart. 1776

Start with n → subtract 1492 → multiply by 6 → add 282 → end with 1986

42. If a number is multiplied by 3 and added to 9, the result is 54. What is the number? 15

Thinking skills: 41–42. Using reverse reasoning

Mixed Review

Write as an expression. **43.** 13 added to the product of 6 and 9 $(6 \cdot 9) + 13$

Solve and check. **44.** $n - 36 = 17$ 53 **45.** $41 + x = 63$ 22 **46.** $9 = x - 334$ 343

COMPUTER ACTIVITY

This computer program can be used to solve equations of the form $\frac{x}{A} = B$. Remind students to separate their number inputs in line 30 with a comma.

```
10 REM: SOLVE AN EQUATION X/A=B
20 PRINT "CHOOSE WHOLE NUMBERS A AND B, A<>0"
30 INPUT A,B
40 PRINT "THE EQUATION IS X/"A;"=";B
50 PRINT "THE SOLUTION IS X=";A*B
60 END
```

See additional answers.

1. Run the program or show what the run would look like.
2. Change the program to solve equations of the form $Ax = B$.

More Practice

Solve.

1. $x - 37 = 110$ 147

2. $r - 83 = 26$ 109

3. $m + 74 = 231$ 157

4. $t - 462 = 19$ 481

5. $w - 396 = 47$ 443

6. $215 = r + 86$ 129

7. $k - 59 = 317$ 376

8. $x + 34 = 416$ 382

9. $t + 21 = 387$ 366

10. $a + 68 = 156$ 88

11. $z - 87 = 216$ 303

12. $y + 87 = 206$ 119

13. $a + 67 = 914$ 847

14. $m - 53 = 422$ 475

15. $c + 23 = 111$ 88

16. $n - 45 = 74$ 119

17. $\frac{a}{7} = 56$ 392

18. $16w = 288$ 18

19. $9c = 207$ 23

20. $12h = 312$ 26

21. $\frac{m}{8} = 24$ 192

22. $276 = 23k$ 12

23. $\frac{r}{12} = 406$ 4872

24. $\frac{z}{7} = 116$ 812

25. $n \cdot 11 = 561$ 51

26. $\frac{y}{27} = 13$ 351

27. $242 = \frac{t}{8}$ 1936

28. $19w = 475$ 25

29. $\frac{x}{27} = 15$ 405

30. $\frac{t}{9} = 63$ 567

31. $27m = 216$ 8

32. $104 = \frac{a}{12}$ 1248

33. $16 = \frac{n}{21}$ 336

34. $t + 38 = 415$ 377

35. $a - 35 = 180$ 215

36. $12 \cdot c = 384$ 32

37. $z + 19 = 111$ 92

38. $26x = 910$ 35

39. $\frac{r}{64} = 11$ 704

40. $y - 908 = 21$ 929

41. $w + 87 = 126$ 39

42. $13 = \frac{h}{19}$ 247

43. $43k = 731$ 17

44. $m - 88 = 316$ 404

45. $\frac{t}{30} = 26$ 780

46. $c - 86 = 104$ 190

47. $n + 77 = 254$ 177

48. $36a = 936$ 26

49. $191 = x - 29$ 220

50. $z + 92 = 647$ 555

51. $930 = 15 \cdot y$ 62

52. $\frac{w}{42} = 15$ 630

53. $21 \cdot t = 693$ 33

54. $387 = a - 26$ 413

55. $r + 49 = 513$ 464

56. $\frac{k}{8} = 404$ 3232

57. $760 = 8m$ 95

58. $\frac{n}{5} = 160$ 800

59. $t - 57 = 487$ 544

60. $x + 81 = 360$ 279

61. $z - 109 = 26$ 135

62. $1395 = 31 \cdot h$ 45

63. $\frac{w}{9} = 18$ 162

64. $r + 75 = 463$ 388

65. $\frac{y}{16} = 25$ 400

66. $v + 67 = 104$ 37

67. $486 = 18n$ 27

68. $476 = c - 738$ 1214

PROBLEM SOLVING: APPLICATIONS

2-7 Keeping Physically Fit

Objective: To solve word problems involving whole numbers.
ALL: 1–9

A calorie is a heat unit that can measure the food energy taken in and used by the body. Average calorie amounts used by activities are shown in the table at right. Calorie needs are shown below.

To keep your weight the same take in 15 calories per pound of your weight, per day.

To gain a pound take in 3500 extra calories.

To lose a pound use 3500 extra calories.

Calorie Use

Activity	Calories per minute
Running	14
Bicycling	11
Swimming	9
Tennis	7
Bowling	5

Problems

Use the information given above to solve. Decide whether to use **pencil and paper, mental math, estimation,** or a **calculator** to find the answer. Use each of these techniques at least once.

1. How many calories would you use by swimming for 30 minutes? 270

2. Would you use more than or less than than 460 calories by bicycling for 45 minutes? more

3. How many calories would you use by running for 1 hour and 5 minutes? 910

4. How many minutes would you have to run to use 350 calories? 25

5. How many more calories would you use by running than by playing tennis for 55 minutes? 385

6. How many calories are needed daily to keep the weight of a 137-pound adult the same? 2055

7. An athlete weighs 129 pounds. Will she gain or lose weight if she eats 15 calories per pound of weight and uses a total of 2000 calories each day? lose

8. During one month, Enrique took in 31,576 fewer calories than he needed to keep his weight the same. How many pounds (to the nearest pound) did he lose? 9

9. ***Data Search*** How many calories do you need each day to maintain your weight? Answers will vary.

What's Your Decision?

Suppose you want to lose 10 pounds by either running or swimming. How will you do it and how long will it take? Answers will vary with student preference.

2-8 Evaluating Formulas

Objective: To evaluate formulas
MIN: 1–9, 16; REG: 1–16; MAX: 1–17; ALL: Mixed Review

A **formula** is an equation that shows a relationship between two or more variables. The formula $\boldsymbol{W = 5H - 190}$ expresses the relationship between a person's height in inches ($\boldsymbol{H}$) and a predicted weight in pounds ($\boldsymbol{W}$).

Example 1

Evaluate the formula $W = 5H - 190$ for $H = 60$.

Solution $W = 5H - 190$
$W = 5(60) - 190$ Replace the variable H with 60.
$W = 110$ Evaluate the resulting numerical expression.

Practice Evaluate the formula above for the heights given.
a. $H = 65$ **135** **b.** $H = 68$ **150**

Example 2

Evaluate the following formula to find Dave's arm strength rating. He weighs 110 lbs., is 63 inches tall, and did 12 dips and 10 pullups. Formula: $A = S\left(\frac{W}{10} + H - 60\right)$. A is arm strength, S is sum of dips and pullups a person can do, W is weight in pounds, and H is height in inches.

Solution

$$A = S\left(\frac{W}{10} + H - 60\right)$$

$$A = 22\left(\frac{110}{10} + 63 - 60\right)$$ Replace the variable S with 22, W with 110, and H with 63. Evaluate.

$$A = 22(11 + 3)$$

$$A = 22(14) = 308$$

Jeff's arm strength rating is 308.

Practice Evaluate the formula above for the values given.
Estrelita: 16 dips, 12 pullups, $W = 110$ lbs., $H = 64$ in. **420**

Oral Exercises

Evaluate mentally. **1.** $95 - ab$ for $a = 9$, $b = 7$ **32** **2.** $r + 23$ for $r = 45$ **68**

3. $A + B$ for $A = 199, B = 78$ 277

4. $\frac{q}{7}$ for $q = 154$ 22

5. $x - y$ for $x = 200, y = 99$ 101

6. $5x$ for $x = 15$ 75

Exercises

A Evaluate the formula for the values given.

1. $W = 5H - 190$ for $H = 66$ 140
2. $S = 16t + 64$, for $t = 12$ 256
3. $P = S - 38$ for $S = 124$ 86
4. $D = 16r - 57$ for $r = 6$ 39
5. $P = 2L + 76$ for $L = 98$ 272
6. $A = 3rr$ for $r = 9$ 243

7. Formula: $D = r \cdot t$ where D is distance in miles, r is rate in mi/hour, and t is time of ride in hours. Find how far Jenny rode her bicycle in 3 hours at 17 mi/hour. 51 miles

8. Formula: $W = r \cdot t + e$ where W is total wages, r is hourly rate of pay, t is time worked in hours, and e is overtime pay. Find how much Brian earned working 37 hours at \$6 an hour, and earning \$58 in overtime pay. \$280

9. Formula: $S = p - d$. Find the sale price of a television when the regular price (p) is \$325 and the discount (d) is \$48. \$277

B Copy and complete the table for each formula.

Formula: $T = t - \frac{a}{300}$

Temperature in degrees Celsius (T) at an altitude in meters (a) when the ground temperature in degrees Celsius is (t).

	10.	11.	12.	13.	14.
a	900	1200	2,400	4,800	6,000
t	20	20	20	20	20
T	17	16	12	4	0

15. Formula: $W = 5000d(d + 1)$ where W is weight in thousands of pounds that a rope with diameter (d) in inches will hold without breaking. Make a table showing values of W for $d = 1, 2, 3, 4$, and 5.

d	1	2	3	4	5
W	10,000	30,000	60,000	100,000	150,000

C Extending Thinking Skills

16. Test the formula $S = 3F - 24$ where S = shoe size and F = length of foot in inches. Make a table to show your data. Is the formula accurate? Answers will vary.

17. The formula for finding the total points scored in a basketball game is $P = 2g + f$, in which P is the total points scored, f is the number of free throws and g is the number of field goals. Write a formula for finding the number of field goals when the total points scored and the number of free throws are known. $g = \frac{P - f}{2}$

Thinking skills: 16. Testing a conjecture 17. Generalizing

Mixed Review

Solve and check. **18.** $m \div 19 = 11$ $m = 209$ **19.** $23t = 437$ $t = 19$

Write as an algebraic expression. **20.** The sum of 37 and a number $n + 37$

2-9 Estimating

Objective: To use rounding or compatible numbers to estimate with whole numbers.
MIN: 1–27; REG: 1–27 even, 28–33; MAX: 1–27 even, 28–35; ALL: Mixed Review

An airline pilot or navigator might estimate the time for a trip, the speed of the plane, or the total number of miles flown.

You have used rounding to estimate sums and differences. You can also use rounding to estimate products and quotients.

Example 1

Estimate the value of the variable by rounding. $807 \cdot 19 = y$

Solution $807 \cdot 19$ Round 807 to 800. Round 19 to 20.

$800 \cdot 20 = 16{,}000$

The estimated value for y is 16,000.

Practice Estimate the value of the variable by rounding.

a. $876 \cdot 79 = n$ 72,000 **b.** $x = 2{,}449 \div 79$ 30

When rounding to estimate a quotient does not result in an easy division, you can **choose a compatible number** that makes the division easier.

Example 2

Estimate the value of the variable by choosing a compatible number.
$1{,}419 \div 28 = c$

Solution $1{,}400 \div 30$ Round the numbers.

$1{,}500 \div 30 = 50$ Choose a number compatible with 30.

The estimated value for c is 50.

Practice Estimate. **a.** $738 \div 9 = t$ $720 \div 9 = 80$ **b.** $m = 1{,}694 \div 37$ $1{,}600 \div 40 = 40$

Oral Exercises 8. 300

Choose a number compatible with the divisor and estimate the quotient.

1. $1{,}100 \div 60$ 20 **2.** $2{,}300 \div 40$ 60 **3.** $1{,}300 \div 60$ 20 **4.** $4{,}600 \div 90$ 50

5. $5{,}700 \div 80$ 70 **6.** $6{,}400 \div 70$ 90 **7.** $19{,}000 \div 40$ 500 **8.** $170{,}000 \div 60$ 3,000

Exercises

A Estimate the value of the variable by rounding.

1. $28 \cdot 82 = x$ 2,400 **2.** $93 \cdot 57 = y$ 5,400 **3.** $n = 78 \cdot 19$ 1,600

4. $p = 45 \cdot 39$ 2,000 **5.** $786 \cdot 43 = a$ 32,000 **6.** $318 \cdot 18 = t$ 6,000

7. $b = 54 \cdot 995$ 50,000 **8.** $n = 513 \cdot 289$ 150,000 **9.** $89 \div 29 = c$ 3

10. $243 \div 57 = z$ 4 **11.** $r = 348 \div 73$ 5 **12.** $s = 724 \div 89$ 8

13. $1{,}599 \div 43 = p$ 40 **14.** $a = 2{,}779 \div 68$ 40 **15.** $5{,}589 \div 69 = t$ 80

16. $z = 8{,}068 \div 85$ 90 **17.** $792 \cdot 564 = n$ 480,000 **18.** $3{,}984 \div 832 = n$ 5

Estimate the value of the variable by choosing a compatible number.

19. $342 \div 7 = s$ 50 **20.** $532 \div 6 = t$ 90 **21.** $n = 653 \div 8$ 80

22. $c = 437 \div 5$ 90 **23.** $294 \div 58 = y$ 5 **24.** $d = 372 \div 55$ 6

25. $553 \div 79 = b$ 7 **26.** $2{,}438 \div 53 = c$ 50 **27.** $2{,}652 \div 68 = d$ 40

B Estimate, then compute the value of the variable in Exercises 28–30.

28. $p = 58 \cdot 94 \cdot 75$ est. 432,000; 408,900 **29.** $z = (1848 \div 56) \cdot 98$ est. 3000; 3234 **30.** $x = (684 \cdot 423) \div 12$ est. 28,000; 24,111

31. A store sold 96 regular tires at an average price of \$56 a tire and 28 snow tires at an average price of \$79 per tire. Estimate the total sales. \$8400

32. The total weekly salary for 12 supervisors was \$6,696. Their total weekly overtime pay was \$492. Estimate a supervisor's average weekly earnings. \$600.

33. Estimate, then find, the number of flight hours logged by a pilot who made six 895-mile trips at an average speed of 618 mi/hour. est. 9; 8.68932 h

C Extending Thinking Skills

34. Copy the division problem at right and fill in the missing digits.

```
     ■ ■
64)■ ■ ■
   ■ ■
   ■ ■ ■
   ■ ■ 2
       3
```

```
     13
64)835
   64
   195
   192
     3
```

35. Mrs. Ward rounded her monthly salary to the nearest hundred and arrived at an estimate of \$40,800 for her yearly salary. Describe her actual monthly salary as accurately as possible. ≥ 3350 and < 3450

Thinking skills: 34. Using logic 35. Estimating/Using reverse reasoning

Mixed Review

Solve and check. **36.** $169 = \frac{b}{13}$ 2197 **37.** $61 = x - 24$ 85 **38.** $19a = 304$ 16

Evaluate the formula $M = 6w + 12$ for: **39.** $w = 9$ 66 **40.** $w = 3$ 30

PROBLEM SOLVING: STRATEGIES

2-10 Draw a Picture

Objective: To solve nonroutine problems, using the strategy Draw a Picture and other strategies learned so far.
MIN: 1–2; REG: 1–5; MAX: 1–9

The problem-solving strategy called **Draw a Picture** can often help you better understand a problem situation and give you a start toward a solution.

Problem The head of a tropical fish is $\frac{1}{3}$ as long as its midsection. Its tail is as long as its head and midsection combined. The total length of the fish is 48 cm. How long is each part of the fish?

It is helpful to draw and label a picture to show the conditions of the problem.

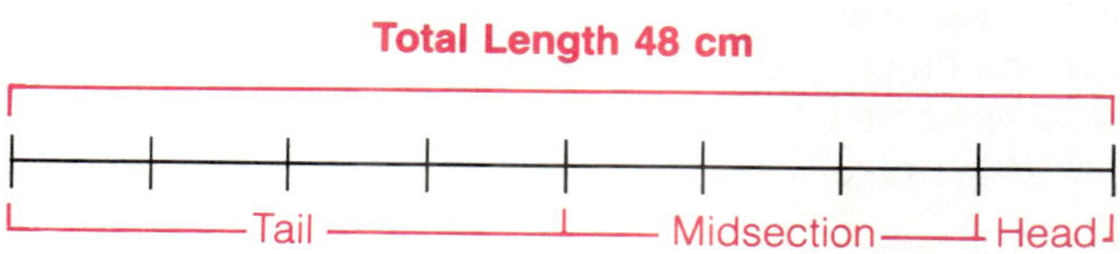

The picture shows that 8 equal segments make up the 48 cm. Each segment must be 6 cm long, so the head is 6 cm long, the midsection is 18 cm long, and the tail is 24 cm long.

Problem-Solving Strategies
Choose the Operations
Guess, Check, Revise
Draw a Picture

This chart shows the strategies introduced so far.

Problems

Solve.

1. The tail of a salamander is three times as long as its midsection. Its head is $\frac{1}{2}$ as long as its midsection. If the total length of the salamander is 27 cm, how long is its tail? **18 cm**

2. Joliet is 564 km from Canton. A bus started at Canton and drove 248 km towards Joliet. A jeep started at Joliet and drive 216 km towards Canton. How far apart were the two vehicles at the end of these trips? **100 km**

3. Alfredo correctly answered 22 questions on a test for a score of 84. There were 15 questions worth 3 points and 11 questions worth 5 points. How many questions of each type did he answer correctly? 13 3-point; 9 5-point

4. A special rubber ball is dropped from the top of a wall 16 feet high. Each time it hits the ground it bounces half as high as its previous high point. If the ball is caught just as it bounces 1 foot high, how far has it traveled? 45 ft

5. Kiyoko scored her test by multiplying the number of problems correct by 3 and subtracting 1 for each of the 16 problems she missed. She answered twice as many problems correctly as the number she missed. What was the total number of problems on the test? What was her score? Total problems, 48; score, 80.

6. Sam's class took twice as long as Mark's class to build a homecoming float. Mark's class took 8 hours longer than Randy's class. The total number of hours spent by all classes was 60. How long did it take each class to build its float? Randy's class = 9, Mark's class = 17, Sam's class = 34

7. Mr. Ashby sold 37 small carved wooden whistles for $5 each, and his last 5 large whistles for $25 each. How many more small whistles will he have to sell to reach a sales total of $500? 38

8. An explorer astronaut completed a 152-km trip around a crater on the moon in 4 hours of traveling time in a moon buggy. She plans to make a 114-km trip at the same speed around another crater and stop for 1 hour along the way to explore on foot. How long will this take? 4 hours

9. A construction engineer wants to make a tunnel through a rock that is 10 meters thick. Each hour the drilling tool goes in 5m, but as the engineer rests the rock caves in and the drilling tool slides back 4m. At this rate, how many hours will it take the engineer to break through the rock? 6 hours

Enrichment

Functions

A **function** can be thought of as a rule that pairs each member of one set, called the **domain**, with one and only one member of another set, called the **range**.

The "Function Machine" below illustrates the idea, using numbers as elements. The set of "input" numbers is the domain. The set of "output" numbers is the range.

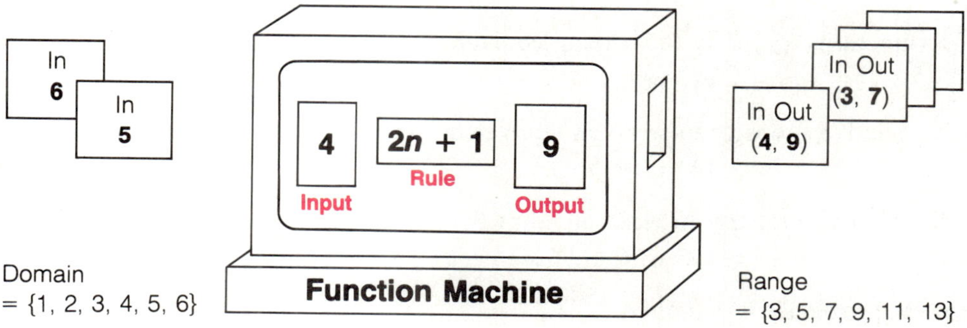

In the rule $f(n) = 2n + 1$, the function is named by the letter f and the input number is represented by the variable n. The output number is represented by either $f(n)$ or $2n + 1$. Using this rule for the input number 4, the output number, $f(n)$, would be found as follows: $f(4) = 2(4) + 1$.

$f(n)$ is read "f of n." Stress that f is not a variable and that n is.

You may use a table to describe the function:

Rule $f(n) = 2n + 1$

Input	1	2	3	4	5	6
Output	3	5	7	9	11	13

Or you may list the pairs as
$f = \{(1, 3), (2, 5), (3, 7), (4, 9), (5, 11), (6, 13)\}$.

1. Domain: $\{1, 2, 3, 4, 5\}$, Function rule: $f(n) = 3n + 4$. List the set of pairs for the function f. $\{(1, 7), (2, 10), (3, 13), (4, 16), (5, 19)\}$

2. Function Rule: $g(x) = \frac{x}{2} + 1$. Find $g(24)$, $g(100)$, $g(498)$, and $g(2)$. 13, 51, 250, 2

3. Function Rule: $h(t) = t \cdot t \cdot t$. Find $h(2)$, $h(4)$, $h(5)$, and $h(10)$. 8, 64, 125, 1000

4. Function Rule: $f(n) = 5n - 3$. Find the value for n if $f(n) = 17$. 4

5. Domain: $\{1, 2, 3, 4, 5, 6, 7, 8\}$, Rule: $f(b) = (b + 3)(b + 3)$. Use a table to show the function. See additional answers.

6. Function: $\{(2, 2), (4, 3), (6, 4), (8, 5), (10, 6)\}$. Give the domain, range, and function rule. See additional answers.

Chapter 2 Review

2-1 Simplify.

1. $64 \div (4 + 4)$ 8 **2.** $6 + [(2 + 1) \div 3]$ 7 **3.** $2 \cdot 6 + 2 \cdot 5$ 22

Evaluate.

4. $\frac{3x}{y}$ for $x = 8$ $\frac{24}{y}$ **5.** $z(c + 4)$ for $z = 6$ 6c + 24 **6.** $6(x - y)$ for $x = 7, y = 5$ 12

2-2 Write as an algebraic expression.

7. twice x 2x **8.** m divided by 4 $m \div 4$ **9.** b multiplied by 5 5b

2-3 Use the commutative property to write an equivalent expression.

10. $f(4)$ 4(f) **11.** ab ba **12.** $14x$ x(14)

Use the associative property to write an equivalent expression.

13. $12(3p)$ $(12 \cdot 3)p$ **14.** $(7t)k$ 7(tk) **15.** $(v \cdot 3)6$ $v(3 \cdot 6)$

Use the distributive property to write an equivalent expression.

16. $4(h + 6)$ 4h + 24 **17.** $w(7 + 1)$ $w \cdot 7 + w$ **18.** $(9 + n)5$ $45 + (n \cdot 5)$

2-4 Use the basic properties to simplify.

19. $3 + (17 + r)$ 20 + r **20.** $9(3m)$ 27m **21.** $12(2x)$ 24x

22. $y + 8y$ 9y **23.** $4a + 7a$ 11a **24.** $18p + 12p$ 30p

2-5 Solve.

25. $6n = 36$ 6 **26.** $\frac{40}{x} = 8$ 5 **27.** $\frac{y}{7} = 5$ 35

2-6 Solve and check.

28. $5x = 75$ 15 **29.** $255 = 5v$ 51 **30.** $\frac{c}{8} = 15$ 120 **31.** $\frac{d}{8} = 23$ 184

2-7 Solve.

32. If Andy uses 9 calories per minute while swimming, how many calories will he use if he swims for 30 minutes? 270

2-8 Solve.

33. Formula: $D = rt$. D is distance in miles, r is rate in miles per hour, and t is time in hours. How far did Brad drive in 6 hours at an average speed of 50 miles per hour? 300 miles

2-9 Use rounding to estimate the value of the variable. Choose compatible numbers as needed.

34. $415 \cdot 42 = t$ 16,000 **35.** $6129 \div 28 = p$ 200 **36.** $1466 \div 32 = r$ 50

Chapter 2 Test

Simplify.

1. $72 \div (6 + 2)$ 9
2. $9 + [(3 + 7) \div 2]$ 14
3. $3 \cdot 8 + 7 \cdot 4$ 52

Evaluate.

4. $9y$ for $y = 5$ 45
5. $\frac{2c}{d}$ for $c = 8$ $16 \div d$
6. $v(6 + m)$ for $v = 6$ $36 + 6m$
7. $\frac{6m}{4}$ for $m = 8$ 12
8. $5(16 - x)$ for $x = 8$ 40
9. $3(a - b)$ for $a = 8, b = 6$ 6

Write as an algebraic expression.

10. twice f $2f$
11. 6 divided by r $6 \div r$
12. 13 less than twice a number $2n - 13$

Use the commutative property to write an equivalent expression.

13. $7w$ $w(7)$
14. $c(8)$ $8c$
15. np pn

Use the associative property to write an equivalent expression.

16. $(2x)y$ $2(xy)$
17. $15(2v)$ $(15 \cdot 2)v$ or $30v$
18. $(b \cdot 4)9$ $b(4 \cdot 9)$ or $b36$

Use the distributive property to write an equivalent expression.

19. $r(2 + 5)$ $2r + 5r$
20. $7(e + 7)$ $7e + 7 \cdot 7$ or $7e + 49$
21. $(4 + a)8$ $4 \cdot 8 + a(8)$ or $32 + a(8)$

Use the basic properties to simplify.

22. $7(8p)$ $56p$
23. $5(6k)$ $30k$
24. $21(11s)$ $231s$
25. $w + 6w$ $7w$
26. $2z + 3z$ $5z$
27. $15n + 20n$ $35n$

Solve.

28. $8n = 32$ 4
29. $\frac{56}{y} = 8$ 7
30. $\frac{a}{3} = 9$ 27

Solve and check.

31. $6z = 78$ 13
32. $220 = 11n$ 20
33. $\frac{r}{7} = 12$ 84
34. $\frac{b}{6} = 19$ 114

Solve.

35. If you use 11 calories per minute bicycling, how many calories would you use on a 30-minute bicycle ride? 330

36. Formula: $A = lw$; A is the area of a rectangle in square centimeters, l is the length of the rectangle, and w is the width of the rectangle. What is the area of a rectangle with width 17 centimeters and length 19 centimeters? 323 sq. centimeters

Use rounding to estimate the value of the variable. Choose compatible numbers as needed.

37. $394 \cdot 28 = m$ 12,000
38. $5776 \div 19 = c$ 290 or 300
39. $8122 \div 23 = h$ 400

Cumulative Review

Solve.

1. $35 - x = 18$ 17
2. $b + 28 = 42$ 14
3. $45 + c = 82$ 37
4. $v - 21 = 50$ 71
5. $48 = z - 27$ 75
6. $39 = 13 + t$ 26

Round to the nearest thousand and estimate the value for each variable.

7. $7{,}154 + 1{,}738 = a$ 9000
8. $n = 3{,}606 - 908$ 3000
9. $3{,}847 + 2{,}929 = b$ 7000

Round to the nearest hundred and estimate the value for each variable.

10. $f = 576 + 114$ 700
11. $703 - 515 = x$ 200
12. $2639 - 482 = g$ 2100

Use rounding to estimate the value of each variable.

13. $216 \cdot 82 = s$ 16,000
14. $3844 \cdot 17 = c$ 80,000
15. $8639 \div 27 = m$ 300

Use front-end estimation to estimate the value of the variable.

16. $v = 221 + 592 + 485$ 1300
17. $4348 + 4369 + 2947 = p$ 12,000

Simplify.

18. $49 \div (5 + 2)$ 7
19. $7 + [(4 + 4) \div 4]$ 9
20. $(2 \cdot 9) + (3 \cdot 4)$ 30

Use the commutative property to write an equivalent expression.

21. $f + 4$ $4 + f$
22. $17 + h$ $h + 17$
23. $c + 981$ $981 + c$
24. $6t$ $t(6)$
25. $z(4)$ $4z$
26. uk ku

Use the associative property to write an equivalent expression. 29. $(m + 16b) + 580$

27. $(a + 44) + 235$ $a + 279$
28. $101 + (555 + w)$ $656 + w$
29. $m + (16b + 580)$
30. $(4e)g$ $4(e \cdot g)$
31. $22(2s)$ $44s$
32. $(y \cdot 8)3$ $y(24)$

Use the distributive property to write an equivalent expression.

33. $k(1 + 7)$ $k + 7k$
34. $2(e + 9)$ $2e + 18$
35. $(3 + p)6$ $18 + 6p$

Use the associative property for multiplication to simplify.

36. $2 \cdot (7 \cdot n)$ $14n$
37. $9(6y)$ $54y$
38. $50(10j)$ $500j$

Use the distributive property to simplify by combining like terms.

39. $v + 5v$ $6v$
40. $8k + 2k$ $10k$
41. $12h + 33h$ $45h$

Chapter 3
Integers

3-1 Integers

Objectives: To give integers for points on the number line, to find the absolute value of an integer, and to compare integers.
MIN: 1–25, 38–40; REG, MAX: 1–25 even, 26–41; ALL: Mixed Review

Positive and negative numbers can be used to describe situations that are opposites of each other, such as profit and loss, temperatures above and below zero, or yards gained and lost.

The complete set of **integers** consists of the positive integers, the negative integers, and 0. On the number line, the **positive integers** are to the right of 0, and the **negative integers** are to the left of 0.

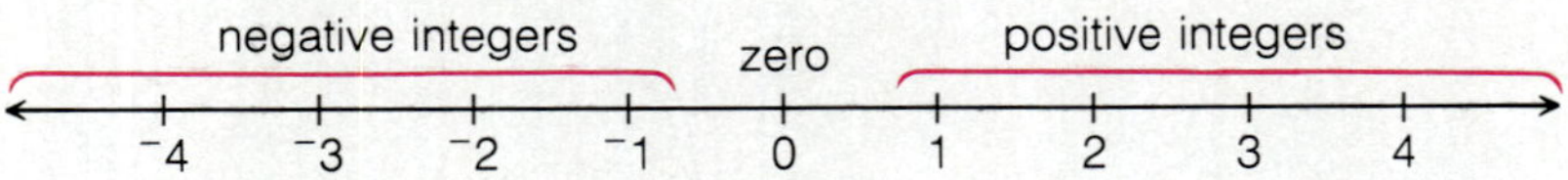

The integer **4** is read as "positive four," or "four." The integer **$^-4$** is read as "negative four." Numbers such as 4 and $^-4$ that are the same distance from 0 on the number line but on opposite sides of it, are called **opposites**. The opposite of 0 is 0. We use a minus sign that is *not* raised to indicate "the opposite of."

The *opposite of 4* is written as -4

The *opposite of* $^-4$ is written as $-(^-4)$.

Example 1

Give the integers for points *A* and *B* on this number line.

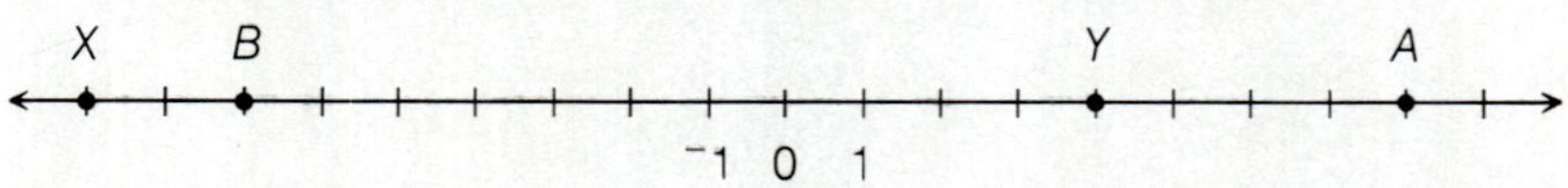

Solution *A*: 8 Point *A* is 8 units *to the right* of 0 on the number line.

B: $^-7$ Point *B* is 7 units *to the left* of 0 on the number line.

Practice Give the integers for points *X* and *Y* on the number line above. **X: $^-9$, Y: 4**

The **absolute value** of an integer is the number of units the integer is from 0 on the number line.

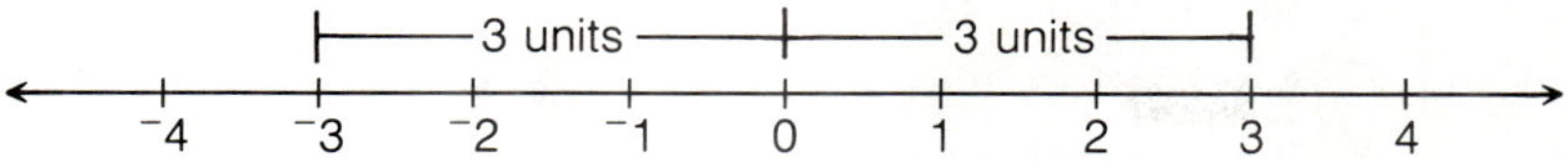

The absolute value of 3 is written $|3|$.
Since 3 is 3 units from 0, $|3| = 3$.
Since $^{-}3$ is 3 units from 0, $|^{-}3| = 3$.

Whether a number is positive or negative, its absolute value is always a positive number.

Example 2

Find the absolute value. $|^{-}7|$

Solution $|^{-}7| = 7$ $^{-}7$ is 7 units from 0 on the number line.

Practice

Find the absolute value. **a.** $|9|$ **9** **b.** $|^{-}85|$ **85**

As with whole numbers, the integer farther to the right on the number line is the greater number. The symbols > (greater than) and < (less than) express inequalities.

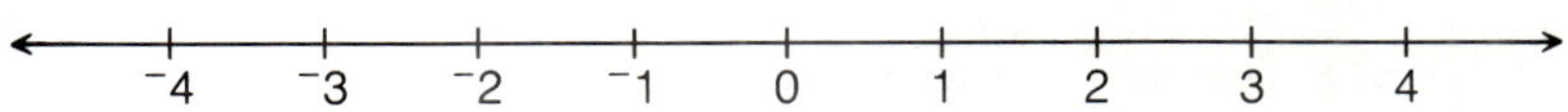

$2 > {}^{-}4$ 2 is **greater than** $^{-}4$.
$^{-}4 < 2$ $^{-}4$ is **less than** 2.

To order a set of integers, you need to compare each integer to the others in turn.

Example 3

Use the symbols < or > to order from least to greatest. $^{-}8, 5, {}^{-}2$

Solution $^{-}8 < {}^{-}2 < 5$ $^{-}8$ is farthest to the left on the number line, $^{-}2$ is next, and 5 is farthest to the right, so $^{-}8 < {}^{-}2$ and $^{-}2 < 5$.

Practice Use the symbols < or > to order from least to greatest.

a. $6, {}^{-}8, {}^{-}16$ **$^{-}16 < {}^{-}8 < 6$**

b. $|^{-}6|, 9, 0, {}^{-}3$ **$^{-}3 < 0 < |^{-}6| < 9$**

Oral Exercises

Give the integer suggested by the situation. Give the opposite of that integer.

1. A gain of \$9 9, −9
2. 900 ft below sea level
3. 4° below zero $^-4$, 4
4. Decrease by 7 kg $^-7$, 7
5. 6 sec before blastoff
6. Lose 4 points $^-4$, 4
7. Gain 7 yards 7, −7
8. 5 strokes under par $^-5$, 5
9. Spent \$50 $^-50$, 50
10. 6 flights up 6, −6
11. Put in 8 liters 8, −8
12. 4 floors down $^-4$, 4

2. $^-900$, 900
5. $^-6$, 6

Exercises

A

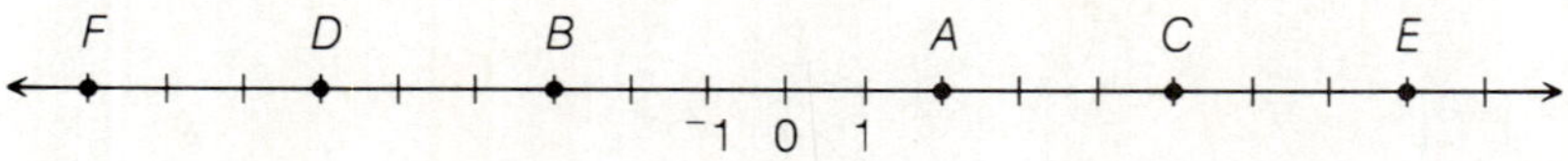

Give the integer for each point on the number line.

1. *A* 2
2. *B* $^-3$
3. *C* 5
4. *D* $^-6$
5. *E* 8
6. *F* $^-9$

Give the absolute value.

7. $|5|$ 5
8. $|^-9|$ 9
9. $|123|$ 123
10. $|^-15|$ 15
11. $|-36|$ 36
12. $|0|$ 0
13. $|199|$ 199
14. $|^-152|$ 152
15. $|^-1|$ 1
16. $|1000|$ 1000

Use the symbols $<$ or $>$ to order from least to greatest.

17. 2, $^-21$ $-21 < 2$
18. $^-7$, 0, 7 $-7 < 0 < 7$
19. $^-1$, $^-6$, 4 $-6 < -1 < 4$
20. $^-5$, $^-3$, 6
21. $^-1$, $^-6$, $^-9$, $^-2$
22. 6, $^-4$, $^-8$
23. $^-23$, 8, $^-24$
24. $^-2$, $|^-3|$, $^-1$, $|4|$
25. $^-13$, 12, $|^-18|$, 6, $^-4$

B For Exercises 26–29, give the integer for the point on the number line.

26. A point 27 units to the right of 0 27
27. A point 124 units to the left of 0 −124
28. A point 20 units to the right of $^-2$ 18
29. A point 16 units to the left of 4 −12
30. What integer is the opposite of $|^-8|$? −8
31. Give the integer represented by $-(^-8)$. 8
32. Find 2 values for x that make the equation $|x| = 5$ true. $x = 5, x = -5$

Complete the following.

33. The opposite of a positive integer is a <u>negative</u> integer.
34. The opposite of a negative integer is a <u>positive</u> integer.
35. The integer <u>0</u> is neither positive nor negative.
36. The absolute value of a nonzero integer is always a <u>positive</u> integer.
37. The opposite of the opposite of a negative integer is a <u>negative</u> integer.

20. $^-5 < ^-3 < 6$
21. $^-9 < ^-6 < ^-2 < ^-1$
22. $^-8 < ^-4 < 6$
23. $^-24 < ^-23 < 8$
24. $^-2 < ^-1 < |^-3| < |4|$
25. $^-13 < ^-4 < 6 < 12 < |^-18|$

C Extending Thinking Skills

Thinking skills: 38–40. Finding a pattern. 41. Using space visualization

Find the next 3 numbers in each pattern by thinking about the number line.

38. 16, 9, 2, __, __, __ **39.** $^-13$, 2, 17, 32, 47, 62 **40.** $^-8$, $^-4$, 0, 4, 8, 12

38. $^-5$, $^-12$, $^-19$

41. If you were to fold a piece of notebook paper in half twice and make a straight cut across the corner with only folded edges, what would the cut-off piece look like when unfolded? a square

Mixed Review

Solve and check. **42.** $\frac{x}{221} = 16$ 3536 **43.** $56 = m - 133$ 189 **44.** $24w = 3144$ 131

Estimate the value of the variable. **45.** $t = \frac{(83 \cdot 28)}{17}$ 120

Simplify. **46.** $3n - n$ 2n **47.** $11(9x)$ 99x **48.** $26a + 7a$ 33a **49.** $\frac{t}{t}$ 1

Evaluate. **50.** $5 \cdot 4 - \frac{36}{9}$ 16 **51.** $\frac{12 + 6}{10 - 7}$ 6 **52.** $5 \cdot 1 \cdot 1$ 5

Basic Property Update

The basic properties for whole number operations are also true for integer operations.

Identity Properties

For every integer **a**,

$a \cdot 1 = 1 \cdot a = a$ $a + 0 = 0 + a = a$

Commutative Properties

For all integers **a** and **b**,

$a + b = b + a$ $a \cdot b = b \cdot a$

Associative Properties

For all integers **a** and **b**,

$(a + b) + c = a + (b + c)$ $(a \cdot b) \cdot c = a \cdot (b \cdot c)$

Distributive Property

For all integers **a, b,** and **c,**

$a(b + c) = ab + ac$

Other properties of integers that follow from the basic properties are:

$a - 0 = a$ $a - a = 0$ $a \cdot 0 = a$

$0 \div a = 0$ $a \div 1 = a$ $a \div a = 1$

3-2 Adding Integers

Objective: To add integers.
MIN: 1–24; REG: 1–24 even, 25–28; MAX: 1–24 even, 25–30; ALL: Mixed Review
An exploratory lesson opener can be found on page 546.

You can use arrows on the number line to find integer sums. Adding a *positive integer* is shown by an arrow to the *right*. Adding a *negative integer* is shown by an arrow to the *left*. The number line below shows how to find the sum $^{-}4 + 6$.

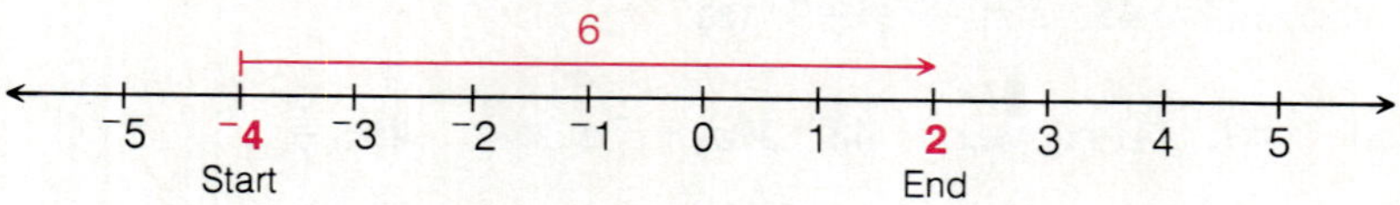

Start at the first number, $^{-}4$. To add positive 6, move to the right 6 units. The ending point, 2, is the sum. $^{-}4 + 6 = 2$.

On the number line, you can see that $4 + {}^{-}4 = 0$.

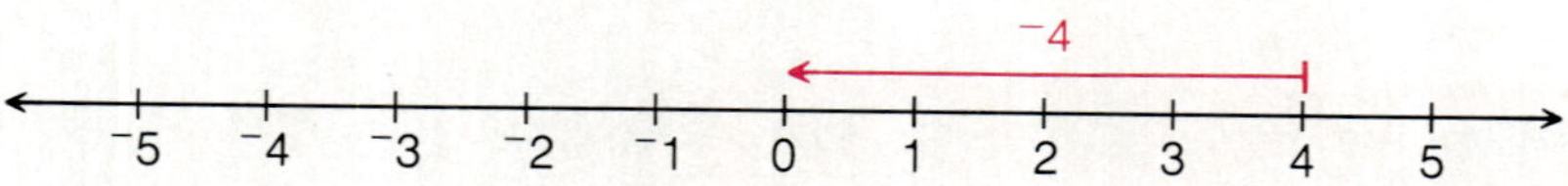

This suggests the following property for adding integers.

Inverse Property of Addition (Opposites Property)

The sum of any integer and its opposite is zero.

For every integer $\boldsymbol{a}$, $\quad \boldsymbol{a + (-a) = 0} \quad$ and $\quad \boldsymbol{-a + a = 0}$

This property and addition on the number line lead to the following rule for finding the sum of any two integers.

Adding Integers

To add integers with like signs,

- Add their absolute values
- Give the sum the sign of the integer addends.

To add integers with unlike signs,

- Find the difference of their absolute values.
- Give the difference the sign of the integer addend with the greater absolute value. If the difference is zero, omit the sign.

Example 1

Find the sum. $^-2 + {}^-3$

Solution $^-2 + {}^-3 = {}^-5$ $|^-2| + |^-3| = 5$. Since both original addends are negative, the sum is negative, $^-5$.

Practice Find the sum.

a. $^-14 + {}^-16$ $^-30$ **b.** $35 + 29$ 64 **c.** $^-34 + {}^-26$ $^-60$

Example 2

Find the sum. $^-6 + 4$

Solution $^-6 + 4 = {}^-2$ $|^-6| - |4| = 2$. Since $|^-6|$ is larger than $|4|$, the sum is negative, $^-2$.

Practice Find the sum.

a. $^-4 + 6$ 2 **b.** $25 + {}^-9$ 16 **c.** $18 + {}^-43$ $^-25$

Example 3

Evaluate. $n + 8$ for $n = {}^-24$

Solution $n + 8$

$^-24 + 8$ Replace n with $^-24$.

$= {}^-16$

Practice Evaluate.

a. $^-9 + x$ for $x = {}^-11$ -20 **b.** $a + b$ for $a = {}^-24$ and $b = 36$ 12

Oral Exercises

Tell whether the sum will be positive, negative, or zero.

1. $9 + 5$ + **2.** $^-6 + {}^-7$ − **3.** $^-8 + {}^-5$ − **4.** $8 + 7$ +

5. $^-9 + {}^-7$ − **6.** $^-7 + 3$ − **7.** $^-6 + 5$ − **8.** $^-7 + 12$ +

9. $5 + {}^-14$ − **10.** $9 + {}^-19$ − **11.** $^-8 + 8$ 0 **12.** $^-4 + 6$ +

Exercises

A Find each sum.

1. $^-5 + (^-7)$ $^-12$ **2.** $9 + 4$ 13 **3.** $^-8 + (^-7)$ -15 **4.** $^-7 + (^-6)$ $^-13$

5. $12 + (^-8)$ 4 **6.** $16 + (^-9)$ 7 **7.** $23 + (^-4)$ 19 **8.** $13 + (^-6)$ 7

9. $^-3 + 11$ 8 **10.** $^-8 + 14$ 6 **11.** $^-5 + 13$ 8 **12.** $^-9 + 25$ 16

13. $^-10 + 4$ -6 **14.** $^-3 + (^-9)$ $^-12$ **15.** $^-6 + 7$ 1 **16.** $26 + 8$ 34

17. $3 + (^-12)$ -9 **18.** $17 + (^-8)$ 9 **19.** $^-24 + 0$ -24 **20.** $^-26 + (^-23)$ $^-49$

Evaluate each expression.

21. $n + (^{-}5)$ for $n = {}^{-}7$ $^{-}12$

22. $^{-}13 + x$ for $x = 35$ 22

23. $r + (^{-}45)$ for $r = 71$ 26

24. $x + (^{-}24)$ for $x = 16$ −8

B Find each sum.

25. $25 + |^{-}12 + {}^{-}15|$ 52

26. $(43 + {}^{-}29) + (^{-}43 + 29)$ 0

27. On the first down, Jerry's football team gained 6 yards. On the second down it lost 14 yards. On the third down it gained 16 yards. If it lost 13 yards on the fourth down, what was its total gain or loss? −5

28. The temperature was recorded at 15° Celsius. It then rose 8°, fell 6°, rose 12°, and fell 23°. What was the temperature after these changes? 6°C

C *Extending Thinking Skills*

Thinking skills: 29. Computing mentally
30. Formulating a proof

29. Add mentally. $^{-}57 + 46 + 178 + (^{-}45) + 199 + 58 + (^{-}178)$ 201

30. This proof uses basic properties to show that $^{-}5 + 8 = 3$.

$^{-}5 + 8$	
$= {}^{-}5 + (5 + 3)$	Substitute 5 + 3 for 8.
$= (^{-}5 + 5) + 3$	Associative Property.
$= 0 + 3$	Additive Inverse Property.
$= 3$	Zero Property.

Write a proof for $9 + {}^{-}4 = 5$.

$9 + {}^{-}4 = (5 + 4) + {}^{-}4$ Substitute 5 + 4 for 9
$= 5 + (4 + {}^{-}4)$ Associative Prop.
$= 5 + 0$ Additive Inverse Prop.
$= 5$ Zero Prop.

Mixed Review

Solve mentally and check. **31.** $4m = 24$ 6 **32.** $\frac{c}{5} = 7$ 35 **33.** $n - 2 = 18$ 20

Order from least to greatest. Use inequality symbols. **34.** $^{-}6, 4, 0, 1, {}^{-}1$
$^{-}6 < {}^{-}1 < 0 < 1 < 4$

NUMBERS TO ALGEBRA

In algebra, $-n$, which means "the opposite of n," is negative when n is replaced with a positive integer, but it is positive when n is replaced with a negative integer.

$-n$	Replace n with 5	$-n$	Replace n with $^{-}5$
-5		$-{}^{-}5$	
$^{-}5$	The result is a negative integer.	5	The result is a positive integer.

Complete each statement.

1. If $n = 8$, then $-n =$ <u>−8</u>.

2. If $n = {}^{-}9$, then $-n =$ <u>9</u>.

3. If $n = {}^{-}24$, then $-n =$ <u>24</u>.

4. If $n = 45$, then $-n =$ <u>−45</u>.

3-3 Subtracting Integers

Objective: To subtract integers.
MIN: 1–39, 51–52; REG, MAX: 1–39 even, 40–52; ALL: Mixed Review

These number lines show that subtracting an integer and adding the opposite of the integer produce the same result.

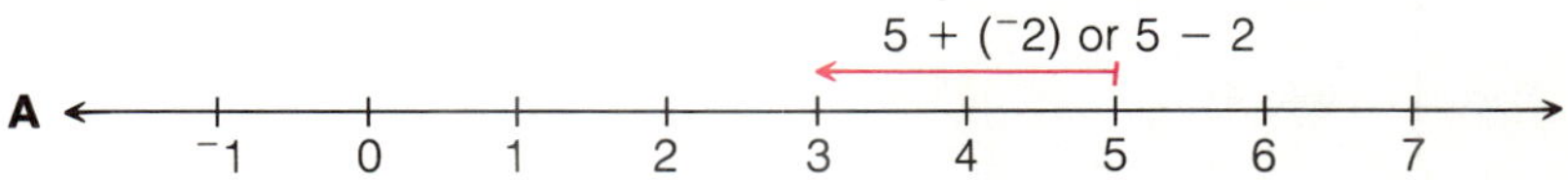

Adding $^-2$ is the same as subtracting 2.

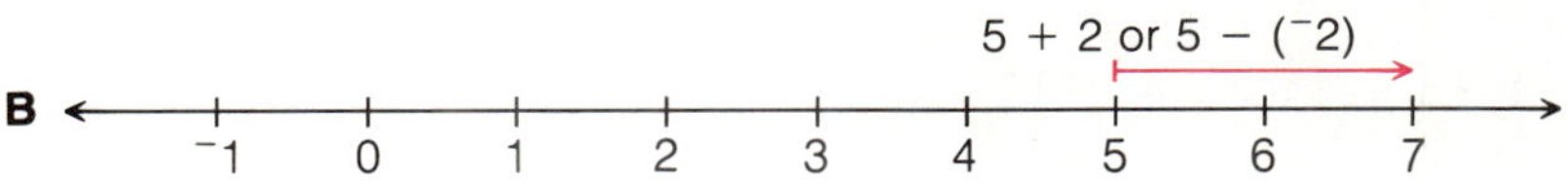

Adding 2 is the same as subtracting $^-2$.

Note that the arrow for $5 - (^-2)$ should go in the opposite direction of the arrow for $5 + (^-2)$.

This suggests the following rule.

> ***Subtracting Integers***
>
> To subtract an integer, add its opposite.
>
> For all integers **a** and ***b***, $\quad \boldsymbol{a - b = a + (-b)}$

You learned earlier in this chapter that -5 represents "the opposite of five" and that $^-5$ (with raised sign) represents the integer "negative five." Since -5 and $^-5$ name the same number, either can be used to represent that integer. Beginning with the following examples, a minus sign that is not raised will always be used to represent a negative integer.

Example 1

Subtract. $\quad -5 - 8$

Solution $\quad -5 - 8$

$= -5 + (-8)$ Subtracting 8 is the same as adding -8.

$= -13$

Practice Subtract. **a.** $2 - 7$ **−5** **b.** $-4 - 6$ **−10** **c.** $-8 - 3$ **−11** **d.** $15 - 9$ **6**

Example 2

Subtract. $7 - (-4)$

Solution $7 - (-4)$

$= 7 + 4$ Subtracting -4 is the same as adding 4.

$= 11$

Practice Subtract. **a.** $9 - (-3)$ 12 **b.** $4 - (-10)$ 14 **c.** $-9 - (-2)$ −7

Example 3

Evaluate. $-8 - n$ for $n = -5$

Solution $-8 - (-5)$ Replace n with -5.

$= -8 + 5$

$= -3$

Practice Evaluate. **a.** $x - (-7)$ for $x = -13$ −6 **b.** $-9 - b$ for $b = -15$ 6

Oral Exercises

Give an addition expression that has the same answer as the subtraction expression.

1. $8 - 4$ $8 + (-4)$ **2.** $3 - 9$ $3 + (-9)$ **3.** $-2 - 7$ $-2 + (-7)$ **4.** $-11 - 6$ $-11 + (-6)$

5. $-8 - 8$ $-8 + (-8)$ **6.** $-5 - 6$ $-5 + (-6)$ **7.** $3 - 6$ $3 + (-6)$ **8.** $-4 - 7$ $-4 + (-7)$

Exercises

A Subtract.

1. $9 - 3$ 6 **2.** $8 - 12$ −4 **3.** $4 - 10$ −6 **4.** $9 - 13$ −4

5. $1 - 7$ −6 **6.** $-3 - 11$ −14 **7.** $-6 - 8$ −14 **8.** $-12 - 5$ −17

9. $-8 - 14$ −22 **10.** $-20 - 6$ −26 **11.** $4 - (-9)$ 13 **12.** $6 - (-15)$ 21

13. $17 - (-8)$ 25 **14.** $24 - (-10)$ 34 **15.** $7 - (-7)$ 14 **16.** $-3 - (-12)$ 9

17. $-8 - (-16)$ 8 **18.** $-14 - (-7)$ −7 **19.** $-18 - (-12)$ −6 **20.** $-50 - (-30)$ −20

21. $7 - 15$ −8 **22.** $-8 - 11$ −19 **23.** $6 - (-8)$ 14 **24.** $-9 - (-13)$ 4

25. $45 - 94$ −49 **26.** $-34 - (-58)$ 24 **27.** $57 - (-24)$ 81 **28.** $-75 - 48$ −123

Evaluate.

29. $n - 9$ for $n = -6$ −15 **30.** $-7 - n$ for $n = 14$ −21 **31.** $12 + n$ for $n = -18$ −6

32. $n + (-6)$ for $n = 11$ 5 **33.** $n - (-11)$ for $n = 5$ 16 **34.** $n - (-13)$ for $n = 9$ 22

35. $-7 + n$ for $n -6$ −13 **36.** $-16 - n$ for $n = -10$ −6 **37.** $-10 + n$ for $n = 12$ 2

38. $n - (-4)$ for $n = -12$, for $n = 16$, and for $n = -53$ −8, 20, −49

39. $6 - r$ for $r = 23$, $r = -9$, and $r = -87$. −17, 15, 93

B Simplify.

40. $5 - (-6) - 20$ −9

41. $-8 - (-2) - (-7)$ 1

42. $-13 - (-7) + (-5)$ −11

43. $(-5 - 3) + (-9) - (-15)$ −2

44. $-6 + (-9 - 12)$ −27

45. $7 - (-8 - 13) + (-5)$ 23

46. Evaluate the formula P = I − E to find the Profit (P) when Income (I) = \$85,654 and Expenses (E) = \$92,472. P = −\$6818

47. Replace a and b with integers to show that $a - b = b - a$ is not true for all integers. 3 − 2 = 1, but 2 − 3 = −1.

48. Replace a, b, and c with integers to show that $(a - b) - c = a - (b - c)$ is not true for all integers. (1 − 2) − 3 = −4, but 1 − (2 − 3) = 2.

49. Death Valley is 282 ft below sea level. Mt. Whitney is 14,494 ft above sea level. How much higher is the elevation of Mt. Whitney than the elevation of Death Valley? 14,776 ft

50. One day the nation's high temperature was 75°F. The low temperature was −8°F. What is the difference between these temperatures? 83°F

C Extending Thinking Skills

51. In a magic square, the sum of the numbers in each row, column, and diagonal is the same. Find the missing numbers in this magic square, assuming the magic sum is −2.

4	−6	−1	1
−7	7	2	−4
3	−3	−8	6
−2	0	5	−5

52. An elevator went up 6 floors, down 9 floors, down 12 more floors, up 8 floors, and down 4 floors. It stopped on the 43rd floor. On what floor did it start? The 54th floor

Thinking skills: 51. Reasoning 52. Using reverse reasoning

Mixed Review

Solve and check. **53.** $8y = 216$ 27 **54.** $t + 163 = 204$ 41

Find the sum. **55.** $-5 + 3$ −2 **56.** $5 + 3$ 8 **57.** $5 + (-3)$ 2

58. $-5 + (-3)$ −8

CALCULATOR ACTIVITY

The change-sign key [+/−] allows you to show a negative integer by changing the displayed number to its opposite.

To add 86 + (−123): 86 [+] 123 [+/−] [=]

To subtract −67 − (−239): 67 [+/−] [−] 239 [+/−] [=]

Add or subtract, using the change-sign key.

1. $258 + (-689)$ −431

2. $-9674 + 6739$ −2935

3. $-783 - (-975)$ 192

4. $-7386 + 2978$ −4408

5. $4547 - (-8964)$ 13,511

6. $-435 + 848$ 413

More Practice

Find the sum.

1. $13 + 9$ 22
2. $28 + (-9)$ 19
3. $-76 + (-84)$ −160
4. $-37 + 19$ −18
5. $64 + (-9)$ 55
6. $21 + 107$ 128
7. $-6 + (-37)$ −43
8. $-26 + 13$ −13
9. $0 + (-11)$ −11
10. $64 + 136$ 200
11. $-57 + 84$ 27
12. $111 + 99$ 210
13. $-6 + (-6)$ −12
14. $6 + (-6)$ 0
15. $-36 + 271$ 235
16. $-28 + (-57)$ −85
17. $-14 + 7$ −7
18. $-21 + (-108)$ −129
19. $147 + 239$ 386
20. $65 + (-19)$ 46

Find the difference.

21. $-5 - 3$ −8
22. $15 - 3$ 12
23. $21 - 38$ −17
24. $-6 - (-31)$ 25
25. $19 - (-21)$ 40
26. $-49 - 51$ −100
27. $-5 - (-3)$ −2
28. $26 - 57$ −31
29. $-12 - 4$ −16
30. $43 - 62$ −19
31. $-9 - (-6)$ −3
32. $25 - (-14)$ 39
33. $180 - 97$ 83
34. $-34 - (-37)$ 3
35. $74 - 105$ −31
36. $-34 - 62$ −96
37. $62 - 87$ −25
38. $91 - (-45)$ 136
39. $-81 - 14$ −95
40. $504 - 367$ 137

Find the sum or difference.

41. $67 - 71$ −4
42. $46 + (-13)$ 33
43. $-19 + 27$ 8
44. $-41 - 16$ −57
45. $-41 + 16$ −25
46. $219 - 167$ 52
47. $50 + (-61)$ −11
48. $29 - 40$ −11
49. $-35 - (-17)$ −18
50. $-64 + 12$ −52
51. $-24 - 12$ −36
52. $-240 - (-13)$ −227
53. $42 - 384$ −342
54. $-27 - (-16)$ −11
55. $-30 - (-42)$ 12
56. $120 + (-49)$ 71
57. $27 + (-62)$ −35
58. $51 - 17$ 34
59. $6 - 31$ −25
60. $-32 + 19$ −13
61. $106 - 112$ −6
62. $102 + (-31)$ 71
63. $19 - (-32)$ 51
64. $19 + (-40)$ −21
65. $-6 + 30$ 24
66. $28 - 35$ −7
67. $-9 + (-11)$ −20
68. $64 - (-16)$ 80
69. $37 - (-42)$ 79
70. $-54 + 17$ −37
71. $-116 + 24$ −92
72. $46 - 59$ −13
73. $-23 + (-66)$ −89
74. $15 - 102$ −87
75. $83 - 117$ −34
76. $-30 + (-47)$ −77
77. $27 - 16$ 11
78. $-47 - 14$ −61
79. $92 + (-18)$ 74
80. $143 + (-51)$ 92
81. $-27 - 6$ −33
82. $-16 + 39$ 23
83. $64 + (-128)$ −64
84. $138 - (-12)$ 150
85. $-30 + 19$ −11
86. $240 - 192$ 48
87. $457 + 393$ 850
88. $19 - 30$ −11
89. $-26 - 17$ −43
90. $-6 + 40$ 34
91. $-319 - (-67)$ −252
92. $-126 + (-12)$ −138
93. $104 + (-35)$ 69
94. $48 - (-52)$ 100
95. $-16 - 12$ −28
96. $-41 + 93$ 52

3-4 Solving Integer Equations: Addition and Subtraction

Objective: To use addition and subtraction to solve equations involving integers.
MIN: 1–30, 37–38; REG, MAX: 1–30 even, 31–37; ALL: Mixed Review

To solve integer equations such as $t + 34 = 28$ or $x - 18 = -35$, you need to get the variable by itself on one side of the equation. The following steps show you how to use the ideas of inverse operations or the additive inverse property and the properties of equality to do this.

If the temperature increased 34°C to reach a high of 28°C, you could solve the equation $t + 34 = 28$ to find the original temperature.

Solving Equations Using Addition and Subtraction

1. Decide which operation (addition or subtraction) has been applied to the variable.

2. Use the inverse operation or use the additive inverse property, adding or subtracting the same number on both sides of the equation.

Examples 1 and 2 show how to use inverse operations and the same steps you used with whole number equations to solve integer equations. Example 3 shows how to use the additive inverse property to solve integer equations.

Example 1

Solve and check. $t + 34 = 28$

Solution $t + 34 = 28$ — You need to get the variable by itself on one side.

$t + 34 - 34 = 28 - 34$ — To undo adding 34, subtract 34 from both sides of the equation so they remain equal. (Subtraction property of equality.)

$t = -6$

Check $-6 + 34 \stackrel{?}{=} 28$ — Replace t with -6 in $t + 34 = 28$.

$28 = 28$ ✓ — The solution is -6.

Practice Solve and check.

a. $x + 19 = -47$ **−66** **b.** $-42 = y + 76$ **−118** **c.** $z + (-35) = (-164)$ **−129**

Example 2

Solve and check. $x - 18 = -35$

Solution $x - 18 = -35$

$x - 18 + 18 = -35 + 18$ To undo subtracting 18, add 18 to both sides so they remain equal. (Addition property of equality.)

$x = -17$

Check $-17 - 18 \stackrel{?}{=} -35$ Replace x with -17 in $x - 18 = -35$.

$-35 = -35$ ✓ The solution is -17.

Practice Solve and check.

a. $b - (-15) = 43$ 28 **b.** $n - 29 = -45$ −16 **c.** $y - (-37) = -64$ −101

Since subtracting an integer is the same as adding the opposite integer, an equation that can be solved by subtracting can also be solved by using the additive inverse property.

Example 3

Solve and check. $n + (-25) = 46$

Solution $n + (-25) = 46$

$n + (-25) + 25 = 46 + 25$ To undo adding -25, use the additive inverse property and add the opposite of -25, or 25. $-25 + 25 = 0$.

$n = 71$

Check $71 + (-25) \stackrel{?}{=} 46$ Replace n with 71 in $n + (-25) = 46$.

$46 = 46$ ✓ The solution is 71.

Practice Solve and check. **a.** $a + (-17) = -75$ −58 **b.** $43 = p + (-26)$ 69

Oral Exercises

To solve, what integer would you add to or subtract from each side?

1. $x + 19 = -48$ **2.** $n - 28 = -56$ **3.** $-379 = x + 85$ **4.** $78 = b - (-45)$

5. $c - 56 = -198$ **6.** $b + (-37) = 84$ **7.** $z + 53 = -53$ **8.** $0 = n - 67$

1. subtract 19, add −19 2. add 28, subtract −28 3. subtract 85, add −85 4. add −45, subtract 45
5. add 56, subtract −56 6. add 37, subtract −37 7. add −53, subtract 53 8. add 67, subtract −67

Exercises

A Solve and check.

1. $n + 27 = -84$ −111 **2.** $x + 97 = 42$ −55 **3.** $a + 157 = -96$ −253

4. $y + 79 = -32$ −111 **5.** $c + 76 = 154$ 78 **6.** $n + 67 = -282$ −349

7. $x + 279 = 194$ −85 **8.** $b + 68 = -253$ −321 **9.** $y + (-36) = 273$ 309

10. $a + (-285) = -643$ −358 **11.** $c + 947 = -765$ −1712 **12.** $x + (-258) = 1752$ 2010

13. $x - 47 = -63$ −16 **14.** $y - 86 = 175$ 261 **15.** $x - 86 = 59$ 145

16. $y - 86 = -254$ −168 **17.** $c - 47 = -195$ −148 **18.** $n - 58 = -127$ −69

19. $c - (-56) = 498$ 442 **20.** $b - (-76) = 485$ 409 **21.** $n - 226 = -46$ 180

22. $y - (-75) = 60$ −15 **23.** $x - 158 = 42$ 200 **24.** $c - 63 = -441$ −378

25. $-278 = n + (-69)$ **26.** $-263 = n - 45$ −218 **27.** $-274 = a + 37$ −311

28. $142 = y - (-349)$ **29.** $x + 367 = 294$ −73 **30.** $-217 = y - 87$ −130

B Solve and check. 25. −209 28. −207 31. 581 32. 23 33. −121

31. $x + (-74 + 39) = 546$ **32.** $a - (-72 + 78) = 17$ **33.** $(-49 + 86) + y = -84$

34. Write and solve an equation with sum −75, one addend −16, and the other addend represented by the variable n. $n + (-16) = -75; n = -59$

35. Write and solve an equation in which 78 is subtracted from a number represented by the variable y to produce the difference −29. $y - 78 = -29; y = 49$

C Extending Thinking Skills

36. When $n + n + n + n$ is evaluated, the result is 9 less than n. Find n. $n = -3$

37. If 5 cards are moved from stack A to stack B, both stacks will contain 12 cards. How many cards were in each stack to begin with? A = 17 B = 7

Thinking skills: 36. Testing a conjecture 37. Reasoning/Testing a conjecture

Mixed Review

Add or subtract. **38.** $7 - (-2)$ 9 **39.** $-9 - (-2)$ −7 **40.** $-4 + (-1)$ −5

Write as an algebraic expression. **41.** twice the sum of 7 and a number $2(7 + n)$

MENTAL MATH

You can find the sum of any number of positive integers in order, starting with 1, by multiplying the last number by the number that follows it and dividing by 2. Check the equations below to see whether they are true. Fill in the blank to complete the generalization.

$$1 + 2 = 2(3) \div 2$$
$$1 + 2 + 3 = 3(4) \div 2$$
$$1 + 2 + 3 + 4 = 4(5) \div 2$$
$$1 + 2 + 3 + 4 + 5 = 5(6) \div 2$$
$$1 + 2 + 3 + 4 + 5 + \ldots + n = \underline{\quad ? \quad}$$

$n(n + 1) \div 2$

Find each sum mentally.

1. $1 + 2 + 3 + 4 + 5 + 6 + 7$ 28 **2.** $1 + 2 + 3 + 4 + 5 + 6 + 7 + 8 + 9$ 45

3. The positive integers from 1 through 20. 210

4. The positive integers from 1 through 99. 4950

PROBLEM SOLVING: APPLICATIONS

3-5 Keeping Checkbook Records

Objective: To solve problems involving checkbook records. ALL: 1–9

A **checking account** at a bank or savings and loan company allows a customer to write checks to pay bills or get needed cash. To keep money in the account, the customer makes deposits.

A deposit of money to an account is called a **credit** (+). A payment (check) or withdrawal is called a **debit** (−). The **balance** is the amount of money in the account at a given point. The balance changes after each withdrawal, payment, or deposit. If an account's debits are greater than its credits, the balance that remains is less than zero and the account has been **overdrawn**. The customer must keep an accurate record of all account **transactions**, or activities.

NUMBER	DATE	DESCRIPTION OF TRANSACTION	PAYMENT/ DEBIT (−)		✓ T	(−) FEE (IF ANY)	DEPOSIT/ CREDIT (+)		BALANCE	
									$ 125	00
126	2/9	Terry's Service Station	$ 79	00						
	2/9	Deposit from Salary					215	00		
127	2/11	Record Land	48	00						
128	2/14	Cancer Research Donation	125	00						
	2/15	Transfer from Savings					575	00		
129	2/15	Food Mart	56	00						
130	2/17	Snapit Camera Shop	279	00						

Problems

Solve problems 1–4, using the transaction record above. For problems 5–8, decide whether to use **pencil and paper, mental math, estimation,** or a **calculator** to find the answer. Use each of these techniques at least once.

1. The amount of money in the account before check number 126 is shown in the balance column. What is the balance after check number 126 is written? $46

2. What is the new balance following the deposit on 2/9? $261

3. What is the new balance following: check number 128? $88 the deposit on 2/15? $663

4. What is the new balance following: check number 129? $607 check number 130? Check your answers by writing credits as positive integers and debits as negative integers and finding their sum. $328

5. A customer wrote a check for $79, wrote a check for $156, made a deposit of $275, and wrote a check for $398. If the balance before these transactions was $517, what was the balance after the transactions? $159

6. Decide whether or not this checking account has been overdrawn: beginning balance: $1353; check for $798; check for $206; deposit of $449; check for $747. Not overdrawn. Balance is $51.

7. A checking account with a beginning balance of $150 had debits of $1250 and credits of $900. By how much was it overdrawn? Overdrawn by $200

8. A checking account had a beginning balance of $435. Deposits were made in the amount of $1196. Checks were written in the amount of $1952. Three of these checks were written when the account was overdrawn. The bank fee for overdrawn checks is $12 per check. How much must be deposited to pay the fee and bring the balance up to $400? $757

9. **Data Search** Find a record form for a checking account. Make up a beginning balance, then 3 checks and 2 deposits. Fill out the record form for the transactions and find the ending balance. Answers will vary. Photocopies of checking account record forms could be used.

What's Your Decision?

At the National Bank you can choose between the two types of non-interest checking accounts shown below.

Regular Account		Zero Balance Account
Account Balance	**Fee for Writing Checks**	
Below $100	$6/month	Annual fee of $24, plus $0.20 for each check.
$100–$200	$5/month	
$200–$300	$4/month	
$300–$400	$3/month	
Over $400	No charge	

You want to open a checking account, and expect to write no more than 10 checks per month. You usually keep a little over $100 in your account. Would you choose a Regular Account or a Zero Balance Account? Why? Allow students to change conditions or assumptions if you wish. Answers will vary.

3-6 Multiplying Integers

Objective: To multiply integers.
MIN: 1–28, 49; REG: 1–49 even; MAX: 1–50 even; ALL: Mixed Review
An exploratory lesson opener can be found on page 547.

You can use repeated addition to show that the product of a positive integer and a negative integer is a negative integer.

$$3(-4) = (-4) + (-4) + (-4) = -12$$

The following statement uses the fact that 3 is the opposite of -3 to suggest that the product of two negative integers is a positive integer:

$$\text{Since } 3(-4) = \mathbf{-12}, \text{ then } -3(-4) = \mathbf{12}$$

The equation pattern below, in which the product increases by 4 each time, also suggests that $-3(-4) = 12$.

$3(-4) = -12$	**You can use basic properties to prove this.**
$2(-4) = -8$	
$1(-4) = -4$	**$-3(-4) + 3(-4) = (-3 + 3)(-4)$ (distributive property)**
$0(-4) = 0$	**$= 0(-4)$ (additive inverse property)**
$-1(-4) = 4$	**$= 0$ (zero property of multiplication)**
$-2(-4) = 8$	**Since their sum is 0, $-3(-4)$ and $3(-4)$ are opposites.**
$-3(-4) = 12$	**So, since $3(-4)$ is -12, $-3(-4)$ must be 12.**

These ideas support a procedure for multiplying integers.

Multiplying Integers

Find the product of the absolute values. Then use the following rules for determining the sign.

- The product of two positive integers is positive.
- The product of two negative integers is positive.
- The product of a positive integer and a negative integer is negative.

Example 1

Find the product. $-6(-4)$

Solution $-6(-4) = 24$ The product of two negative integers is a positive integer.

Practice Find the product. **a.** $9(6)$ **54** **b.** $-7(-8)$ **56** **c.** $-9(-5)$ **45**

Example 2

Find the product. $-8(5)$

Solution $-8(5) = -40$ A positive times a negative integer is a negative integer.

Practice Find the product. **a.** $-5(4)$ −20 **b.** $6(-8)$ −48 **c.** $9(-4)$ −36 **d.** $-7(3)$ −21

Oral Exercises

Tell whether the product is positive or negative.

1. $6(3)$ + **2.** $-7(4)$ − **3.** $-6(-9)$ + **4.** $5(-9)$ −

5. $-8(-3)$ + **6.** $-2(-17)$ + **7.** $-12(15)$ − **8.** $24(-45)$ −

9. $-19(-78)$ + **10.** $28(-14)$ − **11.** $10(-6)$ − **12.** $8(8)$ +

Exercises

A Find the product.

1. $9(7)$ 63 **2.** $-5(-3)$ 15 **3.** $-4(-8)$ 32 **4.** $8(6)$ 48

5. $-9(-4)$ 36 **6.** $-6(-8)$ 48 **7.** $-7(-2)$ 14 **8.** $-3(-8)$ 24

9. $7(8)$ 56 **10.** $-6(-7)$ 42 **11.** $9(-5)$ −45 **12.** $-8(7)$ −56

13. $-7(0)$ 0 **14.** $-1(8)$ −8 **15.** $-9(9)$ −81 **16.** $6(-5)$ −30

17. $-3(-6)$ 18 **18.** $-7(7)$ −49 **19.** $8(-6)$ −48 **20.** $-8(-9)$ 72

21. $-10(-46)$ 460 **22.** $-12(35)$ −420 **23.** $-24(-52)$ 1248 **24.** $54(-23)$ −1242

25. $-76(-98)$ 7448 **26.** $86(-45)$ −3870 **27.** $124(-36)$ −4464 **28.** $-246(-35)$ 8610

B

29. Find the product: $2(-3)(-4)(-5)$. Is the number of negative factors odd or even? Is the product positive or negative? −120; odd; negative

30. Find the product: $(-5)(-6)(-2)(-1)$. Is the number of negative factors odd or even? Is the product positive or negative? 60; even; positive

31. Consider your answers to exercises 29 and 30, try other examples as needed, and complete the following:

a. The product of an odd number of negative factors is __?__ (positive, negative). negative

b. The product of an even number of negative factors is __?__ (positive, negative). positive

Find the product.

32. $-5(-4)(6)$ 120 **33.** $7(-4)(6)$ −168

34. $-36(-25)(-4)(-2)$ 7200 **35.** $-36(-25)(-4)(3)$ −10800

36. $4(-6)(-5)(-23)(-1)$ 2760 **37.** $7(-6)(-5)(-3)(-1)(-8)$ −5040

Evaluate each expression.

38. $-3n$ for $n = 47$ -141

39. $15(z + 6)$ for $z = -23$ -255

40. $-12pq$ for $p = 32$ and $q = -9$ 3456

41. $4a - 3b$ for $a = -5$, $b = 7$ -41

42. Show that $2b + 3b$ has the same value as $5b$ when $b = 4$, $b = -8$, and $b = -34$. What property does this show? **See additional answers.**

43. Show that $-9(a + 25)$ has the same value as $-9a + (-9)(25)$ when $a = -9$, $a = 12$, and $a = -24$. **See additional answers.**

44. The formula $s = -0.4y + 1020$ has been used to predict the record time (s, in seconds) for the mile run in a chosen year (y). Evaluate the formula for $y = 1990$. **$s = 224$ seconds**

45. The air temperature decreases 7°C for each kilometer of increase in altitude. If the temperature was 0°C at a certain altitude, what would the temperature be outside an airplane flying 4 km above this altitude? **−28°C**

Use the change-sign key to find each product. Check your answer with an estimate.

Example: $-578(-637)$ 578 [+/-] [×] 637 [+/-] [=]

46. $-956(498)$ **−476,088**

47. $-853(-32)(-46)$ **−1,255,616**

48. $535(-726)(-43)$ **16,701,630**

C Extending Thinking Skills

49. Find the pattern for each sequence. Give the next 3 integers in the sequence.

a. 3, −6, 6, −12, 9, −18, __, __, __.

b. 6, 2, −2, −6, __, __, __.

a. 12, −24, 15
b. −10, −14, −18

50. Mindy tripled the temperature reading on a cold day and the result was 6 less than the original temperature. How could this be possible? What was the original temperature? **−3°**

Thinking skills 49. Finding a pattern 50. Testing a conjecture

Mixed Review

Solve and check. **51.** $r + 23 = -36$ -59 **52.** $m + 19 = 12$ -7 **53.** $25t = 1075$ 43

Estimate the value of the variable. **54.** $503 \div 9 = a$ 50

MENTAL MATH

Because of the commutative and associative properties, you can choose any pair from among 3 factors a, b, and c to multiply first. That is, $(ab)c = (bc)a = (ac)b$. Choosing **compatible numbers** makes it easier to find products mentally.

Choose compatible numbers to find each product mentally.

1. 4(687)(250) **687,000**

2. 68(50)(20) **68,000**

3. 20(579)(−5) **−57,900**

4. −674(385)(0) **0**

5. 96(4)(25) **9,600**

6. 250(127)(40) **1,270,000**

3-7 Dividing Integers

Objective: To find integer quotients.
MIN: 1–24, 39–40; REG: 1–42 even, 43–44; MAX: 1–42 even, 43–45; ALL: Mixed Review

The relationship between multiplication and division can be used to find the quotient of two integers.

factor		factor		product		product		factor		factor
4	·	**6**	=	**24**	→	**24**	÷	**6**	=	**4**
4	·	**(−6)**	=	**−24**	→	**−24**	÷	**−6**	=	**4**
−4	·	**6**	=	**−24**	→	**−24**	÷	**6**	=	**−4**
−4	·	**(−6)**	=	**24**	→	**24**	÷	**−6**	=	**−4**

The equations above suggest the following procedure.

Dividing Integers

To divide integers, find the quotient of their absolute values. Then use the following rule for determining the sign:

- The quotient of two positive integers is positive.
- The quotient of two negative integers is positive.
- The quotient of a positive and a negative integer is negative.

Example 1

Find the quotient. $-35 \div (-7)$

Solution $-35 \div (-7) = 5$ The quotient of two negative integers is positive. This checks, since $5(-7) = -35$.

Practice Find the quotient. **a.** $30 \div 6$ 5 **b.** $-45 \div (-9)$ 5 **c.** $\frac{-72}{-8}$ 9

Example 2

Find the quotient. $-32 \div 4$

Solution $-32 \div 4 = -8$ The quotient of two integers with unlike signs is negative. This checks, since $-8(4) = -32$.

Practice Find the quotient. **a.** $-48 \div 8$ −6 **b.** $\frac{20}{-4}$ −5 **c.** $40 \div (-5)$ −8

Oral Exercises

Check each quotient by multiplying. Is it correct?

1. $35 \div (-5) = -7$ yes **2.** $-18 \div 6 = 3$ no **3.** $24 \div 8 = 3$ yes

4. $-42 \div (-7) = 6$ yes **5.** $48 \div (-6) = -8$ yes **6.** $0 \div (-12) = 0$ yes

Exercises

A Find the quotient.

1. $56 \div 8$ 7 **2.** $-24 \div (-8)$ 3 **3.** $-20 \div (-4)$ 5 **4.** $81 \div 9$ 9

5. $-36 \div 4$ −9 **6.** $42 \div (-7)$ −6 **7.** $-64 \div 8$ −8 **8.** $72 \div (-9)$ −8

9. $45 \div (-5)$ −9 **10.** $36 \div 6$ 6 **11.** $-48 \div (-8)$ 6 **12.** $-90 \div 10$ −9

13. $9 \div (-1)$ −9 **14.** $0 \div (-12)$ 0 **15.** $-66 \div (-11)$ 6 **16.** $30 \div (-5)$ −6

17. $-54 \div 3$ −18 **18.** $-60 \div (-5)$ 12 **19.** $52 \div (-4)$ −13 **20.** $-51 \div (-3)$ 17

21. $\frac{125}{-5}$ −25 **22.** $\frac{-168}{7}$ −24 **23.** $\frac{-96}{12}$ −8 **24.** $\frac{-75}{-15}$ 5

B Simplify.

25. $(-6)(4) \div (-8)$ 3 **26.** $-40 \div (-8) - 9$ −4 **27.** $-56 \div (2 - 10)$ 7

28. $-18 \div (-3 + 9)$ −3 **29.** $\frac{(-4)(-4)}{2}$ 8 **30.** $\frac{-8 - 19}{-6 - 3}$ 3

Evaluate.

31. $n \div (-8)$ for $n = 48$ −6 **32.** $(-9 + a) \div (-6)$ for $a = 63$ −9

33. Does $a \div (b + c)$ have the same value as $a \div b + a \div c$ when $a = -48$, $b = 12$, and $c = -4$? no

34. When an integer is divided by 9, the quotient is −6. What is the integer? −54

35. The net change in the price of a stock over a three-month period was −27. What was the average change per month? −9

Find the quotient. Check by estimating.

36. $5{,}239{,}374 \div (-6958)$ −753 **37.** $\frac{-1024(-75)}{-32}$ −2400 **38.** $\frac{-94 + 1886}{-108 + 164}$ 32

C *Extending Thinking Skills*

39. Find integers a and b for which $a + b = 6$ and $a \div b = -4$. $a = 8$, $b = -2$

40. Half an integer is five more than the integer. What is the integer? −10

41. The temperature rose at the rate of 3°F per hour to reach 0°F at noon. At what earlier time had the temperature been −21°F? 5 a.m.

Thinking skills: 39, 40. Testing a conjecture 41. Using reverse reasoning

Mixed Review

Solve and check. **42.** $y - 63 = 14$ 77 **43.** $w - (-4) = 61$ 57 **44.** $24r = 384$ 16

Simplify. **45.** $12(9c)$ 108c **46.** $112w + 13w$ 125w **47.** $(9 + t) + 4$ 13 + t

3-8 Solving Integer Equations: Multiplication and Division

Objective: To use multiplication and division to solve equations involving integers.
MIN: 1–28; REG, MAX: 1–28 even, 29–38; ALL: Mixed Review

To solve equations such as $7d = -91$ or $\frac{n}{-9} = 78$, you can use the same steps you used with whole number equations.

- Decide what operation (multiplication or division) has been applied to the variable.
- Use the inverse operation, multiplying or dividing both sides of the equation by the same number.

If quick-freezing for 7 minutes lowers the temperature 91°C, you can solve the equation 7d = −91 to find the temperature change per minute.

Example 1

Solve and check. $-6n = 78$

Solution

$$-6n = 78$$

$$\frac{-6n}{-6} = \frac{78}{-6}$$ To undo multiplying by 6, divide by 6 on both sides.

$$n = -13$$

Check

$$-6(-13) \stackrel{?}{=} 78$$ Replace n with −13 in $-6n = 78$.

$$78 = 78 \checkmark$$ The solution is −13. **The temperature change is −13°C per minute.**

Practice Solve and check. **a.** $-4x = 68$ **−17** **b.** $-182 = 7n$ **−26**

Example 2

Solve and check. $\frac{n}{-4} = 17$

Solution

$$\frac{n}{-4} = 17$$ To undo dividing by −4, multiply by −4 on both sides.

$$\frac{n}{-4}(-4) = 17(-4)$$

$$n = -68$$

Check

$$\frac{-68}{-4} \stackrel{?}{=} 17$$ Replace n with −68 in $\frac{n}{-4} = 17$.

$$17 = 17 \checkmark$$ The solution is −68.

Practice Solve and check. **a.** $\frac{a}{6} = -72$ **−432** **b.** $25 = \frac{x}{-8}$ **−200**

The following property of −1 is useful when solving equations such as $-n = 24$.

Property of −1

The product of −1 and a number is the opposite of the number.

For each number n, $\mathbf{-1n = -n}$ and $\mathbf{-n = -1n}$

Example 3

Solve and check. $-n = 24$

Solution

$$-n = 24$$

$$-1n = 24 \qquad -n = -1n$$

$$\frac{-1n}{-1} = \frac{24}{-1} \qquad \text{Divide (or multiply) each side by } -1.$$

$$n = -24$$

Check

$$-(-24) \stackrel{?}{=} 24 \qquad \text{Replace } n \text{ with 24 in } -n = 24.$$

$$24 = 24 \checkmark \qquad \text{The solution is } -24.$$

Practice Solve and check. **a.** $-x = -15$ 15 **b.** $36 = -b$ −36

Oral Exercises 2. yes 3. no 5. yes

Check. Is the given number a solution to the equation?

1. $-3c = 18,\ c = 6?$ no
2. $5n = -40,\ n = -8?$
3. $-63 = -9x,\ x = -7?$
4. $\frac{y}{-4} = 7,\ y = -28?$ yes
5. $8 = \frac{b}{-3} = b = -24?$
6. $\frac{x}{-9} = -9,\ x = 81?$ yes

Tell what operation you would use to solve the equation.

7. $36c = -324$ div.
8. $-b = 29$ div.
9. $-272 = -34x$ div.
10. $24r = -144$ div.
11. $-169 = \frac{z}{-13}$ mult.
12. $\frac{y}{-57} = 29$ mult.

Exercises 8. −7 16. 702 20. −1085

A Solve and check.

1. $-8y = 96$ −12
2. $2b = -86$ −43
3. $-7n = -105$ 15
4. $-b = 72$ −72
5. $-4z = 92$ −23
6. $8c = -272$ −34
7. $-14a = -168$ 12
8. $-23y = 161$
9. $\frac{n}{17} = -9$ −153
10. $\frac{b}{-26} = 7$ −182
11. $\frac{r}{-31} = -6$ 186
12. $\frac{t}{36} = -8$ −288
13. $\frac{a}{18} = -15$ −270
14. $\frac{a}{-29} = 13$ −377
15. $\frac{x}{46} = -11$ −506
16. $\frac{y}{-54} = -13$
17. $-282 = 47b$ −6
18. $47 = \frac{y}{-16}$ −752
19. $-882 = 14n$ −63
20. $31 = \frac{b}{-35}$

21. $-28n = -1288$ 46 **22.** $-2226 = 42t$ −53 **23.** $\frac{c}{-66} = 77$ −5082 **24.** $-49 = \frac{x}{-34}$ 1666

25. $-16c = 256$ −16 **26.** $\frac{b}{-9} = -347$ 3123 **27.** $-576 = 16p$ −36 **28.** $127 = \frac{r}{-85}$ −10,795

B Solve and check.

29. $(-16 + 25)b = 162$ 18 **30.** $\frac{x}{(-12)(-16)} = -8$ −1536 **31.** $(235 - 421) = -31d$ 6

32. Write and solve an equation in which the product is −518, one factor is −7, and the other factor is represented by the variable n. $-7n = -518, n = 74$

33. A football team lost an average of 12 yards (−12) a play for a total loss of 72 yards (−72). How many plays were run? 6

34. When cooked vegetables were put in a freezer, the temperature dropped an average of 19°C (−19) each hour for 6 hours. If the cooking temperature was 108°C, what was the temperature after the 6-hour drop? −6°C

C Extending Thinking Skills

35. $7a = -224$ $8a = -512$ $9a = -1152$

35. Find the solutions to these equations and look for a pattern. Write the next 3 equations in the pattern. $2a = -2$, $3a = -6$, $4a = -16$, $5a = -40$, $6a = -96$

Use estimation and a calculator to find two solutions to each.

36. $n \cdot n = 1369$ 37, −37 **37.** $n \cdot n = 4624$ 68, −68 **38.** $n \cdot n = 18{,}769$ 137, −137

Mixed Review

Thinking skills: 35. Finding a pattern 36–38. Testing a conjecture/Reasoning

Solve and check. **39.** $30x = 210$ 7 **40.** $t - 26 = 14$ 40 **41.** $m \div 12 = 18$ 216

Multiply. **42.** $9(-2)$ −18 **43.** $-9(-12)$ 108 **44.** $11(-1)$ −11 **45.** $-1(-11)$ 11

COMPUTER ACTIVITY

The following program can help you use a "guess, check, revise" procedure to solve the equation $(n \cdot n) - 3n - 18 = 0$.

```
10 PRINT "GUESS A SOLUTION TO THIS EQUATION"
20 PRINT "NXN-3XN-18=0":INPUT N
30 IF N*N-3*N-18=0 THEN 80
40 PRINT "NO, NXN-3XN-18="N*N-3*N-18
50 PRINT "TRY AGAIN? TYPE YES OR NO.":INPUT T$
60 IF T$="YES" THEN 20
70 PRINT "BYE!":GOTO 90
80 PRINT "YOU'VE FOUND A SOLUTION! GREAT JOB!"
90 END
```

1. Use the program to find two solutions to the equation. 6, −3

2. Change the program to help find solutions to the equation $n \cdot n + n - 12 = 0$.
See additional answers.

More Practice

Solve and check.

1. $-2m = 144$ -72	**2.** $26 + c = 11$ -15	**3.** $t - 16 = 4$ 20	**4.** $\frac{r}{4} = -28$ -112
5. $\frac{x}{-5} = -19$ 95	**6.** $m - 28 = 42$ 70	**7.** $41 = a + (-19)$ 60	**8.** $15n = -540$ -36
9. $k - (-6) = 11$ 5	**10.** $\frac{t}{9} = -16$ -144	**11.** $-12a = 636$ -53	**12.** $r + 6 = 0$ -6
13. $97 + m = -6$ -103	**14.** $350 = -14c$ -25	**15.** $\frac{h}{-3} = -27$ 81	**16.** $x - 6 = -8$ -2
17. $12t = 144$ 12	**18.** $16 = n - 3$ 19	**19.** $w + 30 = 4$ -26	**20.** $\frac{a}{16} = -23$ -368
21. $\frac{m}{-7} = 46$ -322	**22.** $t + 16 = 100$ 84	**23.** $r - 5 = 0$ 5	**24.** $-6k = -594$ 99
25. $c - 6 = -4$ 2	**26.** $\frac{z}{9} = -42$ -378	**27.** $19y = -304$ -16	**28.** $17 = n + 12$ 5
29. $x + 31 = 16$ -15	**30.** $11c = -297$ -27	**31.** $\frac{w}{32} = 9$ 288	**32.** $t - (-2) = 0$ -2
33. $\frac{m}{-6} = -29$ 174	**34.** $z + 26 = 18$ -8	**35.** $h - 0 = -4$ -4	**36.** $-5n = 1035$ -207
37. $24a = -792$ -33	**38.** $99 + h = 67$ -32	**39.** $m - 6 = 21$ 27	**40.** $\frac{t}{11} = -20$ -220
41. $992 = -31n$ -32	**42.** $14 = z - 10$ 24	**43.** $-36 = \frac{r}{6}$ -216	**44.** $46 = 7 + k$ 39
45. $14 = \frac{k}{-9}$ -126	**46.** $21n = 882$ 42	**47.** $(-2) + r = -3$ -1	**48.** $m - 6 = 103$ 109
49. $36 + n = -9$ -45	**50.** $k - 12 = -9$ 3	**51.** $\frac{k}{-3} = 67$ -201	**52.** $14t = -1260$ -90
53. $-26r = -910$ 35	**54.** $-19 = \frac{m}{-21}$ 399	**55.** $t - 12 = 36$ 48	**56.** $n + 3 = 0$ -3
57. $a + 67 = 24$ -43	**58.** $-9w = 1116$ -124	**59.** $47 = m - 14$ 61	**60.** $\frac{z}{21} = -37$ -777
61. $r - 14 = -37$ -23	**62.** $\frac{a}{9} = -54$ -486	**63.** $r + 19 = 101$ 82	**64.** $8k = 520$ 65
65. $z - 39 = -6$ 33	**66.** $t + 91 = 17$ -74	**67.** $6a = -432$ -72	**68.** $\frac{z}{-13} = 25$ -325
69. $\frac{c}{27} = -11$ -297	**70.** $w - (-6) = 12$ 6	**71.** $40 = x + 100$ -60	**72.** $-5m = 575$ -115
73. $984 = 8r$ 123	**74.** $-34 = \frac{t}{-12}$ 408	**75.** $c - 19 = 40$ 59	**76.** $n + 12 = -31$ -43

PROBLEM SOLVING: WRITING EQUATIONS

3-9 Translating Sentences into Equations

Objective: To translate verbal statements into equations.
MIN: 1–21; REG: 1–21 even, 22–25; MAX: 1–21 even, 22–27; ALL: Mixed Review

To translate a verbal statement into an equation, read the verbal statement carefully, thinking about the meaning of each phrase. Then decide what the variable will represent. Notice how the following statement about numbers translates directly to an equation.

Verbal Statement **A number** increased by 5 gives the sum 23

Equation $n + 5 = 23$

Sometimes the translation is less direct.

Verbal Statement $6 less than the **cost of the hat** is $17

Equation $h - 6 = 17$

Example 1

Write an equation. A number increased by 8 is 23.

Solution Let n = a number — First decide what the variable represents.

$n + 8 = 23$ — "a number increased by 8" translates to $n + 8$. "is" translates to "=."

Practice Write an equation. **a.** A number decreased by 9 gives 54. $x - 9 = 54$
b. The sum of a number and 47 is −112. $y + 47 = -112$

Example 2

Write an equation. The $45 baseball glove costs 5 times as much as a ball.

Solution Let b = cost of the ball

$45 = 5b$ — "5 times as much" translates to $5b$.

Practice Write an equation. **a.** 55 times the number of hours is 220. $55h = 220$
b. The number of people divided by 9 gives 12 teams. $n \div 9 = 12$

Oral Exercises

Give an algebraic expression for the phrase.

1. Three times a number $3x$
2. 6 more than a number $x + 6$
3. A number divided by -17 $\frac{x}{-17}$
4. The product of 9 and a number $9x$
5. A number decreased by 5 $x - 5$
6. The quotient of a number and 8 $\frac{x}{8}$

Exercises

A Write an equation.

1. 17 less than a number is 101. $c - 17 = 101$
2. The quotient of a number and -8 is 216. $\frac{r}{-8} = 2$
3. 35 more than a number is 372. $z + 35 = 372$
4. A number decreased by 13 gives 87. $x - 13 = 87$
5. 56 more than a number is 104. $x + 56 = 104$
6. The difference of a number and 9 is 47. $b - 9 =$
7. A number n divided by 25 is -12. $\frac{n}{25} = -12$
8. 57 added to a number gives a total of 123. $p + 57$
9. 9 times a number is -171. $9y = -171$
10. The sum of a number and -73 is 145. $x + (-73)$
11. 17 less than a number is 15. $x - 17 = 15$
12. A number increased by 45 gives 77. $n + 45 = 77$
13. Six times the number of days is 91. $6n = 91$
14. The price decreased by \$29 is \$258. $x - 29 = 258$
15. \$9 more than the total restaurant bill is \$72. $x + 9 = 72$
16. A number increased by 56 produces a total of 124. $b + 56 = 124$
17. 34 multiplied by a number produces the product 272. $34t = 272$
18. 28 is the result of multiplying -4 by a number. $-4x = 28$
19. The difference between the cost of the car and \$6,000 is \$59,820. $x - 6000 = 59{,}820$
20. 369 items divided into boxes of 3 gives the number of boxes. $p = \frac{369}{3}$
21. The present temperature increased by 19 degrees gives a reading of 56 degrees. $t + 19 = 56$

B Write a verbal statement that describes each equation. See additional answers.

22. $x + 9 = 23$ 23. $\frac{p}{-4} = 9$ 24. $-8c = 72$ 25. $16 - n = 7$

C Extending Thinking Skills

26. The sum of 5 times x and 7 times y is 110. Translate into an equation. Find integers x and y that make the equation true. $5x + 7y = 110$; (8, 10)

27. Write and solve an equation for "the sum of 3 page numbers in a row in a book is 828." $n + (n + 1) + (n + 2) = 828$ $n = 275$

Thinking skills: 26, 27. Generalizing/Testing a conjecture

Mixed Review

Solve and check. **28.** $m + 19 = -31$ -50 **29.** $14c = 434$ 31 **30.** $\frac{z}{6} = -6$ -36

Divide. **31.** $\frac{56}{-8}$ -7 **32.** $\frac{-18}{-6}$ 3 **33.** $\frac{-169}{13}$ -13 **34.** $\frac{-324}{-9}$ 36

3-10 Make a Table, Look for a Pattern

Objective: To use the strategies Make a Table, Look for a Pattern, and other strategies learned so far to solve nonroutine problems.
MIN: 1–2; REG: 1–4; MAX: 1–7

Problem-solving strategies called **Make a Table** and **Look for a Pattern** are helpful when solving problems involving numerical relationships. Consider the following problem.

Problem A secret agent was hired for a special assignment that would take exactly 14 days. He could choose to be paid one of the following two ways. Under payment plan A, he would receive $6000 for the job. Under payment plan B, his employer would put $1 into a safe for the first day and increase the amount in the safe to $2 the second day, $4 the third day, $8 the fourth day, and so on, doubling the amount in the safe each day. At the end of the assignment, the agent could claim the contents of the safe. Which payment plan should the agent choose?

To find the amount for payment plan B, you can use the data in the problem to begin a **table**. The red numbers show the data given in the problem. Look for a **pattern** to help extend the table and provide new information.

Notice that the numbers in the second column are found by multiplying 1 less factor of two than the number for the day. On day **4** there were **3** factors of 2, or 2(2)(2), dollars in the safe. So on day **14** there would be **13** factors of 2, or 8192 dollars in the safe.

Day	Number of dollars in the safe
1	**1**
2	**2** ← 2
3	**4** ← 2(2)
4	**8** ← 2(2)(2)
5	16
6	32
⋮	⋮
14	?

Payment plan A: $6000. Payment plan B: $8192. The agent should choose plan B.

Problem-Solving Strategies

Choose the Operations	**Make a Table**
Guess, Check, Revise	**Look for a Pattern**
Draw a Picture	

This chart shows the problem-solving strategies introduced so far.

Problems

Solve.

1. Nina started a computer users club. On the first day, she was the only member. Each day after that, one more member joined than on the previous day. What was the membership of the club after 30 days? 465 members

2. A generous millionaire had an unusual plan for giving away her money. Beginning on her birthday, she would give away $1 the first day, $3 the second day, $5 the third day, and so on. How much money would she have given away after 100 days? $10,000

3. Scientists use a radio telescope to send the following sequence of "beeps" into outer space: 1, 1, 2, 3, 5, 8, 13, 21, 34. They hope that intelligent life will receive these signals and send return signals that continue the sequence. What are the next 5 numbers of "beeps" that intelligent life would send back? 55, 89, 144, 233, 377

4. A sandwich shop has 3-legged stools and 4-legged chairs at its tables. Altogether there are 31 seats and 104 legs. How many of the seats are stools and how many are chairs? 20 stools, 11 chairs

5. Emilio said, "A 1-year-old dog is 7 'dog years' old. My dog is 6 years old in regular years. If you change my age into dog years, I am 49 dog years older than my dog's age in dog years. How old am I?" 13

6. Suppose it takes newborn rabbits two months to mature and produce a new pair of rabbits. After that, they produce a new pair of rabbits on the first day of each month. The new pairs of rabbits grow and reproduce at the same rate. If you started on January 1 with a pair of newborn rabbits and no rabbits died, how many pairs of rabbits would you have on July 1? 13 pairs

7. George had 12 cherry trees. On January 1, 1780, he cut down 2 of the trees. On December 31, 1780, he planted 1 tree. Each year on the same dates he did the same things. On what date did he first have no cherry trees in his yard? January 1, 1790

Enrichment

Sequences

A **sequence** is a set of numbers in a particular order. The numbers in a sequence are called **terms** of the sequence. The table below shows the sequence **1, 3, 5, 7, 9, . . .** The numbers in the top row show the *order* of the terms. The *first* term is 1, the *second* term is 3, the *third* term is 5, and so on. The general rule for finding a particular term of the sequence is called the **rule for the *n*th term**.

Number of the term	1st	2nd	3rd	4th	5th	6th	7th	8th	. . .	*n*th
Terms of the Sequence	1	3	5	7	9	11	13	15	. . .	$2n - 1$

Rule for the *n*th term.

To find the 9th, 10th, and 11th terms of the sequence above, you would replace the variable in the rule for the *n*th term by the number of the term you want to find.

Rule $2n - 1$: $2(9) - 1 = 17$, $2(10) - 1 = 19$, $2(11) - 1 = 21$

The **9**th, **10**th, and **11**th terms of the sequence are **17, 19, 21.**

1. −1, −3, −6, −10, −15

1. The rule for the *n*th term of a sequence is $\frac{n(n+1)}{-2}$. Write the first five terms.

2. The rule for the *n*th term of a sequence is $n(n) - 1$. Give the 8th term of the sequence. Give the 25th term of the sequence. 63, 624

3. Find the rule for the *n*th term of this sequence: 1, 4, 9, 16, 25, 36, 49 . . . $n(n)$

4. Find the rule for the *n*th term of this sequence: −3, −7, −11, −15, −19, −23, −27, . . . $1 - 4(n)$

5. The *n*th term of a sequence is 0 when *n* is even and 1 when *n* is odd. Write the first eight terms of this sequence. 1, 0, 1, 0, 1, 0, 1, 0

6. Write the next five terms in the **Fibonacci sequence**: 1, 1, 2, 3, 5, 8, 13, 21, 34, . . . 55, 89, 144, 233, 377

7. Square any term of the sequence in exercise 6. Then find the product of the term preceding and the term following that term. What do you discover?

8. Estimate how many of the first 25 Fibonacci numbers are even. Extend the sequence to check your guess. There are 8 even Fibonacci numbers in the first 25.

7. The square of any term is equal to the product of the term preceding it and the term following it plus or minus one.

Chapter 3 Review

3-1 Find the absolute value.

1. $|6|$ 6 **2.** $|-3|$ 3 **3.** $|-17|$ 17

Use the inequality symbols. Order from least to greatest.

4. $-8, 2, -2$ $-8 < -2 < 2$ **5.** $-7, -1, 0$ $-7 < -1 < 0$ **6.** $0, -8, -5$ $-8 < -5 < 0$

3-2 Find the sum.

7. $-3 + (-3)$ -6 **8.** $-7 + 4$ -3 **9.** $5 + (-2)$ 3

10. $-20 + (-4)$ -24 **11.** $-15 + 7$ -8 **12.** $12 + (-6)$ 6

Evaluate.

13. $s + (-4)$ for $s = -12$ -16 **14.** $-17 + k$ for $k = 9$ -8

3-3 Subtract.

15. $-7 - 6$ **16.** $5 - 8$ -3 **17.** $-14 - 3$ **18.** $8 - (-3)$ **19.** $6 - 13$ -7

15. -13 17. -17 18. 11

Evaluate.

20. $v - 5$ for $v = 2$ -3 **21.** $-3 - n$ for $n = 9$ -12

3-4 Solve and check.

22. $r + 42 = -7$ -49 **23.** $s - 17 = -34$ -17 **24.** $t - 66 = 15$ 81

25. $-37 = z + 56$ -93 **26.** $v - (-28) = -77$ -105 **27.** $w + (-15) = 56$ 71

Solve.

28. A checking account with a beginning balance of \$120 has credits of \$600 and debits of \$944. By how much is it overdrawn? \$224

3-6 Find the product.

29. $-1(-9)$ 9 **30.** $7(8)$ 56 **31.** $-5(-5)$ 25 **32.** $6(-8)$ -48 **33.** $-4(2)$ -8

3-7 Find the quotient.

34. $-54 \div 6$ -9 **35.** $-81 \div (-9)$ 9 **36.** $28 \div (-4)$ -7

37. $16 \div (-4)$ -4 **38.** $-6 \div (-2)$ 3 **39.** $72 \div (-3)$ -24

3-8 Solve and check.

40. $-81 = 3n$ **41.** $-9y = 27$ **42.** $7 = \frac{r}{-7}$ **43.** $\frac{a}{4} = -6$ **44.** $91 = -r$

40. -27
41. -3
42. -49
43. -24
44. -91

3-9 Write an equation.

45. 10 multiplied by a number n produces the product 110. $10n = 110$

46. The sum of a number and -15 is 60. $n + -15 = 60$

Chapter 3 Test

Find the absolute value.

1. $|23|$ 23
2. $|-8|$ 8
3. $|0|$ 0

Use inequality symbols. Order from least to greatest.

4. 7, −5, 8 $-5 < 7 < 8$
5. −3, −9, 1 $-9 < -3 < 1$
6. −2, 0, −3 $-3 < -2 < 0$

Find the sum.

7. $-3 + 8$ 5
8. $7 + (-1)$ 6
9. $-2 + (-3)$ −5
10. $-1 + (-1)$ −2
11. $-8 + 3$ −5
12. $13 + (-5)$ 8

Evaluate.

13. $m + (-11)$ for $m = 5$ −6
14. $20 + r$ for $r = -7$ 13

Subtract.

15. $1 - 6$ −5
16. $-13 - 2$
17. $12 - 14$
18. $16 - (-5)$
19. $-8 - 7$

16. −15
17. −2
18. 21
19. −15

Evaluate.

20. $-6 - n$ for $n = 7$ −13
21. $-4 - n$ for $n = -7$ 3

Solve and check.

22. $k + 35 = -50$ −85
23. $m - 90 = -10$ 80
24. $73 = p - 2$ 75
25. $-48 = n + 35$ −83
26. $p - (-15) = 43$ 28
27. $s + (-18) = -35$ −17

Solve.

28. By how much is Fred's checking account overdrawn? He started with a balance of \$176, wrote a check for \$140, then wrote a check for \$90. \$54

Find the product.

29. $-5(-4)$ 20
30. $6(8)$ 48
31. $-3(-1)$ 3
32. $9(-2)$ −18
33. $-7(4)$ −28

Find the quotient.

34. $-18 \div 9$ −2
35. $-48 \div (-6)$ 8
36. $24 \div (-3)$ −8
37. $25 \div (-5)$ −5
38. $-9 \div (-3)$ 3
39. $64 \div (-4)$ −16

Solve and check.

40. $-84 = 4s$
41. $-6x = 36$
42. $\frac{t}{5} = -40$
43. $7 = \frac{m}{-8}$
44. $-c = -22$

40. −21
41. −6
42. −200
43. −56
44. 22

Write an equation.

45. 24 less than a number x is 12. $x - 24 = 12$
46. 49 is the result of multiplying −7 by a number m. $-7m = 49$

Cumulative Review

Write an algebraic expression for each phrase.

1. g added to 9 $g + 9$ **2.** k decreased by 3 $k - 3$ **3.** the sum of v and 10 3. $v + 10$

4. half of m $\frac{m}{2}$ **5.** y times 7 $7y$ **6.** 9 divided by w $\frac{9}{w}$

Write an expression for each question.

Wendy has 16 records. How many will she have if

7. she gives away r records? $16 - r$ **8.** she buys b more records? $16 + b$

Simplify.

9. $28 \div (3 + 4)$ 4 **10.** $2 + [(7 + 2) \div 3]$ 5 **11.** $4 \cdot 5 + 3 \cdot 9$ 47

Evaluate each algebraic expression.

12. $s + 2$ for $s = 9$ 11

13. $(4 + 8) - c$ for $c = 10$ 2

14. $f - g$ for $f = 5$, $g = 1$ 4

15. $8k$ for $k = 8$ 64

16. $\frac{2x}{z}$ for $x = 8$, $z = 4$ 4

17. $n(m + 6)$ for $n = 2$, $m = 3$ 18

18. $\frac{7w}{2}$ for $w = 4$ 14

19. $3(10 - c)$ for $c = 5$ 15

20. $y + (-10)$ for $y = 2$ -8

21. $35 + b$ for $b = -10$ 25

22. $-4 - t$ for $t = 5$ -9

23. $-7 - x$ for $x = 3$ -10

Use the commutative property to write an equivalent expression.

24. $y(6)$ $6y$ **25.** pq qp **26.** $12 + z$ $z + 12$

Use the associative property to write an equivalent expression.

27. $20(8p)$ $(20 \cdot 8)p$ **28.** $(5f)n$ $5(fn)$ **29.** $(y + 2) + 9$ $y + (2 + 9)$

Use the distributive property to write an equivalent expression.

30. $3(c + 4)$ $3c + 3 \cdot 4$, $3c + 12$ **31.** $a(8 + 9)$ $8a + 9a$ **32.** $(7 + b)8$ $7 \cdot 8 + 8b$, $56 + 8b$

Use the associative property for multiplication to simplify.

33. $2(5 \cdot t)$ $10t$ **34.** $6(4k)$ $24k$ **35.** $15(2d)$ $30d$

Use the distributive property to simplify by combining like terms.

36. $g + 7g$ $8g$ **37.** $3v + 3v$ $6v$ **38.** $25p + 50p$ $75p$

Compute.

39. $3 - 7$ -4 **40.** $-12 \cdot 8$ -96 **41.** $18 \div (-9)$ -2

Chapter 4
Decimals

4-1 Decimals and Place Value

Objective: To understand decimal notation and place value; to compare and order decimals.
MIN: 1–18, 31–32; REG, MAX: 1–18 even, 19–32; ALL: Mixed Review

An exploratory lesson opener can be found on page 547.

Numbers are represented in the decimal system by using the digits 0 through 9 and grouping by tens. The value of each digit in a decimal depends on its place. The chart below shows the **place value** names for the decimals 24.5 and 6.263.

thousands	hundreds	tens	ones	decimal point	tenths	hundredths	thousandths	
1000	**100**	**10**	**1**		$\frac{1}{10}$	$\frac{1}{100}$	$\frac{1}{1000}$	
		2	4	.	5			twenty-four and five tenths
			6	.	2	6	3	six and two hundred sixty-three thousandths

Example 1

Write $\mathbf{3}(100) + \mathbf{1}(10) + \mathbf{3}(1) + \mathbf{4}\left(\frac{1}{10}\right) + \mathbf{7}\left(\frac{1}{100}\right)$ as a decimal.

Solution 313.47

Practice Write as a decimal.

a. $4(100) + 7(10) + 8(1) + 3\left(\frac{1}{10}\right) + 6\left(\frac{1}{100}\right)$ **478.36**

b. $9\left(\frac{1}{100}\right) + 3\left(\frac{1}{1000}\right)$ **0.093**

Decimals can be used to name points on a number line. The number line shown below is divided into ten equal spaces between 0 and 1. Point *A*, which is at the end of the seventh space, represents the number seven tenths and can be named by the decimal 0.7.

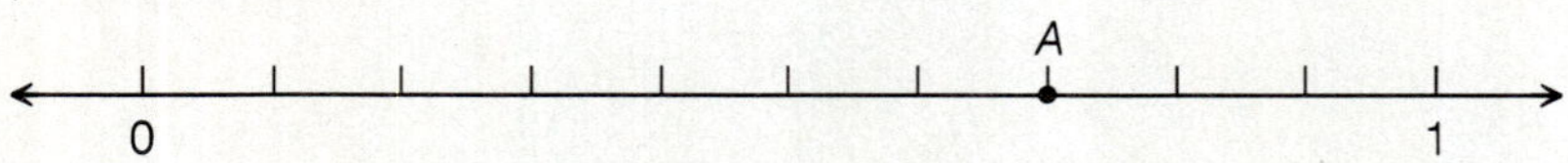

$\frac{7}{10} = 0.7$

Positive numbers represented by decimals are shown to the right of zero on the number line, and negative numbers represented by decimals are shown to the left of zero.

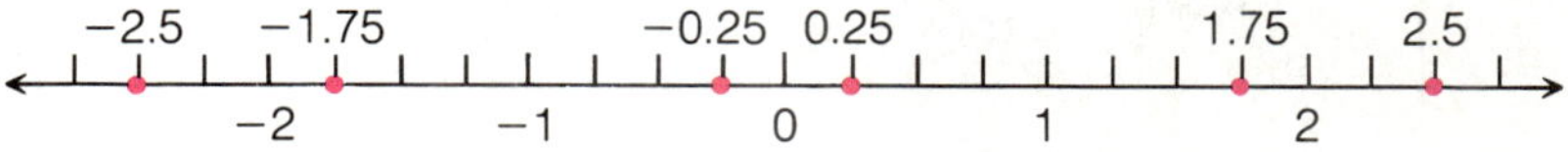

Example 2

Give the decimal names for points A and B.

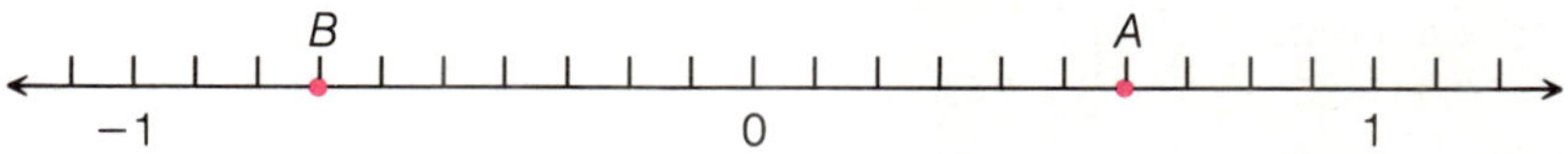

Solution point $A = 0.6$ Point A is at the end of the sixth of 10 equal spaces between 0 and 1.

point $B = -0.7$ Point B is at the end of the seventh of 10 equal spaces between 0 and −1.

Practice

Give the decimal name for point A.

a.

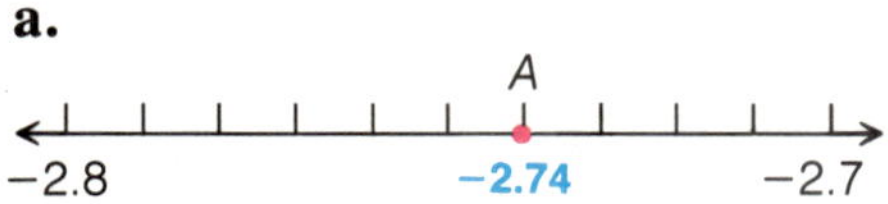

b.

The number farther to the right on the number line is the greater number. To compare decimals without a number line, start on the left, find the first place in which the digits are not equal, and compare the digits. **To compare two negative decimals, first compare their absolute values. The two decimals have the opposite order of their absolute values.**

Example 3

Write >, <, or = for □. 19.31 □ 19.23

Solution $19.31 > 19.23$ Start on the left. The tenths place is the first place in which the digits are not equal. Since 3 > 2, 19.31 > 19.23.

Practice

Write >, <, or = for each □.

a. 47.14 [>] 47.13 **b.** −7.003 [>] −7.03 **c.** 18.06 [=] 18.060

Oral Exercises

1. 8 and 3 tenths
2. 45 and 8 tenths
(etc.)

Read each decimal.

1. 8.3 **2.** 45.8 **3.** 390.1 **4.** 7.64 **5.** 35.18
6. 123.15 **7.** 3014.39 **8.** 0.034 **9.** 0.803 **10.** 493.319

Exercises 1. 540.76 2. 4337.5 3. 748.07 4. 7480.56 5. 578.3 6. 400.04

A Write as a decimal.

1. $5(100) + 4(10) + 7\left(\frac{1}{10}\right) + 6\left(\frac{1}{100}\right)$ **2.** $3(100) + 4(1000) + 3(10) + 7(1) + 5\left(\frac{1}{10}\right)$

3. $7(100) + 8(1) + 7\left(\frac{1}{100}\right) + 4(10)$ **4.** $4(100) + 5\left(\frac{1}{10}\right) + 8(10) + 6\left(\frac{1}{100}\right) + 7(1000)$

5. $8(1) + 7(10) + 5(100) + 3\left(\frac{1}{10}\right)$ **6.** $4\left(\frac{1}{100}\right) + 0(10) + 0(1) + 4(100)$

Give the decimal name for point A.

7. 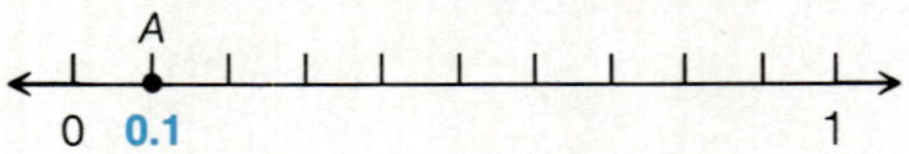

8. 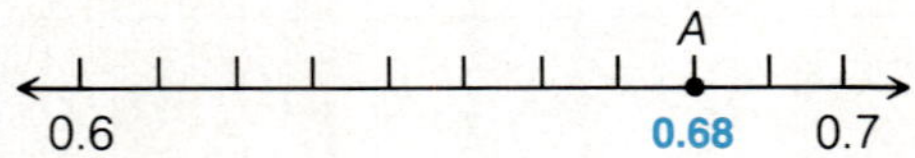

9. 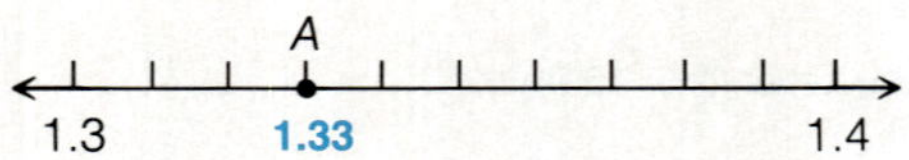

10. 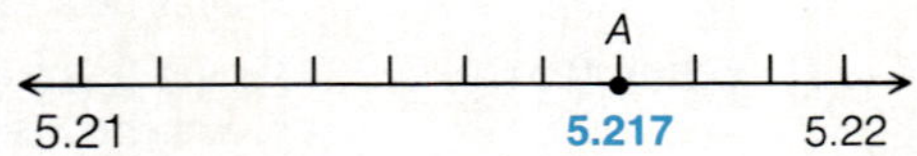

11. 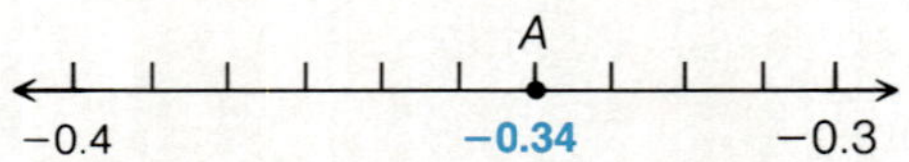

12. 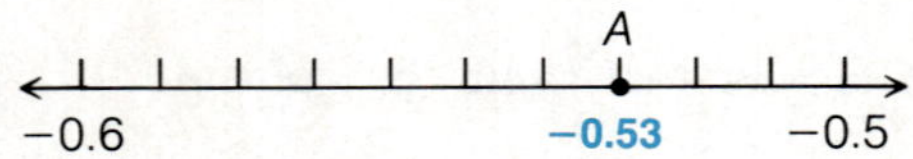

Write $>$, $<$, or $=$ for each $\square$.

13. $6.93 \boxed{=} 6.930$ **14.** $1.01 \boxed{<} 1.101$ **15.** $-4.658 \boxed{>} -4.685$

16. $4.1 \boxed{=} 4.100$ **17.** $7.001 \boxed{<} 7.010$ **18.** $-14.3 \boxed{<} -14.29$

B Write the numbers in order from least to greatest. See additional answers.

19. 3.7, 3.07, 3.069 **20.** 0.004, 0.039, 0.0041

21. 6.101, 6.010, 6.0101 **22.** 4.01, 4, 4.001

23. −5.404, −5.044, −5.040 **24.** −0.001, −0.101, −0.011, −0.01

Write two decimal replacements for x that make the sentence true. Answers will vary.

25. $0.4 < x < 0.5$ **26.** $8.12 < x < 8.13$ **27.** $4.06 > x > 4.05$

28. $12.3 > x > 12.03$ **29.** $9.1 < x < 9.11$ **30.** $-7.6 > x > -7.601$

Have students read $0.4 < x < 0.5$ as "four tenths is less than x is less than 5 tenths."

C Extending Thinking Skills

31. Use only the digits 0 and 1 to write four decimals that are > 0 and < 1. Answers will vary.

32. Use only the digits 5 and 6 to write four decimals that are > 5 and < 6. Answers will vary.

Thinking skills: 31–32. Generalizing

Mixed Review

Solve and check. **33.** $\frac{x}{13} = -17$ −221 **34.** $-24n = -360$ 15 **35.** $16t = -400$ −25

Evaluate for $x = 6$. **36.** $9x - 2$ 52 **37.** $\frac{-36}{x}$ −6 **38.** $-6x - 2$ −38

4-2 Estimating Using Rounding

Objective: To round decimals and estimate decimal sums and differences.
MIN: 1–30, 36–37; REG: 1–30 even, 31–37; MAX: 1–30 even, 31–39; ALL: Mixed Review

This computerized gasoline pump records gallon amounts to the third decimal place. The attendant rounds the number of gallons before writing it on a receipt. The procedure for rounding decimals is given below.

Rounding Decimals

Look at the digit to the right of the place to which you are rounding.

- If it is less than 5, do not change the digit in the place to which you are rounding.
- If it is 5 or more, add 1 to the digit in the place to which you are rounding.

Replace the digits to the right with zeros, or drop them if they are to the right of the decimal point.

Example 1

Round 4.679 to the nearest hundredth.

Solution 4.68 4.679 Since 9 > 5, round 7 up to 8. Drop the digits to the right of the 8.

Practice Round.

a. 7.087 to the nearest hundredth. 7.09

b. 9.783 to the nearest whole number. 10

c. 0.925 to the nearest tenth. 0.9

Example 2

Round 593.827 to the nearest tenth.

Solution 593.8 593.827 Since 2 < 5, leave the 8 unchanged and drop the digits to the right of the tenths place.

Practice Round.

a. 73.484 to the nearest tenth. 73.5 **b.** 0.253 to the nearest hundredth. 0.25

c. 4.82 to the nearest tenth. 4.8

You can use rounding with decimals to make estimates.

Estimating Using Rounding

- Round each decimal.
- Compute with the rounded decimals.

Example 3

Estimate the value of n by rounding. $n = 505.3 + 294.8$

Solution $500 + 300 = 800$ Round 505.3 to 500. Round 294.8 to 300.

The estimated value of n is 800.

Practice Estimate the value of x by rounding.

a. $x = 21.71 + 11.29$ $22 + 11 = 33$

b. $x = 352.92 - 147.19$ $350 - 150 = 200$

Oral Exercises

To round to the indicated place, would you increase the digit in that place or leave it unchanged?

1. 3.578, tenths
2. 5.294, hundredths
3. 32.925, tenths
4. 59.139, hundreds
5. 23.598, whole number
6. 258.37, tenths

1. incr. 2. leave 3. leave 4. incr. 5. incr. 6. incr.

Exercises

A Round to the nearest whole number.

1. 146.73 147
2. 93.48 93
3. 126.55 127
4. 14.73 15
5. 34.61 35
6. 8.499 8
7. 19.501 20
8. 999.99 1000

Round to the nearest tenth.

9. 23.78 23.8
10. 1.725 1.7
11. 42.73 42.7
12. 76.39 76.4
13. 8.294 8.3
14. 39.149 39.1
15. 17.0051 17.0
16. 4.286 4.3

Round to the nearest hundredth.

17. 4.5869 4.59
18. 5.497 5.50
19. 0.2848 0.28
20. 3.0089 3.01
21. 9.896 9.90
22. 0.999 1.00
23. 8.8944 8.89
24. 2.09839 2.10

Estimate the value of the variable by rounding.

25. $n = 218.6 + 101.2$ 320
26. $x = 24.5 - 10.3$ 15
27. $y = 18.603 + 21.2218$ 40
28. $z = 125.3 + 34.8$ 160

Possible answers are given for 25–30.

29. $s = 724.91 + 575.36$ 1300

30. $x = 84.78 - 42.32$ 43

B 31–33 Answers will vary. Examples: 31. 23.46 32. 4.233

31. Find four numbers that, when rounded to the nearest tenth, result in 23.5.

32. Find four numbers that, when rounded to the nearest hundredth, result in 4.23.

33. Use front-end estimation to estimate the sum.
$5.53 + 9.48 + 7.24 + 8.77$ 31

34. Use estimation to decide which is the better buy: 5 bars of soap for \$1.98 or 3 bars of the same soap for \$1.10. 3 for \$1.10

35. Keiko and Sara need to buy 80 hot dogs for a class picnic. They can buy packages of 8 hot dogs for \$1.05 or packages of 10 hot dogs for \$1.53. Use estimation to decide which is the better buy. 8 for \$1.05

C Extending Thinking Skills

Thinking skill: 36–38. Finding a pattern

Look for a pattern. Give the next three numbers.

36. 1, 1, 2, 3, 5, 8, 13, 21, . . . 34, 55, 89

37. 0.001, 0.1, 0.101, 0.201, 0.302, 0.503, 0.805, 1.308, . . . 2.113, 3.421, 5.534

38. −0.03, 0.1, −0.13, 0.26, −0.49, 0.88, −1.63, 3.00, . . . −5.51, 10.14, −18.65

39. 0.001, 0.11, 0.111, 0.221, 0.332, 0.553, 0.885, . . . 1.438, 2.323, 3.761

Mixed Review

Find the value of the variable. **40.** $m = 2\,\frac{17 \cdot 62}{9 - 77}$ −31

Solve and check. **41.** $-36m = 648$ −18 **42.** $t - 26 = -8$ 18 **43.** $\frac{w}{-9} = 14$ −126

Simplify. **44.** $6t - 8t$ −2t **45.** $6(-12)m$ −72m **46.** $-9 + x + 3$ x − 6

COMPUTER ACTIVITY

This program rounds a decimal to the nearest hundredth.

```
10 REM: ROUNDING DECIMALS
20 PRINT "TYPE A DECIMAL THAT HAS THREE OR MORE DECIMAL
   PLACES"
30 INPUT N
40 PRINT N; " ROUNDED TO HUNDREDTHS IS ";
50 PRINT INT(N*100+.5)/100
60 END
```

1. Type and run this program.
2. Change the program so it will round decimals to the nearest tenth. See additional answers.
3. Change the program so it will round decimals to the nearest thousandth. See additional answers.

4-3 Adding and Subtracting with Decimals

Objective: To compute decimal sums and differences.
MIN: 1–22, 32; REG: 1–22 even, 23–32; MAX: 1–22 even, 23–34; ALL: Mixed Review

An 800-meter relay team had individual times of 21.38 seconds, 22.81 seconds, 22.93 seconds, and 21.12 seconds. The time for the team is the sum of the four decimals.

To add or subtract with decimals, align the decimal points and add or subtract as with whole numbers. Place the decimal point in the sum or difference, in line with the other decimal points.

Example 1

Add. $21.83 + 22.81 + 22.93 + 21.12$

Solution

$$\begin{array}{r} 21.83 \\ 22.81 \\ 22.93 \\ +\ 21.12 \\ \hline 88.69 \end{array}$$

Write the numbers vertically, aligning the decimal points, and add.

Place the decimal point in the answer. **The time for the relay team is 88.69 seconds.**

Practice Add. **a.** $34.28 + 259.3$ **293.58** **b.** $46.08 - 2.485$ **43.595**

Example 2

Add. $-5.3 + 2.7$

Solution $-5.3 + 2.7 = -2.6$ Find the difference in absolute values. Since -5.3 has the larger absolute value, the sum is negative.

Practice Add. **a.** $-17.4 + 75.3$ **57.9** **b.** $-3.8 + (7.9 - 3.7)$ **0.4**

Example 3

Evaluate. $v - 7.62$ for $v = 6.1$.

Solution $v - 7.62$

$$6.1 - 7.62$$
$$= 6.1 + (-7.62)$$
$$= -1.52$$

Practice Evaluate.

a. $w - 7.65 + 4.8$ for $w = -2.1$ −4.95 **b.** $z - 3.9 + 9.31$ for $z = 8.63$ 14.04

Oral Exercises

Place the decimal point and read the answer. 1. nine and nine tenths (etc.)

1. 3.2 + 6.7 = 99 9.9

2. 9.51 + 1.3 = 1081 10.81

3. 63.7 − 1.2 = 625 62.5

4. 123.4 + 6.3 = 1297 129.7

5. 13.56 − 6.4 = 716 7.16

6. 406.5 − 25.6 = 3809 380.9

Exercises

A Add or subtract.

1. 35.82 + 12.3 48.12

2. 9.6 + 49.2 58.8

3. 92.3 − 21.5 70.8

4. 7.9 + (−3.8) 4.1

5. 9.73 + (−3.7) 6.03

6. 2.81 + (9.73 − 3.7) 8.84

7. 254.87 − 73.62 181.25

8. −48.52 + 73.36 24.84

9. −38.15 + 8.24 + 17.28

10. 35.9 − 24.8 11.1

11. 71.05 − 39.23 31.82

12. 4.98 + (2.89 + 0.82) 8.69

13. 973.2 − 48.38 924.82

14. 21.03 − (−19.128) 40.158

15. 398.237 − 59.83 338.407

16. 129.36 − 3.295 126.065

17. 79.24 − (−42.48) 121.72

18. 8.083 − 3.591 4.492

9. −12.63

Evaluate. 20. −121.36

19. $75.92 - y$ for $y = 35.1926$ 40.7274

20. $(u - 45.2) + 21.04$ for $u = -97.2$

21. $(45.23 + v) + 38.3$ for $v = -17.5$ 66.03

22. $x - 93.2 + 17.3$ for $x = -0.36$ −76.26

B

23. Evaluate $a - (b - c)$ and $(a - b) - c$ for $a = 12.5$, $b = 5.3$, and $c = 1.4$. Are the two expressions equivalent? no

24. A cross-country course has an uphill section of 0.82 km, a flat section of 1.3 km, a downhill section of 1.1 km, and a wooded section of 1.05 km. How long is the course? 4.27 km

25. A company that makes stereo components had profits for the year, in millions, of \$5.62 on speakers and \$4.38 on turntables. It had losses of \$0.21 on headphones and \$1.03 on amplifiers. What was its net profit? \$8.76 million

Use a calculator with a memory key to evaluate each expression.

26. $x - 2.73$ for $x = 7.9$, $x = 9.32$, and $x = 17.21$ 5.17, 6.59, 14.48

27. $19.8 - x$ for $x = 3.91$, $x = 11.82$, and $x = 6.5$ 15.89, 7.98, 13.3

28. $(x + 5.2) - 3.1$ for $x = 19.6$, $x = 5.3$, and $x = 4.7$ 21.7, 7.4, 6.8

29. $(12.5 - x) + 3.8$ for $x = 5.1$, $x = 7.92$, and $x = 3.3$ 11.2, 8.38, 13

30. Use clustering to estimate the sum. Clustering is introduced on page 51.
$45.68 + 56.79 + 49.53 + 52.75$ 200

31. Explain how you can use compensation to find this difference mentally.
$68.73 - 25.99$ Subtract 26 from 68.73, then add 0.01.

C Extending Thinking Skills

Thinking skills: 32–34. Reasoning

Copy each problem and find the missing digits.

32.
```
   6■.7     61.7
 - 46.■     46.9
   14.8
```

33.
```
   6■■.92   615.92
 - 231.■5   231.15
   384.77
```

34. A business executive can fly from one city to another and then return by train for a total travel time of 7.5 hours. To fly both ways would take a total of 2.5 hours. How long would it take to go both ways by train? 12.5 hours

Mixed Review

Solve and check. **35.** $m - 36 = -15$ 21 **36.** $33r = -594$ −18 **37.** $\frac{x}{-26} = 18$ −468

Evaluate for $n = 3$. **38.** $6n \div 9$ 2 **39.** $-3n + 11$ 2 **40.** $12n \div (-6)$ −6

NUMBERS TO ALGEBRA

The Numbers to Algebra section in Chapter 1 gave the generalization that if a, b, c, and d represent whole numbers, then $(a + b) + (c + d) = (a + c) + (b + d)$. You can use this idea to simplify both numerical and algebraic expressions.

Numbers	Algebra
$(17 + 14) + (3 + 16)$	$(2a + 3b) + (5a + 6b)$
$= (17 + 3) + (14 + 16)$	$= (2a + 5a) + (3b + 6b)$
$= 20 + 30$	$= 7a + 9b$
$= 50$	

Use the idea above to simplify each expression.

1. $(3x + 4y) + (4x + 2y)$
2. $(4a + 3b) + (7a + 6b)$
3. $(7u + 3n) + (4u + 2n)$
4. $(3e + 5f) + (9e + 2f)$

1. $7x + 6y$ 2. $11a + 9b$ 3. $11u + 5n$ 4. $12e + 7f$

4-4 Solving Decimal Equations: Using Addition and Subtraction

Objective: To solve decimal equations involving addition and subtraction.
MIN: 1–18; REG: 1–18 even, 19–25; MAX: 1–18 even, 19–27; ALL: Mixed Review

To solve decimal equations involving addition and subtraction, you can use the same steps you used to solve integer equations.

Example 1

Solve and check. $x + 0.21 = 7.32$

Solution $x + 0.21 = 7.32$ — You need to get the variable by itself on one side.

$x + 0.21 - 0.21 = 7.32 - 0.21$ — To undo adding 0.21, subtract 0.21 from both sides.

$x = 7.11$

Check $7.11 + 0.21 \stackrel{?}{=} 7.32$ — Replace x with 7.11 in $x + 0.21 = 7.32$.

$7.32 = 7.32$ ✓ — The solution is 7.11.

Practice Solve and check. **To solve a, it may help to think of adding to both sides of the equation the additive inverse of 2.39.**

a. $x + 2.39 = -14.85$ **−17.24**

b. $s + 4.287 = 6.24$ **1.953**

Example 2

Solve and check. $x - 0.01 = 4.35$

Solution $x - 0.01 = 4.35$ — You need to get the variable by itself on one side.

$x - 0.01 + 0.01 = 4.35 + 0.01$ — To undo subtracting 0.01, add 0.01 to both sides.

$x = 4.36$

Check $4.36 - 0.01 \stackrel{?}{=} 4.35$ — Replace x with 4.36 in $x - 0.01 = 4.35$.

$4.35 = 4.35$ ✓ — The solution is 4.36.

Practice Solve and check.

a. $x - 2.39 = 14.85$ **17.24**

b. $x - 5.92 = -2.813$ **3.107**

Oral Exercises

Tell what operation you would use to solve the equation.

1. $x - 0.78 = 3.7$ +
2. $x + 8.92 = 13.5$ −
3. $y + 19.3 = 31$ −
4. $x + 25.7 = 14.3$ −
5. $x + 9.6 = 31.7$ −
6. $z - 2.84 = 51.82$ +
7. $s - 0.05 = 0.12$ +
8. $u - 1.24 = 4.82$ +
9. $w + 0.342 = -6.82$ −

Exercises 10. −28.44 15. 150.12 18. 40.82 21. −130.32

A Solve and check.

1. $y + 0.05 = 7.95$ 7.9
2. $z + 17.5 = 46.82$ 29.32
3. $u + 0.99 = 5.72$ 4.73
4. $x - 12.3 = -6.28$ 6.02
5. $x + 29.6 = 142.8$ 113.2
6. $x - 29.6 = -14.3$ 15.3
7. $s + 19.4 = 46.7$ 27.3
8. $s - 79.4 = -46.7$ 32.7
9. $y - 17.9 = 583.6$ 601.5
10. $u + 1.53 = -26.91$
11. $v - 23.7 = -7.41$ 16.29
12. $z + 0.5 = -5.03$ −5.53
13. $x - 23.4 = 345.61$ 369.01
14. $x - 5.05 = -4.38$ 0.67
15. $x - 4.32 = 145.8$
16. $38.06 + x = 91.5$ 53.44
17. $z - 2.04 = 48.13$ 50.17
18. $-14.95 + t = 25.87$

B Solve and check.

19. $-84.02 + z = 46.3$ 130.32
20. $12.3 + x = -6.28$ −18.58
21. $-46.3 = z + 84.02$
22. $(x - 5.2) + 0.47 = 25.84$ 30.57
23. $32.7 + x + 25.2 = 124.8$ 66.9
24. Write and solve an equation with sum of 35.82, one addend of 13.08, and the other addend represented by the variable y. $13.08 + y = 35.82, y = 22.74$
25. Write and solve an equation with sum of 17.36, one addend of 8.24, and the other addend represented by the variable x. $17.36 = 8.24 + x, x = 9.12$

C Extending Thinking Skills

26. Find values for x and y that solve both equations: $x + y = 1.5$ and $x - y = 0.5$. $x = 1, y = 0.5$
27. The numbers in each row, column, and diagonal of this magic square have the same sum. Write equations for finding the values of w, x, y, and z.

1.6	0.2	x	1.3
w	1.1	1.0	0.8
0.9	0.7	0.6	y
0.4	z	1.5	0.1

$1.6 + 0.2 + x + 1.3 = 3.4$
or $x + 1.0 + 0.6 + 1.5 = 3.4$
$w + 1.1 + 1.0 + 0.8 = 3.4$
or $1.6 + w + 0.9 + 0.4 = 3.4$
$0.9 + 0.7 + 0.6 + y = 3.4$
or $1.3 + 0.8 + y + 0.1 = 3.4$
$0.4 + z + 1.5 + 0.1 = 3.4$
or $0.2 + 1.1 + 0.7 + z = 3.4$

Thinking skills:
26. Testing a conjecture
27. Reasoning

Mixed Review

Compute. **28.** $120 - (65 \cdot 12) \div 15$ 68 **29.** $15 + 8 \cdot 6 - 12$ 51

Solve and check. **30.** $16m = -384$ −24 **31.** $r - 16 = -4$ 12

MENTAL MATH

You can "break apart" numbers to add or subtract decimals mentally. For example, to find 9.45 + 11.55, think "9 + 11 is 20 and 0.45 + 0.55 is 1. The sum is 21."

Find each sum or difference by breaking apart the numbers.

1. $7.35 + 9.65$ 17
2. $23.95 + 9.05$ 33
3. $45.85 - 5.40$ 40.45
4. $96.89 - 46.80$ 50.09

4-5 Multiplying with Decimals

Objective: To estimate decimal products by rounding; to compute decimal products.
MIN: 1–18; REG: 1–18 even, 19–32; MAX: 1–18 even, 19–34; ALL: Mixed Review
An exploratory lesson opener can be found on page 548.

Casey is a mountain climber. While planning an expedition and buying supplies, she chose a type of rope priced at \$1.89 per meter. She wanted to buy 19.5 meters of rope, but wasn't sure she had enough money to pay for it. You can estimate the total price by rounding each decimal to the nearest whole number.

Example 1

Estimate the product by rounding. $19.5 \times \$1.89$

Solution

$19.5 \times \$1.89$

$\approx 20 \times 2$ — Round 19.5 to 20 and round 1.89 to 2. Multiply.

$= 40.00$

$19.5 \times \$1.89$ is about \$40.00.

Practice Estimate the product by rounding. **Possible answers are given.**

a. 9.8×12.1 120

b. 19.2×43.9 800

c. 186.2×5.3 1000

d. 493.6×39.4 20,000

Sometimes you need an exact answer rather than an estimate. The steps for multiplying decimals are given below. The *sign* of the product of two decimals is determined in the same way as is the sign for an integer product. The product of two decimals with like signs is positive. The product of two decimals with unlike signs is negative.

Multiplying with Decimals

- Multiply as with whole numbers.
- Place the decimal point in the product so that it has the same number of decimal places as the total number of decimal places in the factors.
- Determine the sign of the product, as with integers.

Example 2

Multiply. $1.02 \times (-0.36)$

Solution

$$\begin{array}{r} 1.02 \\ \times\ 0.36 \\ \hline 612 \\ 306 \\ \hline 0.3672 \end{array}$$

Encourage students to check their calculation with an estimate.

0.3672 Since the sum of the decimal places in the factors is 4, the product has 4 decimal places.

-0.3672 The product of a positive and a negative number is negative.

Practice Multiply. **a.** 4.12×2.1 **b.** $-7.3 \times (-0.02)$ a. 8.652 b. 0.146

Sometimes you need to write zeros to the left of the calculated number in order to place the decimal point.

Example 3

Multiply. 4.3×0.02

Solution

$$\begin{array}{r} 4.3 \\ \times\ 0.02 \\ \hline 0.086 \end{array}$$

Write a zero to the left of the calculated number to give the product 3 decimal places.

Practice Multiply. **a.** -3.7×0.003 **b.** $23.1 \times (-0.002)$ a. −0.0111 b. −0.0462

Oral Exercises

Place the decimal point mentally and read the answer.

1. $3.46 \times 57.9 = 200334$ 200.334

2. $146.2 \times 0.013 = 19006$ 1.9006

3. $12 \times 17.32 = 20784$ 207.84

4. $29.83 \times 0.0035 = 0104405$ 0.104405

5. $9.18 \times 1.018 = 934524$ 9.34524

6. $93.12 \times 33.4 = 3110208$ 3110.208

Exercises

A Estimate the product by rounding. Possible answers are given.

1. 19.8×5.25 100

2. 8.1×4.5 40

3. 3.9×0.75 4

4. 25.2×9.8 250

5. 399.6×20.03 8,000

6. 996.8×37.5 40,000

Multiply.

7. 18.9×0.01 0.189

8. $4.01 \times (-0.32)$ −1.2832

9. $-34.6 \times (-0.1)$ 3.46

10. 16.32×1.03 16.8096

11. -4.38×0.08 −0.3504

12. 196.3×0.01 1.963

13. -84.6×-3.05 258.03

14. 0.193×100 19.3

15. 21.3×0.481 10.2453

16. 471.32×0.001 0.47132

17. 5280×-0.01 −52.8

18. -75.2×100 −7520

B Estimate the product, then compute.

19. 23.2×1.08 25.056 20. 0.515×0.02 0.0103 21. 7.3×0.003 0.0219

22. $2.3 + (3.2 \times 5.3)$ 19.26 23. $4.2(1.3 + 6.7)$ 33.6 24. $5.9(2.1 + 8.3)$ 61.36

Evaluate for $u = 2.3$.

25. $5.2u - 3.1$ 8.86 26. $9.4u + 32.1$ 53.72 27. $3.1u + 8.3$ 15.43

The relationship between temperature in degrees Fahrenheit (°F) and temperature in degrees Celsius (°C) is given by the formula $F = 1.8C + 32$. Evaluate the formula to find F for each value of C.

28. $C = 0°$ 32°F 29. $C = 17°$ 62.6°F 30. $C = 37°$ 98.6°F

31. The hottest temperature ever recorded in North America was 56.7°C, in Death Valley. Use the formula above to find the temperature in degrees Fahrenheit. 134.06°F

32. A record was set by a driver in a three-wheeled car that averaged 157.19 miles per gallon while using 2.871 gallons of fuel. How many miles (to the nearest tenth) did the car travel? 451.3 miles

C Extending Thinking Skills

33. Leslie, Beth, Rafael, and Jason are joining Casey on a mountain climb. In how many different orders can the five climbers tie themselves to the safety rope if Casey is always the lead member of the party? 24

34. A mountain climb has four routes from the base camp to the first campsite, two routes from there to the second campsite, and one route from there to the peak. How many different routes are there to the peak? 8

Thinking skills: 33, 34. Organizing

Mixed Review

Simplify. 35. $46c - 3c$ 43c 36. $35 + (m + 6)$ $m + 41$ 37. $7(-9k)$ $-63k$

Evaluate for $a = 6, b = 4, c = 2$. 38. $a + b - c$ 8 39. $2(a - c) + b$ 12

Round to the nearest tenth and estimate the value of n. 40. $n + 0.196 = 0.662$ 0.5

Order from least to greatest. Use inequality symbols. 41. 1.04, 1.4, 0.14 $0.14 < 1.04 < 1.4$

MENTAL MATH

When multiplying mentally with decimals, it is often helpful to use the distributive property to "break apart" numbers. To find 3×2.3, think "3×2 is 6 and 3×0.3 is 0.9, a total of 6.9."

Use the distributive property to find each product mentally.

1. 3×9.25 27.75 2. 4×5.25 21 3. 5×3.02 15.1

4. 2×7.05 14.1 5. 4×3.6 14.4 6. 3×4.2 12.6

PROBLEM SOLVING: APPLICATIONS

4-6 The Metric System

Objective: To solve problems involving metric units.
ALL: 1–11

The metric system of measurement uses the meter (m) as the basic unit of length, the liter (L) as the basic unit of capacity, and the gram (g) as the basic unit of mass.

A series of prefixes is used with the basic units. The table below shows how each prefix corresponds to a decimal place value.

kilo-	hecto-	deka-	--	deci-	centi-	milli-
1000 thousands	100 hundreds	10 tens	1 ones	0.1 tenths	0.01 hundredths	0.001 thousandths

When a prefix is attached to one of the basic units, a new unit is created. Look at the table below. Multiply a unit by 0.1 to obtain the unit on its right. Multiply a unit by 10 to obtain the unit on its left.

← larger **Metric Units of Length** smaller →						
kilometer (km)	**hecto**meter (hm)	**deka**meter (dkm)	meter (m)	**deci**meter (dm)	**centi**meter (cm)	**milli**meter (mm)
1000 m	100 m	10 m	1 m	0.1 m	0.01 m	0.001 m

The basic units of length, capacity, and mass are related to one another.

A cube with edges the *length* of one **decimeter** has the *capacity* of one **liter**. The *mass* of one liter of water is 1 **kilogram**.

Problems

You might wish to have students choose from among the techniques pencil & paper, mental math, estimation, and calculator. Have them choose the most appropriate technique for each problem.

Solve.

1. How many kilometers long is a 1500-meter race? 1.5 km

2. A man's waist size might be 90 cm. How many meters is that? 0.9 m

3. If the width of the palm of your hand is 0.8 dm, how wide is it in centimeters? 8 cm

4. A football field is about 90 m long. How long is it in decimeters? In dekameters? 900 decimeters 9 dekameters

5. A penny is about 18 mm in diameter. How many centimeters is that? **1.8 cm**

6. A building is 12.34 meters long. How long is it in centimeters? **1234 cm**

7. A stack of five pennies is about 1 cm high. Approximately how many millimeters thick is one penny? **2 mm**

8. If a person's stride is 50 cm long, how many strides equal one kilometer? **2,000**

9. How many millimeters are there in one kilometer? **1,000,000**

10. A thimble has a capacity of about 2 ml. What is the approximate mass of the water contained by a full thimble? (1 liter = 1 kilogram) **2 grams**

11. ***Data Search*** Weigh a milk carton full of water to find the mass of the water in grams. Approximately how many liters does the milk carton hold? **answers will vary**

What's Your Decision?

To find the mass of the water in a swimming pool, would you prefer to know the number of liters or the number of gallons it holds? **Liters; volume-to-mass conversions are simpler.**

One dm^3 has a capacity of 1 liter.

4·7 Dividing with Decimals

Objective: To estimate decimal quotients by rounding and choosing compatible numbers; to compute decimal products.
MIN: 1–21; REG: 1–21 even, 22–28; MAX: 1–21 even, 22–30; ALL: Mixed Review

A university charges \$5789.40 for tuition, fees, room, and board for the school year. A student is permitted 12 monthly payments. What is the amount of the payment each month? You can estimate an answer to this question. When estimating, use the symbol $\approx$, which means "is approximately equal to."

Example 1

Round and choose compatible numbers to estimate.
$5789.40 \div 12$

Solution $5789.40 \div 12$

$\approx 5800 \div 12$ Round 5789.40 to 5800.

$\approx 5500 \div 11$ Change 5800 to 5500 and 12 to 11 to get compatible numbers for division.

≈ 500 $5789.40 \div 12 \approx 500$.

Practice Round and choose compatible numbers to estimate.
a. $25.20 \div 8.10$ 3 **b.** $46.30 \div 5.91$ 8

You will often need an exact answer rather than an estimate. When the divisor is a whole number, place the decimal point in the quotient directly above the decimal point in the dividend.

Example 2

Divide. $13.5 \div 12$

Solution

```
    1.125
12)13.500
   12
    15
    12
     30
     24
      60
      60
       0
```

Since the divisor is a whole number, place the decimal point in the quotient directly above the decimal point of the dividend.

Annex zeros to the dividend as needed in the division process.

Practice Divide. **a.** $382.5 \div 18$ **b.** $223.2 \div 24$ a. 21.25 b. 9.3

When the decimal point is moved one place to the right in both dividend and divisor, the same quotient results.

$$2.0\overline{)6.0}\;=\;3. \qquad 20.\overline{)60.}\;=\;3.$$

This supports the following procedure for dividing with decimals.

Dividing with a Decimal Divisor

- Move the decimal point the same number of places to the right in both divisor and dividend until the divisor is a whole number.
- Divide.
- Determine the sign as you would for division with integers.

Example 3

Divide. $17.25 \div 1.5$

Solution

$$\begin{array}{r} 11.5 \\ 1.5\overline{)17.25} \\ 15 \\ \hline 22 \\ 15 \\ \hline 75 \end{array}$$

Since the divisor is a decimal, move the decimal point until the divisor is a whole number. Move the decimal point one place to the right in both divisor and dividend. Place the decimal point in the quotient.

Practice Divide.

a. $262.2 \div 2.3$ 114 **b.** $1.353 \div 4.1$ 0.33 **c.** $33.37 \div 4.7$ 7.1 **d.** $187.59 \div 3.9$ 48.1

Sometimes you need to annex zeros in the quotient when placing the decimal point.

Example 4

Divide. $-0.0483 \div 2.1$

Solution

$$\begin{array}{r} 0.023 \\ 2.1\overline{)0.0483} \\ 42 \\ \hline 63 \\ 63 \\ \hline \end{array}$$

Move the decimal point in both the divisor and the dividend and place the decimal point one place to the right. In the quotient, annex a zero.

-0.023 The quotient is negative because the divisor and the dividend have unlike signs.

Practice Divide.

a. $0.0045 \div 1.8$ 0.0025 **b.** $0.00496 \div 0.0031$ 1.6 **c.** $0.0315 \div 0.7$ 0.045 **d.** $0.00338 \div 2.6$ 0.0013

Oral Exercises

Place the decimal point in the quotient mentally and read it.

1. $29.61 \div 9 = 329$ 3.29
2. $16.3 \div -5 = -326$ −3.26
3. $104.4 \div 6 = 174$ 17.4
4. $4.816 \div 43 = 112$ .112
5. $129.93 \div 61 = 213$ 2.13
6. $135.47 \div 7.13 = 19$ 19.

Exercises

A Round to the nearest whole number and choose compatible numbers to estimate.

1. $23.72 \div 4.9$ 5
2. $37.2 \div 5.78$ 6
3. $121.4 \div 40.3$ 3
4. $65.7 \div 8.48$ 8
5. $76.8 \div 26.1$ 3
6. $82.1 \div 9.8$ 8

Divide.

7. $33.58 \div 73$ 0.46
8. $-0.4611 \div 53$ −0.0087
9. $6.201 \div 53$ 0.117
10. $1.6758 \div 21$ 0.0798
11. $129.89 \div 31$ 4.19
12. $87.99 \div (-21)$ −4.19
13. $4.42 \div 1.3$ 3.4
14. $-51.3 \div (-1.9)$ 27
15. $2.88 \div 0.18$ 16
16. $10.4 \div (-0.02)$ −520
17. $109.5 \div 0.05$ 2190
18. $-131.58 \div (-8.6)$ 15.3
19. $3.24 \div 2.7$ 1.2
20. $46.08 \div 12.8$ 3.6
21. $27.12 \div 2.4$ 11.3

B Estimate an answer, then divide.

22. $(4.63 + 25.5) \div 2.3$ 13.1
23. $(24.7 \div 6.5) + 4.53$ 8.33
24. $(1.3 + 1.82) \div 2$ 1.56

Evaluate.

25. $x \div 7.5$ for $x = 12$ 1.6
26. $x \div 2.25$ for $x = 72$ 32
27. $x \div 3.5$ for $x = 17.5$ 5

Solve.

28. An insurance policy has an annual premium of \$317.96 and there is an annual service charge of \$2.50 if it is billed quarterly. Estimate the quarterly payments. \$80.00

C Extending Thinking Skills

Find the pattern in the quotients and write one more equation.

29. $1.21 \div .11 =$ _?_ 11
$12.321 \div 1.11 =$ _?_ 11.1
$123.4321 \div 11.11 =$ _?_ 11.11
$1234.54321 \div 111.11 = 11.111$

30. $0.05 \div .99 =$ _?_ .050505
$0.08 \div .99 =$ _?_ .080808
$0.123 \div .999 =$ _?_ .123123
Answers will vary.

Mixed Review

Thinking skill: 29–30. Finding a pattern

Solve and check. 31. $n + 0.65 = 1.7$ 1.05 32. $r - 0.83 = -1.26$ −0.43

Evaluate for $m = 1.63$. 33. $m - 19.09$ 34. $1.27 - m$ −0.36 35. $3.2 + m$ 4.83

Round to the nearest tenth and estimate the value of x. 36. $9.02 + x = 1.49$ −7.5

Write as an algebraic expression. 37. 0.95 added to a number $n + 0.95$

33. −17.46

4-8 Writing Equations to Solve Problems

Objective: To solve word problems by writing equations.
MIN: 1–6; REG: 1–6 even, 7–9; MAX: 1–6 even, 7–11; ALL: Mixed Review

You can use the Problem-Solving Checklist as a guide for solving problems when your plan involves the strategy **Write an Equation**.

Problem-Solving Checklist: Writing Equations

Understand the Question/Find the Needed Data
What do you need to find? Can you show the data in the problem?

Plan What to Do
Can you use a variable to represent an unknown number?
Can you represent other conditions in terms of the variable?
What is equal in the problem? Can you write and solve an equation?

Find the Answer/Check Back
Does the solution of the equation check?
What is the answer to the question in the problem?
Does the answer seem reasonable?

Example

Find the solution by writing and solving an equation. Check your answer.

Karla was running in a 26 km marathon. A friend held up a sign saying, "Only 8 km more to run." How far has Karla run?

Solution

You need to find how many kilometers Karla has run so far.

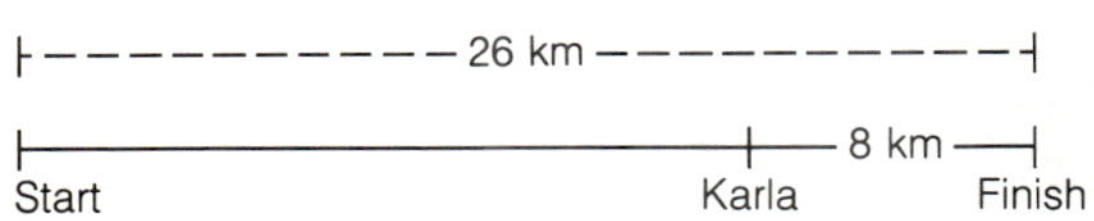

Draw a picture to show the data.

Let d = the distance Karla has run. — Choose a variable to represent what you want to find.

$$d + 8 = 26$$
$$d + 8 - 8 = 26 - 8$$
$$d = 18$$

The distance Karla has run plus 8 km is equal to the total distance of the race. Write and solve an equation.

Check $18 + 8 \stackrel{?}{=} 26$ — Check the equation solution.

$26 = 26$ ✓ — The equation solution checks

Karla has run 18 km. — 18 km and 8 km is 26 km. The answer is reasonable.

Practice Find the solution by writing and solving an equation. Students who sell 25 yearbooks win a radio. Karen needs to sell 6 more yearbooks to win a radio. How many has she sold so far? 19

Oral Exercises

Answer **a** through **e** for each problem.

a. What are you trying to find?

b. What are the important data?

c. What will the variable represent?

d. What is equal in the problem?

e. What is the equation?

1. A salesman sold \$3440 worth of merchandise. This was half his goal for the month. What was his goal? 1e. $g \div 2 = 3440$

2. After Mr. and Mrs. Wong gave their children 45 rare coins, they had 129 left in their collection. How many coins were in their original collection? 2e. $c - 45 = 129$

3. Lupe can start driving in 3 years. If the driving age is 16, how old is Lupe now? 3e. $a + 3 = 16$

4. Phil saves \$2 from his allowance each week. How many weeks will he have to save to buy a \$36 camera? 4e. $2x = 36$

5. Robin scored twice as many points in tonight's game as she did in last night's game. Tonight she scored 22 points. How many did she score last night? 5e. $2x = 22$

Exercises

A Find the solution by writing and solving an equation.

1. A movie was set in a year 16 years before the year 2005. In what year was it set? $m = 2005 - 16$; 1989

2. A theater sold \$789 worth of tickets. Each ticket cost \$3. How many tickets did the theater sell? $3c = 789$; 263 tickets

3. Mr. Bolt sold 17 fewer cars than Ms. Johnson sold. Mr. Bolt sold 53 cars. How many cars did Ms. Johnson sell? $j - 17 = 53$; 70 cars

4. A sweater costs \$24.75. Dana needs \$8.50 more to buy the sweater. How much money does Dana have now? $d + 8.50 = 24.75$; she has \$16.25

5. A runner's time was 4.52 seconds more than the school's record. The runner's time was 29.32 seconds. What was the school's record? $r + 4.52 = 29.32$; record was 24.80 seconds

6. Barry spent \$3.75 of his allowance. He had \$4.25 left. How much was his allowance? $a - 3.75 = 4.25$; \$8.00

B

7. Jan made two dresses, using 6 m of fabric altogether. Each dress took 0.15 of a bolt of fabric. How many meters of fabric were on one bolt?
$0.15r = 6 \div 2$; 20 meters

8. Elaine earns \$4.25 an hour as a part-time clerk at a grocery store. Last week, she earned \$85.75, which included a \$5 bonus. How many hours did she work last week? $4.25h = 85.75 - 5$; 19 hours

9. A number of people signed up for a recreation program. When 17 more enrolled the next day, 15 groups of 8 people could be formed. How many people signed up the first day? $n + 17 = 15 \times 8$; 103 people

C Extending Thinking Skills

Write a word problem that could be solved using the equation.

10. $x + 15 = 43$ **11.** $5x = 115$ answers will vary

Thinking skill: Formulating a problem

Mixed Review

Solve and check. **12.** $t - 1.863 = -2.1$ -0.237 **13.** $m + 3.6 = 8$ 4.4

Evaluate for $a = 1.5$, $b = 2.0$. **14.** $2a - b$ 1 **15.** ab 3 **16.** $a + b - 8$ -4.5

Round to the nearest whole number and estimate the product. **17.** 9.2(0.89) 9

CALCULATOR ACTIVITY

Some calculators have a [K] key that can be used for repeated multiplication by the same factor. The key sequence 3 [×] [K] makes 3 the repeating factor, as shown in the example below.

	Display		Display		Display
3 [×] [K] 4 [=]	12	6 [=]	18	9 [=]	27

This key sequence completes the problems $3 \cdot 4$, $3 \cdot 6$, and $3 \cdot 9$, with the answers displayed after each press of the [=] key.

A store owner is raising all prices by the factor 1.05. Use the constant multiplier key to complete this table of new prices. Round prices to the nearest cent, if necessary.

Current price	New price
\$ 53.95	\$56.65
1. 72.85	\$ 76.49
2. 92.60	\$ 97.23
3. 126.98	\$133.33
4. 361.89	\$379.98
5. 455.99	\$478.79

More Practice

Evaluate.

1. 12.35×0.13 1.6055	**2.** $2.73 + 9.08$ 11.81	**3.** $\frac{0.232}{0.02}$ 11.6	**4.** $12.62 - 1.57$ 11.05
5. $0.67 + 1.93$ 2.6	**6.** 6.24×3.45 21.528	**7.** $8.03 - 9.99$ −1.96	**8.** $\frac{-4.29}{1.65}$ −2.6
9. $\frac{14.3}{-0.55}$ −26	**10.** $15.62 - 13.77$ 1.85	**11.** $4.8 + 0.673$ 5.473	**12.** 9.05×4.4 39.82
13. $1.66 - 0.84$ 0.82	**14.** $\frac{8.25}{3.75}$ 2.2	**15.** -12.05×4.2 −50.61	**16.** $-11.4 + 0.391$ −11.009
17. $10.6 + 1.04$ 11.64	**18.** -2.6×1.03 −2.678	**19.** $11.62 - 7.92$ 3.7	**20.** $\frac{15.21}{0.65}$ 23.4
21. 6.05×3.2 19.36	**22.** $\frac{5.2}{-1.25}$ −4.16	**23.** $8.11 + 6.84$ 14.95	**24.** $44.1 - 37.6$ 6.5
25. $27.04 - 1.62$ 25.42	**26.** $26.1 + 13.9$ 40	**27.** $\frac{23.1}{-6.6}$ −3.5	**28.** 1.75×0.35 0.6125
29. $\frac{9.45}{4.5}$ 2.1	**30.** $16.74 - 3.62$ 13.12	**31.** 16.2×11.5 186.3	**32.** $49.02 + 0.199$ 49.219
33. 11.6×-5.2 −60.32	**34.** $12.62 - 18.04$ −5.42	**35.** $6.25 + 7.75$ 14	**36.** $\frac{4.08}{2.4}$ 1.7
37. $11.6 + 4.3$ 15.9	**38.** $\frac{-45}{(-0.09)}$ 500	**39.** $29.41 - 27.6$ 1.81	**40.** 0.75×9.4 7.05
41. $\frac{-61.56}{8.1}$ −7.6	**42.** $13.11 + 17.49$ 30.6	**43.** 6.1×0.75 4.575	**44.** $3.62 - 1.99$ 1.63
45. $41.35 - 9.16$ 32.19	**46.** -19.25×16.4 −315.7	**47.** $\frac{1.69}{1.3}$ 1.3	**48.** $8.62 + 1.38$ 10
49. $12.62 + 8.17$ 20.79	**50.** $\frac{-2.07}{-0.23}$ 9	**51.** $7.92 - 1.09$ 6.83	**52.** -9.2×6.45 −59.34
53. $6.34 - 4.76$ 1.58	**54.** $14.02 + 6.973$ 54. 20.993	**55.** 84×7.14 599.76	**56.** $\frac{67.32}{1.02}$ 66
57. $\frac{47.3}{-5.5}$ −8.6	**58.** $8.52 - 6.66$ 1.86	**59.** 10.25×2.4 24.6	**60.** $47.9 + 61.43$ 109.33
61. $1.904 + 0.685$ 2.589	**62.** 1.335×2.5 3.3375	**63.** $\frac{33.063}{103}$ 0.321	**64.** $14.11 - 16.8$ −2.69
65. -4.28×1.6 −6.848	**66.** $21.6 - 14.03$ 7.57	**67.** $1.673 + 0.499$ 2.172	**68.** $\frac{2.25}{-4.5}$ −0.5
69. $3.49 - 6.004$ −2.514	**70.** $16.37 + 9.642$ 26.012	**71.** $\frac{-0.196}{98}$ −0.002	**72.** 19.5×30.8 600.6

4-9 Solving Decimal Equations: Using Multiplication and Division

Objective: To solve decimal equations involving multiplication and division.
MIN: 1–15; REG: 1–15 even, 16–26; MAX: 1–15 even, 16–28; ALL: Mixed Review

To solve an equation in which the variable has been multiplied by a decimal, divide both sides of the equation by that decimal.

Example 1

Note that the procedure given here is the same as the method for solving integer equations.

Solve and check. $0.12x = 60$

Solution

$$0.12x = 60$$

To undo multiplying by 0.12, divide by 0.12 on both sides.

$$\frac{0.12x}{\mathbf{0.12}} = \frac{60}{\mathbf{0.12}}$$

$$x = 60 \div 0.12$$

$$x = 500$$

Check

$$0.12(500) \stackrel{?}{=} 60$$

Replace x with 500 in $0.12x = 60$.

$$60 = 60 \checkmark$$

The solution is 500.

Practice Solve and check.

a. $1.5x = 225$ **150** **b.** $0.05x = -46$ **−920**

To solve an equation in which the variable has been divided by a decimal, multiply both sides of the equation by that decimal.

Example 2

Solve and check. $\frac{x}{2.15} = -1.4$

Solution

$$\frac{x}{2.15} = -1.4$$

$$\frac{x}{2.15}(\mathbf{2.15}) = -1.4(\mathbf{2.15})$$

To undo dividing by 2.15, multiply by 2.15 on both sides.

$$x = -3.01$$

Unlike signs result in a negative product.

Check

$$\frac{-3.01}{2.15} \stackrel{?}{=} -1.4$$

Replace x with −3.01 in $\frac{x}{2.15} = -1.4$.

$$-1.4 = -1.4 \checkmark$$

The solution is −3.01.

Practice Solve and check.

a. $\frac{x}{0.14} = 5.2$ **0.728** **b.** $\frac{z}{-3.2} = 5.8$ **−18.56**

Oral Exercises

To solve, would you multiply or divide in the first step?

1. $\frac{x}{2.1} = 5.98$ ×
2. $-8.21y = 5.7$ ÷
3. $23.5z = 8.3$ ÷
4. $19.6 = 3.5w$ ÷
5. $\frac{u}{7.1} = -83.5$ ×
6. $-19.7 = \frac{z}{-3.9}$ ×

Exercises

A Solve and check.

1. $2.1x = 11.13$ 5.3
2. $0.15z = 0.24$ 1.6
3. $1.41y = -9.87$ −7
4. $\frac{w}{6.5} = 7.2$ 46.8
5. $\frac{z}{8.3} = -2.3$ −19.09
6. $\frac{w}{3.2} = 48$ 153.6
7. $-4.5x = 67.5$ −15
8. $\frac{y}{1.7} = -3$ −5.1
9. $3.6z = -43.2$ −12
10. $-0.002u = -576.4$ 288200
11. $1.005x = 20.1$ 20
12. $\frac{x}{-0.09} = 81$ −7.29
13. $\frac{u}{4.98} = -1.2$ −5.976
14. $\frac{-x}{0.55} = 0.2$ −0.11
15. $\frac{y}{-5.4} = 3.2$ −17.28

B Solve and check. 17. −18 18. 0.3125 22. 3.1 23. 113.82 24. 157.48

16. $-321.3 = 5.1x$ −63
17. $-306.9 = 17.05x$
18. $21 = 67.2z$
19. $\frac{-u}{12.5} = -1.4$ 17.5
20. $-0.91 = \frac{u}{1.2}$ −1.092
21. $4.3 = \frac{x}{3.7}$ 15.91
22. $(1.4 + 3.6)x = 15.5$
23. $\frac{x}{4.2} = 5.8 + 21.3$
24. $\frac{z}{8.5 - 2.3} = 25.4$

Write and solve an equation.

25. A number is multiplied by 1.2 and results in the product 264. 1.2x = 264; x = 220
26. A number is divided by 3.4 and results in the quotient 25.3. x ÷ 3.4 = 25.3; x = 86.02

C Extending Thinking Skills

Use inverse operations to find the missing number in each flow chart.

27. [Start with ?] → [multiply by 1.2] → [Add 4.3] → [end with 7.9] 3

28. [Start with 2.3] → [Add 4.6] → [Multiply by ?] → [end with 36.57] 5.3

Mixed Review Thinking skill: Using reverse reasoning

Evaluate for $x = 1.5$, $y = 3.6$. **29.** $2x - 2y$ −4.2 **30.** xy 5.4 **31.** $2x + y$ 6.6

Solve and check. **32.** $m + 1.04 = 3.1$ 2.06 **33.** $-12k = 144$ −12

4-10 Simplify the Problem

Objective: To solve nonroutine problems, using the strategy "Simplify the Problem" and other strategies learned so far.
MIN: 1–2; REG: 1–4; MAX: 1–6

Solving some problems can be quite difficult if you try to find the solution using the numbers given in the problem. The strategy **Simplify the Problem** can help. To simplify, substitute smaller numbers. Then solve the simplified problem, and use your solution to it to help you solve the original problem.

Problem Ms. Corelli needs to install phone lines for a political convention. She has to install lines connecting each committee chairperson's desk with each of the others' desks. Ten desks are to be connected in all. A separate phone line is needed for each connection so people won't listen in on conversations. How many phone lines should Ms. Corelli install?

To solve, try **simplifying** the problem. Suppose there were only 2 desks, rather than 10, that needed phone lines. You could draw a **picture**.

If there were 2 desks, 1 line would be needed.

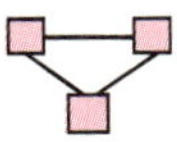
With 3 desks, 3 lines would be needed.

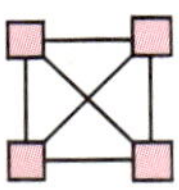
With 4 desks, 6 lines would be needed.

Recording the results from these simpler problems in a **table** helps you see a **pattern**. The difference in the number of lines needed increases by 1 each time.

Number of desks	2	3	4	5	6	7	8	9	10
Lines needed	1	3	6	10	15	21	28	36	45

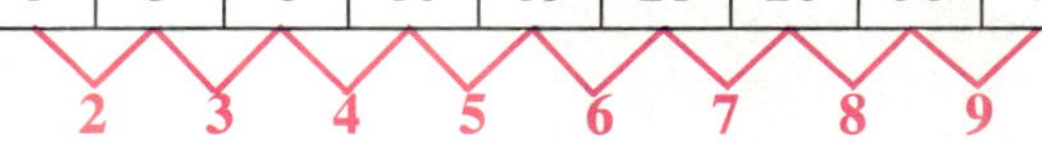

A total of 45 lines is needed to connect the 10 desks.

This chart shows the strategies presented so far. The strategy "Write an Equation" was introduced on page 121.

Problem-Solving Strategies	
Choose the Operations	**Look for a Pattern**
Guess, Check, Revise	**Write an Equation**
Draw a Picture	**Simplify the Problem**
Make a Table	

Problems

Solve.

1. A major company gave a large financial gift to a university. The plan was that $1 million would be given to the university the first year. In each of the following years, the university would receive an amount that was double the amount it received the previous year. What was the total amount of the gift through 12 years? **$4,095,000,000**

2. A grocer made a display with boxes arranged as in the illustration below. Each row had one less box than the row below it and the top row had one box. If 12 boxes were placed on the bottom row, how many were used all together? **78**

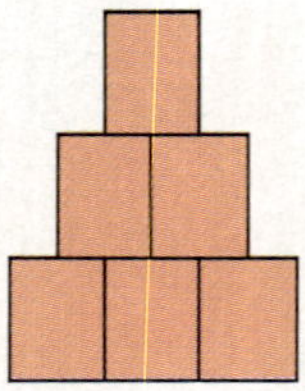

3. A 100 km bicycle race was held on roads near a large city. Markers were placed at the starting line of the race, at the finish line, and at every 10 km in between. How many markers were used for this race? **11 markers**

4. In a recent basketball game, the Giants beat the Lions by 25 points. The coach of the Lions said, "If we had scored twice as many points, we would have beaten the Giants by 14 points." What was the final score of the game? **Giants 64, Lions 39**

5. A boat was having engine trouble. Each day, it could go forward 20 miles. At night, it had to rest its engines and would drift back 5 miles. At this rate, how many days would it take the boat to get 100 miles from its starting point? **7 days**

6. A light bulb inspector found an alarmingly high number of defective bulbs in a recent shipment. Of the 25 bulbs she selected to test, she found 7 bulbs to be defective. How many might she expect to be defective out of the shipment of 275 bulbs? **77 bulbs**

Enrichment

The Binary System and the Computer

A microcomputer that has an 8-bit microprocessor processes information that is coded in strings of 8 digits of zeros and ones.

A digital computer uses only two digits, 0 and 1, to code information. Numbers are represented by a place value system called the **binary system**.

As in the decimal system, each digit in the binary system has a place value. The first seven place values are shown in red below.

Binary Place Value	64	32	16	8	4	2	1
Binary Number	1	1	0	1	1	0	1

The base 10 number represented by the binary number 1101101 is found by multiplying the digit in each place by its place value and adding together all the values.

$$\begin{aligned}(1101101)_2 &= 1(64) + 1(32) + 0(16) + 1(8) + 1(4) + 0(2) + 1(1)\\ &= 109\end{aligned}$$

Find the base 10 number represented by each binary number.

1. 1011 11 **2.** 1101 13 **3.** 110111 55 **4.** 100011 35

Chapter 4 Review

4-1 Write as a decimal.

1. $6(10) + 3(1) + 8\left(\frac{1}{10}\right)$ 63.8

2. $2\left(\frac{1}{1000}\right) + 5(1) + 7\left(\frac{1}{100}\right)$ 5.072

Write >, <, or = for each □.

3. -4.23 □ -4.28 >

4. 1.01 □ 1.010 =

5. 0.51 □ 0.514 <

4-2 Estimate the value of x by rounding. **Possible answers are given.**

6. $x = 94.27 - 13.66$ 80

7. $x = 801.7 + 200.1$ 1000

8. $x = 47.27 + 15.25$ 62

9. $x = 62.85 - 10.13$ 50

4-3 Add or subtract.

10. $7.13 + 1.3$ 8.43

11. $67.19 + 7.17 + 11.11$ 85.47

12. $-3.87 + 9.75$ 5.88

13. $61.3 + (-17.4) + 2.93$ 46.83

Evaluate.

14. $u - 16.45$ for $u = -12.3$ −28.75

15. $93.5 - t$ for $t = 6.8$ 86.7

4-4 Solve and check.

16. $c - 6.29 = -4.83$ 1.46

17. $e + 16.14 = -12.05$ −28.19

18. $m - 8.35 = 21.06$ 29.41

19. $-13.67 + n = 32$ 45.67

4-5 Estimate the product by rounding. **Possible answers are given.**

20. 25.1×7.6 200

21. 3.12×96.78 300

22. 299.6×0.8 300

Find the product.

23. 7.17×3.2 22.944

24. 19.8×-0.07 −1.386

25. -3.8×-9.2 34.96

26. -1462×-0.001 1.462

4-6 Solve.

27. A fence is 17.5 meters long. How many centimeters long is the fence? 1750

28. Hector's cat weighs 4 kilograms. How many grams does the cat weigh? 4000

4-7 Solve and check.

29. $4.4m = 92.4$ 21

30. $-0.03x = 5.4$ −180

31. $1.4s = 12.32$ 8.8

32. $\frac{c}{0.14} = 7.1$ .994

33. $\frac{f}{-3.3} = 1.1$ −3.63

34. $\frac{h}{6.1} = -2.3$ −14.03

4-8 Solve by writing an equation.

35. Ann got a $10.75 discount on a sweater with an original price of $37.90. How much did she pay for the sweater? $p = 37.90 - 10.75$; $27.15

36. A scout troop collected 88.6 kilograms of newspaper on Saturday and 97.4 kilograms on Sunday. How many kilograms did they collect in the two days? $88.6 + 97.4 = t$; 186

Chapter 4 Test

Write as a decimal.

1. $1(10) + 5(1) + 3\left(\frac{1}{10}\right)$ 15.3

2. $4(100) + 6\left(\frac{1}{100}\right) + 8\left(\frac{1}{10}\right)$ 400.86

Write >, <, or = for each □.

3. 290 □ 7.29 >

4. 6.01 □ 6.1 <

5. −3.25 □ −3.45 >

Estimate the value of x by rounding. Possible answers are given.

6. $33.89 - 19.25 = x$ 15

7. $x = 415.2 + 75.5$ 500

8. $x = 391.37 - 106.93$ 300

9. $59.82 + 16.05 = x$ 76

Add or subtract.

10. $32.83 + 12.93$ 45.76

11. $3.14 + 23.36$ 26.5

12. $45.2 + (-2.2)$ 43

13. $-0.33 + 24.81 + 2.66$ 27.14

14. $88.2 - 35.5$ 52.7

15. $-2.76 - 67.21$ −69.97

Solve and check.

16. $f + 17.6 = -12.9$ −30.5

17. $g - 6.43 = 29.62$ 36.05

18. $38.7 + z = 25.86$ −12.84

19. $18.06 + w = -7$ −25.06

Estimate the product by rounding. Possible answers are given.

20. 4.38×6.1 24

21. 19.6×9.8 200

22. 1.02×0.98 1

Find the product.

23. 16.4×0.02 0.328

24. -71.43×-0.01 0.7143

25. -9.4×8.5 −79.9

26. 144.7×-0.1 −14.47

Solve.

27. A container with a volume of 1 cubic centimeter has a capacity of 1 milliliter. What is the capacity of a container with a volume of 75 cubic centimeters? 75 milliliters

28. Lisa ran a 500 meter race. How many kilometers is the race? 0.5 kilometers

Solve and check.

29. $-0.5z = 3.15$ −6.3

30. $1.2p = -5.76$ −4.8

31. $0.18v = 3.24$ 18

32. $\frac{a}{0.8} = -4.4$ −3.52

33. $\frac{s}{-1.9} = -2.7$ 5.13

34. $\frac{w}{3.8} = 4.2$ 15.96

Solve by writing an equation.

35. Jorge made a deposit of \$16.50 on a \$46.50 jacket. How much does he still owe? $16.50 + m = 46.50$. He owes \$30.00.

36. Sophie had 24.7 meters of fencing. She bought a 50-meter roll of fencing. How much fencing does she have now? $f = 24.7 + 50$. She has 74.7 meters.

Cumulative Review

Write as a numerical expression.

1. 7 less than 18 $18 - 7$ **2.** 15 more than 32 $32 + 15$ **3.** 12 increased by 8 $12 + 8$

Write as an algebraic expression.

4. m less than 7 $7 - m$ **5.** c decreased by 9 $c - 9$ **6.** a number added to t 6. $t + n$

Solve the equation for the replacement set given.

7. $c + 4 = 12$ {6, 7, 8, 9} 8 **8.** $z - 8 = 8$ {13, 14, 15, 16} 16

9. $10 - n = 3$ {6, 7, 8, 9} 7 **10.** $4 + g = 14$ {7, 8, 9, 10} 10

Evaluate each algebraic expression.

11. $d + 1$ for $d = 6$ 7 **12.** $h - i$ for $h = 7, i = 0$ 7

13. $g + (-10)$ for $g = 3$ -7 **14.** $35 + t$ for $t = -10$ 25

15. $-4 - c$ for $c = 3$ -7 **16.** $-5 - z$ for $z = 2$ -7

Find the sum or difference.

17. $-2 + 6$ 4 **18.** $-6 + 9$ 3 **19.** $-3 + (-8)$ -11

20. $-2 + (-3)$ -5 **21.** $-6 + 19$ 13 **22.** $12 + (-6)$ 6

23. $5 - 3$ 2 **24.** $-11 - 1$ -12 **25.** $19 - 10$ 9

Solve and check.

26. $x + 22 = 88$ 66 **27.** $g - 31 = 17$ 48 **28.** $300 + k = 414$ 114

29. $506 = 472 - v$ -34 **30.** $383 = 209 - j$ -174 **31.** $365 = h + 100$ 265

32. $p + 45 = -25$ -70 **33.** $c - 77 = -30$ 47 **34.** $88 = k - 1$ 89

Solve.

35. Clyde read 4 more books this month than he read last month. If he read 10 books this month, how many books did he read last month? 6

36. Marquita gave some of her t-shirts away. If she had 14 t-shirts to begin with and has 8 now, how many did she give away? 6

37. A checking account has a beginning balance of \$405. How much money is in the account after a check is written for \$175 and a deposit is made for \$613? \$843

38. By how much is this checking account overdrawn? Beginning balance: \$206; check for \$310; check for \$22. \$126

Chapter 5
Number Theory

5-1 Multiples and Factors

Objective: To write multiples and factors of a number.
MIN: 1–35; REG: 1–35 even, 36–41; MAX: 1–35 even, 36–43; ALL: Mixed Review
An exploratory lesson opener can be found on page 548.

The product of two whole numbers is a **multiple** of each of the whole numbers. The multiples of a number can be found by multiplying the number by 0, 1, 2, 3, 4, and so on.

$$0 \times 2 = \boldsymbol{0}, \quad 1 \times 2 = \boldsymbol{2}, \quad 2 \times 2 = \boldsymbol{4}, \quad 3 \times 2 = \boldsymbol{6}, \ldots$$

0, 2, 4, 6, . . . are multiples of **2**. All whole numbers that are multiples of 2 are called **even numbers**. All whole numbers that are not multiples of 2 are called **odd numbers**.

Example 1

Encourage students to practice skip-counting by 2s, 3s, 4s . . . 8s, and 9s to produce multiples of these numbers.

Write the first five nonzero multiples of 6.

Solution 6, 12, 18, 24, 30 Multiply 6 by 1, 2, 3, 4, and 5.

Practice

a. Write the first three nonzero multiples of 7. **7, 14, 21**

b. Write the first five nonzero multiples of 9. **9, 18, 27, 36, 45**

On some calculators, you can enter a number and press [+] [=] [=] [=] . . . *to display its multiples.*

When two or more whole numbers are multiplied to form a product, each is called a **factor** of the product.

factors of 20 → 4 × 5 = 20 ← The product of 4 times 5

When a whole number is divided by one of its nonzero factors, the quotient is a whole number and the remainder is zero. Thus, you can find out if one number is a factor of another by dividing. If the quotient is a whole number with a remainder of 0, then the divisor and the quotient are each factors of the dividend.

Example 2

Is 9 a factor of 153?

Solution $153 \div 9 = 17$ Since $153 \div 9 = 17$, with a remainder of 0,

9 is a factor of 153. then $17 \times 9 = 153$, and 9 and 17 are factors of 153.

17 is also a factor of 153.

Practice

a. Is 13 a factor of 106? **No**

b. Is 7 a factor of 84? **Yes**

To find all the factors of a whole number, divide the number by each whole number, beginning with 1. Stop dividing when the factors begin to repeat.

Example 3

Find all the factors of 12.

Solution The factors of 12 are 1, 2, 3, 4, 6, and 12.

$12 \div 1 = 12$, so **1** and **12** are factors of 12.
$12 \div 2 = 6$, so **2** and **6** are factors of 12.
$12 \div 3 = 4$, so **3** and **4** are factors of 12.
$12 \div 4 = 3$ produces no new factors. The factors 3 and 4 repeat. Stop.

Practice

a. Find all the factors of 18. **1, 2, 3, 6, 9, 18**

b. Find all the factors of 32. **1, 2, 4, 8, 16, 32**

Oral Exercises

Is the statement true or false?

1. 25 is a multiple of 5. **T**

2. 24 is a factor of 7. **F**

3. 4 is a multiple of 16. **F**

4. 28 is a multiple of 4. **T**

5. 5 is a factor of 25. **T**

6. 32 is a multiple of 9. **F**

7. 7 is a multiple of 35. **F**

8. 21 is a multiple of 3. **T**

9. 8 is a multiple of 64 **F**

10. 40 is a multiple of 5. **T**

Exercises

A Give the first four nonzero multiples of each number. **See additional answers**

1. 3	**2.** 5	**3.** 7	**4.** 2	**5.** 6	**6.** 8
7. 9	**8.** 10	**9.** 4	**10.** 11	**11.** 20	**12.** 100

Is the first number a factor of the second?

13. 4, 20 **Yes**	**14.** 5, 32 **No**	**15.** 3, 37 **No**	**16.** 8, 40 **Yes**	**17.** 9, 63 **Yes**
18. 6, 54 **Yes**	**19.** 8, 70 **No**	**20.** 7, 42 **Yes**	**21.** 4, 96 **Yes**	**22.** 5, 75 **Yes**
23. 3, 48 **Yes**	**24.** 2, 124 **Yes**	**25.** 3, 252 **Yes**	**26.** 4, 256 **Yes**	**27.** 9, 576 **Yes**

Find all the factors of the given number. **See additional answers.**

28. 16	**29.** 20	**30.** 28	**31.** 84
32. 40	**33.** 60	**34.** 34	**35.** 105

B

36. Solve the equations. Use the results to list all the factors of 72.

a. $1m = 72$ 72 b. $2n = 72$ 36 c. $3p = 72$ 24

d. $4q = 72$ 18 e. $6r = 72$ 12 f. $8s = 72$ 9

36. 1, 2, 3, 4, 6, 8, 9, 12, 18, 24, 36, 72

37. Solve the equation $18b = 1638$. Then give two 2-digit factors of 1638. 91. 18, 91

38. Find a number that has only two factors. 2, 3, or any prime

39. Find four numbers that have only three factors. Describe these numbers. 4, 9, 25, 49; squares

40. If n is any whole number, which expressions represent even numbers? a, c, f

a. $2n$ b. $3n$ c. $4n$ d. $2n + 1$ e. $2n - 1$ f. $2n + 2$ g. $n + 2$

41. Describe five ways in which twenty-four cars can be parked in rows in a parking lot, with the same number in each row.
For example, 2 rows of 12, 3 rows of 8, 4 rows of 6, 6 rows of 4, 8 rows of 3.

C Extending Thinking Skills

Thinking skills: 42. Generalizing 43. Formulating a proof

42. Complete these generalizations about even (E) and odd (O) numbers.

a. E + E = E b. O + O = E c. E + O = O

d. E × E = E e. O × O = O f. E × O = E

43. The number 6 is called a **perfect** number because the sum of its factors other than itself is 6. Show that 28 is a perfect number.
Factors: 1, 2, 4, 7, 14, 28. 1 + 2 + 4 + 7 + 14 = 28

Mixed Review

Solve and check. 44. $-9x = 59.4$ −6.6 45. $m \div 3.7 = 3.92$ 14.504

Evaluate for $h = -9.45$, $p = 4.5$. 46. $(h \div p) + 2$ −0.1 47. $2p - h + 1.75$ 20.2

CALCULATOR ACTIVITY

You can use a calculator with a memory to find the factors of a given whole number quickly. First enter the number and push [M] or [M+] to record it in the calculator's memory. Then try each divisor 2, 3, 4, 5, . . . in order, as shown below.

This key causes the number in memory to be displayed. → [MR] [÷] 2 (trial divisor) [=]

When the answer is a whole number, the trial divisor and the whole number are factors of the number in the memory.

Find the factors of each number.

1. 78 2. 90 3. 105 4. 210

1. 1, 2, 3, 6, 13, 26, 39, 78
2. 1, 2, 3, 5, 6, 9, 10, 15, 18, 30, 45, 90
3. 1, 3, 5, 7, 15, 21, 35, 105
4. 1, 2, 3, 5, 6, 7, 10, 14, 15, 21, 30, 35, 42, 70, 105, 210

5-2 Divisibility

Objective: To use divisibility rules.
MIN: 1–32; REG: 1–32 even, 33–46; MAX: 1–32 even, 33–49; ALL: Mixed Review

A whole number is **divisible** by another whole number if, when it is divided by that number, the result is a whole number quotient with a remainder of zero. For example, **21** is divisible by ***3*** because **21** ÷ ***3*** gives the quotient 7 with the remainder 0. A whole number is divisible by each of its factors.

Every number has 1 as a factor. The divisibility rules below are shortcuts for determining whether a whole number has 2, 3, or 5 as a factor.

Divisibility by 2, 3, and 5

- A number is **divisible by 2** if its ones digit is even (0, 2, 4, 6, or 8).
- A number is **divisible by 3** if the sum of its digits is divisible by 3.
- A number is **divisible by 5** if its ones digit is 0 or 5.

The examples below show how these rules are applied.

Example 1

Is 3,764 divisible by 2?

Solution 3,764 is divisible by 2. Since the ones digit (4) is even, the number is divisible by 2.

Practice Which numbers are divisible by 2? **a, c, d, f**

a. 58 **b.** 27 **c.** 36 **d.** 80 **e.** 653 **f.** 7254

Example 2

Is 560 divisible by 5?

Solution 560 is divisible by 5. Since 560 ends in **0**, it is divisible by 5.

Practice Which numbers are divisible by 5? **b, c**

a. 204 **b.** 375 **c.** 800 **d.** 5,509

Example 3

Is 837 divisible by 3?

Solution 837 is divisible by 3. 8 + 3 + 7 = 18. Since the sum of the digits is divisible by 3, 837 is divisible by 3.

Practice Which numbers are divisible by 3? a, b

a. 81 **b.** 243 **c.** 953 **d.** 9,613

Oral Exercises

Is the digit odd or even?

1. 6 even **2.** 0 even **3.** 5 odd **4.** 8 even **5.** 3 odd

Find the sum of the digits of the number.

6. 49 13 **7.** 357 15 **8.** 986 23 **9.** 5,768 26 **10.** 57,643 25

Exercises

A State whether the number is divisible by 2.

1. 45 No **2.** 36 Yes **3.** 74 Yes **4.** 67 No
5. 48 Yes **6.** 170 Yes **7.** 299 No **8.** 594 Yes

State whether the number is divisible by 5.

9. 58 No **10.** 75 Yes **11.** 60 Yes **12.** 125 Yes
13. 702 No **14.** 66,345 Yes **15.** 100,000 Yes **16.** 400,551 No

State whether the number is divisible by 3.

17. 48 Yes **18.** 63 Yes **19.** 98 No **20.** 173 No
21. 147 Yes **22.** 273 Yes **23.** 2,706 Yes **24.** 3,195 Yes
25. 2,897 No **26.** 54,623 No **27.** 81,659 No **28.** 97,497 Yes
29. 43,248 Yes **30.** 386,322 Yes **31.** 872,724 Yes **32.** 497,333 No

B A whole number is **divisible by 6** if it is divisible by 2 *and* by 3. State whether each number is divisible by 6.

33. 468 Yes **34.** 2,678 No **35.** 42,672 Yes **36.** 54,987 No

A whole number is **divisible by 9** if the sum of its digits is divisible by 9. State whether each number is divisible by 9.

37. 81 Yes **38.** 342 Yes **39.** 5,670 Yes **40.** 8,769 No

A whole number is **divisible by 4** if its last 2 digits are divisible by 4. State whether each number is divisible by 4.

41. 532 Yes **42.** 9,366 No **43.** 24,847 No **44.** 57,936 Yes

A number is also divisible by 4 if twice its tens digit plus its ones digit is divisible by 4.

45. Write a rule for divisibility by 10. A number is divisible by 10 if its last digit is 0.

46. Fill in the missing hundreds and ones digits in 3,▮4▮ so that the number will be:

a. divisible by 2 b. divisible by 3 c. divisible by 5

see page bottom

C Extending Thinking Skills

47. Find the smallest number that is divisible by each of the numbers 1 through 9. 2520

48. Give an example to show that a number that is divisible by both 2 and 4 is not necessarily divisible by 8. Answers may vary. Example: 12 is divisible by 2 and 4, but not by 8.

49. A number is divisible by 1, 5, and 7. It is not an odd number, and it is less than 100. Find the number. Find three other divisors of the number. 70. Divisors: 2, 10, 14, 35, 70

Thinking skills: 47. Reasoning 48. Seeking a counterexample 49. Using logic

Mixed Review

Write <, >, or = for each □. 50. 1.01 □ 10.1 < 51. −6.25 □ −7.19 >

52. −0.01 □ 0.001 < 53. 0.14 □ 0.14 = 54. 4.031 □ 4.013 >

Evaluate. 55. $k - 61.73$, for $k = 16.937$ −44.793 56. $m - 10.62$, for $m = 10.637$ .017

57. $(17 - r) + 26.1$, for $r = 16.937$ 26.163 58. $\frac{11.776}{x}$, for $x = 5.12$ 2.3

Solve and check. 59. $r - 9.05 = 6.21$ 15.26 60. $-1.3y = 1.69$ −1.3

Use the variable n to write an equation for each statement.

61. 11 more than twice a number gives 27. $2n + 11 = 27$

62. Half of a number equals the product of 3 and 4 $\frac{n}{2} = 3 \cdot 4$

MENTAL MATH

A way of deciding mentally whether a number is divisible by 3 is to "cast out multiples of 3" before adding the digits. Study the following mental steps for deciding whether the number 356,937 is divisible by 3.

Throw out all the digits that are multiples of 3.	→	Add the remaining digits.	→	Continue to add digits and throw out multiples of 3.
~~3~~5~~6~~,~~93~~7		5 + 7 = 12		1 + 2 = ~~3~~

If all the digits in the original number are thrown out, or if the final digit is thrown out, the number is divisible by 3.

Cast out multiples of 3 to decide mentally whether each number is divisible by 3.

1. 919,536 Yes 2. 837,654 Yes 3. 273,649 No

4. 943,761 Yes 5. 6,589,335 Yes 6. 74,239,436 No

46. Answers will vary, but must meet these requirements:
 a. ones digit must be 0, 2, 4, 6, or 8.
 b. sum of hundreds and ones digits must be 2, 5, 8, 11, 14, or 17.
 c. ones digit must be 0 or 5.

PROBLEM SOLVING: WRITING EQUATIONS

5-3 Writing Equations to Solve Problems

Objective: To solve word problems by writing and solving equations.
MIN: 1–5, 9–10; REG, MAX: 1–5 even, 6–10; ALL: Mixed Review

As shown in Chapter 4, the Problem-Solving Checklist can be a guide for solving problems when your plan is to write and solve an equation.

Problem-Solving Checklist: Writing Equations

Understand the Question/Find the Needed Data
What do you need to find? Can you show the data in the problem?

Plan What to Do
Can you use a variable to represent missing data?
Can you represent other conditions in terms of the variable?
What is equal in the problem? Can you write and solve an equation?

Find the Answer/Check Back
Does the equation solution check?
What is the answer to the question in the problem?
Does the answer seem reasonable?

Example

Find the solution by writing and solving an equation. Gloria had $51, which was just enough to pay for school yearbooks for herself and 2 friends. How much did a yearbook cost?

Solution Let c = cost of a yearbook — You want to find the cost of a single yearbook. Choose a variable to represent this.

$$3c = 51$$

Look for what is equal. $51 equals 3 times the cost of a yearbook.

$$\frac{3c}{3} = \frac{51}{3}$$

Write and solve an equation.

$$c = 17$$

Check $3(17) \stackrel{?}{=} 51$

$$51 = 51 \checkmark$$

The equation solution checks.

Each yearbook costs $17.

Estimate: three $20 yearbooks would cost $60. The answer is reasonable.

Practice Find the solution by writing and solving an equation. Todd divided the price of a stereo system by six to find the amount of one of six equal payments. The payment was \$84. What was the total price? $\frac{s}{6} = 84$ **\$504**

Exercises

A Solve by writing an equation. Check your answer.

1. Tim picked up a pizza for himself and three friends. Later, when they divided the cost of the pizza equally, each of them paid \$2.41. What was the total cost of the pizza? $\frac{c}{4} = 2.41$ **\$9.64**

2. A bandleader earns \$48 more per performance than the saxophonist. If the bandleader earns \$125 for a performance, how much does the saxophonist earn? $s + 48 = 125$ **\$77**

3. The even number 146 is twice an odd number. What is that odd number? $2r = 146$ **The number is 73.**

4. Sixteen years ago, Sylvia was 26 years old. How old is Sylvia now? $s - 16 = 26$ **She is 42.**

5. A class sold a total of 453 tickets to a play. This was three times the number they had sold on a certain day. How many did they sell on that day? $3t = 453$ **They sold 151.**

B

6. To make a 20-percent profit on an item, a store manager must sell the item for 1.2 times the amount he paid. On a compact-disc player that sells for \$354, the manager makes a 20-percent profit. Did the manager pay more or less than \$300 for it? $354 = 1.2x$ $x = 295$ **Less than \$300.**

7. Mrs. Luzinski bought 12 video cassettes in packages of 4. She multiplied the cost of 1 cassette to find a total cost of \$84. How much did each cassette cost? $12c = 84$ **Each cost \$7.**

8. An egg farm shipped out 8064 eggs on Friday. The eggs were packed in boxes of 1 dozen. How many boxes were used? The boxes were packed in crates that held a dozen boxes each. How many crates were used? $b \cdot 12 = 8064$ **672 boxes.** $c \cdot 12 = 672$ **56 crates.**

C Extending Thinking Skills

9. Write a word problem that could be solved using the equation $3x = 78$. **Answers will vary.**

10. Write a word problem that could be solved using the equation $x - 26 = 19$. **Answers will vary. Thinking skills: 9, 10. Formulating a problem**

Mixed Review

Evaluate for $n = 9$. **11.** $n - 11$ **−2** **12.** $-n - 11$ **−20** **13.** $n + 11$ **20**

14. $-n + 11$ **2** **15.** $n - n$ **0** **16.** $2n$ **18** **17.** $\frac{n}{-3}$ **−3** **18.** $6 - 2n$ **−12**

Solve and check. **19.** $-3r = 8.22$ **−2.74** **20.** $\frac{t}{1.6} = 32.2$ **51.52**

21. $m - 19.26 = 0.004$ **19.264** **22.** $\frac{a}{-9.2} = -6.15$ **56.58** **23.** $-6.1 = \frac{x}{7.5}$ **−45.75**

5-4 Prime and Composite Numbers

Objective: To decide whether a number is prime or composite.
MIN: 1–20, 27; REG: 1–20 even, 21–24, 27; MAX: 1–20 even, 21–27; ALL: Mixed Review
An exploratory lesson opener can be found on page 548.

The number 5 has exactly two factors, 1 and 5. Any whole number that has exactly two factors, 1 and itself, is called a **prime number**.

Any whole number greater than 1 that has more than two factors is called a **composite number**. The number 1 has only one factor, so it is neither prime nor composite. The table below shows whether each number from 1 to 12 is prime, composite, or neither. The first five prime numbers are 2, 3, 5, 7, and 11.

Scientists have sent signals of prime number sequences into space in hopes of communicating with life on other planets.

Number	Factors	Type
1	1	neither
2	1, 2	prime
3	1, 3	prime
4	1, 2, 4	composite
5	1, 5	prime
6	1, 2, 3, 6	composite
7	1, 7	prime
8	1, 2, 4, 8	composite
9	1, 3, 9	composite
10	1, 2, 5, 10	composite
11	1, 11	prime
12	1, 2, 3, 4, 6, 12	composite

To decide whether a number is prime or composite, find and count its factors.

Example

State whether the number 14 is prime or composite.

Solution Factors of 14: 1, 2, 7, 14. Since 14 has more than 2 factors,
14 is a composite number. it is composite.

Practice State whether the number is prime or composite. **a.** 21 com. **b.** 17 prime **c.** 39 com.

Oral Exercises

Give all the factors. Is the number prime or composite?

1. 16	2. 13	3. 15	4. 24
$16 = 1 \times 16$ $16 = 2 \times 8$ $16 = 4 \times 4$ 1, 2, 4, 8, 16 com.	$13 = 1 \times 13$ 1, 13 prime	$15 = 1 \times 15$ $15 = 3 \times 5$ 1, 3, 5, 15 com.	$24 = 1 \times 24$ $24 = 2 \times 12$ $24 = 3 \times 8$ $24 = 4 \times 6$

1, 2, 3, 4, 6, 8, 12, 24 com.

Exercises

A State whether the given number is prime or composite. If it is composite, give a factor of the number other than itself and 1. Hint: Divisibility rules may help.

1. 6 com. 2, 3 **2.** 7 prime **3.** 9 com. 3 **4.** 14 com. 2, 7
5. 18 com. 2, 3, etc. **6.** 22 com. 2, 11 **7.** 26 com. 2, 13 **8.** 32 com. 2, 4, etc.
9. 40 com. 2, 4, etc. **10.** 33 com. 3, 11 **11.** 34 com. 2, 17 **12.** 45 com. 3, 5, etc.
13. 2 prime **14.** 49 com. 7 **15.** 29 prime **16.** 41 prime
17. 47 prime **18.** 57 com. 3, 19 **19.** 77 com. 7, 11 **20.** 51 com. 3, 17

B

21. List all prime numbers less than 25. 2, 3, 5, 7, 11, 13, 17, 19, 23

22. Find the smallest number for n that gives a composite value for $6n + 1$. 4

23. Find the smallest number for n that gives a composite value for $6n - 1$. 6

24. Evaluate the formula $P = (n \cdot n) - n + 11$ for $n =$ each of the whole numbers 1 through 11. For which value of n is P not prime?
For 11, P is not prime.

C Extending Thinking Skills

25. Consecutive primes, such as 3 and 5, that have a difference of 2 are called twin primes. Find three more pairs of twin primes. 5 and 7, 11 and 13, 17 and 19 etc.

26. Continue this pattern until you arrive at a number that is not a prime.
11, 13, 17, 23, . . . 31, 41, 53, 67, 83, 101, 121

27. A mathematician named Goldbach claimed that every even number greater than 2 is the sum of two prime numbers. No one has proven this true, but no one has proven it false! Show that it is true for the even numbers 4 through 30. See additional answers.

Thinking skills: 25. Reasoning 26. Finding a pattern 27. Formulating a proof

Mixed Review

Solve and check. **28.** $12.3 = r + 6.72$ 5.58 **29.** $n \div 0.35 = 1.75$ 0.6125 **30.** $2.4c = 4.08$ 1.7
31. $t + 0.8 = 1.1$ 0.3 **32.** $-0.09y = -45$ 500 **33.** $m \div 3.2 = -6.05$ −19.36

5-5 Powers and Exponents

Objectives: To write, interpret, and multiply numbers using exponential notation.
MIN: 1–40; REG: 1–50 even, 51–53; MAX: 1–50 even, 51–55; ALL: Mixed Review

A product in which the factors are identical is called a **power** of that factor. $32 = 2 \cdot 2 \cdot 2 \cdot 2 \cdot 2$, so 32 is the fifth power of 2. **Exponents** are used to write powers in short form. The exponent indicates the number of times the **base** is used as a factor.

$$2 \cdot 2 \cdot 2 \cdot 2 \cdot 2 = 2^5$$ ← **Exponent** (5); ← **Base** (2)

We read this as "two to the fifth power."

Other examples in which exponents are used to show powers are given below.

4^2 is 4 to the **second power** (or 4 squared) and means $4 \cdot 4$, or 16.

10^3 is 10 to the **third power** (or 10 cubed) and means $10 \cdot 10 \cdot 10$, or 1000.

3^4 is 3 to the **fourth power** and means $3 \cdot 3 \cdot 3 \cdot 3$, or 81.

a^6 is a to the **sixth power** and means $a \cdot a \cdot a \cdot a \cdot a \cdot a$.

A number raised to the first power is that number. For example, $10^1 = 10$.

To show the factors of a number expressed using exponents, you write the number in **expanded form**, also called factored form.

Example 1

Write in expanded form. Simplify if possible. 10^6

Solution $10^6 = 10 \cdot 10 \cdot 10 \cdot 10 \cdot 10 \cdot 10$ The exponent, 6, indicates that 10 is a factor 6 times.

$= 1,000,000$

Practice Write in expanded form and simplify.

a. 3^4 **b.** $(-2)^4$ **c.** a^3 $a \cdot a \cdot a$

a. $3 \cdot 3 \cdot 3 \cdot 3 = 81$ b. $(-2)(-2)(-2)(-2) = 16$

Example 2

Write using exponents. $7 \cdot 7 \cdot 7 \cdot 7$

Solution 7^4 The base, 7, is a factor 4 times.

Practice Write using exponents.

a. $10 \cdot 10 \cdot 10$ 10^3 **b.** $-2(-2)$ $(-2)^2$ **c.** $bbbb$ b^4

To find the product of numbers expressed using exponents, you can simply add exponents to find the total number of factors in the product. **Remind students that an exponent tells the number of times a base is used as a factor.**

$$2^4 \cdot 2^3 = \underbrace{(2 \cdot 2 \cdot 2 \cdot 2)}_{4 \text{ factors}} \cdot \underbrace{(2 \cdot 2 \cdot 2)}_{3 \text{ factors}} = \underbrace{2^7}_{7 \text{ factors}}$$

4 factors + 3 factors = 7 factors

This leads to a rule for *multiplying* numbers with like bases.

The fact that $a^0 = 1$ is discussed in Chapter 6.

Multiplying Powers with Like Bases

To multiply two or more powers with like bases, first add the exponents. Use this sum as the exponent together with the original base to express the product.

For any number **a**, and whole numbers **m** and **n**, $a^m \cdot a^n = a^{m+n}$

Example 3

Multiply. Give the answer in exponent form. $3^2 \cdot 3^4$

Solution $3^2 \cdot 3^4 = 3^6$ To multiply powers with like bases, add the exponents. $2 + 4 = 6$.

Practice Multiply. Give the answer in exponent form.

a. $(-5)^2 \cdot (-5)^4$ **$(-5)^6$** **b.** $10^2 \cdot 10^5$ **10^7** **c.** $x^2 \cdot x^3$ **x^5**

The rule above works only when two powers with the *same bases* are *multiplied*. It does not apply when multiplying powers with unlike bases or when *adding* powers with like bases.

Example 4

Simplify. $3^2 + 3^4$

Solution $3^2 + 3^4 = 9 + 81$ Find $3^2 = 3(3) = 9$, and find $3^4 = 3(3)(3)(3) = 81$.

$= 90$ Add. **Since 90 is not a power of three, the answer cannot be simply expressed using exponents.**

Practice Simplify. **a.** $10^2 + 10^3$ **b.** $(-5)^2 + (-5)^4$ **a. $100 + 1000 = 1100$** **b. $25 + 625 = 650$**

Oral Exercises

Name the base and exponent, then read the expression. **1. "ten to the fourth power" (etc.)**

1. 10^4 **b10, e4**
2. 9^5 **b9, e5**
3. 3^{17} **b3, e17**
4. 5^7 **b5, e7**
5. 11^4 **b11, e4**
6. 7^6 **b7, e6**
7. 2^8 **b2, e8**
8. $(-4)^{19}$ **b−4, e19**
9. 12^3 **b12, e3**
10. $(-8)^6$ **b−8, e6**
11. 8^{12} **b8, e12**
12. $(-3)^2$ **b−3, e2**

Exercises

A Write in expanded form. Simplify if possible. **See additional answers.**

1. 10^5 **2.** 5^4 **3.** 7^3 **4.** 12^5 **5.** 9^3

6. $(-10)^6$ **7.** 8^2 **8.** $(-6)^3$ **9.** n^4 **10.** b^5

Write using exponents. **12.** $(-3)^5$

11. $5 \cdot 5 \cdot 5 \cdot 5$ 5^4 **12.** $(-3)(-3)(-3)(-3)(-3)$ **13.** $10 \cdot 10 \cdot 10 \cdot 10$ 10^4

14. $2 \cdot 2 \cdot 2 \cdot 2 \cdot 2$ 2^5 **15.** $(9)(9)(9)(9)$ 9^4 **16.** $yyyyyy$ y^6

Multiply. Give answers in exponent form. **20.** $(-5)^4$ **22.** $(-3)^{17}$

17. $3^2 \cdot 3^4$ 3^6 **18.** $2^4 \cdot 2^6$ 2^{10} **19.** $10^7 \cdot 10^3$ 10^{10} **20.** $(-5)^1 \cdot (-5)^3$

21. $10^5 \cdot 10^9$ 10^{14} **22.** $(-3)^8 \cdot (-3)^9$ **23.** $7^{23} \cdot 7^4$ 7^{27} **24.** $4^5 \cdot 4^8$ 4^{13}

25. $2^3 \cdot 2^8$ 2^{11} **26.** $5^5 \cdot 5^2$ 5^7 **27.** $3^8 \cdot 3^6$ 3^{14} **28.** $10^5 \cdot 10^8 \cdot 10^3$

29. $p^3 \cdot p^1$ p^4 **30.** $r^4 \cdot r^3$ r^7 **31.** $c^3 \cdot c^5 \cdot c^7$ c^{15} **32.** $a^2 \cdot a^4$ a^6

28. 10^{16}

Add.

33. $2^4 + 2^5$ 48 **34.** $5^2 + 5^3$ 150 **35.** $10^2 + 10^1$ 110 **36.** $(-3)^4 + (-3)^2$ 90

Simplify.

37. $2^3 + 2^2$ $8 + 4 = 12$ **38.** $(-3)^2 \cdot (-3)^3$ $(-3)^5 = -243$ **39.** $5^1 \cdot 5^2$ $5^3 = 125$ **40.** $4^2 + 4^3$ $16 + 64 = 80$

B Write using exponents.

41. 100 10^2 **42.** 1,000 10^3 **43.** 10,000 10^4 **44.** 100,000 10^5

Evaluate the expression.

45. $y^5 \cdot y^3$ for $y = 4$ 65,536 **46.** $10^a \cdot 10^b$ for $a = 2, b = 3$ 100,000

47. $5^x \cdot 5^y$ for $x = 4, y = 3$ 78,125 **48.** $10^a + 10^b$ for $a = 2, b = 3$ 1,100

49. $10^a + b$ for $a = 2, b = 23$ 123 **50.** $4^p \cdot 4^q \cdot 4^r$ for $p = 3, q = 5, r = 2$ 1,048,576

Complete the equations in 51. Then answer the questions in 52–53.

51. $(-2)^1 = ?$ -2 $(-2)^2 = ?$ 4 $(-2)^3 = ?$ -8 $(-2)^4 = ?$ 16 $(-2)^5 = ?$ -32 $(-2)^6 = ?$ 64

52. When a negative integer is raised to an odd power, is the result a positive or a negative integer? negative

53. When a negative integer is raised to an even power, is the result a positive or a negative integer? positive

C Extending Thinking Skills

Thinking skills: 54. Finding a pattern 55. Reasoning

54. Jesse suggested to his parents the following plan for his monthly allowance. He would get 1 penny on the first day, 2 pennies on the second day, 4 pennies on the third day, 8 pennies on the fourth day, and so on. If Jesse's parents agree to the plan, how much should he get on the 30th day? 2^{29} pennies, or $5,368,709.12

55. A *googol* and a *googolplex* are very large numbers that are defined as follows.

$$10^{100} = \text{one googol} \qquad 10^{\text{googol}} = \text{one googolplex}$$

a. One googol is one followed by how many zeros? 100

b. One googolplex is 1 followed by how many zeros? A googol, or 10^{100}.

Mixed Review

Evaluate for $a = 2.5$, $b = 3$, $c = 3.25$. **56.** abc 24.375 **57.** $a - c$ −0.75
58. $3a - 2b + c$ 4.75 **59.** $bc - a$ 7.25 **60.** $bc \div (-a)$ −3.9 **61.** $2a - b$ 2
Solve. **62.** $10.4 + x = 8.7$ −1.7 **63.** $9.8(x) = 22.54$ 2.3
64. $396.49 - x = 177.5$ 218.99 **65.** $18.4x = 86.48$ 4.7 **66.** $\frac{x}{24.64} = 250.3$ 6,167.392

ESTIMATION

To gain a feeling for certain large number amounts, it helps to compare them to powers of ten. Choose the power of ten, 10^4, 10^5, 10^6, 10^7, 10^8, or 10^9, that you would estimate is closest to each of the following amounts. Use an atlas or other reference book to check your estimates.

1. the population of the city in which you live, or the nearest large city.

2. the population of the United States.

3. the population of the world.

4. the population of the largest city in the world. Answers will vary.

5-6 Prime Factorization

Objective: To write the prime factorization of a number.
MIN: 1–29; REG: 1–29 even, 30–39; MAX: 1–29 even, 30–42; ALL: Mixed Review

A factor tree shows a number as a product of prime factors. This factor tree shows that $24 = 2 \cdot 2 \cdot 2 \cdot 3$, or $2^3 \cdot 3$.

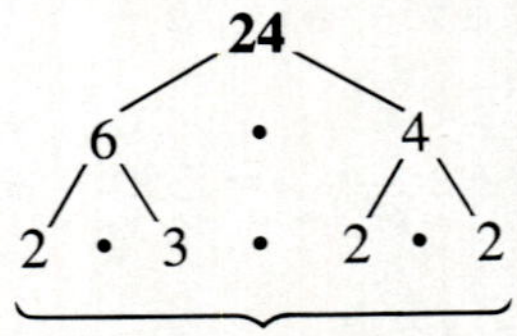

Prime factorization of 24

This expression of 24 as a product of prime factors is called the **prime factorization** of 24. We say that 24 has been **factored completely**. A factor tree for 24 that begins with $8 \cdot 3$ will have the same prime factors appearing in the bottom row, but perhaps in a different order. This suggests the following theorem.

Unique Factorization Theorem

Every composite number can be expressed as the product of prime numbers in only one way, except for the order of the factors.

Also known as the Fundamental Theorem of Arithmetic.

Example 1

Make a factor tree to find the prime factorization of 60.

Solution

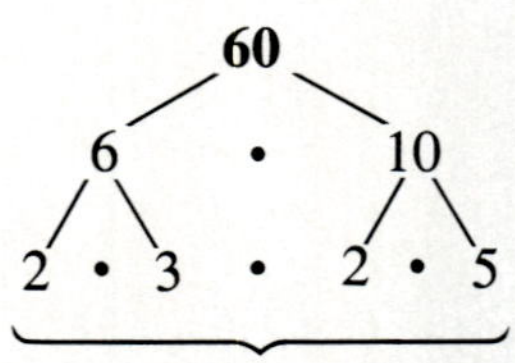

For the first row, look for any two numbers with product 60.

Note that trees could also be made with 3 × 15 or 2 × 30 in the first row.

The prime factorization of 60 is $2 \cdot 2 \cdot 3 \cdot 5$.

Practice Make a factor tree to find the prime factorization of each number.

a. 40 **b.** 54

a.

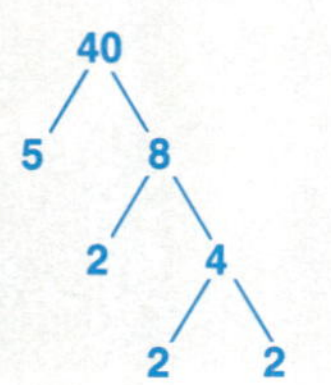

$2 \cdot 2 \cdot 2 \cdot 5$

b.

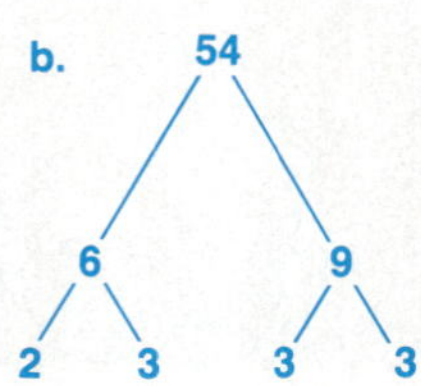

$2 \cdot 3 \cdot 3 \cdot 3$

Example 2

Write the prime factorization of 84, using exponents.

Solution $84 = 2 \cdot 42 = 2 \cdot 6 \cdot 7 = 2 \cdot 2 \cdot 3 \cdot 7 = 2^2 \cdot 3 \cdot 7$

First find any 2 factors. Then continue to find factors until all are prime.

Practice Write the prime factorization of each, using exponents.

a. 28 $2^2 \cdot 7$ **b.** 72 $2^3 \cdot 3^2$

Another way of finding the prime factorization of a number is to use repeated division.

Example 3

Use repeated division to find the prime factorization of 60.

Solution

```
2|60
  2|30
    3|15
      5|5
        1
```

Beginning with 2, check whether each prime is a factor of the number. If it is, divide the number by the prime. Continue until you arrive at a quotient of 1.

$60 = 2 \cdot 2 \cdot 3 \cdot 5$

The product of the divisors is the prime factorization of the number.

Practice Use repeated division to find the prime factorization of each.

a. 42 $2 \cdot 3 \cdot 7$ **b.** 315 $3 \cdot 3 \cdot 5 \cdot 7$

Oral Exercises

Give each number as the product of two factors. 5. $2 \cdot 6$, $3 \cdot 4$ 8. $2 \cdot 8$, $4 \cdot 4$

1. 6 $2 \cdot 3$ **2.** 8 $2 \cdot 4$ **3.** 4 $2 \cdot 4$ **4.** 9 $3 \cdot 3$ **5.** 12

6. 15 $3 \cdot 5$ **7.** 14 $2 \cdot 7$ **8.** 16 **9.** 21 $3 \cdot 7$ **10.** 26 $2 \cdot 13$

Exercises

A Copy and complete each factor tree to find the prime factorization of each number.

1. 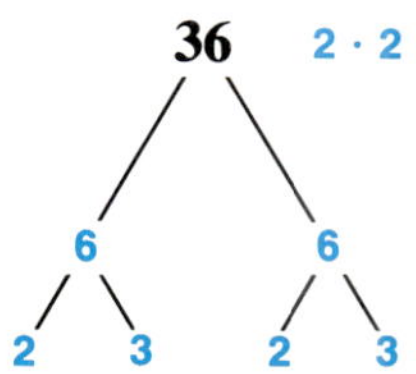

$2 \cdot 2 \cdot 3 \cdot 3$

2. 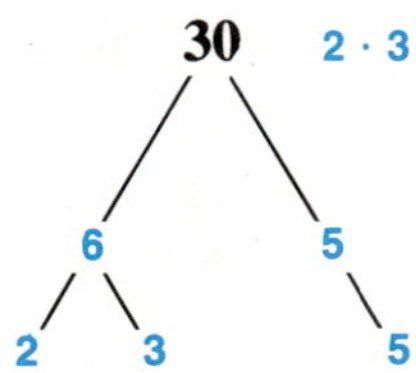

$2 \cdot 3 \cdot 5$

3. 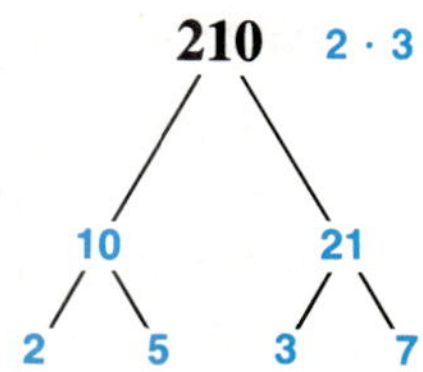

$2 \cdot 3 \cdot 5 \cdot 7$

Make a factor tree to find the prime factorization of each number. See additional answers.

4. 32 **5.** 48 **6.** 63 **7.** 140 **8.** 150

Use exponents to show the prime factorization of each number.

9. 12 $2^2 \cdot 3$ **10.** 18 $2 \cdot 3^2$ **11.** 36 $2^2 \cdot 3^2$ **12.** 45 $3^2 \cdot 5$

13. 507 $3 \cdot 13^2$ **14.** 75 $3 \cdot 5^2$ **15.** 90 $2 \cdot 3^2 \cdot 5$ **16.** 84 $2^2 \cdot 3 \cdot 7$

17. 240 $2^4 \cdot 3 \cdot 5$ **18.** 126 $2 \cdot 3^2 \cdot 7$ **19.** 330 $2 \cdot 3 \cdot 5 \cdot 11$ **20.** 288 $2^5 \cdot 3^2$

21. 225 $3^2 \cdot 5^2$ **22.** 236 $2^2 \cdot 59$ **23.** 462 $2 \cdot 3 \cdot 7 \cdot 11$ **24.** 585 $3^2 \cdot 5 \cdot 13$

Use repeated division to find the prime factorization of each number.

25. 297 $3 \cdot 3 \cdot 3 \cdot 11$ **26.** 108 $2 \cdot 2 \cdot 3 \cdot 3 \cdot 3$ **27.** 216 $2 \cdot 2 \cdot 2 \cdot 3 \cdot 3 \cdot 3$ **28.** 625 $5 \cdot 5 \cdot 5 \cdot 5$ **29.** 9,282 $2 \cdot 3 \cdot 7 \cdot 13 \cdot 17$

B Solve the equation to complete the prime factorization.

30. $546 = 2 \cdot 3 \cdot 7 \cdot n$ 13 **31.** $385 = n \cdot 7 \cdot 11$ 5

32. $285 = 3 \cdot 5 \cdot n$ 19 **33.** $2093 = 7 \cdot 13 \cdot n$ 23

Solve the equation to find the number whose prime factorization is given.

34. $3^2 \cdot 5^3 = n$ 1125 **35.** $2^2 \cdot 3^4 \cdot 7 = x$ 2268

36. $z = 5^3 \cdot 7^2 \cdot 13$ 79,625 **37.** $t = 2^4 \cdot 5^2 \cdot 7^3$ 137,200

Use a calculator for exercises 38 and 39.

38. Evaluate the expression $n \cdot 7 \cdot 11 \cdot 13$ for the following values of n: $n = 263$, $n = 389$, $n = 59$. Do you see a shortcut for evaluating such an expression? Use the shortcut to evaluate the expression for $n = 974$.
263,263 389,389 59,059 Shortcut: Repeat n in the thousands and units. 974,974

39. Rima's teacher asked her to replace a with her age in the expression $a \cdot 3 \cdot 37 \cdot 91$, and then evaluate the expression. If Rima is 14 years old, what is the value of the expression? Evaluate the expression for $a = 23$, $a = 47$, $a = 75$, $a =$ your age. What do you discover?
141,414; 232,323; 474,747; 757,575. Pattern: repeats the age

C Extending Thinking Skills

40. Jeffrey's teacher told him that her age was a 2-digit number that was equal to twice the product of its digits. How old was Jeffrey's teacher? 36

41. Find the smallest number that has six different primes in its prime factorization. $2 \cdot 3 \cdot 5 \cdot 7 \cdot 11 \cdot 13 = 30{,}030$

42. Look for a pattern in the prime factorization of each of the four numbers in the sequence. Give the next two numbers. 2, 6, 30, 210, __, __. 2310, 30,030

Thinking skills: 40. Testing a conjecture 41. Reasoning 42. Finding a pattern

Mixed Review

Use the variable n to write an equation for each statement.

43. 9 times the sum of 12 and a number gives 135. $9(12 + n) = 135$

44. 266 is 56 more than the product of a number and 15. $266 = 56 + 15n$

45. A number added to 36 gives the product of 5 and 4. $36 + n = 20$

Solve and check. **46.** $36n = -558$ -15.5 **47.** $c - 19.65 = -17.3$ 2.35

5-7 Greatest Common Factor

Objective: To find the greatest common factor of numbers.
MIN: 1–32; REG: 1–32 even, 33–40; MAX: 1–32 even, 33–42; ALL: Mixed Review

The **Greatest Common Factor (GCF)** of two whole numbers is the greatest whole number that is a factor of *both* the numbers.

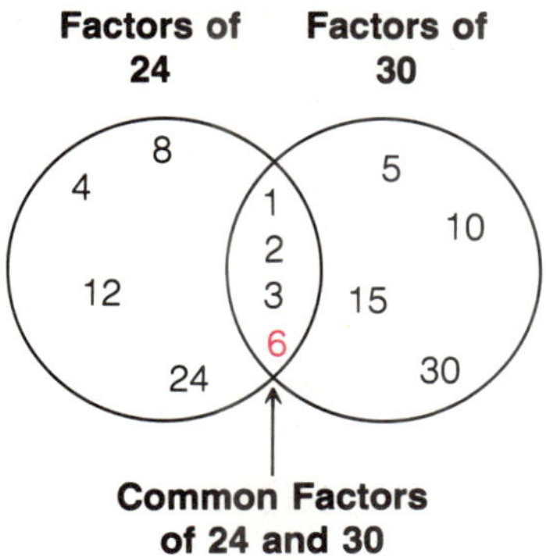

Factors of 24	**Factors of 30**
1, 2, 3, 4, 6, 8, 12, 24	1, 2, 3, 5, 6, 10, 15, 30

Common Factors of 24 and 30 1, 2, 3, 6

Greatest Common Factor of 24 and 30 6

In the example above, the factors of both numbers were listed to find the GCF. A method often used for finding the GCF of two whole numbers is to list the factors of the smaller number only, then look for the largest of these factors that is also a factor of the larger number.

Example 1

List factors of the smaller number to find the GCF of 18 and 24.

Solution Factors of 18: 18, 9, 6, 3, 2, 1 — Neither 18 nor 9 is a factor of 24.
6 is the GCF of 18 and 24. — 6 is the largest factor of 18 that is also a factor of 24.

Practice List factors of the smaller number to find the GCF of each pair.

a. 18, 27 **b.** 20, 12

a. 1, 2, 3, 6, 9, 18. GCF = 9
b. 1, 2, 3, 4, 6, 12. GCF = 4

Another way of finding the GCF is to use the prime factorization of the numbers. This method is useful in work with algebraic expressions. Factors that occur in the prime factorizations of both numbers are included as factors in the GCF.

Example 2

Use prime factorization to find the GCF of 36 and 60.

Solution $36 = \mathbf{2} \times \mathbf{2} \times \mathbf{3} \times 3$ Write the prime factorizations of 36 and 60.

$60 = \mathbf{2} \times \mathbf{2} \times \mathbf{3} \times 5$

GCF of 36 and 60 $= 2 \times 2 \times 3 = 12$ The GCF is the product of all the common prime factors.

Practice Use prime factorization to find the GCF. **a.** 12, 16 **b.** 24, 30

a. $12 = 2 \cdot 2 \cdot 3$, $16 = 2 \cdot 2 \cdot 2 \cdot 2$ GCF = 4
b. $24 = 2 \cdot 2 \cdot 2 \cdot 3$, $30 = 2 \cdot 3 \cdot 5$ GCF = 6

Oral Exercises

Give the GCF of the two numbers. The set of factors of each number is given.

1. 12: {1, 2, 3, 4, 6, 12} 4
20: {1, 2, 4, 5, 10, 20}

2. 18: {1, 2, 3, 6, 9, 18} 6
24: {1, 2, 3, 4, 6, 8, 12, 24}

3. 12: {1, 2, 3, 4, 6, 12} 4
16: {1, 2, 4, 8, 16}

4. 20: {1, 2, 4, 5, 10, 20} 4
28: {1, 2, 4, 7, 14, 28}

5. 21: {1, 2, 3, 7, 21} 7
56: {1, 2, 4, 7, 28, 56}

6. 70: {1, 2, 5, 7, 10, 14, 35, 70} 14
42: {1, 2, 3, 6, 7, 14, 21, 42}

Note that any two numbers share 1 as a common factor.

Exercises

A List factors of the smaller number to find the GCF of each pair.

1. 8, 27 1
2. 18, 30 6
3. 16, 24 8
4. 21, 28 7
5. 12, 36 12
6. 18, 27 9
7. 48, 64 16
8. 36, 54 18
9. 24, 40 8
10. 30, 45 15
11. 10, 21 1
12. 16, 40 8
13. 27, 36 9
14. 18, 45 9
15. 32, 40 8
16. 18, 24 6

Use prime factorization to find the GCF of each pair.

17. 30, 105 15
18. 42, 44 2
19. 36, 54 18
20. 60, 126 6
21. 45, 60 15
22. 28, 42 14
23. 48, 72 24
24. 26, 51 1
25. 63, 84 21
26. 90, 189 9
27. 56, 90 2
28. 84, 108 12
29. 144, 216 72
30. 136, 162 2
31. 130, 182 26
32. 154, 192 2

B Use prime factorization to find the GCF of each set of numbers.

33. 12, 18, 24 6
34. 28, 42, 90 2
35. 30, 36, 48 6
36. 16, 28, 40 4
37. 32, 56, 72 8
38. 30, 45, 60 15

39. Two numbers whose GCF is 1 are said to be **relatively prime**. Are these number pairs relatively prime?

a. 9, 10 yes **b.** 20, 27 yes **c.** 165, 182 yes

40. Carver Junior High School has 210 students in 8th grade and 180 students in 9th grade. The principal wants to divide the students into classes that are all the same size but keep the grades separate. What is the largest possible class size that could be used? 30 students

C Extending Thinking Skills

41. The number n is between 60 and 70. The GCF of n and 27 is 9. Find n. 63

42. A rectangular meeting room in a hotel is 66 feet by 78 feet. The room is to be covered with square pieces of carpet that are all the same size. If the square carpet pieces cannot be cut, what is the largest square piece that can be used? 6′ × 6′

Thinking skills: 41. Testing a conjecture 42. Reasoning

Mixed Review

Simplify. **43.** $t(6 - 4)$ 2t **44.** $6m + 9m$ 15m **45.** $n \div (-n)$ −1

46. $6(2c)$ 12c **47.** $2(3x) - 4$ 6x − 4 **48.** $9(y + 2y)$ 27y **49.** $n - n$ 0

Evaluate for $n = 6$. **50.** $2(4n)$ 48 **51.** $3n + 2n$ 30 **52.** $n(0)$ 0

Solve and check. **53.** $2x + 3x = 45$ 9 **54.** $-1.5c = 30$ −20

NUMBERS TO ALGEBRA

Prime factorization is used to find the GCF of two algebraic expressions in the same way it is used to find the GCF of two whole numbers.

Numbers	Algebra
▪ Find the GCF of 108 and 120. $108 = 2 \cdot 2 \cdot 3 \cdot 3 \cdot 3$ $120 = 2 \cdot 2 \cdot 2 \cdot 3 \cdot 5$ Since 2 is a common factor twice and 3 is a common factor once, the CGF is $2^2 \cdot 3$, or 12.	▪ Find the GCF of x^2y^3 and x^3yz. $x^2y^3 = x \cdot x \cdot y \cdot y \cdot y$ $x^3yz = x \cdot x \cdot x \cdot y \cdot z$ Since x is a common factor twice and y is a common factor once, the GCF is x^2y.

Find the GCF of each pair of expressions.

1. xy^3 and x^3y xy **2.** $x^2y^5z^2$ and x^3y^4 x^2y^4 **3.** $12x^3y^2$ and $18xy^3$ $6xy^2$

4. xy^2z^3 and x^2y^3z xy^2z **5.** $16x^2z^3$ and $24x^3z^2$ $8x^2z^2$ **6.** $27ab^3$ and $45a^2b$ $9ab$

PROBLEM SOLVING: APPLICATIONS

5-8 Driving a Car

Objective: To solve word problems involving braking and stopping distances.
ALL: 1–12

Most states require that anyone applying for a driver's license take a written examination. Questions about reaction, braking, and stopping distances are often included in such an exam. The information in the graph below, taken from a driving manual, is important in road safety.

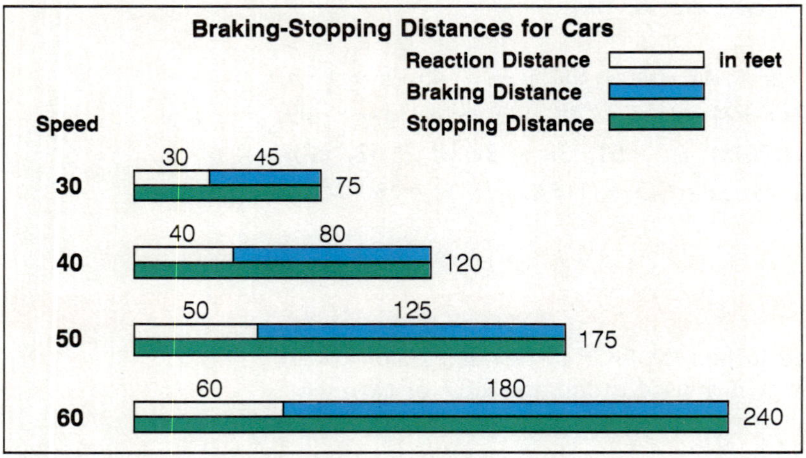

In the graph, the Stopping Distance is the sum of the Reaction Distance and the Braking Distance. The *Reaction Distance* is the number of feet the car travels *after* the driver decides to stop and *before* he or she applies the brakes.

Problems

You might wish to have students choose from among the techniques pencil and paper, mental math, estimation, and calculator. Have them choose the most appropriate technique for each problem.

Use the graph above to solve problems 1–4.

1. If a car is traveling at a speed of 40 mi/h and the driver sees an animal in the road, the reaction distance before the brakes are applied is the same number of feet as the speed. After the driver applies the brakes, how far will the car go before it stops? **80 ft**

2. What is the reaction distance when a car is traveling 60 mi/h? What are the braking and stopping distances? **reaction 60 ft braking 180 ft stopping 240 ft**

3. How much farther is the stopping distance at 60 mi/h than at 50 mi/h? **65 ft**

4. Estimate the amount by which the stopping distance increases when a car's speed doubles from 30 to 60 mi/h. Is the increase about two, three, four, or five times the original speed? About 3 times.

Use the following formulas as needed to solve problems 5–12.

The ***Reaction Distance,*** **R**, in feet has been determined to be about the same as the speed the car is traveling in miles per hour.

The ***Braking Distance,*** **B**, is the distance in feet traveled after the brakes are applied:

$$\mathbf{B = 0.05R^2}$$

The ***Stopping Distance,*** **S**, is the Reaction Distance plus the Braking Distance:

$$\mathbf{S = R + 0.05R^2}$$

5. What is the braking distance when a car is traveling 70 mi/h? 245 ft

6. What is the stopping distance when a car is traveling 70 mi/h? 315 ft

7. Find the braking and stopping distances for the speed of 55 mi/h. Braking–151.25 ft Stopping–206.25 ft

8. Some states have suggested that the speed limit be raised to 65 mi/h. How does the stopping distance at this speed compare with that at 55 mi/h? Stopping distance 276.25, 70 ft more

9. If the driver of a car traveling 45 mi/h sees a truck stopped in the road ahead and immediately applies the brakes, how far will the car travel before coming to a stop? 146.25 ft

Tell students to use the idea of GCF to solve problems 10 and 11.

10. A driver education instructor wanted to place 72 sophomores and 120 juniors in driver education classes so that each class would be the same size and would be as large as possible, but the grades would be kept separate. How many students will be in each class, and how many classes will be needed? 24 students in 8 classes

11. A highway planner wants to place rest stops at equal distances along an interstate highway connecting three cities. The distance from the first to the second city is 475 miles. The distance from the second to the third city is 285 miles. There is to be a rest stop in each of the cities, and the distance between the rest stops is to be as great as possible. How far apart will the rest stops be? 95 miles

12. **Data Search** Find three different speed limits posted on streets in your community and calculate the stopping distance for each speed. Answers will vary.

What's Your Decision?

How many feet of space should you keep between your car and the car in front of you for every 10 mi/h of speed? Use the data from the graph on page 154 to help you decide. Give reasons for your decision. Answers will vary. Approximately 40 ft

5-9 Least Common Multiple

Objective: To find the least common multiple of numbers.
MIN: 1–28; REG: 1–28 even, 29–45; MAX: 1–28 even, 29–47; ALL: Mixed Review

The **Least Common Multiple (LCM)** of two whole numbers is the smallest *nonzero* whole number that is a multiple of *both* of the whole numbers.

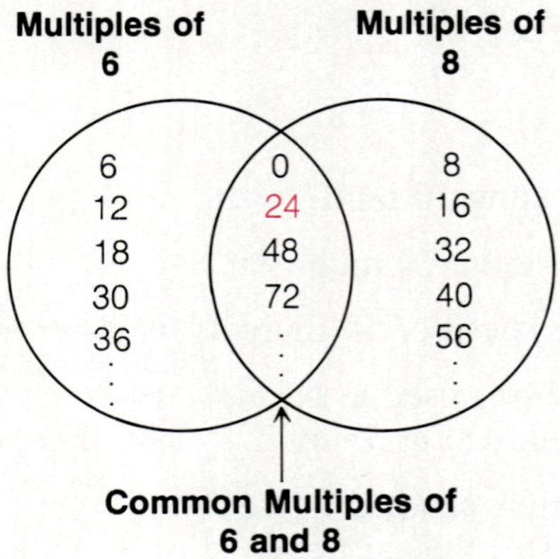

Multiples of 6
0, 6, 12, 18, 24, 30, 36, . . .

Multiples of 8
0, 8, 16, 24, 32, 40, 48, . . .

Common Multiples of 6 and 8 0, 24, 48, 72, . . .

Least Common Multiple of 6 and 8 24

In the example above, multiples of both numbers were listed to find the LCM. A method often used to find the LCM of two numbers is to list multiples of the larger number, then look for the smallest of these that is also a multiple of the smaller number.

Example 1

List multiples of the larger number to find the LCM of 9 and 12.

Solution 12, 24, 36 — Start a list of the nonzero multiples of the greater number, 12.

The LCM of 9 and 12 is 36 — Find the lowest number in the list that is also a multiple of the lesser number, 9.

Practice List multiples of the greater number to find the LCM of each pair.

a. 6, 9 **b.** 8, 12 a. 9, *18*, . . . LCM = 18
b. 12, *24*, . . . LCM = 24

You can use the prime factorizations of two numbers to find the LCM of the numbers. This method is useful in work with algebraic expressions. The factors in the LCM are only those factors needed so that the prime factorization of each number occurs in the LCM.

Example 2

Use prime factorizations to find the LCM of 42 and 60.

Solution $42 = \mathbf{2 \times 3 \times 7}$

$60 = 2 \times 2 \times 3 \times 5$

$\text{LCM}(42, 60) = \mathbf{2 \times 3 \times 7} \times 2 \times 5$ Write the prime factorization of the lesser number. Then include factors from the greater number as needed to include the prime factorization of the greater number in the LCM.

$= 420$

Practice Use prime factorization to find the LCM of each pair. **a.** 12, 15 **b.** 24, 30
a. $2 \cdot 2 \cdot 3 \cdot 5 = 60$ b. $2 \cdot 2 \cdot 2 \cdot 3 \cdot 5 = 120$

Oral Exercises

Give five nonzero multiples of each number.

1. 4 **2.** 5 **3.** 8 **4.** 6 **5.** 7
6. 10 **7.** 12 **8.** 9 **9.** 15 **10.** 11

1. 4, 8, 12, 16, 20 2. 5, 10, 15, 20, 25 3. 8, 16, 24, 32, 40 4. 6, 12, 18, 24, 30 5. 7, 14, 21, 28, 35
6. 10, 20, 30, 40, 50 7. 12, 24, 36, 48, 60 8. 9, 18, 27, 36, 45 9. 15, 30, 45, 60, 75 10. 11, 22, 33, 44, 55

Exercises

A List multiples of the larger number to find the LCM of each pair.

1. 8, 10 40 **2.** 4, 6 12 **3.** 6, 10 30 **4.** 6, 8 24
5. 8, 12 24 **6.** 6, 15 30 **7.** 9, 15 45 **8.** 12, 16 48
9. 6, 21 42 **10.** 9, 5 45 **11.** 10, 15 30 **12.** 8, 20 40
13. 6, 14 42 **14.** 3, 4 12 **15.** 4, 10 20 **16.** 8, 18 72

Use prime factorization to find the LCM of each pair.

17. 20, 25 100 **18.** 8, 18 72 **19.** 30, 70 210 **20.** 18, 60 180
21. 28, 40 280 **22.** 27, 36 108 **23.** 42, 70 210 **24.** 20, 24 120
25. 63, 90 630 **26.** 50, 75 150 **27.** 60, 80 240 **28.** 72, 84 504

B Find the LCM of the three numbers given.

29. 4, 6, 15 60 **30.** 8, 10, 12 120 **31.** 9, 12, 15 180 **32.** 8, 15, 18 360
33. 6, 12, 18 36 **34.** 30, 42, 48 1680 **35.** 45, 63, 75 1575 **36.** 30, 40, 60 120
37. 28, 40, 56 280 **38.** 27, 36, 45 540 **39.** 27, 45, 60 540 **40.** 28, 35, 56 280

41. What is the shortest length of ribbon that can be cut either into a whole number of 24-cm pieces or into a whole number of 30-cm pieces? 120 cm

42. Suppose two race cars pass the flag at the same time and each continues to travel at a constant rate of speed. If the first car travels around the track in 24 seconds and the second car goes around in 27 seconds, how long will it be before they are at the flag at the same time again? 216 seconds

Using a calculator, list multiples and find the LCM of the numbers given.

43. 108, 135 540 **44.** 84, 126 252 **45.** 60, 126, 210 1,260

C *Extending Thinking Skills*

46. Mr. Lint goes to the laundromat every 10th day. Mr. Tidy goes every 6th day and Ms. Suds goes every 9th day. If they all go on January 1, on what date will they all go there on the same day again? (Assume that it is a leap year.) April 1

47. Carlos said, "You can always find the LCM of two numbers by dividing the product of the numbers by their GCF. You can also find the GCF of two numbers by dividing the product of the numbers by their LCM." Test his idea with these pairs of numbers:

a. $\frac{48}{2} = 24$, $\frac{48}{24} = 2$
b. $\frac{150}{5} = 30$, $\frac{150}{30} = 5$
c. $\frac{216}{6} = 36$, $\frac{216}{36} = 6$

a. 6, 8 **b.** 10, 15 **c.** 12, 18 Is Carlos' generalization true for these three cases? yes

Thinking skills: 46. Reasoning/Finding a pattern 47. Generalizing/Seeking a counterexample

Mixed Review

Write in exponent form.

48. $x \cdot x \cdot x \cdot x$ x^4 **49.** $(-2)(-2)(-2)$ $(-2)^3$

Solve and check.

50. $-12.5n = 50$ -4 **51.** $t - 16.45 = 2.37$ 18.82

COMPUTER ACTIVITY

This program finds the LCM of two numbers.

```
10 PRINT"TO FIND THE LCM"
20 PRINT "OF A AND B WHERE A>B"
30 PRINT "TYPE IN NUMBERS FOR A AND B"
40 INPUT A,B
50 FOR X=1 TO B
60 LET N = A*X
70 IF N/B=INT(N/B) THEN 90
80 NEXT X
90 PRINT "LCM(";A;",";B;")=";N
100 END
```

1. Run the program to find the LCM of 84 and 126. 252

2. Use the generalization suggested in exercise 47 to make additions to the program so that it will also find the GCF of the two numbers.
95 PRINT "GCF(";A;",";B;") = ";A*B/N

PROBLEM SOLVING: STRATEGIES

5-10 Make an Organized List

Objective: To solve nonroutine problems, using the strategy Make an Organized List and other strategies learned so far.
MIN: 1–2; REG: 1–4; MAX: 1–6

A strategy called **Make an Organized List** is helpful for solving problems that involve finding all possible ways to accomplish something. Making an organized list helps determine when all possible ways have been found. Consider the problem below.

Problem A computer club's members use two letters followed by three numbers to form identification codes. They use the letters C and I and the numbers 2, 3, and 4. They allow repetition of the letters, but not of the numbers. How many members can there be before they must change their method of making identification codes?

To solve this problem, you might make an organized list to show all the different identification codes that could be formed. There are four possible letter combinations, CI, IC, II, and CC. You can use these to list all possibilities systematically.

CI-234	IC-234	II-234	CC-234	Notice that the numbers starting with 2 are listed first.
CI-243	IC-243	II-243	CC-243	
CI-324	IC-324	II-324	CC-324	then the numbers starting with 3,
CI-342	IC-342	II-342	CC-342	
CI-423	IC-423	II-423	CC-423	and finally, the numbers starting with 4.
CI-432	IC-432	II-432	CC-432	

Twenty-four different identification numbers can be made from the letters C and I and the numbers 2, 3, and 4. After 24 members join the club, the system for forming ID numbers will have to be expanded.

Problem-Solving Strategies

Choose the Operations	**Look for a Pattern**
Guess, Check, and Revise	**Write an Equation**
Draw a Picture	**Simplify the Problem**
Make a Table	**Make an Organized List**

This chart shows the strategies presented so far.

Problems

Solve.

1. How many different arrangements of 3 letters can be made from the letters in the word MATH if no repetition of letters is allowed? How many of these are actual words? **24 arrangements. MAT, HAT, TAM, HAM are words.**

2. A baseball coach listed the possible batting orders for his first four batters, Allen, Burge, Cotton, and Denby. His only requirement was that Cotton could not bat immediately after Denby. How many different batting order choices did the coach have? **18**

3. The oil well on Mr. Greeley's property produces 19.7 barrels of oil each day. If a barrel contains 42 gallons and oil is worth $33 a barrel, how much money does the oil well yield per year? **$237,286.50**

4. When Tim puts his coins in groups of 2, 3, or 4, he always has one coin left over. He can put the coins in groups of 5 with none left over. What is the lowest number of coins Tim can have? **25**

5. Jeral's chess club has 9 members. They are planning to have a holiday tournament in which every member plays every other member just once. How many games would be played? **36**

6. A school with 900 students has exactly 900 lockers. The principal, who used to be a math teacher, met the students outside the building on the first day of school, and described the following weird plan. The first student is to enter the school and open all the lockers. The second student will then follow the first and close every even-numbered locker. The third student will follow and reverse every third locker by closing open lockers and opening closed lockers. The fourth student will reverse every fourth locker, and so on until all 900 students have walked past and reversed lockers. The students who can predict which lockers will remain open will get first choice of lockers. Which lockers will remain open? **The squares–1, 4, 9, 16, 25 . . . 900. They alone have an odd number of factors. (Squares of 1–30.)**

Enrichment

More About Prime Numbers

Euclid, who lived about 300 B.C., proved that there is an infinite number of prime numbers. Today, mathematicians still study the properties of prime numbers, and continue to find larger and larger primes. By studying a list of prime numbers, you can find some interesting patterns and relationships. The table below shows the first 99 primes.

Note that the Sieve of Eratosthenes (see Teaching Suggestions for 5–4) can be used to generate the primes in this table.

The First 99 Prime Numbers										
	0	**1**	**2**	**3**	**4**	**5**	**6**	**7**	**8**	**9**
0		2	3	5	7	11	13	17	19	23
1	29	31	37	41	43	47	53	59	61	67
2	71	73	79	83	89	97	101	103	107	109
3	113	127	131	137	139	149	151	157	163	167
4	173	179	181	191	193	197	199	211	223	227
5	229	233	239	241	251	257	263	269	271	277
6	281	283	293	307	311	313	317	331	337	347
7	349	353	359	367	373	379	383	389	397	401
8	409	419	421	431	433	439	443	449	457	461
9	463	467	479	487	491	499	503	509	521	523

To use the table, notice that the 56th prime, for example, is found in the row labeled "5" and in the column labeled "6." It is 263.

1. What is the 23rd prime? the 50th prime? the 89th prime? the 99th prime? **83, 229, 461, 523**

2. Where are these in the sequence of primes? 59, 113, 227, 379, 521
17th, 30th, 49th, 75th, 98th

3. Count the primes between 0 and 100, 100 and 200, 200 and 300, 300 and 400, and 400 and 500. Do you think the primes get fewer and fewer as the whole numbers get larger? **yes**

4. Consecutive primes with a difference of 2, such as 3 and 5, are called twin primes. How many pairs of twin primes are there in the first 99 primes? **25**

5. How many groups of three consecutive odd numbers are there in the first 99 primes? **1**

6. Look at the ones digit of the 2- and 3-digit primes. What do you discover? **It is 1, 3, 7, or 9.**

7. If you square a whole number and add 1, will you get a prime over half the time? **no**

8. Prime numbers 17 and 71 are "reversal primes." What other reversal primes can you find in the table? **13, 31 37, 73 79, 97 113, 311**

Chapter 5 Review

5-1 Give the first five nonzero multiples of each number.

1. 5 5, 10, 15, 20, 25 **2.** 10 10, 20, 30, 40, 50 **3.** 70 70, 140, 210, 280, 350

Find all the factors of the given number.

4. 16 1, 2, 4, 8, 16 **5.** 32 1, 2, 4, 8, 16, 32 **6.** 85 1, 5, 17, 85

5-2 Tell whether the number is divisible by 2, by 3, and by 5.

7. 286 2 **8.** 207 3 **9.** 732 2, 3

10. 1,345 5 **11.** 9,940 2, 5 **12.** 6,532 2

5-3 Find the solution by writing and solving an equation.

13. The even number 218 is twice an odd number. What is the odd number? $218 = 2x$, $x = 109$

14. Roberto needs 144 bottles of juice for a party. If juice comes 12 bottles to a case, how many cases does he need? $12c = 144$, $c = 12$

5-4 State whether the number is prime or composite.

15. 39 composite **16.** 24 composite **17.** 55 composite

5-5 Write in expanded form. Simplify if possible.

18. 8^3 $8 \cdot 8 \cdot 8 = 512$ **19.** $(-6)^4$ $-6(-6)(-6)(-6) = 1{,}296$ **20.** s^6 $s \cdot s \cdot s \cdot s \cdot s \cdot s$

Write using exponents.

21. $6 \cdot 6 \cdot 6 \cdot 6 \cdot 6$ 6^5 **22.** $a \cdot a \cdot a$ a^3 **23.** $-2(-2)(-2)(-2)$ $(-2)^4$

Multiply. Give the answers in exponent form.

24. $8^2 \cdot 8^1$ 8^3 **25.** $(-3)^4 \cdot (-3)^2$ $(-3)^6$ **26.** $x^3 \cdot x^2$ x^5

Simplify.

27. $3^3 + 3^2$ 36 **28.** $(-10)^4 + (-10)^1$ 9990

5-6 Use exponents to show the prime factorization of each number.

29. 36 $2^2 \cdot 3^2$ **30.** 75 $3 \cdot 5^2$ **31.** 150 $2 \cdot 3 \cdot 5^2$ **32.** 378 $2 \cdot 3^3 \cdot 7$

5-7 Find the Greatest Common Factor (GCF) of each pair.

33. 15, 21 3 **34.** 39, 52 13 **35.** 60, 144 12

5-8

36. Use the formula $D = r \cdot t$ to find the rate (r) at which Greg traveled when he went the distance (D) of 100 miles in the time (t) of $2\frac{1}{2}$ hours. 40 mi/h

5-9 Find the Least Common Multiple (LCM) of each pair.

37. 8, 12 24 **38.** 16, 24 48 **39.** 18, 32 288

Chapter 5 Test

Give the first five nonzero multiples of each number.

1. 3 3, 6, 9, 12, 15
2. 8 8, 16, 24, 32, 40
3. 40 40, 80, 120, 160, 200

Find all the factors of the given number.

4. 12 1, 2, 3, 4, 6, 12
5. 28 1, 2, 4, 7, 14, 28
6. 48 1, 2, 3, 4, 6, 8, 12, 16, 24, 48

Tell whether the number is divisible by 2, by 3, and by 5.

7. 6,372 2, 3
8. 4,715 5
9. 2,596 2
10. 9,360 2, 3, 5
11. 6,580 2, 5
12. 3,505 5

Find the solution by writing and solving an equation.

13. Sandy has \$130 more in her savings account than she has in her checking account. If she has \$571 in her savings account, how much does she have in her checking account? $c + 130 = 571, c = \$441$

14. Diane paid \$24 for a case of 12 bottles of orange juice. What was the cost of each bottle of orange juice? $12b = 24, b = \$2$

State whether the number is prime or composite.

15. 23 prime
16. 56 composite
17. 77 composite

Write in expanded form. Simplify if possible.

18. 7^4 $7 \cdot 7 \cdot 7 \cdot 7 = 2401$
19. $(-2)^5$ $-2(-2)(-2)(-2)(-2) = -32$
20. r^7 $r \cdot r \cdot r \cdot r \cdot r \cdot r \cdot r$

Write each of these products in exponent form.

21. $10 \cdot 10$ 10^2
22. $b \cdot b \cdot b \cdot b \cdot b \cdot b$ b^6
23. $-4(-4)(-4)$ $(-4)^3$

Multiply. Give the answer in exponent form.

24. $3^3 \cdot 3^3$ 3^6
25. $(-2)^5 \cdot (-2)^3$ $(-2)^8$
26. $n^1 \cdot n^4$ n^5

Simplify.

27. $2^4 + 2^1$ 18
28. $(-3)^2 + (-3)^3$ -18

Use exponents to show the prime factorization of each number.

29. 48 $2^4 \cdot 3$
30. 80 $2^4 \cdot 5$
31. 105 $3 \cdot 5 \cdot 7$
32. 280 $2^3 \cdot 5 \cdot 7$

Find the Greatest Common Factor (GCF) of each pair.

33. 16, 20 4
34. 42, 60 6
35. 55, 121 11

36. Use the formula $D = r \cdot t$ to find the time (t) it took for Brad to go a distance (D) of 125 miles at a rate (r) of 50 mi/h. $2\frac{1}{2}$ hours

Find the Least Common Multiple (LCM) of each pair.

37. 6, 15 30
38. 8, 20 40
39. 18, 30 90

Cumulative Review

Evaluate.

1. $5y$, for $y = 5$ **25**

2. $\frac{3a}{b}$, for $a = 6$, $b = 9$ **2**

3. $\frac{3p}{4}$, for $p = 8$ **6**

4. $3(10 - x)$, for $x = 7$ **9**

Write an equation for each verbal statement.

5. 18 less than a number m is 20. $m - 18 = 20$

6. 50 is the result of multiplying -5 by a number f. $-5f = 50$

Solve.

7. If you use 11 calories per minute bicycling, how many calories would you use on a 20-minute bicycle ride? **220**

8. Sue wants to use 450 calories by swimming. How long will she have to swim if she uses 9 calories each minute? **50 minutes**

The formula for the area of a rectangle is $A = lw$; A = area in square centimeters, l = length of the rectangle, w = width of the rectangle. Use the formula to solve the following problem.

9. What is the area of a rectangle with width 25 centimeters and length 30 centimeters? **750 sq. cm**

Find each product.

10. $-6(-9)$ **54**

11. $7(2)$ **14**

12. $-5(-8)$ **40**

13. $3(-1)$ **−3**

14. $-4(4)$ **−16**

15. $6(-3)$ **−18**

Find each quotient.

16. $-42 \div 6$ **−7**

17. $-24 \div (-8)$ **3**

18. $45 \div (-5)$ **−9**

19. $12 \div (-4)$ **−3**

20. $-6 \div (-3)$ **2**

21. $72 \div (-8)$ **−9**

Solve and check.

22. $8z = 104$ **13**

23. $204 = 17x$ **12**

24. $\frac{y}{9} = 16$ **144**

25. $-77 = 7t$ **−11**

26. $-3x = 75$ **−25**

27. $\frac{j}{6} = -17$ **−102**

Find each product or quotient.

28. 4.42×1.09 **4.8178**

29. -5.5×2.1 **−11.55**

30. $-99.44 \div 11$ **−9.04**

31. $-3.84 \div (-1.2)$ **3.2**

Solve and check.

32. $-0.5g = 2.60$ **−5.2**

33. $1.5z = -6.75$ **−4.5**

34. $\frac{s}{0.7} = -3.1$ **−2.17**

35. $\frac{p}{-2.2} = -3.4$ **7.48**

Chapter 6

Rational Numbers: Addition and Subtraction

6-1 Fractions and Equivalent Fractions

Objectives: To write equivalent fractions and determine whether fractions are equivalent; to reduce fractions to lowest terms.
MIN: 1–44, 62; REG: 1–44 even, 45–60, 62; MAX: 1–60 even, 61–63; ALL: Mixed Review

The symbol $\frac{a}{b}$, where a and b are whole numbers and $b \neq 0$, is called a **fraction**. The number above the bar is the **numerator** and the number below the bar is the **denominator**. A fraction can describe a region or a set. Fractions that show the same amount are called **equivalent fractions**.

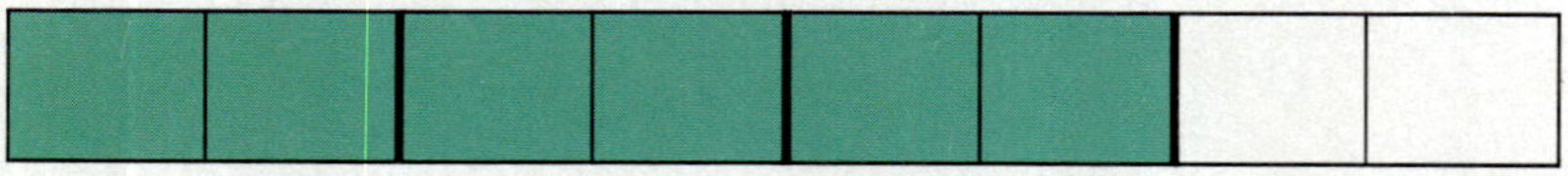

This region is divided into equal parts. The fractions $\frac{3}{4}$ and $\frac{6}{8}$ name the same part of the region.

Notice that $\frac{3}{4} = \frac{3 \cdot 2}{4 \cdot 2} = \frac{6}{8}$ and $\frac{6}{8} = \frac{6 \div 2}{8 \div 2} = \frac{3}{4}$.

Property of Equivalent Fractions

Multiplying or dividing both the numerator and denominator of a fraction by the same nonzero integer results in an equivalent fraction.

For all numbers **a, b,** and **c,** **($b \neq 0, c \neq 0$)**

$\frac{a}{b} = \frac{a \cdot c}{b \cdot c}$ and $\frac{a}{b} = \frac{a \div c}{b \div c}$

You might have students produce sequences of equivalent fractions by multiplying both numerator and denominator by 2, 3, 4, 5, and so on.

Example 1

Write an equivalent fraction by replacing the variable with a whole number.

a. $\frac{3}{8} = \frac{x}{24}$ **b.** $\frac{20}{25} = \frac{y}{5}$

Solution **a.** $\frac{3}{8} = \frac{9}{24}$ Since $8 \times 3 = 24$, multiply the numerator by 3.

b. $\frac{20}{25} = \frac{4}{5}$ Since $25 \div 5 = 5$, divide the numerator by 5.

Practice Write an equivalent fraction by replacing the variable with a whole number.

a. $\frac{3}{5} = \frac{x}{60}$ **36** **b.** $\frac{7}{15} = \frac{y}{75}$ **35**

If two fractions are equivalent, their **cross products** are equal. And if the cross products are equal, the fractions are equivalent.

$$5 \cdot 36 = 180 \qquad 12 \cdot 15 = 180$$

$$\frac{5}{12} = \frac{15}{36}$$

Example 2

Check cross products to decide whether $\frac{15}{25}$ and $\frac{5}{8}$ are equivalent.

Solution

$$\frac{15}{25} \stackrel{?}{=} \frac{5}{8}$$

$15 \cdot 8 \quad 25 \cdot 5$ Find the cross products.

$$120 \neq 125$$

$$\frac{15}{25} \neq \frac{5}{8}$$

Practice Check cross products to decide whether each pair of fractions is equivalent.

a. $\frac{3}{5}, \frac{12}{20}$ **yes** **b.** $\frac{7}{12}, \frac{28}{48}$ **yes**

A fraction is in **lowest terms** when the only common factor of the numerator and denominator is 1. One way to reduce a fraction to lowest terms is to divide the numerator and denominator by the greatest common factor. Another is to factor the numerator and denominator into prime factors and divide common factors.

A third way is to divide both the numerator and denominator repeatedly by common factors until the only common factor is 1.

Example 3

Reduce to lowest terms. $\frac{24}{36}$

Solution

Method 1

$$\frac{24}{36} = \frac{24 \div 12}{36 \div 12} = \frac{2}{3}$$

The GCF of 24 and 36 is 12. Divide both the numerator and the denominator by 12.

Method 2

$$\frac{24}{36} = \frac{2 \cdot 2 \cdot 2 \cdot 3}{2 \cdot 2 \cdot 3 \cdot 3}$$

Factor 24 and 36 into prime factors.

$$= \frac{\cancel{2} \cdot \cancel{2} \cdot 2 \cdot \cancel{3}}{\cancel{2} \cdot \cancel{2} \cdot 3 \cdot \cancel{3}} = \frac{2}{3}$$

Divide common factors.

Practice Reduce to lowest terms. **a.** $\frac{18}{36}$ $\frac{1}{2}$ **b.** $\frac{70}{105}$ $\frac{2}{3}$

Oral Exercises

Give two equivalent fractions for each figure.

1. 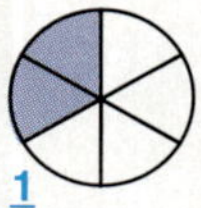 $\frac{2}{6}, \frac{1}{3}$

2.  $\frac{4}{10}, \frac{2}{5}$

3.  $\frac{8}{12}, \frac{4}{6}, \frac{2}{3}$

Exercises

A Write an equivalent fraction by replacing the variable with a whole number.

1. $\frac{7}{15} = \frac{x}{30}$ 14
2. $\frac{1}{2} = \frac{y}{36}$ 18
3. $\frac{18}{30} = \frac{t}{5}$ 3
4. $\frac{4}{9} = \frac{d}{27}$ 12
5. $\frac{16}{64} = \frac{g}{4}$ 1
6. $\frac{2}{3} = \frac{x}{54}$ 36
7. $\frac{5}{6} = \frac{m}{60}$ 50
8. $\frac{56}{80} = \frac{n}{10}$ 7
9. $\frac{13}{15} = \frac{x}{75}$ 65
10. $\frac{5}{18} = \frac{y}{90}$ 25
11. $\frac{49}{140} = \frac{a}{20}$ 7
12. $\frac{156}{240} = \frac{b}{60}$ 39
13. $\frac{7}{12} = \frac{k}{60}$ 35
14. $\frac{8}{9} = \frac{h}{108}$ 96
15. $\frac{72}{150} = \frac{x}{25}$ 12
16. $\frac{25}{60} = \frac{y}{240}$ 100

Check cross products to decide which pairs of fractions are equivalent.

17. $\frac{2}{3}, \frac{4}{9}$ no
18. $\frac{3}{5}, \frac{9}{15}$ yes
19. $\frac{4}{7}, \frac{8}{14}$ yes
20. $\frac{4}{10}, \frac{6}{15}$ yes
21. $\frac{4}{14}, \frac{6}{21}$ yes
22. $\frac{5}{7}, \frac{35}{49}$ yes
23. $\frac{12}{54}, \frac{2}{9}$ yes
24. $\frac{7}{12}, \frac{3}{4}$ no

Reduce each fraction to lowest terms.

25. $\frac{40}{52}$ $\frac{10}{13}$
26. $\frac{24}{60}$ $\frac{2}{5}$
27. $\frac{48}{84}$ $\frac{4}{7}$
28. $\frac{66}{102}$ $\frac{11}{17}$
29. $\frac{68}{76}$ $\frac{17}{19}$
30. $\frac{108}{156}$ $\frac{9}{13}$
31. $\frac{36}{54}$ $\frac{2}{3}$
32. $\frac{78}{112}$ $\frac{39}{56}$
33. $\frac{52}{56}$ $\frac{13}{14}$
34. $\frac{105}{126}$ $\frac{5}{6}$
35. $\frac{66}{378}$ $\frac{11}{63}$
36. $\frac{231}{351}$ $\frac{77}{117}$
37. $\frac{9}{21}$ $\frac{3}{7}$
38. $\frac{6}{54}$ $\frac{1}{9}$
39. $\frac{21}{35}$ $\frac{3}{5}$
40. $\frac{14}{56}$ $\frac{1}{4}$
41. $\frac{42}{90}$ $\frac{7}{15}$
42. $\frac{15}{21}$ $\frac{5}{7}$
43. $\frac{63}{81}$ $\frac{7}{9}$
44. $\frac{88}{121}$ $\frac{8}{11}$

B Give two pairs of values for x and y that will make equivalent fractions. **Answers will vary.**

45. $\frac{3}{8}, \frac{x}{y}$
46. $\frac{7}{12}, \frac{x}{y}$
47. $\frac{20}{32}, \frac{x}{y}$
48. $\frac{18}{36}, \frac{x}{y}$
49. $\frac{5}{16}, \frac{x}{y}$
50. $\frac{3}{30}, \frac{x}{y}$
51. $\frac{24}{60}, \frac{x}{y}$
52. $\frac{1}{23}, \frac{x}{y}$

Check cross products to decide which pairs of fractions are equivalent.

53. $\frac{12}{25}, \frac{60}{125}$ yes
54. $\frac{7}{9}, \frac{63}{88}$ no
55. $\frac{9}{13}, \frac{45}{72}$ no
56. $\frac{6}{15}, \frac{42}{105}$ yes
57. $\frac{91}{124}, \frac{7}{8}$ no
58. $\frac{3}{18}, \frac{15}{80}$ no
59. $\frac{24}{36}, \frac{144}{216}$ yes
60. $\frac{35}{64}, \frac{245}{428}$ no

C Extending Thinking Skills 62. $\frac{2}{4} = \frac{4}{8} = \frac{8}{16}$ and $\frac{2}{8} = \frac{4}{16}$ are possible answers.

61. Is the generalization $\frac{a}{b} = \frac{a + c}{b + c}$ true or false? False. $\frac{3}{5} \neq \frac{3 + 2}{5 + 2}$

62. Find all the ways in which the numbers 2, 4, 8, and 16 can be placed in the squares at right to complete a true statement.

$$\frac{\square}{\square} = \frac{\square}{\square}$$

63. Find all the ways in which any four of the numbers 2, 3, 4, 5, 6, 7, 8, 9, and 10 can be placed in the squares to make a true statement. See additional answers.

Thinking skills: 61. Seeking a counterexample 62–63. Testing a conjecture

Mixed Review

Simplify. **64.** $9c - 16c$ $-7c$ **65.** $4m + 2m - m$ $5m$ **66.** $7x + 5 + 4x$ $11x + 5$

67. $4(3r) - 10 + 2r$ $14r - 10$ **68.** $6(y + 4y) + 3y$ $33y$ **69.** $t + t + t$ $3t$

70. $p^2 \cdot p^2 \cdot p^4$ p^8 **71.** $a \cdot a \cdot a$ a^3 **72.** $3^5 \cdot 3^{11}$ 3^{16}

Find the greatest common factor (GCF). **73.** 25, 80 5 **74.** 12, 21 3

Solve and check. **75.** $14.7 = 9.35 - c$ -5.35 **76.** $1.5n = -6$ -4

Evaluate for $y = 1.6$. **77.** $y^2 + 4$ 6.56 **78.** $y^2(y + 2)$ 9.216

NUMBERS TO ALGEBRA

You can reduce fractional expressions in algebra in the same way you reduce fractions in arithmetic.

Numbers	Algebra
$\frac{6}{15} = \frac{2 \cdot \cancel{3}}{5 \cdot \cancel{3}} = \frac{2}{5}$	$\frac{2a\cancel{n}}{5b\cancel{n}} = \frac{2a\cancel{n}}{5b\cancel{n}} = \frac{2a}{5b}$
$\frac{15}{21} = \frac{\cancel{3} \cdot 5}{7 \cdot \cancel{3}} = \frac{5}{7}$	$\frac{6x}{15xy} = \frac{2 \cdot \cancel{3} \cdot \cancel{x}}{\cancel{3} \cdot 5 \cdot \cancel{x} \cdot y} = \frac{2}{5y}$
$\frac{30}{70} = \frac{\cancel{2} \cdot 3 \cdot \cancel{5}}{\cancel{2} \cdot \cancel{5} \cdot 7} = \frac{3}{7}$	$\frac{4m\cancel{n}}{9\cancel{n}} = \frac{2 \cdot 2 \cdot m \cdot \cancel{n}}{3 \cdot 3 \cdot \cancel{n}} = \frac{4m}{9}$

Reduce to lowest terms. The variables a, b, and c represent distinct whole numbers, not including 0.

1. $\frac{2a}{3a}$ $\frac{2}{3}$ **2.** $\frac{5ab}{25a}$ $\frac{b}{5}$ **3.** $\frac{21a}{24b}$ $\frac{7a}{8b}$

4. $\frac{15ab}{48b}$ $\frac{5a}{16}$ **5.** $\frac{abc}{2bc}$ $\frac{a}{2}$ **6.** $\frac{46ab}{48bc}$ $\frac{23a}{24c}$

7. $\frac{96a}{24bc}$ $\frac{4a}{bc}$ **8.** $\frac{18c}{12ac}$ $\frac{3}{2a}$ **9.** $\frac{15a}{35a}$ $\frac{3}{7}$

10. $\frac{51ab}{102bc}$ $\frac{a}{2c}$ **11.** $\frac{24c}{70ac}$ $\frac{12}{35a}$ **12.** $\frac{17a^2}{34ab}$ $\frac{a}{2b}$

6-2 Improper Fractions and Mixed Numbers

Objectives: To write improper fractions as integers or mixed numbers; to write mixed numbers as improper fractions.
MIN: 1–32; REG: 1–32 even, 33–39; MAX: 1–32 even, 33–40; ALL: Mixed Review

A tailor uses about $4\frac{1}{2}$ yards of fabric to make a suit.

A fraction is a **proper fraction** if its numerator is less than its denominator. If its numerator is greater than or equal to its denominator, it is an **improper fraction**. For example, $\frac{4}{5}$ is a proper fraction, but $\frac{5}{4}$ is an improper fraction. When an improper fraction is written as an integer and a fraction, it is called a mixed numeral or **mixed number**. As shown below, the improper fraction $\frac{5}{4}$ is the same as the mixed number $1\frac{1}{4}$.

$\frac{5}{4} = 1\frac{1}{4}$

To change an improper fraction to a mixed number, use the fact that a fraction $\frac{a}{b}$ can be interpreted as $a \div b$.

Example 1

Write as an integer or mixed number. $\frac{19}{7}$

Solution

$$\begin{array}{r} 2 \\ 7\overline{)19} \\ \underline{14} \\ 5 \end{array}$$

$\frac{19}{7}$ means $19 \div 7$.

$\frac{19}{7} = 2\frac{5}{7}$ Write the quotient as the integer. Write the remainder over the divisor as the proper fraction.

Practice Write each as an integer or a mixed number.

a. $\frac{11}{3}$ $3\frac{2}{3}$ **b.** $\frac{24}{4}$ 6 **c.** $\frac{36}{7}$ $5\frac{1}{7}$

Example 2

Write as an improper fraction in lowest terms. $3\frac{1}{7}$

Solution

$3\frac{1}{7}$

$3 \cdot 7 + 1 = 22$ There are $3 \cdot 7$ or 21 sevenths in the 3 whole units. 21 sevenths plus 1 seventh equal 22 sevenths.

$3\frac{1}{7} = \frac{22}{7}$

Practice Write each as an improper fraction in lowest terms.

a. $4\frac{3}{4}$ $\frac{19}{4}$ **b.** $5\frac{1}{3}$ $\frac{16}{3}$ **c.** $7\frac{3}{10}$ $\frac{73}{10}$

Oral Exercises

Give an improper fraction and a mixed number for each.

1. 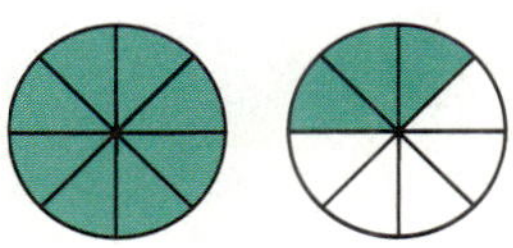

2. 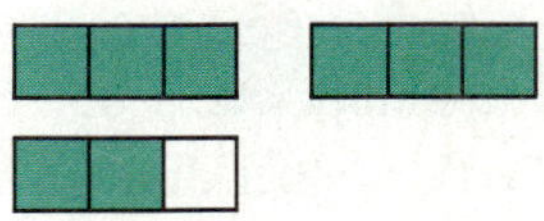

3. 

1. $\frac{11}{8}$, $1\frac{3}{8}$ 2. $\frac{8}{3}$, $2\frac{2}{3}$ 3. $\frac{12}{5}$, $2\frac{2}{5}$

Exercises

A Write each as an integer or a mixed number.

1. $\frac{25}{6}$ $4\frac{1}{6}$ **2.** $\frac{41}{7}$ $5\frac{6}{7}$ **3.** $\frac{51}{3}$ 17 **4.** $\frac{17}{11}$ $1\frac{6}{11}$

5. $\frac{91}{11}$ $8\frac{3}{11}$ **6.** $\frac{238}{10}$ $23\frac{4}{5}$ **7.** $\frac{121}{11}$ 11 **8.** $\frac{215}{15}$ $14\frac{1}{3}$

9. $\frac{342}{100}$ $3\frac{21}{50}$ **10.** $\frac{75}{8}$ $9\frac{3}{8}$ **11.** $\frac{80}{16}$ 5 **12.** $\frac{230}{25}$ $9\frac{1}{5}$

13. $\frac{290}{25}$ $11\frac{3}{5}$ **14.** $\frac{75}{12}$ $6\frac{1}{4}$ **15.** $\frac{35}{10}$ $3\frac{1}{2}$ **16.** $\frac{440}{100}$ $4\frac{2}{5}$

Write each as an improper fraction in lowest terms.

17. $3\frac{7}{8}$ $\frac{31}{8}$ **18.** $4\frac{2}{3}$ $\frac{14}{3}$ **19.** $2\frac{9}{16}$ $\frac{41}{16}$ **20.** $3\frac{6}{15}$ $\frac{17}{5}$

21. $5\frac{7}{9}$ $\frac{52}{9}$ **22.** $7\frac{9}{30}$ $\frac{73}{10}$ **23.** $8\frac{2}{3}$ $\frac{26}{3}$ **24.** $9\frac{7}{11}$ $\frac{106}{11}$

25. $13\frac{7}{10}$ $\frac{137}{10}$ **26.** $15\frac{3}{50}$ $\frac{753}{50}$ **27.** $11\frac{3}{20}$ $\frac{223}{20}$ **28.** $21\frac{7}{8}$ $\frac{175}{8}$

29. $5\frac{3}{51}$ $\frac{86}{17}$ **30.** $7\frac{53}{100}$ $\frac{753}{100}$ **31.** $9\frac{71}{80}$ $\frac{791}{80}$ **32.** $5\frac{31}{40}$ $\frac{231}{40}$

B Evaluate each expression. Write as a mixed number.

33. $\frac{a}{b}$ for $a = 23$ and $b = 5$ $4\frac{3}{5}$

34. $\frac{u}{v}$ for $u = 27$ and $v = 11$ $2\frac{5}{11}$

35. $\frac{x}{y}$ for $x = 73$ and $y = 17$ $4\frac{5}{17}$

36. $\frac{n}{m}$ for $n = 41$ and $m = 12$ $3\frac{5}{12}$

37. $\frac{2x}{y}$ for $x = 33$ and $y = 5$ $13\frac{1}{5}$

38. $\frac{3a}{2b}$ for $a = 13$ and $b = 7$ $2\frac{11}{14}$

C Extending Thinking Skills

39. $\frac{3}{2} = 1\frac{1}{2}; \frac{4}{3} = 1\frac{1}{3}; \frac{7}{4} = 1\frac{3}{4}; \frac{7}{3} = 2\frac{1}{3}; \frac{7}{2} = 3\frac{1}{2}$

39. Find how many ways the numbers 1, 2, 3, 4, and 7 can be placed in the boxes to make a true statement. Numbers may be repeated more than once. The fraction part of the mixed number should be less than 1.

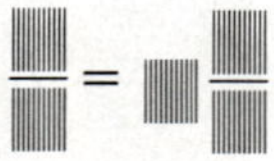

40. Write a generalization for changing any mixed number $a\frac{b}{c}$ to an improper fraction. $a\frac{b}{c} = \frac{ac + b}{c}$

Thinking skills: 39. Testing a conjecture 40. Generalizing

Mixed Review

Give the prime factorization of each. **41.** 30 $2 \cdot 3 \cdot 5$ **42.** 27 $3 \cdot 3 \cdot 3$

Evaluate for $a = 1.5$, $b = 2.75$, $c = 3$. **43.** abc 12.375 **44.** $a(c + b)$ 8.625

Tell whether the number is prime or composite. **45.** 53 prime **46.** 15 composite

Write and solve an equation. **47.** A number divided by 5 gives 8. $n \div 5 = 8, n = 40$

Solve and check. **48.** $120 - z = 64.37$ 55.63 **49.** $6m = -36.138$ -6.023

COMPUTER ACTIVITY

This computer program will reduce the fraction $\frac{A}{B}$ to lowest terms.

```
5 PRINT "TYPE IN THE NUMERATOR AND THE DENOMINATOR OF
YOUR FRACTION SEPARATED BY A COMMA."
10 INPUT N,D
20 T = ABS(N):B=ABS(D)
30 X=INT(T/B):R=T-X*B
40 IF R=0 THEN GOTO 60
50 T=B:B=R: GOTO 30
60 PRINT "THE REDUCED FRACTION IS ";N/B;"/";D/B
70 END
```

Run the program for each fraction.

a. $\frac{2,945}{32,395}$ $\frac{1}{11}$ **b.** $\frac{12,369}{11,994,447}$ $\frac{4,123}{3,998,149}$ **c.** $\frac{1,351,350}{11,781}$ $\frac{1,950}{17}$

6-3 Rational Numbers

Objectives: To write the opposite of a rational number; to graph rational numbers on a number line.
MIN: 1–26, 39; REG: 1–26 even, 27–39; MAX: 1–26 even, 27–40; ALL: Mixed Review

A number that can be expressed in the fractional form $\frac{a}{b}$, where a and b are integers and $b \neq 0$, is called a **rational number**. Each rational number corresponds to *one* set of equivalent fractions and names one point on the number line. For example, in the set $\{\frac{2}{3}, \frac{4}{6}, \frac{6}{9}, \frac{8}{12}, \ldots\}$ each fraction names the same rational number.

The number line below shows points for some rational numbers expressed as fractions or mixed numbers.

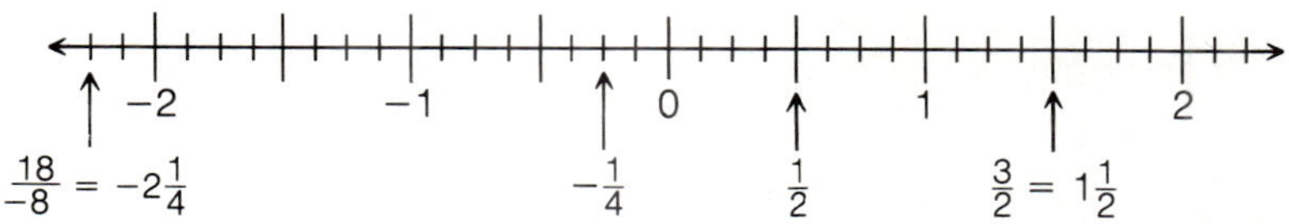

Whole numbers, integers, and certain decimals are rational numbers because they can be expressed as fractions. For example:

$$4 = \frac{4}{1} \qquad -5 = \frac{-5}{1} \qquad 0.3 = \frac{3}{10}$$

The opposite of a negative rational number is a positive rational number. The opposite of a positive rational number is a negative rational number. There are three ways to show a negative rational number.

Example 1

Write the opposite of $\frac{3}{5}$ in three different ways.

Solution

$-\frac{3}{5}, \frac{-3}{5}, \frac{3}{-5}$ The negative sign can precede the entire fraction, precede the numerator, or precede the denominator.

Practice Write the opposite of each in three different ways.

a. $\frac{5}{6}$ **b.** $\frac{7}{3}$ **c.** $\frac{11}{12}$ **d.** $\frac{15}{9}$

a. $-\frac{5}{6}, \frac{-5}{6}, \frac{5}{-6}$ b. $-\frac{7}{3}, \frac{-7}{3}, \frac{7}{-3}$
c. $-\frac{11}{12}, \frac{-11}{12}, \frac{11}{-12}$ d. $-\frac{15}{9}, \frac{-15}{9}, \frac{15}{-9}$

Example 2

Graph the following rational numbers.

$-\frac{1}{4}, \frac{3}{4}, -\frac{6}{4}, \frac{6}{4}$

Solution

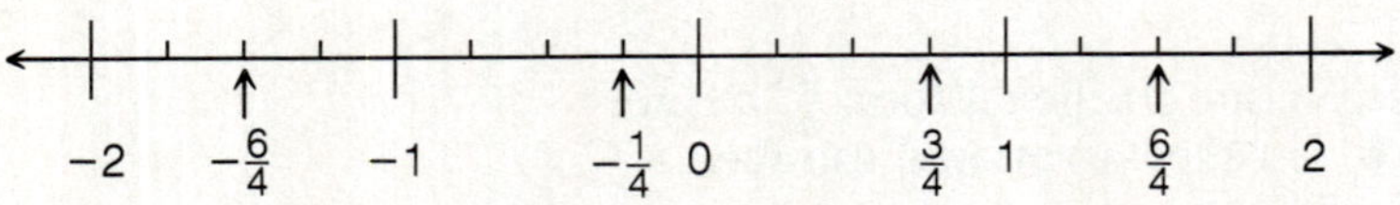

Divide the number line into fourths between integers.

Practice Graph the following rational numbers.

$-\frac{3}{8}, -2, \frac{5}{8}, \frac{1}{8}, \frac{9}{8}$

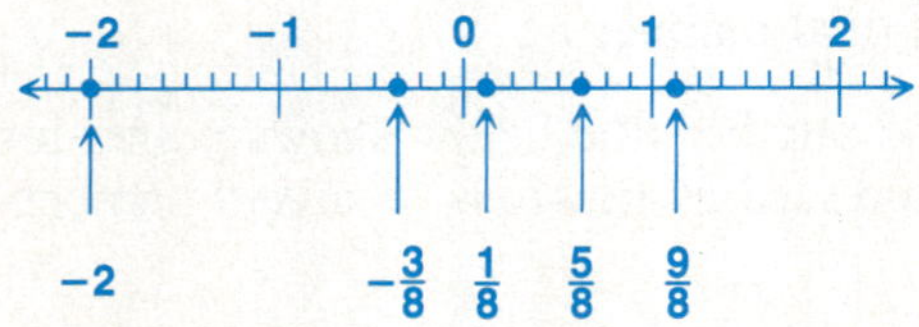

Oral Exercises

Give the rational number for each point.

1.

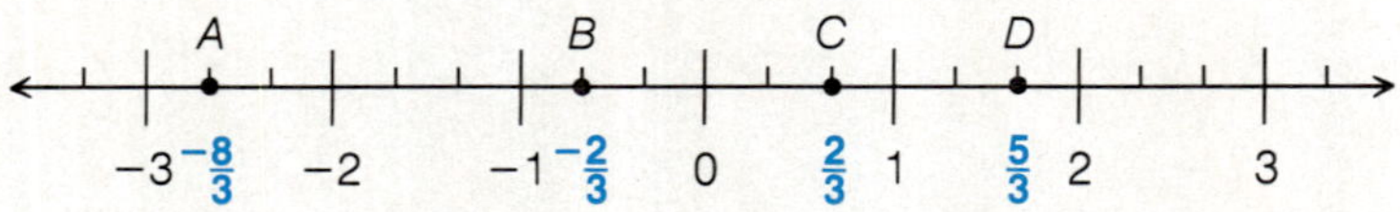

2.

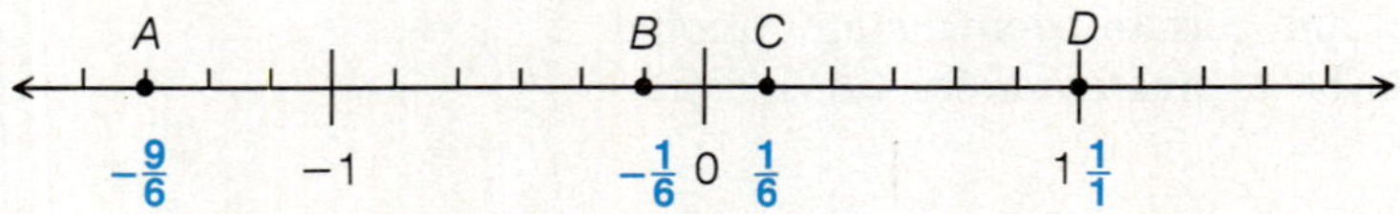

3.

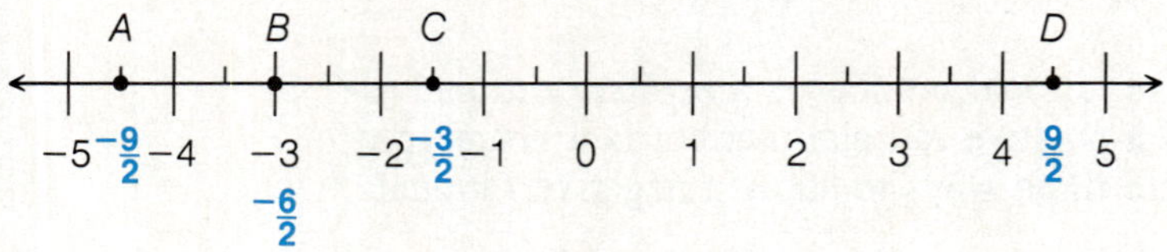

Exercises

A Write the opposite of each in three different ways.

1. $\frac{3}{8}$ $-\frac{3}{8}, \frac{-3}{8}, \frac{3}{-8}$ **2.** $\frac{2}{9}$ $-\frac{2}{9}, \frac{-2}{9}, \frac{2}{-9}$ **3.** $\frac{11}{16}$ $-\frac{11}{16}, \frac{-11}{16}, \frac{11}{-16}$ **4.** $\frac{5}{6}$ $-\frac{5}{6}, \frac{5}{-6}, \frac{-5}{6}$

5. $\frac{5}{2}$ $-\frac{5}{2}, \frac{-5}{2}, \frac{5}{-2}$ **6.** $\frac{4}{9}$ $-\frac{4}{9}, \frac{-4}{9}, \frac{4}{-9}$ **7.** $\frac{3}{2}$ $-\frac{3}{2}, \frac{-3}{2}, \frac{3}{-2}$ **8.** $\frac{1}{10}$ $-\frac{1}{10}, \frac{-1}{10}, \frac{1}{-10}$

9. $\frac{3}{7}$ $-\frac{3}{7}, \frac{-3}{7}, \frac{3}{-7}$ **10.** $\frac{12}{5}$ $-\frac{12}{5}, \frac{-12}{5}, \frac{12}{-5}$ **11.** $\frac{3}{16}$ $-\frac{3}{16}, \frac{-3}{16}, \frac{3}{-16}$ **12.** $\frac{6}{5}$ $-\frac{6}{5}, \frac{-6}{5}, \frac{6}{-5}$

13. $\frac{9}{12}$ $-\frac{9}{12}, \frac{-9}{12}, \frac{9}{-12}$ **14.** $\frac{23}{100}$ **15.** $\frac{18}{24}$ $-\frac{18}{24}, \frac{-18}{24}, \frac{18}{-24}$ **16.** $\frac{75}{125}$

14. $-\frac{23}{100}, \frac{-23}{100}, \frac{23}{-100}$ 16. $-\frac{75}{125}, \frac{-75}{125}, \frac{75}{-125}$

Graph the rational numbers. See additional answers.

17. $1\frac{2}{3}, \frac{1}{3}, \frac{2}{3}, 1\frac{1}{3}$

18. $1\frac{2}{5}, \frac{2}{5}, \frac{3}{5}, 2$

19. $-\frac{1}{4}, -1\frac{3}{4}, \frac{4}{4}, -1$

20. $\frac{1}{6}, -\frac{4}{6}, -\frac{5}{6}, \frac{5}{6}$

21. $3\frac{1}{2}, 2, \frac{1}{2}, 1\frac{1}{2}, \frac{5}{2}$

22. $\frac{2}{7}, 1\frac{1}{7}, \frac{4}{7}, \frac{12}{7}$

23. $-\frac{2}{4}, \frac{3}{4}, -\frac{5}{4}, -2, \frac{1}{4}$

24. $1\frac{1}{3}, -\frac{5}{3}, -2, \frac{2}{3}, \frac{8}{3}$

25. $\frac{1}{8}, -\frac{3}{8}, -\frac{5}{8}, \frac{6}{8}, \frac{7}{8}$

26. $\frac{1}{5}, 0, 1, \frac{10}{5}, \frac{4}{5}, \frac{2}{5}$

B Write the opposite of each expression.

27. $\frac{3}{x}$ $-\frac{3}{x}$

28. $-\frac{2}{y}$ $\frac{2}{y}$

29. $-\frac{h}{2}$ $\frac{h}{2}$

30. $\frac{t}{7}$ $-\frac{t}{7}$

31. $-\frac{6}{n}$ $\frac{6}{n}$

32. $\frac{y}{10}$ $-\frac{y}{10}$

33. $-\frac{12}{m}$ $\frac{12}{m}$

34. $-\frac{x}{100}$ $\frac{x}{100}$

Evaluate each expression for the values given. 36. $\frac{10}{7}$ or $1\frac{3}{7}$

35. $\frac{x + y}{x}$, $x = 5$, $y = -2$ $\frac{3}{5}$

36. $\frac{a + b}{c + d}$, $a = 4$, $b = 6$, $c = -5$, $d = 12$

37. $\frac{2a + 1}{b}$, $a = 7$, $b = 2$ $\frac{15}{2}$ or $7\frac{1}{2}$

38. $\frac{5x - 2}{3y}$, $x = -2$, $y = 4$ -1

C *Extending Thinking Skills*

Find the pattern and complete each sequence.

39. $\frac{1}{3}, -\frac{1}{5}, \frac{1}{7}, -\frac{1}{9}, \frac{1}{11}$, —, —, — $-\frac{1}{13}, \frac{1}{15}, -\frac{1}{17}$

40. $-\frac{1}{4}, -\frac{1}{2}, -\frac{3}{4}, -1, -1\frac{1}{4}$, —, —, — $-1\frac{1}{2}, -1\frac{3}{4}, -2$

Thinking skill: 39–40. Finding a pattern.

Mixed Review

Simplify. 41. $(3m + 2n) + (m + 6n)$ $4m + 8n$ 42. $7(m + 2) + (m + 3)$ $8m + 17$

Solve and check. 43. $2a = -7.562$ -3.781 44. $r + 2.695 = -8.305$ -11

ESTIMATION

You can think about a number line to estimate what integer is closest to a given fraction or mixed number.

Estimate.

1. $5\frac{3}{8}$ 5

2. $-1\frac{7}{8}$ -2

3. $-\frac{1}{3}$ 0

4. $1\frac{1}{5}$ 1

5. $-4\frac{1}{4}$ -4

6. $-4\frac{5}{6}$ -5

6-4 Comparing and Ordering Rational Numbers

Objective: To compare and order rational numbers.
MIN: 1–32; REG: 1–32 even, 33–46; MAX: 1–32 even, 33–48; ALL: Mixed Review
An exploratory lesson opener can be found on page 549.

You can use a number line to compare rational numbers. On the number line below, -2 is to the left of $-1\frac{1}{2}$; $-\frac{5}{8}$ is to the left of $-\frac{1}{8}$; and $1\frac{7}{8}$ is to the right of $1\frac{1}{8}$.

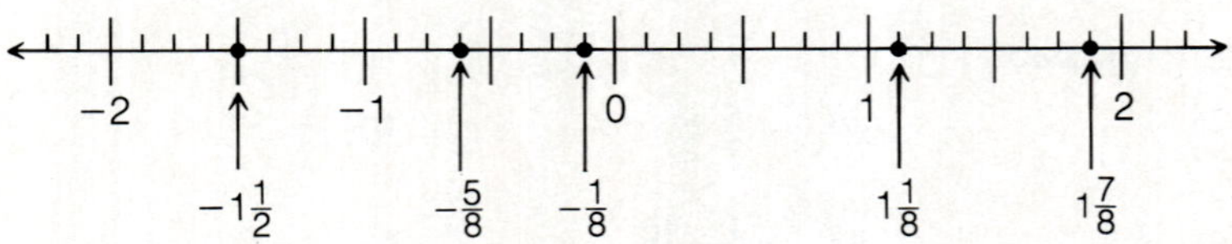

You can show these relationships as:

$$-2 < -1\tfrac{1}{2} \qquad -\tfrac{5}{8} < -\tfrac{1}{8} \qquad 1\tfrac{7}{8} > 1\tfrac{1}{8}$$

When two fractions have the same sign, you can compare them by changing them to equivalent fractions with like denominators or to decimals.

When two fractions are both negative, you can compare them most easily by changing them to equivalent fractions with like denominators.

Example 1

Write >, <, or = for □. Use equivalent fractions to decide.

$\frac{5}{6} \square \frac{7}{8}$

Solution

$\frac{5}{6} = \frac{20}{24}$ Change $\frac{5}{6}$ and $\frac{7}{8}$ to fractions with like denominators. 24 is the least common multiple of 6 and 8.

$\frac{7}{8} = \frac{21}{24}$

$\frac{5}{6} < \frac{7}{8}$ Since $20 < 21$, $\frac{20}{24} < \frac{21}{24}$, so $\frac{5}{6} < \frac{7}{8}$.

Practice Write <, >, or = for each □.

a. $\frac{3}{5} \square \frac{4}{7}$ > **b.** $-\frac{5}{12} \square -\frac{3}{8}$ <

You can use the fact that $\frac{a}{b}$ means $a \div b$ to change a rational number to a decimal. Since some decimals are repeating or very long, you may choose to round the decimal.

Example 2

Write $<$, $>$, or $=$ for $\square$. Use decimals rounded to the nearest thousandth to decide. $\frac{4}{9} \square \frac{3}{8}$

Solution

$$\begin{array}{r} 0.4444 \\ 9\overline{)4.0000} \\ \underline{3\ 6} \\ 40 \\ \underline{36} \\ 40 \\ \underline{36} \\ 40 \\ \underline{36} \\ 4 \end{array} \qquad \begin{array}{r} 0.375 \\ 8\overline{)3.000} \\ \underline{2\ 4} \\ 60 \\ \underline{56} \\ 40 \\ \underline{40} \\ 0 \end{array}$$

Annex zeros and divide to the ten thousandths place if necessary. Round to the nearest thousandth. $\frac{4}{9} = 0.444$ and $\frac{3}{8} = 0.375$.

$\frac{4}{9} > \frac{3}{8}$ 0.444 > 0.375, so $\frac{4}{9} > \frac{3}{8}$.

Practice Write $<$, $>$, or $=$ for $\square$. Use decimals rounded to the nearest thousandth to decide. **a.** $\frac{5}{8} \square \frac{7}{12}$ ($>$) **b.** $-\frac{4}{5} \square -\frac{5}{7}$ ($<$)

When two fractions are equivalent, their cross products are equal. You can use this fact to compare positive rational numbers.

Comparing Positive Rational Numbers

To compare two positive rational numbers, compare their cross products. For integers **a, b, c,** and **d,**

$\frac{a}{b} < \frac{c}{d}$ if $ad < bc$ and $\frac{a}{b} > \frac{c}{d}$ if $ad > bc$

Example 3

Write $<$, $>$, or $=$ for $\square$. Use cross products to decide.

$\frac{5}{9} \square \frac{4}{7}$

Solution $\frac{5}{9} \square \frac{4}{7}$

$5 \cdot 7 \quad 9 \cdot 4$ First find the cross products. The rational numbers compare as their cross products compare.

$35 < 36$

$\frac{5}{9} < \frac{4}{7}$ 35 < 36, so $\frac{5}{9} < \frac{4}{7}$.

Practice Write $<$, $>$, or $=$ for each $\square$. Use cross products to decide.

a. $\frac{5}{6} \square \frac{4}{7}$ ($>$) **b.** $\frac{9}{13} \square \frac{8}{15}$ ($>$)

Oral Exercises

Tell which rational number is greater.

1. $\frac{5}{7}, \frac{6}{7}$ $\frac{6}{7}$ 2. $-\frac{4}{18}, -\frac{5}{18}$ $-\frac{4}{18}$ 3. $\frac{5}{9}, \frac{4}{9}$ $\frac{5}{9}$ 4. $-\frac{15}{12}, -\frac{11}{12}$ $-\frac{11}{12}$ 5. $-\frac{5}{2}, -\frac{1}{2}$ $-\frac{1}{2}$

Exercises

A Write <, >, or = for each □. Use equivalent fractions to decide.

1. $\frac{2}{5} \boxed{<} \frac{3}{7}$ 2. $-\frac{13}{18} \boxed{<} -\frac{2}{3}$ 3. $\frac{3}{4} \boxed{>} \frac{5}{8}$ 4. $-\frac{3}{8} \boxed{<} -\frac{5}{17}$

5. $\frac{5}{13} \boxed{<} \frac{6}{15}$ 6. $\frac{35}{51} \boxed{>} \frac{2}{3}$ 7. $\frac{25}{54} \boxed{>} \frac{4}{9}$ 8. $\frac{29}{32} \boxed{>} \frac{7}{8}$

Write <, >, or = for each □. Use decimals rounded to the nearest thousandth to decide.

9. $-\frac{5}{9} \boxed{>} -\frac{7}{12}$ 10. $\frac{4}{5} \boxed{>} \frac{2}{3}$ 11. $\frac{4}{9} \boxed{>} \frac{3}{7}$ 12. $\frac{17}{30} \boxed{<} \frac{3}{5}$

13. $\frac{3}{5} \boxed{<} \frac{7}{10}$ 14. $-\frac{5}{6} \boxed{<} -\frac{6}{8}$ 15. $\frac{11}{12} \boxed{>} \frac{13}{15}$ 16. $\frac{7}{9} \boxed{<} \frac{8}{10}$

17. $-\frac{5}{8} \boxed{>} -\frac{7}{10}$ 18. $\frac{2}{3} \boxed{<} \frac{8}{11}$ 19. $\frac{9}{16} \boxed{>} \frac{8}{15}$ 20. $-\frac{3}{7} \boxed{>} -\frac{9}{16}$

Write <, >, or = for each □. Use cross products to decide.

21. $\frac{3}{10} \boxed{<} \frac{1}{3}$ 22. $\frac{24}{2} \boxed{=} \frac{12}{1}$ 23. $\frac{13}{12} \boxed{<} \frac{8}{7}$ 24. $\frac{6}{5} \boxed{<} \frac{15}{12}$

25. $\frac{17}{20} \boxed{>} \frac{9}{11}$ 26. $\frac{9}{24} \boxed{<} \frac{2}{5}$ 27. $\frac{14}{25} \boxed{>} \frac{16}{30}$ 28. $\frac{34}{60} \boxed{>} \frac{17}{32}$

29. $\frac{5}{8} \boxed{>} \frac{4}{9}$ 30. $\frac{9}{13} \boxed{>} \frac{4}{7}$ 31. $\frac{13}{15} \boxed{<} \frac{7}{8}$ 32. $\frac{5}{9} \boxed{<} \frac{3}{5}$

B Write in order from least to greatest.

33. $\frac{1}{5}, \frac{1}{3}, \frac{1}{4}$ $\frac{1}{5}, \frac{1}{4}, \frac{1}{3}$ 34. $-\frac{3}{4}, -\frac{7}{8}, -\frac{5}{7}$ $-\frac{7}{8}, -\frac{3}{4}, -\frac{5}{7}$ 35. $\frac{4}{9}, \frac{6}{12}, \frac{2}{3}$ $\frac{4}{9}, \frac{6}{12}, \frac{2}{3}$

36. $\frac{2}{5}, \frac{7}{15}, \frac{10}{20}$ $\frac{2}{5}, \frac{7}{15}, \frac{10}{20}$ 37. $\frac{5}{6}, \frac{6}{7}, \frac{3}{8}$ $\frac{3}{8}, \frac{5}{6}, \frac{6}{7}$ 38. $-\frac{5}{12}, -\frac{6}{15}, -\frac{3}{4}$ 38. $-\frac{3}{4}, -\frac{5}{12}, -$

Write <, >, or = for each □. The variable stands for a positive integer.

39. $\frac{x}{5} \boxed{>} \frac{x}{7}$ 40. $\frac{y}{24} \boxed{<} \frac{y}{12}$ 41. $\frac{j}{8} \boxed{=} \frac{j}{8}$ 42. $\frac{t}{4} \boxed{<} t$

43. $\frac{2d}{6} \boxed{<} \frac{3d}{6}$ 44. $\frac{5}{a} \boxed{>} \frac{3}{a}$ 45. $\frac{4}{2x} \boxed{>} \frac{4}{3x}$ 46. $\frac{2a}{3b} \boxed{=} \frac{6a}{9b}$

C Extending Thinking Skills

47. Write in order from least to greatest. $\frac{5}{6}, \frac{5}{8}, \frac{5}{3}, \frac{5}{2}, \frac{5}{9}, \frac{5}{5}, \frac{5}{12}, \frac{5}{4}$ $\frac{5}{12}, \frac{5}{9}, \frac{5}{8}, \frac{5}{6}, \frac{5}{5}, \frac{5}{4}, \frac{5}{3}, \frac{5}{2}$

48. Make an organized list of all rational numbers less than 1 whose numerators and denominators are selected from the numbers 2, 3, 4, 5, 6, 7, 8, 9, and 10.

See additional answers. Thinking skills: 47. Generalizing 48. Organizing

Mixed Review

Evaluate for $x = 6$. **49.** $x^2 + 19$ 55 **50.** $x(x + 2)$ 48 **51.** $0.5x$ 3

Solve and check. **52.** $-1.35m = -8.37$ 6.2 **53.** $r - 16.28 = -11.06$ 5.22

The basic properties for integers also apply to rational numbers. The multiplicative inverse property will be introduced in Chapter 7. The properties below hold for all rational numbers $\frac{a}{b}$, $\frac{c}{d}$, and $\frac{e}{f}$, where $b \neq 0$, $d \neq 0$, and $f \neq 0$.

Basic Property Update

Commutative Properties

$\frac{a}{b} + \frac{c}{d} = \frac{c}{d} + \frac{a}{b}$ $\quad$ $\frac{a}{b} \cdot \frac{c}{d} = \frac{c}{d} \cdot \frac{a}{b}$

Associative Properties

$\left(\frac{a}{b} + \frac{c}{d}\right) + \frac{e}{f} = \frac{a}{b} + \left(\frac{c}{d} + \frac{e}{f}\right)$ $\quad$ $\left(\frac{a}{b} \cdot \frac{c}{d}\right) \cdot \frac{e}{f} = \frac{a}{b} \cdot \left(\frac{c}{d} \cdot \frac{e}{f}\right)$

Identity Properties

$\frac{a}{b} + 0 = 0 + \frac{a}{b} = \frac{a}{b}$ $\quad$ $\frac{a}{b} \cdot 1 = 1 \cdot \frac{a}{b} = \frac{a}{b}$

Additive Inverse Property

$\frac{a}{b} + \left(-\frac{a}{b}\right) = -\frac{a}{b} + \frac{a}{b} = 0$

Distributive Property

$\frac{a}{b}\left(\frac{c}{d} + \frac{e}{f}\right) = \frac{a}{b} \cdot \frac{c}{d} + \frac{a}{b} \cdot \frac{e}{f}$

CALCULATOR ACTIVITY

You can use a calculator to find the decimal form of a rational number by dividing the numerator by the denominator. To find the decimal form of $\frac{3}{4}$, calculate $3 \div 4$.

		Display
Rational number: $\frac{3}{4}$	3 [÷] 4 [=]	0.75
Rational number: $-\frac{1}{6}$	1 [+/−] [÷] 6 [=]	−0.1666667

Use a calculator to find the decimal form of each rational number.

1. $\frac{4}{5}$ 0.8 **2.** $-\frac{7}{8}$ −0.875 **3.** $\frac{13}{16}$ 0.8125 **4.** $\frac{17}{20}$ 0.85 **5.** $-\frac{7}{25}$ −0.28

6-5 Adding and Subtracting with Like Denominators

Objective: To add and subtract rational numbers with like denominators.
MIN: 1–27; REG: 1–27 even, 28–48; MAX: 1–27 even, 28–51; ALL: Mixed Review

The number line below suggests that $a \cdot \frac{1}{b} = \frac{a}{b}$ when $b \neq 0$.

$$\frac{1}{7} + \frac{1}{7} + \frac{1}{7} + \frac{1}{7} + \frac{1}{7} = 5 \cdot \frac{1}{7} = \frac{5}{7}$$

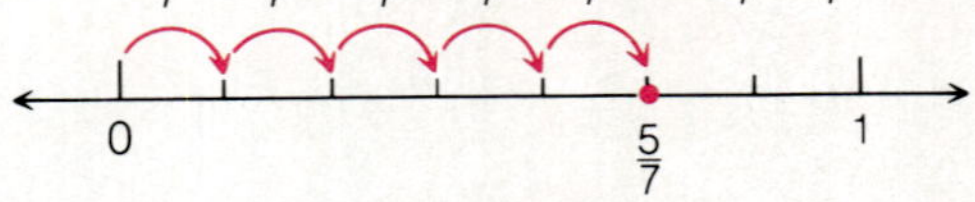

This can be used to explain a method for adding and subtracting rational numbers with like denominators.

$$-\frac{2}{5} + \frac{4}{5} = -2 \cdot \frac{1}{5} + 4 \cdot \frac{1}{5} \qquad a \cdot \frac{1}{b} = \frac{a}{b}$$

$$= (-2 + 4) \cdot \frac{1}{5} \qquad \text{Distributive property}$$

$$= \frac{-2 + 4}{5} \qquad a \cdot \frac{1}{b} = \frac{a}{b}$$

$$= \frac{2}{5}$$

Adding and Subtracting with Like Denominators

To add or subtract rational numbers with like denominators, add or subtract the numerators. Write the sum or difference over the common denominator.

For all integers **a, b,** and **c,** where **c ≠ 0,**

$$\frac{a}{c} + \frac{b}{c} = \frac{a + b}{c} \qquad \frac{a}{c} - \frac{b}{c} = \frac{a - b}{c}$$

Example 1

Subtract. Reduce to lowest terms. $\frac{4}{9} - \frac{7}{9}$

Solution $\frac{4}{9} - \frac{7}{9} = \frac{4 - 7}{9}$ Subtract the numerators. Write the difference over the common denominator.

$= -\frac{3}{9} = -\frac{1}{3}$ Reduce to lowest terms.

Practice Add or subtract. Reduce to lowest terms.

a. $\frac{5}{12} + \frac{4}{12}$ $\frac{3}{4}$ **b.** $-\frac{2}{7} + \frac{4}{7}$ $\frac{2}{7}$ **c.** $-\frac{7}{18} - \frac{13}{18}$ $-1\frac{1}{9}$

Example 2

Subtract. Reduce to lowest terms. $\frac{13}{25} - \left(-\frac{7}{25}\right)$

Solution

$$\frac{13}{25} - \left(-\frac{7}{25}\right) = \frac{13}{25} + \frac{7}{25} \quad \text{Subtracting } -\frac{7}{25} \text{ is the same as adding } \frac{7}{25}.$$

$$= \frac{13 + 7}{25}$$

$$= \frac{20}{25} = \frac{4}{5}$$

Practice Subtract. Reduce to lowest terms.

a. $\frac{5}{18} - \left(-\frac{4}{18}\right)$ $\frac{1}{2}$ **b.** $-\frac{8}{12} - \left(-\frac{5}{12}\right)$ $-\frac{1}{4}$

Oral Exercises

Name the property (commutative, associative, identity, or inverse) shown.

1. $\frac{5}{6} + \left(-\frac{5}{6}\right) = 0$ inv.

2. $-\frac{3}{4} + \left(-\frac{1}{4}\right) = -\frac{1}{4} + \left(-\frac{3}{4}\right)$ commut.

3. $\frac{12}{10} = 0 + \frac{12}{10}$ ident.

4. $1\frac{3}{4} + (-3) = -3 + 1\frac{3}{4}$ commut.

5. $0 + \frac{1}{6} = \frac{1}{6}$ ident.

6. $\frac{4}{5} + \left(\frac{2}{5} + \frac{3}{5}\right) = \left(\frac{4}{5} + \frac{2}{5}\right) + \frac{3}{5}$ assoc.

7. $-\frac{3}{2} + \frac{3}{2} = 0$ inv.

8. $\left(-\frac{4}{8} + \frac{1}{8}\right) + \frac{3}{8} = -\frac{4}{8} + \left(\frac{1}{8} + \frac{3}{8}\right)$ assoc.

Exercises

A Add or subtract. Reduce to lowest terms.

1. $\frac{3}{8} + \frac{1}{8}$ $\frac{1}{2}$

2. $-\frac{2}{7} + \frac{3}{7}$ $\frac{1}{7}$

3. $\frac{5}{11} - \frac{2}{11}$ $\frac{3}{11}$

4. $\frac{7}{16} - \frac{3}{16}$ $\frac{1}{4}$

5. $\frac{7}{8} - \frac{9}{8}$ $-\frac{1}{4}$

6. $\frac{11}{21} - \frac{5}{21}$ $\frac{2}{7}$

7. $-\frac{7}{11} + \frac{9}{11}$ $\frac{2}{11}$

8. $\frac{23}{27} + \frac{13}{27}$ $1\frac{1}{3}$

9. $\frac{48}{13} - \frac{22}{13}$ 2

10. $\frac{12}{7} - \frac{5}{7}$ 1

11. $\frac{7}{19} - \frac{11}{19}$ $-\frac{4}{19}$

12. $\frac{3}{13} - \frac{9}{13}$ $-\frac{6}{13}$

13. $-\frac{2}{15} - \left(-\frac{13}{15}\right)$ $\frac{11}{15}$

14. $\frac{12}{8} - \left(-\frac{4}{8}\right)$ 2

15. $-\frac{18}{25} - \frac{37}{25}$ $-2\frac{1}{5}$

16. $-\frac{14}{15} - \frac{1}{15}$ -1

17. $\frac{24}{15} - \left(-\frac{14}{15}\right)$ $2\frac{8}{15}$

18. $-\frac{9}{10} - \left(-\frac{9}{10}\right)$ 0

19. $-\frac{18}{24} - \frac{12}{24}$ $-1\frac{1}{4}$

20. $\frac{27}{60} - \left(-\frac{5}{60}\right)$ $\frac{8}{15}$

21. $\frac{3}{16} - \left(-\frac{12}{16}\right)$ $\frac{15}{16}$

22. $\frac{35}{50} - \frac{45}{50}$ $-\frac{1}{5}$

23. $-\frac{21}{24} - \left(-\frac{5}{24}\right)$ $-\frac{2}{3}$

24. $\frac{16}{35} - \frac{25}{35}$ $-\frac{9}{35}$

25. $-\frac{15}{60} + \frac{35}{60}$ $\frac{1}{3}$

26. $\frac{8}{24} + \frac{16}{24}$ 1

27. $\frac{18}{36} - \left(-\frac{12}{36}\right)$ $\frac{5}{6}$

B Evaluate each expression. Reduce to lowest terms.

28. $a - \frac{3}{8}$ when $a = \frac{7}{8}$ $\frac{1}{2}$

29. $x + \frac{1}{9}$ when $x = \frac{4}{9}$ $\frac{5}{9}$

30. $s + \frac{27}{2}$ when $s = \frac{11}{2}$ 19

31. $y - \frac{9}{14}$ when $y = \frac{13}{14}$ $\frac{2}{7}$

32. $x + y$ when $x = \frac{7}{19}$ and $y = \frac{8}{19}$ $\frac{15}{19}$

33. $u - v$ when $u = \frac{19}{24}$ and $v = \frac{5}{24}$ $\frac{7}{12}$

Solve. Reduce to lowest terms.

34. $\frac{5}{8} + \left(-\frac{7}{8}\right) - \frac{12}{8}$ $-1\frac{3}{4}$

35. $-\frac{5}{24} - \left(-\frac{13}{24}\right) - \left(-\frac{6}{24}\right)$ $\frac{7}{12}$

36. $\frac{45}{60} - \frac{15}{60} - \left(-\frac{75}{60}\right)$ $1\frac{3}{4}$

37. $-\frac{11}{12} - \frac{13}{12} - \left(-\frac{36}{12}\right)$ 1

Write each expression as a fraction. Example: $\frac{12}{m} - \frac{7}{m} = \frac{12-7}{m} = \frac{5}{m}$

38. $\frac{5}{y} - \frac{7}{y}$ $-\frac{2}{y}$

39. $\frac{6}{z} + \frac{5}{z}$ $\frac{11}{z}$

40. $\frac{12}{s} - \frac{7}{s}$ $\frac{5}{s}$

41. $\frac{27}{x} - \frac{15}{x}$ $\frac{12}{x}$

42. $\frac{123}{t} + \frac{35}{t}$ $\frac{158}{t}$

43. $\frac{67}{u} - \frac{25}{u}$ $\frac{42}{u}$

"Break apart" numbers to find each sum mentally.
Example: To find $5\frac{1}{4} + \frac{3}{4}$, think "$\frac{1}{4} + \frac{3}{4}$ is 1, and 5 + 1 is 6."

44. $6\frac{1}{3} + \frac{2}{3}$ 7

45. $7\frac{3}{8} + \frac{1}{8}$ $7\frac{1}{2}$

46. $26\frac{1}{5} + 4\frac{3}{5}$ $30\frac{4}{5}$

47. The total length of a bicycle race course will be $\frac{5}{8}$ mile, and $\frac{3}{8}$ mile is completed so far. What part of the course is not yet completed?

48. A rancher repaired three sections of fencing. The first section was $\frac{1}{10}$ mile long, the second was $\frac{3}{10}$ mile long, and the third was $\frac{4}{10}$ mile long. What was the total length of the fencing the rancher repaired? $\frac{4}{5}$ mi

47. $\frac{5}{8} - \frac{3}{8} = \frac{2}{8} = \frac{1}{4}$ mi (or $\frac{2}{5}$ of the course)

C *Extending Thinking Skills*

49. Find all the whole number values of t that make the equation true.
$\frac{19}{t} + \frac{5}{t} =$ a whole number $t = 1, 2, 3, 4, 6, 8, 12, 24$

50. Find two rational numbers with a sum of 0 and a difference of $\frac{1}{2}$. $\frac{1}{4}, -\frac{1}{4}$

51. Find two rational numbers with a sum of $\frac{1}{2}$ and a difference of 0. $\frac{1}{4}, \frac{1}{4}$

Thinking skills: 49. Generalizing 50–51. Testing a conjecture

Mixed Review

Solve and check. **52.** $-27a = 334.8$ -12.4 **53.** $r + 0.414 = 2.396$ 1.982

Find the greatest common factor (GCF). **54.** 32, 96, 48 16

6-6 Adding and Subtracting with Unlike Denominators

Objective: To add and subtract rational numbers with unlike denominators
MIN: 1–24, 31; REG: 1–24 even, 25–31; MAX: 1–24 even, 25–32; ALL: Mixed Review

Wrestlers keep accurate records of weight gains and losses. If Jim gained $\frac{5}{8}$ pound one week and $\frac{2}{3}$ pound the next week, he could add $\frac{5}{8}$ and $\frac{2}{3}$ to find his total gain for the two weeks.

To add or subtract rational numbers with unlike denominators, first change them to equivalent fractions with like denominators. This is usually the least common multiple of the denominators, called the **least common denominator**. Add or subtract the numerators and write the sum or difference over the common denominator.

Example 1

Add. Reduce to lowest terms. $\frac{5}{8} + \frac{2}{3}$

Solution

$\frac{5}{8} + \frac{2}{3} = \frac{15}{24} + \frac{16}{24}$ Change each to a fraction with a denominator of 24.

$= \frac{15 + 16}{24}$

$= \frac{31}{24}$

$= 1\frac{7}{24}$

Practice Add. Reduce to lowest terms.

a. $\frac{4}{5} + \frac{5}{12}$ $1\frac{13}{60}$ **b.** $\frac{3}{10} + \frac{6}{15}$ $\frac{7}{10}$

Example 2

Subtract. Reduce to lowest terms. $-\frac{5}{6} - \frac{1}{2}$

Solution

$$-\frac{5}{6} - \frac{1}{2} = -\frac{5}{6} - \frac{3}{6}$$ 6 is the least common multiple of 6 and 2.

$$= \frac{-5 - 3}{6}$$

$$= -\frac{8}{6}$$

$$= -1\frac{2}{6} = -1\frac{1}{3}$$ Reduce to lowest terms.

Practice Subtract. Reduce to lowest terms.

a. $\frac{5}{6} - \left(-\frac{1}{8}\right)$ $\frac{23}{24}$ **b.** $-\frac{2}{5} - \frac{3}{6}$ $-\frac{9}{10}$

Oral Exercises

Give the least common denominator for each pair of fractions.

1. $\frac{1}{2}, \frac{3}{4}$ 4 **2.** $\frac{2}{3}, \frac{1}{6}$ 6 **3.** $\frac{1}{5}, \frac{3}{10}$ 10 **4.** $\frac{1}{2}, \frac{2}{3}$ 6 **5.** $\frac{1}{3}, \frac{1}{4}$ 12

6. $\frac{2}{3}, \frac{3}{5}$ 15 **7.** $\frac{1}{4}, \frac{3}{8}$ 8 **8.** $\frac{4}{5}, \frac{7}{25}$ 25 **9.** $\frac{5}{6}, \frac{1}{3}$ 6 **10.** $\frac{1}{2}, \frac{7}{10}$ 10

Exercises

A Add or subtract. Reduce to lowest terms.

1. $\frac{1}{2} + \frac{3}{8}$ $\frac{7}{8}$ **2.** $\frac{2}{3} - \frac{1}{4}$ $\frac{5}{12}$ **3.** $\frac{1}{6} - \frac{3}{4}$ $-\frac{7}{12}$

4. $\frac{7}{8} - \frac{1}{4}$ $\frac{5}{8}$ **5.** $-\frac{1}{2} + \frac{1}{5}$ $-\frac{3}{10}$ **6.** $-\frac{2}{5} - \frac{3}{4}$ $-1\frac{3}{20}$

7. $-\frac{7}{10} - \frac{1}{2}$ $-1\frac{1}{5}$ **8.** $\frac{2}{3} - \frac{1}{12}$ $\frac{7}{12}$ **9.** $\frac{3}{4} + \frac{5}{6}$ $1\frac{7}{12}$

10. $-\frac{8}{16} + \frac{3}{4}$ $\frac{1}{4}$ **11.** $\frac{3}{4} - \frac{1}{5}$ $\frac{11}{20}$ **12.** $\frac{5}{6} + \frac{7}{10}$ $1\frac{8}{15}$

13. $\frac{5}{8} - \left(-\frac{5}{6}\right)$ $1\frac{11}{24}$ **14.** $\frac{5}{18} - \left(-\frac{3}{6}\right)$ $\frac{7}{9}$ **15.** $-\frac{11}{12} - \left(-\frac{5}{3}\right)$ $\frac{3}{4}$

16. $-\frac{12}{8} - \frac{2}{3}$ $-2\frac{1}{6}$ **17.** $-\frac{7}{10} - \left(-\frac{29}{100}\right)$ $-\frac{41}{100}$ **18.** $\frac{14}{8} - \left(-\frac{6}{4}\right)$ $3\frac{1}{4}$

19. $\frac{7}{8} - \left(-\frac{5}{6}\right)$ $1\frac{17}{24}$ **20.** $-\frac{5}{8} + \frac{5}{12}$ $-\frac{5}{24}$ **21.** $-\frac{7}{12} - \left(-\frac{4}{9}\right)$ $-\frac{5}{36}$

22. $\frac{4}{5} - \left(-\frac{3}{8}\right)$ $1\frac{7}{40}$ **23.** $\frac{3}{4} - \frac{4}{7}$ $\frac{5}{28}$ **24.** $-\frac{1}{8} - \left(-\frac{5}{12}\right)$ $\frac{7}{24}$

B Add. Reduce to lowest terms.

25. $-\frac{4}{9} + \frac{2}{3} - \frac{4}{9}$ $-\frac{2}{9}$

26. $-\frac{5}{6} - \frac{3}{4} + \frac{1}{2}$ $-1\frac{1}{12}$

27. $-\frac{2}{3} + \frac{7}{9} + \frac{1}{2}$ $\frac{11}{18}$

28. $-\frac{4}{6} - \left(-\frac{7}{12}\right) - \frac{3}{8}$ $-\frac{11}{24}$

29. The high jump record used to be 5 ft $1\frac{1}{4}$ in. Today the record was broken by $\frac{1}{2}$ in. What is the new high jump record? 5 ft $1\frac{3}{4}$ in.

30. A wrestler lost $\frac{1}{2}$ lb during the morning and another $\frac{3}{4}$ lb during the afternoon. How much weight in all did the wrestler lose that day? $1\frac{1}{4}$ lb

C Extending Thinking Skills

31. Place the digits 2, 3, 4, and 5 in the boxes to make the equation true. $-\frac{\square}{\square} + \frac{\square}{\square} = \frac{7}{20}$ ($-\frac{2}{5} + \frac{3}{4}$)

32. Complete the following proof by supplying the missing reasons.

$$\frac{a}{b} + \frac{c}{d} = \frac{a \cdot d}{b \cdot d} + \frac{c \cdot b}{d \cdot b}$$ property of equivalent fractions

$$= \frac{a \cdot d}{b \cdot d} + \frac{c \cdot b}{b \cdot d}$$ why? commutative

$$= a \cdot d \cdot \frac{1}{b \cdot d} + (c \cdot b) \cdot \frac{1}{b \cdot d}$$ $\frac{a}{b} = b \cdot \frac{1}{b}$

$$= (a \cdot d + c \cdot b) \cdot \frac{1}{b \cdot d}$$ why? distributive

$$= \frac{a \cdot d + c \cdot b}{b \cdot d}$$ why? $\frac{a}{b} = a \cdot \frac{1}{b}$

Testing a conjecture 32. Proving

Mixed Review

Find all the factors of each number. **33.** 25 1, 5, 25 **34.** 21 1, 3, 7, 21

Give the absolute value and opposite of each. **35.** −7 7, 7 **36.** 19 19, −19

Estimate, then find the value of the variable. **37.** $M = 6.5(68 \div 17)$ est. ~28, $M = 26$

Solve and check. **38.** $19.8 = -4c$ −4.95 **39.** $-11.62 + n = -49.33$ −37.71

ESTIMATION

You can substitute compatible numbers to estimate the sum or difference of rational numbers by changing them to approximate equivalent fractions. For example, to estimate the sum $\frac{1}{4} + \frac{7}{8}$, think: $\frac{7}{8}$ is about $\frac{3}{4}$, so $\frac{1}{4} + \frac{7}{8}$ is about $\frac{1}{4} + \frac{3}{4}$, or 1.

Estimate each sum or difference.

1. $\frac{7}{8} + \frac{12}{13}$ 2 **2.** $\frac{3}{7} + \frac{5}{8}$ 1 **3.** $\frac{11}{12} - \frac{3}{4}$ $\frac{1}{4}$ **4.** $\frac{14}{15} + \frac{1}{2}$ $1\frac{1}{2}$

6-7 Adding and Subtracting with Mixed Numbers

Objective: To add and subtract mixed numbers.
MIN: 1–18; REG 1–18 even, 19–29; MAX: 1–18 even, 19–30; ALL: Mixed Review
An exploratory lesson opener can be found on page 549.

A builder estimated that $4\frac{1}{2}$ cubic yards of concrete were needed for a job. Only $2\frac{2}{3}$ cubic yards were used. You could subtract to check the estimate.

The rules for adding and subtracting rational numbers with like and unlike denominators apply to mixed numbers.

Example 1

Subtract. $\begin{array}{r} 4\frac{1}{2} \\ -2\frac{2}{3} \\ \hline \end{array}$

Solution

$$\begin{array}{rcccl} 4\frac{1}{2} & = & 4\frac{3}{6} & = & 3\frac{9}{6} \\ -2\frac{2}{3} & = & 2\frac{4}{6} & = & 2\frac{4}{6} \\ \hline & & & & 1\frac{5}{6} \end{array}$$

Find equivalent fractions with like denominators and rename $4\frac{3}{6}$ as $3\frac{9}{6}$.
Subtract the fractions, and then subtract the whole numbers.

Practice Add or subtract. Reduce to lowest terms if necessary.

a. $\begin{array}{r} 3\frac{1}{4} \\ -1\frac{1}{2} \\ \hline \end{array}$ **b.** $\begin{array}{r} 3\frac{4}{5} \\ +5\frac{5}{8} \\ \hline \end{array}$ **c.** $\begin{array}{r} 4\frac{3}{5} \\ +2\frac{3}{5} \\ \hline \end{array}$

a. $1\frac{3}{4}$ b. $9\frac{17}{40}$ c. $7\frac{1}{5}$

Sometimes it helps to change mixed numbers to improper fractions before adding or subtracting.

Example 2

Add and reduce to lowest terms. $-2\frac{2}{3} + 1\frac{1}{6}$

Solution $-2\frac{2}{3} + 1\frac{1}{6} = -\frac{8}{3} + \frac{7}{6}$ Write each mixed number as an improper fraction.

$= -\frac{16}{6} + \frac{7}{6}$ 6 is the least common denominator.

$= \frac{-16 + 7}{6}$

$= -\frac{9}{6} = -1\frac{3}{6} = -1\frac{1}{2}$

Practice Add or subtract. Reduce to lowest terms.

a. $1\frac{1}{2} + \frac{3}{4}$ $2\frac{1}{4}$ **b.** $-3\frac{1}{3} - \left(-4\frac{2}{5}\right)$ $1\frac{1}{15}$ **c.** $2\frac{3}{8} + 1\frac{7}{8}$ $4\frac{1}{4}$

Oral Exercises

Change each mixed number to an improper fraction mentally.

1. $1\frac{3}{5}$ $\frac{8}{5}$ **2.** $-4\frac{1}{2}$ $-\frac{9}{2}$ **3.** $-5\frac{3}{4}$ $-\frac{23}{4}$ **4.** 4 $\frac{4}{1}$ **5.** $6\frac{2}{3}$ $\frac{20}{3}$

6. $-3\frac{1}{10}$ $-\frac{31}{10}$ **7.** $1\frac{7}{12}$ $\frac{19}{12}$ **8.** $2\frac{5}{6}$ $\frac{17}{6}$ **9.** $-3\frac{1}{9}$ $-\frac{28}{9}$ **10.** $-7\frac{5}{8}$ $-\frac{61}{8}$

Exercises

A Add or subtract. Reduce to lowest terms.

1. $1\frac{2}{3} - \frac{3}{5}$ $1\frac{1}{15}$ **2.** $2\frac{2}{5} + 1\frac{1}{5}$ $3\frac{3}{5}$ **3.** $2\frac{3}{7} + \frac{1}{3}$ $2\frac{16}{21}$

4. $3\frac{1}{6} - \frac{2}{3}$ $2\frac{1}{2}$ **5.** $3\frac{3}{5} + \frac{5}{12}$ $4\frac{1}{60}$ **6.** $3\frac{7}{8} + 2\frac{1}{3}$ $6\frac{5}{24}$

7. $3\frac{3}{8} + 2\frac{5}{8}$ 6 **8.** $5\frac{5}{12} - 3\frac{2}{3}$ $1\frac{3}{4}$ **9.** $5\frac{3}{4} + 6\frac{1}{3}$ $12\frac{1}{12}$

10. $-3\frac{5}{6} - 2\frac{1}{8}$ $-5\frac{23}{24}$ **11.** $-4\frac{3}{10} + 7\frac{3}{4}$ $3\frac{9}{20}$ **12.** $-1\frac{2}{9} - 1\frac{5}{12}$ $-2\frac{23}{36}$

13. $-4 - 2\frac{4}{7}$ $-6\frac{4}{7}$ **14.** $-4\frac{5}{6} + 7$ $2\frac{1}{6}$ **15.** $-1\frac{5}{12} - 4\frac{5}{16}$ $-5\frac{35}{48}$

16. $2\frac{3}{4} - 1\frac{7}{8}$ $\frac{7}{8}$ **17.** $-6\frac{1}{4} + 4\frac{1}{2}$ $-1\frac{3}{4}$ **18.** $5\frac{3}{8} + 2\frac{3}{4}$ $8\frac{1}{8}$

B For Exercises 19–24, estimate, using rounding to the nearest whole number or a **front-end** technique. For example, to find $7\frac{3}{4} + 2\frac{1}{3}$, think: $7\frac{3}{4}$ is closest to 8, $2\frac{1}{3}$ is closest to 2. Estimate: 10. Or think: 7 + 2 is 9, $\frac{3}{4} + \frac{1}{3}$ is about 1. Estimate: 10.

19. $3\frac{4}{5} - 1\frac{1}{3}$ 3 **20.** $12\frac{5}{8} - \left(-3\frac{2}{5}\right)$ 16 **21.** $-7\frac{1}{2} + 3\frac{1}{4}$ −4

22. $\frac{1}{5} - 4\frac{7}{12} + 6\frac{4}{9}$ 2 **23.** $-5\frac{5}{6} - \left(-\frac{7}{9}\right) + 12$ 7 **24.** $6\frac{2}{3} - 25 - \left(-12\frac{9}{10}\right)$ −5

Estimate each sum using clustering. For example, to find $3\frac{3}{4} + 4\frac{1}{5} + 3\frac{7}{8}$, think "all the numbers cluster around 4, and $3 \cdot 4 = 12$."

25. $4\frac{5}{6} + 5\frac{3}{8} + 4\frac{9}{10}$ 15 **26.** $6\frac{3}{4} + 7\frac{1}{6} + 6\frac{7}{8}$ 21 **27.** $12\frac{1}{5} + 11\frac{7}{8} + 11\frac{9}{10}$ 36

28. A school's record for a race was 4 minutes $23\frac{1}{2}$ seconds. A runner ran this race in 4 minutes $35\frac{3}{10}$ seconds. By how many seconds did the runner miss tying the record? $11\frac{4}{5}$ seconds

29. A carpenter cut $5\frac{1}{2}$ inches off one end of a 6-foot-long piece of wood and $3\frac{3}{5}$ inches off the other end. How long was the wood after these two pieces were cut off? $5'2\frac{9}{10}''$

C Extending Thinking Skills

30. Write the rational numbers $\frac{1}{2}$, 1, $1\frac{1}{2}$, 2, $2\frac{1}{2}$, and 3 in the circles so that the sum along each side of the triangle is $4\frac{1}{2}$.

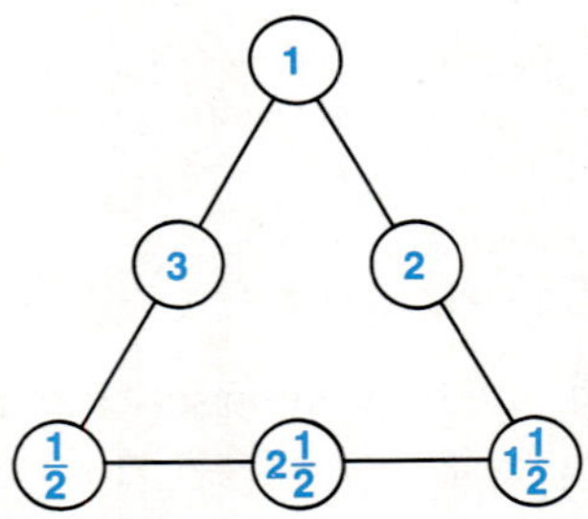

Thinking skill: 30. Testing a conjecture

Mixed Review

Find the least common multiple (LCM). **31.** 6, 8, 10, 12 120

Evaluate for $y = -3$. **32.** $y(y - 4)$ 21 **33.** $y^2 + 3y$ 0

34. $9 - 6y$ 27

MENTAL MATH

You can choose compatible numbers to add or subtract rational numbers mentally. For example, in the equation $2\frac{1}{4} + 3\frac{2}{3} + 4\frac{3}{4}$ you can first add $2\frac{1}{4}$ and $4\frac{3}{4}$ to get 7 and then add 7 and $3\frac{2}{3}$ to get $10\frac{2}{3}$.

Choose compatible numbers to add or subtract mentally.

1. $-5\frac{2}{5} + 7 - 1\frac{3}{5}$ 0 **2.** $6\frac{2}{3} + 4\frac{1}{2} - 2\frac{2}{3}$ $8\frac{1}{2}$ **3.** $3\frac{4}{9} + 4\frac{5}{6} + 1\frac{5}{9}$ $9\frac{5}{6}$

4. $10\frac{5}{8} - 2\frac{3}{8} - 3$ $5\frac{1}{4}$ **5.** $-4\frac{4}{5} + 5\frac{5}{12} - \frac{1}{5}$ $\frac{5}{12}$ **6.** $-5\frac{6}{7} - 7\frac{1}{3} - 2\frac{1}{7}$ $-15\frac{1}{3}$

More Practice

Add or subtract. Reduce to lowest terms.

1. $\frac{1}{2} + \frac{3}{8}$ $\frac{7}{8}$
2. $\frac{7}{10} + \left(-\frac{7}{10}\right)$ 0
3. $-\frac{2}{9} + \frac{4}{9}$ $\frac{2}{9}$
4. $\frac{3}{4} - \frac{1}{5}$ $\frac{11}{20}$
5. $\frac{4}{5} + \left(-\frac{1}{5}\right)$ $\frac{3}{5}$
6. $1\frac{2}{7} - \frac{3}{7}$ $\frac{6}{7}$
7. $\frac{3}{4} + \frac{1}{2}$ $1\frac{1}{4}$
8. $1\frac{5}{12} - \frac{3}{4}$ $\frac{2}{3}$
9. $2\frac{3}{5} - \frac{2}{3}$ $1\frac{14}{15}$
10. $\frac{1}{2} - \frac{2}{3}$ $-\frac{1}{6}$
11. $\frac{9}{10} - \frac{1}{2}$ $\frac{2}{5}$
12. $-\frac{5}{12} + \frac{3}{4}$ $\frac{1}{3}$
13. $\frac{3}{10} + \left(-\frac{1}{6}\right)$ $\frac{2}{15}$
14. $\frac{4}{9} + 1\frac{2}{3}$ $2\frac{1}{9}$
15. $4\frac{1}{2} - 1\frac{2}{3}$ $2\frac{5}{6}$
16. $\frac{2}{3} - \left(-\frac{4}{5}\right)$ $1\frac{7}{15}$
17. $\frac{5}{6} + \frac{1}{8}$ $\frac{23}{24}$
18. $1\frac{1}{2} + \frac{1}{4}$ $1\frac{3}{4}$
19. $\frac{3}{4} - 3\frac{1}{12}$ $-2\frac{1}{3}$
20. $4\frac{3}{10} - 3\frac{1}{2}$ $\frac{4}{5}$
21. $\frac{2}{5} - \frac{3}{8}$ $\frac{1}{40}$
22. $\frac{8}{9} - \frac{7}{10}$ $\frac{17}{90}$
23. $1\frac{1}{2} + \frac{1}{5}$ $1\frac{7}{10}$
24. $\frac{9}{10} + \left(-\frac{1}{2}\right)$ $\frac{2}{5}$
25. $1\frac{3}{7} - \frac{6}{7}$ $\frac{4}{7}$
26. $1\frac{7}{12} - 2\frac{1}{4}$ $-\frac{2}{3}$
27. $\frac{3}{4} - \frac{4}{7}$ $\frac{5}{28}$
28. $2\frac{1}{3} - \frac{3}{4}$ $1\frac{7}{12}$
29. $\frac{9}{25} + \frac{1}{15}$ $\frac{32}{75}$
30. $\frac{9}{10} + 2\frac{1}{3}$ $3\frac{7}{30}$
31. $4\frac{1}{3} - 2\frac{5}{6}$ $1\frac{1}{2}$
32. $3\frac{1}{9} + \frac{3}{4}$ $3\frac{31}{36}$
33. $1\frac{3}{4} - \frac{5}{8}$ $1\frac{1}{8}$
34. $\frac{2}{7} + \left(-\frac{3}{5}\right)$ $-\frac{11}{35}$
35. $5\frac{2}{3} - 4\frac{7}{8}$ $\frac{19}{24}$
36. $\frac{5}{16} + \left(-\frac{1}{4}\right)$ $\frac{1}{16}$
37. $-\frac{3}{7} + 1\frac{3}{7}$ 1
38. $6\frac{13}{15} - 5$ $1\frac{13}{15}$
39. $3\frac{3}{5} - \frac{2}{3}$ $2\frac{14}{15}$
40. $3\frac{3}{4} - 5\frac{1}{4}$ $-1\frac{1}{2}$
41. $3\frac{5}{9} - \left(-2\frac{3}{5}\right)$ $6\frac{7}{45}$
42. $\frac{3}{8} + 2\frac{1}{2}$ $2\frac{7}{8}$
43. $2\frac{1}{2} + \left(-3\frac{2}{7}\right)$ $-\frac{11}{14}$
44. $\frac{17}{22} - \frac{3}{11}$ $\frac{1}{2}$
45. $-\frac{1}{4} + \frac{5}{6}$ $\frac{7}{12}$
46. $4\frac{1}{3} - 1\frac{1}{3}$ 3
47. $1\frac{2}{3} - \frac{7}{9}$ $\frac{8}{9}$
48. $\frac{3}{4} + \frac{4}{5}$ $1\frac{11}{20}$

PROBLEM SOLVING: APPLICATIONS

6-8 The Stock Market

Objective: To solve problems involving stock prices.
ALL: 1–10

When you buy **shares** of **stock** from a corporation, you become a part owner of the corporation, and are called a shareholder or a stockholder. Anyone can buy stock, but most people have a stockbroker help them buy it. Stock prices are published each day in newspapers. A stock's price is given as a fractional part of a dollar. For example, $19\frac{5}{8}$ means 19 and $\frac{5}{8}$ dollars, or \$19.625.

Below is a typical newspaper stock report. The company names are given as abbreviations under the word "stock."

Today's Stock Report

(1) High	(1) Low	(2) Stock	(3) High	(3) Low	(4) Last	(5) Chg.
19 3/4	9 1/2	**AWARD**	16 1/4	14	15 5/8	+2 1/4
28	10 7/8	**ConnEt**	16 1/8	13	14 1/8	−1 5/8
62 1/4	47	**NATS**	59 7/8	56 1/2	57 3/4	+1 7/8
45 1/2	22	**PODCo**	33 1/2	29 7/8	31	−3/4
21 1/4	14 7/8	**RICH**	19 3/4	18 3/8	19 5/8	+1/2

(1) High/Low–The highest and lowest selling prices of each share for the year.
(2) Stock–The name of the company.
(3) High/Low–The highest and lowest selling prices of each share for this day.
(4) Last–The price of the last share sold this day. Also called "closing price."
(5) Chg. (Change)–The difference between today's closing price and yesterday's closing price.

Problem Find the price of the last RICH Corporation share sold yesterday.

Solution $19\frac{5}{8} - \frac{1}{2} = 19\frac{1}{8}$ The change (Chg.) was an increase (+) of $\frac{1}{2}$. To find yesterday's last price, subtract $\frac{1}{2}$ from today's last price.

Problems

You might wish to have students choose from among the techniques pencil and paper, mental math, estimation, and a calculator. Have them choose the most appropriate technique for each problem.

Solve.

1. PODCo shares went down today compared to yesterday. What was the last price for a share yesterday? **31 3/4**

2. AWARD shares went up today compared to yesterday. What was the last price for a share yesterday? **13 3/8**

3. What was the closing price of NATS shares yesterday? 55 7/8

4. What was the closing price of ConnEt shares yesterday? 15 3/4

5. What is the difference between the yearly high and low of PODCo shares? 23 1/2

6. What is the difference in the yearly high and low prices for RICH shares? 6 3/8

7. How much would it cost to buy 100 shares of RICH stock at today's high price? $1975

8. About how much would it cost to buy 1500 shares of ConnEt stock at today's high price? $24,000

9. Suppose you owned one share in each of the companies listed on page 190. Did the total value of your stock increase or decrease yesterday? By how much? increase 2 1/4

10. ***Data Search*** Choose any five stocks from a newspaper report. Find the closing price of each stock for the day previous to the day of the report. Answers will vary.

What's Your Decision?

Suppose you have $2,000 to invest in stocks. You want to buy at least 20 shares of each stock and you want at least 3 different stocks. Use estimation to decide which of the stocks in the table on page 190 and how many shares of each you would choose for your investment.

6-9 Solving Equations: Using Addition and Subtraction

Objective: to solve equations involving addition and subtraction of rational numbers.
MIN: 1–27; REG: 1–27 even, 28–36; MAX: 1–27 even, 28–38; ALL: Mixed Review

The steps you have learned for solving equations involving whole numbers, decimals, and integers apply to equations involving all rational numbers.

- Decide which operation (addition or subtraction) has been applied to the variable.
- Use the inverse operation or use the additive inverse property, adding or subtracting the same number on both sides of the equation.

To solve an equation such as $x - \frac{4}{5} = -\frac{3}{10}$, you need to get the variable by itself.

Example 1

Solve and check. $x - \frac{4}{5} = -\frac{3}{10}$

Solution $x - \frac{4}{5} = -\frac{3}{10}$

$$x - \frac{4}{5} + \frac{4}{5} = -\frac{3}{10} + \frac{4}{5}$$ Addition property of equality.

$$x = -\frac{3}{10} + \frac{4}{5}$$

$$= -\frac{3}{10} + \frac{8}{10}$$

$$x = \frac{5}{10} = \frac{1}{2}$$

Check $\frac{1}{2} - \frac{4}{5} \stackrel{?}{=} -\frac{3}{10}$ Replace x with $\frac{1}{2}$ in $x - \frac{4}{5} = -\frac{3}{10}$.

$$\frac{5}{10} - \frac{8}{10} \stackrel{?}{=} -\frac{3}{10}$$

$$-\frac{3}{10} = -\frac{3}{10} \checkmark$$ The solution is $\frac{1}{2}$.

Practice Solve and check.

a. $x + \frac{7}{12} = \frac{1}{4}$ $-\frac{1}{3}$

b. $y - \frac{5}{6} = -\frac{3}{4}$ $\frac{1}{12}$

c. $-\frac{1}{4} = t - \frac{7}{8}$ $\frac{5}{8}$

d. $-\frac{4}{5} + m = \frac{2}{3}$ $1\frac{7}{15}$

Example 2

Solve and check. $\frac{5}{6} + a = \frac{4}{9}$

Solution

$$\frac{5}{6} + a = \frac{4}{9}$$

$$-\frac{5}{6} + \frac{5}{6} + a = -\frac{5}{6} + \frac{4}{9}$$ Add the inverse to both sides.

$$a = -\frac{5}{6} + \frac{4}{9}$$

$$= -\frac{15}{18} + \frac{8}{18}$$

$$a = -\frac{7}{18}$$

Check

$$\frac{5}{6} + \left(-\frac{7}{18}\right) \stackrel{?}{=} \frac{4}{9}$$ Replace a with $-\frac{7}{18}$ in $\frac{5}{6} + a = \frac{4}{9}$.

$$\frac{15}{18} + \left(-\frac{7}{18}\right) \stackrel{?}{=} \frac{8}{18}$$

$$\frac{8}{18} = \frac{8}{18} \checkmark$$ The solution is $-\frac{7}{18}$.

Practice Solve and check. **a.** $f + \frac{5}{12} = -\frac{3}{8}$ $-\frac{19}{24}$ **b.** $\frac{5}{9} = y - \left(-\frac{4}{5}\right)$ $-\frac{11}{45}$

Oral Exercises

To solve, would you add or subtract in the first step?

1. $n - \frac{6}{8} = -\frac{2}{3}$ + **2.** $y - \frac{5}{2} = -\frac{1}{2}$ + **3.** $\frac{3}{4} + b = -\frac{3}{5}$ −

4. $x - \frac{5}{6} = \frac{7}{12}$ + **5.** $-\frac{2}{3} + a = -6$ + **6.** $p + \frac{3}{4} = \frac{8}{9}$ −

7. $t - \frac{7}{10} = -\frac{4}{5}$ + **8.** $m - \frac{3}{10} = -\frac{6}{8}$ + **9.** $\frac{1}{5} + x = -\frac{3}{8}$ −

Exercises

A Solve and check.

1. $x - \frac{2}{3} = \frac{4}{9}$ $1\frac{1}{9}$ **2.** $t - \frac{4}{3} = -\frac{5}{7}$ $\frac{13}{21}$ **3.** $m - \frac{3}{5} = -\frac{5}{6}$ $-\frac{7}{30}$

4. $-\frac{6}{18} + x = \frac{4}{3}$ $1\frac{2}{3}$ **5.** $p - \frac{2}{5} = -\frac{5}{6}$ $-\frac{13}{30}$ **6.** $y - \frac{5}{3} = \frac{5}{12}$ $2\frac{1}{12}$

7. $a - \frac{3}{4} = -\frac{5}{8}$ $\frac{1}{8}$ **8.** $y - \frac{1}{2} = -\frac{7}{10}$ $-\frac{1}{5}$ **9.** $-\frac{3}{20} + b = -\frac{2}{5}$ $-\frac{1}{4}$

10. $x - \frac{7}{8} = -\frac{5}{12}$ $\frac{11}{24}$ **11.** $p - \frac{14}{25} = -\frac{3}{2}$ $-\frac{47}{50}$ **12.** $z - \frac{10}{30} = \frac{2}{5}$ $\frac{11}{15}$

13. $d - \frac{3}{5} = -\frac{1}{10}$ $\frac{1}{2}$

14. $n - \frac{3}{10} = -\frac{2}{50}$ $\frac{13}{50}$

15. $k - \frac{6}{15} = -\frac{12}{9}$ $-\frac{14}{15}$

16. $x + \frac{4}{7} = -\frac{2}{5}$ $-\frac{34}{35}$

17. $\frac{1}{5} + y = -\frac{1}{8}$ $-\frac{13}{40}$

18. $-\frac{6}{24} = \frac{2}{3} + h$ $-\frac{11}{12}$

19. $h + \frac{5}{6} = \frac{7}{8}$ $\frac{1}{24}$

20. $\frac{7}{16} + x = -\frac{3}{8}$ $-\frac{13}{16}$

21. $\frac{6}{5} = \frac{2}{3} + x$ $\frac{8}{15}$

22. $\frac{9}{16} + y = -\frac{13}{15}$ $-1\frac{103}{240}$

23. $\frac{3}{7} = h + \frac{4}{5}$ $-\frac{13}{35}$

24. $y + \frac{12}{25} = \frac{3}{10}$ $-\frac{9}{50}$

25. $\frac{9}{10} + z = -16\frac{3}{8}$ $-17\frac{11}{40}$

26. $-\frac{5}{16} = \frac{5}{6} + k$ $-1\frac{7}{48}$

27. $-\frac{9}{10} = \frac{7}{12} + x$ $-1\frac{29}{60}$

B Solve and check.

28. $x + \left(\frac{5}{8} - \frac{5}{6}\right) = \frac{15}{24}$ $\frac{5}{6}$

29. $\left(-\frac{6}{18} - \frac{3}{4}\right) + y = \frac{2}{9}$ $1\frac{11}{36}$

30. $-\frac{4}{15} = \left(-\frac{1}{5} + \frac{1}{2}\right) + h$ $-\frac{17}{30}$

31. $\left[-\frac{2}{3} - \left(-\frac{5}{6}\right)\right] + k = -\frac{5}{12}$ $-\frac{7}{12}$

32. $\left(\frac{4}{7} - \frac{1}{5}\right) + x = \frac{11}{5}$ $1\frac{29}{35}$

33. $-\frac{13}{24} = \left(-\frac{5}{12} + \frac{3}{8}\right) + d$ $-\frac{1}{2}$

34. Write two different equations with the solution $\frac{3}{5}$.

35. Write an equation for which the solution $-\frac{7}{12}$ is found by subtracting.

36. Write an equation for which the solution $\frac{11}{16}$ is found by adding.

34–36 Answers will vary. Check by solving.

C *Extending Thinking Skills*

37. Find 3 rational numbers, each with a numerator of 1, whose sum is $\frac{13}{16}$. $\frac{1}{2}, \frac{1}{4}, \frac{1}{16}$

38. Chris's wardrobe is made up of 6 different pairs of trousers, half as many shirts as pairs of trousers, and 2 sweaters, one green and one gray. How many different combinations of trousers, shirt and sweater does Chris have with these items? 36

Thinking skills: 37. Testing a conjecture 38. Organizing

Mixed Review

Solve and check. **39.** $-36c = 66.6$ -1.85 **40.** $r \div 16 = -0.575$ -9.2

Find all the factors of the number. **41.** 19 1, 19 **42.** 9 1, 3, 9 **43.** 12 1, 2, 3, 4, 6, 12

MENTAL MATH

You can use compensation to find certain differences of rational numbers mentally. For example, to find $5 - 3\frac{7}{8}$, think "$5 - 4$ is 1. $\frac{1}{8}$ too much has been subtracted, so it must be added back on. $1 + \frac{1}{8}$ is $1\frac{1}{8}$."

Use compensation to subtract mentally.

1. $8 - 4\frac{3}{4}$ $3\frac{1}{4}$

2. $9 - 6\frac{4}{5}$ $2\frac{1}{5}$

3. $7\frac{1}{10} - 3\frac{9}{10}$ $3\frac{1}{5}$

4. $8\frac{1}{3} - 2\frac{2}{3}$ $5\frac{2}{3}$

6-10 Deciding What the Variable Represents

Objective: To practice solving word problems by writing and solving equations.
MIN: 1–5; REG: 1–5 even, 6–8; MAX: 1–5 even, 6–10; ALL: Mixed Review

You can use the Problem-Solving Checklist on page 121 to help you solve problems. When your plan involves writing and solving an equation, keep these questions in mind.

These steps are part of the "Plan" phase of the Checklist on p. 22.

- Can you use a variable to represent an unknown number?
- Can you represent other conditions in terms of the variable?
- What is equal?
- Can you write and solve an equation?

Example

The width of a steel rod is $\frac{11}{16}$ inches. The designer says this rod is $\frac{3}{8}$ inch wider than the rod he had originally designed. What was the width of the rod originally designed?

Solution Let x = width of the original rod — Let the variable stand for the width of the original rod.

$x + \frac{3}{8}$ = width of final rod — The width of the final rod is $\frac{3}{8}$ inch wider, or $x + \frac{3}{8}$.

$\frac{11}{16}$ = width of final rod

$x + \frac{3}{8} = \frac{11}{16}$ — The two expressions for the width of the rod must be equal.

$$x + \frac{3}{8} + \left(-\frac{3}{8}\right) = \frac{11}{16} + \left(-\frac{3}{8}\right)$$

$$x = \frac{5}{16}$$

The original rod was $\frac{5}{16}$ in. wide.

Check $\frac{5}{16} + \frac{3}{8} = \frac{5}{16} + \frac{6}{16} = \frac{11}{16}$ — $\frac{5}{16}$ is less than $\frac{11}{16}$. The answer seems reasonable.

Exercises

A Solve by writing an equation. Check.

1. After cutting $\frac{1}{8}$ inch off the end of a copper pipe, a plumber had the exact length needed to connect a water line. The exact length needed was $\frac{3}{4}$ inch. What was the length of the original piece of pipe? **$p - \frac{1}{8} = \frac{3}{4}$ $\frac{7}{8}$ inch.**

2. The price of a stereo system increased twice in the last month. The second increase was \$25, for a total increase of \$45.50. What was the first increase? $i + 25 = 45.50$ \$20.50

3. The total length of a race track for bicycles is $\frac{5}{8}$ mile. The first $\frac{1}{5}$ mile is very hilly and the rest is flat. How much of the course is flat? $r + \frac{1}{5} = \frac{5}{8}$ length $\frac{17}{40}$ mi

4. Kim grew a lot during the first 6 months of last year. During the second 6 months she grew only $\frac{1}{8}$ inch. She grew $5\frac{1}{2}$ inches in all last year. How much did she grow during the first 6 months? $x + \frac{1}{8} = 5\frac{1}{2}$ She grew $5\frac{3}{8}$ in.

5. Al had 48 tapes in his music collection. His older sister Carrie gave him some more tapes. Now he has a total of 73 tapes. How many tapes did Carrie give to Al? $48 + t = 73$ She gave him 25.

B

6. Rick sold 7 tickets to the talent show during the first week of sales and 5 tickets during the second week of sales. Each ticket cost the same. He turned in a total of \$15. What was the cost of each ticket? $7x + 5x = 15$ each cost \$1.25

7. A wire had to be wrapped in a plastic coating. The wire was $\frac{7}{12}$ inch wide. Two layers of plastic had to be put over the wire so the total thickness of the plastic would equal the thickness of the wire. The first layer of plastic was $\frac{1}{2}$ inch thick. How thick was the second layer? $\frac{1}{2} + x = \frac{7}{12}$ second layer: $\frac{1}{12}$ in. thick

8. A machinist usually makes a certain part by welding together two metal strips, one $\frac{1}{8}$ inch thick and the other $\frac{1}{3}$ inch thick. She has a metal strip $\frac{1}{4}$ inch thick that she can use to make this part. How thick a strip must she weld on to make the part the correct thickness? $\frac{1}{8} + \frac{1}{3} = \frac{1}{4} + s$ weld on $\frac{5}{24}''$ thick piece

C Extending Thinking Skills

9. Write a word problem that would be solved using the equation $x + \frac{1}{2} = \frac{5}{8}$.
Answers will vary.

10. Write a word problem that would be solved using the equation $25 - x = 64$.
Answers will vary. Thinking skills: 9–10. Formulating a problem

Mixed Review

Give the prime factorization of each. **11.** 49 $7 \cdot 7$ **12.** 34 $2 \cdot 17$ **13.** 90 $2 \cdot 3 \cdot 3 \cdot 5$

Evaluate for $n = 4$. **14.** $5(n + 2) + 12$ 42 **15.** $n(6 - n) + 7$ 15

Write an algebraic expression for the phrase.

16. 14 times the difference of a number and 5. $14(n - 5)$

17. The sum of 64 and a number, divided by 12. $(64 + n) \div 12$

Simplify. **18.** $x \cdot y \cdot x \cdot y$ x^2y^2 **19.** $4m + 3m$ $7m$ **20.** $c + c + c$ $3c$

Give the least common multiple (LCM). **21.** 3, 6, 7, 4 84 **22.** 9, 12 36

6-11 Use Logical Reasoning

Objective: To solve nonroutine problems, using Logical Reasoning and other strategies learned so far.
MIN: 1–2; REG: 1–4; MAX: 1–6

Some problems must be solved by understanding the given relationships among facts and using known facts and relationships to make conclusions. The problem-solving strategy **Use Logical Reasoning** is a name for this process.

Problem Wally, Kim, José, and Rosalie each like one sport. One plays basketball every day, one plays tennis on weekends, one likes squash, and one coaches a soccer team. Wally's sister likes tennis. Wally hates playing basketball. Kim likes coaching. Which sport does each one prefer?

You can solve this problem by recording the given information in a chart and making conclusions based on it. The charts below show the reasoning you might go through to solve this problem.

	B	T	S	C
W	no	no		
K				
J		no		
R				

Wally's sister plays tennis so Wally and José do not play tennis. Wally does not play basketball.

	B	T	S	C
W	no	no		no
K	no	no	no	yes
J		no		no
R				no

If Kim coaches then no one else coaches.

	B	T	S	C
W	no	no	yes	no
K	no	no	no	yes
J	yes	no		no
R		yes		no

Wally must play squash and Rosalie must play tennis. José has to play basketball.

Problem-Solving Strategies

Choose the Operations	**Write an Equation**
Guess, Check, Revise	**Simplify the Problem**
Draw a Picture	**Make an Organized List**
Make a Table	**Use Logical Reasoning**
Look for a Pattern	

This chart lists the strategies presented so far.

Problems

Solve. Ned-Jefferson Mary-Stovertown Steve-Millerville Jack-Newville

1. Ned, Mary, Steve, and Jack live in the towns Millerville, Jefferson, Newville, and Stovertown. None lives in a city that has the same first letter as his or her name. Neither Jack nor Ned has ever been to Millerville. Mary has spent all of her life in Stovertown. Which person lives in which town?

2. Two men and two women each play a different sport. Their names are Elena, Connie, Barry, and Miguel. One plays baseball, one plays soccer, one plays basketball, and one plays football. Connie's brother is part of this group; he plays basketball and he is married to the one who plays soccer. The one who plays football doesn't have any brothers or sisters. Both Miguel and the one who plays baseball are single. Who plays each game? Elena-soccer Barry-basketball Connie-baseball Miguel-football

3. Ricky and his friends are sitting at a large round table playing a game. Ricky's mother gave them a box of 25 oranges to share. The box was passed around the table, and each person took 1 orange until there were no more oranges left. Ricky took the first orange and the last orange and he may have taken more than the first and last orange. How many people could have been sitting at the table? 3, 4, 6, 8, 12, or 24.

4. A large circus tent is set up with 12 poles. The poles are arranged in a circle with a separate rope connecting each pair of poles. How many ropes are needed to connect these 12 poles? 66

5. The varsity baseball team won 5 out of every 7 games it played this year. It lost 24 games. How many games did it win? won 60

6. Walker, Franklin, and King Schools have students in an all-star band. There are 150 band members. Walker school has 300 students and 30 are in the band. Franklin School has 400 students and 60 are in the band. How many students from King School are in the band? 60

Enrichment

Using a Calculator to Add or Subtract Fractions with Unlike Denominators

You can use a calculator to add or subtract fractions with unlike denominators. The examples show how the calculator solution corresponds to the worked-out solution.

Worked-out Solution

$$\frac{2}{3} + \frac{4}{5} = \frac{2}{3} \cdot \frac{5}{5} + \frac{4}{5} \cdot \frac{3}{3}$$
$$= \frac{2 \cdot 5 + 4 \cdot 3}{3 \cdot 5}$$
$$= \frac{22}{15} = 1\frac{7}{15}$$

The method shown here involves multiplying each fraction addend by 1 in the form $\frac{a}{a}$, where a is the denominator of the other addend.

Calculator Solution

$$\overset{n_1}{\underset{d_1}{\frac{2}{3}}} + \overset{n_2}{\underset{d_2}{\frac{4}{5}}}$$

To find the numerator: **Display**

2 [×] 5 [+] 3 [×] 4 [=] 22 $n_1 \cdot d_2 + n_2 \cdot d_1$

To find the denominator: **Display**

3 [×] 5 [=] 15 $d_1 \cdot d_2$

Solution $\frac{22}{15} = 1\frac{7}{15}$

Use a calculator to add or subtract.

1. $\frac{12}{13} + \frac{4}{5}$ $1\frac{47}{65}$
2. $\frac{5}{9} + \frac{11}{12}$ $1\frac{17}{36}$
3. $\frac{32}{60} - \frac{5}{24}$ $\frac{13}{40}$

4. $\frac{7}{10} + \frac{24}{50}$ $1\frac{9}{50}$
5. $\frac{7}{8} - \frac{3}{16}$ $\frac{11}{16}$
6. $\frac{23}{24} - \frac{11}{15}$ $\frac{9}{40}$

7. $-\frac{4}{15} - \frac{8}{12}$ $-\frac{14}{15}$
8. $-\frac{33}{40} + \frac{17}{20}$ $\frac{1}{40}$
9. $\frac{13}{15} - \frac{3}{10}$ $\frac{17}{30}$

Use a calculator to solve each equation.

10. $x - \frac{7}{9} = \frac{5}{12}$ $1\frac{7}{36}$
11. $x + \frac{13}{20} = \frac{7}{18}$ $-\frac{47}{180}$
12. $\frac{5}{16} + x = \frac{17}{25}$ $\frac{147}{400}$

13. $x + \frac{5}{8} = -\frac{4}{5}$ $-1\frac{17}{40}$
14. $x - \frac{1}{10} = \frac{1}{20}$ $\frac{3}{20}$
15. $-\frac{7}{12} + x = -\frac{3}{10}$ $\frac{17}{60}$

Instructions are given for calculators programmed to follow order of operations rules. For calculators not programmed in this way, use [=] and [M] keys as necessary.

Chapter 6 Review

6-1 Write an equivalent fraction by replacing the variable with a whole number.

1. $\frac{5}{6} = \frac{f}{72}$ 60 **2.** $\frac{k}{4} = \frac{13}{52}$ 1 **3.** $-\frac{4}{15} = -\frac{r}{75}$ 20

Reduce each fraction to lowest terms.

4. $\frac{27}{36}$ $\frac{3}{4}$ **5.** $-\frac{42}{91}$ $-\frac{6}{13}$ **6.** $\frac{144}{156}$ $\frac{12}{13}$

6-2 Write each as a mixed number.

7. $\frac{19}{3}$ $6\frac{1}{3}$ **8.** $-\frac{123}{11}$ $-11\frac{2}{11}$ **9.** $\frac{187}{30}$ $6\frac{7}{30}$

Write each as an improper fraction in lowest terms.

10. $5\frac{1}{7}$ $\frac{36}{7}$ **11.** $-8\frac{2}{5}$ $-\frac{42}{5}$ **12.** $4\frac{6}{10}$ $\frac{23}{5}$

6-3 **13.** Graph $-\frac{3}{8}$, $\frac{5}{8}$, -1, and $-1\frac{3}{4}$ on a number line.

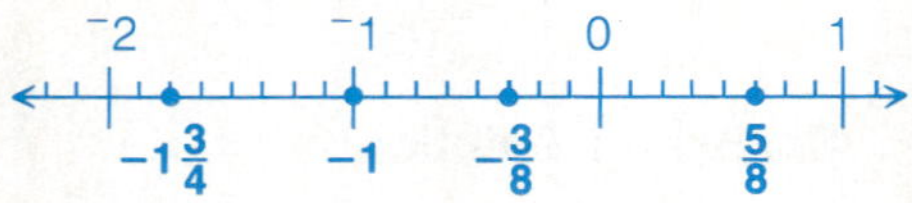

6-4 Write <, >, or = for each □.

14. $\frac{2}{5} \square \frac{6}{15}$ = **15.** $-\frac{5}{6} \square -\frac{11}{12}$ > **16.** $\frac{9}{13} \square \frac{2}{3}$ > **17.** $-\frac{2}{9} \square -\frac{15}{45}$ >

6-5 Add or subtract. Reduce to lowest terms.

18. $\frac{5}{12} + \frac{5}{12}$ $\frac{5}{6}$ **19.** $-\frac{7}{9} + \frac{5}{9}$ $-\frac{2}{9}$ **20.** $-\frac{13}{25} - \left(-\frac{4}{25}\right)$ $-\frac{9}{25}$

6-6

21. $\frac{1}{6} + \frac{1}{2}$ $\frac{2}{3}$ **22.** $-\frac{8}{15} - \left(-\frac{1}{5}\right)$ $-\frac{1}{3}$ **23.** $-\frac{7}{8} - \left(-\frac{1}{4}\right)$ $-\frac{5}{8}$

6-7

24. $2\frac{1}{3} + 1\frac{3}{4}$ $4\frac{1}{12}$ **25.** $7\frac{1}{6} - 3\frac{1}{2}$ $3\frac{2}{3}$ **26.** $8\frac{1}{5} - \left(-4\frac{1}{2}\right)$ $12\frac{7}{10}$

6-8 Solve.

27. Mr. Greyson had $4\frac{1}{2}$ pounds of apples. He used $2\frac{3}{4}$ pounds to make a pie. How many pounds did he have left? $1\frac{3}{4}$ pounds

28. Henry and Steve hiked $4\frac{3}{8}$ miles, stopped for an hour, and then hiked $5\frac{1}{4}$ miles farther. How far did they hike? $9\frac{5}{8}$ miles

6-9 Solve and check.

29. $y - \frac{1}{8} = \frac{1}{2}$ $\frac{5}{8}$ **30.** $x + \frac{3}{5} = -\frac{1}{10}$ $-\frac{7}{10}$ **31.** $\frac{7}{12} = c - \frac{2}{3}$ $1\frac{1}{4}$

6-10

32. Find the solution by writing and solving an equation.

Mr. MacGregor bought 3 yards of string. After tying up his tomato plants, he had $1\frac{3}{8}$ yards of string left over. How much string did he use? $x + 1\frac{3}{8} = 3$ $1\frac{5}{8}$ yards.

Chapter 6 Test

Write an equivalent fraction by replacing the variable with a whole number.

1. $\frac{6}{7} = \frac{v}{77}$ 66 **2.** $\frac{c}{9} = \frac{49}{63}$ 7 **3.** $-\frac{16}{17} = -\frac{t}{102}$ 96

Reduce each fraction to lowest terms.

4. $\frac{56}{84}$ $\frac{2}{3}$ **5.** $-\frac{12}{90}$ $-\frac{2}{15}$ **6.** $\frac{124}{172}$ $\frac{31}{43}$

Write each as a mixed number.

7. $\frac{38}{6}$ $6\frac{1}{3}$ **8.** $-\frac{53}{3}$ $-17\frac{2}{3}$ **9.** $\frac{288}{100}$ $2\frac{22}{25}$

Write each as an improper fraction in lowest terms.

10. $2\frac{5}{7}$ $\frac{19}{7}$ **11.** $-4\frac{3}{11}$ $-\frac{47}{11}$ **12.** $9\frac{7}{20}$ $\frac{187}{20}$

13. Graph $\frac{3}{8}$, $-\frac{5}{8}$, -1, and $1\frac{1}{2}$ on a number line.

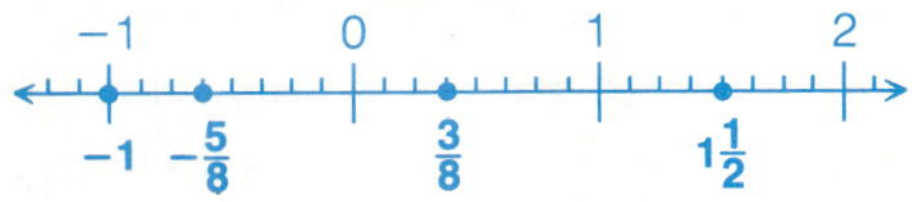

Write <, >, or = for each □.

14. $\frac{5}{9} \square \frac{2}{3}$ < **15.** $-\frac{1}{3} \square -\frac{4}{7}$ > **16.** $\frac{8}{7} \square \frac{13}{11}$ < **17.** $-\frac{9}{24} \square -\frac{12}{36}$ <

Add or subtract. Reduce to lowest terms.

18. $\frac{6}{11} + \frac{4}{11}$ $\frac{10}{11}$ **19.** $-\frac{2}{5} + \frac{4}{5}$ $\frac{2}{5}$ **20.** $-\frac{15}{24} - \left(-\frac{7}{24}\right)$ $-\frac{1}{3}$

21. $\frac{5}{6} + \frac{2}{3}$ $1\frac{1}{2}$ **22.** $-\frac{7}{12} - \frac{1}{4}$ $-\frac{5}{6}$ **23.** $-\frac{7}{9} - \left(-\frac{1}{3}\right)$ $-\frac{4}{9}$

24. $6\frac{1}{2} + 1\frac{3}{8}$ $7\frac{7}{8}$ **25.** $8\frac{1}{4} - 1\frac{1}{2}$ $6\frac{3}{4}$ **26.** $9\frac{5}{6} - \left(-2\frac{1}{3}\right)$ $12\frac{1}{6}$

Solve.

27. Mr. Cole had $8\frac{1}{2}$ pounds of flour. He gave a neighbor $2\frac{3}{4}$ pounds. How much did he have left? $5\frac{3}{4}$ lb

28. Becky and Linnea rode their bikes $5\frac{3}{4}$ miles, stopped for lunch, then rode $2\frac{3}{8}$ miles more. How far did they ride? $8\frac{1}{8}$ miles

Solve and check.

29. $p - 1\frac{1}{5} = 1\frac{4}{15}$ $2\frac{7}{15}$ **30.** $z + \frac{3}{4} = -\frac{1}{6}$ $-\frac{11}{12}$ **31.** $\frac{5}{14} = a - \left(-\frac{1}{7}\right)$ $\frac{3}{14}$

32. Find the solution by writing and solving an equation.

Mrs. Chung used $2\frac{1}{2}$ gallons of paint on her bedroom and hallway. If her hallway took $\frac{2}{3}$ gallon of paint, how much did her bedroom take?
$2\frac{1}{2} = \frac{2}{3} + x$ or $x = 2\frac{1}{2} - \frac{2}{3}$ $1\frac{5}{6}$ gallons

Cumulative Review

Write as a decimal.

1. $4(10) + 7(1) + 8\left(\frac{1}{10}\right)$ **47.8**
2. $2(100) + 9\left(\frac{1}{100}\right) + 6\left(\frac{1}{10}\right)$ **200.69**
3. $5(1) + 7\left(\frac{1}{100}\right) + 1\left(\frac{1}{10}\right)$ **5.17**
4. $3\left(\frac{1}{10}\right) + 7\left(\frac{1}{100}\right) + 9\left(\frac{1}{1000}\right)$ **0.379**

Write $<$, $>$, or $=$ for each $\square$.

5. $4.430 \square 4.43$ $=$
6. $4.01 \square 4.1$ $<$
7. $-2.73 \square -3.12$ $>$
8. $6 \square -4$ $>$
9. $-2 \square -10$ $>$
10. $-5 \square 0$ $<$

Round.

11. 83 to the nearest ten **80**
12. 291 to the nearest hundred **300**
13. 7,486 to the nearest thousand **7,000**
14. 24,576 to the nearest ten thousand **20,000**

Round to the nearest ten and estimate the value of each variable.

15. $38 + 21 = a$ **60**
16. $x = 286 - 57$ **230**
17. $12 + 54 = p$ **60**

Round to the nearest hundred and estimate the value of each variable.

18. $m = 475 + 391$ **900**
19. $402 - 174 = w$ **200**
20. $3927 - 301 = f$ **3600**

Estimate the value of x by rounding. **Possible answers are given.**

21. $51.32 - 38.11 = x$ **10**
22. $x = 295.2 + 48.8$ **350**
23. $r = 67.59 - 16.63$ **50**
24. $47.92 + 50.05 = r$ **100**

Estimate the product by rounding. **Possible answers are given.**

25. 3.73×7.2 **28**
26. 7.2×6.8 **49**
27. 1.08×6.9 **7**

Give the first five nonzero multiples of each number.

28. 4 **4, 8, 12, 16, 20**
29. 7 **7, 14, 21, 28, 35**
30. 30 **30, 60, 90, 120, 150**

Find all the factors of the given number.

31. 32 **1, 2, 4, 8, 16, 32**
32. 42 **1, 2, 3, 6, 7, 14, 21, 42**
33. 60 **1, 2, 3, 4, 5, 6, 10, 12, 15, 20, 30, 60**

State whether the number is divisible by 2.

34. 48 **yes**
35. 30 **yes**
36. 81 **no**

State whether the number is divisible by 3.

37. 38 **no**
38. 450 **yes**
39. 522 **yes**

State whether the number is divisible by 5.

40. 15 **yes**
41. 510 **yes**
42. 301 **no**

Chapter 7
Rational Numbers: Multiplication and Division

7-1 Multiplying Rational Numbers

Objective: To multiply rational numbers expressed as fractions and mixed numbers.
MIN: 1–27 even; REG: 28–50; MAX: 1–50 even, 51–53; ALL: Mixed Review

In the auditorium seating chart below, the front section contains $\frac{1}{2}$ of the total number of seats. The left section contains $\frac{1}{3}$ of the total number. The *front left* section contains $\frac{1}{2}$ of $\frac{1}{3}$, or $\frac{1}{6}$ of the total number.

Back Left	Back Center	Back Right
Front Left	Front Center	Front Right

This suggests that for all integers a and b, where ($a \neq 0$, $b \neq 0$),

$$\frac{1}{a} \cdot \frac{1}{b} = \frac{1}{ab}$$

This rule, together with properties introduced earlier, can be used to explain the steps in multiplying two rational numbers.

$$\frac{2}{3} \cdot \frac{3}{4} = \left(2 \cdot \frac{1}{3}\right)\left(3 \cdot \frac{1}{4}\right) \qquad a \cdot \frac{1}{b} = \frac{a}{b}$$

$$= (2 \cdot 3)\left(\frac{1}{3} \cdot \frac{1}{4}\right) \qquad \text{Commutative and associative properties}$$

$$= 6 \cdot \frac{1}{12} \qquad \frac{1}{a} \cdot \frac{1}{b} = \frac{1}{ab}$$

$$= \frac{6}{12} \qquad a \cdot \frac{1}{b} = \frac{a}{b}$$

This suggests the following rule. **Remind students that since *a*, *b*, *c* and *d* are integers, the rules for determining the sign of an integer product apply here.**

Multiplying Rational Numbers

To multiply two rational numbers, multiply the numerators and multiply the denominators.

For all rational numbers $\frac{a}{b}$ and $\frac{c}{d}$ ($b \neq 0$, $d \neq 0$),

$$\frac{a}{b} \cdot \frac{c}{d} = \frac{a \cdot c}{b \cdot d}$$

Example 1

Multiply. Reduce to lowest terms. $\frac{2}{5}\left(\frac{5}{12}\right)$

Solution

$\frac{2}{5}\left(\frac{5}{12}\right) = \frac{2(5)}{5(12)}$ Multiply numerators; multiply denominators.

$= \frac{10}{60} = \frac{1}{6}$ You may wish to have students simplify by cancelling before they multiply: $\frac{\overset{1}{\cancel{2}}}{\underset{1}{\cancel{5}}} \cdot \frac{\overset{1}{\cancel{5}}}{\underset{6}{\cancel{12}}} = \frac{1}{6}$.

Practice Multiply. Reduce to lowest terms.

a. $-\frac{5}{6}\left(-\frac{3}{8}\right)$ $\frac{5}{16}$ **b.** $\frac{2}{3}\left(\frac{8}{9}\right)$ $\frac{16}{27}$

When multiplying mixed numbers, first change them to improper fractions.

Example 2

Multiply. Reduce to lowest terms. $-1\frac{1}{5}\left(-3\frac{1}{2}\right)$

Solution

$-1\frac{1}{5}\left(-3\frac{1}{2}\right) = \frac{-6}{5}\left(\frac{-7}{2}\right)$ Change mixed numbers to improper fractions.

$= \frac{-6(-7)}{5(2)}$

$= \frac{42}{10}$ When both factors are negative, the product is positive.

$= 4\frac{2}{10} = 4\frac{1}{5}$

Practice Multiply. Reduce to lowest terms.

a. $2\frac{2}{3}\left(1\frac{1}{5}\right)$ $3\frac{1}{5}$ **b.** $3\frac{1}{12}(-2)$ $-6\frac{1}{6}$

Oral Exercises

Give each mixed number as an improper fraction.

1. $3\frac{1}{2}$ $\frac{7}{2}$ **2.** $-4\frac{2}{3}$ $-\frac{14}{3}$ **3.** $1\frac{4}{5}$ $\frac{9}{5}$ **4.** $2\frac{1}{8}$ $\frac{17}{8}$

5. $10\frac{1}{10}$ $\frac{101}{10}$ **6.** $4\frac{3}{4}$ $\frac{19}{4}$ **7.** $-1\frac{5}{8}$ $-\frac{13}{8}$ **8.** $6\frac{2}{3}$ $\frac{20}{3}$

State whether the product is positive or negative. Do not compute.

9. $-\frac{7}{8}\left(\frac{1}{2}\right)$ − **10.** $\frac{2}{5}\left(\frac{9}{10}\right)$ + **11.** $-\frac{3}{5}\left(-\frac{2}{3}\right)$ +

12. $4\frac{1}{2}\left(-\frac{3}{4}\right)$ − **13.** $-\frac{5}{6}\left(-\frac{7}{9}\right)$ + **14.** $\frac{4}{9}\left(\frac{35}{100}\right)$ +

Exercises

A Multiply. Reduce to lowest terms.

1. $\frac{3}{5}\left(\frac{3}{4}\right)$ $\frac{9}{20}$
2. $\frac{3}{5}\left(\frac{5}{7}\right)$ $\frac{3}{7}$
3. $\frac{1}{2}\left(\frac{5}{16}\right)$ $\frac{5}{32}$
4. $-\frac{8}{16}\left(\frac{1}{2}\right)$ $-\frac{1}{4}$
5. $-\frac{4}{3}\left(\frac{5}{8}\right)$ $-\frac{5}{6}$
6. $-\frac{6}{4}\left(-\frac{3}{5}\right)$ $\frac{9}{10}$
7. $\frac{14}{6}\left(-\frac{3}{10}\right)$ $-\frac{7}{10}$
8. $\frac{5}{3}\left(\frac{15}{8}\right)$ $\frac{25}{8} = 3\frac{1}{8}$
9. $\frac{7}{12}\left(\frac{4}{3}\right)$ $\frac{7}{9}$
10. $-\frac{3}{8}\left(-\frac{2}{3}\right)$ $\frac{1}{4}$
11. $-\frac{5}{2}\left(-\frac{5}{10}\right)$ $\frac{5}{4} = 1\frac{1}{4}$
12. $\frac{12}{25}\left(\frac{2}{10}\right)$ $\frac{12}{125}$
13. $\frac{4}{12}\left(\frac{15}{24}\right)$ $\frac{5}{24}$
14. $\frac{24}{26}\left(-\frac{3}{5}\right)$ $-\frac{36}{65}$
15. $-\frac{6}{7}\left(-\frac{14}{10}\right)$ $\frac{6}{5} = 1\frac{1}{5}$
16. $-3\frac{1}{10}\left(1\frac{1}{3}\right)$ $-4\frac{2}{15}$
17. $6\frac{1}{2}\left(2\frac{2}{8}\right)$ $14\frac{5}{8}$
18. $4\frac{1}{4}\left(1\frac{2}{5}\right)$ $5\frac{19}{20}$
19. $\frac{6}{2}\left(2\frac{1}{4}\right)$ $6\frac{3}{4}$
20. $-8\frac{1}{8}\left(-\frac{3}{4}\right)$ $6\frac{3}{32}$
21. $-4\frac{5}{6}\left(2\frac{2}{3}\right)$ $-12\frac{8}{9}$
22. $8\frac{7}{9}\left(-5\frac{2}{5}\right)$ $-47\frac{2}{5}$
23. $2\frac{3}{4}\left(8\frac{4}{7}\right)$ $23\frac{4}{7}$
24. $-1\frac{1}{10}\left(3\frac{2}{10}\right)$ $-3\frac{13}{25}$
25. $1\frac{3}{4}\left(-\frac{4}{5}\right)$ $-1\frac{2}{5}$
26. $-2\frac{3}{4}\left(-2\frac{1}{4}\right)$ $6\frac{3}{16}$
27. $5\left(4\frac{2}{5}\right)$ 22

B Find the value of each expression. Reduce to lowest terms.

28. $\frac{2}{3}b$ for $b = 1\frac{1}{4}$ $\frac{5}{6}$
29. $-\frac{4}{3}a$ for $a = 12$ -16
30. $1\frac{1}{5}y - \frac{3}{10}$ for $y = -1\frac{3}{5}$ $-2\frac{11}{50}$
31. $\frac{1}{8}x + 2$ for $x = 3\frac{1}{2}$ $2\frac{7}{16}$
32. $1\frac{1}{2}h$ for $h = -\frac{5}{6}$ $-1\frac{1}{4}$
33. $\frac{2}{3}x + \frac{1}{4}$ for $x = -\frac{5}{6}$ $-\frac{11}{36}$

Find the value of x. Reduce to lowest terms.

34. $x = \frac{2}{3}\left(\frac{3}{5} + \frac{6}{15}\right)$ $\frac{2}{3}$
35. $x = \left(\frac{1}{2} + \frac{3}{4}\right)\frac{2}{5}$ $\frac{1}{2}$
36. $x = -\frac{18}{24}(9 - 15)$ $4\frac{1}{2}$
37. $x = -\frac{4}{5}(-5 + 8)$ $-2\frac{2}{5}$
38. $x = \frac{4}{5}\left(\frac{4}{5} + \frac{3}{4}\right)$ $1\frac{6}{25}$
39. $x = \frac{3}{8}(6 - 7)$ $-\frac{3}{8}$

Solve.

40. A cookbook recommends roasting a turkey at a low temperature $\frac{3}{4}$ hour for each pound. How long should you cook a $10\frac{1}{2}$ pound turkey? $7\frac{7}{8}$ hr

41. Each section of a fence is $6\frac{1}{2}$ ft long. How long is this fence if it has 8 sections? 52 ft

42. A certain steel bar weighs $2\frac{1}{2}$ pounds per foot. What would be the weight of a piece $3\frac{3}{4}$ ft long? $9\frac{3}{8}$ lb

Estimate each product by rounding to the nearest whole number.

43. $3\frac{4}{5}\left(7\frac{1}{8}\right)$ 28 **44.** $5\frac{1}{6}\left(6\frac{9}{10}\right)$ 35 **45.** $7\frac{3}{4}\left(4\frac{1}{3}\right)$ 32 **46.** $8\frac{1}{7}\left(9\frac{7}{8}\right)$ 80

Estimate each product by substituting compatible numbers.
For example, to estimate $\frac{1}{3}(25)$, think "$\frac{1}{3}$ of 24 is 8."

47. $\frac{1}{5}(39)$ 8 **48.** $\frac{1}{4}(26)$ 7 **49.** $\frac{2}{3}(17)$ 12 **50.** $\frac{1}{10}(79)$ 8

C Extending Thinking Skills

51. To find the value of x in the equations below, look for a pattern. Then write two similar equations with the same solution. $x = 48$

$$\frac{1}{3}x = 16, \frac{1}{4}x = 12, \frac{1}{6}x = 8$$ Answers will vary. Example: $\frac{1}{12}x = 4$ $\frac{1}{8}x = 6$

52. Look for a pattern to find the next two numbers in this sequence.

$$\frac{3}{2}, -\frac{3}{5}, \frac{6}{25}, -\frac{12}{125}, \text{—}, \text{—}$$ $\frac{24}{625}, \frac{-48}{3125}$

53. Marie's age is 4 more than $2\frac{1}{2}$ times her brother's age. If Marie is 24, how old is her brother? 8

Thinking skills: 51. Testing a conjecture 52. Finding a pattern 53. Testing a conjecture

Mixed Review

Simplify. **54.** $(5m + 12) + (4m - 6)$ $9m + 6$ **55.** $c + t + t + t + c$ $2c + 3t$

Find the greatest common factor (GCF). **56.** 9, 42, 51 3 **57.** 4, 9 1

Reduce to lowest terms. **58.** $\frac{16}{100}$ $\frac{4}{25}$ **59.** $-\frac{12}{60}$ $-\frac{1}{5}$ **60.** $\frac{27}{51}$ $\frac{9}{17}$

Write each mixed number as an improper fraction.

61. $-6\frac{4}{5}$ $-\frac{34}{5}$ **62.** $12\frac{1}{3}$ $\frac{37}{3}$ **63.** $5\frac{7}{16}$ $\frac{87}{16}$ **64.** $-2\frac{7}{8}$ $-\frac{23}{8}$

Solve and check. **65.** $19.5x = -1.17$ -0.06 **66.** $a - 1.06 = -0.98$ 0.08

67. $n - \frac{1}{2} = \frac{3}{8}$ $\frac{7}{8}$ **68.** $y + \frac{3}{14} = \frac{5}{8}$ $\frac{23}{56}$ **69.** $-\frac{1}{3} = z - \frac{3}{5}$ $\frac{4}{15}$

MENTAL MATH

You can use the distributive property and "break apart numbers" to find the products of rational numbers mentally. For example, to find $8 \cdot 3\frac{1}{4}$, think: "8 times 3 plus 8 times $\frac{1}{4}$ equals 24 plus 2, or 26."

Use the distributive property to find each product mentally.

1. $12 \cdot 2\frac{1}{3}$ 28 **2.** $-16 \cdot 1\frac{1}{4}$ -20 **3.** $20 \cdot \left(-2\frac{3}{10}\right)$ -46 **4.** $\frac{1}{2} \cdot 8\frac{1}{3}$ $4\frac{1}{6}$

5. $8\frac{1}{2} \cdot 4$ 34 **6.** $-100 \cdot 1\frac{1}{10}$ -110 **7.** $50 \cdot 2\frac{1}{50}$ 101 **8.** $\frac{1}{4} \cdot 4\frac{1}{2}$ $1\frac{1}{8}$

7-2 Dividing Rational Numbers

Objective: To divide rational numbers expressed as fractions and mixed numbers.
MIN: 1–28; REG: 1–28 even, 29–33; MAX: 1–28 even, 29–35; ALL: Mixed Review

An exploratory lesson opener can be found on page 549.

Two numbers are called **multiplicative inverses** or **reciprocals** of each other if their product is 1. The multiplicative inverse of 3 is $\frac{1}{3}$ since $3 \times \frac{1}{3} = 1$. The multiplicative inverse of $-\frac{3}{4}$ is $-\frac{4}{3}$ since $-\frac{3}{4}\left(-\frac{4}{3}\right) = 1$.

Inverse Property of Multiplication

The product of a rational number and its multiplicative inverse (or reciprocal) is 1.

For every rational number $\frac{a}{b}$ **($a \neq 0, b \neq 0$),**

$$\frac{a}{b} \cdot \frac{b}{a} = \frac{b}{a} \cdot \frac{a}{b} = 1$$

We can use this property to develop a rule for dividing two rational numbers.

$$\frac{4}{5} \div \frac{2}{3} = \frac{\frac{4}{5}}{\frac{2}{3}}$$ $\frac{a}{b} \div \frac{c}{d}$ is the same as $\frac{\frac{a}{b}}{\frac{c}{d}}$.

$$\frac{4}{5} \div \frac{2}{3} = \frac{\frac{4}{5} \cdot \frac{3}{2}}{\frac{2}{3} \cdot \frac{3}{2}}$$ $\frac{a}{b} = \frac{a \cdot c}{b \cdot c}$.

$$\frac{4}{5} \div \frac{2}{3} = \frac{\frac{4}{5} \cdot \frac{3}{2}}{1}$$ Inverse property of multiplication.

$$\frac{4}{5} \div \frac{2}{3} = \frac{4}{5} \cdot \frac{3}{2}$$

Dividing Rational Numbers

To divide by a rational number, multiply by its inverse. For rational numbers $\frac{a}{b}$ and $\frac{c}{d}$ **($b \neq 0, c \neq 0, d \neq 0$),**

$$\frac{a}{b} \div \frac{c}{d} = \frac{a}{b} \cdot \frac{d}{c}$$

The rule for determining the sign of an integer quotient applies to rational numbers.

Example 1

Divide. Reduce to lowest terms. $\frac{1}{4} \div \frac{2}{6}$

Solution

$$\frac{1}{4} \div \frac{2}{6} = \frac{1}{4} \cdot \frac{6}{2}$$ Dividing by $\frac{2}{6}$ is the same as multiplying by $\frac{6}{2}$.

$$= \frac{6}{8} = \frac{3}{4}$$

Practice Divide. Reduce to lowest terms. **a.** $-\frac{2}{3} \div \frac{1}{6}$ -4 **b.** $\frac{5}{24} \div 5$ $\frac{1}{24}$

Example 2

Divide. Reduce to lowest terms. $-1\frac{3}{4} \div 2\frac{1}{2}$

Solution

$$-1\frac{3}{4} \div 2\frac{1}{2} = -\frac{7}{4} \div \frac{5}{2}$$ Change mixed numbers to improper fractions.

$$= -\frac{7}{4} \cdot \frac{2}{5}$$ Dividing by $\frac{5}{2}$ is the same as multiplying by $\frac{2}{5}$.

$$= -\frac{14}{20} = -\frac{7}{10}$$

Practice Divide. Reduce to lowest terms. **a.** $8 \div \left(-3\frac{3}{4}\right)$ $-2\frac{2}{15}$ **b.** $3\frac{1}{8} \div 2\frac{1}{12}$ $1\frac{1}{2}$

Oral Exercises

Express each as a multiplication problem. Do not solve.

1. $\frac{3}{4} \div \frac{2}{5}$ $\frac{3}{4}\left(\frac{5}{2}\right)$ **2.** $\frac{7}{9} \div \frac{4}{5}$ $\frac{7}{9}\left(\frac{5}{4}\right)$ **3.** $\frac{6}{5} \div \left(-\frac{8}{3}\right)$ $\frac{6}{5}\left(-\frac{3}{8}\right)$ **4.** $\frac{5}{8} \div \frac{2}{3}$ $\frac{5}{8}\left(\frac{3}{2}\right)$

5. $-\frac{3}{4} \div 5$ $-\frac{3}{4}\left(\frac{1}{5}\right)$ **6.** $12 \div \left(-\frac{1}{2}\right)$ $\frac{12}{1}\left(-\frac{2}{1}\right)$ **7.** $\frac{5}{6} \div \frac{1}{3}$ $\frac{5}{6}\left(\frac{3}{1}\right)$ **8.** $\frac{3}{10} \div \left(-\frac{15}{100}\right)$ $\frac{3}{10}\left(-\frac{100}{15}\right)$

Exercises

A Divide. Reduce to lowest terms.

1. $\frac{3}{5} \div \frac{1}{5}$ 3 **2.** $\frac{1}{2} \div \frac{3}{4}$ $\frac{2}{3}$ **3.** $\frac{5}{8} \div \frac{7}{8}$ $\frac{5}{7}$ **4.** $\frac{1}{9} \div \left(-\frac{4}{3}\right)$ $-\frac{1}{12}$

5. $-\frac{5}{8} \div \frac{3}{2}$ $-\frac{5}{12}$ **6.** $\frac{5}{9} \div \left(-\frac{7}{3}\right)$ $-\frac{5}{21}$ **7.** $-\frac{3}{10} \div \frac{8}{5}$ $-\frac{3}{16}$ **8.** $\frac{9}{12} \div \frac{5}{6}$ $\frac{9}{10}$

9. $-\frac{5}{9} \div \left(-\frac{1}{3}\right)$ $1\frac{2}{3}$ **10.** $\frac{5}{12} \div \frac{2}{3}$ $\frac{5}{8}$ **11.** $\frac{5}{4} \div \left(-\frac{5}{8}\right)$ -2 **12.** $-\frac{3}{8} \div 5$ $-\frac{3}{40}$

13. $\frac{13}{20} \div \frac{7}{10}$ $\frac{13}{14}$ **14.** $\frac{1}{10} \div \frac{3}{12}$ $\frac{2}{5}$ **15.** $\frac{18}{24} \div \frac{1}{8}$ 6 **16.** $1\frac{1}{2} \div \left(-\frac{3}{4}\right)$ -2

17. $-3\frac{1}{3} \div \left(-\frac{5}{6}\right)$ 4 **18.** $3\frac{1}{3} \div 2\frac{1}{4}$ $1\frac{13}{27}$ **19.** $5\frac{2}{3} \div 1\frac{3}{4}$ $3\frac{5}{21}$ **20.** $2\frac{1}{4} \div (-6)$ $-\frac{3}{8}$

21. $-\frac{4}{5} \div 2\frac{2}{3}$ $-\frac{3}{10}$

22. $-2\frac{1}{2} \div \left(-1\frac{3}{4}\right)$ $1\frac{3}{7}$

23. $\frac{1}{6} \div 3\frac{1}{2}$ $\frac{1}{21}$

24. $\frac{7}{3} \div \left(-1\frac{3}{7}\right)$ $-1\frac{19}{30}$

25. $3\frac{1}{10} \div \frac{1}{100}$ 310

26. $6\frac{1}{4} \div 10\frac{1}{4}$ $\frac{25}{41}$

27. $-18 \div \left(-2\frac{11}{16}\right)$ $6\frac{30}{43}$

28. $6\frac{2}{3} \div 2\frac{3}{4}$ $2\frac{14}{33}$

B Evaluate each expression. Reduce to lowest terms.

29. $\frac{x}{16} - \frac{1}{2}$ for $x = -\frac{3}{8}$ $-\frac{67}{128}$

30. $\frac{-3\frac{1}{3}}{x}$ for $x = \frac{5}{12}$ -8

31. $\frac{5}{x} + \frac{3}{4}$ for $x = 1\frac{1}{2}$ $4\frac{1}{12}$

32. Suppose you walk at a rate of $2\frac{1}{2}$ miles per hour. How long would it take to walk to school and back if the school were $3\frac{1}{2}$ miles from home? $2\frac{4}{5}$ h, or 2 h 48 min.

33. Christopher wants to engrave his name on an ID bracelet. There are $1\frac{3}{4}$ inches of space on the bracelet. He can choose from three sizes of letters: $\frac{1}{8}$ inch wide, $\frac{1}{4}$ inch wide, or $\frac{1}{2}$ inch wide. Which size or sizes of lettering could he use for his name? $\frac{1}{8}$ in.

C Extending Thinking Skills

34. Find all the ways to complete the equation $\frac{x}{y} \cdot \frac{m}{n} = \frac{9}{15}$ where $x < 9$, $m < 9$, $y < 15$, and $n < 15$. Replace each letter with a different whole number.

35. Find two rational numbers with a sum of 1, product of $-\frac{6}{25}$, and quotient of -6. $\frac{6}{5}, -\frac{1}{5}$

Thinking skills: 34. Organizing 35. Testing a conjecture

34.

x	y	m	n
3	5	1	1
1	5	3	1
3	1	1	5
1	1	3	5
3	3	3	5
3	5	3	3
3	5	5	5
5	5	3	5

x	y	m	n
6	10	1	1
1	10	6	1
etc.			

Mixed Review

Solve and check. 36. $a \div 1.75 = -16$ -28 37. $r + 0.9 = -1.3$ -2.2

Find the least common multiple (LCM). 38. 4, 6, 9, 12 36

ESTIMATION

You can substitute compatible numbers to estimate quotients involving rational numbers.

Example: Choose the best estimate. $14\frac{1}{8} \div 7$ <2 or >2?

Think: $14\frac{1}{8}$ is about 14 and $14 \div 7 = 2$. Since $14\frac{1}{8} > 14$, the quotient will be greater than 2.

Use compatible numbers to choose the best estimate.

1. $7\frac{2}{9} \cdot 2\frac{1}{3}$ <14 or >14? >14

2. $\frac{2}{5} \cdot 18$ <18 or >18? <18

3. $12\frac{1}{3} \div 4$ <3 or >3? >3

4. $8\frac{3}{4} \div 2\frac{1}{4}$ <4 or >4? <4

5. $22\frac{1}{3} \div 8$ <3 or >3? <3

6. $24 \div 2\frac{5}{8}$ <8 or >8? >8

7-3 Solving Equations: Using Multiplication and Division

To solve equations using the multiplicative inverse.
MIN: 1–24; REG: 1–24 even, 25–33; MAX: 1–24 even, 25–36; ALL: Mixed Review

You can solve equations such as $\frac{3}{4}x = \frac{5}{8}$ by dividing both sides by $\frac{3}{4}$. It is simpler, however, to multiply by the reciprocal of $\frac{3}{4}$. Keep in mind the following steps:

- Decide which operation (multiplication or division) has been applied to the variable.
- Use the inverse operation or use the multiplicative inverse property, multiplying or dividing by the same number on both sides of the equation.

Both examples use the multiplicative inverse.

Example 1

Solve and check. $\frac{3}{4}x = \frac{5}{8}$

Solution

$$\frac{3}{4}x = \frac{5}{8}$$

$$\frac{4}{3} \cdot \left(\frac{3}{4}x\right) = \frac{4}{3} \cdot \frac{5}{8}$$ Multiply both sides by the reciprocal of $\frac{3}{4}$ so they remain equal.

$$1 \cdot x = \frac{20}{24}$$ $\frac{4}{3} \cdot \frac{3}{4} = 1$. Associative property and inverse property.

$$x = \frac{5}{6}$$ Identity property.

Check

$$\frac{3}{4} \cdot \frac{5}{6} \stackrel{?}{=} \frac{5}{8}$$ Replace x with $\frac{5}{6}$.

$$\frac{15}{24} \stackrel{?}{=} \frac{5}{8}$$

$$\frac{5}{8} = \frac{5}{8} \checkmark$$ The solution is $\frac{5}{6}$.

Practice Solve and check.

a. $\frac{1}{6}f = 4$ 24 **b.** $-\frac{3}{4}y = 1\frac{1}{2}$ -2

c. $6y = \frac{3}{8}$ $\frac{1}{16}$ **d.** $7p = \frac{7}{10}$ $\frac{1}{10}$

Example 2

Solve and check. $-3\frac{1}{2}x = 2\frac{3}{4}$

Solution

$$-3\frac{1}{2}x = 2\frac{3}{4}$$

$$-\frac{7}{2}x = \frac{11}{4}$$ Change mixed numbers to improper fractions.

$$-\frac{2}{7}\left(-\frac{7}{2}x\right) = -\frac{2}{7} \cdot \frac{11}{4}$$ Multiply both sides by the reciprocal of $-\frac{7}{2}$.

$$1 \cdot x = -\frac{22}{28}$$

$$x = -\frac{11}{14}$$

Check

$$-\frac{7}{2}\left(-\frac{11}{14}\right) \stackrel{?}{=} \frac{11}{4}$$ Replace x with $-\frac{11}{14}$ in $-\frac{7}{2}x = \frac{11}{4}$.

$$\frac{77}{28} \stackrel{?}{=} \frac{11}{4}$$

$$\frac{11}{4} = \frac{11}{4}\ \checkmark$$ The solution is $-\frac{11}{14}$.

Practice Solve and check. **a.** $4\frac{1}{3}y = 5$ $1\frac{2}{13}$ **b.** $-2\frac{4}{5} = 1\frac{5}{6}h$ $-1\frac{29}{55}$

Oral Exercises

To solve, what operation would you use in the first step?

1. $\frac{1}{6}y = \frac{2}{3}$ ×
2. $\frac{y}{5} = \frac{1}{2}$ ×
3. $6b = -\frac{3}{5}$ ÷
4. $\frac{5}{6}x = \frac{7}{12}$ ×
5. $\frac{2}{3}a = -6$ ×
6. $p - 3\frac{3}{4} = \frac{8}{9}$ +
7. $\frac{t}{-7} = \frac{4}{5}$ ×
8. $m + \frac{3}{10} = \frac{6}{8}$ −

Exercises

A Solve and check.

1. $4x = \frac{3}{4}$ $\frac{3}{16}$
2. $\frac{t}{4} = \frac{5}{7}$ $2\frac{6}{7}$
3. $\frac{4}{5}m = 5$ $6\frac{1}{4}$
4. $6x = \frac{4}{3}$ $\frac{2}{9}$
5. $\frac{1}{5}y = 1$ 5
6. $\frac{y}{6} = \frac{5}{12}$ $2\frac{1}{2}$
7. $\frac{2}{3}h = -6$ −9
8. $-\frac{1}{2}y = \frac{7}{10}$ $-1\frac{2}{5}$
9. $3b = -\frac{2}{3}$ $-\frac{2}{9}$
10. $-\frac{5}{6}h = -\frac{3}{2}$ $1\frac{4}{5}$
11. $\frac{p}{14} = \frac{3}{2}$ 21
12. $\frac{z}{-10} = \frac{2}{5}$ −4
13. $-\frac{7}{16}x = -\frac{3}{8}$ $\frac{6}{7}$
14. $\frac{9}{16}y = \frac{2}{5}$ $\frac{32}{45}$
15. $\frac{k}{6} = -\frac{12}{9}$ −8
16. $1\frac{2}{3}x = \frac{6}{5}$ $\frac{18}{25}$
17. $-1\frac{3}{8}x = -\frac{3}{4}$ $\frac{6}{11}$
18. $4\frac{3}{5}h = 8$ $1\frac{17}{23}$
19. $4\frac{3}{10} = 2\frac{3}{5}g$ $1\frac{17}{26}$
20. $1\frac{2}{9} = 18h$ $\frac{11}{162}$
21. $3\frac{1}{3}x = -12$ $-3\frac{3}{5}$
22. $-4f = -3\frac{5}{6}$ $\frac{23}{24}$
23. $-\frac{9}{10} = 5\frac{7}{12}x$ $-\frac{54}{335}$
24. $\frac{y}{12} = 2\frac{3}{10}$ $27\frac{3}{5}$

B Solve and check.

25. $\left(\frac{3}{4} + \frac{5}{8}\right)x = 2$ $1\frac{5}{11}$ **26.** $-\frac{7}{12}(-6y) = 4$ $1\frac{1}{7}$ **27.** $\left(\frac{2}{3}x\right)\frac{1}{2} = 5$ 15

28. $\left(5 - 2\frac{3}{4}\right)x = -2\frac{5}{6}$ $-1\frac{7}{27}$ **29.** $\left(\frac{1}{2} + \frac{7}{9} - \frac{1}{2}\right)t = 14$ 18 **30.** $5\frac{1}{2}r = 0$ 0

For Exercises 31–33, solve mentally by choosing **compatible numbers**. For example, to solve $x = \frac{1}{5}\left(\frac{7}{8}\right)(5)$, think "$\frac{1}{5}(5)$ is 1; $1\left(\frac{7}{8}\right)$ is $\frac{7}{8}$.

31. $x = \frac{1}{6}\left(2\frac{1}{4}\right)(6)$ $2\frac{1}{4}$ **32.** $x = \frac{3}{8}\left(\frac{1}{6}\right)(8)$ $\frac{1}{2}$ **33.** $x = 9\left(2\frac{1}{5}\right)\left(\frac{1}{3}\right)$ $6\frac{3}{5}$

C Extending Thinking Skills

34. Find a number for which the following is true: $\frac{1}{5}$ of the number added to 28 is triple the number. 10

35. Mentally find two values of y that make this equation true. $\frac{1}{2}y^2 = 8$. $y = 4, y = -4$

36. Mentally find two values of x that make this equation true. $8x^2 = 2$. $x = \frac{1}{2}, x = -\frac{1}{2}$

Thinking skills: 34. Testing a conjecture 35–36. Computing mentally

Mixed Review

Simplify. **37.** $(4c + 2) + (4c - 2)$ $8c$ **38.** $6(x - 3) + 4x + 2$ $10x - 16$

Solve and check. **39.** $-16.8 = r \div 2$ -33.6 **40.** $z + 29 = 317$ 288

COMPUTER ACTIVITY

You can use a computer program to find decimal or integer solutions to equations. To solve the equation $\frac{3}{4}x = 15$, the program computes:

Remind students to separate their number inputs with a comma in line 20.

$$x = \frac{4 \cdot 15}{3}$$ Multiply by 4 and divide by 3.

```
10  PRINT "SOLVING EQUATIONS OF THE FORM A/B*N=C"
20  INPUT "ENTER A, B, AND C"; A, B, C
30  IF A=0 OR B=0 THEN GOTO 70
40  N=(B*C)/A
50  PRINT "N="; N
60  GOTO 80
70  PRINT "NEITHER A NOR B CAN EQUAL ZERO."
80  PRINT "DO YOU WANT TO CONTINUE (Y/N)?"
90  INPUT A$:IF A$="Y" THEN 20
100 END
```

Use the program to solve each equation for N.

1. $\frac{15}{29}N = 25$ **2.** $\frac{23}{27}N = -21$ **3.** $\frac{63}{82}N = 136$ **4.** $4\frac{2}{3}N = 215$

5. Modify the program to solve equations such as $37N = \frac{3}{7}$.

6. Modify the program to solve equations such as $\frac{2}{3}N = \frac{4}{5}$. 5, 6. See additional answers.

1. 48.333333 2. −24.6521739 3. 177.01587 4. 46.0714286

More Practice

Solve and check. Reduce your answer to lowest terms.

1. $3t = \frac{3}{5}$ $\frac{1}{5}$	**2.** $\frac{r}{5} = \frac{6}{11}$ $2\frac{8}{11}$	**3.** $3\frac{1}{2}x = 4$ $1\frac{1}{7}$
4. $-\frac{k}{9} = \frac{4}{5}$ $-7\frac{1}{5}$	**5.** $2\frac{1}{4}z = \frac{4}{9}$ $\frac{16}{81}$	**6.** $9c = \frac{2}{5}$ $\frac{2}{45}$
7. $6c = \frac{8}{15}$ $\frac{4}{45}$	**8.** $-\frac{2}{3}f = 1\frac{4}{5}$ $-2\frac{7}{10}$	**9.** $\frac{3}{4}m = 6\frac{1}{3}$ $8\frac{4}{9}$
10. $\frac{5}{9}w = \frac{10}{17}$ $1\frac{1}{17}$	**11.** $1\frac{1}{2}x = 5$ $3\frac{1}{3}$	**12.** $1\frac{4}{5}y = \frac{3}{4}$ $\frac{5}{12}$
13. $11t = -\frac{9}{10}$ $-\frac{9}{110}$	**14.** $\frac{6}{11}r = \frac{9}{11}$ $1\frac{1}{2}$	**15.** $\frac{a}{3} = \frac{5}{7}$ $2\frac{1}{7}$
16. $2\frac{3}{5}k = -4$ $-1\frac{7}{13}$	**17.** $3\frac{1}{4}n = -\frac{3}{4}$ $-\frac{3}{13}$	**18.** $-6c = \frac{11}{12}$ $-\frac{11}{72}$
19. $-\frac{a}{2} = -\frac{3}{5}$ $1\frac{1}{5}$	**20.** $6\frac{2}{3}x = 15$ $2\frac{1}{4}$	**21.** $\frac{7}{10}k = 1\frac{3}{4}$ $2\frac{1}{2}$
22. $-3\frac{1}{8}c = \frac{15}{16}$ $-\frac{3}{10}$	**23.** $\frac{m}{4} = -\frac{2}{9}$ $-\frac{8}{9}$	**24.** $4t = -\frac{2}{5}$ $-\frac{1}{10}$
25. $\frac{k}{12} = \frac{7}{8}$ $10\frac{1}{2}$	**26.** $\frac{3}{4} = r + 1\frac{1}{2}$ $-\frac{3}{4}$	**27.** $z - \frac{1}{2} = -\frac{3}{8}$ $\frac{1}{8}$
28. $\frac{5}{6}(m) = -3\frac{3}{4}$ $-4\frac{1}{2}$	**29.** $-\frac{4}{5} + c = 1$ $1\frac{4}{5}$	**30.** $k - \frac{7}{8} = 1\frac{1}{3}$ $2\frac{5}{24}$
31. $12z = \frac{3}{5}$ $\frac{1}{20}$	**32.** $\frac{h}{-3} = \frac{11}{15}$ $-2\frac{1}{5}$	**33.** $\frac{9}{16}z = -4\frac{2}{3}$ $-8\frac{8}{27}$
34. $\frac{11}{12}a = -6$ $-6\frac{6}{11}$	**35.** $c - 4\frac{1}{3} = \frac{2}{5}$ $4\frac{11}{15}$	**36.** $1\frac{5}{6} = a + \frac{3}{5}$ $1\frac{7}{30}$
37. $-5x = 2\frac{1}{8}$ $-\frac{17}{40}$	**38.** $3\frac{1}{3} = z - \left(-\frac{1}{2}\right)$ $2\frac{5}{6}$	**39.** $m + \frac{1}{3} = -\frac{3}{4}$ $-1\frac{1}{12}$
40. $8a = -\frac{5}{6}$ $-\frac{5}{48}$	**41.** $k - 2\frac{4}{5} = \frac{1}{2}$ $3\frac{3}{10}$	**42.** $\frac{2}{3}x = -9$ $-13\frac{1}{2}$
43. $2\frac{1}{6}x = 9\frac{1}{2}$ $4\frac{5}{13}$	**44.** $\frac{3}{5} = t + \frac{1}{2}$ $\frac{1}{10}$	**45.** $\frac{m}{-6} = \frac{8}{15}$ $-3\frac{1}{5}$
46. $y - \frac{5}{12} = -\frac{3}{4}$ $-\frac{1}{3}$	**47.** $-\frac{4}{9} = c + 2\frac{1}{3}$ $-2\frac{7}{9}$	**48.** $6n = \frac{3}{5}$ $\frac{1}{10}$

PROBLEM SOLVING: WRITING EQUATIONS

7-4 Deciding What the Variable Represents

Objective: To select appropriate variables in writing equations to solve word problems.
MIN: 1–5; REG: 1–8; MAX: 1–10; ALL: Mixed Review

You can use the Problem-Solving Checklist on page 121 to help you solve word problems. When your plan involves writing and solving an equation, keep the following questions in mind.

- Can you use a variable to represent an unknown number?
- Can you represent other conditions in terms of the variable?
- What is equal?
- Can you write and solve an equation?

The variable can often represent the number that the question in the problem is asking you to find. Sometimes, however, the variable must be used to represent another unknown in the problem.

Example

A jazz group gave 2 shows in one night. Attendance at the second show was $1\frac{1}{2}$ times the attendance at the first show. If there were 87 people at the second show, what was the total attendance for the two shows?

Solution

Let f = number at the first show $1\frac{1}{2}f$ = number at the second show	Total attendance is the sum of the numbers for both shows. Let the variable stand for the unknown number at the first show. The number at the second show is $1\frac{1}{2}$ times the number at the first show.
$1\frac{1}{2}f = 87$	The number of people at the second show was 87.
$\frac{3}{2}f = 87$	
$\frac{2}{3} \cdot \frac{3}{2}f = \frac{2}{3} \cdot 87$	Multiply both sides by the reciprocal of $\frac{3}{2}$.
$f = 58$	There were 58 people at the first show.
$58 + 87 = 145$	Add the numbers for both shows to answer the question in the problem.

The total attendance was 145.

Exercises

A Solve by writing an equation.

1. In one year, 120 students enrolled at a technical school. This was $\frac{3}{5}$ of the number of people accepted. How many of those accepted did not enroll? $\frac{3}{5}x = 120$ $x = 200$ $200 - 120 = 80$ 80 did not enroll

2. The cost of an adult ticket to a show is $2\frac{1}{2}$ times the cost of a child's ticket. An adult ticket is \$3.75. What would the total ticket cost be for 1 adult and 1 child? $2\frac{1}{2}c = 3\frac{3}{4}$ $c = 1.50$ $3.75 + 1.50 = 5.25$ total cost: \$5.25

3. A cook needed $\frac{1}{3}$ hour cooking time for each pound of turkey. How big was the turkey if it cooked for $5\frac{1}{2}$ hours? $\frac{1}{3}p = 5\frac{1}{2}$; $16\frac{1}{2}$ lb

4. A grocery store manager said that $\frac{1}{24}$ of the number of checks received in May were from a certain bank. The store received 96 checks from that bank. How many checks did it receive from other banks? $\frac{1}{24}c = 96$ $c = 2{,}304$ $2304 - 96 = 2208$ 2208 from other banks

5. To allow for waste and leftovers, a roofer always orders about $\frac{1}{10}$ more shingles than are needed to cover the exact measurements of a roof. The roofer ordered $2\frac{1}{2}$ extra bundles of shingles for a roof. How many bundles were needed for the exact measurements? $\frac{1}{10}s = 2\frac{1}{2}$ $s = 25$ 25 bundles needed

B

6. The odometer on Marta's bicycle read 2375 when she left home. At the end of her trip the odometer read 2453. Marta bicycled for $6\frac{1}{2}$ hours that day. What was her average speed for the trip? $6\frac{1}{2}x = 2453 - 2375$; 12 mi/h

7. If Tim lost $\frac{1}{2}$ lb on Saturday and another $\frac{1}{4}$ lb on Sunday, he would have been down to the wrestling weight his coach wanted for him. His coach wanted him to weigh $105\frac{1}{2}$ lbs. How much did Tim weigh before Saturday? $x - \frac{1}{2} - \frac{1}{4} = 105\frac{1}{2}$; $106\frac{1}{4}$ lbs

8. Fran likes to work 20 hours each week. Last week she worked only $4\frac{1}{2}$ hours on Tuesday, $4\frac{1}{2}$ hours on Thursday, and $2\frac{1}{4}$ hours on Friday. Fran earned \$54. How much would she earn in a 20-hour work week? $\left(4\frac{1}{2} + 4\frac{1}{2} + 2\frac{1}{4}\right)x = 54$ $4.80(20) = 96$; \$96

C Extending Thinking Skills

Write a word problem that could be solved using the equation.

9. $\frac{3}{4}x = 8$ Answers will vary. **10.** $\left(\frac{1}{2} + \frac{1}{4}\right)x = 10$

Thinking skills: 9–10. Formulating a problem

Mixed Review

Solve and check. **11.** $x - \frac{3}{8} = \frac{2}{3}$ $1\frac{1}{24}$ **12.** $\frac{4}{11} = c + \frac{1}{2}$ $-\frac{3}{22}$ **13.** $y + \frac{1}{4} = \frac{11}{16}$ $\frac{7}{16}$

14. $r + 1.03 = -2.67$ -3.7 **15.** $-9.86m = 226.78$ -23 **16.** $z - 14.4 = 12.73$ 27.13

Evaluate for $m = \frac{3}{4}$. **17.** $m + \frac{1}{2}$ $1\frac{1}{4}$ **18.** $\frac{2}{5} - m$ $-\frac{7}{20}$ **19.** $\frac{3}{16} + m$ $\frac{15}{16}$

PROBLEM SOLVING: APPLICATIONS

7-5 Mathematics and Masonry

Objective: To solve applied problems involving multiplication of rational numbers.
All: 1–10

A **mason** is a person who builds with stones, bricks, and concrete. Much of the work a mason does involves computation with rational numbers.

Problems

You might wish to have students choose from among the techniques pencil and paper, mental math, estimation, and calculator. Have them choose the most appropriate technique for each problem.

Solve.

1. Figure 1 shows a hollow brick that is used to line the insides of chimneys. What is the overall length of one of these bricks? 13″

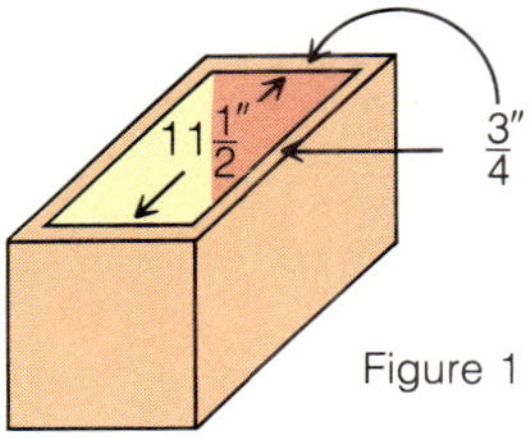

Figure 1

2. The stairway in figure 2 has five risers. What is the rise of these steps? $36\frac{1}{4}''$

Figure 2

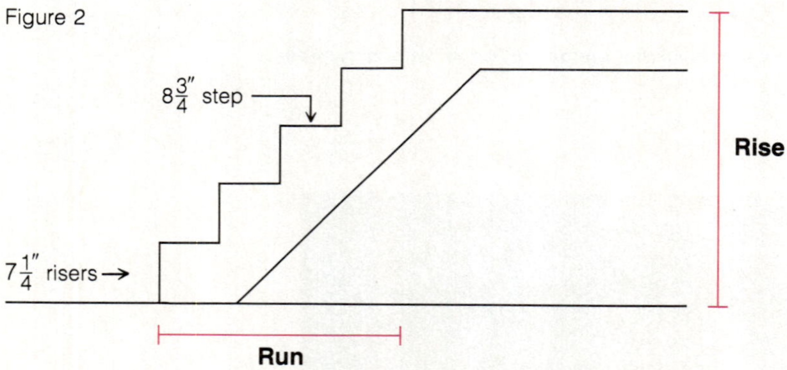

3. The stairway in figure 2 has 4 steps. What is the run of this stairway? 35″

4. A certain stairway has 14 risers. If each riser is $6\frac{5}{8}$ inches high, what is the rise of this stairway? $92\frac{3}{4}''$

5. A standard-size brick weighs $2\frac{3}{4}$ pounds. How much would 36 of these bricks weigh? 99 lb

6. Plans for construction of a patio require 888 square feet of brick. The mason figures that with the size of brick and mortar to be used, about $5\frac{1}{2}$ bricks are needed per square foot. About how many bricks are needed to construct the patio? 4884

7. A mason needs $\frac{5}{8}$ of a cubic yard of mortar to lay 1000 bricks in a wall. About how many cubic yards of mortar would he need to lay 6400 bricks? 4 cu. yds

8. A standard-size brick is $2\frac{1}{2}$ inches thick. What is the height of a wall made of 6 rows of this brick with $\frac{1}{2}$ inch of mortar at its base and $\frac{1}{2}$ inch of mortar between rows? 18″

9. A standard-size brick is $8\frac{1}{4}$ inches long. How long would a wall be if it had 24 bricks side-by-side, with $\frac{3}{8}$ inch of mortar between bricks? $206\frac{5}{8}''$

10. ***Data Search*** Suppose you were going to cover one wall of your classroom with brick. Find the approximate total cost of the brick you would need for the job. Answers will vary.

What's Your Decision?

You have 700 standard-size bricks ($8\frac{1}{4}''$ by $2\frac{1}{2}''$) and plan to build a patio, using $\frac{1}{2}''$ of sand between bricks. You hope to use as many of the bricks as possible. What shape and size patio would you design? Answers will vary.

7-6 Rational Numbers and Repeating Decimals

Objective: To change a rational number to a decimal.
MIN: 1–16, 34–35; REG: 1–16 even, 17–35; MAX: 1–16 even, 17–36; ALL: Mixed Review

You can write $\frac{3}{8}$ as the **terminating decimal** 0.375, because when you divide 3 by 8 the division process ends with a remainder of 0, or terminates. A **repeating decimal** is a decimal with a set of digits that repeats endlessly. For example, you can write $\frac{15}{33}$ as a repeating decimal.

The rational numbers and the irrational numbers make up the set of real numbers. The real numbers are studied again in Chapter 14.

$$\frac{15}{33} = 0.454545\ldots$$

$$= 0.\overline{45}$$ The bar indicates the set of digits that repeats.

A rational number can be expressed as either a terminating or repeating decimal. A decimal that is neither terminating nor repeating, such as 3.121121112 . . . , names an **irrational number**.

Example

Write $\frac{3}{11}$ as a decimal. Use a bar for a repeating decimal.

Solution

$$\begin{array}{r} 0.2727\ldots = 0.\overline{27} \\ 11\overline{)3.0000} \\ \underline{2\,2} \\ 80 \\ \underline{77} \\ 30 \\ \underline{22} \\ 80 \\ \underline{77} \\ 30 \end{array}$$

Annex zeros and continue dividing until the quotient either terminates or begins to repeat.

Practice Write as a decimal. Use a bar for a repeating decimal. **a.** $\frac{2}{9}$ $0.\overline{2}$ **b.** $\frac{1}{6}$ $0.1\overline{6}$

Oral Exercises

State whether each is a terminating decimal or repeating decimal, or names an irrational number. For a repeating decimal, identify the digits that repeat.

1. 0.171717 . . . rep. $\overline{17}$ **2.** 0.236 term. **3.** 3.12121212 . . . rep. $\overline{12}$

4. 3.123456 . . . irrat. **5.** 0.34555 term. **6.** 2.010010001 . . . irrat.

Exercises

A Write as a decimal. Use a bar for a repeating decimal.

1. $\frac{5}{11}$ $0.\overline{45}$ **2.** $\frac{1}{3}$ $0.\overline{3}$ **3.** $\frac{7}{20}$ 0.35 **4.** $\frac{23}{9}$ $2.\overline{5}$

5. $\frac{1}{15}$ $0.0\overline{6}$ **6.** $\frac{3}{4}$ 0.75 **7.** $\frac{27}{16}$ 1.6875 **8.** $\frac{7}{18}$ $0.3\overline{8}$

9. $\frac{2}{3}$ $0.\overline{6}$ **10.** $\frac{5}{8}$ 0.625 **11.** $\frac{4}{9}$ $0.\overline{4}$ **12.** $\frac{7}{16}$ 0.4375

13. $\frac{23}{40}$ 0.575 **14.** $\frac{11}{3}$ $3.\overline{6}$ **15.** $\frac{9}{24}$ 0.375 **16.** $\frac{6}{11}$ $0.\overline{54}$

B Write each as a decimal.

17. $\frac{16}{25}$ 0.64 **18.** $\frac{48}{11}$ $4.\overline{36}$ **19.** $\frac{88}{33}$ $2.\overline{6}$ **20.** $\frac{72}{99}$ $0.\overline{72}$

21. $\frac{51}{85}$ 0.6 **22.** $\frac{17}{32}$ 0.53125 **23.** $\frac{65}{18}$ $3.6\overline{1}$ **24.** $\frac{44}{54}$ $0.\overline{814}$

25. $\frac{5}{12}$ $0.41\overline{6}$ **26.** $\frac{19}{15}$ $1.2\overline{6}$ **27.** $\frac{9}{33}$ $0.\overline{27}$ **28.** $\frac{16}{6}$ $2.\overline{6}$

29. Solve for x: $x = 34.\overline{45} - 1.\overline{45}$ 33 **30.** Solve for x: $x = 23.\overline{8} - 2.3\overline{8}$ 21.5

Write the fractions as decimals to solve. Give an approximate answer if the decimal repeats.

31. A new auditorium has 250 seats. About $\frac{7}{8}$ of the seats have right-hand desks. How many seats have right-hand desks? 219

32. A consumer guide suggests that people should save about $\frac{1}{6}$ of their earnings. How much money should a person who earns \$32,500 a year save? \$5417

33. In a school with 360 graduating students, $\frac{3}{4}$ of the students plan to go to college. How many students plan to go to college? 270

C Extending Thinking Skills

34. Arrange in order from smallest to largest.

$0.44, 0.4, 0.4\overline{3}, 0.4\overline{29}, 0.45, 0.4\overline{5}$ $0.4, 0.4\overline{29}, 0.4\overline{3}, 0.44, 0.45, 0.4\overline{5}$

35. Write a decimal that neither terminates nor repeats. Answers will vary. Ex. 0.1020030004 . . .

36. Find the decimal equivalents for $\frac{1}{11}$, $\frac{2}{11}$, $\frac{3}{11}$, and $\frac{4}{11}$. Look for a pattern in these decimals to give the decimal equivalents for $\frac{5}{11}$, $\frac{6}{11}$, $\frac{7}{11}$, and $\frac{8}{11}$.
$0.\overline{45}, 0.\overline{54}, 0.\overline{63}, 0.\overline{72}$ Thinking skills: 34. Reasoning 35. Generalizing 36. Finding a pattern

Mixed Review

Find the least common multiple (LCM). **37.** 3, 4, 5 60 **38.** 2, 3, 4 12

Solve and check. **39.** $a - 3.97 = 2.43$ 6.4 **40.** $-4c = -16.8$ 4.2

Find the greatest common factor (GCF). **41.** 8, 16, 34 2

7-7 More About Exponents

Objectives: To find the quotient of two numbers expressed with exponents; to understand negative exponents.
MIN: 1–28, 45; REG: 1–28 even, 29–45; MAX: 1–28 even, 29–48; ALL: Mixed Review

You have used the rule $a^m \cdot a^n = a^{m+n}$ to multiply powers with the same base.

The following suggests a rule for simplifying expressions in the form $\frac{a^m}{a^n}$.

$$\frac{4^5}{4^2} = \frac{\overset{1}{\cancel{4}} \cdot \overset{1}{\cancel{4}} \cdot 4 \cdot 4 \cdot 4}{\underset{1}{\cancel{4}} \cdot \underset{1}{\cancel{4}}} = 4 \cdot 4 \cdot 4 = 4^3$$

Dividing Powers with Like Bases

To find the quotient of two numbers in exponential form with the same base, subtract the exponent of the denominator from the exponent of the numerator. Write this difference as the exponent of the base.

For all values of ***a*** except 0 and for all numbers ***m*** and ***n***, $\frac{a^m}{a^n} = a^{m-n}$

Example 1

Simplify. Write the expression with exponents. $\frac{2^5}{2^3}$

Solution

$\frac{2^5}{2^3} = 2^{5-3} = 2^2$ Since the base is the same, subtract the exponents.

Practice Simplify. Write the expression with exponents.

a. $\frac{5^6}{5^3}$ 5^3 **b.** $\frac{(-3)^7}{(-3)^4}$ $(-3)^3$ **c.** $\frac{10^4}{10}$ 10^3

Example 2

Simplify. Write the expression with exponents. $\frac{x^6}{x^2}$

Solution

$\frac{x^6}{x^2} = x^{6-2} = x^4$ Since the base is the same, subtract the exponents.

Practice Simplify. Write the expression with exponents.

a. $\frac{y^8}{y^3}$ y^5 **b.** $\frac{m^5}{m}$ m^4 **c.** $\frac{z^3}{z^2}$ z

You can use the rule given on page 221 to simplify the expression $\frac{5^2}{5^4}$:

$$\frac{5^2}{5^4} = 5^{2-4} = 5^{-2}$$

You could also write the following: $\frac{5^2}{5^4} = \frac{\overset{1}{\cancel{5}} \cdot \overset{1}{\cancel{5}}}{\underset{1}{\cancel{5}} \cdot \underset{1}{\cancel{5}} \cdot 5 \cdot 5} = \frac{1}{5 \cdot 5} = \frac{1}{5^2}$

This shows that $\frac{1}{5^2}$ is the same as 5^{-2}. In general, you can use a **negative exponent** to write $\boldsymbol{a^{-m} = \frac{1}{a^m}}$.

Example 3

Write the expression 3^{-2} without exponents.

Solution

$3^{-2} = \frac{1}{3^2} = \frac{1}{9}$ Since the exponent is negative, you can rewrite the expression as $\frac{1}{3^2}$, and $3^2 = 9$.

Practice Write each expression without exponents. **a.** 5^{-3} $\frac{1}{125}$ **b.** 4^{-1} $\frac{1}{4}$

This rule for dividing powers with like bases also shows that any number to the zero power, $\boldsymbol{a^0}$, is **1**. $\frac{a^m}{a^m} = a^{m-m} = a^0 = 1$

Oral Exercises

State each using exponents.

1. $3 \cdot 3 \cdot 3 \cdot 3$ **2.** $\frac{1}{2 \cdot 2 \cdot 2 \cdot 2 \cdot 2}$ **3.** $\frac{1}{5 \cdot 5 \cdot 5}$ **4.** $4 \cdot 4 \cdot 2 \cdot 2 \cdot 2$

1. 3^4 2. 2^{-5} 3. 5^{-3} 4. $4^2 \cdot 2^3$

Exercises

A Simplify. Write the expression with exponents.

1. $\frac{3^4}{3}$ 3^3 **2.** $\frac{5^4}{5^2}$ 5^2 **3.** $\frac{(-4)^3}{(-4)}$ $(-4)^2$ **4.** $\frac{2^6}{2^5}$ 2

5. $\frac{10^5}{10}$ 10^4 **6.** $\frac{(-4)^7}{(-4)^5}$ $(-4)^2$ **7.** $\frac{8^5}{8}$ 8^4 **8.** $\frac{(-2)^5}{(-2)}$ $(-2)^4$

9. $\frac{6^2}{6}$ 6 **10.** $\frac{(-3)^8}{(-3)^5}$ $(-3)^3$ **11.** $\frac{x^5}{x^4}$ x **12.** $\frac{y^6}{y^5}$ y

13. $\frac{t^3}{t}$ t^2 **14.** $\frac{r^6}{r^4}$ r^2 **15.** $\frac{g^5}{g^4}$ g **16.** $\frac{y^4}{y}$ y^3

17. $\frac{x^6}{x}$ x^5 **18.** $\frac{m^4}{m^2}$ m^2 **19.** $\frac{n^6}{n^4}$ n^2 **20.** $\frac{s^7}{s^2}$ s^5

Write the expression without exponents.

21. 4^{-2} $\frac{1}{16}$ **22.** 3^{-3} $\frac{1}{27}$ **23.** $(-2)^{-4}$ $\frac{1}{16}$ **24.** $(-3)^3$ -27

25. 10^{-4} $\frac{1}{10{,}000}$ **26.** $(-2)^2$ 4 **27.** $(-5)^{-2}$ $\frac{1}{25}$ **28.** 2^6 64

B Simplify. Write the expression with exponents.

29. $3^2 \cdot 3^{-5}$ 3^{-3} **30.** $4 \cdot 4^5 \cdot 4^{-3}$ 4^3 **31.** $x \cdot x^{-3} \cdot x^3$ x

32. $\frac{4}{4^4}$ 4^{-3} **33.** $\frac{3^2 \cdot 3^4}{3^5}$ 3 **34.** $\frac{x^5}{x^2 \cdot x}$ x^2

35. $(-3)^{-4} \cdot (-3)^{-1}$ $(-3)^{-5}$ **36.** $a^{-3} \cdot a^{-4} \cdot a^9$ a^2 **37.** $5^{-2} \cdot 5^{-3} \cdot 5^4$ 5^{-1}

38. $\frac{z}{z^2 \cdot z^2}$ z^{-3} **39.** $\frac{(-2)^4(-2)^2}{(-2)^4}$ $(-2)^2$ **40.** $\frac{5^3 \cdot 5^2}{5^7}$ 5^{-2}

41. What does n equal in $3^n = 27$? 3 **42.** What does n equal in $(-4)^n = \frac{1}{-64}$? −3

43. Compute $4^{-3} \cdot 4^5$. Then compute $4^5 \cdot 4^{-3}$. Is $a^n \cdot a^m = a^m \cdot a^n$ true for all values of a, n, and m? Why? Name a property that explains the equality. See additional answers.

44. Write $(4x^2)^3$ as an expression with one exponent.
$(4x^2)^3 = (4x^2)(4x^2)(4x^2) = 4 \cdot 4 \cdot 4 \cdot x^2 \cdot x^2 \cdot x^2 = 64x^6$

C *Extending Thinking Skills*

Express each as a whole number without exponents.

45. $\frac{1}{3^{-2}}$ 9 **46.** $\frac{1}{4(4^{-2})}$ 4 **47.** $\frac{2^{-2}}{2^{-4}}$ 4

48. A poll taker interviewed only 1 person on Monday. On each of the following days, she interviewed 2 more people than on the previous day. How many interviews had she completed by the end of the day on Saturday? 36

Thinking skills: 45–47. Generalizing 48. Finding a pattern

Mixed Review

Evaluate for $y = -1.2$. **49.** $-1.5y$ 1.8 **50.** $y + 9.35$ 8.15 **51.** $6.3 - y$ 7.5

Solve and check. **52.** $r \div 16 = -3$ −48 **53.** $c + 7 = -1.4$ −8.4

NUMBERS TO ALGEBRA

You can simplify algebraic expressions as you did numerical expressions.

Numbers	Algebra
$\frac{2^4}{3^3} \cdot \frac{3^2}{2^2} = \frac{2 \cdot 2 \cdot 2 \cdot 2 \cdot 3 \cdot 3}{2 \cdot 2 \cdot 3 \cdot 3 \cdot 3} = \frac{2^2}{3}$	$\frac{2x^3}{5y} \cdot \frac{6y^2}{4x^4} = \frac{2 \cdot x \cdot x \cdot x \cdot 2 \cdot 3 \cdot y \cdot y}{5 \cdot y \cdot 2 \cdot 2 \cdot x \cdot x \cdot x \cdot x} = \frac{3y}{5x}$

Simplify.

1. $\frac{x^2}{y^4} \cdot \frac{y^6}{x}$ xy^2 **2.** $\frac{4a^3}{5b^5} \cdot \frac{3b}{6a^2}$ $\frac{2a}{5b^4}$ **3.** $\frac{m^4}{n^5} \cdot \frac{6n^7}{5m^5}$ $\frac{6n^2}{5m}$

4. $\frac{3x^3}{5y^2} \cdot \frac{y^3}{12x^3}$ $\frac{y}{20}$ **5.** $xy^2 \cdot x^{-3}y^4$ $\frac{y^6}{x^2}$ **6.** $\frac{ab^5}{c} \cdot \frac{a^3c^4}{b^2}$ $a^4b^3c^3$

7-8 Scientific Notation

Objective: To change numbers expressed in scientific notation into standard form, and vice-versa.
MIN: 1–24; REG: 1–24 even, 25–31; MAX: 1–24 even, 25–34; ALL: Mixed Review

An exploratory lesson opener can be found on page 550.

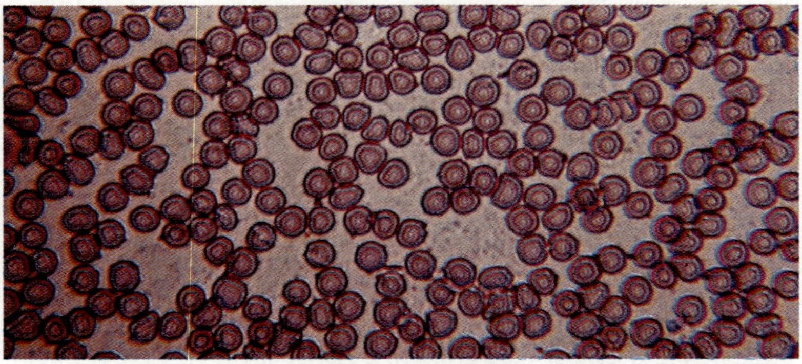

The human body replaces 2.0×10^{11} red blood cells every day. A cell in the body might be as small as 3×10^{-4} inches in diameter.

To simplify work with very large or very small numbers, you can use **scientific notation**. A number written as the product of a power of 10 and a number greater than or equal to 1 but less than 10 is expressed in scientific notation. The number 3.45×10^3 is in scientific notation. The numbers 12.3×10^2 and 124.5 are not in scientific notation.

Example 1

Write 3.23×10^4 in decimal form.

Solution

32,300 Multiplying 3.23 by 10^4 moves the decimal point 4 places to the right.

Practice Write each in decimal form.

a. 1.8×10^{-4} 0.00018 **b.** 6.556×10^2 655.6 **c.** 4×10^{-2} 0.04

Example 2

Write 1,234,000 in scientific notation.

Solution

1.234×10^6 Move the decimal point 6 places to the left. Multiply by 10^6.

Practice Write each in scientific notation.

a. 4567 4.567×10^3 **b.** 234,000 2.34×10^5 **c.** 50,000,000 5×10^7

Example 3

Write 0.000345 in scientific notation.

Solution

3.45×10^{-4} Move the decimal point 4 places to the right. Multiply by 10^{-4}.

Practice Write each in scientific notation.

a. 0.0206 2.06×10^{-2} **b.** 0.000008 8×10^{-6} **c.** 0.2004 2.004×10^{-1}

Oral Exercises

Tell where the decimal should be placed to express each number in scientific notation.

1. 32,500 3.25 **2.** 35 3.5 **3.** 0.005 5
4. 0.6 6 **5.** 770 7.7 **6.** 82.5 8.25
7. 430,000 4.3 **8.** 18.6 1.86 **9.** 0.000050 5
10. 0.09 9 **11.** 0.0072 7.2 **12.** 25,000 2.5

Exercises

A Write each in decimal form. 1–24. See additional answers.

1. 3.5×10^4 **2.** 6.2×10^{-2} **3.** 4.05×10^5 **4.** 7.0×10^{-5}

Write each in scientific notation.

5. 135 **6.** 23,000 **7.** 345,000 **8.** 8.4
9. 1240 **10.** 650,000 **11.** 4,550,000 **12.** 600
13. 99,000 **14.** 1,000,000 **15.** 0.078 **16.** 0.4
17. 0.000677 **18.** 0.0055 **19.** 0.000001 **20.** 0.05
21. 0.0000405 **22.** 0.000007 **23.** 0.101 **24.** 0.00000003

B Use the rules for multiplying and dividing exponents to write each product or quotient in scientific notation.

25. $(4.0 \times 10^3)(2 \times 10^4)$ 8×10^7 **26.** $(3.2 \times 10^{-2})(5 \times 10^{-4})$ 1.6×10^{-5}

27. $\frac{6 \times 10^5}{2 \times 10^4}$ 3×10^1 **28.** $\frac{5.2 \times 10^5}{1.3 \times 10^3}$ 4×10^2

29. Every 0.4 cubic inch of human blood contains about 5,500,000 red blood cells. Write the number of red blood cells in scientific notation. 5.5×10^6

30. The red blood cell is one of the smallest cells in the human body. It has a diameter of 0.0003 inch. Write this number in scientific notation. 3×10^{-4}

31. Some cells in the human body are so small that 200,000 could be placed on the head of a pin. Write this number in scientific notation. 2×10^5

C Extending Thinking Skills

32. Solve for y. $(7 \times 10^2)y = 6.3 \times 10^6$ 9×10^3

33. Use the digits 1, 2, 3, and 4 and one negative sign to write a number in scientific notation that is as close to 0.01 as possible. 1.34×10^{-2}

34. Chemical X is made in units weighing 2.5×10^{-2} g. To produce one unit of this chemical, 4×10^{-3} g of chemical A and 5.0×10^{-2} g of chemical B are needed. In the production process, 2.5×10^{-3} g of chemical A and 4.2×10^{-2} g of chemical B are burned off. What amount of each unit of chemical X comes from chemicals other than A and B? 1.55×10^{-2}g

Thinking skills: 32 Generalizing 33. Testing a conjecture 34. Reasoning

Mixed Review

Reduce each fraction to lowest terms. **35.** $\frac{27}{135}$ $\frac{1}{5}$ **36.** $\frac{48}{192}$ $\frac{1}{4}$ **37.** $\frac{12}{140}$ $\frac{3}{35}$

Solve and check. **38.** $a - \frac{3}{4} = -\frac{5}{8}$ $\frac{1}{8}$ **39.** $h + \frac{13}{15} = 1\frac{1}{3}$ $\frac{7}{15}$

Simplify. **40.** $z \cdot z \cdot 10 \cdot 4 \cdot z$ $(40)z^3$ **41.** $t + 2t + 3t + 4t$ $10t$

CALCULATOR ACTIVITY

Some calculators allow you to enter numbers in scientific notation. You can use them to work with small or large numbers that would not fit on the calculator's display screen in standard form.

Using a scientific calculator, follow the example below. The EE key tells the calculator you are entering a power of ten. If the number or the exponent is negative, enter the negative sign before you enter the number or the exponent.

Problem in standard notation:
(52,000,000)(230,000,000) = 11,960,000,000,000,000

Problem in scientific notation:
$(5.2 \times 10^7)(2.3 \times 10^8) = 1.196 \times 10^{16}$

5.2 [EE] 7 [×] 2.3 [EE] 8 [=] Display: 1.196 16

Notice that the calculator display shows only 16, rather than 10^{16}, in the product. The answer is written 1.196×10^{16}.

Use a calculator to find each product.

1. $(4.74 \times 10^8)(8.5 \times 10^9)$ 4.029×10^{18}
2. $(6.33 \times 10^7)(1.9 \times 10^7)$ 1.2027×10^{15}
3. (245,600,000)(700,000) 1.7192×10^{14}
4. (1,754)(6,570,000,000,000) 1.1524×10^{16}
5. (1,250,000,000,000)(12,240) 1.53×10^{16}
6. (24,400,000,000)(5,500,000,000) 1.342×10^{20}

PROBLEM SOLVING: STRATEGIES

7-9 Work Backward

Objective: To solve nonroutine problems, using the strategy Work Backward and other strategies learned so far.
MIN: 1–2; REG: 1–4; MAX: 1–6

Sometimes a word problem describes a sequence of actions involving numbers, gives the result, and asks for the number started with. A problem of this type can be solved by using a strategy called **Work Backward**.

Problem On Monday Jeff opened a savings account for his summer earnings and deposited all of his first week's earnings. On Tuesday, he deposited \$25 into the account. He withdrew \$23 on Wednesday to buy tapes and another \$15 on Thursday for other expenses. On Friday, he withdrew half of what was left in the account to buy some clothing. He then had \$12.50 remaining in the account. How much money did he deposit on Monday?

To solve this problem you can start with the amount of money Jeff had at the end and work backward, using the inverse operations. The lists below show the data given in the story and indicate how to work backward.

Data in the Story	Work Backward
Deposited x ↓	\$38
Deposited (added) \$25 ↓	↑ Subtract \$25; \$63 − \$25 = \$38
Withdrew (subtracted) \$23 ↓	↑ Add \$23; \$40 + \$23 = \$63
Withdrew (subtracted) \$15 ↓	↑ Add \$15; \$25 + \$15 = \$40
Withdrew half (divided by 2) ↓	↑ Multiply by 2; \$12.50 × 2 = \$25
Final amount–\$12.50	↑ \$12.50

Jeff deposited \$38 on Monday.

This chart shows the strategies presented so far.

Problem-Solving Strategies

Choose the Operations	Write an Equation
Guess, Check, Revise	Simplify the Problem
Draw a Picture	Make an Organized List
Make a Table	Use Logical Reasoning
Look for a Pattern	Work Backward

Problems

Solve.

1. Ned, Gary, Kris, and Brenda worked at a school car wash on Saturday. One person washed each car. Ned washed twice as many cars as Gary. Gary washed 4 fewer than Kris and Kris washed 5 more than Brenda. If Brenda washed 8 cars, how many cars did these students wash all together? **48; Ned 18, Gary 9, Kris 13**

2. Two jars of chemicals labeled A and B were mixed so that each contained 64 ml at the end. The mixing process involved first pouring from B into A as much liquid as A contained, and finally pouring from A into B as much liquid as B now had. How much liquid was in each jar before they were mixed? **A 48 ml, B 80 ml**

3. Jana spent exactly $1.00 on some snack items at the natural foods store. She bought 11 items on the price list at the right. Which items could she have bought? **See additional answers.**

Snack Items	
Honey Drops 2 for $0.15	Carob Chews 3 for $0.25
Granola Bars $0.10 each	

4. Suppose you have two pails, one that holds 4L of water and one that holds 9L. There are no markings on either pail to indicate quantities. How can you measure out 6L of water using these two pails? **See additional answers.**

5. Felipe lives at the corner of 1st and A Streets. Each day he goes to Matty's house at the corner of 5th and F Streets, traveling only east and north. How many days can he go without having to repeat a route? **126 days**

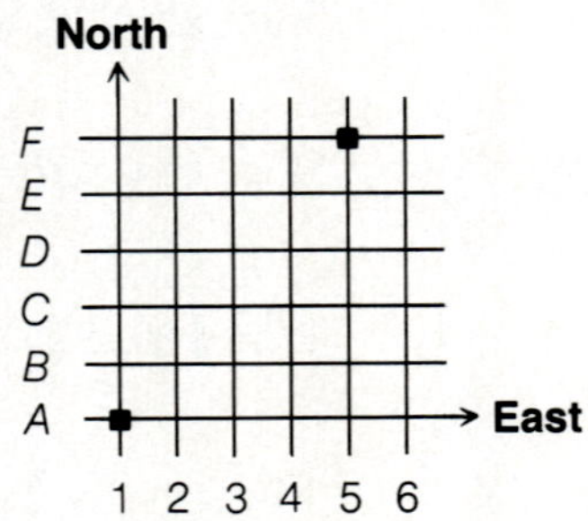

6. A rumor was passed around a town that had a population of 20,550. Each person who heard the rumor passed it on to 4 other people the next day, and then told no one else. One person made up the rumor and told it to 4 people on October 1st. What is the earliest day that everyone in this town might have heard the rumor? **October 8**

Enrichment

Writing a Repeating Decimal as a Fraction

You can express a terminating decimal as a fraction. For example, $1.235 = \frac{1235}{1000}$.

To write a repeating decimal as a fraction, you can begin by multiplying by 10^n, where n is the number of digits in the repeating decimal part.

Write $0.\overline{36}$ as a fraction.

$x = 0.\overline{36}$ Original equation.

$100x = 36.\overline{36}$ Multiply both sides of the original equation by 10^2 since there are **2** digits in **36**.

$$\begin{aligned} 100x &= 36.\overline{36} \\ -\quad x &= 0.\overline{36} \\ \hline 99x &= 36 \end{aligned}$$

Subtract the original equation.

$x = \frac{36}{99} = \frac{4}{11}$ Solve for x. Reduce to lowest terms.

The following example involves a single repeating digit.

Write $1.2\overline{3}$ as a fraction.

$x = 1.233333\ldots$ Original equation.

$10x = 12.33333\ldots$ Multiply both sides of the equation by 10, since only 1 digit repeats.

$$\begin{aligned} 10x &= 12.33333\ldots \\ -\quad x &= 1.23333\ldots \\ \hline 9x &= 11.1 \end{aligned}$$

Subtract the original equation.

$x = \frac{11.1}{9} \cdot \frac{10}{10}$ Solve for x. Multiply by $\frac{10}{10}$ to remove the decimal from the numerator.

$x = \frac{111}{90} = 1\frac{21}{90} = 1\frac{7}{30}$

Write each as a fraction. Reduce to lowest terms.

1. $x = 0.\overline{3}$ $\frac{1}{3}$ **2.** $x = 0.1\overline{6}$ $\frac{1}{6}$ **3.** $x = 0.41\overline{6}$ $\frac{5}{12}$

4. $x = 0.\overline{425}$ $\frac{421}{990}$ **5.** $x = 1.3\overline{8}$ $1\frac{7}{18}$ **6.** $x = 0.5\overline{27}$ $\frac{29}{55}$

7. $x = 1.1\overline{2}$ $1\frac{11}{90}$ **8.** $x = 0.\overline{618}$ $\frac{34}{55}$ **9.** $x = 1.\overline{225}$ $1\frac{225}{999}$

Chapter 7 Review

7-1 Multiply. Reduce to lowest terms.

1. $-\frac{4}{9}\left(-\frac{3}{4}\right)$ $\frac{1}{3}$
2. $\frac{6}{7}\left(-\frac{2}{3}\right)$ $-\frac{4}{7}$
3. $-1\frac{3}{5}\left(3\frac{4}{7}\right)$ $-5\frac{5}{7}$
4. $7\frac{1}{2}\left(2\frac{4}{5}\right)$ 21

7-2 Divide. Reduce to lowest terms.

5. $\frac{7}{12} \div \left(-\frac{4}{3}\right)$ $-\frac{7}{16}$
6. $-\frac{9}{10} \div \left(-\frac{2}{5}\right)$ $2\frac{1}{4}$
7. $4\frac{3}{8} \div 1\frac{3}{4}$ $2\frac{1}{2}$
8. $-4\frac{1}{2} \div 6\frac{3}{4}$ $-\frac{2}{3}$

7-3 Solve and check.

9. $\frac{4}{5}m = \frac{2}{15}$ $\frac{1}{6}$
10. $\frac{f}{-6} = \left(-\frac{2}{3}\right)$ 4
11. $-5\frac{1}{4}s = 2\frac{4}{5}$ $-\frac{8}{15}$
12. $-4\frac{4}{5} = 1\frac{6}{10}n$ -3

7-4 Find the solution by writing and solving an equation.

13. Twenty-one people came to the first electronics class. This was $\frac{7}{8}$ of the number that signed up. How many people signed up? $\frac{7}{8}x = 21$ 24 signed up

7-5 Solve.

14. Mario placed dominoes end-to-end to form a line $37\frac{1}{2}$ inches long. If each domino was $2\frac{1}{2}$ inches long, how many were used? 15

15. If $\frac{1}{4}$ cup of lemon juice is used to make 2 cups of lemonade, how much lemon juice is needed to make 12 cups of lemonade? $1\frac{1}{2}$ cups

7-6 Write each as a decimal. Use a bar for a repeating decimal.

16. $\frac{11}{18}$ $0.6\overline{1}$
17. $\frac{9}{20}$ 0.45
18. $\frac{2}{9}$ $0.\overline{2}$

7-7 Simplify. Write the expression with exponents.

19. $\frac{8^5}{8^2}$ 8^3
20. $\frac{(-5)^8}{(-5)}$ $(-5)^7$
21. $\frac{x^7}{x^5}$ x^2

Write each expression without exponents.

22. 4^{-3} $\frac{1}{64}$
23. $(-6)^{-2}$ $\frac{1}{36}$
24. $\frac{(-8)^3}{(-8)^5}$ $\frac{1}{64}$

7-8 Write each in decimal form.

25. 4.7×10^3 4700
26. 1.9×10^{-2} 0.019
27. 3.7×10^{-5} 0.000037

Write each in scientific notation.

28. 436,000 4.36×10^5
29. 57,000 5.7×10^4
30. 80,000,000 8.0×10^7
31. 0.000722 7.22×10^{-4}
32. 0.000001 1.0×10^{-6}
33. 0.05 5.0×10^{-2}

Chapter 7 Test

Multiply. Reduce to lowest terms.

1. $\left(\frac{5}{12}\right)\left(\frac{3}{8}\right)$ $\frac{5}{32}$

2. $\left(\frac{1}{2}\right)\left(-\frac{4}{5}\right)$ $-\frac{2}{5}$

3. $\left(-1\frac{7}{8}\right)\left(4\frac{2}{5}\right)$ $-8\frac{1}{4}$

4. $\left(-1\frac{1}{5}\right)\left(-2\frac{1}{2}\right)$ 3

Divide. Reduce to lowest terms.

5. $-\frac{8}{9} \div \left(-\frac{4}{5}\right)$ $1\frac{1}{9}$

6. $\frac{3}{10} \div \frac{15}{100}$ 2

7. $6\frac{3}{7} \div \left(-1\frac{1}{4}\right)$ $-5\frac{1}{7}$

8. $-6\frac{1}{4} \div \frac{7}{10}$ $-8\frac{13}{14}$

Solve and check.

9. $-5y = -\frac{2}{3}$ $\frac{2}{15}$

10. $1\frac{1}{2}v = \frac{7}{10}$ $\frac{7}{15}$

11. $1\frac{1}{3} = -12a$ $-\frac{1}{9}$

12. $-6\frac{2}{3} = 3\frac{1}{3}x$ -2

Find the solution by writing and solving an equation.

13. Debbie walked from her house to a bus stop $4\frac{1}{2}$ km away. This was $\frac{1}{3}$ of the total distance to her aunt's house. What was the total distance? $\frac{1}{3}x = 4\frac{1}{2}$ $13\frac{1}{2}$ km

Solve.

14. A stack of 34 identical books is on the teacher's desk. If the stack is $59\frac{1}{2}$ inches high, how thick is each book? $1\frac{3}{4}$ in.

15. A necklace is made of beads that are $\frac{3}{4}$ inch wide. If there are 30 beads with no space between, how long is the necklace? $22\frac{1}{2}$ in.

Write each as a decimal. Use a bar for a repeating decimal.

16. $\frac{21}{40}$ 0.525

17. $\frac{1}{3}$ $0.\overline{3}$

18. $\frac{2}{11}$ $0.\overline{18}$

Simplify. Write each expression with exponents.

19. $\frac{9^8}{9^4}$ 9^4

20. $\frac{(-2)^6}{(-2)^1}$ $(-2)^5$

21. $\frac{p^9}{p^5}$ p^4

Write each expression without exponents.

22. 2^{-4} $\frac{1}{16}$

23. 3^{-3} $\frac{1}{27}$

24. $\frac{(-7)^4}{(-7)^6}$ $\frac{1}{49}$

Write each in decimal form.

25. 1.47×10^6 1,470,000

26. 9.0×10^{-3} 0.009

27. 2.11×10^{-5} 0.0000211

Write each in scientific notation.

28. 329 3.29×10^2

29. 85,000 8.5×10^4

30. 99,000,000 9.9×10^7

31. 0.08 8.0×10^{-2}

32. 0.0000101 1.01×10^{-5}

33. 0.0079 7.9×10^{-3}

Cumulative Review

Evaluate.

1. $49.07 + h$ for $h = 18.47$ 67.54
2. $m + 13.73$ for $m = 3.85$ 17.58
3. $21.8 + v$ for $v = -3.3$ 18.5
4. $p + 17.32 + 1.94$ for $p = -0.54$ 18.72
5. $29.1 - f$ for $f = 80.5$ −51.4
6. $w - 22.71$ for $w = -9.13$ −31.84

Add or subtract. Reduce to lowest terms.

7. $\frac{2}{13} + \frac{4}{13}$ $\frac{6}{13}$
8. $-\frac{6}{8} + \frac{1}{8}$ $-\frac{5}{8}$
9. $-\frac{22}{30} - \left(-\frac{4}{30}\right)$ $-\frac{3}{5}$
10. $\frac{1}{3} + \frac{2}{9}$ $\frac{5}{9}$
11. $-\frac{4}{15} - \frac{1}{5}$ $-\frac{7}{15}$
12. $-\frac{3}{8} - \left(-\frac{1}{4}\right)$ $-\frac{1}{8}$
13. $8\frac{1}{4} + 1\frac{5}{8}$ $9\frac{7}{8}$
14. $4\frac{3}{5} - 2\frac{1}{2}$ $2\frac{1}{10}$
15. $7\frac{1}{3} - \left(-1\frac{1}{2}\right)$ $8\frac{5}{6}$

Solve and check.

16. $m + 23.4 = -16.7$ −40.1
17. $r - 8.12 = 6.16$ 14.28
18. $b - \frac{2}{5} = \frac{3}{10}$ $\frac{7}{10}$
19. $t + \frac{1}{3} = -\frac{5}{6}$ $-1\frac{1}{6}$
20. $\frac{4}{15} = c - \left(-\frac{1}{5}\right)$ $\frac{1}{15}$

Solve by writing an equation.

21. Sally bought a ring on sale for \$25.59. This was \$6.80 less than the original price. What was the original price? $x - 6.80 = 25.59$ \$32.39

22. Raul drove 15 more miles on Tuesday than he drove on Monday. If he drove 46 miles on Tuesday, how many miles did he drive Monday? $m + 15 = 46$ 31 miles

23. Vince has \$210 more in his savings account than he has in his checking account. If he has \$616 in his savings account, how much does he have in his checking account? $c + 210 = 616$ \$406

24. Mrs. Tanaka has 13 envelopes. How many more envelopes does she need if she is sending 56 letters? $13 + e = 56$ she needs 43

Today's Stock Report

High	Low	Stock	High	Low	Last	Chg.
17 1/8	12 5/8	**Zzt**	14 7/8	13 3/4	13 3/4	−1/2
25 1/2	10 1/2	**EEWA**	17 3/8	14 3/8	16 3/4	+1/4

25. What is the difference between the yearly low and today's low for EEWA shares? 3 7/8

26. What was the closing price for Zzt shares yesterday? 14 1/4

Chapter 8
Equations and Inequalities

8-1 Solving Equations: Combined Operations

Objective: To solve equations involving more than one operation.
MIN: 1–24; REG: 1–24 even, 25–31; MAX: 1–24 even, 25–33; ALL: Mixed Review

A grocer could solve the equation $0.2n + 8 = 40$ to find the number of 0.2-kg oranges in a crate that weighs 8 kg when empty and 40 kg when full.

You have learned to solve equations in which one operation has been applied to the variable. To solve equations in which more than one operation has been applied to the variable, you will need to observe the order of operations, then undo them, as shown in the table below.

The properties of equality emphasize the importance of applying the inverse operations to *both* sides of the equation.

Equation	Order of Operations Applied to the Variable	To Undo These Operations
$3n + 5 = 44$	First n was multiplied by 3. Then 5 was added.	Subtract 5 from each side. Then divide each side by 3.
$\frac{2}{3}b - 4 = 4$	First b was multiplied by $\frac{2}{3}$. Then 4 was subtracted.	Add 4 to each side. Then multiply each side by the reciprocal of $\frac{2}{3}$, or $\frac{3}{2}$.
$6(x + 2) = 30$	First 2 was added to x. Then this sum was multiplied by 6.	Divide each side by 6. Then subtract 2 from each side.
$\frac{y - 7}{3} = 8$	First 7 was subtracted from y. Then this difference was divided by 3.	Multiply each side by 3. Then add 7 to each side.

Remind students that when solving equations the objective is to get the variable by itself on one side.

This leads to steps for solving equations with combined operations.

Solving Equations with Combined Operations

1. Identify the order in which the operations have been applied to the variable.
2. Undo the operations in reverse order by applying the inverse operations or the inverse properties on both sides of the equation.

Example 1

Work through the practice problems carefully with students to develop further understanding.

Solve and check. $2n + 8 = 40$

Solution

$$2n + 8 = 40$$

First n was multiplied by 2. Then 8 was added.

$$2n + 8 - 8 = 40 - 8$$
$$2n = 32$$

Since adding 8 was the last operation applied, undo it by subtracting 8 from each side.

$$\frac{2n}{2} = \frac{32}{2}$$

Then divide by 2 to undo multiplying by 2.

$$n = 16$$

Check

$$2(16) + 8 \stackrel{?}{=} 40$$

Replace n with 16 in $2n + 8 = 40$.

$$32 + 8 \stackrel{?}{=} 40$$
$$40 = 40 \checkmark$$

The solution is 16.

Practice Solve and check. **a.** $\frac{3}{4}z + 6 = 18$ **16** **b.** $-y - 5 = 12$ **−17** **c.** $6x + 36 = 144$ **18**

Example 2

Solve and check. $5(x + 7) = 105$

Solution

$$5(x + 7) = 105$$

First 7 was added to x. Then this sum was multiplied by 5.

$$\frac{5(x + 7)}{5} = \frac{105}{5}$$

Divide each side by 5.

$$x + 7 = 21$$
$$x + 7 - 7 = 21 - 7$$

Subtract 7 from each side.

$$x = 14$$

Check

$$5(14 + 7) \stackrel{?}{=} 105$$

Replace x with 14 in $5(x + 7) = 105$.

$$5(21) \stackrel{?}{=} 105$$
$$105 = 105 \checkmark$$

The solution is 14.

Practice Solve and check. **a.** $\frac{(x + 8)}{6} = 7$ **34** **b.** $-6(n - 7) = 96$ **−9**

Oral Exercises

To solve, what operation would you use in the first step?

1. $\frac{x}{2} - 4 = -8$ Add.
2. $\frac{d}{4} + 8 = 18$ Subt.
3. $4r + 16 = 28$ Subt.
4. $9s - 4 = 59$ Add.
5. $\frac{x - 14}{8} = -7$ Mult.
6. $6(x - 2) = 18$ Div.

Exercises

A Solve and check.

1. $3b - 5 = 16$ 7
2. $4r + 16 = -80$ −24
3. $9z - 45 = 45$ 10
4. $-x + 46 = 27$ 19
5. $68 = 5n + 43$ 5
6. $37 = 12p - 23$ 5
7. $\frac{y}{4} - 12 = 8$ 80
8. $\frac{c}{8} + 16 = 23$ 56
9. $14 = \frac{a}{24} - 13$ 648
10. $3x + 7 = -38$ −15
11. $\frac{2}{7}n - 18 = 40$ 203
12. $39 = 4y - 45$ 21
13. $4(c + 6) = 56$ 8
14. $6c - 8 = 42$ $8\frac{1}{3}$
15. $-45 = 9(s + 12)$ −17
16. $45 = \frac{x}{3} + 36$ 27
17. $0.8a + 3.4 = 7.2$ 4.75
18. $16 = -b + 3\frac{1}{4}$ $-12\frac{3}{4}$
19. $\frac{x + 8}{9} = -6$ −62
20. $\frac{p - 7}{12} = -16$ −185
21. $7y - 13 = 50$ 9
22. $\frac{x + 4}{6} = 4$ 20
23. $7(x - 7) = 49$ 14
24. $3x - 6 = 9$ 5

B Solve and check. Before assigning Exercises 25–30, work through some examples with students.

25. $3 = \frac{6x + 18}{9}$ $1\frac{1}{2}$
26. $\frac{14n + 8}{3} = 26$ 5
27. $\frac{12x - 12}{8} = -9$ −5
28. $\frac{3y - 4}{4} = 2$ 4
29. $\frac{x + 6}{3} = 5$ 9
30. $\frac{3b - 5}{2} = 8$ 7
31. The formula for the perimeter of a rectangle is $P = 2l + 2w$. Find the length (l), when the perimeter (P) is 50 and the width (w) is 9. 16

C Extending Thinking Skills

32. A taxi fare included a starting fee of \$1.25 plus \$0.55 for each minute of the trip. The total cost for the trip was \$8.40. How many minutes did the trip take? 13
Encourage students to use a Guess, Check, Revise strategy; no formal equations until 8-3.
33. Complete the generalization for solving each type of equation.
a. If $ax + b = c$, then $x =$ __?__ $\frac{c - b}{a}$
b. If $ax - b = c$, then $x =$ __?__ $\frac{c + b}{a}$

Thinking skills: 32. Testing a conjecture 33. Generalizing

Mixed Review

Evaluate for $n = \frac{2}{3}$. **34.** $n - 1\frac{1}{2}$ $-\frac{5}{6}$ **35.** $\frac{4}{5}n$ $\frac{8}{15}$ **36.** $\frac{3}{5} - n$ $-\frac{1}{15}$

Solve and check. **37.** $0.8a = -2.4$ −3 **38.** $1.02 + c = -0.85$ −1.87

PROBLEM SOLVING: WRITING EQUATIONS

8-2 Translating Sentences Involving Combined Operations

Objective: To translate verbal statements into equations.
MIN: 1–8; REG: 1–11; MAX: 1–8 even, 9–13; ALL: Mixed Review

You have learned to translate a verbal statement suggesting a single operation into an equation. To solve some problems you need to choose a variable and translate a verbal statement that suggests a combination of operations, as shown below.

Verbal Statement: **Tim's \$15 overtime pay** is *\$3 less* than **twice his regular pay**.

Equation: $2p - 3 = 15$

Example

Write an equation. Carmen's class of 25 students has only 3 more than twice the number of students in Yoshi's class.

Solution Let s = number of students in Yoshi's class — Choose what the variable represents

$2s + 3 = 25$ — "Twice the number of students" translates to $2s$. "Three more than twice . . ." translates to $2s + 3$. "Is" indicates equality.

Practice Write an equation.

a. When Julio's age is doubled and increased by 5, the result is Eugene's age, or 39.

b. The \$65 cost of a pair of shoes is only \$4 less than 3 times what shoes cost in 1960.

a. $2J + 5 = 39$ b. $3s - 4 = 65$

Oral Exercises

Give an equation.

1. 36 more than 6 times a number gives 84. $6x + 36 = 84$
2. 14 less than 7 times a number gives 28. $7x - 14 = 28$
3. The difference of 3 times a number and −9 is −9. $3x - (-9) = -9$
4. The quotient of 6 times a number and 9 is 45. $6x \div 9 = 45$

Exercises

A Write an equation.

1. Emiko's weekly salary, \$200, is \$25 more than twice Jane's salary. $2j + 25 = 200$

2. Stefan traveled 300 km, which was 35 km less than half as far as he traveled yesterday. $\frac{1}{2}y - 35 = 300$

3. The 67 students in Gloria's class are 4 more than 3 times the number in Vi's class. $67 = 3v + 4$

4. Nan earned \$56, which was \$8 more than twice Juanita's earnings. $56 = 2j + 8$

5. Ben's 94 km bike trip on Tuesday was 17 km more than half the length of his trip on Monday. $94 = \frac{1}{2}m + 17$

6. The \$564 collected for the band trip is \$36 more than $\frac{1}{4}$ the amount needed.

7. The election vote turnout of 376 was 56 more than $\frac{3}{4}$ of last year's turnout.

8. If Mr. Yamamoto doubles his old salary and adds his age, 46, the total is \$46,000, his new salary. $2s + 46 = \$46{,}000$

6. $564 = \frac{1}{4}n + 36$
7. $376 = \frac{3}{4}e + 56$

B Write a verbal description of each equation. The oral exercises (p. 237) provide examples of verbal descriptions.

9. $2x + 3 = 13$ **10.** $4n - 8 = 7$ **11.** $\frac{1}{4}p + 2 = 9$

9–11. Answers will vary. Example for 9: Two times a number increased by 3 is 13.

C *Extending Thinking Skills*

12. Half an hour ago, it was 3 times as long after noon as it was until midnight. What time is it now? 9:30 p.m.

13. If 4 frogs can catch 4 flies in 4 minutes, how many frogs can catch 100 flies in 100 minutes? 4 frogs

Thinking skills: 12. Testing a conjecture 13. Reasoning

Mixed Review

Solve and check. **14.** $c - 1\frac{2}{3} = \frac{1}{2}$ $2\frac{1}{6}$ **15.** $\frac{2}{5}m = 3\frac{3}{4}$ $9\frac{3}{8}$

Simplify. **16.** $4(m + 4) - 16$ $4m$ **17.** $10(t + 30) + 500$ $10t + 800$

CALCULATOR ACTIVITY

You can use inverse operations and the calculator to solve certain equations quickly.

For example, to solve $3x + 4 = 85$, you might think "The variable x has been multiplied by 3, 4 has been added to this product and the result is 85. Working backward from 85, I'll undo adding 4 by subtracting 4, then undo multiplying by 3 by dividing by 3."

Display

85 [−] 4 [=] [÷] 3 [=] 27 The solution is 27.

Use inverse operations and a calculator to solve.

1. $4x - 17 = 207$ 56 **2.** $\frac{n}{19} + 26 = 39$ 247 **3.** $\frac{p}{15} - 69 = 16$ 1275 **4.** $\frac{5a}{12} + 13 = 23$ 24

PROBLEM SOLVING: WRITING EQUATIONS

8-3 Writing Equations to Solve Problems

Objective: To solve word problems by writing and solving equations.
MIN: 1–5, 9–10; REG, MAX: 1–5 even, 6–10; ALL: Mixed Review

You can use the Problem-Solving Checklist on page 121 to help you solve problems. When your plan involves writing and solving an equation, keep these questions in mind.

- Can you use a variable to represent an unknown number?
- Can you represent other conditions in terms of the variable?
- What is equal?
- Can you write and solve an equation?

Example

Peter bought a 10-speed bike for \$172. He made a down-payment of \$76 and monthly payments of \$24. How many months did it take him to pay for the bike?

Solution

Let m = the number of monthly payments. — Let m represent the unknown number.

$$24m + 76 = 172$$

The total amount paid can be expressed in two equal ways. Write an equation.

$$24m + 76 - 76 = 172 - 76$$
$$24m = 96$$
$$m = 4$$

Check $24(4) + 76 \stackrel{?}{=} 172$

$$172 = 172 \checkmark$$

The equation solution checks.

It took Peter 4 months to pay for the bicycle.

Estimate: \$76 plus 4 payments of \$25 would be \$176. The answer is reasonable.

Oral Exercises

Give an equation.

1. Rachel's present salary, \$25,000, is \$365 more than twice what it was when she began the job. What was her starting salary? $2s + 365 = 25{,}000$

2. Calvin saved half his allowance, earned \$18, and had a total of \$30. What is his allowance? $\frac{1}{2}a + 18 = 30$

3. Mr. Meyer's golf score, 83, was only 5 more than twice his age. How old is Mr. Meyer? $2a + 5 = 83$

4. Sandy added 16 stamps to her collection and then gave half her stamps to her sister. Then she had 29 stamps. How many stamps did she have at the beginning? $\frac{1}{2}(s + 16) = 29$

Exercises

A Solve by writing an equation.

1. Three-fourths of the members of a tennis club signed up in advance for a tournament. On the day of the tournament, 9 more entered, making a total of 84. How many members did the tennis club have? $\frac{3}{4}m + 9 = 84$ 100

2. Jason earned \$6 an hour for a job. He got a bonus of \$75, making his total earnings \$297. How many hours did he work? $6h + 75 = 297$ 37

3. The principal of Parkside School reported that there was an average of 27 students in each classroom. Then 56 more students enrolled in the school, making a new total of 704 students. How many classrooms were there? $27c + 56 = 704$ 24

4. Before giving 12 ounces of flour to a friend, Mrs. Higuera had $\frac{3}{4}$ of a sack of flour. The remaining flour weighed 48 ounces. How many ounces did a full bag weigh? $\frac{3}{4}b - 12 = 48$ 80 ounces

5. The championship basketball team fell short of getting double its opponents's score by only 3 points. The team scored 81 points. How many points did the opponents score? $81 = 2s - 3$ 42

B Solve by writing an equation.

6. A tank that was $\frac{3}{4}$ full of liquid fertilizer had 565 gallons pumped out and 832 gallons put in, leaving 2400 gallons in the tank. What would a full tank hold? $\frac{3}{4}t - 565 + 832 = 2400$ 2844 gallons

7. The temperature dropped by half, decreased by 3°C, and rose 9°C. The thermometer then read 14°C. What was the beginning temperature? $\frac{1}{2}t - 3 + 9 = 14$ 16°C

8. The Ortiz family drove at an average rate of 53 mi/h, stopped for food, then drove 78 miles more, in $1\frac{1}{2}$ hours. Their odometer showed they had traveled a total of 343 miles. What was their total driving time?
$53(t - 1\frac{1}{2}) + 78 = 343$ $6\frac{1}{2}$ hours

C Extending Thinking Skills

9. Write a word problem that could be solved using the equation $3x - 4 = 17$.
Answers will vary.

10. Write a word problem that could be solved using the equation $54 - 2x = 20$.
Thinking skills: 9–10. Formulating a problem.

Mixed Review

Evaluate for $a = 2.5$, $b = 1.4$. **11.** $3(a - b)$ 3.3 **12.** $a(b + 2)$ 8.5

Solve and check. **13.** $-2.4a = 3$ −1.25 **14.** $n + 0.6 = -2.4$ −3

8-4 Simplifying to Solve Equations: Using the Distributive Property

Objective: To simplify equations using the distributive property, then solve them.
MIN: 1–24; REG: 1–24 even, 25–34; MAX: 1–24 even, 26–36; ALL: Mixed Review
An exploratory lesson opener can be found on page 550.

You have learned to use a reversed form of the distributive property to simplify an algebraic expression such as $3x + 4x$ as follows:

Help students see that the reversed form is obtained by reversing the order of the factors and the sides of the equation $a(b + c) = ab + ac$.

$$ba + ca = (b + c)a$$
$$3x + 4x = (3 + 4)x = 7x$$

The distributive property of multiplication over subtraction is stated below.

Distributive Property: Multiplication over Subtraction

For all numbers **a, b,** and **c, a(b − c) = ab − ac**

You can use a reversed form of the distributive property of multiplication over subtraction to simplify an algebraic expression such as $7x - 5x$ as follows:

$$ba - ca = (b - c)a$$
$$7x - 5x = (7 - 5)x = 2x$$

Example 1

Solve and check. $5x + 2x = -161$

Solution

$5x + 2x = -161$

$7x = -161$ Because of the distributive property, $5x + 2x = (5 + 2)x = 7x$.

$\frac{7x}{7} = -\frac{161}{7}$ Divide each side by 7, or use the multiplicative inverse property and multiply by $\frac{1}{7}$.

$x = -23$

Check $5(-23) + 2(-23) \stackrel{?}{=} -161$ Substitute -23 for x in $5x + 2x = -161$.

$-115 + -46 \stackrel{?}{=} -161$

$-161 = -161$ ✓

Practice Solve and check.

a. $-6x + 19x = 962$ 74 **b.** $\frac{2}{5}y + \frac{4}{5}y = 36$ 30

Example 2

Solve and check. $7x - 5x = 86$

Solution $7x - 5x = 86$

$2x = 86$ Because of the distributive property, $7x - 5x = (7 - 5)x = 2x$.

$\frac{2x}{2} = \frac{86}{2}$

$x = 43$

Check $7(43) - 5(43) \stackrel{?}{=} 86$ Substitute 43 for x in $7x - 5x = 86$.

$301 - 215 \stackrel{?}{=} 86$

$86 = 86$ ✓

Practice Solve and check. **a.** $-6n - 4n = 65$ **-6.5** **b.** $72 = 13y - 5y$ **9**

Oral Exercises

Use the distributive property to simplify the equation.

1. $4x + 7x = 77$ **11x = 77** **2.** $9x + 6x = 30$ **15x = 30** **3.** $25x + 75x = 500$ **100x = 500**

4. $45 = 5x + 4x$ **45 = 9x** **5.** $60 = 21x - 6x$ **60 = 15x** **6.** $125 = 65x - 40x$ **125 = 25x**

Exercises

A Solve and check.

1. $5x + 2x = 84$ **12** **2.** $4a + 12a = -48$ **-3** **3.** $9c + 8c = 68$ **4**

4. $12b + 14b = 52$ **2** **5.** $6n + 9n = 75$ **5** **6.** $-3t + 9t = 144$ **24**

7. $24z - 8z = 64$ **4** **8.** $18r - 9r = -108$ **-12** **9.** $-9a - 4a = 78$ **-6**

10. $12y - 3y = 207$ **23** **11.** $25b - 8b = 51$ **3** **12.** $13c - 6c = 98$ **14**

13. $65 = 8p + 5p$ **5** **14.** $126 = -14s + 8s$ **-21** **15.** $26z + 37z = 315$ **5**

16. $21x - 9x = -96$ **-8** **17.** $34n - 26n = 136$ **17** **18.** $82t - 69t = 156$ **12**

19. $96 = 31d - 19d$ **8** **20.** $288 = -a + 13a$ **24** **21.** $735 = 63r - 28r$ **21**

22. $\frac{3}{4}y + \frac{1}{2}y = 24$ **$19\frac{1}{5}$** **23.** $4.6x + 3.4x = 128$ **16** **24.** $\frac{7}{8}n - \frac{3}{4}n = -6$ **-48**

B Solve and check.

25. $-5x + 7x + 9x = -88$ **-8** **26.** $16n - 9n + 12n = 76$ **4**

27. $6.3p + 8.6p - 5.4p = 28.5$ **3** **28.** $-9b + -9b + -9b = -108$ **4**

29. $\frac{3}{2}a + \frac{3}{4}a + \frac{3}{8}a = 105$ **40** **30.** $2\frac{1}{2}c + 5\frac{1}{4}c - 3\frac{1}{3}c = 106$ **24**

31. Evaluate both $(b - c)a$ and $ba - ca$ for $a = -3$, $b = 6$, and $c = 9$. Choose new values for a, b, and c and evaluate again. Did the expressions have the same value in both cases? **Yes**

Write equations to solve Exercises 32–34.

32. Alicia bought some \$6 records and an \$18 record-cleaning kit. The total cost was \$60. How many records did she buy? $6r + 18 = 60$ 7 records

33. Teri bought 8 tickets to the water slide and spent \$4 on refreshments. The total cost was \$10. What was the cost of each ticket? $8t + 4 = 10$ 75¢

34. A family traveled in a car for a number of hours at a speed of 55 mi/h. After stopping for gas, they went 76 miles farther. They traveled a total of 296 miles. How many hours did they travel before stopping for gas?
$55h + 76 = 296$ 4 hours

C Extending Thinking Skills

35. Bill wrote: "For all whole numbers a, b, and c, $a \cdot (b + c) = a + (b \cdot c)$." Check this for $a = 2$, $b = 3$, and $c = 4$, and then for $a = 3$, $b = 5$, and $c = 6$. If you can find one set of values for a, b, and c for which $a \cdot (b + c)$ *does not equal* $a + (b \cdot c)$, you will have found a *counterexample* proving Bill's generalization false! Can you do it? Yes. Answers will vary. One counterexample is $a = 3$, $b = 4$, $c = 5$.

36. Adam, Bob, Carl, and Dave are brothers. Adam is $\frac{1}{2}$ as old as Carl. Carl is 3 years older than Dave. Dave and Bob together are 17 years old. Bob is 8. Who is the youngest? Adam

Thinking skills: 35. Seeking a counterexample 36. Using logic

Mixed Review

Find the greatest common factor (GCF). **37.** 27, 45, 108 9

Evaluate for $m = 3.2$, $n = -0.8$. **38.** $3n - m$ -5.6 **39.** $4(m + n)$ 9.6

COMPUTER ACTIVITY

This program will solve equations of the form $Ax + B = C$. The values for A, B, and C can be positive or negative rational numbers and are entered as decimals.

Be sure students separate A, B, and C by commas.

```
10 PRINT "THIS PROGRAM WILL SOLVE EQUATIONS OF THE FORM
   AX+B=C"
20 INPUT "TYPE IN INTEGER OR DECIMAL VALUES FOR A, B,
   AND C.";A,B,C
30 PRINT "THE EQUATION IS ";A;"X+";B;"=";C
40 PRINT "THE SOLUTION IS X=";(C-B)/A
50 INPUT "DO YOU WANT TO SOLVE ANOTHER ONE?";A$
60 IF LEFT$(A$,1)="Y" THEN GOTO 20
70 END
```

1. Use the program to solve the equation $2x - 1 = 5$. 3

2. Use the program to solve the equation $\frac{1}{4}n + \left(-\frac{1}{2}\right) = \frac{3}{8}$. 3.5

3. Make up some equations and use the program to solve them.

1. This equation can be rewritten as $2x + (-1) = 5$ and the numbers 2, −1, and 5 can be entered for A, B, and C.
2. The decimals 0.25, −0.5, and 0.375 can be entered for A, B, and C.

8-5 More Simplifying to Solve Equations

Objective: To solve equations requiring simplification of expressions and combined operations.
MIN: 1–24, 33–34; REG, MAX: 1–24 even, 25–34; ALL: Mixed Review

You learned earlier that terms such as n and $5n$ are called *like terms* because they have the same variable part, n. The terms $-8x$, $0.4x$, and $\frac{2}{3}x$ are also like terms because their variable part, x, is the same. Terms such as $3a$ and $4b$ have different variables and thus are unlike terms. Terms such as $5y$, $5y^2$, and $5xy$ are also unlike terms, since their variable parts are not identical.

To solve some equations you must first simplify expressions by combining all like terms. In the following examples, the commutative, associative, and distributive properties are used to do this.

Example 1

Solve and check. $9x + 5(x + 7) = -49$

Solution

$9x + 5(x + 7) = -49$	Multiply by 5 first, so you will have like terms to combine.
$9x + 5x + 35 = -49$	Distributive property: $5(x + 7) = 5x + 5 \cdot 7$
$14x + 35 = -49$	Distributive property: $5x + 9x = (5 + 9)x = 14x$.
$14x + 35 + (-35) = -49 + (-35)$	Add -35 to each side.
$14x = -84$	
$\frac{14x}{14} = -\frac{84}{14}$	Divide each side by 14.
$x = -6$	

Check

$9(-6) + 5(-6 + 7) \stackrel{?}{=} -49$	Replace x with -6 in $9x + 5(x + 7) = -49$.
$-54 + 5(1) \stackrel{?}{=} -49$	
$-49 = -49$ ✓	The solution is -6.

Practice Solve and check.

a. $2b + 3(b - 7) = 44$ 13 **b.** $4(2y + 9) + 7y = -24$ −4

You can use the commutative and associative properties for addition together to add any three numbers in any order. You can use the commutative and associative properties of multiplication to multiply any three numbers in any order. In the following example, these ideas are used to simplify expressions.

Example 2

Solve and check. $3(5n) + 14 + 6n = 21$

Solution

$$3(5n) + 14 + 6n = 21$$

$$15n + 14 + 6n = 21$$ Associative property: $3(5n) = (3 \cdot 5)n = 15n$.

$$15n + 6n + 14 = 21$$ Commutative and associative properties: $15n + 14 + 6n = 15n + 6n + 14$.

$$21n + 14 = 21$$ Distributive property: $15n + 6n = (15 + 6)n = 21n$.

$$21n + 14 - 14 = 21 - 14$$

$$21n = 7$$

$$\frac{21n}{21} = \frac{7}{21}$$

$$n = \frac{1}{3}$$

Check

$$3\left(5 \cdot \frac{1}{3}\right) + 14 + 6\left(\frac{1}{3}\right) \stackrel{?}{=} 21$$ Replace n with $\frac{1}{3}$ in $3(5n) + 14 + 6n = 21$.

$$3\left(\frac{5}{3}\right) + 14 + 2 \stackrel{?}{=} 21$$

$$5 + 16 \stackrel{?}{=} 21$$

$$21 = 21\checkmark$$ The solution is $\frac{1}{3}$.

Practice Solve. **a.** $4z + 9 + 3(2z) = 129$ 12 **b.** $-2(7c) - 12 + 5c = 51$ −7

Oral Exercises

Simplify.

1. $3(y + 4)$ 3y + 12 **2.** $4(p - 8)$ 4p − 32 **3.** $2(b + 9) + 3$ 2b + 21

4. $5n + 9 + 3n$ 8n + 9 **5.** $7 + 8x - 2x$ 7 + 6x **6.** $13 + 6c + 12c$ 18c + 13

Exercises

A Solve and check.

1. $5(n + 3) + 5 = -25$ −9

2. $4(x - 3) + 8 = 60$ 16

3. $-8a + 6(a + 7) = 1$ $20\frac{1}{2}$

4. $6c + 4(c + 8) = 48$ 1.6

5. $10z + 5(z - 12) = 0$ 4

6. $7y + 7(y + 3) = -21$ −3

7. $5(4d) + 7 + (-8d) = 88$ $6\frac{3}{4}$

8. $-3(7a) + 17 + 6a = 82$ $-4\frac{1}{3}$

9. $4(3c) + 9 + 8c = 109$ 5

10. $9b + 6(4b) - 12 = 87$ 3

11. $12r - 8 + 5(3r) = 46$ 2

12. $24 + 3(-4s) + 6s = -24$ 8

13. $4t + 3t - 9 = 76$ $12\frac{1}{7}$

14. $-x + 9 + 7x = 99$ 15

15. $5y - 3y + (-4y) = 96$ −48

16. $6(n + 6) + 7n = -55$ −7

17. $5(z + 4) + 6z = 97$ 7

18. $-3(2c) + 4(5c) + 11 = 81$ 5

19. $-9 + 4d + 8d = 93$ $8\frac{1}{2}$

20. $-4(x - 8) + 20 = -13$ $16\frac{1}{4}$

21. $21 + 5(b + 7) = -54$ -22

22. $\frac{4}{5}p - \frac{1}{5}p - 42 = 8$ $83\frac{1}{3}$

23. $61 = \frac{1}{3}(6c) + 3c - 9$ 14

24. $24 = 18 + \frac{1}{2}(s + 6)$ 6

B Solve and check.

25. $3(x + 4) + 5(x - 2) = 66$ 8

26. $8(p - 3) + 3p + 7p = 138$ 9

27. $-3(4 + x) + 7(-3x) = 72$ $-3\frac{1}{2}$

28. $-4(3 + z) + 6z - 12 = -36$ -6

Use a calculator to solve the equations in Exercises 29–30.

29. $678(n + 39) + 457n = 77{,}517$ 45

30. $87(29x) + 43x + 57{,}650 = 1198$ -22

31. If Juana multiplies her age by 6, the result is 156. Mary is 3 times as old as Juana. Write an equation to find Juana's age. Use the solution to find Mary's age.
$6j = 156$ Juana is 26. Mary is 78.

32. In June, a salesman sold 80 cars. This was 8 more than 3 times the number he sold in May. Write an equation to find how many he sold in May. $80 = 3m + 8$ He sold 24.

C Extending Thinking Skills

33. The sum of what three consecutive page numbers in a book is 264? 87, 88, 89

34. An apartment manager used 339 metal digits to number apartments consecutively, beginning with 1. How many apartments did he number? 149

Thinking skills: 33. Testing a conjecture 34. Reasoning

Mixed Review

Write each as a decimal. 35. $\frac{3}{4}$ 0.75 36. $\frac{7}{8}$ 0.875 37. $\frac{3}{5}$ 0.6 38. $\frac{9}{10}$ 0.9

Give the prime factorization of each. 39. 15 $3 \cdot 5$ 40. 17 17 41. 28 $2 \cdot 2 \cdot 7$

NUMBERS TO ALGEBRA

The number examples below suggest an important generalization in algebra.

Numbers	Algebra
$-(9 - 5) = 5 - 9 = -4$ $-(2 - 7) = 7 - 2 = 5$ $-\left(\frac{3}{4} - \frac{1}{4}\right) = \frac{1}{4} - \frac{3}{4} = -\frac{1}{2}$	For all numbers a and b, $-(a - b) = b - a$

Use the generalization above to simplify.

1. $-(5 - x)$ $x - 5$
2. $-(3 - 2n)$ $2n - 3$
3. $-(7 - 5t)$ $5t - 7$
4. $-(4 - 3a) + 7$ $3a + 3$
5. $-(2b - 6) + 9$ $-2b + 15$
6. $-(12a - 4) + a$ $4 - 11a$

8-6 Solving Equations with Variables on Both Sides

Objective: To solve equations in which variables appear on both sides.
MIN: 1–24; REG: 1–24 even, 25–36; MAX: 1–24 even, 25–38; ALL: Mixed Review

You might solve an equation such as $5h = 12 + 3h$ to help you decide whether to rent a tool at \$5/h or at a base price of \$12 plus \$3/h.

You can solve an equation with like terms on both sides by using the properties of equality to get an equation in which the variable appears on only one side.

Example 1

Solve and check. $5h = 12 + 3h$

Solution $5h = 12 + 3h$

$5h - 3h = 12 + 3h - 3h$ Subtract $3h$ from each side so that all terms with a variable are on the same side of the equation.

$2h = 12$

$\frac{2h}{2} = \frac{12}{2}$

$h = 6$

Check $5(6) \stackrel{?}{=} 12 + 3(6)$ Replace h with 6 in $5h = 12 + 3h$.

$30 \stackrel{?}{=} 12 + 18$

$30 = 30 \checkmark$ The solution is 6.

Practice Solve and check.

a. $2x + 72 = 4x$ **36** **b.** $24 + y = 9y$ **3**

Discuss cases such as $45 + x = 6x$ where the variable might better be by itself on the right, rather than on the left.

Example 2

Solve and check. $6x - 2 = 4x + 3$

Solution $6x - 2 = 4x + 3$ You want to get the variables on one side, numbers on the other.

$6x - 2 + 2 = 4x + 3 + 2$ Add 2 to each side.

$6x = 4x + 5$

$6x - 4x = 4x - 4x + 5$ Subtract $4x$ from each side so all terms with a variable will be on the same side.

$2x = 5$

$\frac{2x}{2} = \frac{5}{2}$

$x = \frac{5}{2}$

Check $6\left(\frac{5}{2}\right) - 2 \stackrel{?}{=} 4\left(\frac{5}{2}\right) + 3$ Replace x with $\frac{5}{2}$ in $6x - 2 = 4x + 3$.

$15 - 2 \stackrel{?}{=} 10 + 3$

$13 = 13 \checkmark$

Practice Solve and check. **a.** $3y + 4 = 6y + 2$ $\frac{2}{3}$ **b.** $-7a + 8 = 3a - 2$ 1

Oral Exercises

What would you do to get the variable alone on one side of the equation?

1. $5x = 14 - 2x$ add $2x$
2. $8b = 18 + 2b$ subt. $2b$
3. $7n = 4n + 12$ subt. $4n$
4. $9z = 4z - 25$ subt. $4z$
5. $5r = 7r + 24$ subt. $7r$
6. $3p = 8p - 35$ subt. $8p$
7. $3c + 8 = 7c$ subt. $3c$
8. $4y - 12 = 8y$ subt. $4y$
9. $12n - 18 = 6n$ subt. $12n$

Exercises

A Solve and check.

1. $9x = 26 - 4x$ 2
2. $7n = 15 - 8n$ 1
3. $12a = 48 - 4a$ 3
4. $13b = 27 + 4b$ 3
5. $16c = 42 + 9c$ 6
6. $15y = 72 + 7y$ 9
7. $7r = -5r + 144$ 12
8. $7z = 3z - 52$ -13
9. $180 - 9s = 9s$ 10
10. $4t = 136 + 9t$ $-27\frac{1}{5}$
11. $5x = 30 - x$ 5
12. $54 + 11n = 2n$ -6
13. $121 + 4x = 15x$ 11
14. $36 + 9x = 3x$ -6
15. $84 + 6x = 18x$ 7
16. $-4a + 3 = 2a + 15$ -2
17. $8c - 2 = 5c + 4$ 2
18. $12x + 4 = 9x - 11$ -5
19. $9r + 7 = 4r - 8$ -3
20. $3s + 7 = -5s - 9$ -2
21. $15z - 9 = -3z + 9$ 1
22. $\frac{5}{8}x + 12 = \frac{3}{8}x + 4$ -32
23. $\frac{2}{5}n - 3 = \frac{9}{5}n - 5$ $1\frac{3}{7}$
24. $\frac{5}{6}a + 4 = \frac{2}{3}a + 6$ 12

B Solve and check.

25. $3(x + 4) = -5x - 30$ $-5\frac{1}{4}$ **26.** $5(b - 3) = 7b - 14$ $-\frac{1}{2}$ **27.** $5(r + 6) = -5(r + 3)$ $-4\frac{1}{2}$

28. $9(z - 6) = 4(z + 12)$ $20\frac{2}{5}$ **29.** $6(s + 1) = 4(s + 2)$ 1 **30.** $4(a - 2) = 2(a + 8)$ 12

31. $\frac{1}{3}(x + 6) = \frac{5}{6}x$ 4 **32.** $\frac{1}{2}(y - 16) = \frac{3}{4}y$ −32 **33.** $\frac{3}{8}(n + 24) = \frac{1}{4}n$ −72

Write an equation to solve each problem in Exercises 34–36.

34. Mrs. Greer rented a tool for a base charge of \$15 plus \$3 per hour. The total rental cost was \$33. For how long did she rent the tool? $3a + 15 = 33$ 6 hours

35. Jeff's checking account charges were \$3.00 per month plus \$0.15 per check. Jeff was charged \$7.50 for the month. How many checks did he write? $3 + 0.15x = 7.50$ 30 checks

36. Christina scored 35 points in each round of a game, plus 46 bonus points. Her total score was 256. How many rounds did she play? $35r + 46 = 256$ 6 rounds

C Extending Thinking Skills

37. If one of the daughters in the Biggs family had been a boy, the number of boys and girls would have been equal. If one of the sons had been a girl, there would have been twice as many girls as boys. How many children were in the family? 12

38. Each of three large blocks weighs the same. Each of five small blocks weighs the same. Each large block weighs 3 times as much as a small block. All together, the blocks weigh 112 kg. What does a large block weigh? 24 kg

Thinking skills: 37. Reasoning/Testing a conjecture 38. Reasoning

Mixed Review

Write <, >, or = for each □. **39.** $\frac{5}{9} \square \frac{31}{54}$ < **40.** $-\frac{3}{5} \square -\frac{41}{65}$ > **41.** $\frac{52}{81} \square \frac{2}{3}$ <

Solve and check. **42.** $c - 16 = 57$ 73 **43.** $\frac{3}{4}m = 24$ 32 **44.** $r + \frac{1}{2} = -\frac{2}{5}$ $-\frac{9}{10}$

Simplify. Write the result with exponents. **45.** $\frac{m^5}{m^2}$ m^3 **46.** $\frac{r^3}{r}$ r^2

MENTAL MATH

Use a calculator to check whether or not the following equations are true. Then look for a pattern and use it to find the products below mentally.

$29 \cdot 31 = (30 \cdot 30) - (1 \cdot 1)$ $28 \cdot 32 = (30 \cdot 30) - (2 \cdot 2)$
$39 \cdot 41 = (40 \cdot 40) - (1 \cdot 1)$ $38 \cdot 42 = (40 \cdot 40) - (2 \cdot 2)$
$49 \cdot 51 = (50 \cdot 50) - (1 \cdot 1)$ $48 \cdot 52 = (50 \cdot 50) - (2 \cdot 2)$

Find each product mentally.

1. $59 \cdot 61$ 3599 **2.** $69 \cdot 71$ 4899 **3.** $79 \cdot 81$ 6399 **4.** $58 \cdot 62$ 3596 **5.** $68 \cdot 72$ 4896 **6.** $78 \cdot 82$ 6396

PROBLEM SOLVING: APPLICATIONS

8-7 The Cost of Operating a Car

Objective: To solve word problems involving costs of car operation.
ALL: 1–11

What does it cost to operate a car? The chart below gives ways to estimate the yearly expenses for a car of average size.

Estimated Costs of Owning and Operating a Car

- For a person of average income, the annual cost of owning and operating a car, including financing, might be from 0.10 to 0.15 times his or her annual income.
- The cost of operating a car, including gas, oil, repairs, insurance, and taxes, might be about 17¢ per mile.
- The depreciation, or loss of value, of a car can be estimated for the first three years by subtracting $\frac{20}{100}$ of its original value if it is one year old, $\frac{35}{100}$ if it is two years old, and $\frac{50}{100}$ if it is three years old.

Problems

Solve, using the information above as needed.

1. Estimate the annual cost of operating a car, including financing, for a person earning $28,000 a year. **$2,800–$4,200**

2. Mr. Young's income was $35,000 last year. His car expenses were $4,823. By how much did his car expenses differ from 0.15 of his income? **$427**

3. Sue Brown bought a new car for $8,500. What will its value be after she has owned it for one year? **$6,800**

4. José Lopez drove his car 13,125 miles in one year. What would you expect his expenses, not including depreciation, to be? **$2231.25**

5. Marti Berk earns $32,000 a year. Her car expenses were $4,500. Are her expenses within the range given in the chart for her income? **yes**

6. Tami Yang bought a new sports car for $14,500. What will be the depreciated value of the car after she has driven it for 3 years? **$7,250**

7. Mrs. Tran bought a new car that cost $9,450. Make a table showing the depreciated value of the car after 1, 2, and 3 years. **See additional answers.**

8. Carol Glynn drove her new car 15,000 miles the first year. Estimate her operating expenses, not including depreciation, for that year. **$2,550**

9. Ken Harding calculated that the expense, not including depreciation, for operating his car one year was $3,500. Estimate how far he had driven the car. 20,000 miles

10. Hector Sanchez's new car cost $10,500. He drove it 15,500 miles the first year. His basic costs were: $97 for repairs, maintenance and accessories, $346 for gas and oil, $295 for insurance, and $328 for taxes. How much per mile, including depreciation, did it cost Hector to operate his car during the first year? $\frac{3166}{15500}$, or about 20.4¢

11. *Data Search* Find the current value of a new car you would like to buy. Make a chart showing the depreciated value of the car each year for a 3-year period. Then decide what it should cost you to buy the car used in 3 years. Answers will vary.

What's Your Decision?

Comparing a Standard and Diesel Automobile

- The diesel engine averages 1.5 times as many miles per gallon of fuel as the standard model.
- The diesel engine model costs $500 more than the standard model.
- Regular fuel costs $1.20 per gallon. Diesel fuel costs $1.25 per gallon.

You are trying to decide whether to buy the diesel or the standard model of a new car. You estimate that you will drive 12,000 miles per year and will keep the car for 4 years. The standard model gets an average of 30 miles to a gallon of gas. Which will you buy?
Answers may vary. If driven 41,000 miles, the diesel model would save enough on gas costs to offset the extra cost of the engine.

More Practice

Solve and check.

1. $3a - 5 = 19$ 8
2. $-16 = 3c + 2$ −6
3. $(m + 6) \div 12 = 5$ 54
4. $6n - 9n = 36$ −12
5. $3.6 = 10.5 - 3m$ 2.3
6. $(n + 6) \div 3 = 1.4$ −1.8
7. $(r + 4) \div 5 = 8$ 36
8. $2(a + 1.7) = 6$ 1.3
9. $1.43 = 0.4y - 0.6y$ −7.15
10. $16 - 2n = 10$ 3
11. $6(m - 4) = 24$ 8
12. $19.4 - 0.6c = -16.6$ 60
13. $-9 = 3y + 12$ −7
14. $0.4n = 4 + 0.9n$ −8
15. $(a + 5) \div 0.6 = 10$ 1
16. $3y + 6 = 2y$ −6
17. $4(z + 3) = 4$ −2
18. $(r + 9) \div 4 = 3.5$ 5
19. $2(y + 6) = 12$ 0
20. $4(1 - m) = 6$ −0.5
21. $11c + 2 = 6c + 12$ 2
22. $-15 - c = 4c$ −3
23. $14 + a = 3a + 2$ 6
24. $9m + 4m + 6 = 32$ 2
25. $6(2 - r) = 15$ −0.5
26. $m + 4 = -3m$ −1
27. $1.6n + 3.8n = 54$ 10
28. $-63 = 7(x - 9)$ 0
29. $(c + 3) \div 2 = 5$ 7
30. $2.5(z + 4) = 3.75$ −2.5
31. $5(10 - c) = 125$ −15
32. $14y + 3 = 13.5$ 0.75
33. $3(a + 2) = 4 + 2a$ −2
34. $144 = 12(c - 2)$ 14
35. $6c + 4c = -30$ −3
36. $4x + 2x = 10 + x$ 2
37. $37 = 14m + 93$ −4
38. $0.5(a + 2) = 2.5$ 3
39. $11r + 4 = 9r + 8$ 2
40. $-16 = 4c + 6$ −5.5
41. $13k = 19k + 12$ −2
42. $2(a + 6) + 3a = 32$ 4
43. $11z = 20 - 9z$ 1
44. $15r = 3(r + 28)$ 7
45. $25 = 4(m - 3) - 3$ 10
46. $9c + 6 = 15$ 1
47. $(x + 4) \div 6 = -2$ −16
48. $19h + 4h = 126.5$ 5.5
49. $21 + 4h = 11h$ 3
50. $-6 = 0.75z + 3$ −12
51. $1.2m - 6.5m = 21.2$ −4
52. $26 - 3c = 10c$ 2
53. $4t - 12 = t + 3$ 5
54. $9a - 16 = 5a + 4$ 5
55. $11t - 4 = 40$ 4
56. $6y = 9 + 9y$ −3
57. $(m - 2) \div 7 = 1.5$ 12.5
58. $46 = 9x - 4x$ 9.2
59. $8m = 4m - 24$ −6
60. $6r + 9 = 9r - 27$ 12
61. $8.6n - 5n = 9$ 2.5
62. $19t + 36 = 112$ 4
63. $5x + 6 = 12 + 8x$ −2
64. $(m - 16) \div 2 = 1$ 18
65. $11.5 + 2h = 8.5$ −1.5
66. $6a + 2(a - 4) = 8$ 2
67. $12m - 4 = 38$ 3.5
68. $7c + 11c = 162$ 9
69. $11t - 4.5t = 6.5$ 1
70. $180 = 8r + 7r$ 12
71. $19n - 13n = 4.5$ 0.75
72. $4a + 13 = 5a + 4$ 9
73. $16k + 8k = 312$ 13
74. $5m - 11m = 138$ −23
75. $9t - 2 = 4t + 18$ 4
76. $9 - 14m = -33$ 3
77. $59 - 8w = 3$ 7
78. $11y + 6 = 14y - 27$ 11
79. $12 + 4c = 2c$ −6
80. $(z + 2) \div 2 = -2$ −6
81. $(x - 20) \div 6 = 0.75$ 24.5

8-8 Solving Inequalities

Objective: To solve inequalities involving one operation.
MIN: 1–18; REG: 1–18 even, 19–25; MAX: 1–18 even, 19–27; ALL: Mixed Review
An exploratory lesson opener can be found on page 550.

A booster increases the 4 miles per second orbit speed of a satellite to at least 7 miles per second to send it into outer space. You can write the inequality $4 + b \geq 7$.

An **inequality** is a statement that uses the symbol $>$, $<$, $\geq$ (greater than or equal to), or $\leq$ (less than or equal to) to compare two expressions. You can solve inequalities involving addition or subtraction in the same way you solve equations. The following properties, and similar ones for inequalities involving $<$, are used.

Addition and Subtraction Properties of Inequalities

For all numbers a, b, and c,

if $a > b$, then $a + c > b + c$ and $a - c > b - c$

For all numbers a, b, and c, if $a < b$, then $a + c < b + c$ and $a - c < b - c$

Example 1

Solve. $x - 7 \geq -8$

Solution $x - 7 \geq -8$ Undo subtracting 7 by adding 7 to each side.

$x - 7 + 7 \geq -8 + 7$ Adding to both sides does not change the direction of the inequality sign.

$x \geq -1$

Practice Solve. **a.** $n + 4 \leq 5$ $n \leq 1$ **b.** $1 > x - \frac{1}{2}$ $x < 1\frac{1}{2}$

The display below shows the effects of multiplying or dividing each side of an inequality by a negative number.

$$3 < 4$$
$$-2 \cdot 3 \; ? \; -2 \cdot 4$$
$$-6 > -8$$

When you multiply or divide each side of an inequality by a **negative** number, the inequality sign is **reversed**.

$$6 < 8$$
$$\frac{6}{-2} \; ? \; \frac{8}{-2}$$
$$-3 > -4$$

The following properties and similar ones for inequalities involving < are used when solving inequalities by multiplying or dividing.

Multiplication and Division Properties of Inequalities

For all numbers **a, b,** and **c,** where **c** is positive,

if $a > b$, then $a \cdot c > b \cdot c$ and $\frac{a}{c} > \frac{b}{c}$

For all numbers **a, b,** and **c,** where **c** is negative,

if $a > b$, then $a \cdot c < b \cdot c$ and $\frac{a}{c} < \frac{b}{c}$

For all numbers *a*, *b*, and c, where c is positive, if $a < b$, then $a \cdot c < b \cdot c$ and $\frac{a}{c} < \frac{b}{c}$

For all numbers *a*, *b*, and c, where c is negative, if $a < b$, then $a \cdot c > b \cdot c$ and $\frac{a}{c} > \frac{b}{c}$

Example 2

Solve and check. $-4m > -2$

Solution $-4m > -2$

$-\frac{1}{4}(-4m) < -\frac{1}{4}(-2)$ When you multiply both sides by a negative rational number, you must change the direction of the inequality sign.

$1 \cdot m < \frac{2}{4}$

$m < \frac{1}{2}$

Check To check the computation, replace the inequality sign in $-4m > -2$ with an equal sign and see whether $m = \frac{1}{2}$ is a solution of the resulting equation.

$-4 \cdot 0 \overset{?}{>} -2$ To check whether the inequality sign is correct, see whether a solution to $m < \frac{1}{2}$, such as 0, is also a solution to $-4m > -2$.

$0 > -2$ ✓

Note that it would not be possible to check *every* solution to this type of inequality.

Practice Solve. **a.** $\frac{3}{4}b > 12$ $b > 16$ **b.** $15 < -3p$ $p < -5$

Oral Exercises

To solve, what operation would you use?

1. $x + 9 > 13$ subt. 9 **2.** $b - 4 < 8$ add 4 **3.** $-3n > 12$ div. −3 **4.** $15 < y + 7$ subt. 7

5. $\frac{c}{4} < 3$ mult. 4 **6.** $18 > z - 2$ add 2 **7.** $25 < 5p$ div. 5 **8.** $4 > \frac{n}{5}$ mult. 5

Exercises

A Solve and check.

1. $a - 5 < 1$ $a < 6$
2. $2 > b + 2$ $b < 0$
3. $c - 1 < 4$ $c < 5$
4. $-7z < 35$ $z > -5$
5. $\frac{1}{3}p > 7$ $p > 21$
6. $25 < 5n$ $n > 5$
7. $t + 9 < -15$ $t < -24$
8. $4 < x - 5$ $x > 9$
9. $y + 8 < 2\frac{1}{2}$ $y < -5\frac{1}{2}$
10. $d + 1 > 3$ $d > 2$
11. $\frac{n}{4} < -1$ $n < -4$
12. $\frac{c}{2} > 3$ $c > 6$
13. $s - 9 > 0$ $s > 9$
14. $-8c > 56$ $c < -7$
15. $12 < n + 5$ $n > 7$
16. $\frac{3}{8} < n - \frac{5}{6}$ $n > 1\frac{5}{24}$
17. $2m \leq \frac{5}{6}$ $m \leq \frac{5}{12}$
18. $2\frac{1}{3}b > -6$ $b > -2\frac{4}{7}$

B Solve and check.

Write an inequality for each sentence in Exercises 19–21.

19. Three is less than a number x. $3 < x$
20. A number y is greater than negative two. $y > -2$
21. A number is less than or equal to 5 halves. $x \leq \frac{5}{2}$
22. The minimum speed on a highway is 40 mi/h and the maximum speed is 55 mi/h. Write an inequality that describes the situation when a car is:

a. breaking the speed limit $s > 55$ **b.** going too slow $s < 40$

Solve the problems in Exercises 23–25 by writing and solving inequalities.

Example: Five more than an integer is less than 27. Find the largest integer that meets this condition.

Solution: $n + 5 < 27$ $n < 22$ The largest integer satisfying this inequality is 21.

23. Six less than an integer n is greater than 25. Find the least possible value for n. 32
24. Eight times an integer n is less than 48. Find the greatest possible value for n. 5
25. Louise said she would sell her house only if the sealed bid were more than \$4000 above her cost of \$92,000. Bids are in multiples of \$500. What is the smallest bid that Louise would accept? \$96,500

C Extending Thinking Skills

26. How can you use unmarked 8-liter and 5-liter containers to measure out exactly 7 liters? See additional answers.
27. Solve the inequality $|x - 3| > 5$ for the replacement set $\{-10, -9, \ldots 10\}$
$x = \{-10, -9, -8, -7, -6, -5, -4, -3, 9, 10\}$

Thinking skills: 26, 27. Reasoning

Mixed Review

Evaluate for $x = -0.5$, $y = 0.5$. **28.** $2x^2y$ 0.25 **29.** $6(x + y)$ 0

30. $1.6x + 3.2y$ 0.8 **31.** $y(1 - 2x)$ 1 **32.** $x(y - x)$ −0.5 **33.** $x \div (-y)$ 1

8-9 Solving Inequalities: Combined Operations

Objective: To solve inequalities involving more than one operation or requiring simplification.
MIN: 1–27; REG: 1–27 even, 28–37; MAX: 1–27 even, 28–39; ALL: Mixed Review

You can use the same steps to solve inequalities that involve more than one operation as you used for solving equations with combined operations. Observe the order in which operations are applied to the variable and undo the operations, in reverse order. As with equations, the goal is to get the variable by itself on one side.

Example 1

Solve. $-x + 4 < 6$

Solution $-x + 4 < 6$

$-x + 4 - 4 < 6 - 4$ Subtract 4 from each side. The inequality sign stays the same.

$-x < 2$

$(-1)(-x) > (-1)2$ Multiply each side by -1. Reverse the inequality sign.

$x > -2$

Practice Solve.

a. $\frac{n}{4} - 3 > 2$ $n > 20$ **b.** $12 \leq 2a + 3$ $a \geq 4\frac{1}{2}$

To begin to solve an inequality, you often need to simplify an expression by combining like terms. The basic properties help.

Example 2

Solve. $-4b + 6b \geq 5$

Solution $-4b + 6b \geq 5$

$2b \geq 5$ Distributive property: $-4b + 6b = 2b$.

$\frac{1}{2} \cdot 2b \geq \frac{1}{2} \cdot 5$ Multiply each side by $\frac{1}{2}$. The inequality sign stays the same.

$b \geq \frac{5}{2}$

Practice Solve. **a.** $8r - 5r < -12$ $r < -4$ **b.** $5(x + 2) > 25$ $x > 3$

To solve an inequality with variables on both sides, you can add or subtract a variable expression from each side to get the variable by itself, as with equations.

Example 3

Solve. $5n - 6 < 3n$

Solution $5n - 6 < 3n$

$5n - 6 + 6 < 3n + 6$ Add 6 to each side. The inequality sign stays the same.

$5n < 3n + 6$

$5n - 3n < 3n - 3n + 6$ Subtract $3n$ from each side. The inequality sign stays the same.

$2n < 6$

$\frac{2n}{2} < \frac{6}{2}$ Divide each side by 2. The inequality sign stays the same.

$n < 3$

Practice Solve.

a. $-2b + 5 > 3b$ $b < 1$ **b.** $5 + 6c \leq -8c + 3$ $c \leq -\frac{1}{7}$

Oral Exercises

Is the given number a solution to the inequality?

1. $2z + 3 < 12$; 4 Yes **2.** $\frac{n}{2} - 1 > 20$; 50 Yes **3.** $2b + 3b < 10$; 0 Yes

4. $-p + 2 > 0$; 3 No **5.** $3(s + 1) < 12$; 3 No **6.** $4y - 2y > 9$; 5 Yes

7. $6 + 2n < -1$; -4 Yes **8.** $3c - 2c \geq 0$; -5 No **9.** $5(r - 1) \leq 9$; 3 No

Exercises

A Solve.

1. $2a - 1 > 5$ $a > 3$ **2.** $3x + 2 < -4$ $x < -2$ **3.** $2s + 5 < 3$ $s < -1$

4. $n - 3 > 5$ $n > 8$ **5.** $4r + 2 > -22$ $r > -6$ **6.** $\frac{t}{2} - 8 > 14$ $t > 44$

7. $-5 + 3c > 31$ $c > 12$ **8.** $\frac{n}{3} - 2 < 4$ $n < 18$ **9.** $-5 < -3p + 34$ $p < 13$

10. $4y + 7y > -66$ $y > -6$ **11.** $9x + 3x < 96$ $x < 8$ **12.** $54 < -7c + 4c$ $c < -18$

13. $12n - 4n < 64$ $n < 8$ **14.** $4x + 6x < 100$ $x < 10$ **15.** $15 \leq 14a - 9a$ $a \geq 3$

16. $3(r + 2) > -5$ $r > -3\frac{2}{3}$ **17.** $2(n - 6) \geq 4$ $n \geq 8$ **18.** $12 < 5(c + 8)$ $c > -5\frac{3}{5}$

19. $7x + 3 < 3x$ $x < -\frac{3}{4}$ **20.** $-9b > 26 + 4b$ $b < -2$ **21.** $-24 + 6a > -2a$ $a > 3$

22. $-2s + 3 < 5s - 4$ $s > 1$ **23.** $9x - 2 \geq -12x + 1$ $x \geq \frac{1}{7}$ **24.** $4z - 3 < 10z - 5$ $z > \frac{1}{3}$

25. $\frac{2}{5}x - 3 > 2$ $x > 12\frac{1}{2}$ **26.** $4c + 7 + 3c < -35$ $c < -6$ **27.** $\frac{3}{8}n - 5 > \frac{7}{8}n$ $n < -10$

B Solve and check.

28. $-3x + \frac{1}{2} < \frac{3}{5}$ $x > -\frac{1}{30}$ **29.** $-\frac{x}{2} + \frac{3}{8} > -\frac{1}{8}$ $x < 1$ **30.** $-5y - \frac{5}{9} > -2\frac{1}{3}$ $y < \frac{16}{45}$

31. Write an inequality for "the difference of twice a number and 9 is less than 57."

32. Write an inequality for "12 more than half a number is less than or equal to 30."

33. Write an inequality for "the sum of two thirds of a number and six is less than forty." $\frac{2}{3}n + 6 < 40$ 31. $2n - 9 < 57$ 32. $\frac{1}{2}n + 12 \leq 30$

34. Five more than twice an integer is less than 51. Write and solve an inequality to find the largest integer that meets this condition. $5 + 2n < 51$ largest = 22

35. Six less than half an integer is greater than 15. Write and solve an inequality to find the smallest integer that meets this condition. $\frac{1}{2}n - 6 > 15$ smallest = 43

36. Badminton team A scored 5 less than half of team B's score. Team B scored less than 15 points. What is the greatest score team A could have? 2

37. Mrs. Taka feels that she should pay no more than $15,000 plus $2\frac{1}{2}$ times her yearly income for a house. What is the smallest yearly income she could have if she is to buy a $90,000 house? $2\frac{1}{2}y + 15{,}000 \geq 90{,}000$ $30,000

C *Extending Thinking Skills*

38. A sentence with "and" is true if and only if *both* parts of the sentence are true. Give four solutions to this inequality: $x < 10$ **and** $x > 5$ Answers will vary. Encourage students to give both integer and rational solutions.

39. A sentence with "or" is true if and only if *one or the other or both* parts are true. Give four solutions to this inequality: $x < -2$ **or** $x > 2$ Answers will vary.

Thinking skills: 38, 39. Using logic

Mixed Review

Find the least common multiple (LCM). **40.** 4, 10, 15 60 **41.** 9, 15 45

Simplify. **42.** $p^3 \cdot p^9 \cdot p$ p^{13} **43.** $m^2 \cdot n^3 \cdot m^4 \cdot n$ m^6n^4

ESTIMATION

You can substitute compatible numbers chosen from 0, $\frac{1}{2}$, and 1, to estimate certain sums. For example, $\frac{1}{12}$ is "close to zero," $\frac{5}{9}$ is "a little greater than $\frac{1}{2}$," $\frac{5}{11}$ is "a little less than $\frac{1}{2}$," $\frac{14}{15}$ is "a little less than 1," and $\frac{11}{10}$ is "a little more than 1."

Estimate whether each sum is more or less than the number given.

1. $\frac{5}{6} + \frac{7}{8}$; 2 less **2.** $\frac{3}{8} + \frac{1}{10} + \frac{4}{5}$; $1\frac{1}{2}$ less **3.** $\frac{7}{8} + \frac{8}{9} + \frac{11}{12}$; 3 less

4. $\frac{4}{9} + \frac{5}{8} + \frac{7}{12} + \frac{1}{2}$; 2 more **5.** $5\frac{3}{5} + 7\frac{1}{12} + \frac{11}{10}$; 13 more **6.** $\frac{362}{365} + 1\frac{181}{365}$; $2\frac{1}{2}$ less

PROBLEM SOLVING: STRATEGIES

8-10 Using Several Strategies

Objective: To solve nonroutine problems, using combinations of various strategies.
MIN: 1–2; REG: 1–4; MAX: 1–6

You have learned that more than one strategy can be used when solving a problem. In fact, using several strategies is often helpful. For example, **Simplify the Problem, Draw a Picture, Look for a Pattern,** and **Make a Table** are all used to solve the following problem.

Problem How many different squares are there on a patio made of 6 rows of 6 square tiles?

To solve this problem, you might first **simplify** it by starting with fewer squares and **draw** the following pictures.

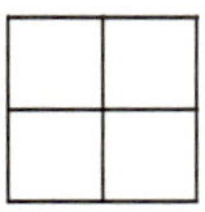

4 1 by 1 squares
1 2 by 2 squares

2 by 2 4 + 1 = 5 squares in all.

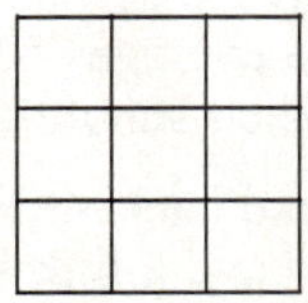

9 1 by 1 squares
4 2 by 2 squares
1 3 by 3 squares

3 by 3 9 + 4 + 1 = 14 squares in all.

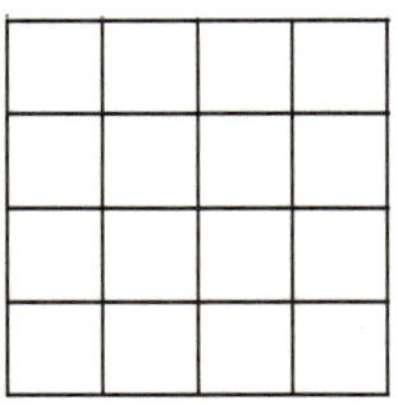

16 1 by 1 squares
9 2 by 2 squares
4 3 by 3 squares
1 4 by 4 squares

4 by 4 16 + 9 + 4 + 1 = 30 squares in all.

Then you could organize the information in a **table** and look for a **pattern**. The total number of squares can be found by squaring the side length and adding this to all the square numbers less than it.

Length of Side	Total Number of Squares
1	1
2	5 = 4 + 1
3	14 = 9 + 4 + 1
4	30 = 16 + 9 + 4 + 1

The total number of different squares on the 6-by-6 square patio is

$$36 + 25 + 16 + 9 + 4 + 1 = 91.$$

Problem-Solving Strategies	
Choose the Operations	**Write an Equation**
Guess, Check, Revise	**Simplify the Problem**
Draw a Picture	**Make an Organized List**
Make a Table	**Use Logical Reasoning**
Look for a Pattern	**Work Backwards**

This chart shows the strategies presented.

Problems

Solve.

1. How many different sizes of squares are there on an ordinary checkerboard of 64 squares? 8

2. What is the largest amount of money you could have in quarters, dimes, nickels, or pennies without being able to make change for $1? $1.19

3. Two people on motorcycles leave state parks that are 600 miles apart and travel toward each other. One cycle averages 55 mi/h. The other averages 45 mi/h. They both start at 11 a.m. When should they expect to meet? 5:00 p.m.

4. You have 5 sections of chain and each section has 3 links. The cost to have a link cut is 10¢. The cost to have a link welded is 20¢. How can you join the sections into one continuous chain for less than $1? See additional answers.

5. A prospector wanted to sell 8 pieces of gold. The buyer said he had weighed them and they were all the same weight. The prospector suspected that one of the pieces was heavier than the other 7. He found the heavier piece by using a balance scale and only two weighings. Describe how he did it. See additional answers.

6. A biologist started with 1 microbe in a special liquid. During each hour, the microbe population became 3 times as large as it was the previous hour. How many microbes were there when the biologist stopped the experiment after 6 hours? 729

Enrichment

Multiplying Binomials

An expression with one term, such as **$5x$**, is called a **monomial**. An expression with two terms, such as **$4x + 3$**, is called a **binomial**. An expression with three terms, such as **$x + xy - 3$**, is called a **trinomial**. A **polynomial** is a monomial or the sum of monomials, such as **$2n^2 + 3np + p + 5$**.

You have used the distributive property to multiply a monomial by a binomial when you simplified expressions such as $5(x + 2)$. You can also use the distributive property in two steps to multiply a binomial times a binomial to find a product such as $(x + 2)(y + 3)$.

To use the distributive property in the first step, replace a with $(x + 2)$, replace b with y, and replace c with 3. Then use the distributive property in the second step to simplify $(x + 2)y$ and $(x + 2)3$ to get the final product.

$$
\begin{aligned}
a\,(b + c) &= a\,b + a\,c \\
&\downarrow \\
(x + 2)(y + 3) &= (x + 2)y + (x + 2)3 \\
&= xy + 2y + x(3) + (2)(3) \\
&= xy + 2y + 3x + 6
\end{aligned}
$$

Notice that each term of the second binomial is multiplied by every term of the first binomial. That is, first y is multiplied by x and then by 2, then 3 is multiplied by x and then by 2. This shortcut is shown by the diagram below.

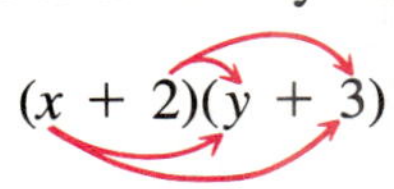

$$(x + 2)(y + 3) = xy + 2y + x(3) + (2 \cdot 3)$$

Students can also multiply by using the mnemonic FOIL (First, Outer, Inner, Last): First xy, Outer $x(3)$, Inner $2y$, Last $(2 \cdot 3)$.

Multiply.

1. $(x + 1)(y + 4)$
   $xy + 4x + y + 4$
2. $(p + 2)(q + 5)$
   $pq + 5p + 2q + 10$
3. $(b + 3)(c + 1)$
   $bc + b + 3c + 3$
4. $(y + 6)(z + 7)$
   $yz + 7y + 6z + 42$
5. $(x + 2)(x + 3)$
   $x^2 + 5x + 6$
6. $(t + 1)(t + 5)$
   $t^2 + 6t + 5$
7. $(s + 6)(s + 4)$
   $s^2 + 10s + 24$
8. $(n + 2)(n + 2)$
   $n^2 + 4n + 4$
9. $(a + 10)(a + 9)$
   $a^2 + 19a + 90$
10. $(t + 1)(t + 7)$
    $t^2 + 8t + 7$
11. $(a + 6)(b + 4)$
    $ab + 6b + 4a + 24$
12. $(n + 3)(m + 8)$
    $nm + 3m + 8n + 24$
13. $(s + 4)(s + 9)$
    $s^2 + 13s + 36$
14. $(y + 5)[y + (-3)]$
    $y^2 + 2y - 15$
15. $[s + (-4)](s + 4)$
    $s^2 - 16$

Chapter 8 Review

8-1 Solve.

1. $5h - 12 = 38$ 10

2. $24 = -c + 5$ −19

3. $\frac{f + 2}{7} = -3$ −23

4. $3(y + 1) = 15$ 4

8-2 Write an equation.

5. When Sally's age is multiplied by 5 and increased by 3, the result is Toshi's age, 33. $5s + 3 = 33$

6. Joaquin's weight, 100 kg, is 2 kg more than twice his sister Abbey's weight. How much does Abbey weigh? $2a + 2 = 100$

8-3 Solve by writing an equation.

7. Alice scored 22 goals this month. That is 2 fewer than 3 times the number of goals she scored last month. How many goals did she score last month? $3g - 2 = 22$ 8 goals

8. Gil had a carton of milk that was $\frac{3}{4}$ full. He gave Mr. Fletcher 16 oz, leaving 8 oz in the carton. How many ounces did the carton hold? $\frac{3}{4}x - 16 = 8$ 32 oz

8-4 Solve and check.

9. $72 = 4p + 8p$ 6

10. $8c + 7c = -120$ −8

11. $108 = 15r - 6r$ 12

12. $25t - 11t = 70$ 5

8-5 Solve and check.

13. $9c + 5(c + 4) = -92$ −8

14. $3(d - 1) + 4d = 67$ 10

15. $12n - 8 + 4(3n) = 88$ 4

16. $6(y - 8) + 3y = 6$ 6

8-6 Solve and check.

17. $5t + 12 = 7t - 4$ 8

18. $144 - 7t = 5t$ 12

19. $3y + 7 = 4y$ 7

20. $-6r + 1 = -20 + r$ 3

8-7 Solve.

21. A class picnic was attended by 4 less than $\frac{2}{3}$ of the total number of students in the class. Forty-six students came to the picnic. What is the total number of students in the class? 75

8-8 Solve and check.

22. $p + 10 < 17$ $p < 7$

23. $45 > -9p$ $p > -5$

8-9 Solve.

24. $15 < 3a + 3$ $a > 4$

25. $-6f + 8f < -20$ $f < -10$

Chapter 8 Test

Solve.

1. $5t - 1 = 14$ 3
2. $-28 = 3n - 1$ −9
3. $\frac{a + 5}{7} = 4$ 23
4. $3(m - 5) = -27$ −4

Write an equation.

5. The \$38 cost of a sweater is \$2 more than 3 times the cost of a tie. $38 = 3t + 2$
6. Brant's age, 37, is 3 less than twice Ricky's age. $2r - 3 = 37$

Solve by writing an equation.

7. Mike's height is 5 inches less than twice his little sister's height. Mike is 71 inches tall. What is his little sister's height? $2s - 5 = 71$ 38 inches
8. Cheryl averaged 28 points per game this year. This is 4 more than $\frac{3}{4}$ of last year's average. How many points per game did she average last year? $\frac{3}{4}n + 4 = 28$ 32 points

Solve and check.

9. $18c + 3c = 84$ 4
10. $-55 = 6y + 5y$ −5
11. $20x - 13x = -56$ −8
12. $135 = 22t - 7t$ 9

Solve and check.

13. $3(m + 4) + 6m = 66$ 6
14. $b + 4(b - 9) = 14$ 10
15. $7c - 8 + 3(2c) = 31$ 3
16. $5(x - 8) - 2x = -16$ 8

Solve and check.

17. $7x - 2 = 5x + 16$ 9
18. $14p = 20p - 18$ 3
19. $35 + 6n = 5n$ −35
20. $20r + 7 = 15r + 22$ 3

Solve.

21. Suzy is paid \$5 per hour for washing cars. Her total earnings one Saturday came to \$39. This included \$4 in tips. How many hours did she work that day? 7

Solve and check.

22. $4 > f - 6$ $f < 10$
23. $6r < -12$ $r < -2$

Solve.

24. $8b - 5b > 9$ $b > 3$
25. $\frac{n}{3} - 4 > 2$ $n > 18$

Cumulative Review

Write in expanded form. Simplify if possible.

1. 9^5 $9 \cdot 9 \cdot 9 \cdot 9 \cdot 9 = 59{,}049$ **2.** $(-6)^4$ $(-6)(-6)(-6)(-6) = 1296$ **3.** t^3 $t \cdot t \cdot t$

Write using exponents.

4. $9 \cdot 9 \cdot 9 \cdot 9$ 9^4 **5.** $ccccccc$ c^7 **6.** $-2(-2)(-2)$ $(-2)^3$

Multiply. Give the answer in exponent form.

7. $4^3 \cdot 4^2$ 4^5 **8.** $(-5)^7 \cdot (-5)^3$ $(-5)^{10}$ **9.** $v^3 \cdot v$ v^4

Evaluate each expression.

10. $3^3 + 3^1$ 30 **11.** $(-4)^2 + (-4)^2$ 32

Give the prime factorization.

12. 72 $2 \cdot 2 \cdot 2 \cdot 3 \cdot 3$ **13.** 25 $5 \cdot 5$

14. 110 $2 \cdot 5 \cdot 11$ **15.** 200 $2 \cdot 2 \cdot 2 \cdot 5 \cdot 5$

Find the Greatest Common Factor (GCF) of each pair of numbers.

16. 12, 20 4 **17.** 42, 48 6 **18.** 45, 120 15

Find the Least Common Multiple (LCM) of each pair of numbers.

19. 2, 15 30 **20.** 8, 12 24 **21.** 12, 30 60

Write each as a decimal. Use a bar for repeating decimals.

22. $\frac{2}{3}$ $0.\overline{6}$ **23.** $\frac{7}{8}$ 0.875 **24.** $\frac{5}{11}$ $0.\overline{45}$

Simplify. Write the expression with exponents.

25. $\frac{5^8}{5^3}$ 5^5 **26.** $\frac{(-7)^7}{(-7)^3}$ $(-7)^4$ **27.** $\frac{r^5}{r^8}$ $\frac{1}{r^3}$ or r^{-3}

Write the expression without exponents.

28. 2^{-3} $\frac{1}{8}$ **29.** $(-3)^{-2}$ $\frac{1}{9}$ **30.** $\frac{(-6)^5}{(-6)^7}$ $\frac{1}{36}$

Write in scientific notation.

31. 414 4.14×10^2 **32.** 33,000 3.3×10^4 **33.** 12,000,000 1.2×10^7

34. 0.05 5.0×10^{-2} **35.** 0.0000225 2.25×10^{-5} **36.** 0.0089 8.9×10^{-3}

Write in decimal form.

37. 2.78×10^4 27,800 **38.** 3.0×10^{-5} 0.00003 **39.** 7.08×10^{-3} 0.00708

Chapter 9
Ratio, Proportion, and Percent

9-1 Ratio

Objective: To write a ratio as a fraction in lowest terms.
MIN: 1–20; REG: 1–20 even, 21–35; MAX: 1–20 even, 21–37; ALL: Mixed Review

A comparison of one number to another is called a **ratio**.

The instrument ratio helps balance the sound of an orchestra.

Instrument Familes in an Orchestra			
Strings 67	Woodwinds 19	Brass 14	Percussion 11

The ratio of brass instruments to percussion instruments is fourteen to eleven. This ratio can be written as

$$\frac{14}{11}, \quad 14:11, \quad \text{or } 14 \text{ to } 11.$$

A ratio is in **lowest terms** when the only common factor of the two numbers being compared is 1.

Example

Write the ratio $\frac{18}{15}$ as a fraction in lowest terms.

Solution

$\frac{18}{15} = \frac{6}{5}$ Divide 18 and 15 by 3, the greatest common factor.

Practice Write each ratio as a fraction in lowest terms.

a. $9:27$ $\frac{1}{3}$ **b.** 35 to 28 $\frac{5}{4}$ **c.** $\frac{36}{24}$ $\frac{3}{2}$

Oral Exercises

Give the ratio. Do not express in lowest terms.

1. cellos to violins **10:34**
2. double basses to harps **9:2**
3. English horns to piccolos **1:1**
4. harps to violins **2:34**
5. violas to cellos **12:10**
6. flutes to harps **2:2**
7. piccolos to bassoons **1:5**
8. saxophones to clarinets **2:6**

Strings	Woodwinds
34 violins 10 cellos 12 violas 9 double basses 2 harps	2 oboes 1 English horn 2 flutes 1 piccolo 5 bassoons 6 clarinets 2 saxophones

Exercises

A Write each ratio as a fraction in lowest terms.

1. $\frac{24}{16}$ $\frac{3}{2}$
2. 8 to 14 $\frac{4}{7}$
3. 16:36 $\frac{4}{9}$
4. 26:10 $\frac{13}{5}$
5. 9 to 24 $\frac{3}{8}$
6. $\frac{3}{33}$ $\frac{1}{11}$
7. 12 to 7 $\frac{12}{7}$
8. 30:18 $\frac{5}{3}$
9. 36:24 $\frac{3}{2}$
10. 7 to 35 $\frac{1}{5}$
11. $\frac{16}{10}$ $\frac{8}{5}$
12. $\frac{8}{64}$ $\frac{1}{8}$
13. 28:7 $\frac{4}{1}$
14. 10 to 90 $\frac{1}{9}$
15. 64 to 36 $\frac{16}{9}$
16. 48:144 $\frac{1}{3}$
17. 10 to 36 $\frac{5}{18}$
18. 24:60 $\frac{2}{5}$
19. 18:36 $\frac{1}{2}$
20. 60 to 200 $\frac{3}{10}$

B Write each ratio as a fraction in lowest terms.

21. 2 harps to 10 cellos $\frac{1}{5}$
22. 6 clarinets : 2 saxophones $\frac{3}{1}$
23. 34 violins : 12 violas $\frac{17}{6}$
24. 9 double basses to 12 violas $\frac{3}{4}$

Write each ratio in two other ways. 28–30 see below.

25. x to y $\frac{x}{y}$, $x:y$
26. y to x $\frac{y}{x}$, $y:x$
27. $3a$ to b $3a:b$, $\frac{3a}{b}$
28. $(s + t)$ to p
29. $(m + n)$ to z
30. $(a + b)$ to $(a - b)$
31. A softball team won 12 games and lost 6. What is its ratio of wins to losses? 2:1
32. A 120-piece marching band has 7 treble drums and 3 bass drums. What is the ratio of these drums to all instruments in the band? 1:12
33. A rock band needed 250 hours of studio work to record an album with 45 minutes of music. What is the ratio of minutes of album music to minutes of work needed? 3:1000
34. A plane that held a total of 225 passengers had 12 passengers in first class and 113 passengers in "coach." What was the ratio of full seats to empty seats? 5:4
35. Six buses took 425 students to a state capital. Four adults were on each bus. What was the ratio of adults to students? 24:425

C Extending Thinking Skills

36. How many girls are in a class of 32 students if the ratio of girls to boys is 3:5? 12
37. A builder used 12 parts sand, 15 parts gravel, 6 parts cement, and 3 parts water in a concrete mix. You can write these relationships as 12:15:6:3. Give the ratio in lowest terms. 4:5:2:1

Thinking Skills: 36. Reasoning 37. Generalizing

Mixed Review

Solve and check. 38. $9y = 3y - 45$ -7.5 39. $2(a + 6) - 4(a + 1) = 4$ 2

40. $z + 45 < 36$ $z < -9$ 41. $18 > -3c$ $c > -6$ 42. $9 < t + 2$ $t > 7$

Write each as a decimal. 43. $3 \div 8$ 0.375 44. $1 \div 20$ 0.05 45. $3 \div 25$ 0.12

Evaluate. Reduce to lowest terms. 46. $\left(\frac{3}{4}\right)\left(\frac{2}{3}\right)$ $\frac{1}{2}$ 47. $\frac{4}{7} + \frac{1}{2}$ $1\frac{1}{14}$ 48. $\frac{3}{5} \div \frac{7}{15}$ $1\frac{2}{7}$

28. $\frac{s+t}{p}$, $s + t:p$ 29. $\frac{m+n}{z}$, $m + n:z$ 30. $a + b:a - b$, $\frac{a+b}{a-b}$

9-2 Proportion

Objective: To determine whether ratios are equal; to solve proportions for missing terms.
MIN: 1–32; REG: 1–32 even, 33–47; MAX: 1–32 even, 33–49; ALL: Mixed Review

The ratios $\frac{10}{18}$ and $\frac{15}{27}$ can both be written in lowest terms as $\frac{5}{9}$. You can write the equation $\frac{10}{18} = \frac{15}{27}$.

An equation stating that two ratios are equal is called a **proportion**. Notice that the cross products shown by the arrows below in the proportion $\frac{10}{18} = \frac{15}{27}$ are equal.

$10 \cdot 27 = 270$ $\qquad$ $18 \cdot 15 = 270$

$$\frac{10}{18} = \frac{15}{27}$$

Property of Proportions

Two ratios are equal if and only if the cross products are equal.

$\frac{a}{b} = \frac{c}{d}$ **if and only if** $ad = bc$ **(**$b \neq 0$ **and** $d \neq 0$**)**

Example 1

Write = or ≠ for the □. Use the property of proportions.

$\frac{28}{21} \square \frac{8}{6}$

Solution

$\frac{28}{21} \quad \square \quad \frac{8}{6}$

$28 \cdot 6 \stackrel{?}{=} 21 \cdot 8$ Find the cross products.

$168 = 168$

$\frac{28}{21} = \frac{8}{6}$ Since the cross products are equal, the ratios are equal.

Practice Write = or ≠ for the □. Use the property of proportions.

a. $\frac{8}{12} \square \frac{12}{15}$ ≠ $\qquad$ **b.** $\frac{8}{3} \square \frac{16}{6}$ =

When one of the numbers in a proportion is not known, you can use the property of proportions to write an equation.

Example 2

Solve and check. $\frac{x}{40} = \frac{3}{5}$

Solution $\frac{x}{40} = \frac{3}{5}$

$5 \cdot x = 40 \cdot 3$ Property of proportions.

$5x = 120$

$\frac{5x}{5} = \frac{120}{5}$ Divide each side by 5.

$x = 24$

Check $\frac{24}{40} \stackrel{?}{=} \frac{3}{5}$ Replace x with 24.

$5 \cdot 24 \stackrel{?}{=} 40 \cdot 3$ Property of proportions.

$120 = 120$ ✓ The solution is 24.

Practice Solve and check. **a.** $\frac{3}{4} = \frac{m}{20}$ 15 **b.** $\frac{2}{7} = \frac{18}{b}$ 63

Example 3

Use a proportion to solve. The ratio of adults to students on a train is 2 to 11. There are 12 adults on the train. How many students are on the train?

Solution

$\frac{2}{11} = \frac{12}{x}$ ← adults / ← students — To write the proportion, set up each ratio in the same way. Both ratios compare adults to students.

$2 \cdot x = 11 \cdot 12$ Set up an equation by finding the cross products. Then solve the equation.

$x = \frac{11 \cdot 12}{2}$

$x = 66$

There are 66 students on the train.

Practice Use a proportion to solve. Chemical C is made up of chemicals A and B in the ratio of 17 to 5. To make a batch of chemical C with 102 units of chemical A, how many units of chemical B are needed? 30

Oral Exercises

State the equation you would solve to find the missing number.

1. $\frac{1}{2} = \frac{x}{18}$ $1 \cdot 18 = 2x$ **2.** $\frac{n}{48} = \frac{3}{4}$ $3 \cdot 48 = 4n$ **3.** $\frac{4}{1} = \frac{24}{t}$ $1 \cdot 24 = 4t$ **4.** $\frac{3}{10} = \frac{y}{40}$ $3 \cdot 40 = 10y$

5. $\frac{x}{52} = \frac{5}{4}$ $52 \cdot 5 = 4x$ **6.** $\frac{16}{n} = \frac{1}{3}$ $16 \cdot 3 = 1n$ **7.** $\frac{4}{5} = \frac{m}{65}$ $4 \cdot 65 = 5m$ **8.** $\frac{6}{9} = \frac{4}{r}$ $9 \cdot 4 = 6r$

Exercises

A Write = or ≠ for each □. Use the property of proportions.

1. $\frac{4}{5} \square \frac{12}{15}$ =
2. $\frac{5}{2} \square \frac{35}{14}$ =
3. $\frac{4}{15} \square \frac{3}{7}$ ≠
4. $\frac{8}{6} \square \frac{28}{22}$ ≠
5. $\frac{9}{16} \square \frac{3}{4}$ ≠
6. $\frac{8}{15} \square \frac{20}{45}$ ≠
7. $\frac{3}{2} \square \frac{12}{8}$ =
8. $\frac{17}{34} \square \frac{1}{2}$ =
9. $\frac{28}{35} \square \frac{4}{5}$ =
10. $\frac{4}{7} \square \frac{32}{56}$ =
11. $\frac{9}{4} \square \frac{63}{28}$ =
12. $\frac{12}{30} \square \frac{10}{25}$ =

Solve and check.

13. $\frac{24}{x} = \frac{4}{3}$ **18**
14. $\frac{5}{3} = \frac{y}{42}$ **70**
15. $\frac{1}{2} = \frac{m}{18}$ **9**
16. $\frac{t}{14} = \frac{5}{2}$ **35**
17. $\frac{12}{27} = \frac{8}{m}$ **18**
18. $\frac{r}{27} = \frac{8}{18}$ **12**
19. $\frac{u}{7} = \frac{22}{14}$ **11**
20. $\frac{12}{30} = \frac{10}{n}$ **25**
21. $\frac{4}{5} = \frac{28}{x}$ **35**
22. $\frac{15}{y} = \frac{10}{8}$ **12**
23. $\frac{63}{144} = \frac{t}{16}$ **7**
24. $\frac{8}{15} = \frac{m}{105}$ **56**
25. $\frac{7}{x} = \frac{4}{9}$ **$15\frac{3}{4}$**
26. $\frac{y}{42} = \frac{15}{18}$ **35**
27. $\frac{100}{m} = \frac{90}{45}$ **50**
28. $\frac{18}{y} = \frac{126}{150}$ **$21\frac{3}{7}$**

Use a proportion to solve Exercises 29–32.

29. Four shovels of sand are used for every 5 shovels of gravel in making cement. How many shovels of sand are needed for 25 shovels of gravel? **20**

30. Bob's snack mix contains both peanuts and pecans in a ratio of 8 to 5. How many grams of pecans does he need if he uses 520 g of peanuts? **325**

31. The ratio of Ann's weight to Tina's weight is 5 : 6. Ann weighs 85 pounds. How much does Tina weigh? **102**

32. The ratio of expenses to income in the Johnsons' business is 5 to 8. What are their expenses for a month in which their income is $9,800? **$6,125**

B Estimate to find two equal ratios from each set. Then write a proportion and use cross products to see whether it is true.

33. $\frac{2}{3}; \frac{10}{18}; \frac{5}{9}$ **$\frac{10}{18} = \frac{5}{9}$**
34. $\frac{48}{60}; \frac{56}{75}; \frac{4}{5}$ **$\frac{48}{60} = \frac{4}{5}$**
35. $\frac{7}{18}; \frac{42}{126}; \frac{77}{198}$ **$\frac{7}{18} = \frac{77}{198}$**
36. $\frac{5}{12}; \frac{25}{60}; \frac{50}{144}$ **$\frac{5}{12} = \frac{25}{60}$**
37. $\frac{24}{125}; \frac{4}{25}; \frac{32}{200}$ **$\frac{4}{25} = \frac{32}{200}$**
38. $\frac{6}{15}; \frac{42}{90}; \frac{84}{180}$ **$\frac{42}{90} = \frac{84}{180}$**
39. $\frac{3}{4}; \frac{18}{24}; \frac{28}{35}$ **$\frac{3}{4} = \frac{18}{24}$**
40. $\frac{5}{8}; \frac{30}{45}; \frac{40}{64}$ **$\frac{5}{8} = \frac{40}{64}$**
41. $\frac{10}{12}; \frac{20}{25}; \frac{15}{18}$ **$\frac{10}{12} = \frac{15}{18}$**

Solve each proportion.

42. $\frac{24}{13} = \frac{x}{91}$ **168**
43. $\frac{32}{x} = \frac{128}{60}$ **15**
44. $\frac{15}{8} = \frac{105}{x}$ **56**

45. The ratio of full seats to the total number of seats on Mr. Chu's flight is 3 to 7. There are 28 empty seats. How many people are on the plane? **21**

46. The ratio of girls to boys in Maria's school is 3:4. The school has 228 boys. How many students does it have all together? **399**

47. Find two values for x and y so that $\frac{x}{12} = \frac{28}{y}$. **Answers will vary. $xy = 336$.**

C Extending Thinking Skills

48. Eric's school system has 9000 pupils and a teacher-pupil ratio of 1:30. How many more or fewer teachers are needed to reduce the teacher-pupil ratio to 1:25? **60 more**

49. It takes 12 minutes to cut a log into 4 pieces. How long will it take to cut a log into 6 pieces? $\frac{12}{3} = \frac{t}{5}$ $t = 20$ **20 minutes**

Thinking skills: 48. Reasoning 49. Using space visualization

Mixed Review

Reduce to lowest terms. **50.** $\frac{70}{98}$ $\frac{5}{7}$ **51.** $\frac{27}{144}$ $\frac{3}{16}$ **52.** $\frac{12}{22}$ $\frac{6}{11}$ **53.** $\frac{40}{64}$ $\frac{5}{8}$

Simplify. **54.** $\frac{n^7}{n^2}$ n^5 **55.** $\frac{x^4}{x^2}$ x^2 **56.** $\frac{(-3)^5}{(-3)^2}$ -27

Solve and check. **57.** $\frac{2}{3}n = \frac{1}{2}$ $\frac{3}{4}$ **58.** $c - \frac{4}{5} = \frac{1}{8}$ $\frac{37}{40}$ **59.** $\frac{1}{2}(t + \frac{1}{3}) = -\frac{1}{6}$ $-\frac{2}{3}$

Estimate each product or quotient. **60.** 385×22 **61.** 2.85×16.3
62. $71.2 \div 12.3$ **63.** $83.7 \div 8.5$ **64.** $608 \div 33$ **65.** $124.7 \div 11.1$

60–65. Possible answers are given. 60. 8,000 61. 48 62. 6 63. 9 64. 20 65. 11

NUMBERS TO ALGEBRA

Study the examples to see how proportions involving numbers are related to proportions in algebra.

Numbers	Algebra	
$\frac{8}{5} = \frac{16}{12 - 2}$	$\frac{5}{3} = \frac{10}{x - 1}$	
$8 \cdot (12 - 2) = 5 \cdot 16$	$5 \cdot (x - 1) = 3 \cdot 10$	Property of proportions.
$8 \cdot 12 - 8 \cdot 2 = 5 \cdot 16$	$5x - 5 = 30$	Distributive property.
$96 - 16 = 80$	$5x = 35$	
$80 = 80$	$x = 7$	

Solve each proportion. Check your solution.

1. $\frac{2}{3} = \frac{a + 1}{12}$ **7** **2.** $\frac{n + 1}{9} = \frac{2}{6}$ **2** **3.** $\frac{4}{3} = \frac{32}{x - 1}$ **25**

4. $\frac{1}{5} = \frac{x + 2}{25}$ **3** **5.** $\frac{x - 3}{6} = \frac{12}{9}$ **11** **6.** $\frac{5}{4} = \frac{y + 1}{8}$ **9**

9-3 Rate

Objective: To simplify rates, and to solve problems involving rates.
MIN: 1–17; REG: 1–17 even, 18–20; MAX: 1–17 even, 18–22; ALL: Mixed Review

A **rate** is a ratio that involves two different units. A rate is usually given as a quantity per unit, such as kilometers per hour (km/h). You can use division to simplify a rate.

Example 1

Simplify. 176 km/2 h

Solution $\frac{176 \text{ km}}{2 \text{ h}} = 88 \text{ km/h}$ 88 km/1 hr can be written as 88km/h.

Practice Simplify each rate. **a.** 87 km/10 L **8.7 km/L** **b.** 750 words/2 min **375 words/min**

Example 2

Solve. A typist types 150 words in two minutes. How many words can he type in 5 minutes?

Solution

$$\frac{x}{5} = \frac{150}{2}$$
$$2x = 150 \cdot 5$$
$$2x = 750$$
$$x = 375$$

He can type 375 words in 5 minutes.

Practice Solve.

How much should Fran get for 3 hours of work at $4.25/hour? **$12.75**

Oral Exercises

Describe a situation in which each rate might be used. **Answers will vary.**

1. km/L **2.** miles/hour **3.** km/hour **4.** revolutions/min
5. beats/min **6.** words/min **7.** dollars/hour **8.** cars/day

Exercises

A Simplify each rate.

1. $56/7 h **$8/h** **2.** 75 km/10 L **7.5 km/L** **3.** 1250 words/5 min **250 w/m**
4. 78 cm/6 s **13 cm/s** **5.** 3,750 km/4 h **937.5 km/h** **6.** 35 days/5 weeks **7 days/wk**

7. 105 people/35 cars 3 people/car

8. 143 players/11 teams 13 players/team

9. 2128 revolutions/32 min 66.5 rev/min

10. $220/40 h $5.50/h

11. 385 min/7 classes 55 min/class

12. 576 students/24 teachers 24 stud./teach.

Solve.

13. A motorist drove 1200 km on 45 L of gasoline. How many liters are needed to drive 500 km? 18.75

14. Apples are on sale at 4 for 60¢. How much will 18 apples cost? $2.70

15. What is the hourly rate of a mechanic who charges $160 for 4 hours of labor? $40

16. A typist can type 275 words in 5 minutes. How many can she type in 12 minutes? 660

17. What is the rate in beats per minute for a pulse rate of 25 beats per 15 seconds? 100 b/min

B

18. Dana works $5\frac{1}{2}$ hours per day at $4.50 per hour. What does she earn in 5 days? $123.75

19. A man bought a crate of apples for $10. The apples cost 90¢/lb and 4 apples equal about 1 pound. About how many apples should the crate contain? 44

20. How long would it take to lay 8 rows of 18 bricks each at a rate of 4 bricks per minute? 36 minutes

C Extending Thinking Skills

21. If $\frac{a}{b} = \frac{c}{d}$ then $\frac{a + b}{b} = \frac{c + d}{d}$. Show why this is true. See additional answers.

22. A recipe calls for two ounces of butter to every 5 tablespoons of flour. A cube of butter is 4 oz, and there are 4 cubes per pound. How many tablespoons of flour are needed for $4\frac{1}{2}$ pounds of butter? 180

Thinking skills: 21. Formulating a proof 22. Reasoning

Mixed Review

Solve and check. 23. $4(m + 2) = 14$ 1.5 24. $3.35 = 1 - t$ −2.35

25. $r + 0.027 = -0.901$ −0.928 26. $(x + 5) \div 6 = 102$ 607 27. $3r = -0.24$ −0.08

ESTIMATION

Use any estimation techniques you have learned to estimate the following rates.

1. 729 km on 25 L of gasoline. How many km per L is this? ≈ 30 km/L
2. 338 words in 5 minutes. What is the typing rate in words per minute? ≈ 70 words/min
3. $4.13 for 5 hours parking. What is the hourly parking rate? ≈ 80¢
4. $37.20 for 9 hours of work. What is the hourly wage? ≈ $4
5. 500 miles in $3\frac{1}{4}$ hours. How many miles per hour is this rate? ≈ 150 mi/h

9-4 Translating Problems Involving Two Expressions

Objective: To practice writing equations for solving word problems involving two expressions.
MIN: 1–9; REG: 1–10; MAX: 1–12; ALL: Mixed Review

Some word problems involve two unknowns. An expression for one of the unknowns can be given in terms of the other. To solve such problems, first decide which unknown the variable will represent, then express the other in terms of the variable.

Example 1

Write an expression for each unknown.

Beverly is 3 years older than Celia.

Celia's age? Beverly's age?

Solution

Let c = Celia's age — Since Beverly's age is given in relation to Celia's, make Celia's age the first unknown.

$c + 3$ = Beverly's age — Then express Beverly's age is terms of Celia's.

Practice Sue has 3 times as many books as Dawn. Write an expression for each unknown.

a. The number Dawn has? d
b. The number Sue has? $3d$

Example 2

Write an equation.

A large bus holds 24 students more than a small bus. The two buses hold 76 students altogether.

Solution

Let s = number a small bus holds — Since the number a large bus holds is given relative to the number a small bus holds, let s represent the number the small bus holds.

$s + 24$ = number a large bus holds — Then express the number the large bus holds in terms of the number the small bus holds.

$s + (s + 24) = 76$ — The two buses together hold 76 students.

Practice Write an equation.

Fred collected twice as many shells as Nick. Together they collected 24. $n + 2n = 24$

Oral Exercises

Let m = the number of math books. Give an algebraic expression for each in terms of m.

1. 7 fewer history books than math books $m - 7$
2. 8 times as many English books as math books. $8m$
3. $\frac{1}{2}$ as many science books as math books $\frac{m}{2}$
4. 12 fewer art books than math books $m - 12$

Exercises

A Write an expression for each unknown.

1. Carl had 7 more than Kim: the number Kim had? k the number Carl had? $k + 7$
2. 3 times as many cars as bikes: the number of bikes? b the number of cars? $3b$
3. $\frac{1}{2}$ as many boys as girls: the number of girls? g the number of boys? $\frac{1}{2}g$
4. Saturday seats cost double weekday seats: the cost of weekday seats? w the cost of Saturday seats? $2w$

Write an equation.

5. There are 8 more boys than girls in the 63-member band this year. $g + (g + 8) = 63$
6. One side of a record album lasts 98 seconds longer than the other. The total length of the album is 1936 seconds. $x + (x + 98) = 1936$
7. The second math quiz was worth half as many points as the first quiz. Both quizzes together were worth 110 points. $x + \frac{x}{2} = 110$
8. A tape cost \$2.10 more than an album. Together they cost \$13.60. $a + (a + 2.10) = 13.60$
9. John scored 25 more points in his second bowling game than his first. He had a total of 297 points for the two games. $p + (p + 25) = 297$

B Write an equation.

10. A theater has twice as many seats in each row of the middle section as in each row of the two side sections. It has a total of 96 seats per row made up of a middle row and two equal side rows. $x + x + 2x = 96$

C Extending Thinking Skills

Write a word problem that would be solved using the equation. Answers will vary.

11. $x + (x + 2) = 18$.
12. $x + 5x = 36$.

Thinking skills: 11–12. Formulating a problem

Mixed Review

Simplify. **13.** $2a + 3(a - 6) - 4a$ $a - 18$ **14.** $-m - 5(2m + 4) - 8m$ $-19m - 20$

Solve and check. **15.** $22 + 9z = 4$ -2 **16.** $2(x + 1) = 0$ -1

17. $3(14 - n) + 7 = 5n + 1$ 6 **18.** $4(c + 3) = 14c - 3$ 1.5 **19.** $9y + 16 = 3 - 4y$ -1

PROBLEM SOLVING: WRITING EQUATIONS

9-5 Solving Problems Involving Two Expressions

Objective: To use equations to solve word problems involving two expressions.
MIN: 1–5; REG: 1–5 even, 6–8; MAX: 1–5 even, 6–10; ALL: Mixed Review

You can write and solve an equation to find the answer to a word problem in which one unknown can be expressed in terms of another. Keep these questions in mind:

- Can you use a variable to represent an unknown number?
- Can you represent other conditions in terms of the variable?
- What is equal?
- Can you write and solve an equation?

Example

A train engine picked up $3\frac{1}{2}$ times as many cars at the second stop as it picked up at its first stop. Then it had a total of 81 cars. How many cars did it pick up at its second stop?

Solution

Let f = number picked up at the first stop.

$3\frac{1}{2}f$ = number picked up at the second stop — The number of cars picked up at the second stop is given relative to the number picked up at the first stop.

$f + 3\frac{1}{2}f = 81$

$4\frac{1}{2}f = 81$

$\frac{9}{2}f = 81$ — Multiply both sides by $\frac{2}{9}$ to get f alone.

$f = 18$ — The train picked up 18 cars at the first stop.

$3\frac{1}{2} \cdot 18 = 63$ — The problem asks for the number picked up at the second stop, so find the value of $3\frac{1}{2}f$ when $f = 18$.

The train picked up 63 cars at its second stop. — $63 < 81$. The answer is reasonable.

Practice Solve. Mrs. Kang has 4 fewer girls than boys in her art club this year. There are 28 students in the art club this year. How many are girls?
$b + (b - 4) = 28$ $b = 16$ $b - 4 = 12$ 12 are girls

Oral Exercises

Give an equation.

1. Sid paid \$4 more for jeans than for a shirt. The total cost was \$38. How much was each? **$s + (s + 4) = 38$**

2. The deep end of a swimming pool is $\frac{1}{4}$ the length of the shallow end. The total length of the pool is 90 m. How long is each section? $s + \frac{1}{4}s = 90$

3. Vicky used 15 times as much fertilizer as grass seed on her lawn. Altogether, she used 32 pounds of material. How many pounds of fertilizer did she use? $15x + x = 32$

Exercises

A Solve by writing an equation.

1. An airplane had 65 more occupied seats than empty seats. It had a total of 211 seats. How many seats were occupied? $x + (x + 65) = 211$ $x = 73$; 138 occupied

2. A box of popcorn cost twice as much as a cup of juice. Lyle bought 1 of each and paid $2.25. What was the price of popcorn? $j + 2j = 2.25$ $j = 75$; popcorn was $1.50

3. A class of 354 graduates has 64 fewer male graduates than female graduates. How many are male and how many are female? $f + (f - 64) = 354$ $f = 209$ 145 male 209 female

4. Of 410 students graduating, the number planning to get a job is $\frac{1}{4}$ as many as the number planning to go to college. How many students are not planning to go to college? $\frac{1}{4}c + c = 410$ $c = 328$; 82 are not.

5. A large bottle of fruit juice contains 4 times as much as the regular size. The total number of ounces in the two sizes is 40. How many ounces are in each size? $4r + r = 40$ $r = 8$; large 32, regular 8

B

6. Gilberto bought 2 shirts and a sweater and paid $48. The sweater cost twice as much as each shirt. How much was the sweater and how much was each shirt?
$2x + 2x = 48$ sweater = $24; shirt = $12

7. Jerry collected twice as many donations as Fred. Tom collected 12 more donations than Fred. Each donation was at least $2. Altogether, the three boys collected 48 donations. How many donations did each boy collect? $f + 2f + (f + 12) = 48$
Fred 9 Jerry 18 Tom 21

8. The weekend rate for a 5-passenger rental car is $\frac{1}{2}$ as much as the weekday rate. Mrs. Cooper rented a car for 3 weekdays and 1 weekend day and paid $171.50. How much more is the cost per day during the week than on the weekend? $3d + \frac{1}{2}d = 171.50$ $24.50 more

C Extending Thinking Skills

9. Write a word problem that could be solved using the equation $x + \frac{1}{2}x = 24$.

10. Write a word problem that could be solved using the equation $x + 2x + 3x = 126$.
Thinking skills: 9–10. Formulating a problem Answers will vary.

Mixed Review

Find the least common multiple (LCM). **11.** 5, 6, 8 120 **12.** 3, 6, 7 42

Evaluate for $n = \frac{2}{3}$. **13.** $\frac{1}{5}n + \frac{1}{2}$ $\frac{19}{30}$ **14.** $n - \frac{2}{7}$ $\frac{8}{21}$ **15.** $\frac{9}{10} - n$ $\frac{7}{30}$ **16.** $\frac{n}{4} + \frac{1}{6}$ $\frac{1}{3}$

More Practice

Write each ratio in two other ways.

1. 6 to x $\frac{6}{x}$, $6:x$
2. m to $5n$ $\frac{m}{5n}$, $m:5n$
3. $(r + 3)$ to t $\frac{r+3}{t}$, $r + 3:t$
4. $(r + s)$ to $(s + t)$ $\frac{r+s}{s+t}$, $(r + s):(s + t)$

Write = or ≠ for each □. Use the property of proportions.

5. $\frac{4}{3} \square \frac{9}{7}$ ≠
6. $\frac{7}{8} \square \frac{42}{48}$ =
7. $\frac{6}{14} \square \frac{33}{77}$ =
8. $\frac{12}{8} \square \frac{26}{18}$ ≠

Simplify each rate.

9. 28 days/7 months 4 days/mo
10. 135 mi/3 h 45 mi/h
11. 440 words/5 minutes 88 words/min

Solve.

12. Farmer Hoff raises chickens. He has 216 Rhode Island Reds, 144 Leghorns, and 180 Bantams. **a.** What is the ratio of Rhode Island Reds to Leghorns? **b.** Of Leghorns to Bantams? **c.** Of Bantams to all other chickens?
12. a. $\frac{r}{l} = \frac{216}{144} = \frac{3}{2}$ b. $\frac{l}{b} = \frac{144}{180} = \frac{4}{5}$ c. $\frac{b}{a} = \frac{180}{360} = \frac{1}{2}$

13. There are 32 students in Mrs. Chu's class. Four students are left-handed. What is the ratio of left-handed students to right-handed students? $\frac{4}{32-4} = \frac{4}{28} = \frac{1}{7}$

14. Acme Industries has 120 employees, including 15 supervisors. What is the ratio of supervisors to all other employees? $\frac{15}{120-15} = \frac{1}{7}$

15. The ratio of Anaya's age to Raphael's age is 7:12. Anaya is 14. How old is Raphael? $\frac{7}{12} = \frac{14}{r}$, 24

16. The ranger at a campground counted tents and camper vans and found that the ratio of tents to vans was 12:17. She counted 84 tents. How many vans did she find? $\frac{12}{17} = \frac{84}{v}$, 119

17. A truck driver drove 371 miles in 7 hours. How far did he drive in one hour? $\frac{371}{7} = 53$; 53 mi/h

18. Ernesto can assemble 3 circuit boards in 10 minutes. How many boards can he assemble in one hour? $\frac{c}{60} = \frac{3}{10}$; 18 boards

19. Tanisha worked 3 hours and was paid \$13.50. At that rate, how long will it take Tanisha to earn \$45.00? $\frac{\$13.50}{3} = \frac{\$45.00}{T}$, 10

20. Nate is 3 years older than Maury. Their combined age is 35. How old is Nate? How old is Maury? $m + (m + 3) = 35$; Nate is 19, Maury is 16

21. Steve sold 12 more tickets than Daniel, and Kyoki sold twice as many as Daniel. They sold 216 tickets in all. How many tickets did each person sell?
$D + (D + 12) + 2D = 216$ Daniel, 51 Steve, 63 Kyoki, 102

9-6 Scale Drawings

Objective: To use scale drawings to find actual dimensions; to reduce actual dimensions to a scale.
MIN: 1–13; REG: 1–13 even, 14–19; MAX: 1–13 even, 14–22; ALL: Mixed Review

In the **scale drawing** below, the dimensions of every object in the drawing are reduced by the same ratio or **scale**. The scale for this drawing is 5 cm to 2 m. This means that an object that has a length of 5 cm in the scale drawing has an actual length of 2 m.

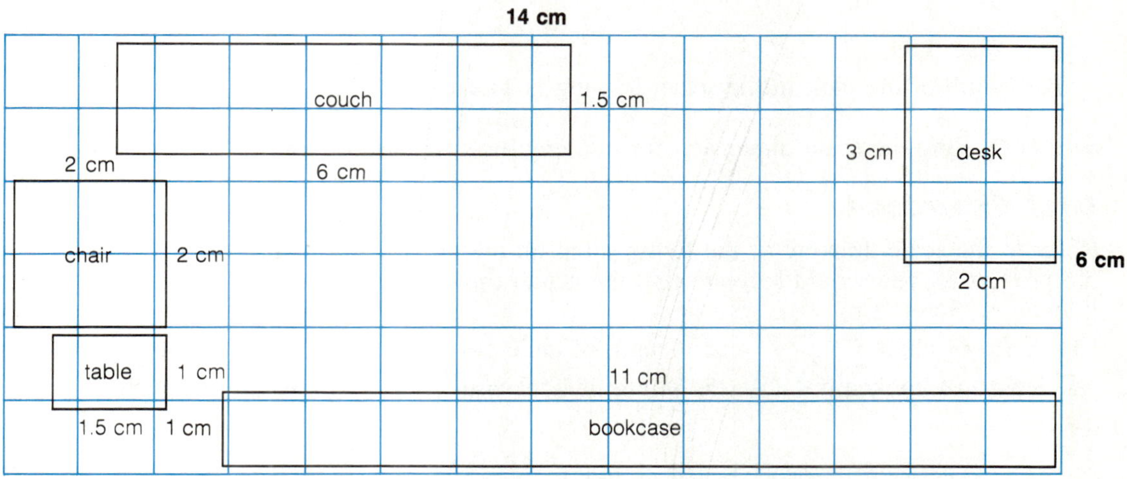

You can use the scale in the drawing to set up a proportion and find the actual dimensions of the objects shown in the picture.

Example 1

Find the actual length of the couch in the scale drawing above.

Solution Let L = actual length of couch

$\frac{5}{2} = \frac{6}{L} \begin{matrix} \leftarrow \text{cm} \\ \leftarrow \text{m} \end{matrix}$ The scale is 5 cm to 2 m. The length of the couch in the drawing is 6 cm, and the actual length of the couch is unknown.

$5 \cdot L = 2 \cdot 6$

$L = 2.4$

The actual length of the couch is 2.4 m.

Practice

Find the actual length and width of the desk in the drawing above. L = 1.2 m W = 0.8 m

You can find the scale dimensions of an object if you know its actual dimensions and the scale.

Example 2

The actual dimensions of a park are 825 m (width) and 1350 m (length). Find the width of this park on a scale drawing with a scale of 1 cm = 75 m.

Solution

Let W = scale drawing width of the park

$$\frac{W}{825} = \frac{1}{75} \begin{matrix} \leftarrow \text{cm} \\ \leftarrow \text{m} \end{matrix}$$ The width of the park in the scale drawing is unknown. The actual width is 825 m.

$75 \cdot W = 1 \cdot 825$

$W = 11$

The width of the park in the scale drawing is 11 cm.

Practice Find the scale dimension for the length of the park. **18 cm**

Oral Exercises

Refer to the scale drawing of the living room on the previous page and give the proportion you would solve to find the actual dimension. The scale for the drawing is 5 cm : 2 m.

1. width of chair $\frac{5}{2} = \frac{2}{x}$
2. length of table $\frac{5}{2} = \frac{1.5}{x}$
3. width of table $\frac{5}{2} = \frac{1}{x}$
4. length of bookcase $\frac{5}{2} = \frac{11}{x}$
5. width of bookcase $\frac{5}{2} = \frac{1}{x}$
6. width of couch $\frac{5}{2} = \frac{1.5}{x}$

Exercises

A Refer to the scale drawing to find the actual straight-line distance between each pair of towns. The scale is 1 cm : 150 km.

1. San Anselmo to Easthaven **600 km**
2. Arcata to Hadleigh **495 km**
3. Easthaven to Miwok **900 km**
4. Hadleigh to San Anselmo **600 km**
5. Arcata to Silver Lake **720 km**
6. Silver Lake to San Anselmo **450 km**
7. Miwok to Silver Lake **555 km**

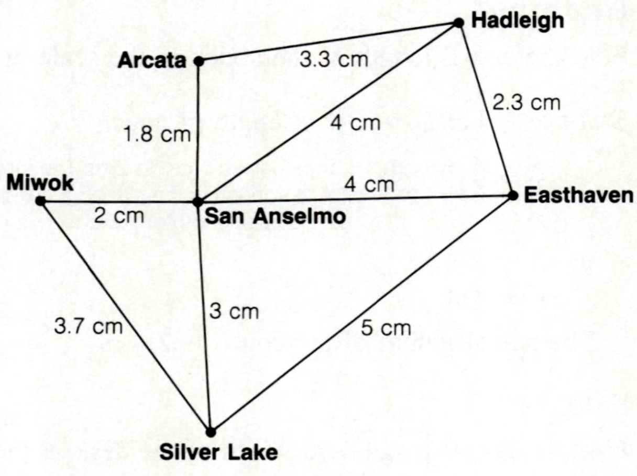

Find the scale dimensions for each if the scale is 5 mm : 2 m.

8. Dining Room: 5 m × 4 m

9. Bedroom: 4 m × 5 m 10 mm × 12.5 mm

10. Bench: 1.5 m × 1 m 3.75 mm × 2.5 mm

11. Table: 1 m × 3 m 2.5 mm × 7.5 mm

12. Dresser: 2 m × 0.5 m 5 mm × 1.25 mm

13. Kitchen table: 1.4 m × 0.66 m

8. 12.5 mm × 10 mm 13. 3.5 mm × 1.65 mm

B The scale for this drawing of a patio is 1 cm : 1.25 m. Measure the dimensions of the patio to the nearest millimeter. Then use a proportion to calculate the actual dimensions.

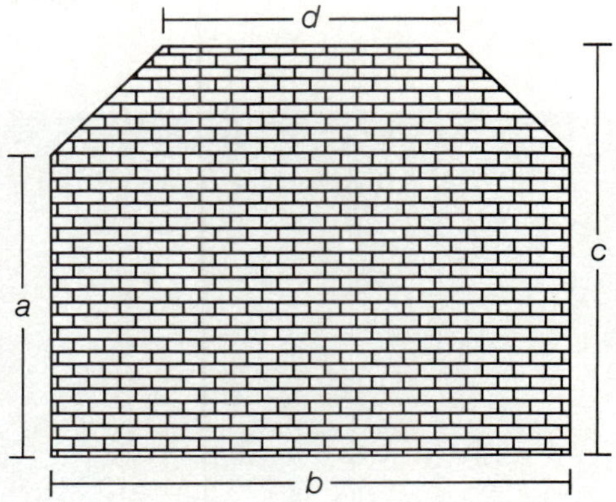

14. dimension a 4 m

15. dimension c 5.375 m

16. dimension b 6.625 m

17. dimension d 3.75 m

Note that students will need to use a metric ruler to complete Exercises 14–17.

18. The distance from A to B on the map is 4.2 cm. The distance from B to C is 1.4 cm. B is on a straight road from A to C. The scale on the map is 1 cm : 125 km. What is the actual distance from A to C? 700 km

19. On a map with scale 2 cm : 25 km, what would be the dimensions of a 145 km by 80 km rectangle? 11.6 cm × 6.4 cm

C Extending Thinking Skills

20. An insect in a picture is $\frac{1}{2}$ inch long and a label says "enlarged 12 times." What is the insect's actual length? $\frac{1}{24}$ inch

21. Charles wants to make a scale drawing to represent a driveway 64 ft long. His paper is $8\frac{1}{2}$ inches square. Which of the following scales could he use? c or d

a. 1 in. = 1 ft **b.** $\frac{1}{4}$ in. = 1 ft **c.** $\frac{1}{8}$ in. = 1 ft **d.** $\frac{1}{16}$ in. = 1 ft

22. The length of a rectangle is 5 m more than the width. The ratio of the length to the width is 5 to 4. What are the dimensions of the rectangle? (Let x = the width of the rectangle.) 25 × 20

Thinking Skills: 20–22. Using space visualization

Mixed Review

Simplify. **23.** $14m + 7m$ $21m$ **24.** $3c + 9 + (-6c) - 12$ $-3c - 3$ **25.** $t + 3t + 8$ $4t + 8$

Evaluate for $a = \frac{1}{2}$, $b = \frac{2}{3}$, $c = \frac{1}{5}$. **26.** $ab + c$ $\frac{8}{15}$ **27.** $a(2c - b)$ $-\frac{2}{15}$

28. $3a - 2b$ $\frac{1}{6}$ **29.** $\frac{ac}{b}$ $\frac{3}{20}$

Simplify. **30.** $\frac{x^5}{x^4}$ x **31.** $\frac{5^4}{5^2}$ 25 **32.** $\frac{x^6}{x^4}$ x^2 **33.** $\frac{7^5}{7}$ 2,401

Write each as an improper fraction. **34.** $9\frac{2}{3}$ $\frac{29}{3}$ **35.** $-6\frac{3}{5}$ $-\frac{33}{5}$

9-7 Determining the Best Buy

Objective: To compare unit prices and determine best buys.
ALL: 1–14

Grocery stores often show unit prices.

The cost of one unit of an item is called the **unit price**. You can find and compare unit prices to determine the better buy. The item with the lower unit price is usually considered the better buy.

A & B Discount	
100 vitamins	2/$1.99
video cassettes	3/$17.50
earrings	2 pair for $5
hand lotion	6 oz/$4.99
cologne	2 oz/$2.99
shampoo	15 oz/$2.99
glass cleaner	32 oz/$1.19

Bob's Super Saver	
100 vitamins	4/$3.75
video cassettes	2/$11
earrings	4 pair for $11
hand lotion	10 oz/$8.10
cologne	3 oz/$3.59
shampoo	24 oz/$5.28
glass cleaner	24 oz/$0.96

Example

Find the unit prices and determine which price advertised above is the better buy for a video cassette.

Solution $\frac{17.50}{3} = 5.8333\ldots \approx 5.84$ Divide the total price by the number of cassettes to find the unit price. Prices are rounded to the next highest cent.

$\frac{11.00}{2} = 5.50$ **Discuss the fact that money is usually rounded to the next highest cent, not the nearest cent.**

The price 2 for $11 is better than the price 3 for $17.99.

Problems

You might wish to have students choose from among the techniques pencil and paper, mental math, estimation, and calculator. Have them choose the most appropriate technique for each problem.

Use data from the advertisements on page 282 to find the unit prices and tell which store has the better buy for each. [Unit prices given here for A & B, then Bob's.]

1. hand lotion 0.84, 0.81; Bob's **2.** shampoo 0.20, 0.22; A & B **3.** cologne 1.50, 1.20; Bob's

4. glass cleaner 0.037, 0.04 A & B **5.** earrings 2.50, 2.75; A & B **6.** vitamins 1.00, 0.94; Bob's

Solve.

7. Jones' store has a price of 3 for $50 on video games. Smith's store is selling 8 video games for $134.50. Which store has the better unit price for video games? Jones, $16.67 each. (Smith $16.82)

8. Soup is priced at 2 cans for $0.75 at one store. At another store, six cans of the same soup cost $1.98. Find the unit prices to determine the better buy. The second, 33¢ to 37.5¢

9. At A & B Discount, apple juice costs $1.26 for the 1.4 L size. At Bob's Super Saver it costs $1.53 for the 1.8 L size. Which store has the better buy on apple juice? Bob's, 0.85 to 0.90 per liter

10. Ground beef is on sale for $1.25/lb. How much could you get for $5? 4 lb

11. Shampoo at A & B Discount is usually $2.99 for 15 oz. Today, the 10 oz. size is on sale for $2.10. Estimate the cost per ounce. About how much more or less per ounce would the smaller size cost? About 1¢ more

12. Squash costs $1.85/kg. Estimate the cost of a squash weighing 0.54 kg. $1

Use the advertisement below to solve.

13. Mrs. Hamilton bought 2 loaves of wheat bread, 3 loaves of rye bread, 12 muffins, and 15 biscuits on sale. How much did she save off the regular price? $1.70

Item	Sale Price	Regular Price
Wheat Bread	2 for $1.25	$0.75 a loaf
Rye Bread	2 for $1.40	$0.90 a loaf
Muffins	6 for $0.75	$0.15 each
Biscuits	6 for $0.54	$0.12 each

14. ***Data Search*** Choose two grocery stores. Find common brands for: $\frac{1}{2}$ gallon whole milk; 20 lbs dog food; 3 cans frozen orange juice; 2 loaves whole wheat bread; 1 whole chicken. Which store has the better buy for each? Answers will vary.

What's Your Decision?

You have $10 to spend on items in the advertisements on page 282. What would you buy and from which store? Answers will vary.

9-8 Percent

Objective: To write a percent as a fraction in lowest terms; to use a proportion to write a fraction as a percent.
MIN: 1–40; REG: 1–40 even, 41–44; MAX: 1–40 even, 41–47; ALL: Mixed Review

An exploratory lesson opener can be found on page 551.

The ratio of a number to 100 is called a **percent**. The word percent means *per one hundred*, and is represented by the symbol %. You can write a percent as a fraction.

Example 1

Write 68% as a fraction in lowest terms.

Solution

$68\% = \frac{68}{100}$ Write the number in front of the % sign over 100.

$= \frac{17}{25}$

Practice Write each as a fraction in lowest terms.

a. 55% $\frac{11}{20}$ **b.** 4% $\frac{1}{25}$ **c.** 21% $\frac{21}{100}$

Example 2

Write $12\frac{1}{2}\%$ as a fraction in lowest terms.

Solution

$12\frac{1}{2}\% = \frac{12\frac{1}{2}}{100}$ Write the number in front of the % sign over 100.

$= 12\frac{1}{2} \div 100$ $\frac{a}{b}$ is the same as $a \div b$.

$= \frac{25}{2} \div 100$

$= \frac{25}{2} \times \frac{1}{100}$

$= \frac{25}{200}$

$= \frac{1}{8}$ $12\frac{1}{2}\% = \frac{1}{8}$.

Practice Write each as a fraction in lowest terms.

a. $5\frac{1}{2}\%$ $\frac{11}{200}$ **b.** $2\frac{3}{4}\%$ $\frac{11}{400}$ **c.** $6\frac{2}{3}\%$ $\frac{1}{15}$

You can use a proportion to change a fraction to percent.

Example 3

Use a proportion to write $\frac{3}{8}$ as a percent.

Solution

$\frac{3}{8} = \frac{x}{100}$ — 3 is to 8 as what is to 100? Change $\frac{3}{8}$ to a fraction with a denominator of 100.

$100 \cdot 3 = 8 \cdot x$ — Property of proportions. **Visualizing $\frac{\text{part}}{\text{whole}} = \frac{\text{percent}}{100}$ may help students remember how to set up proportions.**

$300 = 8x$

$\frac{300}{8} = \frac{8x}{8}$

$37\frac{1}{2} = x$

$\frac{3}{8}$ is equal to $37\frac{1}{2}\%$ — The numerator over 100 is the percent. Write the numerator in front of the % sign.

Practice Use a proportion to write each as a percent. **a.** $\frac{24}{25}$ **96%** **b.** $\frac{5}{6}$ **$83\frac{1}{3}\%$**

Oral Exercises

Give each as a percent.

1. $\frac{37}{100}$ **37%** **2.** 88 : 100 **88%** **3.** 53 to 100 **53%** **4.** 40 out of 100 **40%**

5. 1 : 100 **1%** **6.** $\frac{25}{100}$ **25%** **7.** $\frac{65}{100}$ **65%** **8.** 36 out of 100 **36%**

9. $\frac{1}{100}$ **1%** **10.** 99 : 100 **99%** **11.** $\frac{100}{100}$ **100%** **12.** $\frac{1\frac{3}{4}}{100}$ **$1\frac{3}{4}\%$**

Exercises

A Write each as a fraction in lowest terms.

1. 45% $\frac{9}{20}$ **2.** 70% $\frac{7}{10}$ **3.** 5% $\frac{1}{20}$ **4.** 75% $\frac{3}{4}$

5. 20% $\frac{1}{5}$ **6.** 50% $\frac{1}{2}$ **7.** 10% $\frac{1}{10}$ **8.** 35% $\frac{7}{20}$

9. 74% $\frac{37}{50}$ **10.** 80% $\frac{4}{5}$ **11.** $1\frac{1}{2}\%$ $\frac{3}{200}$ **12.** 100% $\frac{1}{1}$

13. 8% $\frac{2}{25}$ **14.** $10\frac{1}{4}\%$ $\frac{41}{400}$ **15.** $5\frac{3}{4}\%$ $\frac{23}{400}$ **16.** $12\frac{1}{2}\%$ $\frac{1}{8}$

17. 63% $\frac{63}{100}$ **18.** $25\frac{1}{5}\%$ $\frac{63}{250}$ **19.** $3\frac{1}{3}\%$ $\frac{1}{30}$ **20.** 48% $\frac{12}{25}$

21. $9\frac{3}{4}\%$ $\frac{39}{400}$ **22.** 91% $\frac{91}{100}$ **23.** $14\frac{9}{10}\%$ $\frac{149}{1000}$ **24.** $1\frac{1}{10}\%$ $\frac{11}{1000}$

Use a proportion to write each as a percent.

25. $\frac{3}{4}$ **75%** **26.** $\frac{1}{2}$ **50%** **27.** $\frac{1}{6}$ **$16\frac{2}{3}\%$** **28.** $\frac{4}{40}$ **10%**

29. $\frac{1}{3}$ $33\frac{1}{3}\%$

30. $\frac{50}{50}$ 100%

31. $\frac{4}{5}$ 80%

32. $\frac{3}{10}$ 30%

33. $\frac{3}{25}$ 12%

34. $\frac{11}{20}$ 55%

35. $\frac{5}{8}$ $62\frac{1}{2}\%$

36. $\frac{1}{20}$ 5%

37. $\frac{5}{6}$ $83\frac{1}{3}\%$

38. $\frac{18}{25}$ 72%

39. $\frac{3}{16}$ $18\frac{3}{4}\%$

40. $\frac{7}{8}$ $87\frac{1}{2}\%$

B Solve. **Note that Chapter 10 emphasizes percent application problems.**

41. If the ratio $2x:100$ is equivalent to 68%, what is x? 34

42. In a survey of 500 mothers with children under age 18, about 300 worked outside the home. What percentage of those surveyed was this? about 60%

43. The Better Business Bureau in Greenville had 85 buyer complaints one year. Twenty of these involved automobile businesses. What percentage of the complaints involved automobile businesses? $23\frac{9}{17}\%$

44. What percentage of the whole numbers from 1 to 100 are prime numbers? What percentage are composite numbers? 25% prime, 74% composite

C Extending Thinking Skills

Use estimation to decide which test each student took if the math test had 30 questions and the English test had 50 questions.

45. Wanda had 24 correct and scored 80%. math

46. Danny missed 10 and scored 80%. English

47. James scored 90%. If he had answered 5 more questions correctly, he would have scored 100%. English

Thinking skills: 45–47. Estimating

Mixed Review

Solve and check. **48.** $3m + 13 = 1 + m$ −6 **49.** $4(3 - t) = 3(1 - t)$ 9

50. $-32 = 16a$ −2 **51.** $6(c + 4.5) = 36.6$ 1.6 **52.** $14k - 21 = 0$ 1.5

ESTIMATION

Study the percents to which these commonly-used fractions correspond.

$\frac{1}{5} = 20\%$ $\frac{1}{4} = 25\%$ $\frac{1}{3} = 33\frac{1}{3}\%$

$\frac{1}{2} = 50\%$ $\frac{2}{3} = 66\frac{2}{3}\%$ $\frac{3}{4} = 75\%$

Substitute compatible numbers to estimate the equivalent percent for each.

1. $\frac{24}{49}$ $\approx 50\%$

2. $\frac{29}{41}$ $\approx 75\%$

3. $\frac{9}{28}$ $\approx 33\frac{1}{3}\%$

4. $\frac{5}{16}$ $\approx 33\frac{1}{3}\%$

5. $\frac{13}{18}$ $\approx 66\frac{2}{3}\%$

6. $\frac{5}{24}$ $\approx 20\%$

9-9 Percents Greater Than 100 and Less Than 1

Objective: To write fractions as percents greater than 100 and less than 1; to write such percents as fractions.
MIN: 1–40; REG: 1–40 even, 41–45; MAX: 1–40 even, 41–47; ALL: Mixed Review

An exploratory lesson opener can be found on page 551.

A sales clerk had to sell 5 computers to meet his sales goal. He actually sold 6 computers. You could use a proportion to find what percentage of the goal he made.

You have worked with percents between 1 and 100. Percents can also be greater than 100 or less than 1.

Example 1

Use a proportion to write $\frac{6}{5}$ as a percent.

Solution

$\frac{6}{5} = \frac{x}{100}$ Change $\frac{6}{5}$ to a fraction with a denominator of 100.

$100 \cdot 6 = 5 \cdot x$

$600 = 5x$

$\frac{600}{5} = x$

$120 = x$

$\frac{6}{5}$ is equal to 120% Notice that the fraction $\frac{6}{5}$ is greater than 1, and the percentage is greater than 100%.

Practice Use a proportion to write each fraction as a percent.

a. $\frac{5}{4}$ 125% **b.** $1\frac{3}{20}$ 115%

Example 2

Use a proportion to write $\frac{1}{200}$ as a percent.

Solution

$\frac{1}{200} = \frac{x}{100}$ Change $\frac{1}{200}$ to a fraction with a denominator of 100.

$100 = 200x$

$\frac{100}{200} = \frac{200x}{200}$

$\frac{1}{2} = x$

$\frac{1}{200}$ is equal to $\frac{1}{2}\%$ Since percent means per 100, we can write $\frac{\frac{1}{2}}{100}$ as $\frac{1}{2}\%$.

Practice Use a proportion to write each fraction as a percent.

a. $\frac{3}{400}$ $\frac{3}{4}\%$ **b.** $\frac{2}{250}$ $\frac{4}{5}\%$

Example 3

Write 150% as a fraction or mixed number in lowest terms.

Solution

$150\% = \frac{150}{100}$ Write the number in front of the % sign over 100.

$= \frac{3}{2}$

$= 1\frac{1}{2}$

Practice Write each percent as a fraction or mixed number in lowest terms. **a.** 225% $\frac{9}{4}$ **b.** 164% $\frac{41}{25}$

Example 4

Write $\frac{1}{4}\%$ as a fraction or mixed number in lowest terms.

Solution

$\frac{1}{4}\% = \frac{\frac{1}{4}}{100}$ Write the number in front of the % sign over 100.

$= \frac{1}{4} \times \frac{1}{100}$ $\frac{a}{b} = a \cdot \frac{1}{b}$

$= \frac{1}{400}$

Practice Write each percent as a fraction or mixed number in lowest terms.

a. $\frac{1}{2}\%$ $\frac{1}{200}$ **b.** $\frac{1}{8}\%$ $\frac{1}{800}$

Oral Exercises

Give each ratio as a percent. Which are greater than 100%? Which are less than 100%?

1. $\frac{23}{100}$ 23%
2. $\frac{125}{100}$ 125%
3. $\frac{2}{3}:100$ $\frac{2}{3}\%$
4. $\frac{9}{100}$ 9%
5. $\frac{\frac{3}{4}}{100}$ $\frac{3}{4}\%$
6. 200 : 100 200%
7. $\frac{1}{2}$ to 100 $\frac{1}{2}\%$
8. $\frac{105}{100}$ 105%
9. $\frac{1}{5}:100$ $\frac{1}{5}\%$
10. 300 to 100 300%
11. $\frac{85}{100}$ 85%
12. $\frac{9}{10}$ to 100 $\frac{9}{10}\%$
13. $1\frac{1}{2}$ to 100 $1\frac{1}{2}\%$
14. $\frac{500}{100}$ 500%
15. $1\frac{5}{8}:100$ $1\frac{5}{8}\%$
16. $\frac{\frac{1}{10}}{100}$ $\frac{1}{10}\%$

Exercises

A Use a proportion to write each fraction as a percent.

1. $\frac{175}{100}$ 175%
2. $\frac{3}{2}$ 150%
3. $2\frac{1}{4}$ 225%
4. $\frac{5}{3}$ $166\frac{2}{3}\%$
5. $\frac{6}{3}$ 200%
6. $2\frac{1}{2}$ 250%
7. $\frac{4}{1}$ 400%
8. $3\frac{1}{5}$ 320%
9. $\frac{8}{6}$ $133\frac{1}{3}\%$
10. $1\frac{1}{10}$ 110%
11. $1\frac{3}{8}$ $137\frac{1}{2}\%$
12. $\frac{26}{20}$ 130%
13. $\frac{1}{500}$ $\frac{1}{5}\%$
14. $\frac{1}{125}$ $\frac{4}{5}\%$
15. $\frac{3}{1000}$ $\frac{3}{10}\%$
16. $\frac{1}{400}$ $\frac{1}{4}\%$
17. $\frac{8}{1000}$ $\frac{4}{5}\%$
18. $\frac{1}{160}$ $\frac{5}{8}\%$
19. $\frac{3}{500}$ $\frac{3}{5}\%$
20. $\frac{4}{600}$ $\frac{2}{3}\%$
21. $2\frac{1}{5}$ 220%
22. $\frac{1}{300}$ $\frac{1}{3}\%$
23. $\frac{3}{400}$ $\frac{3}{4}\%$
24. $\frac{5}{1000}$ $\frac{1}{2}\%$

Write each percent as a fraction, mixed number, or whole number in lowest terms.

25. 110% $1\frac{1}{10}$
26. 250% $2\frac{1}{2}$
27. 300% 3
28. 108% $1\frac{2}{25}$
29. 175% $1\frac{3}{4}$
30. 500% 5
31. 200% 2
32. 1000% 10
33. $\frac{1}{5}\%$ $\frac{1}{500}$
34. $\frac{3}{4}\%$ $\frac{3}{400}$
35. $\frac{3}{8}\%$ $\frac{3}{800}$
36. $1\frac{1}{4}\%$ $\frac{1}{80}$
37. 225% $2\frac{1}{4}$
38. $\frac{7}{10}\%$ $\frac{7}{1000}$
39. $2\frac{1}{2}\%$ $\frac{1}{40}$
40. $1\frac{5}{8}\%$ $\frac{13}{800}$

B

41. Bike shop employees set a goal of selling 200 new bikes in a 3-month period. They sold 75 bikes the first month, 130 bikes the second month, and 125 the third month. What percentage of their goal did they reach? 165%

42. A survey of 250 schools showed that 10 schools had more than 30 computers. What percentage of the schools surveyed had more than 30 computers? 4%

43. The Marshalls budgeted \$225 a month for clothing. Last month they spent \$75 for clothing. What percentage of the budgeted amount was spent? $33\frac{1}{3}\%$

44. Does $x\% + y\% = (x + y)\%$? Use number replacements for x and y to show why the two expressions are or are not equivalent. yes

45. Does $\frac{x\%}{y\%} = \left(\frac{x}{y}\right)\%$? Use number replacements for x and y to show why the two expressions are or are not equivalent. no; example: $x = 20$, $y = 40$; $0.5 \neq 0.005$

C Extending Thinking Skills

46. Within a 10 cm by 10 cm region on a sheet of paper, draw a rectangle with a perimeter of 20 cm and an area that is 16% of the 10 cm by 10 cm region.
8 × 2 units

47. This large cube is made up of 27 smaller cubes. If the large cube was painted red on all 6 sides, what percentage of the cubes would be painted red on exactly 1 side? 2 sides? $22\frac{2}{9}\%$ one side, $44\frac{4}{9}\%$ two sides

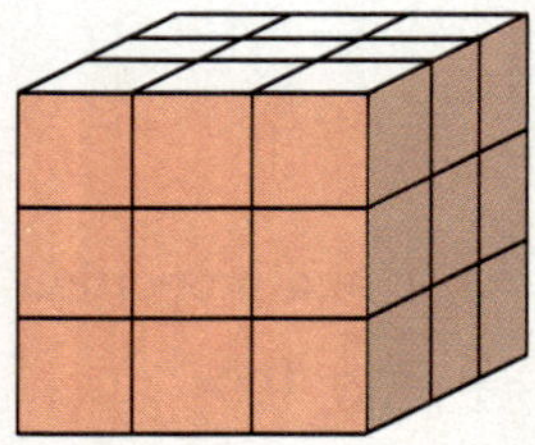

Thinking Skills: 46. Testing a conjecture 47. Using space visualization

Mixed Review

Find the greatest common factor (GCF). **48.** 8, 64, 92 4 **49.** 6, 21 3

Write each as a decimal. Use a bar for a repeating decimal.

50. $\frac{2}{3}$ $0.\overline{6}$ **51.** $\frac{5}{6}$ $0.8\overline{3}$ **52.** $\frac{7}{8}$ 0.875 **53.** $\frac{3}{10}$ 0.3 **54.** $\frac{8}{11}$ $0.\overline{72}$

Write each as a mixed number. **55.** $\frac{26}{3}$ $8\frac{2}{3}$ **56.** $\frac{211}{16}$ $13\frac{3}{16}$ **57.** $\frac{67}{8}$ $8\frac{3}{8}$

COMPUTER ACTIVITY

This program will write a fraction in the form $\frac{A}{B}$ as a percent.

```
10 PRINT "CHANGE A/B TO A PERCENT."
20 PRINT "WHAT IS A?"
30 INPUT A
40 PRINT "WHAT IS B?"
50 INPUT B
60 PRINT A;"/";B;" IS ";(A/B) * 100;"%."
70 INPUT "DO YOU WANT TO TRY AGAIN? (Y OR N)?";A$
80 IF A$ = "Y" THEN GOTO 10
90 END
```

Use the program to change each to a percent.

1. $\frac{5}{6}$ 83.3333% **2.** $\frac{7}{25}$ 28% **3.** $\frac{65}{80}$ 81.25%

4. $\frac{4}{3}$ 133.3333% **5.** $\frac{12}{48}$ 25% **6.** $\frac{18}{5}$ 360%

9-10 Percent and Decimals

Objective: To write percents as decimals and decimals as percents.
MIN: 1–48; REG: 1–48 even, 49–54; MAX: 1–48 even, 49–58; ALL: Mixed Review

When you solve problems involving percent, you will often be working with percents expressed as fractions or percents expressed as decimals. You have learned how to change percents to fractions and fractions to percents. The following examples show how to change a percent to a decimal and a decimal to a percent.

Example 1

Write $25\frac{1}{2}\%$ as a decimal.

Solution $25\frac{1}{2}\% = 25.5\%$

$= \frac{25.5}{100}$ Write as a fraction with a denominator of 100.

$= 0.255$ Change the fraction to a decimal.

Practice Write each as a decimal.

a. 5% 0.05 **b.** $8\frac{1}{4}\%$ 0.0825 **c.** 125% 1.25

Example 2

Write 0.35 as a percent.

Solution $0.35 = \frac{35}{100}$ Change the decimal to a fraction.

$= 35\%$

Practice Write each as a percent.

a. 0.60 60% **b.** 0.085 $8\frac{1}{2}\%$ **c.** 2.14 214%

Here are three illustrations of a percent expressed as a decimal.

Percent		Decimal
68%	$\frac{68}{100}$	0.68
1.8%	$\frac{1.8}{100} \cdot \frac{10}{10} = \frac{18}{1000}$	0.018
12.75%	$\frac{12.75}{100} \cdot \frac{100}{100} = \frac{1275}{10000}$	0.1275

The chart on page 291 suggests shortcuts for changing a percent to a decimal or a decimal to a percent. To change a *percent to a decimal*, move the decimal point in the percent two places to the left and drop the percent symbol. To change a *decimal to a percent*, move the decimal point in the decimal two places to the right and add the percent symbol.

Oral Exercises

Is the equation true or false?

1. $67\% = \frac{67}{100}$ T
2. $41\% = 41$ F
3. $0.78 = 7.8\%$ F
4. $0.07 = 7\%$ T
5. $\frac{42.5}{100} = 42\frac{1}{2}\%$ T
6. $3\% = 0.3$ F
7. $100\% = 1$ T
8. $0.62 = 62\%$ T
9. $90\% = 0.90$ T
10. $425\% = 0.425$ F
11. $1\% = 0.01$ T
12. $105\% = 1.05$ T

Exercises

A Write each percent as a decimal.

1. 40% 0.4
2. 10% 0.1
3. 5% 0.05
4. 65% 0.65
5. 2% 0.02
6. 15% 0.15
7. 150% 1.5
8. 1% 0.01
9. 90% 0.9
10. 13% 0.13
11. 100% 1
12. 25% 0.25
13. $4\frac{1}{2}\%$ 0.045
14. $\frac{1}{4}\%$ 0.0025
15. 200% 2
16. 80% 0.8
17. 12% 0.12
18. $10\frac{1}{2}\%$ 0.105
19. 34% 0.34
20. $2\frac{1}{4}\%$ 0.0225
21. $\frac{3}{4}\%$ 0.0075
22. $12\frac{3}{4}\%$ 0.1275
23. $5\frac{5}{8}\%$ 0.05625
24. $75\frac{3}{4}\%$ 0.7575

Write each decimal as a percent.

25. 0.62 62%
26. 0.55 55%
27. 0.05 5%
28. 0.75 75%
29. 0.8 80%
30. 1.25 125%
31. 0.001 $\frac{1}{10}\%$
32. 0.95 95%
33. 0.015 $1\frac{1}{2}\%$
34. 3.05 305%
35. 0.125 $12\frac{1}{2}\%$
36. 0.508 $50\frac{4}{5}\%$
37. 1.5 150%
38. 0.245 $24\frac{1}{2}\%$
39. 0.005 $\frac{1}{2}\%$
40. 0.57 57%
41. 0.465 $46\frac{1}{2}\%$
42. 2.4 240%
43. 0.075 $7\frac{1}{2}\%$
44. 3 300%
45. 4.38 438%
46. 2.456 $245\frac{3}{5}\%$
47. 0.006 $\frac{3}{5}\%$
48. 0.06 6%

B Solve.

49. A batting average is the number of hits divided by the number of times at bat, expressed as a decimal rounded to the thousandths place. In 1941, Ted Williams got a hit 40.6% of his times at bat. What was Williams' batting average in 1941? 0.406

50. In 1927, Babe Ruth hit 60 home runs to set a major league record that held for 34 years. In that same year, Ruth's batting average was 0.356. What percentage of his times at bat in 1927 did Ruth get a hit? 35.6%

51. In 1985, Willie McGee had 216 hits in 612 at-bats. What was his batting average? 0.353

52. In 1985, Dwight Gooden won 24 games and lost 4. What percentage of these games did he win? 85.7%

53. Change $x\%$ to a decimal where x is a 1-digit number. $x(0.01)$

54. Change y to a percent where y is a whole number. $y(100)$

C Extending Thinking Skills

Use estimation to find a value for x. Then use a calculator to find the exact value for x.

55. $\frac{x}{16} = 75\%$ 12

56. $\frac{x}{20} = 55\%$ 11

57. $\frac{x}{90} = 34\frac{4}{9}\%$ 31

58. $\frac{x}{500} = \frac{3}{5}\%$ 3

Thinking skill: 55–58 Estimating.

Mixed Review

Solve and check. **59.** $19(m + 3) = 76$ 1 **60.** $2(r - 5) = 11r + 8$ −2

61. $-15 = 3(9 + m) - 3$ −13 **62.** $14 < 3r + 2$ $4 < r$ **63.** $-4c + 2c > -16$ $c < 8$

Evaluate for $a = 3$, $b = 4$. **64.** $3(a - b)$ −3 **65.** $b(3a + 2b)$ 68

66. $a \div (2b)$ 0.375 **67.** $-a(5 - b)$ −3 **68.** $b(b - a)$ 4 **69.** $6(-a)(-b)$ 72

CALCULATOR ACTIVITY

You can change a fraction to a percent by first changing the fraction to a decimal and then changing the decimal to a percent.

Example: Change $\frac{7}{8}$ to a percent.

Fraction		Display	
$\frac{7}{8}$	7 [÷] 8 [=]	0.875	$0.875 = 87.5\%$

Note that with a calculator having a percent key, you can push: 7 [÷] 8 [%] [=] to get answer, 87.5.

Use a calculator to change each fraction to a percent.

1. $\frac{3}{4}$ 75% **2.** $\frac{3}{10}$ 30% **3.** $\frac{1}{6}$ $16.\overline{6}\%$ **4.** $\frac{7}{25}$ 28%

5. $\frac{4}{3}$ $133.\overline{3}\%$ **6.** $\frac{3}{16}$ 18.75% **7.** $\frac{3}{20}$ 15% **8.** $\frac{12}{5}$ 240%

More Practice

Solve.

1. Feldman's Bakery sells 12 dozen wheat rolls and 10 dozen rye rolls every day. What is the ratio of wheat rolls to rye rolls? $\frac{6}{5}$

2. The ratio of boys to girls on the Blue Jays soccer team is 5 : 4. There are 15 boys on the team. How many team members are girls? $\frac{5}{4} = \frac{15}{g}$ 12

3. Cindy swims 36 laps of the pool in 45 minutes. At this rate, how many laps will she swim in an hour? $\frac{36}{45} = \frac{x}{60}$ 48

4. Jim is two inches taller than Carlos. Their combined height is 146 inches. How tall is Carlos? $c + (c + 2) = 146$ 72 inches

5. The ratio of home-team fans to visiting-team fans at the basketball game is $\frac{5}{3}$. 1250 fans are cheering for the home team. How many are cheering for the visiting team? $\frac{5}{3} = \frac{1250}{x}$ 750

6. Anita's sunflower plant sprouted and grew 7 feet tall in just four weeks. How many inches did it grow each day? $\frac{7 \times 12}{28} = \frac{x}{1}$ 3 inches

7. The Varsity Theater has 220 regular seats and 80 balcony seats. What is the ratio of balcony seats to total seats? $\frac{80}{300} = \frac{4}{15}$

8. Angelo scored 45 more points on his second math quiz than on his first. His combined score was 149. What did he score on the first quiz? $x + (x + 45) = 149$ 52

9. The students in Miss Oliveira's class compared lunches. They found that 13 had peanut butter sandwiches, 6 had tuna sandwiches, and 2 had egg salad sandwiches. What was the ratio of tuna sandwiches to all sandwiches? $\frac{6}{13 + 6 + 2} = \frac{2}{7}$

10. What is the hourly rate of a typist who charges \$150 for 12 hours of labor? $150 \div 12 = r$ \$12.50

11. A double yogurt dish cost \$0.95 more than a single dish. Together they cost \$3.15. What did each cost? $x + (x + 0.95) = 3.15$ single \$1.10 double \$2.05

12. The ratio of Mr. Kovac's age to his son's age is $\frac{5}{2}$. Mr. Kovac's son is eighteen. How old is Mr. Kovac? $\frac{5}{2} = \frac{x}{18}$ 45

13. On Saturday, 96 gallons of green and white paint were sold at O'Hara's House Paints. White paint outsold green paint by 28 gallons. How many gallons of white and how many gallons of green were sold that day?
$g + (g + 28) = 96$ 62 gallons white; 34 gallons green

PROBLEM SOLVING: STRATEGIES

9-11 Multiple Solutions

Objective: To solve nonroutine problems, finding more than one way to solve each.
MIN: 1–2; REG: 1–4; MAX: 1–6

You have learned that many problems can be solved by using a combination of strategies. You might first simplify the problem, then draw a picture, make a table, and finally look for a pattern. For many problems, a solution can be found in several different ways. Two people can use completely different strategies but find the same correct solution. In the following example, a problem is solved two ways.

Problem Ms. Malito saves coupons from a service station to get a gift. The station gives 5-point coupons and 3-point coupons. She has exactly 22 coupons, for a total of 86 points. She has fewer than 15 of each type of coupon. How many of each does she have?

Solution 1

One way to solve this problem is to use the **Guess, Check, Revise** strategy. You could guess a number for each type of coupon so that the sum of the two numbers is 22, and check whether they would total 86 points. If they did not, you would revise your guess and continue the process until you found the right combination. The process might be like this.

Try 14 for 3-point coupons and 8 for 5-point coupons: $14 \cdot 3 = 42$, $8 \cdot 5 = 40$, $42 + 40 = 82$ Too low.

Try 12 for 3-point coupons and 10 for 5-point coupons: $12 \cdot 3 = 36$, $10 \cdot 5 = 50$, $36 + 50 = 86$ Correct!

Solution 2

Another way to solve this problem is to **Make a Table** and look for the correct combination.
Logically, if Ms. Malito has <15 of each coupon, she must have >7 of each. Discuss.

3-point coupons	**Number**	1	2	3	4	5	6	7	8	9	10	11	**12**	13	14
	Points	3	6	9	12	15	18	21	24	27	30	33	**36**	39	42
5-point coupons	**Number**	1	2	3	4	5	6	7	8	9	**10**	11	12	13	14
	Points	5	10	15	20	25	30	35	40	45	**50**	55	60	65	70

Problem-Solving Strategies	
Choose the Operations	**Write an Equation**
Guess, Check, Revise	**Simplify the Problem**
Draw a Picture	**Make an Organized List**
Make a Table	**Use Logical Reasoning**
Look for a Pattern	**Work Backwards**

This chart shows the strategies presented in previous chapters.

Problems

Find two different ways to solve each problem.

1. Jim Westly grows pine trees on a small farm. Twenty of his trees were killed in a very cold winter. That spring he bought the same number of new trees as had survived last winter. Later, he sold all of his trees to 6 customers, each of whom bought 15 trees. How many trees did he originally have? **65**

2. Mrs. Ungar is 28 years old and her daughter, Heather, is 6 years old. How old will Heather be when she is half as old as her mother? **22**

Solve.

3. Fred drove 25 km farther than Tammy. If she had driven twice as far as she did, Fred would have driven only 10 km more than Tammy. How far did each drive? **Fred 40, Tammy 15**

4. Cesar has $50 to buy tickets to a concert. Seats in the front section cost $7 and seats in the back section cost $5. Cesar bought 8 tickets and spent exactly $50. How many of each kind did he buy? **3 at $5 and 5 at $7**

5. A neighborhood was given 3 numbers to use for 3-digit prefixes on telephone numbers. The numbers were 3, 0, and 9. The only guideline was that the 0 could not be the first number. How many prefixes did the neighborhood have to choose from? **18**

6. An airline reported that on flights from city A to city B, only 2 out of every 5 seats were filled. Each plane holds 250 passengers and each ticket costs $125. How much more money per flight would the airline make if it had every seat filled with a paying passenger? **$18,750**

Enrichment

A Golden Rectangle

Early Greek architects, painters, and sculptors identified what they believed to be the rectangle most pleasing to the human eye. They called it the **Golden Rectangle**. The ratio of the length l to the width w of this rectangle, $\frac{l}{w}$, was called the **Golden Ratio**. The Greeks discovered that the following proportion holds only for a Golden Rectangle:

The actual Golden Ratio, written as a decimal, is 1.618. Students might average the ratios they find to estimate the Golden Ratio.

$$\frac{l}{w} = \frac{l + w}{l}$$

1. Each of the rectangles below is a Golden Rectangle. Measure the length and width of each, to the nearest millimeter. Find the ratio $\frac{l}{w}$ for each. Express each ratio as a decimal, and use them to estimate the Golden Ratio as a decimal. **Note that students will need to use metric rulers. You might wish to have them use calculators to find the ratios.**

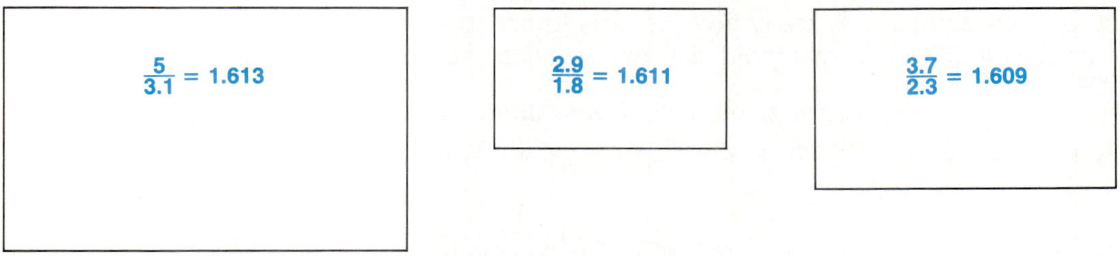

2. Measure the length and width of each rectangle below to the nearest millimeter and compute the ratio $\frac{l}{w}$ for each. For which rectangle is this ratio closest to the Golden Ratio? Is this rectangle most pleasing to your eye?

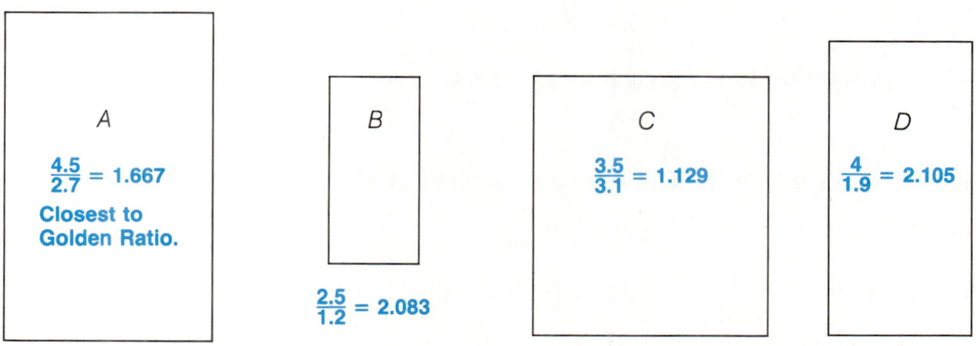

Chapter 9 Review

9-1 Write each ratio as a fraction in lowest terms.

1. 9:21 3/7 **2.** 48 to 30 8/5 **3.** 70/15 14/3

9-2 Solve and check.

4. $\frac{y}{40} = \frac{5}{8}$ 25 **5.** $\frac{4}{7} = \frac{n}{35}$ 20 **6.** $\frac{1}{6} = \frac{5}{k}$ 30

9-3 Simplify the rate.

7. 114 km/3 h 38 km/h **8.** $120/6 days $20/day **9.** 78 kg/6 days 13 kg/day

Solve.

10. Ginger walked 20 km in 4 hours. How many kilometers per hour is that? 5 km/hour

9-4 Write an equation.

11. Last year Kelly read 3 times as many mysteries as science fiction books. She read 44 of the two types combined. How many did she read of each type? $s + 3s = 44$

9-5 Solve by writing an equation.

12. Sayako invited $\frac{1}{2}$ as many boys as girls to her party. She sent out 24 invitations in all. How many did she send to boys and how many to girls? $\frac{1}{2}g + g = 24$ 8 boys, 16 girls

9-6 Find the scale dimensions of each. The scale is 1 cm = 0.5 m.

13. Counter: 2 m × 0.5 m 4 cm × 1 cm **14.** Closet: 1.5 m × 0.5 m 3 cm × 1 cm

9-7 Solve.

15. Which is the better buy, 8 square feet of wrapping paper for $2.25 or 5 square feet of wrapping paper for $1.30? 5 ft² for $1.30

9-8 Write each as a fraction in lowest terms.

16. 14% $\frac{7}{50}$ **17.** 7% $\frac{7}{100}$ **18.** $7\frac{1}{2}\%$ $\frac{3}{40}$

Use a proportion to write each as a percent.

19. $\frac{3}{4}$ 75% **20.** $\frac{9}{25}$ 36% **21.** $\frac{1}{8}$ 12.5%

9-9 Use a proportion to write each as a percent.

22. $\frac{225}{100}$ 225% **23.** $\frac{8}{5}$ 160% **24.** $\frac{3}{400}$ 0.0075% or $\frac{3}{4}\%$

Write each percent as a fraction or mixed number in lowest terms.

25. 275% $2\frac{3}{4}$ **26.** $\frac{2}{5}\%$ $\frac{1}{250}$ **27.** $2\frac{3}{4}\%$ $\frac{11}{400}$

9-10 Write as a decimal. **28.** 4% 0.04 **29.** $8\frac{1}{2}\%$ 0.085

Write as a percent. **30.** 81 8100% **31.** 0.075 $7\frac{1}{2}\%$

Chapter 9 Test

Write each ratio as a fraction in lowest terms.

1. 8 : 32 $\frac{1}{4}$
2. 12 to 15 $\frac{4}{5}$
3. $\frac{72}{27}$ $\frac{8}{3}$

Solve and check.

4. $\frac{a}{42} = \frac{3}{7}$ 18
5. $\frac{5}{6} = \frac{m}{30}$ 25
6. $\frac{3}{4} = \frac{15}{c}$ 20

Simplify the rate.

7. 225 km/3 h 75 km/h
8. \$500/4 days \$125/day
9. 186 beats/3 minutes 62 beats/min

Solve.

10. Mr. Bertolini charged \$60 for 4 hours labor. What is his hourly rate? \$15

Write an equation.

11. Geri's father is 4 times Geri's age. The sum of their ages is 45. $4g + g = 45$

Solve by writing an equation.

12. Peter has taken piano lessons for half as many years as Laura has. The sum of the number of years they have both taken lessons is 12. How many years has each taken lessons? $l + \frac{1}{2}l = 12$ Laura 8 years, Peter 4 years

Find the scale dimensions of each. The scale is 1 cm = 0.5 m.

13. Desk: 1.5 m × 1 m 3 cm × 2 cm

14. Patio: 3.5 m × 3 m 7 cm × 6 cm

Solve.

15. Orange juice costs \$2.69 for the 64-ounce size and \$1.27 for the 32-ounce size. Which size is the better buy? 32 oz.

Write each as a fraction in lowest terms.

16. 22% $\frac{11}{50}$
17. 1% $\frac{1}{100}$
18. $4\frac{1}{2}$% $\frac{9}{200}$

Use a proportion to write each as a percent.

19. $\frac{1}{4}$ 25%
20. $\frac{17}{20}$ 85%
21. $\frac{3}{8}$ $37\frac{1}{2}$%
22. $\frac{450}{100}$ 450%
23. $\frac{12}{5}$ 240%
24. $\frac{25}{500}$ 5%

Write each percent as a fraction or mixed number in lowest terms.

25. 325% $3\frac{1}{4}$
26. $\frac{3}{4}$% $\frac{3}{400}$
27. $2\frac{1}{2}$% $\frac{1}{40}$

Write as a decimal.

28. 9% 0.09
29. $12\frac{1}{5}$% 0.122

Write as a percent.

30. 0.20 20%
31. 0.025 $2\frac{1}{2}$%

Cumulative Review

Write an equivalent fraction by replacing the variable with a whole number.

1. $\frac{7}{8} = \frac{b}{96}$ $\frac{84}{96}$ **2.** $\frac{y}{9} = \frac{56}{63}$ $\frac{8}{9}$ **3.** $-\frac{15}{22} = -\frac{f}{110}$ $-\frac{75}{110}$

Reduce to lowest terms.

4. $\frac{52}{84}$ $\frac{13}{21}$ **5.** $-\frac{16}{88}$ $-\frac{2}{11}$ **6.** $\frac{124}{160}$ $\frac{31}{40}$

Write each as a mixed number.

7. $\frac{41}{6}$ $6\frac{5}{6}$ **8.** $\frac{-73}{3}$ $-24\frac{1}{3}$ **9.** $\frac{149}{100}$ $1\frac{49}{100}$

Write each as an improper fraction in lowest terms.

10. $3\frac{1}{7}$ $\frac{22}{7}$ **11.** $-5\frac{3}{12}$ $-\frac{21}{4}$ **12.** $4\frac{6}{15}$ $\frac{22}{5}$

Write <, >, or = for each □.

13. $\frac{7}{9}$ □ $\frac{3}{5}$ > **14.** $-\frac{2}{3}$ □ $-\frac{5}{6}$ > **15.** $\frac{4}{7}$ □ $\frac{10}{11}$ <

16. $-\frac{9}{14}$ □ $-\frac{15}{36}$ < **17.** $\frac{20}{11}$ □ $\frac{14}{5}$ < **18.** $\frac{12}{60}$ □ $\frac{9}{45}$ =

Solve. Reduce to lowest terms.

19. $\frac{1}{2}\left(\frac{4}{5}\right)$ $\frac{2}{5}$ **20.** $\frac{2}{5}\left(-\frac{3}{8}\right)$ $-\frac{3}{20}$

21. $-1\frac{5}{7}\left(4\frac{2}{3}\right)$ -8 **22.** $-2\frac{4}{5}\left(-3\frac{1}{2}\right)$ $9\frac{4}{5}$

23. $-\frac{5}{9} \div \left(-\frac{2}{3}\right)$ $\frac{5}{6}$ **24.** $\frac{7}{20} \div \frac{84}{100}$ $\frac{5}{12}$

Solve.

25. $6y - 3 = 15$ 3 **26.** $-22 = 3n - 1$ -7

27. $\frac{m + 4}{5} = 2$ 6 **28.** $3(d - 7) = -27$ -2

Write an equation.

29. The \$72 cost of a jacket is \$6 more than 6 times the cost of a tie. $72 = 6t + 6$

Solve by writing an equation.

30. Bill's height is 4 inches more than twice his cousin's height. His cousin is 32 inches tall. How tall is Bill? $b = 2(32) + 4$, $b = 68$

Chapter 10
Using Percent

10-1 Finding a Percent of a Number

Objective: To find a percent of a number.
MIN: 1–27; REG: 1–27 even, 28–40; MAX: 1–27 even, 28–44; ALL: Mixed Review
An exploratory lesson opener can be found on page 551.

A customer gave a waiter 15% of the total bill as a tip. The total bill was $60. You can use a proportion or an equation to find the amount of the tip.

You can find a percent of a number by using a proportion or an equation.

You may wish to teach your preferred method. The exercises are appropriate for practice of either the proportion or the equation method.

Example 1

Use a proportion to solve. 15% of 60

Solution

$\frac{15}{100} = \frac{n}{60}$ You are looking for a number that has the same ratio to 60 as 15 has to 100.

$100 \cdot n = 15 \cdot 60$ Property of proportions.

$\frac{100n}{100} = \frac{15 \cdot 60}{100}$ Divide both sides by 100 to get the variable by itself.

$n = 9$

15% of 60 is 9. **The tip was $9.**

Practice Use a proportion to solve.

a. 5% of 120 **6** **b.** 150% of 60 **90** **c.** 12% of 30 $\frac{18}{5}$

When the percent involves a fraction, you can most easily find the percent of a number using an equation.

Example 2

Use an equation to solve. $12\frac{1}{2}\%$ of 25

Solution $0.125 \times 25 = n$ Think: $12\frac{1}{2}\%$ of 25 is what number?

$$0.125 \times 25 = n$$

$$3.125 = n$$

$12\frac{1}{2}\%$ of 25 is 3.125.

Practice Use an equation to solve.

a. 60% of 130 78 **b.** $4\frac{1}{2}\%$ of 75 3.375 **c.** 120% of 85 102

Oral Exercises

Give each percent as a decimal and as a fraction in lowest terms.

1. 25% $0.25, \frac{1}{4}$ **2.** 50% $0.5, \frac{1}{2}$ **3.** 75% $0.75, \frac{3}{4}$ **4.** $66\frac{2}{3}\%$ $0.\overline{6}, \frac{2}{3}$ **5.** 1% $0.01, \frac{1}{100}$

6. 100% 1, 1 **7.** 10% $0.1, \frac{1}{10}$ **8.** $33\frac{1}{3}\%$ $0.\overline{3}, \frac{1}{3}$ **9.** 90% $0.9, \frac{9}{10}$ **10.** 80% $0.8, \frac{4}{5}$

Exercises

A Use a proportion to solve.

1. 20% of 60 12 **2.** 9% of 360 32.4 **3.** 15% of 160 24 **4.** 4% of 60 2.4

5. 90% of 50 45 **6.** 70% of 8 5.6 **7.** 3% of 180 5.4 **8.** 25% of 18 4.5

9. 28% of 84 23.52 **10.** 30% of 412 123.6 **11.** 12% of 25 3 **12.** 25% of 55 13.75

Use an equation to solve.

13. 40% of 20 8 **14.** $12\frac{1}{2}\%$ of 56 7 **15.** $\frac{1}{2}\%$ of 500 2.5

16. 35% of 40 14 **17.** 0.3% of 126 0.378 **18.** 2.5% of 3 0.075

19. 15% of 80 12 **20.** 250% of 100 250 **21.** 12% of 125 15

Use a proportion or an equation to solve.

22. 18% of 54 9.72 **23.** 9% of 24.5 2.205 **24.** 250% of 20 50

25. $\frac{1}{4}\%$ of 148 0.37 **26.** 4.5% of 60 2.7 **27.** 0.5% of 12 0.06

B Evaluate each expression.

28. 34% of n, $n = 85$ 28.9 **29.** $x\%$ of 150, $x = 5\frac{1}{2}$ 8.25 **30.** 150% of h, $h = 20$ 30

Choose compatible numbers to give an estimate for each.
Example: 48% of \$82. Think: "50% of 80 is 40, so 48% of \$82 is about \$40."

31. 20% of \$505 \$100 **32.** 5% of \$95 \$5 **33.** 78% of \$310 \$240

34. A high school had 700 students in its graduating class. Out of this graduating class, 65% of the students went to college. How many students went to college? 455

35. A salesperson made \$45,500 last year. Her expenses were $67\frac{1}{2}\%$ of her income. How much money did she have left after expenses? \$14,787.50

36. Ruben took 24 shots in last night's basketball game. His shooting percentage was $37\frac{1}{2}\%$. Each basket was worth 2 points. How many points did Ruben score? 18

Some tips are easy to calculate mentally. To find 15% of \$64, you can "break apart" 15% as follows: 10% of \$64 is \$6.40. 5% of \$64 is half of \$6.40, or \$3.20. The tip is \$6.40 plus \$3.20, or \$9.60. Calculate a 15% tip mentally for each bill amount.

37. \$42 \$6.30 **38.** \$15 \$2.25 **39.** \$35 \$5.25 **40.** \$18 \$2.70

C *Extending Thinking Skills*

Estimate the percentage of each square that is shaded.

41.

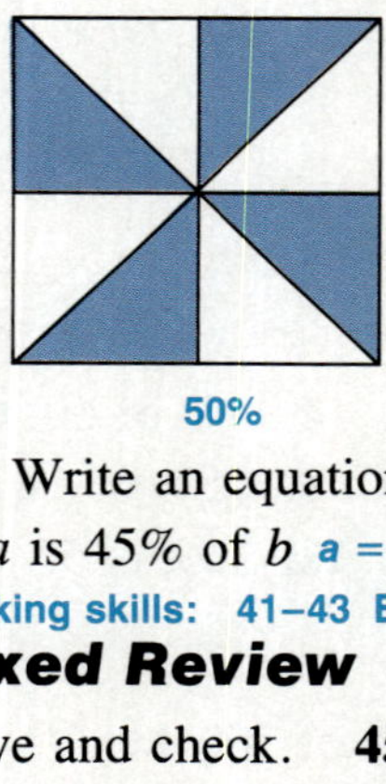

50%

42.

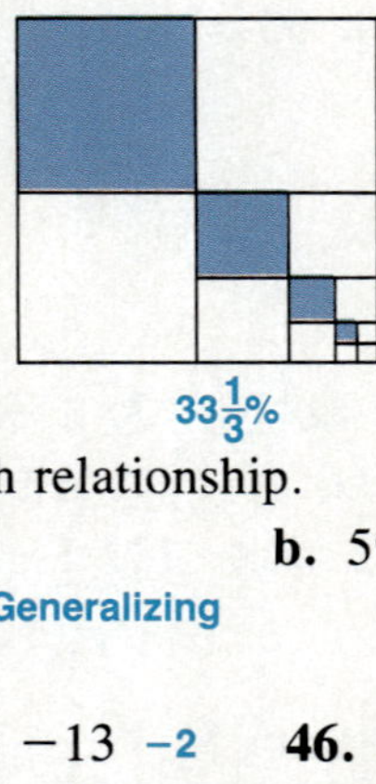

$33\frac{1}{3}\%$

43.

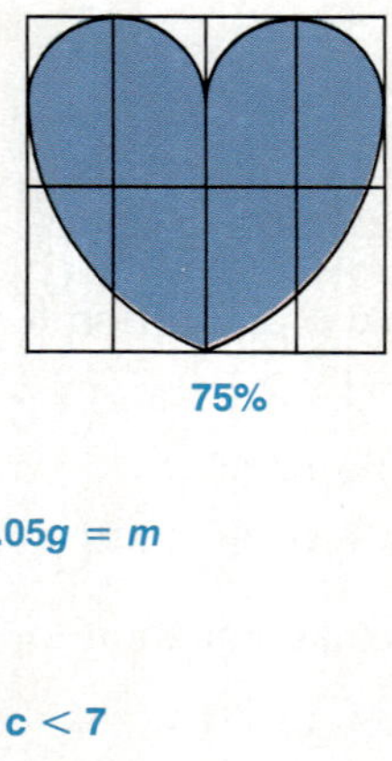

75%

44. Write an equation to show each relationship.

a. a is 45% of b $a = 0.45b$

b. 5% of g is m $0.05g = m$

Thinking skills: 41–43 Estimating 44. Generalizing

Mixed Review

Solve and check. **45.** $9 + 11c = -13$ −2 **46.** $21 - 3c > 0$ $c < 7$

47. $9z + 36 = -9$ −5 **48.** $4(m + 6) + 7 = 3(5m + 2) - 8$ 3

MENTAL MATH

You can substitute compatible numbers and change a percent to a fraction to compute a percent of a number mentally. For example, to find $66\frac{2}{3}\%$ of 12, you can think, "$66\frac{2}{3}\%$ is the same as $\frac{2}{3}$, and $\frac{2}{3}$ of 12 is 8."

Compute mentally.

1. 50% of 90 45 **2.** $33\frac{1}{3}\%$ of 30 10 **3.** 200% of 35 70

4. 25% of 48 12 **5.** 75% of 20 15 **6.** 10% of \$80 \$8

10-2 Finding What Percent One Number is of Another

Objective: To find what percent one number is of another.
MIN: 1–28; REG: 1–28 even, 29–37; MAX: 1–28 even, 29–40; ALL: Mixed Review

An exploratory lesson opener can be found on page 552.

Sometimes you may want to find what percent one number is of another. For instance, if you worked at a record store and wanted to compare the number of records and tapes sold of each type of music, you could find what percent each number is of the total number sold.

	Country	Rock	Classical	Jazz	Total Sold
Records	20	78	17	35	150
Tapes	10	66	19	24	120

The chart shows that 24 jazz tapes were sold out of a total of 120 tapes. You can use a proportion to find what percent of the tapes sold were jazz tapes.

Example 1 shows proportion method; Example 2 shows equation method. The exercises are appropriate for practice of either method. You may wish to teach the one you prefer.

Example 1

Use a proportion to solve. 24 is what percent of 120?

Solution $\frac{n}{100} = \frac{24}{120}$ You want to find a number that has the same ratio to 100 as 24 has to 120.

$120 \cdot n = 100 \cdot 24$ Property of proportions.

$\frac{120n}{120} = \frac{100 \cdot 24}{120}$

$n = 20$

24 is 20% of 120. $\frac{20}{100}$ is the same as 20%.

Practice Use a proportion to solve.

a. 30 is what percent of 150? **20%**

b. What percent of 120 is 40? **$33\frac{1}{3}\%$**

The chart above shows that 78 out of the 150 records sold were rock music. The following example shows how to use an equation to find what percent of the records sold were rock.

Example 2

Use an equation to solve. 78 is what percent of 150?

Solution $78 = p \cdot 150$ Think: 78 is what percent of 150?

$\frac{78}{150} = \frac{p \cdot 150}{150}$ $78 = p \cdot 150$

$0.52 = p$ Change $\frac{78}{150}$ to a decimal by dividing 78 by 150.

78 is 52% of 150. $0.52 = \frac{52}{100}$ or 52%.

Practice Use an equation to solve.

a. 61 is what percent of 120? $50\frac{5}{6}\%$

b. What percent of 150 is 20? $13\frac{1}{3}\%$

Oral Exercises

Tell whether each statement is true or false. If it is false, tell why.

1. 1 out of 10 is 10%. T

2. 70 out of 150 is greater than 50%. F

3. 60 out of 120 is 50%. T

4. 90 out of 75 is greater than 100%. T

5. 40 out of 50 is less than 30%. F

6. 2 out of 100 is greater than 1%. T

Exercises

A Use a proportion to solve.

1. What percent of 20 is 3? 15%

2. 25 is what percent of 125? 20%

3. 27 out of 45 is what percent? 60%

4. 3 is what percent of 24? $12\frac{1}{2}\%$

5. What percent of 85 is 68? 80%

6. 11 is what percent of 20? 55%

7. 6 is what percent of 80? $7\frac{1}{2}\%$

8. 4 is what percent of 40? 10%

9. What percent of 125 is 50? 40%

10. 18 is what percent of 60? 30%

Use an equation to solve.

11. What percent of 40 is 8? 20%

12. 60 out of 150 is what percent? 40%

13. What percent of 24 is 8? $33\frac{1}{3}\%$

14. 18 out of 108 is what percent? $16\frac{2}{3}\%$

15. 100 is what percent of 300? $33\frac{1}{3}\%$

16. 39 is what percent of 50? 78%

17. 12 is what percent of 54? $22\frac{2}{9}\%$

18. What percent of 40 is 13? $32\frac{1}{2}\%$

19. What percent of 70 is 42? 60%

20. 16 is what percent of 80? 20%

Use a proportion or an equation to solve.

21. What percent of 50 is 40? 80%

22. 17 is what percent of 68? 25%

23. 25 is what percent of 1000? $2\frac{1}{2}\%$

24. 1 is what percent of 200? $\frac{1}{2}\%$

25. 8 is what percent of 64? $12\frac{1}{2}\%$

26. What percent of 50 is 15? 30%

27. What percent of 65 is 13? 20%

28. 6 is what percent of 15? 40%

B Find each percent in Exercises 29–33. Round to the nearest tenth of a percent.

Example: 24 is what percent of 120?

Solution: $\frac{x}{100} = \frac{24}{120} \longrightarrow x = \frac{24}{120} \cdot 100 = 20$ 24 is 20% of 120.

29. What percent of 2 is 0.5? 25%

30. What percent of 48 is 18? 37.5%

31. 48 is what percent of 108? 44.4%

32. What percent of 5 is 1.5? 30%

33. Mr. Perez paid \$3 tax on a \$60 hotel bill. What percent of \$60 was the tax? 5%

34. Of 200 grocery shoppers surveyed, 92 did not have a regular shopping day. What percentage is this? 46%

35. Out of 125 students, 65 received an A or a B on the last math test. What percentage of the students received an A or a B? 52%

36. A math test has 75 total possible points. What percent correct is 68 points? $90\frac{2}{3}\%$

37. An English test has a total of 120 points. A score of 70% is passing. Dana got 80 points. Did she pass? no

38. Of 150 people who went to a play, 40 thought it was excellent, 60 thought it was average, and 50 thought it was poor. What percentage of the people thought the play was better than average? $26\frac{2}{3}\%$

C Extending Thinking Skills

39. Find x and y: x is 50% of y and $x + y = 75$. 25, 50

40. Find x and y: x is $33\frac{1}{3}\%$ of y and $3x + y = 120$. 20, 60

Thinking skills: 39–40. Testing a conjecture

Mixed Review

Write as a decimal. **41.** 25% 0.25 **42.** 80% 0.8 **43.** 124% 1.24 **44.** 19% 0.19

Solve and check. **45.** $6(m + 3) = 0$ −3 **46.** $4(9r + 7r) = 256$ 4

ESTIMATION

You can estimate what percent one number is of another by using known fractions. For example, to solve "21 is what percent of 80?" think, "Since $\frac{20}{80}$ is $\frac{1}{4}$ or 25%, $\frac{21}{80}$ is about 25% of 80." Estimate the percent.

1. 25 is what percent of 75? 33%
2. 14 is what percent of 30? 50%
3. 62 is what percent of 90? 67%
4. 76 is what percent of 50? 150%

10-3 Finding a Number When a Percent of it is Known

Ojective: To find a number when a percent of it is known.
MIN: 1–26; REG: 1–26 even, 27–37; MAX: 1–26 even, 27–40

Packages of computer disks are on sale for 25% off the original price. This is a savings of \$15 per package. To find the original price you can use either a proportion or an equation.

Example 1 You may wish to teach your preferred method. The exercises are appropriate for practice of either.

Use a proportion to solve. 15 is 25% of what number?

Solution $\frac{25}{100} = \frac{15}{n}$ 15 has the same ratio with what number as 25 has with 100?

$25 \cdot n = 100 \cdot 15$

$\frac{25n}{25} = \frac{100 \cdot 15}{25}$

$n = 60$

15 is 25% of 60.

Practice Use a proportion. Round to the nearest tenth if necessary.

a. 45% of what number is 18? 40 **b.** 24 is 12% of what number? 200

Example 2

Use an equation to solve. 25% of what number is 15?

Solution $0.25n = 15$ Think: 25% of what number is 15? ($0.25 \cdot n = 15$)

$\frac{0.25n}{0.25} = \frac{15}{0.25}$

$n = 60$

25% of 60 is 15.

Practice Use an equation to solve. Round to the nearest tenth if necessary.

a. 22 is 35% of what number? 62.9 **b.** 4.5% of what number is 9? 200

Oral Exercises

State the equation you would use for each problem.

1. 60 is 75% of what number? $60 = 0.75n$
2. 40% of what number is 90? $0.40n = 90$
3. 30% of what number is 15? $0.30n = 15$
4. 15 is 150% of what number? $15 = 1.5n$
5. 125% of what number is 85? $1.25n = 85$
6. 40 is 35% of what number? $40 = 0.35n$

Exercises

A Use a proportion to solve. Round to the nearest tenth if necessary.

1. 75% of what number is 24? 32

2. 20 is 4% of what number? 500

3. 40 is 25% of what number? 160

4. 15 is 6% of what number? 250

5. 5 is 1% of what number? 500

6. 56% of what number is 28? 50

7. 52 is 4% of what number? 1300

8. 150% of what number is 12? 8

Use an equation to solve. Round to the nearest tenth if necessary.

9. 60 is 20% of what number? 300

10. 75% of what number is 120? 160

11. 85 is 30% of what number? 283.3

12. 7 is 23% of what number? 30.4

13. 5 is 17% of what number? 29.4

14. 15% of what number is 12? 80

15. 4% of what number is 56? 1400

16. 60 is 75% of what number? 80

Use a proportion or an equation to solve. Round to the nearest tenth if necessary.

17. $4\frac{1}{2}$% of what number is 12? 266.7

18. 25% of what number is $10\frac{1}{2}$? 42

19. 4 is 1% of what number? 400

20. 26 is 50% of what number? 52

21. 2.25 is $12\frac{1}{2}$% of what number? 18

22. 250% of what number is $\frac{3}{4}$? 0.3

23. 150% of what number is 12? 8

24. 80% of what number is 28? 35

25. 25 is 40% of what number? 62.5

26. 15% of what number is 26? 173.3

B Use a calculator for Exercises 27–32. Check the reasonableness of your answer with an estimate.

27. 85% of what number is 350? 411.76

28. 28.75 is 75% of what number? 38.33

29. $12\frac{1}{2}$% of what number is 95.5? 764

30. 775 is $25\frac{1}{4}$% of what number? 3069.31

31. 130 is 85% of what number? 152.94

32. 120% of what number is 25.25? 21.04

Solve.

33. The choir at a local high school is the largest in the state. The choir has 192 members. This is 24% of all the students in the school. What is the total number of students in the school? 800 students

34. A car race was stopped after 425 miles had been completed. The race was considered official, since 85% of the race had been completed. What was the length of the original race? 500 miles

35. There are 120 students involved in athletic programs at Washington School this year. This number is 150% of the students involved last year. How many students were in the athletic programs last year? 80 students

36. One day 20% of the students in a school went on a field trip. Only 240 students were left in the school. How many students went on the field trip? 60 students

37. So far, 135 cubic yards of concrete have been poured for new tennis courts. The job foreman says 75% of the pouring is now completed. What is the total number of cubic yards of concrete that will be used for this job? 180 yd^3

C Extending Thinking Skills

38. What is 45% of $3x$ if 30% of x is 45? 202.5

39. A jar has 100 ml of water in it and is 20% full. How much water will be in the jar when it is 80% full? 400 ml

40. Look at the 10 × 10 block below. Imagine stacking blocks like it and then painting only the outside faces of the stack red. How many blocks would need to be stacked so that $53\frac{1}{3}\%$ of the 1-by-1 blocks had only one face painted red?

Thinking skills: 38. Reasoning 39. Visualizing 40. Using space visualization

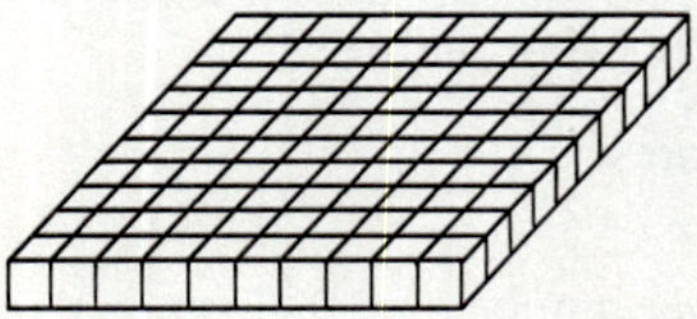

3 10 × 10 blocks (300 blocks)
160 with one red face
(64 on top, 64 on bottom,
8 on each of the other 4 sides)

Mixed Review

Write each fraction in lowest terms, then write as a percent.

41. $\frac{6}{8}$ $\frac{3}{4}$, 75% **42.** $\frac{27}{15}$ $\frac{9}{5}$, 180% **43.** $\frac{60}{12}$ $\frac{5}{1}$, 500% **44.** $\frac{45}{75}$ $\frac{3}{5}$, 60% **45.** $\frac{18}{4}$ $\frac{9}{2}$, 450%

Solve and check. **46.** $\frac{3}{4}x + \frac{1}{3} = \frac{5}{6}$ $\frac{2}{3}$ **47.** $\frac{2}{5}r - \frac{3}{4} = \frac{1}{2}$ $\frac{25}{8}$ **48.** $-\frac{5}{6} = \frac{2}{3}m + \frac{4}{9}$ $-\frac{23}{12}$

CALCULATOR ACTIVITY

You can use the [%] key on a calculator to solve percent problems.

On some calculators, you must push [=] after [%] to get the answer.

Examples:

		Display
a. Find 50% of 300.	50 [×] 300 [%]	150
b. What percent of 50 is 20?	20 [÷] 50 [%]	40 (40%)
c. 30% of what number is 24?	24 [÷] 30 [%]	80

Use the percent key to solve.

1. Find 30% of 64. 19.2

2. What percent of 80 is 24? 30%

3. 15% of what number is 30? 200

4. 35 is what percent of 80? 43.75%

5. 45% of what number is 54? 120

6. 85% of 12 is what number? 10.2

10-4 Practice Solving Problems

Objective: To solve word problems by writing equations in which one unknown is represented in terms of another.
MIN: 1–5; REG: 1–8; MAX: 1–10; ALL: Mixed Review

You can use the Problem-Solving Checklist on page 121 to help you solve problems. When your plan involves writing an equation, keep these questions in mind.

- Can you use a variable to represent an unknown number?
- Can you represent other conditions in terms of the variable?
- What is equal?
- Can you write and solve an equation?

Example

Solve by writing an equation. The cost of a small pizza is 75% of the cost of a medium pizza. What is the cost of each size pizza if the total cost for 1 medium and 1 small pizza is \$15.05?

Solution Let m = cost of medium pizza

$0.75m$ = cost of small pizza

Since the cost of the small pizza is given relative to the cost of the medium pizza, let m represent the cost of the medium pizza.

$$m + 0.75m = 15.05$$
$$1.75m = 15.05$$
$$m = 8.6$$

A medium pizza costs \$8.60.

$$0.75m = (0.75)(8.6) = 6.45$$

A small pizza costs \$6.45.

Practice Solve by writing an equation. Linda has 1 more than half as many paper route customers as George. Linda has 35 customers. How many does George have? **$35 = \frac{1}{2}x + 1$, $x = 68$ customers**

Exercises

A Solve by writing an equation.

1. Donna has 3 times as many books as Gloria. Together they have 76 books. How many does Gloria have? **$g + 3g = 76$ 19 books**

2. Mr. Lee saved \$24.00 when he bought a suit on sale at 15% off the regular price. What was the regular price of the suit? **$0.15x = 24$ \$160**

3. A delivery person has 5 more than twice the number of morning customers on his afternoon route. He has 45 customers in the afternoon. How many morning customers does he have? $2x + 5 = 45$ 20

4. Rick bought 1 record at the regular price and another for 50% of that price. The total price was \$12.24. What was the cost of each record?

5. Karen works as a waitress. One week she worked for 20 hours and got \$64.50 in tips. Her total wages and tips came to \$146.50. What is her hourly wage? $20w + 64.50 = 146.50$ \$4.10

B 4. $p + \frac{1}{2}p = 12.24$ \$8.16 regular \$4.08 50% off

6. Martha bought a pair of jeans on sale. The sale price was 50% of the original price. She paid with a \$20 bill and received \$4.25 in change. If there was no tax on the jeans, what was the original price? $0.5x = 20 - 4.25$ The jeans were \$31.50

7. A salesperson said she would take 10% off the price of a sweater. What was the original price of the sweater if the reduced price was \$24.30? $24.30 + 0.1s = s$ \$27

8. Pat was given a 15% raise. She works 20 hours per week. Her earnings after the raise were \$103.50 per week. How much money did she make per week before her raise? $20(1.15x) = 103.50$ $x = 4.50$ $4.50 \times 20 = 90$ \$90

C Extending Thinking Skills

Answers will vary.

9. Write a word problem that could be solved using the equation $0.05x = \$16.80$.

10. Write a word problem that could be solved using the equation $2x + 5 = 25$.

Thinking skills: 9–10. Formulating a problem

Mixed Review

Evaluate for $a = 2$, $b = 3$, $c = 2$. **11.** b^2c^2 36 **12.** $2ac - 3b$ −1

13. $2b - a$ 4 **14.** $5(c - b) + 3a$ 1 **15.** $10b + ca$ 34

Write without exponents. **16.** 5^3 125 **17.** 4^{-3} $\frac{1}{64}$ **18.** $\frac{(-6)^4}{(-6)^6}$ $\frac{1}{36}$

10-5 Percent of Increase and Decrease

Objective: To find percents of increase and decrease.
MIN: 1–16; REG: 1–16 even, 17–24; MAX: 1–16 even, 17–26; ALL: Mixed Review

Ms. Hernandez earned $500 per week last year. This year she will earn $530 per week. You can use a proportion to find the percent of increase of Ms. Hernandez's salary.

To find the percent of increase or decrease:

- First subtract to find the amount of increase or decrease.
- Then find what percent of the original price that amount is.

Example 1

Find the percent of increase. Original amount = $500; new amount = $530

Solution

Amount of increase = $530 − 500   amount of increase = new amount − original amount
= $30

Percent of increase

$\frac{x}{100} = \frac{30}{500}$ To find the percent of increase you need to find what percent 30 is of 500.

$500x = 100 \cdot 30$

$x = 6$

The percent of increase is 6%.

Practice Find the percent of increase.
Original amount = 40; new amount = 45 $12\frac{1}{2}\%$

Example 2

Find the percent of decrease. Original amount = 60; new amount = 45.

Solution

Amount of decrease = 60 − 45   amount of decrease = original amount − new amount

= 15

Percent of decrease

$\frac{x}{100} = \frac{15}{60}$ To find the percent of decrease, you need to find what percent 15 is of 60.

$60x = 100 \cdot 15$

$x = 25$

The percent of decrease is 25%

Practice Find the percent of decrease.
Original amount = 125; new amount = 65 **48%**

Oral Exercises

State whether the change would be an increase or a decrease.

1. Original price: \$65
   Sale price: \$59 **D**
2. Last year's cost: \$100
   This year's cost: \$112 **I**
3. During the sale: \$75
   After the sale: \$85 **I**
4. 1986 price: \$85
   1987 price: \$105 **I**
5. Last year's income: \$420
   This year's income: \$375 **D**
6. Price 5 years ago: \$12
   Price today: \$8.50 **D**

Exercises

A Find the percent of increase. Round to the nearest tenth if necessary.

1. Original amount = \$20
   New amount = \$30 **50%**
2. Original amount = \$50
   New amount = \$75 **50%**
3. Original amount = \$120
   New amount = \$150 **25%**
4. Original amount = \$300
   New amount = \$360 **20%**
5. Original amount = \$140
   New amount = \$168 **20%**
6. Original amount = \$345
   New amount = \$483 **40%**
7. Original amount = \$525
   New amount = \$609 **16%**
8. Original amount = \$185
   New amount = \$296 **60%**

Find the percent of decrease. Round to the nearest tenth if necessary.

9. Original amount = \$60
   New amount = \$15 **75%**
10. Original amount = \$85
    New amount = \$34 **60%**
11. Original amount = \$80
    New amount = \$45 **43.8%**
12. Original amount = \$78
    New amount = \$39 **50%**

13. Original amount = \$190
New amount = \$80 57.9%

14. Original amount = \$135
New amount = \$120 11.1%

15. Original amount = \$48
New amount = \$32 33.3%

16. Original amount = \$95
New amount = \$85 10.5%

B For Exercises 17–19, use the formula below and a calculator to find the percent of increase or decrease. Round each percent to the nearest tenth if necessary.

$$\text{percent of increase or decrease} = \frac{\text{amount of increase or decrease}}{\text{original amount}}$$

17. Original amount = \$750; new amount = \$1000 33.3%
18. Original amount = \$1250; new amount = \$1000 20%
19. Original amount = \$2995; new amount = \$6500 117%
20. Last year 30 students took a drafting class. This year 35 students are taking the class. What is the percentage of increase in enrollment? $16\frac{2}{3}\%$
21. Kevin weighed 45 kg last month. This month he weighs 42 kg. By what percentage did his weight decrease? $6\frac{2}{3}\%$
22. A calculator cost \$25 five years ago. Today the price of the same calculator has decreased by 80%. What is the cost of the calculator today? \$5
23. A gold ring cost \$125 five years ago. There was a 420% increase in the price in the last 5 years. What is the price of that ring today? \$650 (\$125 + \$525 increase)
24. A computer had 128 K of internal memory when it was first produced. Now its internal memory has been increased by 300%. What is its amount of internal memory now? 512 K

C Extending Thinking Skills

25. Last month a clothing store manager decreased prices on all stock by 10%. This month she increased all prices by 10%. What would you pay for a coat this month that had cost \$75 before prices were decreased last month? \$74.25
26. The original price of an item, increased by a certain percentage, is \$310. Decreased by the same percentage, it is \$190. What is the original price? What is the percentage of increase or decrease? \$250, 24%

Thinking skills: 25. Reasoning 26. Testing a conjecture

Mixed Review

Write each as a decimal. Use a bar for a repeating decimal.

27. $\frac{4}{5}$ 0.8 28. $\frac{6}{11}$ $0.\overline{54}$ 29. $\frac{8}{3}$ $2.\overline{6}$ 30. $\frac{3}{8}$ 0.375 31. $\frac{5}{16}$ 0.3125

Reduce to lowest terms. 32. $\frac{16}{64}$ $\frac{1}{4}$ 33. $\frac{35}{112}$ $\frac{5}{16}$ 34. $\frac{52}{78}$ $\frac{2}{3}$ 35. $\frac{9}{114}$ $\frac{3}{38}$

Solve and check. 36. $26a + 2(a + 3) = 146$ 5 37. $14r + 3 = 2r - 21$ −2

38. $42 - 13t = 9 - 2t$ 3 39. $6(a - 7) + 2 = 50a + 4$ −1

10-6 Discount, Sale Price, and Commission

Objective: To find percents of discounts, sale prices, and commissions.
MIN: 1–16; REG: 1–16 even, 17–21; MAX: 1–16 even, 17–23; ALL: Mixed Review
You may want to have students use calculators for this lesson.

When you buy an item on sale, you receive a discount from the original price. The **discount** is the amount subtracted from the **regular price**. The **sale price** of an item is the regular price less the discount. A **commission** is the amount a salesperson receives for making a sale. A commission is usually a percentage of the sale price.

BOAT SALE
REGULAR PRICE $8500 15% Discount Today Only!!!!

Example 1

Find the discount and sale price for the boat in the advertisement above.

Solution

Discount = 15% of $8500
= 0.15 × 8500
Discount = $1275

You can find the discount by multiplying the original price by the discount percent.

Sale price = $8500 − $1275
Sale price = $7225

You can find the sale price by subtracting the amount of the discount from the original price.

Practice Find the discount and sale price for each. Round to the nearest cent, if necessary.

a. Regular price = $2250
Discount = 20%

b. Regular price = $875
Discount = 12%

a. discount = $450 sale price = $1800 b. discount = $105 sale price = $770

Example 2

The sale price of a boat is $7225. What is the amount of commission for the sale of the boat if the commission is 1.5%? Round to the nearest cent.

Solution

Commission = 1.5% of $7225

You can find the commission by multiplying the sale price by the commission percent.

= 0.015 × 7225
Commission = $108.38

Round 108.375 to the nearest cent.

Practice Find the commission on each. Round to the nearest cent.

a. Sale price = $600
Commission = 2% **$12**

b. Sale price = $1250
Commission = 2.5% **$31.25**

Oral Exercises

State whether each is true or false.

1. A discount is an amount subtracted from the original amount. T
2. The commission is the regular price less the discount. F
3. The percent of discount tells the amount of commission. F
4. The sale price is less than the regular price. T
5. The sale price is the regular price less the discount. T

Exercises

A Find the discount and sale price for each. Round to the nearest cent if necessary.

1. Regular price = \$10
   Discount = 25% **\$2.50, \$7.50**
2. Regular price = \$5
   Discount = 10% **\$0.50, \$4.50**
3. Regular price = \$12
   Discount = 2.75% **\$0.33, \$11.67**
4. Regular price = \$30
   Discount = 15% **\$4.50, \$25.50**
5. Regular price = \$260
   Discount = 12.5% **\$32.50, \$227.50**
6. Regular price = \$9.85
   Discount = 4% **\$0.39, \$9.46**
7. Regular price = \$689.95
   Discount = 15% **\$103.49, \$586.46**
8. Regular price = \$249.95
   Discount = 10% **\$25, \$224.95**
9. Regular price = \$4570
   Discount = 12.5% **\$571.25, \$3998.75**
10. Regular price = \$25,750
    Discount = 6% **\$1545, \$24,205**

Find the commission. Round to the nearest cent if necessary.

11. Sale price = \$45
    Commission = 3% **\$1.35**
12. Sale price = \$125
    Commission = 2% **\$2.50**
13. Sale price = \$12.95
    Commission = $5\frac{1}{2}\%$ **\$0.71**
14. Sale price = \$58.60
    Commission = $1\frac{3}{4}\%$ **\$1.03**
15. Sale price = \$1375
    Commission = 12% **\$165**
16. Sale price = \$4575
    Commission = 5% **\$228.75**

B Use a calculator to help solve Exercises 17–19.

17. How much money would you save if you bought both the screen house and the tent canopy on sale? **\$23.39**

18. What is the percentage of discount on the tent canopy? **30.05%**

19. What is the percentage of discount on a tent with a sale price of \$64.95 and regular cost of \$84.95? **23.5%**

Camping Gear
Screen House REG. \$79.95 Today—18% OFF!!
Tent Canopy REG. \$29.95 Today—Save \$9.00

Find the discount, sale price, and commission rounded to the nearest cent.

20. Regular price = \$169
Discount = 30%
Commission = 4%
\$50.70, \$118.30, \$4.73

21. Regular price = \$3,560
Discount = 10%
Conmission = 6%
\$356, \$3204, \$192.24

C *Extending Thinking Skills*

22. Mr. Lehr sold 75 pairs of shoes each week for 10 weeks. Ms. Lorenzo sold 1 pair of shoes the first week, 2 pairs the second week, 4 pairs the third week, and so on, doubling the number sold each week. Which salesperson sold the most shoes by the end of the 10 weeks? Ms. Lorenzo

23. A store had a 20% discount on every item in stock. Would a customer be better off if the 15% sales tax were applied before or after the discount? After.

Thinking skills: 22. Finding a pattern 23. Reasoning

Mixed Review

Write each as a fraction in lowest terms. **24.** 35% $\frac{7}{20}$ **25.** 40% $\frac{2}{5}$ **26.** 125% $\frac{5}{4}$

Write each as a decimal. **27.** 25% 0.25 **28.** 140% 1.4 **29.** 11% 0.11 **30.** 60% 0.6

Evaluate for $a = 6$, $b = 3$. **31.** $\frac{2}{3}(a + b)$ 6 **32.** $\frac{a}{b}$ 2 **33.** $4(a - b)$ 12

Solve and check. **34.** $2(t + 2) < 16$ $t < 6$ **35.** $-4r < 12$ $r > -3$

COMPUTER ACTIVITY

The program below will compute the discount (D) and the sale price (S) of an item when you input the list price (L) and the percentage of discount (P).

```
10  PRINT "FIND THE SALE PRICE."
20  PRINT "ENTER THE LIST PRICE, L, AND"
30  PRINT "THE PERCENT OF DISCOUNT, D, (IN DECIMAL
    FORM)."
40  INPUT L,P
50  PRINT "THE LIST PRICE IS $";L
60  PRINT "THE PERCENT OF DISCOUNT IS ";P
70  D=P*L
80  S=L-D
90  PRINT "THE DISCOUNT IS $";D
100 PRINT "THE SALE PRICE IS $";S
110 END
```

Use this program to find the discount and sale price for each item.

1. List price, \$375; discount, 10%
\$37.50, \$337.50

2. List price, \$2250; discount, 30%
\$675.00, \$1575.00

3. List price, \$9479; discount, 5%
\$473.95, \$9005.05

4. List price, \$359; discount, $12\frac{1}{2}\%$
\$44.88, \$314.12

Determine which would cost less.

5. A \$95 radio with an 8% discount or a \$110 radio with a 20% discount.
\$95 radio (\$87.40 to \$88.00)

More Practice

Solve.

1. Marilyn deposits 15% of her paycheck into her savings account each week. Her paycheck this week is $215. How much should she deposit into savings? **$32.25**

2. Out of 30 students, 12 had perfect scores on the spelling test last week. What percentage of the students had perfect scores? **40%**

3. The library at Bayview School has 4,257 nonfiction books. This is 60% of all the books in the library. What is the total number of books in the library? **7095**

4. Steve sold 72 records today. He sold 75% of them in the afternoon. How many records did Steve sell in the morning. **18**

5. Of the 900 students at Harrison Middle School, 360 are in sixth grade, 225 are in seventh, and 315 are in eighth. What percentage of the students are in sixth grade? in seventh? in eighth? **6th 40%, 7th 25%, 8th 35%**

6. McNally's Snack Mix contains peanuts, pecans and almonds. The mix is 35% peanuts. How many ounces of peanuts are in a 12-ounce can? **4.2**

7. At the Round Hill Apartments, 70 percent of the apartments are rented to families with children. If 35 are rented to families with children, how many apartments are there in all? **50**

8. Last season the Cooley Cougars won 12 football games. This year they won 8 games. What was the percentage of decrease in games won? **$33\frac{1}{3}$%**

9. Jennifer went to the store to buy 48 different items. She found and bought 75% of the items on her list. How many items did she buy? **36**

10. Sol saved $24 by buying his tape player on sale for 40% off the regular price. What was the regular price of the tape player? **$60**

11. This year, 455 people went to the Pine City Fourth of July picnic. This number is 140% of the number that went last year. How many people went to the picnic last year? **325**

PROBLEM SOLVING: APPLICATIONS

10-7 Finding Sales Tax

Objective: To solve problems involving sales tax percentages.
ALL: 1–13

In many places you pay a sales tax when you buy an item. The sales tax is a percentage of the regular price of the item. The tables below show tax rates in certain states and cities for a recent year. If you live in a state that has a sales tax and a city that has a sales tax, both taxes are added to the price of an item you buy.

STATE SALES AND USE TAXES			
STATE	**% RATE**	**STATE**	**% RATE**
California	4.75	S. Dakota	4
Colorado	3	Texas	4.125
Missouri	4.225	Virginia	3
New York	4	Washington	6.5

CITY SALES TAXES			
CITY	**% RATE**	**CITY**	**% RATE**
Anaheim, CA	1.25	New York, NY	4.25
Berkeley, CA	1.75	Rapid City, SD	2
Boulder, CO	2.15	San Antonio, TX	1
Jefferson City, MO	1	Seattle, WA	0.925

Problem Use the tax rate table to find the total cost, including state tax, of a \$500 television set bought in Virginia.

Solution

Let T = amount of tax paid

$T = \$500 \times 0.03$ Tax = selling price × tax percent.

$T = \$15$

Let C = total cost of the television set

$C = \$500 + 15$ Total cost = selling price + tax.

$C = \$515$

The total cost of the television set is \$515.

Problems

You might wish to have students choose from among the techniques pencil and paper, mental math, estimation, and calculator. Have them choose the most appropriate technique for each problem.

Use the tax rate tables to solve. Round to the nearest cent if necessary.

1. Celia bought a car in Texas for \$1200. How much state tax did she pay? \$49.50

2. What would the New York state sales tax be on a \$500 video cassette recorder? **\$20**

3. A family in Colorado bought a \$12,500 camper. How much was the state sales tax? **\$375**

4. A man in Anaheim, CA, bought a new suit for \$125. How much city sales tax did he pay? **\$1.56**

5. In the state of Washington, what would the total cost be for a car priced at \$7500? **\$7987.50**

6. A tape deck was on sale in Seattle, WA, for \$379. What is the city tax on this tape deck? **\$3.51**

7. Mr. Montes paid \$10,000 for a boat. This price was 5% less than the original price. Mr. Montes lived in San Antonio, TX. How much city tax would he pay on the boat? **\$100**

8. A refrigerator originally priced at \$875 was on sale for \$696.99 in Berkeley, CA. What was the city tax on this refrigerator? **\$12.20**

To find the total state and city tax on an item, add the two percentage rates. An item bought in Berkeley, CA, would have a city tax of 1.75% plus a state tax rate of 4.75%. The total tax on the item would be 6.5%.

9. Ms. Flores bought a portable television on sale for \$175 in San Antonio, TX. What was the total tax she paid on the item? **\$8.97**

10. Ms. Harrison bought 4 new tires for a truck. Each tire was priced at \$50. She bought the tires in Boulder, CO. What was the approximate total cost for all of the tires, including tax? **\$210.30**

11. The Cho family in Seattle, WA, bought 3 coats that cost \$125, \$65, and \$50. What was the total cost for the coats, including tax? **\$257.82**

12. Mr. Sanders of Anaheim, CA, was given \$1,250 off the price of a \$10,450 van. What was the total price he paid for the van, including tax? **\$9752**

13. ***Data Search*** Find the total cost, including all taxes, of purchasing a \$200 stereo in your home town. If there is no state or local sales tax where you live, use the tax rate for a city or state near you. **answers will vary**

What's Your Decision?

Suppose you could choose to live in any of the cities listed in the chart on page 320. You estimate that you will spend \$5000 a year on taxable items. Which city would you consider if one of the factors important to you is the amount of sales tax you would pay in that city and state?
San Antonio, TX, has the lowest sales tax of the cities listed (5.125%).

10-8 Simple Interest

Objective: To find simple interest amounts.
MIN: 1–25; REG: 1–25 even, 26–35; MAX: 1–25 even, 26–37; ALL: Mixed Review
You may want to have students use calculators for this lesson.

Interest is a charge for the use of money. When you borrow money, you pay interest for the use of the money. When you place money in a savings account, the bank pays you interest for the use of your money. To find **simple interest**, you multiply the amount of money borrowed or saved by the interest rate and multiply that amount by the time the money is used.

Formula: Simple Interest

$I = Prt$; I = simple interest
P = principal (amount borrowed or saved)
r = interest rate
t = time

Example 1

Find the interest charged and the total amount. Round to the nearest cent.
\$300 at 1.5% per month for 6 months

Solution $I = Prt$ Interest = principal × rate × time
$= 300(0.015)(6)$ P = \$300, r = 1.5% per month, t = 6 months
$= 27$

The amount of interest is \$27.

$A = P + I$ Total amount = principal + interest
$= 300 + 27$
$= 327$

The total amount of money is \$327.

Practice Find the interest and the total amount. Round to the nearest cent.

a. \$250 at 10% per year for 3 years. **\$75, \$325**
b. \$75 at 5% per month for 6 months. **\$22.50, \$97.50**

An interest rate is given as a percentage for a unit of time such as 5% per year or $1\frac{1}{2}$% per month. When you use the formula for simple interest, the unit of time for the interest rate (r) and the time that the money is earning interest (t) must be the same.

Example 2

Find the interest and the total amount. Round to the nearest cent.
$600 at 12% per year for 6 months.

Solution $P = \$600$

$r = 12\%$ per year

$t = 6 \text{ mos.} = \frac{1}{2} \text{ year} = 0.5 \text{ year}$ — *r* and *t* must be for the same unit of time. Change months to years.

$I = Prt$

$= 600(0.12)(0.5)$

$= 36$

The amount of interest is $36.

$A = P + I$

$= 600 + 36$

$= 636$

The total amount of money is $636.

Practice Find the interest and the total amount. Round to the nearest cent.

a. $125 at 1% per month for 1 year. **$15, $140**

b. $2000 at 18% per year for 6 months. **$180, $2180**

Oral Exercises

Replace each variable to make a true statement.

1. $0.75 = x\%$ **75%**
2. 1 year $= y$ months **12**
3. 4 months $= d$ years $\frac{1}{3}$
4. $0.45 = t\%$ **45**
5. $0.2 = m\%$ **20%**
6. 3 months $= n$ years $\frac{1}{4}$
7. 18 months $= d$ years $1\frac{1}{2}$
8. $0.05 = x\%$ **5%**
9. $0.085 = b\%$ **8.5%**
10. 3 years $= t$ months **36**

Exercises

A Find the interest charged and the total amount paid. Round to the nearest cent.

1. $120 at 1% per month for 3 months **$3.60, $123.60**
2. $95 at 18% per year for 2 years **$34.20, $129.20**
3. $400 at 1% per month for 6 months **$24, $424**
4. $300 at 18% per year for 2 years **$108, $408**
5. $500 at 2% per month for 10 months **$100, $600**
6. $2500 at 12% per year for 2 years **$600, $3100**
7. $375 at 1.5% per month for 18 months **$101.25, $476.25**

8. \$2500 at 2% per month for 12 months **\$600, \$3100**
9. \$425 at 1.5% per month for 18 months **\$114.75, \$539.75**
10. \$750 at 1% per month for 24 months **\$180, \$930**
11. \$695 at $1\frac{1}{4}$% per month for 14 months **\$121.63, \$816.63**
12. \$200 at $1\frac{1}{2}$% per month for 6 months **\$18, \$218**
13. \$400 at 2% per month for 18 months **\$144, \$544**
14. \$750 at 10% per year for 3 years **\$225, \$975**
15. \$320 at 6% per year for 4 years **\$76.80, \$396.80**
16. \$1200 at 1% per month for 9 months **\$108, \$1,308**
17. \$700 at 2% per month for 2 years **\$336, \$1,036**
18. \$1000 at 12% per year for 4 months **\$40, \$1,040**
19. \$10,000 at 10% per year for 18 months **\$1,500, \$11,500**
20. \$100 at 18% per year for 30 months **\$45, \$145**
21. \$300 at 15% per year for 9 months **\$33.75, \$333.75**
22. \$400 at 1% per month for 2 years **\$96, \$496**
23. \$90 at 8% per year for 21 months **\$12.60, \$102.60**
24. \$845 at $1\frac{1}{2}$% per month for 3 years **\$456.30, \$1301.30**
25. \$2000 at $1\frac{1}{2}$% per year for 3 months **\$7.50, \$2007.50**

B Complete the table for a monthly finance charge of 1.5%. Always round the finance charge to the next whole cent.

Example: October \$1500 − 200 = \$1300 new amount

\$1300(0.015) = \$19.50 finance charge

\$1300 + 19.50 = \$1319.50 balance

	Month	Balance	Payment	New Amt.	Finance Chg.	Balance
	Oct.	\$1500	\$200	\$1300	\$19.50	\$1319.50
26.	Nov.	\$1319.50	\$200	\$1119.50	**\$16.80**	\$1136.30
27.	Dec.	\$1136.30	\$200	**\$936.30**	**\$14.05**	**\$950.35**
	Jan.	\$950.35	\$200	\$750.35	\$11.26	\$761.61
28.	Feb.	**\$761.61**	\$200	**\$561.61**	**\$8.43**	**\$570.04**
29.	Mar.	**\$570.04**	\$200	**\$370.04**	**\$5.56**	**\$375.60**
30.	Apr.	\$375.60	\$200	**\$175.60**	**\$2.64**	**\$178.24**
31.	May	\$178.24	**\$178.24**	-0-	-0-	-0-
32.	TOTAL				**\$78.24**	

In Exercises 33–35, write an equation and solve for the missing value. Round to the nearest tenth of a percent if necessary.

33. $I = \$8.44$; $P = \$75$; $t = 7\frac{1}{2}$ months $8.44 = 75 \times r \times 7\frac{1}{2}, r = 0.015$ or $1\frac{1}{2}\%$ per month

34. $I = \$12.50$; $r = 1\%$ per month; $t = 1$ year $12.50 = 0.01(12)P$; \$104.17

35. $I = \$27$; $P = \$450$; $r = 18\%$ per year $27 = 450 \times 0.18 \times T, T = 4$ months

C Extending Thinking Skills

36. Write the digits 1, 2, 3, 4, or 5 in the boxes to obtain a simple interest amount of approximately \$13. Use each digit only once.

$$I = Prt$$

$$I = (▮▮)(0.▮▮)(▮)$$ or $13 \cdot 0.25 \cdot 4$ / $25 \cdot 0.13 \cdot 4$

37. Lucky Larry found an amazing investment plan. For every \$2.50 invested, he had a 360% rate of increase. When he closed his plan, he had a total of \$138. How much did he invest? \$38.33

Thinking skills: 36, 37. Testing a conjecture

Mixed Review

Solve and check. **38.** $\frac{6}{5} = \frac{x}{2}$ 2.4 **39.** $\frac{n}{15} = \frac{64}{80}$ 12 **40.** $\frac{27}{9} = \frac{54}{t}$ 18

Write an expression for each if there are half as many trucks as cars.

41. Number of cars c **42.** Number of trucks $\frac{c}{2}$

Write an expression for each if $\frac{1}{3}$ of the trucks are red.

43. Number of trucks t **44.** Number of red trucks $\frac{1}{3}t$

NUMBERS TO ALGEBRA

You can use the skills you learned for solving equations to write a related formula for $I = Prt$.

Numbers	Algebra
Solve $15 = 100(0.05)t$ for t,	Solve $I = Prt$ for t.
$\frac{15}{100(0.05)} = t$	$\frac{I}{Pr} = t$

1. $t = \frac{D}{r}$
2. $l = \frac{V}{WH}$
3. $c = \frac{F - 32}{1.8}$
4. $A = \frac{p - b - c}{2}$

Write a related formula.

1. Solve $D = rt$ for t.

2. Solve $V = LWH$ for L.

3. Solve $F = 1.8C + 32$ for C.

4. Solve $p = 2A + B + C$ for A.

10-9 Making Circle Graphs

Objective: To make circle graphs.
MIN: 1–7; REG: 1–8; MAX: 1–9; ALL: Mixed Review

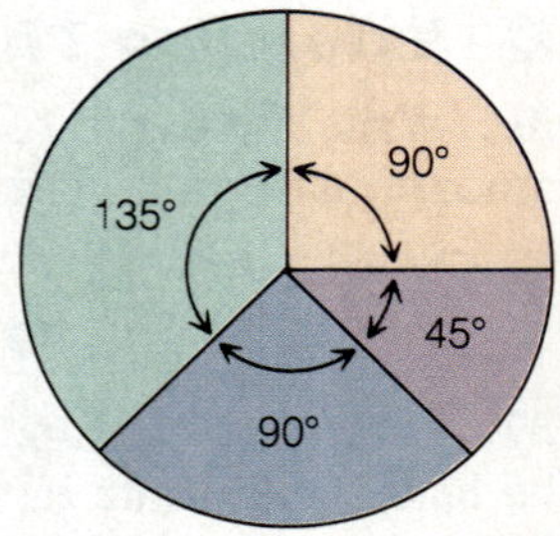

A **circle graph** is useful for picturing a total amount that is divided into parts. Each part of the circle graph is called a **sector**. A circle graph shows the relationships of the parts to each other and to the total.

The circle graph at right has four sectors. The measure of the angle for each sector is shown. The sum of the angles of the sectors of a circle equals 360°. To make sectors, you need to draw angles. You can use a protractor to draw angles for sectors.

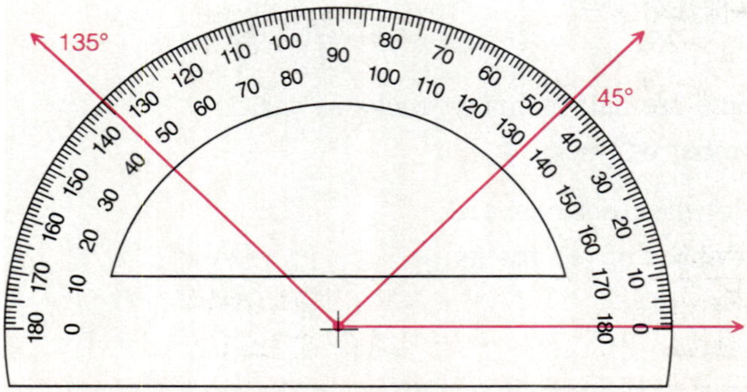

To make a circle graph, follow these steps:

1. Find the total for the data.
2. Express each part as a percent of the total.
3. Find the degrees for each sector.
4. Draw and label the sectors.
5. Title the graph.

Example

Make a circle graph to show the data. Denise's college expenses: \$1500 tuition; \$2250 room and board; \$1250 miscellaneous.

Solution

$1500 + 2250 + 1250 = 5000$ The circle represents the total expenses.

Tuition: $\frac{1500}{5000} = 0.30 = 30\%$

Room/Board: $\frac{2250}{5000} = 0.45 = 45\%$

Miscellaneous: $\frac{1250}{5000} = 0.25 = 25\%$

Find what percent of the whole each item represents.

Tuition: $0.30 \times 360 = 108°$
Room/Board: $0.45 \times 360 = 162°$
Miscellaneous: $0.25 \times 360 = 90°$

Every circle is made up of 360°. Find what part of the 360° each sector represents.

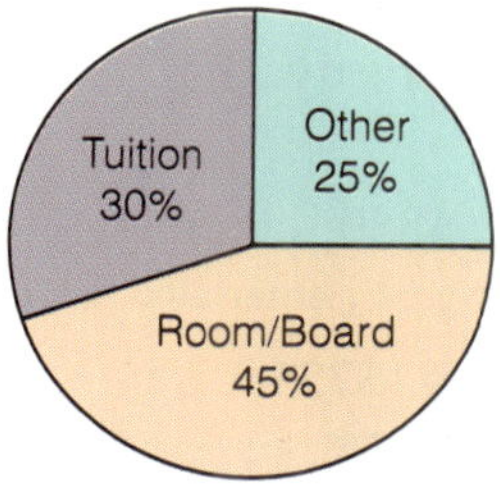

Denise's College Expenses

Use a compass to draw a circle. Use a protractor to draw each angle. Label each sector.

Give the graph a title that clearly indicates what the graph shows.

Practice Make a circle graph. Plans of seniors at Cooper High: Four-year college, 100; work, 55; two-year college, 30; armed services, 10; undecided, 5.

4-year college 180°
work 99°
2-year college 54°
armed service 18°
undecided 9°

Oral Exercises

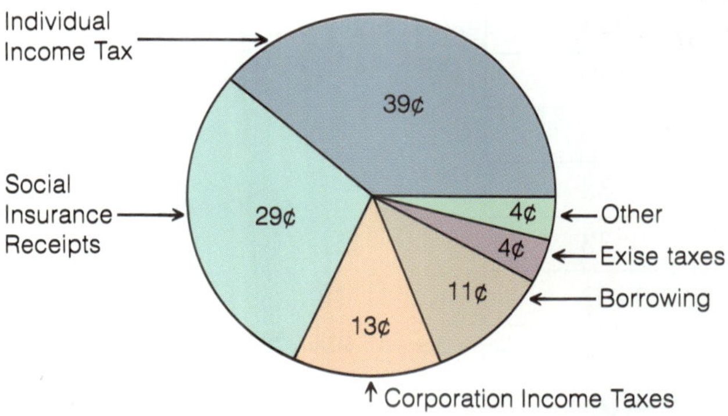

Federal Income for a Given Year (per dollar)

Answer each question about the circle graph.

1. What is the total amount represented in this graph? **\$1 of revenue**
2. Do any of the sectors show more than 180°? **no**
3. What is the greatest source of federal income? **Individual Income Tax**
4. Which two sources of income together represent more than 50%?
5. Is 39¢ the same as 39% in this graph? **yes**

4. indiv. income tax + soc. ins. receipts *or* indiv. income tax + corp. inc. tax

Exercises

A Make a circle graph to show the data.

1. Movies in town over the last six months: adventure, 50%; science fiction, 20%; children's, 15%; comedy, 10%; other, 5%. See additional answers.

2. After-school jobs: restaurants and food stands, 45%; stores, 35%; private homes, 10%; other, 10%. restaur. 162°; stores 126°; homes 36°; other 36°

3. Transportation to school: walk, 45%; bus, 30%; bicycle, 17%; car, 8%. 3. walk 162° bus 108° bicycle 61° car 29°

4. Rainfall, in centimeters: summer, 30 cm; fall, 15 cm; winter, 30 cm; spring, 45 cm. summer 90°; fall 45°; winter 90°; spring 135°

5. Students' favorite records: popular, 105; soundtrack, 42; comedy, 38; classical, 15. popular 189°; soundtrack 76°; comedy 68°; classical 27°

6. Ms. Bradley's expenses for one month: rent, \$424; food, \$350; clothing, \$150; entertainment, \$50; other, \$25. rent 153°; food 126°; clothing 54°; entertain. 18°; other 9°

7. Norman's car expenses for one year: payments, \$2520; gas, \$725; insurance, \$450; repairs, \$210; registration, \$75. payments 228°; gas 66°; ins. 41°; repairs 19°; reg. 7°

B

8. Make a circle graph to show the monthly electricity use.

	Present Reading (kWh)	Previous Reading (kWh)
Jan–Feb	6781	4993
Mar–Apr	8305	6781
May–Jun	9738	8305
Jul–Aug	0538	9738
Sep–Oct	1163	0538
Nov–Dec	2403	1163

To find the number of kilowatt hours (kWh) used, subtract the previous reading from the present reading (except when the meter passes 9999 and starts at 0000).

Jan–Feb 87°; Mar–Apr 74°; May–Jun 70°; Jul–Aug 39°; Sep–Oct 30°; Nov–Dec 60°

C Extending Thinking Skills

9. Make a circle graph, using the numbers 1–20 and the categories below.

a. numbers with 1 or 2 factors 162°

b. numbers with 3 or 4 factors 126°

c. numbers with 5 or 6 factors 72°

Thinking skill: 9. Organizing

Mixed Review

Write each as a fraction in lowest terms. **10.** 8% $\frac{2}{25}$ **11.** 35% $\frac{7}{20}$ **12.** 48% $\frac{12}{25}$

Solve and check. **13.** $4t - 19 = -7$ 3 **14.** $11(t + 3) = 18t - 2$ 5

PROBLEM SOLVING: STRATEGIES

10-10 Problems without Solutions

Objective: To solve nonroutine problems; to recognize problems without solutions.
MIN: 1–3; REG: 1–5; MAX: 1–7

For some problems, no solution is possible. In the following example, the strategy **Make an Organized List** is used to determine that there is no solution.

Problem Nancy said she lost 72¢ behind the sofa. Although she wasn't sure what coins she had lost, Nancy knew she did not have a half-dollar and she thought she had lost 5 coins. What 5 coins did Nancy lose that totaled 72¢?

You can use an organized list to make sure you have checked all possibilities. Possible coins to consider are quarters, dimes, nickels, and pennies. Since she had 72¢, she must have had 2 pennies. This reduces the problem to finding 3 coins that total 70¢. The organized list below shows that there are no combinations of quarters, dimes, nickels, and pennies that give 72¢ in 5 coins.

Quarters	Dimes	Nickels	Pennies	Total
3	0	0	2	77¢
2	1	0	2	62¢
1	2	0	2	47¢
1	1	1	2	42¢
1	0	2	2	37¢

If she did not have a half-dollar, Nancy could not have lost 5 coins that totaled 72¢.

Problem-Solving Strategies

Choose the Operations	**Write an Equation**
Guess, Check, Revise	**Simplify the Problem**
Draw a Picture	**Make an Organized List**
Make a Table	**Use Logical Reasoning**
Look for a Pattern	**Work Backwards**

This chart shows the strategies presented in previous chapters.

Problems

Solve. If the problem has no solution, show why.

Sandwiches prices include tax	
Hamburger	$1.25
Chicken	$1.65
Pizza	$1.10
Roast Beef	$1.45

1. Ms. Banfield bought 2 different sandwiches from this menu and paid for them with a $5 bill. The clerk said the change was $2.35. Which sandwiches did Ms. Banfield buy? no solution

2. Nicole was in a walkathon for charity. She got 25 people to pledge $0.15 for each kilometer she walked and 45 people to pledge $0.25 for each kilometer she walked. She walked 35 km in the walkathon. How much money did she collect in pledges? $525

3. Thirty-five people signed up for a 25-km run. The race organizer wants to give each runner a different 4-digit number by using the digits 3, 4, 5, and 6 so that each digit is used once in each number. How many more people could sign up for the race and each be given a different number? 0 (only 24 possible)

4. Kara found 17 coins in an old pocketbook. She counted a total of $1.15. How many coins of each kind did she find?
2Q, 2D, 8N, 5P; or 10D, 2N, 5P; or 1Q, 2D, 14N; or 3Q, 2D, 2N, 10P

5. Will works for the school yearbook. He sets the page numbers for the yearbook by hand. If each digit has to be set separately, and there are 250 pages in the yearbook, how many digits must Will set? 642 (9 × 90 × 2 = 151 × 3)

6. A gardener bought 200 m of wire fencing to fit exactly around the outside of a rectangular flower bed. The length of the flower bed is 12 m more than the width. What is the length and width of this flower bed? 44m × 56 m

7. Jeremy makes 3-legged stools and 4-legged stools. He has 48 legs for stools. He wants to make 12 stools, some of each kind. Can he do this and use all the legs? No.

Enrichment

Compound Interest

The formula $I = Prt$ can be used to find **compound interest**. When interest is **compounded**, the amount of interest earned in a given time period is added to the principal. Interest is then earned on the new principal. Interest may be compounded annually (every 1 year), semiannually (every $\frac{1}{2}$ year), quarterly (every $\frac{1}{4}$ year), monthly (every $\frac{1}{12}$ year), or daily. The example below shows how to use a calculator to find compound interest.

Suppose $5000 is deposited at an interest rate of 10% per year, compounded semiannually. What will the principal be after 1 year?

First $\frac{1}{2}$ Year

		Display	
Store 5000 in memory.	5000 [M+]	5000	Principal.
Multiply the principal by the interest rate, 0.1, and the time period, 0.5 year, to find the interest.	5000 [×] 0.1 [×] 0.5 [=]	250	Interest.
Add the interest to the principal amount stored in the memory to get the new principal.	250 [+] [MR] [=]	5250	New principal.

Second $\frac{1}{2}$ Year

Replace 5000 with 5250 in memory.	[M+]		
Find the interest for the second half-year, using the new principal.	5250 [×] 0.1 [×] 0.5 [=]	262.5	Interest.
Add the interest to the principal.	262.50 [+] [MR] [=]	5512.5	Principal at the end of the year.

The amount after the first six months is $5250. The amount after the full year is $5512.50.

Use a calculator and the formula $I = Prt$ to solve.

1. Amount deposited: $150. Rate of interest: 5% per year compounded semiannually. What will the principal be after 2 years? $165.57

2. Amount deposited: $3500. Rate of interest: 6% per year, compounded quarterly. What will the principal be after 2 years? $3942.72

Chapter 10 Review

10-1 Use a proportion or an equation to solve.

1. 40% of 135 54
2. 5% of 240 12
3. 2.5% of 90 2.25

10-2 Use a proportion or an equation to solve.

4. What percent of 60 is 12? 20%
5. 42 is what percent of 105? 40%
6. What percent of 120 is 90? 75%
7. 7 out of 28 is what percent? 25%

10-3 Use a proportion or an equation to solve. Round to the nearest tenth if necessary.

8. 35 is 20% of what number? 175
9. 50% of what number is 51? 102
10. 60% of what number is 30? 50
11. 7 is 1% of what number? 700

10-4 Solve by writing an equation.

12. A skirt costs 80% as much as a sweater. What is the cost of each if the total cost for the skirt and sweater is $45? sweater $25, skirt $20

10-5 Find the percent of increase or decrease.

13. Original amount = $80; new amount = $96 20% increase
14. Original amount = $140; new amount = $196 40% increase
15. Original amount = $150; new amount = $105 30% decrease

10-6 Solve.

16. When Rosa eats at a restaurant, she leaves a tip that is 15% of her bill. How much tip will she leave if the bill is $12.00? $1.80

10-7 Find the discount and the sale price for each.

17. Regular price = $25
    Discount = 20% $5, $20
18. Regular price = $1,750
    Discount = 4% $70, $1,680

Find the commission. Round to the nearest cent if necessary.

19. Sale price = $95
    Commission = 3% $2.85
20. Sale price = $705.20
    Commission = 1.5% $10.58

10-8 Find the interest and the total amount. Round to the nearest cent if necessary.

21. $300 at 8% per year for 2 years $48, $348
22. $160 at 1.5% per month for 7 months $16.80, $176.80

10-9

23. Make a circle graph to show the data. Library books checked out: mystery, 30; fiction, 65; nonfiction, 105; children's, 60; other, 40.
Mystery 36° Fiction 78° Nonfiction 126° Children's 72° Other 48°

Chapter 10 Test

Use a proportion or an equation to solve.

1. 25% of 64 16
2. 110% of 50 55
3. 12.5% of 40 5

Use a proportion or an equation to solve.

4. What percent of 25 is 7? 28%
5. 26 is what percent of 65? 40%
6. What percent of 152 is 76? 50%
7. 15 out of 150 is what percent? 10%

Use a proportion or an equation to solve. Round to the nearest tenth if necessary.

8. 3 is 5% of what number? 60
9. 90% of what number is 63? 70
10. 40% of what number is 34? 85
11. 66 is 75% of what number? 88

Solve by writing an equation.

12. James saves \$17.40 when he bought a fish tank at 30% off the original price. What was the original price of the fish tank? $0.3f = 17.40$, $f = \$58$

Find the percent of increase or decrease.

13. Original amount = \$150; new amount = \$180 20% increase
14. Original amount = \$70; new amount = \$133 90% increase
15. Original amount = \$225; new amount = \$45 80% decrease
16. Original amount = \$650; new amount = \$455 30% decrease

Solve.

17. Pam teaches karate, and 35% of her students are children. If Pam has 180 students, how many are children? 63

Find the discount and the sale price for each. Round to the nearest cent if necessary.

18. Regular price = \$50
    Discount = 40% \$20, \$30
19. Regular price = \$3,550
    Discount = 6% \$213, \$3,337

Find the commission. Round to the nearest cent if necessary.

20. Sale price = \$70 \$2.80
    Commission = 4%
21. Sale price = \$73.90 \$5.91
    Commission = 8%

Find the interest and the total amount. Round to the nearest cent if necessary.

22. \$350 at 15% per year for 6 years \$315, \$665
23. \$700 at 2% per month for 3 months \$42, \$742

24. Make a circle graph to show the data. Greeting cards sold: birthday, 25%; anniversary, 15%; graduation, 10%; special occasion, 30%; other, 20%.
Birthday = 90° Anniv = 54° Graduation = 36° Special Occasion = 108° Other = 72°

Cumulative Review

Solve. Reduce to lowest terms.

1. $4\frac{1}{8} \div \left(-1\frac{1}{4}\right)$ $-3\frac{3}{10}$

2. $-3\frac{3}{4} \div 5\frac{5}{8}$ $-\frac{2}{3}$

3. $-6x = -\frac{3}{4}$ $\frac{1}{8}$

4. $\frac{2}{3}v = \frac{5}{12}$ $\frac{5}{8}$

5. $2\frac{1}{4} = -12a$ $-\frac{3}{16}$

6. $-5\frac{5}{6} = 2\frac{1}{2}x$ $-2\frac{1}{3}$

Solve by writing an equation.

7. Susanne ran the same distance each day for 5 days. She ran a total of 17 km. How far did she run each day? $5x = 17$ 3.4 km

8. 21 people attended math class on Friday. This is $\frac{7}{8}$ of the number of people signed up for the class. How many people are signed up for the class? $\frac{7}{8} \times p = 21, p = 24$

Solve.

9. $12c + 4c = 80$ 5

10. $-48 = 7v + 5v$ -4

11. $22x - 17x = -65$ -13

12. $150 = 30t - 5t$ 6

13. $7b - 2b = 145$ 29

14. $-96 = 12z - 4z$ -12

15. $3(x - 2) + x = 9$ $3\frac{3}{4}$

16. $2y + 4(y + 5) = 80$ 10

17. $2x + 12(x + 2) = 94$ 5

18. $-56 = -4n + 2(n + 7)$ 35

19. $5(d + 4) + 7d = 68$ 4

20. $c + 3(c - 4) = 16$ 7

Find the actual dimensions of the garden. Scale is 2 mm = 0.5 m.

21. Length (l) 80 m

22. Width (w) 100 m

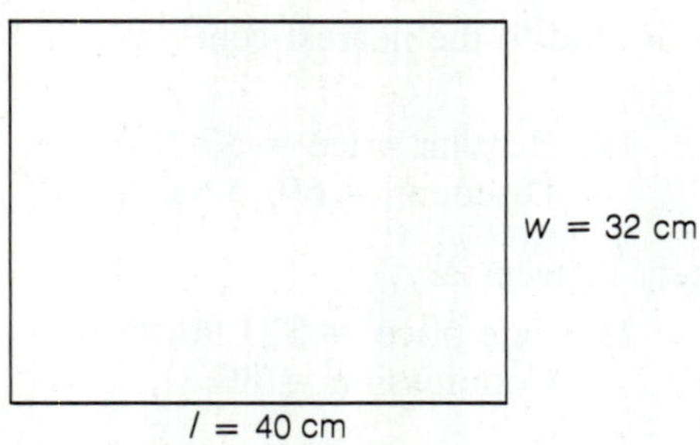

Use the scale of 1 cm = 0.5 m to find the scale dimensions of each.

23. Gate: 1.5 m × 2 m 3 cm × 4 cm

24. Fence: 7.5 m × 1.5 m 15 cm × 3 cm

Solve.

25. Blueberries cost $2.69 for the 16-ounce size and $1.42 for the 8-ounce size. Which size is the better buy? 16 oz

Chapter 11
Equations in Geometry

11-1 Basic Figures

Objective: To recognize and identify basic geometric figures.
MIN: 1–8; REG: 1–8 even, 9–13; MAX: 1–8 even, 9–14; ALL: Mixed Review

The three basic geometric figures are point, line, and plane.

	Figure	Name	Symbol
A **point** is the simplest geometric figure. It shows a location.		Point *A*	*A*
Given any two points, there is only one **line** through the two points.		Line *AB*	$\overleftrightarrow{AB}$
Three points that are not on the same line determine a **plane**.	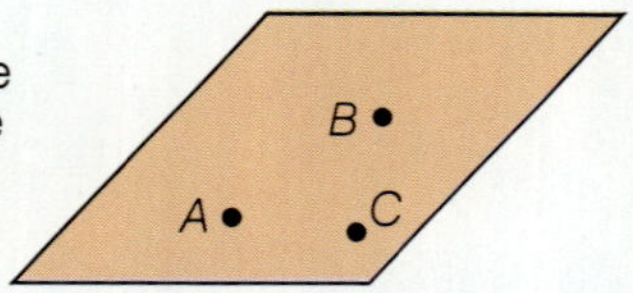	Plane *ABC*	*ABC*

The arrows at the end of a line show that the line goes on endlessly in two directions.

A plane is flat and continues in all directions.

Segments and rays are parts of a line.

	Figure	Name	Symbol
A **segment** *PQ* includes endpoints *P* and *Q* and all points between.		segment *PQ*	$\overline{PQ}$
A **ray** *AB* extends in one direction from endpoint *A*.	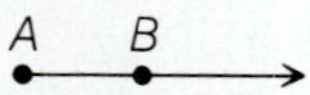	ray *AB*	$\overrightarrow{AB}$

Example

Name a point, a line, a plane, a segment, and a ray on the figure.

Solution W
$\overleftrightarrow{WX}$
WXZ
$\overline{XW}$
$\overrightarrow{YX}$

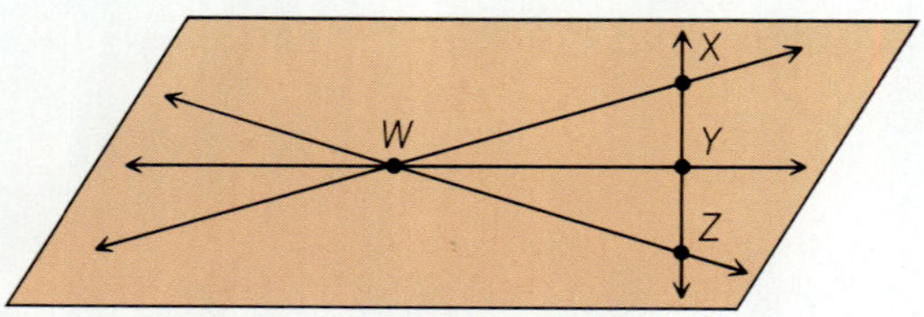

Practice:
Points: X, Y, Z
Lines: $\overleftrightarrow{WY}$, $\overleftrightarrow{WZ}$, $\overleftrightarrow{XZ}$
Segments: $\overline{YW}$, $\overline{ZW}$, $\overline{XY}$, $\overline{YZ}$, $\overline{XZ}$
Rays: $\overrightarrow{YW}$, $\overrightarrow{YZ}$, $\overrightarrow{WX}$, $\overrightarrow{XW}$, $\overrightarrow{WY}$, $\overrightarrow{WZ}$, $\overrightarrow{ZW}$, $\overrightarrow{XZ}$, $\overrightarrow{ZX}$

Practice Name another point, line, segment, and ray on the figure.

Oral Exercises

State whether the object suggests a point, a ray, a line, or a plane.

1. tip of an ice pick point
2. top of a desk plane
3. the tip of a scissors point
4. a flat ceiling plane
5. a tightly-stretched string line
6. a laser beam ray

Exercises

A Use the figure at right for Exercises 1–8.

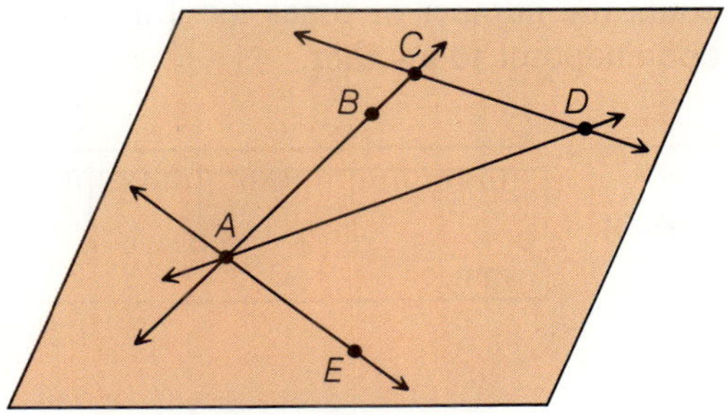

1. Name three points. *A, B, C, D, E*
2. Name three lines. $\overleftrightarrow{AC}$, $\overleftrightarrow{AB}$, $\overleftrightarrow{CD}$, $\overleftrightarrow{DA}$, $\overleftrightarrow{AE}$, $\overleftrightarrow{BC}$
3. Name a plane. *ACD, AED, BCD,* etc.
4. Name two segments with endpoint *C*.
5. Name four segments with endpoint *A*.
6. Name two lines that pass through point *C*.
7. Name two rays with endpoint *D*.
8. Name two rays with endpoint *B*.

B Use the figure at right for Exercises 9–13.

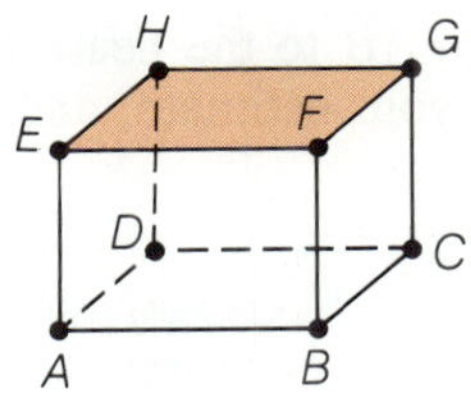

9. Name three segments with endpoint *A*.
10. Name two planes that include $\overline{EF}$.
11. Name two planes that include $\overline{CG}$.
12. Name twelve different segments.
13. Name six different planes.

13. *AEF, FBC, CGH, HDA, FGH, ABC* (note: *AEF* could be named *ABE*, etc.)

4. $\overline{BC}$, $\overline{AC}$, $\overline{CD}$
5. $\overline{AE}$, $\overline{AB}$, $\overline{AC}$, $\overline{AD}$
6. $\overleftrightarrow{AC}$, $\overleftrightarrow{DC}$, $\overleftrightarrow{BC}$
7. $\overrightarrow{DC}$, $\overrightarrow{DA}$
8. $\overrightarrow{BC}$, $\overrightarrow{BA}$
9. $\overline{AE}$, $\overline{AB}$, $\overline{AD}$
10. *EFH, EFB*
11. *CGF, CGH*
12. $\overline{AB}$, $\overline{BC}$, $\overline{CD}$, $\overline{DA}$, $\overline{EA}$, $\overline{FB}$, $\overline{GC}$, $\overline{HD}$, $\overline{EF}$, $\overline{FG}$, $\overline{GH}$, $\overline{HE}$

C Extending Thinking Skills

14. Copy and complete the third and fourth figures below. Then draw the next two figures to find the next three numbers in the pattern.

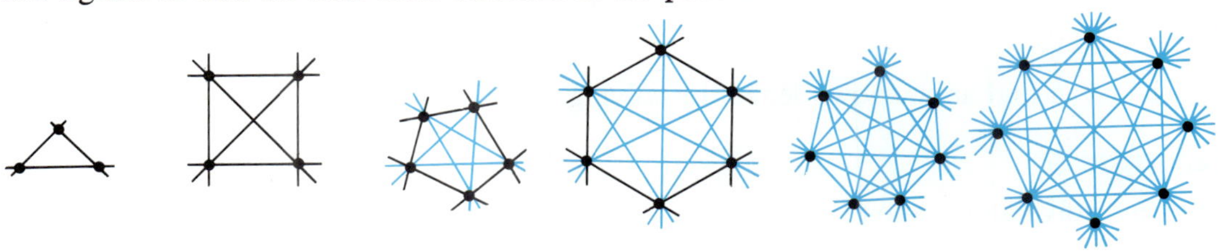

3 points	4 points	5 points	6 points	7 points	8 points
3 lines	6 lines	10 lines	15 _?_ lines	21 _?_ lines	28 _?_ lines

Thinking skills: 14. Finding a pattern

Mixed Review

Solve and check. **15.** $2(m - 9) + 5m = 6 + m$ 4 **16.** $29.6 - t = -3.9$ 33.5

17. $2x < 5$ $x < 2.5$

18. $3x - 6 > 15$ $x > 7$

19. $4x + 2 > 10$ $x > 2$

20. $-2x + 1 < 3$ $x > -1$

11-2 Length and Perimeter

Objective: To estimate and measure lengths of segments; to write equations to find perimeters and unknown lengths.
MIN: 1–16; REG: 1–16 even, 17–20; MAX: 1–16 even, 17–23; ALL: Mixed Review

To find the length of a segment, choose a unit of length and count the number of times the unit can be laid end-to-end from one endpoint to another.

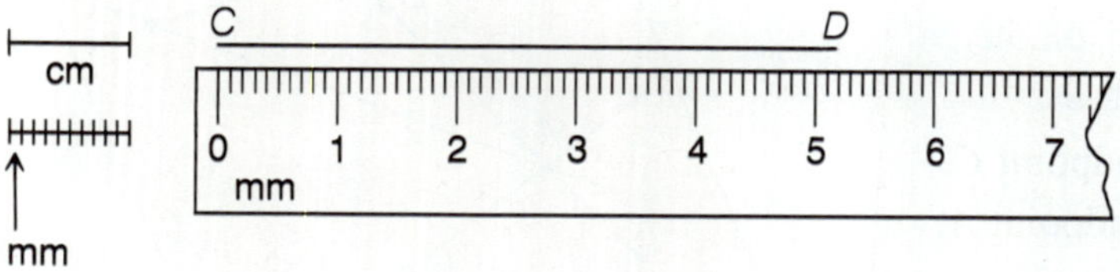

The length of $\overline{CD}$ is equal to 5.2 cm.
We write $CD = 5.2$ cm.

To give students a sense of unit lengths, discuss metric measurements of common items–e.g., a soup can is 10 cm high.

Example 1

Estimate the length of $\overline{AB}$ to the nearest centimeter. Use a ruler to check your estimate, and give the length.

Solution

$AB \approx 4$ cm — Mentally picture one centimeter to estimate how many centimeters long $\overline{AB}$ is.

Check $AB = 4.3$ cm — The estimate is about right.

Practice Estimate the lengths. Use a ruler to check, and give the length.

a. $CD = \underline{?}$ cm = **3 cm**
b. $DE = \underline{?}$ cm = **2 cm**

You can often find an unknown length by solving an equation.

Example 2

Write and solve an equation to find the length of $\overline{CD}$.

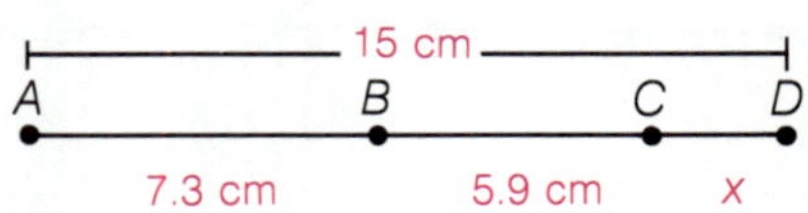

Solution Let $x = CD$. Then — Let x be the unknown length.

$$7.3 + 5.9 + x = 15$$
$$13.2 + x = 15$$
$$-13.2 + 13.2 + x = -13.2 + 15$$
$$x = 1.8$$

$CD = 1.8$ cm

Practice Write and solve an equation to find the length of $\overline{BC}$.

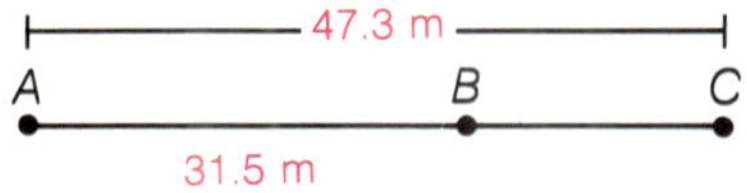

31.5 + x = 47.3 BC = 15.8 m

The total distance around a figure is called its **perimeter**. The perimeter of a rectangle is equal to twice its length plus twice its width. The formula is written as $P = 2l + 2w$.

You can find the perimeter of a nonrectangular figure by adding the lengths of the sides.

Example 3

Find the perimeter.

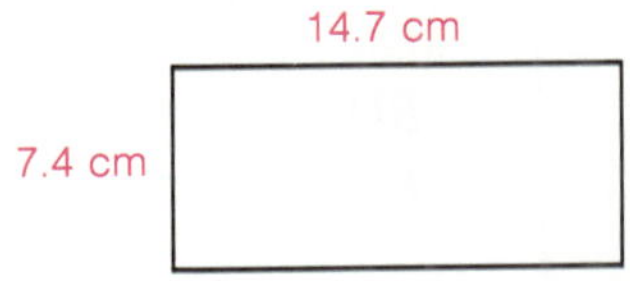

Solution
$$\begin{aligned} P &= 2l + 2w \\ &= 2(14.7) + 2(7.4) \\ &= 29.4 + 14.8 \\ &= 44.2 \end{aligned}$$

The perimeter is 44.2 cm.

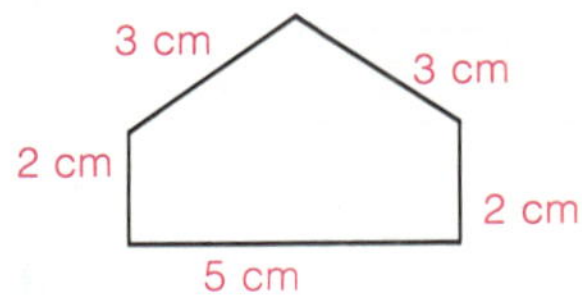

Practice Find the perimeter.
15 cm

When you know the perimeter of a figure but not the length of one side, you can write and solve an equation to find the unknown length.

Example 4

Write and solve an equation to find the length of $\overline{AB}$.

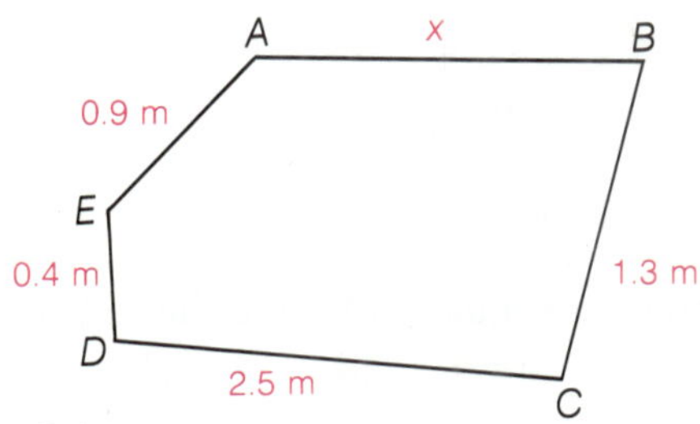

Perimeter = 7.2 m

Solution Let $x = AB$.

$$\begin{aligned} x + 1.3 + 2.5 + 0.4 + 0.9 &= 7.2 \\ x + 5.1 &= 7.2 \\ x + 5.1 - 5.1 &= 7.2 - 5.1 \\ x &= 2.1 \end{aligned}$$

$AB = 2.1$ m

Practice

Write and solve an equation to find the unknown length.
x + 5.1 + 3.7 + 3.9 = 17
x = 4.3

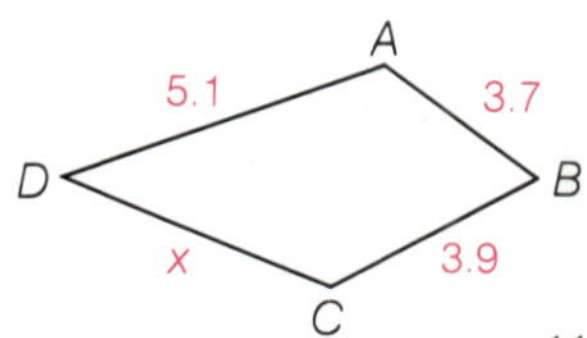

Perimeter = 17 cm

Oral Exercises

Give an equation that is suggested by each figure.

1.

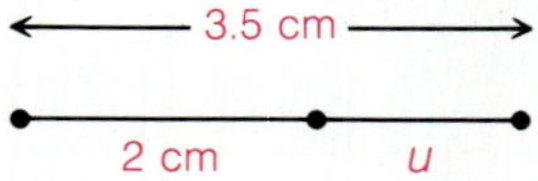

$u + 2 = 3.5$

2.

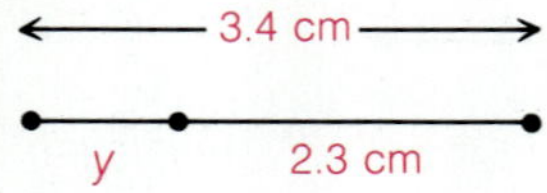

$Y + 2.3 = 3.4$

3.

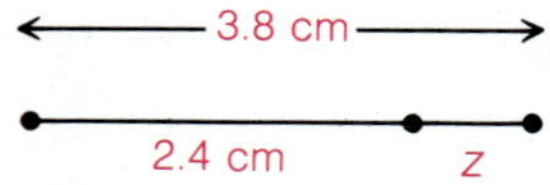

$2.4 + z = 3.8$

Exercises

A Estimate the length of each segment to the nearest centimeter. Use a ruler to check your measure, and give the length.

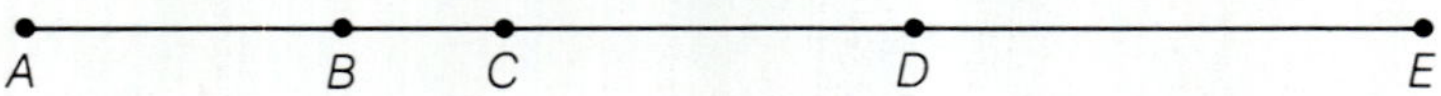

1. $AC = \underline{?}$ 3.2 cm

2. $BD = \underline{?}$ 3.8 cm

3. $CE = \underline{?}$ 6.1 cm

4. $BC = \underline{?}$ 1.1 cm

5. $BE = \underline{?}$ 7.2 cm

6. $AE = \underline{?}$ 9.3 cm

Write and solve an equation to find each unknown length. 7–16. See additional answers.

7.

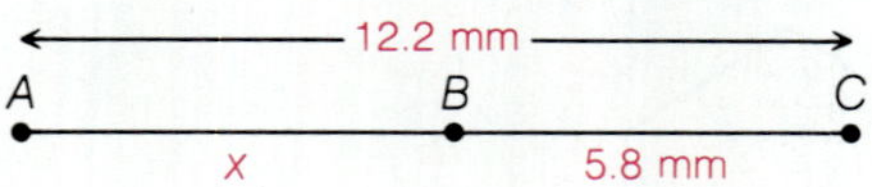

8.

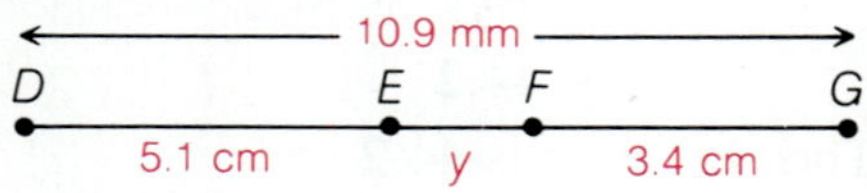

9.

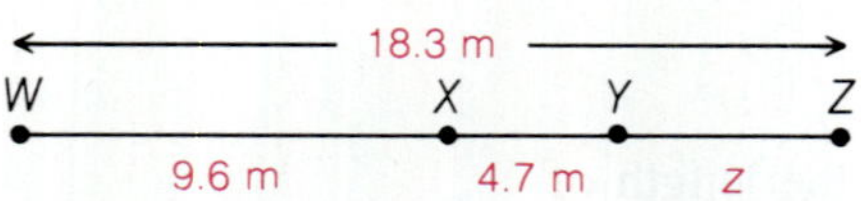

10.

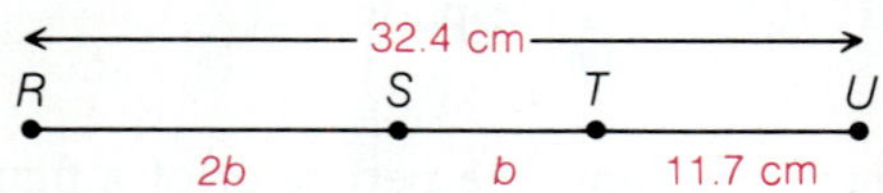

Find the perimeter.

11.

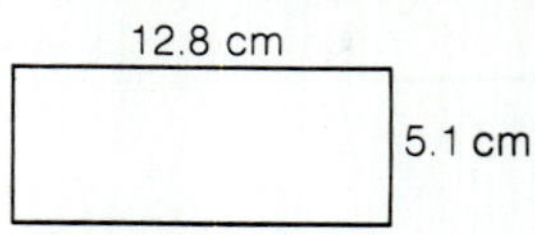

12.

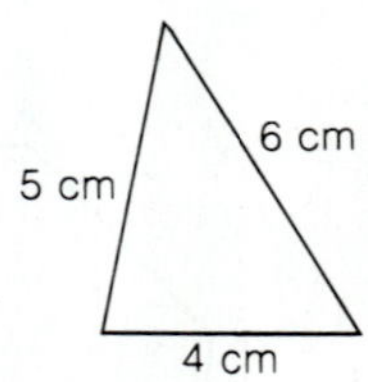

13.

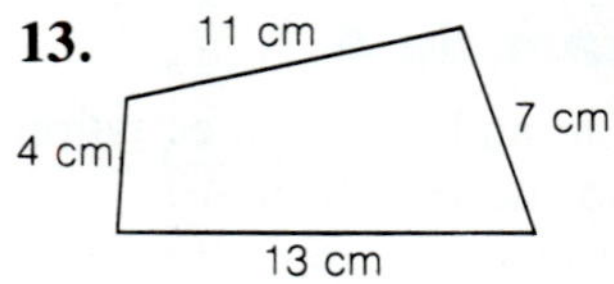

Write and solve an equation to find the unknown length.

14.

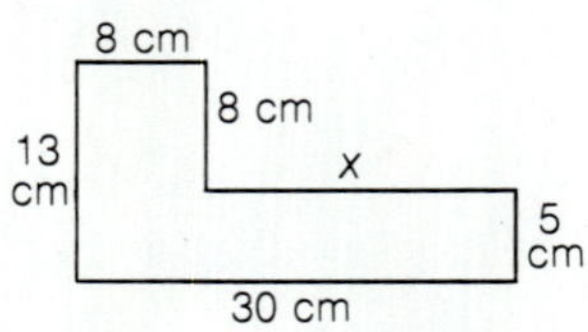

Perimeter = 86

15.

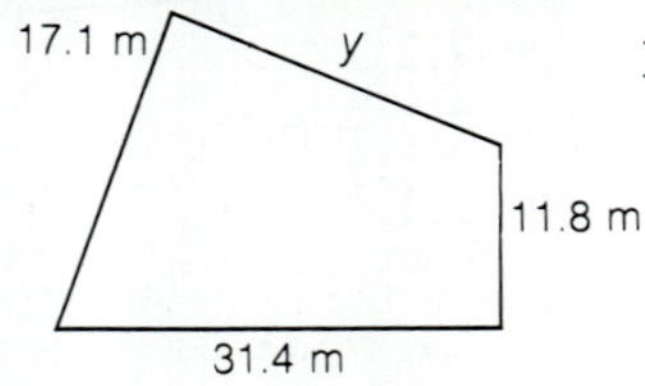

Perimeter = 76.6

16.

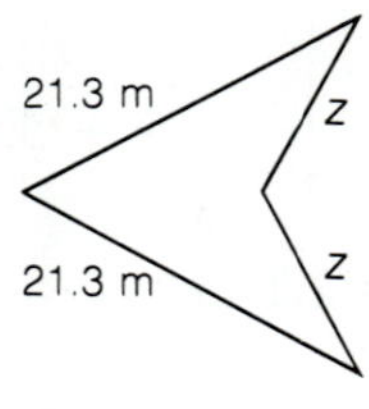

Perimeter = 68.6

B

17. The perimeter of the triangle at right is 14 cm. Write and solve an equation to find the lengths of the sides. **14 = x + x + 1.5x**
x = 4, sides are 4 cm, 4 cm, 6 cm

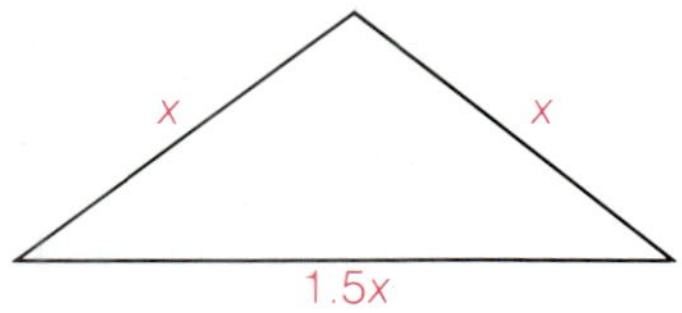

18. An 8.9 km cross-country course has an 0.8 km section that is flat, a 2.3 km section on a slight grade, a section on a steep uphill grade, a 4.5 km section on a slight downhill grade, and an 0.8 km section on rough ground. Write and solve an equation to find the length of the steep uphill grade.
0.8 + 2.3 + g + 4.5 + 0.8 = 8.9   g = 0.5 km

19. A rectangle with a perimeter of 96 cm has a length 3 times its width. Write and solve an equation to find the lengths of the sides.
2(3x) + 2x = 96   x = 12   3x = 36

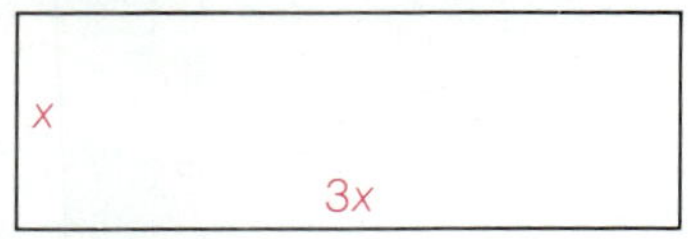

20. A field has sides of length 519.3 m, 391.4 m, and 791.6 m. Its perimeter is 2141.5 m. Write and solve an equation to find the length of the fourth side.
s + 519.3 + 391.4 + 791.6 = 2,141.5   s = 439.2 m

C Extending Thinking Skills **Thinking skills: 21–23. Using space visualization; reasoning**

Write and solve equations to find x and y in each figure below. Then find the perimeter of the figure. Assume all angles are right angles. **See additional answers.**

21.

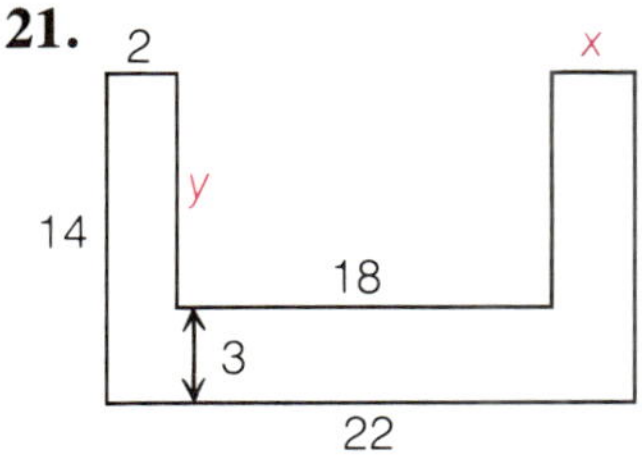

22.

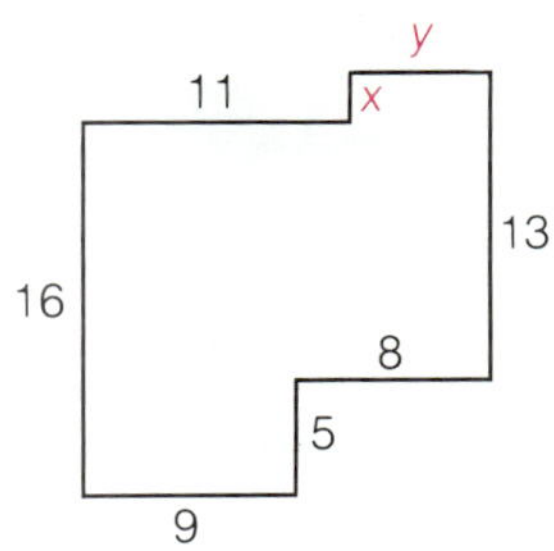

23.

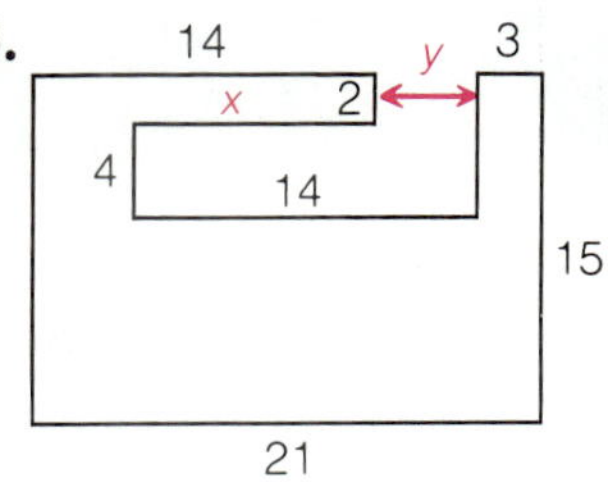

Mixed Review

Write as a decimal. **24.** $9\frac{1}{4}$ **9.25** **25.** $\frac{6}{5}$ **1.2** **26.** $-\frac{3}{10}$ **−0.3** **27.** $2\frac{5}{8}$ **2.625**

Write as a mixed number. **28.** $\frac{11}{3}$ $3\frac{2}{3}$ **29.** $\frac{14}{5}$ $2\frac{4}{5}$ **30.** $\frac{9}{4}$ $2\frac{1}{4}$

Evaluate for $a = 2$, $b = 1.5$, $c = 1.8$ **31.** $2a - bc$ **1.3** **32.** $-a(b - c)$ **0.6**

Write each percent as a fraction or mixed number in lowest terms.
33. 140% $1\frac{2}{5}$ **34.** 250% $2\frac{1}{2}$ **35.** 108% $1\frac{2}{25}$ **36.** 70% $\frac{7}{10}$ **37.** 37% $\frac{37}{100}$

Write an equation to find each. **38.** 60% of 12 **x = 0.6(12), 7.2** **39.** 0.5% of 60 **x = 0.005(60), 0.3**
40. 16 is what percent of 40? **40%** **41.** 27 is what percent of 18? **150%**

11-3 Angles and Angle Measures

Objective: To measure angles; to classify angles as right, acute or obtuse; to write equations to find angle measures.
MIN: 1–16; REG: 1–16 even, 17–22, MAX: 1–16 even, 17–24; ALL: Mixed Review

A navigator uses angles and angle measures to plot a ship's course.

Figure	Name	Symbol
An **angle** is two rays with a common endpoint called the **vertex**. Each ray is called a side of the angle. 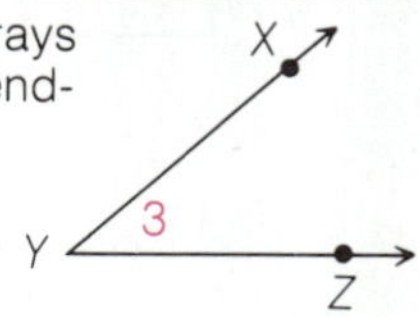	angle *XYZ* or angle *Y* or angle 3	$\angle XYZ$ or $\angle Y$ or $\angle 3$

A **protractor** is used to measure angles. The unit of angle measure is the **degree** (°). To measure an angle, place the arrow of the protractor at the vertex of the angle and the baseline of the protractor along one side of the angle. Then read the number of degrees from the other side of the angle. In the drawing at the right, the measure of $\angle ABC$ is 50°. We write: $m\angle ABC = 50°$.

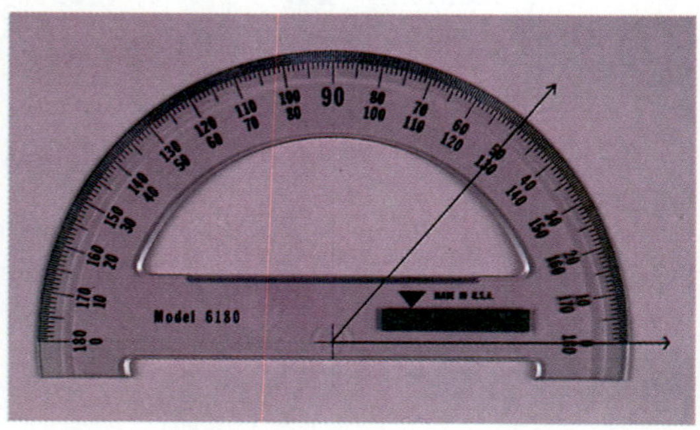

Angles are classified into three groups according to their measures. The measure of an **acute angle** is less than 90°. The measure of a **right angle** is 90°. The measure of an **obtuse angle** is greater than 90°.

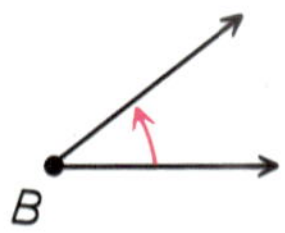

Acute Angle
$m\angle B < 90°$

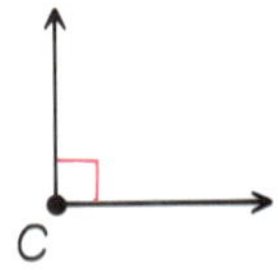

Right Angle
$m\angle C = 90°$

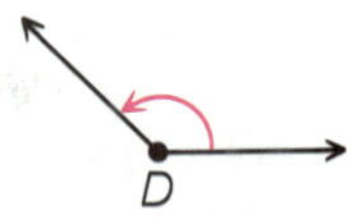

Obtuse Angle
$m\angle D > 90°$

Example 1

In the figure, name: **a.** one acute angle **b.** one obtuse angle

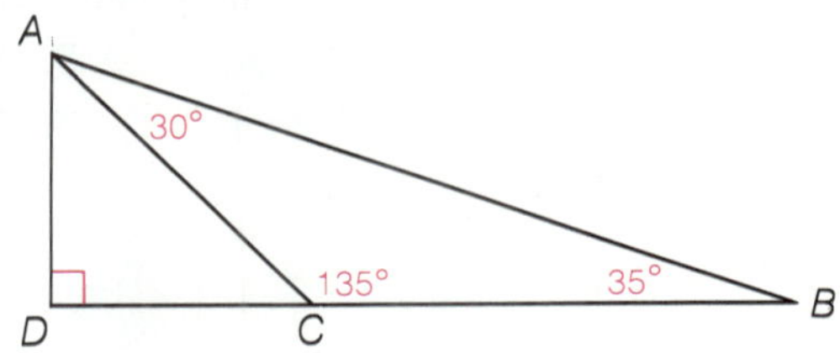

Solution

a. $\angle CAB$ **b.** $\angle ACB$

Practice In the figure, name:

a. two other acute angles $\angle CBA$, $\angle DAB$, $\angle DAC$, $\angle DCA$

b. a right angle $\angle ADC$

Example 2

Write and solve an equation to find $m\angle ABD$.

Discuss the fact that $m\angle ABC = m\angle ABD + m\angle DBC$.

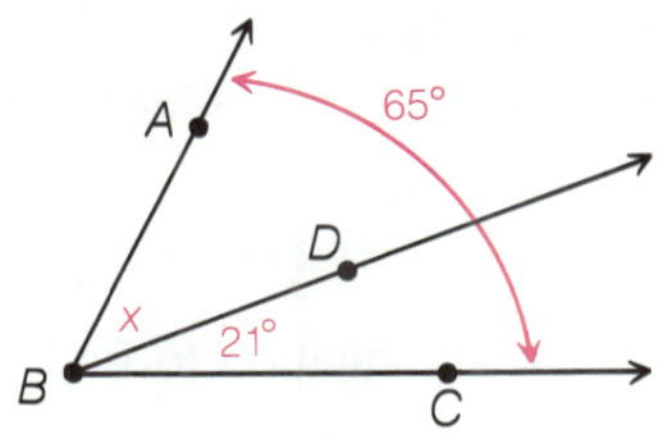

Solution

$$m\angle ABC = m\angle ABD + m\angle DBC$$
$$65 = x + 21$$
$$65 - 21 = x + 21 - 21$$
$$44 = x$$
$$m\angle ABD = 44°$$

Practice For the figure above, write and solve an equation to find each.

a. $m\angle DBC$ if $m\angle ABD$ is 28° **b.** $m\angle ABD$ if $m\angle DBC$ is 49°

a. $28 + x = 65$ 37°
b. $x + 49 = 65$ 16°

Two angles are **complementary** if the sum of their measures is 90°. Two angles are **supplementary** if the sum of their measures is 180°.

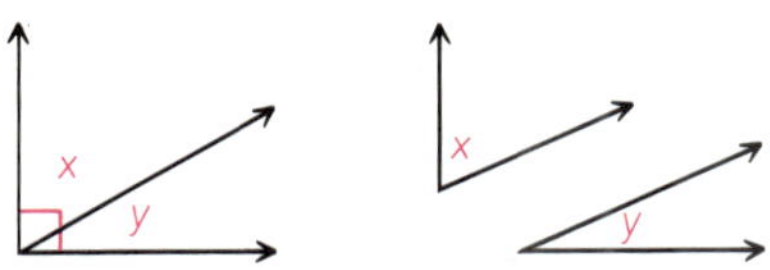

Complementary angles
$x + y = 90$

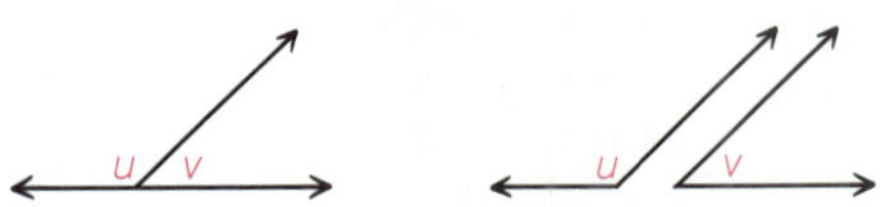

Supplementary angles
$u + v = 180$

Example 3

Write and solve an equation to find the complement of a 49° angle.

Solution $x + 49 = 90$ Let x = the measure of the complement. The sum of the measures of an angle and its complement is 90°.

$x + 49 - 49 = 90 - 49$

$x = 41$ The complement of a 49° angle is a 41° angle.

Practice Write and solve an equation to find:

a. the supplement of a 119° angle
119 + x = 180 61°

b. the complement of a 48° angle
48 + x = 90 42°

Oral Exercises

State whether an angle with the given measure is acute, right, or obtuse.

1. 10° **acute** **2.** 90° **right** **3.** 175° **obtuse** **4.** 1° **acute** **5.** 95° **obtuse**
6. 80° **acute** **7.** 18° **acute** **8.** 100° **obtuse** **9.** 89° **acute** **10.** 130° **obtuse**

Exercises

A Use the figure at right for Exercises 1–5.

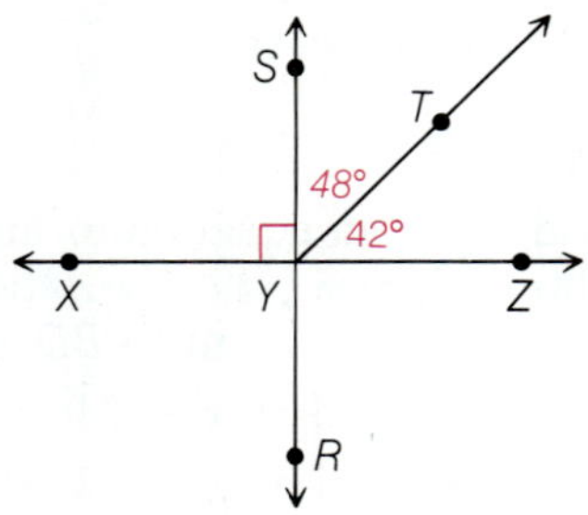

1. Name an acute angle. **∠TYZ, ∠TYS, etc.**
2. Name an obtuse angle. **∠XYT, ∠RYT, etc.**
3. Name two right angles. **∠SYZ, ∠XYS, ∠XYR, ∠RYZ**
4. Name a pair of complementary angles. **∠SYT, ∠TYZ**
5. Name a pair of supplementary angles. **∠XYS, ∠SYZ ∠XYT, ∠TYZ**

Write and solve an equation to find:

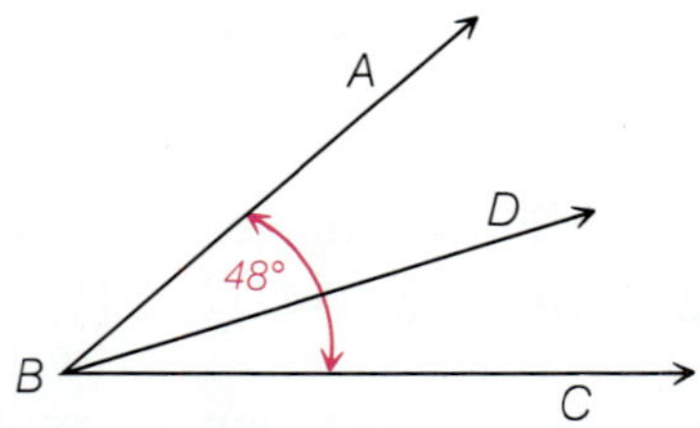

6. m∠ABD if m∠CBD = 15° **x + 15 = 48 33°**
7. m∠CBD if m∠ABD = 31° **x + 31 = 48 17°**
8. m∠ABD if m∠CBD = 21° **x + 21 = 48 27°**
9. m∠CBD if m∠ABD = 35° **x + 35 = 48 13°**
10. m∠ABD if m∠CBD = 19.2° **x + 19.2 = 48 28.8°**
11. m∠CBD if m∠ABD = 27.5° **x + 27.5 = 48 20.5°**

Write and solve an equation to find: **See below.**

12. the complement of a 23° angle
13. the complement of a 53° angle
14. the supplement of a 123° angle
15. the supplement of a 153° angle
16. the supplement of a 23° angle **x + 23 = 180; 157°**
17. the complement of a 17° angle **x + 17 = 90; 73°**

B

18. If m∠A = x, what is the measure of the complement of ∠A? **90 − x**
19. If m∠B = x, what is the measure of the supplement of ∠B? **180 − x**

12. x + 23 = 90 67° 13. x + 53 = 90 37° 14. x + 123 = 180 57° 15. x + 153 = 180 27°

20. Draw four angles of different measures. Estimate their measures. Check your estimates with a protractor. **Answers will vary.**

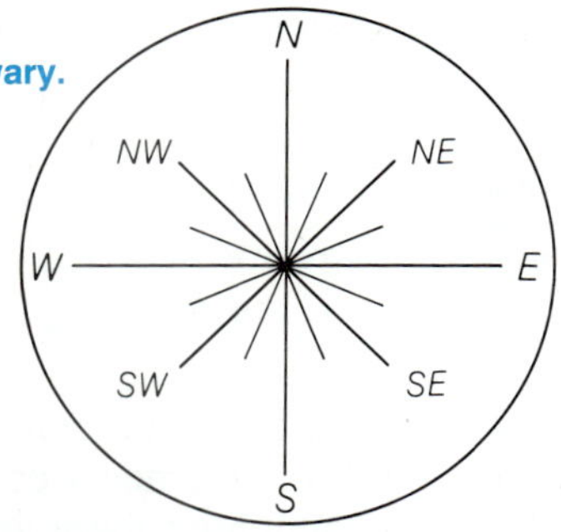

21. What is the measure of the angle between east and northeast? **45°**

22. What is the measure of the angle between northeast and north-northeast? **22.5° Students may need to be told that north-northeast is the direction between north and northeast.**

23. Two supplementary angles have measures of $2x$ and $3x$. Write and solve an equation to find the measure of the angles. **$2x + 3x = 180$ angles: 72°, 108°**

24. Two complementary angles have measures of y and $5y$. Write and solve an equation to find the measure of angle y. **$y + 5y = 90$ angles: 15°, 75°**

C *Extending Thinking Skills*

25. A ship travels in a direction 12° east of due north until it reaches a lighthouse. It then turns 30° to the left. How many degrees from due north is the ship now traveling? **18° west of due north**

26. After a ship travels due east for several hours, it turns 21 degrees to the left. How many degrees east of north is the ship now traveling? **69°**

Thinking skills: 25, 26. Reasoning

Mixed Review

Write as a fraction in lowest terms. **27.** 40% $\frac{2}{5}$ **28.** 35% $\frac{7}{20}$

Solve and check. **29.** $4.5n + 16.3 = 21.7$ **1.2** **30.** $-6.2m + 18.3 = 6m$ **1.5**

ESTIMATION

Use the 60°, 45°, and 30° angles to estimate the measures of the angles below.

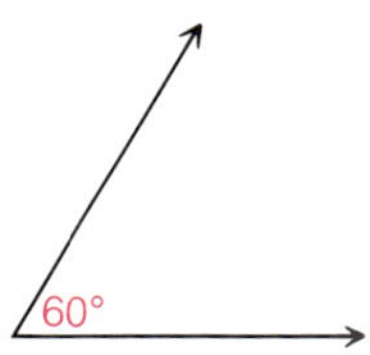

$\frac{2}{3}$ of a right angle

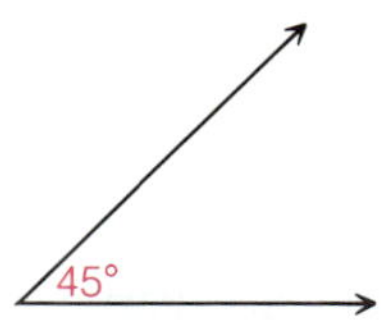

$\frac{1}{2}$ of a right angle

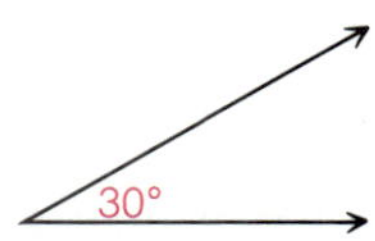

$\frac{1}{3}$ of a right angle

1.

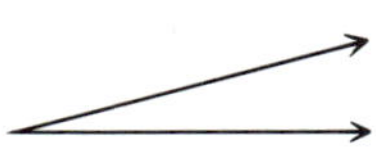

15°

2.

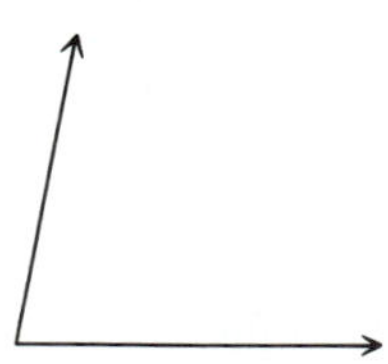

80°

3.

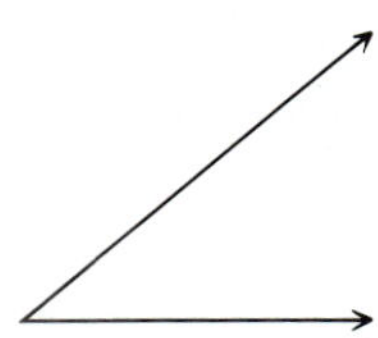

50°

11-4 Parallel and Perpendicular Lines

Objective: To identify lines as parallel or perpendicular; to find measures of vertical and corresponding angles.
MIN: 1–8; REG: 1–8 even, 9–11; MAX: 1–8 even, 9–13; ALL: Mixed Review
An exploratory lesson opener can be found on page 552.

Two lines in the same plane are **parallel** if they have no points in common. Two segments or rays in the same plane are parallel if the lines containing them are parallel. Lines in the same plane that are not parallel are called **intersecting** lines. Lines that intersect to form right angles are called **perpendicular** lines.

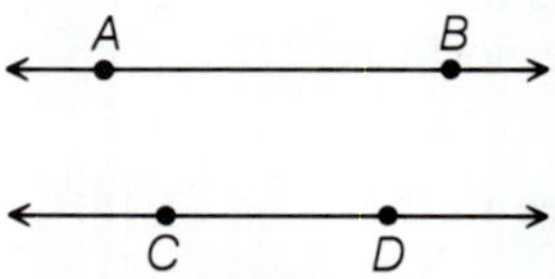

parallel lines

$\overleftrightarrow{AB}$ is parallel to $\overleftrightarrow{CD}$
$\overleftrightarrow{AB} \parallel \overleftrightarrow{CD}$

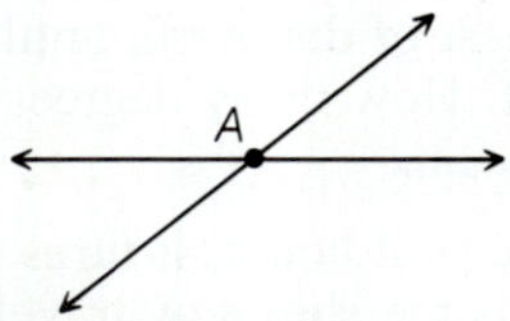

lines intersect at *A*

A is a point of intersection

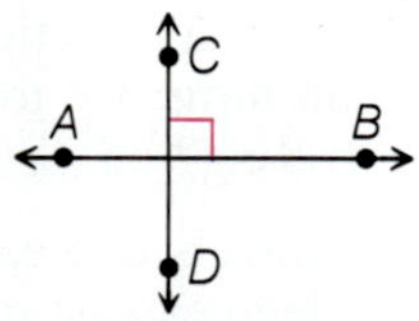

perpendicular lines

$\overleftrightarrow{AB}$ is perpendicular to $\overleftrightarrow{CD}$
$\overleftrightarrow{AB} \perp \overleftrightarrow{CD}$

Two intersecting lines form two pairs of **vertical angles**. In the figure at right, ∠2 and ∠4 are vertical angles, and ∠1 and ∠3 are vertical angles.

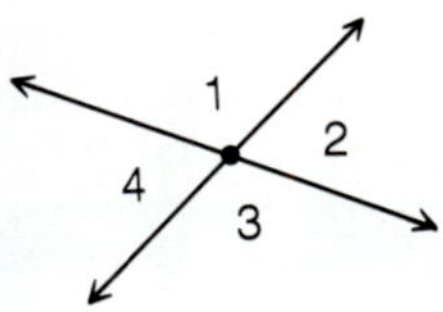

$m\angle 1 = m\angle 3$
$m\angle 2 = m\angle 4$

Measure of Vertical Angles

Vertical angles have the same measure.

When lines intersect, you can sometimes use vertical angles to find the measure of an angle not given.

Example 1

Find m∠1 on the figure at right.

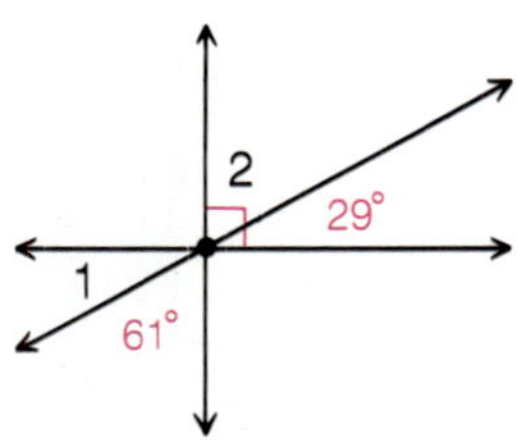

Solution $m\angle 1 = 29°$ ∠1 is vertical to an angle of 29°. Vertical angles have the same measure.

Practice Find m∠2 on the figure at right.
61°

A line that intersects two other lines is called a **transversal**. When a transversal intersects a pair of parallel lines, the pairs of **corresponding angles** formed have the same measure.

Corresponding angles in this figure are ∠1 and ∠5, ∠2 and ∠6, ∠3 and ∠7, and ∠4 and ∠8.

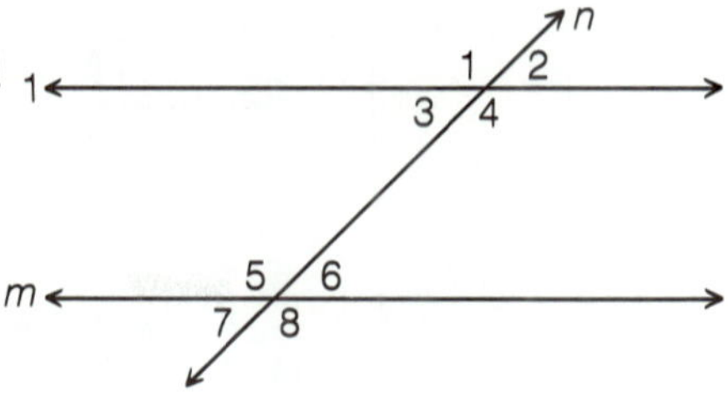

m∠1 = m∠5
m∠2 = m∠6
m∠3 = m∠7
m∠4 = m∠8

Measures of Corresponding Angles

Where a transversal intersects a pair of parallel lines, corresponding angles have the same measure.

Example 2

Lines *a* and *b* are parallel. Find the measures of ∠1, ∠2, and ∠3.

Solution

$m\angle 1 + 121 = 180$	∠1 is supplementary to the angle
$m\angle 1 = 59°$	measuring 121°
$m\angle 2 = 121°$	Vertical angles have equal measures.
$m\angle 3 = 59°$	Corresponding angles have equal measures.

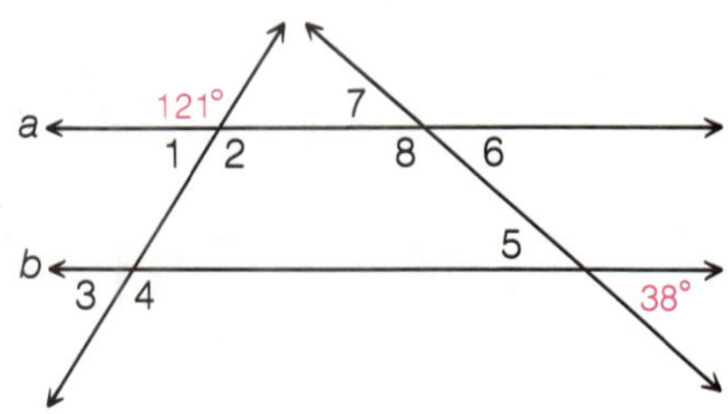

Practice Find the measures of ∠5, ∠6, ∠7, and ∠8.
m∠5 = 38°, m∠6 = 38°, m∠7 = 38°, m∠8 = 142°

Oral Exercises

1. Name a pair of parallel lines. ***m* ∥ *n* *p* ∥ *q***
2. Name a pair of perpendicular lines ***m* ⊥ *r***
3. Name an angle that is vertical to ∠4 **∠6**
4. Name an angle corresponding to ∠3 **∠6**
5. Name an angle corresponding to ∠5 **∠7**

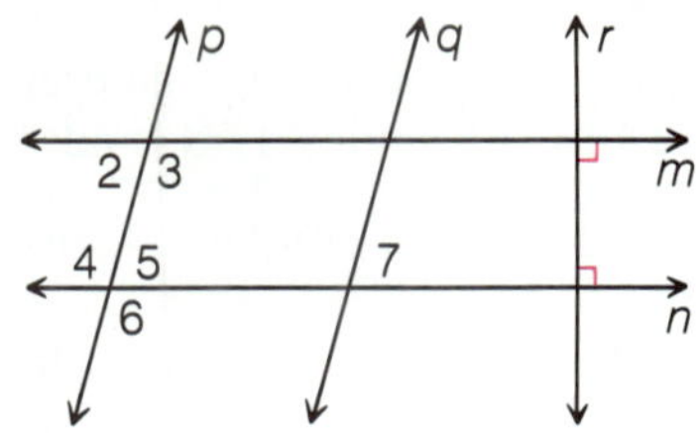

m ∥ *n*
p ∥ *q*

Exercises

A Find the measures of ∠1, ∠2, ∠3, and ∠4.

1. m∠1 = 52°
m∠2 = 38°
2. m∠1 = 23°
m∠2 = 67°
3. m∠1 = 90°
m∠2 = 40°

1.

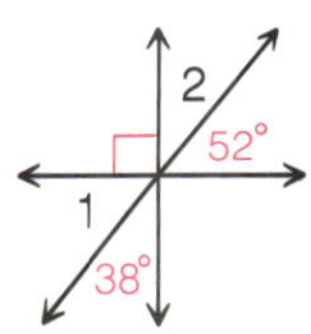

2.

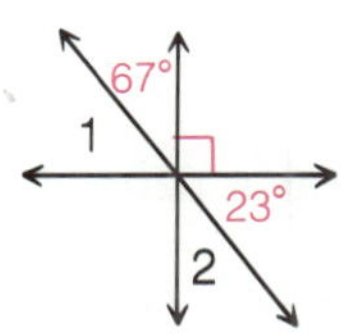

3.

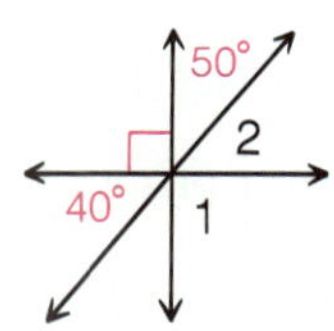

4. 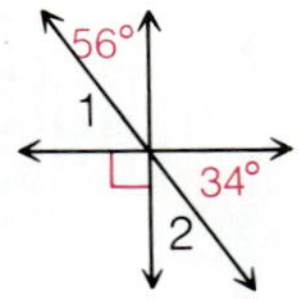

4. m∠1 = 34°
m∠2 = 56°

5. 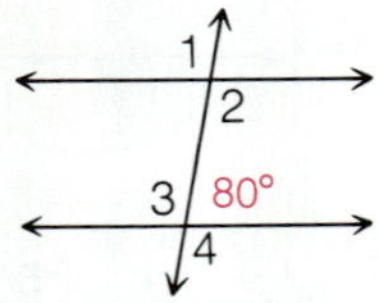

5. m∠1 = m∠2 = m∠3 = m∠4 = 100°

6. 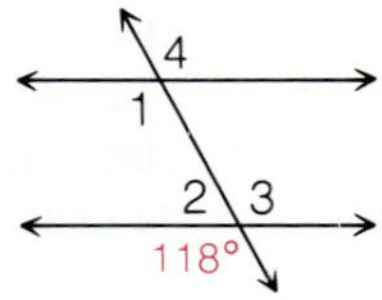

6. m∠1 = 118°
m∠2 = 62°
m∠3 = 118°
m∠4 = 118°

7. 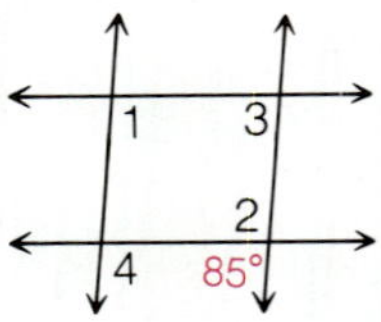

m∠1 = 95°
m∠2 = 95°
m∠3 = 85°
m∠4 = 95°

8. 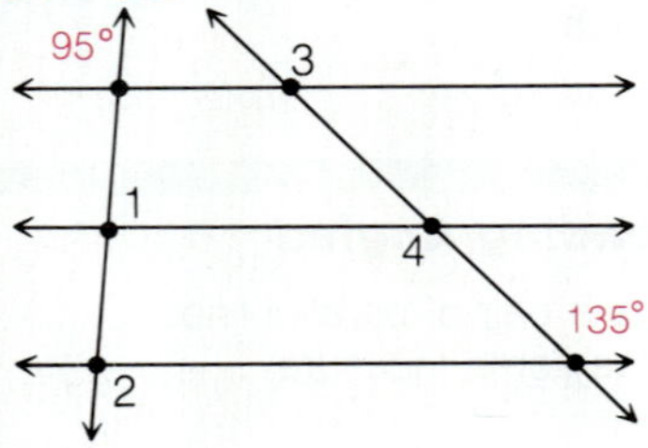

m∠1 = 85°
m∠2 = 95°
m∠3 = 135°
m∠4 = 135°

B Write and solve an equation to find the measure of ∠*ABC*.

9. 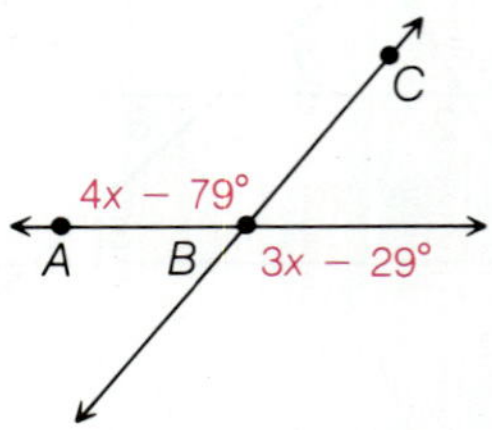

4x − 79 = 3x − 29; x = 50
m∠ABC = 121°

10. 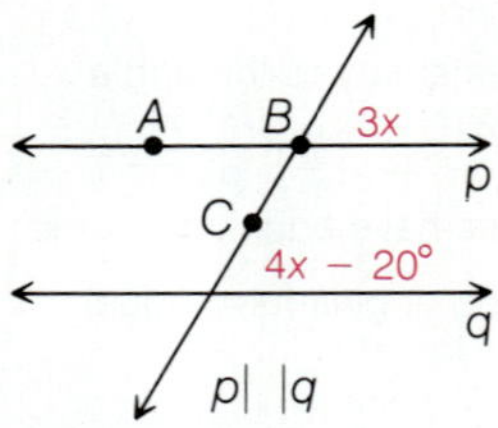

3x = 4x − 20; x = 20°
m∠ABC = 60°

11. 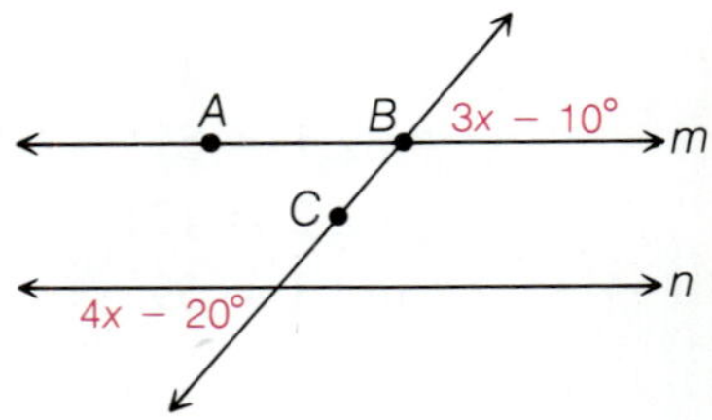

3x − 10 = 4x − 20; x = 10°
m∠ABC = 20°

C Extending Thinking Skills

12. In the plan for a fence gate at right, what are the measures of angles 1, 3, and 4?
12. m∠1 = m∠4 = 45°; m∠3 = 135°.

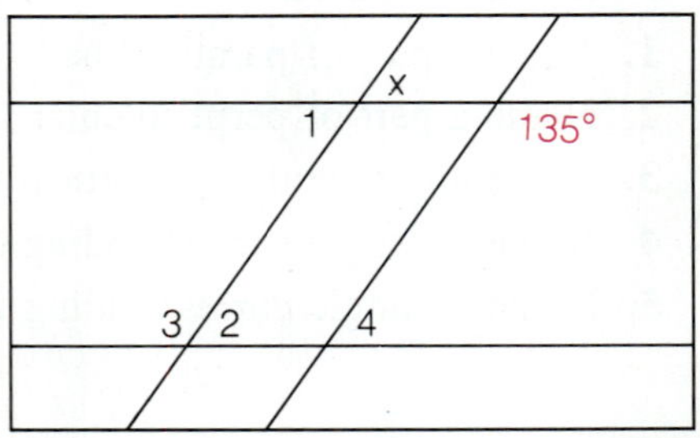

13. Angles 1 and 2 are called **alternate interior angles**.
m∠1 = *m*∠*x* (why?) Vertical angles
m∠*x* = m∠2 (why?) Corresponding angles
Formulate a generalization about alternate interior angles: If 2 lines are parallel, alternate interior angles are _?_. If 2 lines are parallel, alternate interior angles are equal. Thinking skill: 12, 13. Reasoning

Mixed Review

Use a proportion to write as a percent. **14.** $\frac{4}{5}$ 80% **15.** $\frac{1}{8}$ 12.5%

16. What percent of 16 is 12? 75% **17.** What percent of 45 is 27? 60%

11-5 Triangles

Objective: To classify triangles according to type, using specific names; to write equations to find angle measures in triangles.
MIN: 1–10; REG: 1–10 even, 11–14; MAX: 1–10 even, 11–15; ALL: Mixed Review

A **triangle** is a figure in a plane, made up of three segments meeting at endpoints. In the triangle at right, $\overline{AB}$, $\overline{BC}$, and $\overline{AC}$ are called the **sides**. Points A, B, and C are called **vertices**.

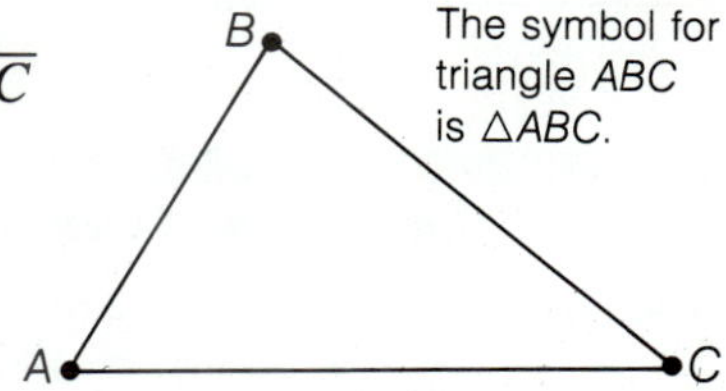

The symbol for triangle ABC is $\triangle ABC$.

Triangles are classified by the measures of their angles or the lengths of their sides.

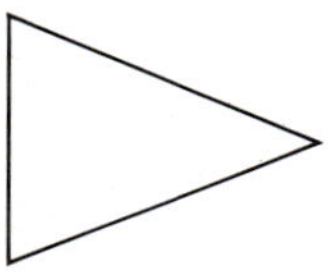

Acute
all acute angles

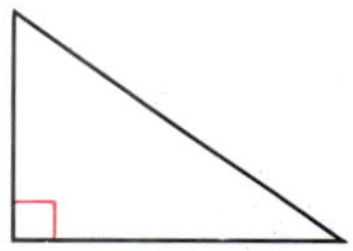

Right
one right angle

Obtuse
one obtuse angle

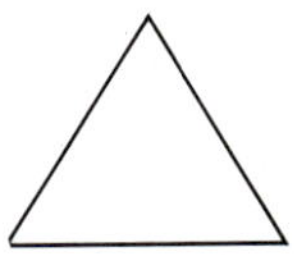

Equilateral
all sides equal in length

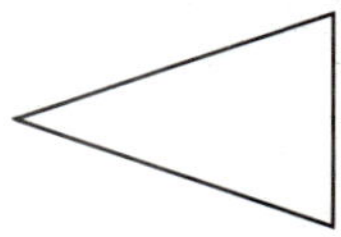

Isosceles
at least two sides equal in length

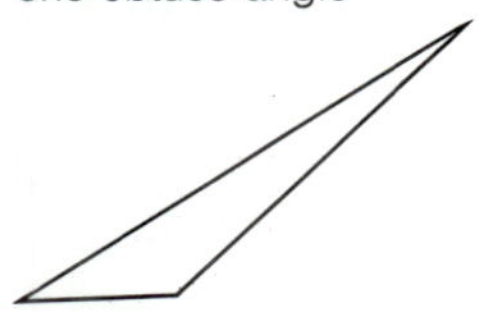

Scalene
no two sides equal in length

Example 1

Classify $\triangle ABC$ and $\triangle DEF$ by the measures of the angles and the lengths of the sides.

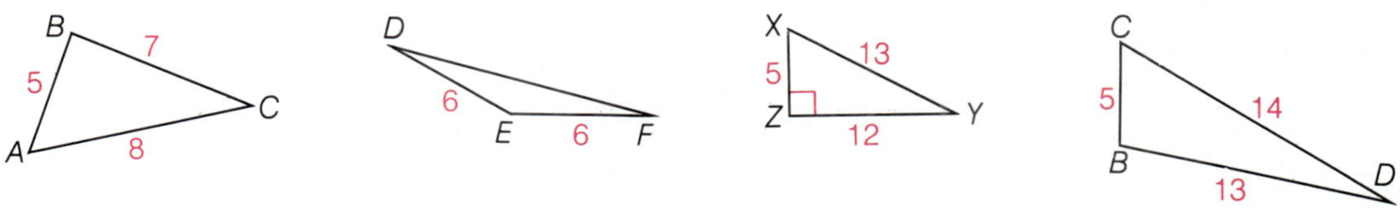

Solution

$\triangle ABC$ is acute and scalene. All angles are acute and no two sides are equal in length.

$\triangle DEF$ is obtuse and isosceles. Angle E is obtuse; two sides are equal in length.

Practice Classify $\triangle XYZ$ and $\triangle BCD$ above by the measures of their angles and the lengths of the sides.
$\triangle XYZ$ is right and scalene. $\triangle BCD$ is obtuse and scalene.

The picture below suggests an important property of triangles.

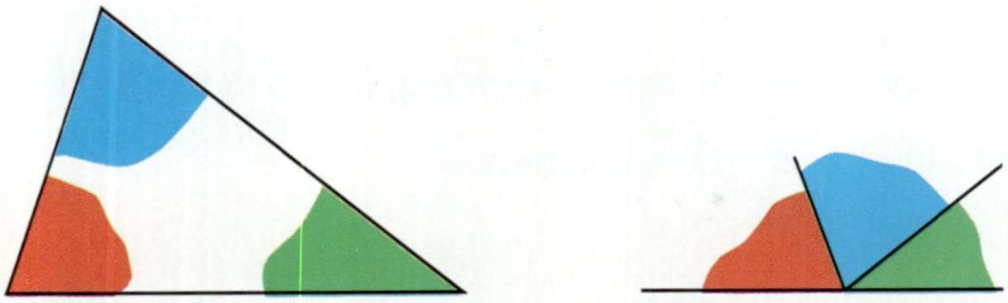

Have students cut out paper triangles, tear off points, and arrange as shown.

Sum of the Angle Measures of a Triangle

The sum of the angle measures in any triangle is 180°.

$m\angle A + m\angle B + m\angle C = 180°$

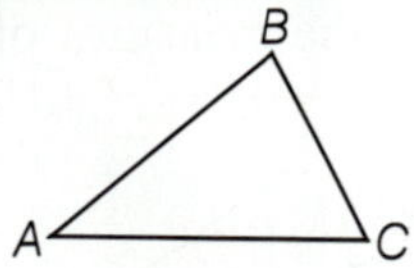

Example 2

Write and solve an equation to find m∠A.

Solution $x + 30 + 130 = 180$

$x + 160 = 180$

$x = 20$

$m\angle A = 20°$.

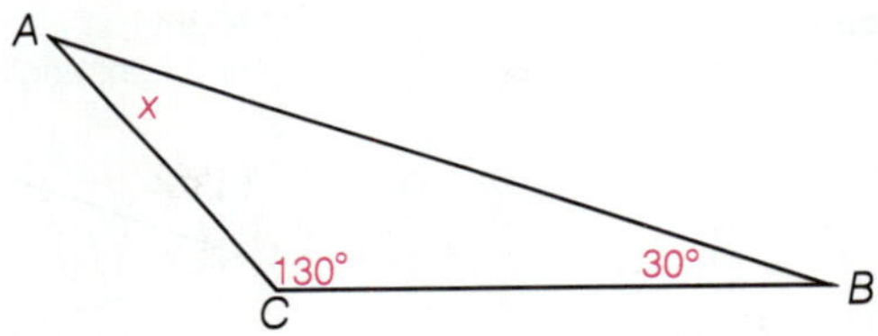

Practice Two angles of a triangle have measures 37° and 92°. Write and solve an equation to find the measure of the third angle. **$x + 37 + 92 = 180$ $x = 51$**

Oral Exercises

On the figure at right, name the following:

1. An obtuse triangle.
2. Three right triangles.
3. Two acute triangles.

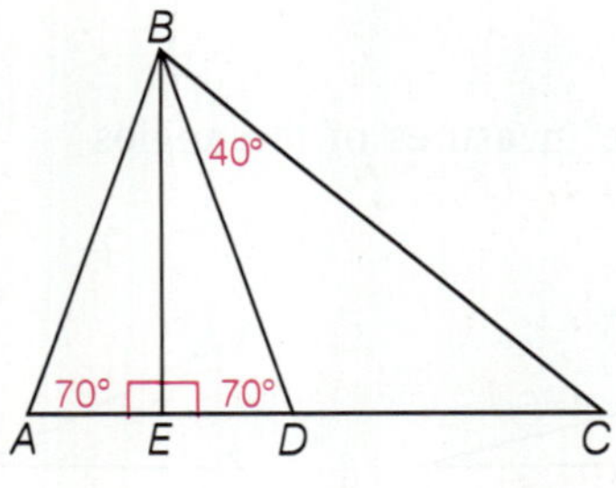

1. △BDC
2. △BEA, △BED, △BEC
3. △BAD, △BAC

Exercises

A Classify each triangle by the measures of its angles and the lengths of its sides.

1. 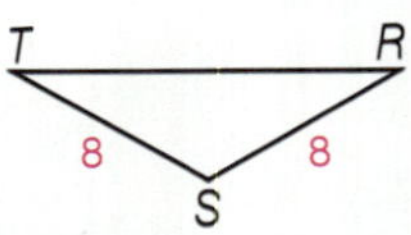

obtuse isosceles

2. 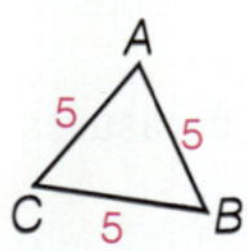

acute equilateral

3. 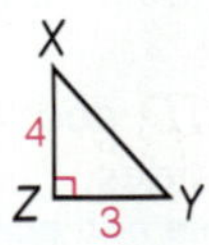

right scalene

4. 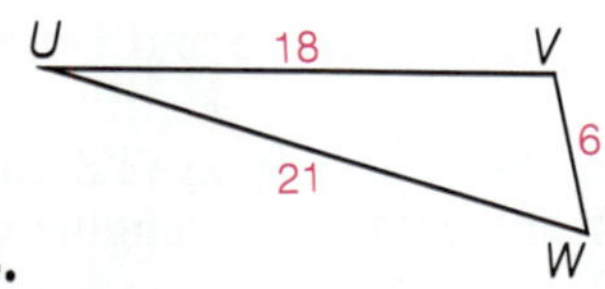

obtuse scalene

Write and solve an equation to find m∠A. See additional answers.

5. 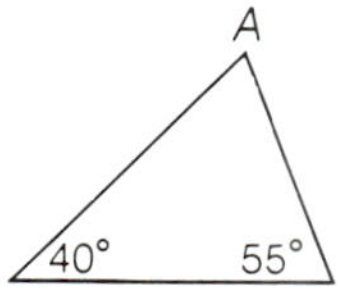

6. 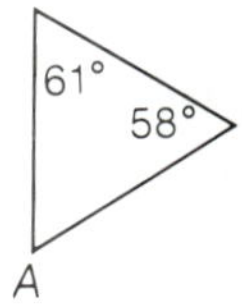

7. 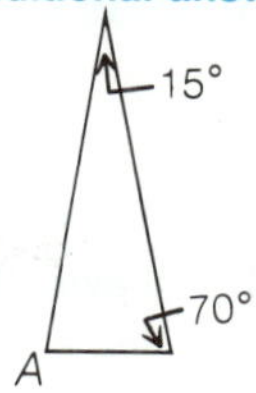

8. 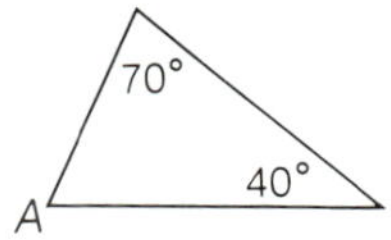

Write and solve an equation to find the measure of the third angle.

9. Two angles of a triangle have measures of 70° and 32°. $x + 70 + 32 = 180$; 78°

10. Two angles of a triangle have measures of 119° and 25°. $x + 119 + 25 = 180$; 36°

B Write and solve an equation to find the measure of each angle.

11. 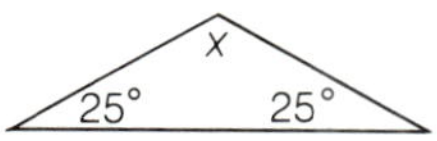

$x + 25 + 25 = 180$; 130°

12. 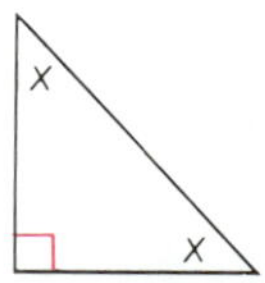

$x + x + 90 = 180$; 45°

13. 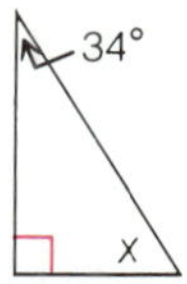

$x + 90 + 34 = 180$; 56°

14. 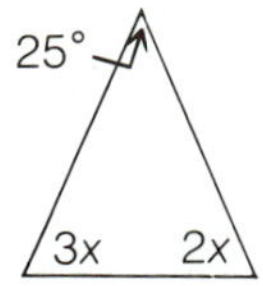

$3x + 2x + 25 = 180$; 62°, 93°

C Extending Thinking Skills

Find the pattern and use the variable n to complete the following: $x = \underline{\ ?\ }$

15. 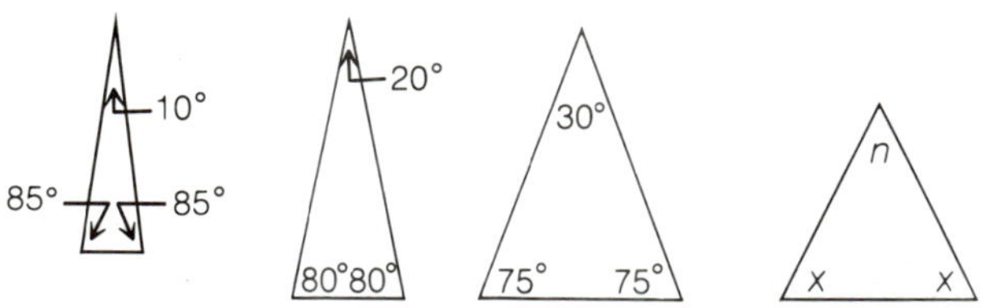

$x = \dfrac{180 - n}{2}$

Thinking skill: 16. Finding a pattern; generalizing

Mixed Review

Solve and check. **16.** $-16c > 32$ $c < -2$ **17.** $4m + 6 < 2m + 12$ $m < 3$

18. What percent of 20 is 4? 20% **19.** What percent of 20 is 17? 85%

CALCULATOR ACTIVITY

You can use the key sequence below to solve an equation such as $x + 45 + 73 = 180$.

								Display
45	[+]	73	[=]	[+/−]	[+]	180	[=]	62

Use a calculator to solve.

1. $x + 48 + 23 = 180$ 109 **2.** $x + 56 + 112 = 180$ 12 **3.** $x + 84 + 27 = 180$ 69

11-6 Polygons

Objectives: To classify polygons according to type, using specific names; to write and solve equations to find the measure of an angle in a polygon.
MIN: 1–8; REG: 1–8 even, 9–13; MAX: 1–8 even, 9–14; ALL: Mixed Review

A **polygon** is a plane figure formed by three or more segments that intersect only at their endpoints so that exactly two segments meet at each endpoint. A polygon is named according to the number of its sides.

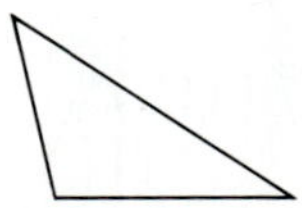

triangle

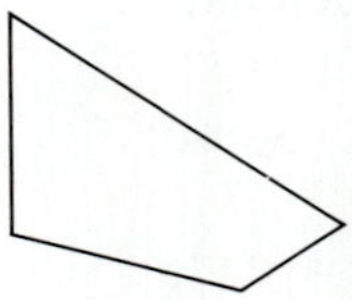

quadrilateral

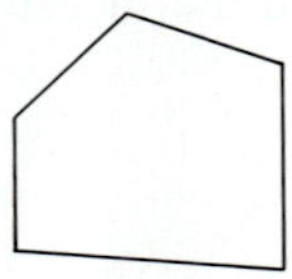

pentagon

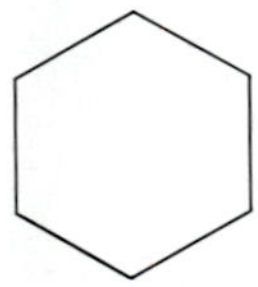

hexagon

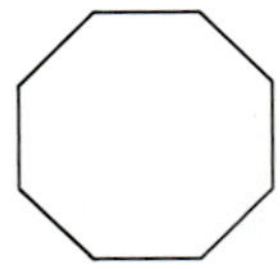

octagon

In a **regular polygon**, all sides and all angles have the same measure. The hexagon and octagon above are regular polygons.

A **quadrilateral** is a polygon with four sides. Some quadrilaterals have specific names.

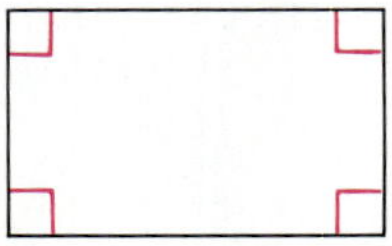

rectangle
4 right angles

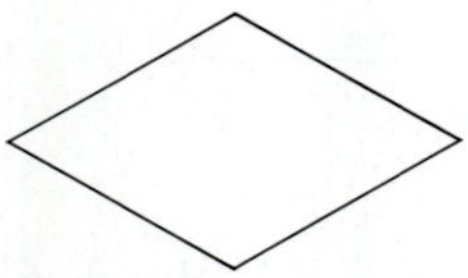

rhombus
all sides
the same length

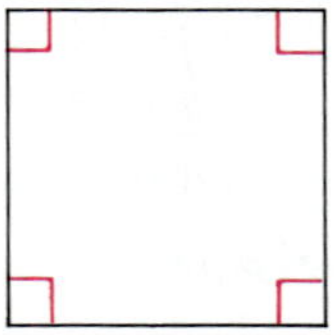

square
4 right angles;
all sides the same length

parallelogram
2 pairs of
parallel sides

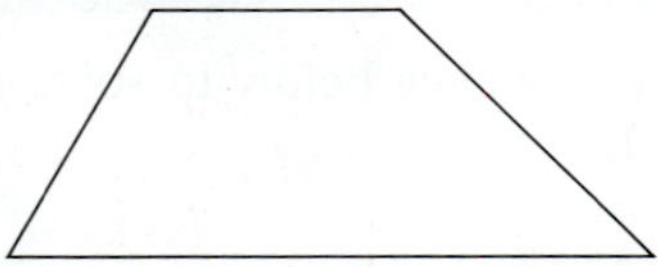

trapezoid
1 pair of
parallel sides

Example 1

Name each polygon and indicate whether it is regular. Use specific names for quadrilaterals.

a.

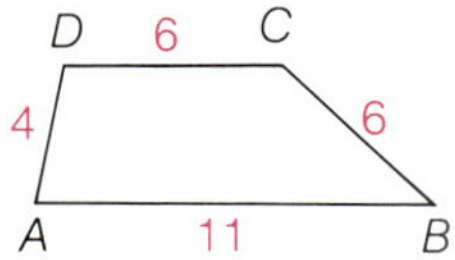

b.

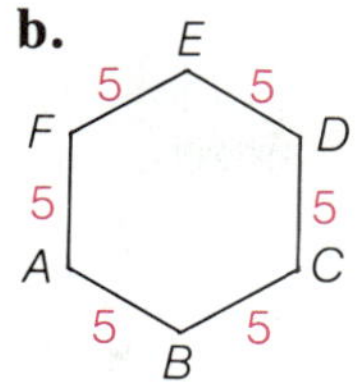

Solution

a. *ABCD* is a trapezoid. Quadrilateral with only one pair of parallel sides.

b. *ABCDEF* is a regular hexagon. All angles and all sides are equal in measure. It is regular.

Practice Name each polygon and indicate whether it is regular. Use specific names for quadrilaterals.

a.

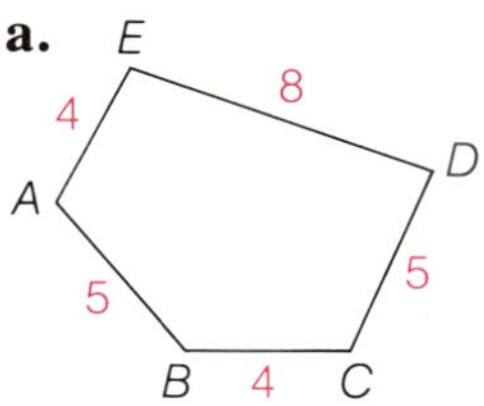

a. pentagon, not regular

b.

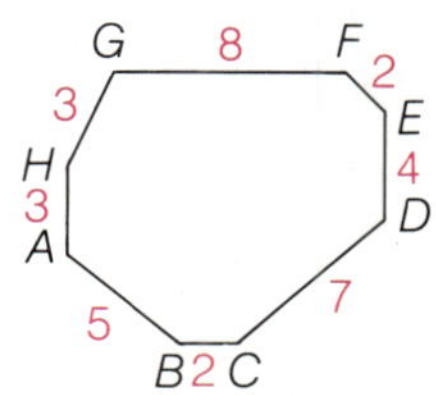

b. octagon, not regular

The figure below shows an important property of quadrilaterals.

Sum of the Angle Measures of a Quadrilateral

The sum of the measures of the angles in any quadrilateral is 360°.

$m\angle A + m\angle B + m\angle C + m\angle D = 360°$

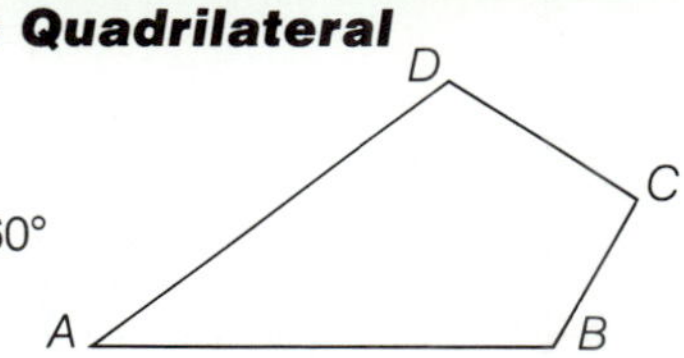

Example 2

Write and solve an equation to find m∠*A*.

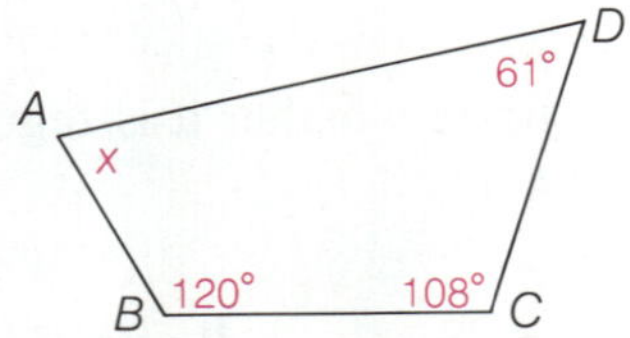

Solution

$$x + 120 + 108 + 61 = 360$$ The sum of the measures of the angles in a quadrilateral is 360°.

$$x + 289 = 360$$ Combine terms.

$$x + 289 - 289 = 360 - 289$$

$$x = 71$$

m ∠*A* = 71°.

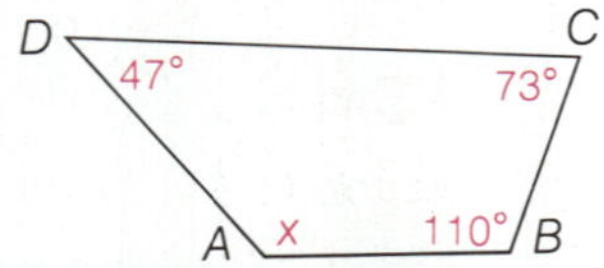

Practice Write and solve an equation to find m∠*A*.
x + 47 + 73 + 110 = 360 m∠A = 130°

Oral Exercises **1. pentagon 2. quadrilateral 3. hexagon 4. triangle 5. octagon**

Name a polygon that has:

1. 5 sides **2.** 4 sides **3.** 6 sides **4.** 3 sides **5.** 8 sides

6. When is a polygon a regular polygon? **When sides have equal lengths and angles have equal measures.**

7. What is another name for a regular triangle. **equilateral triangle**

8. What is another name for a regular quadrilateral? **square**

Exercises

A Name each polygon and indicate whether it is regular. Use any special names for quadrilaterals.

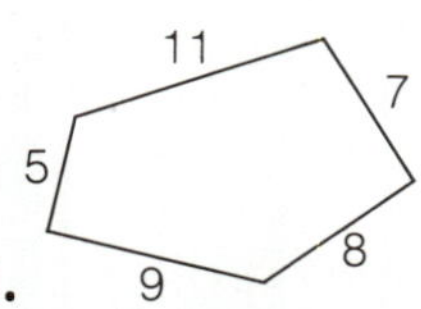

1. **pentagon, not reg.**

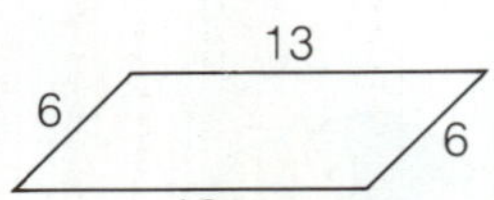

2. **parallelogram, not reg.**

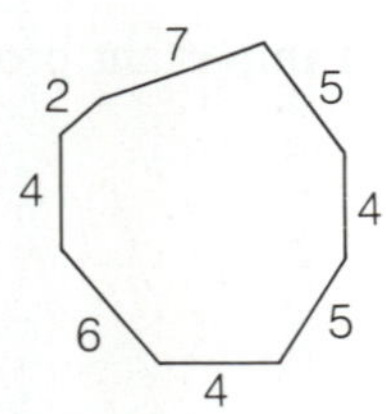

3. **octagon, not reg.**

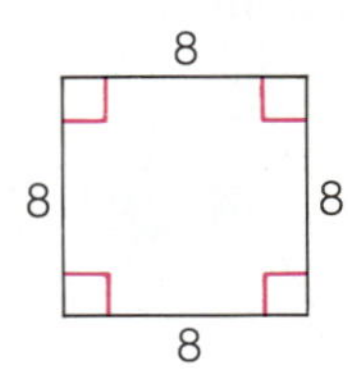

4. **square, reg.**

Write and solve an equation to find m∠*A*.

5.

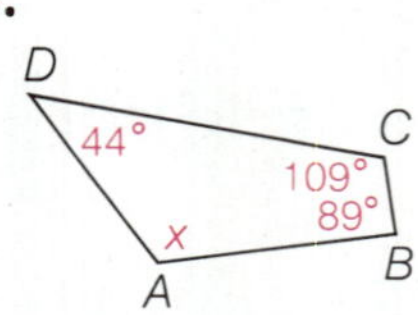

6.

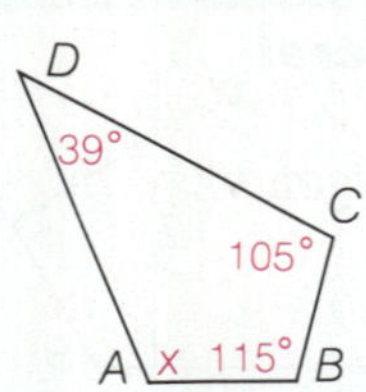

7.

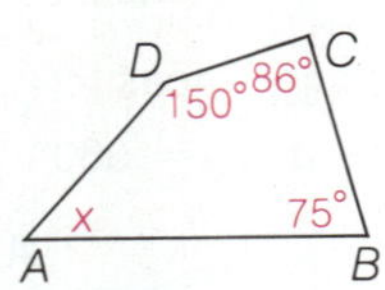

8.

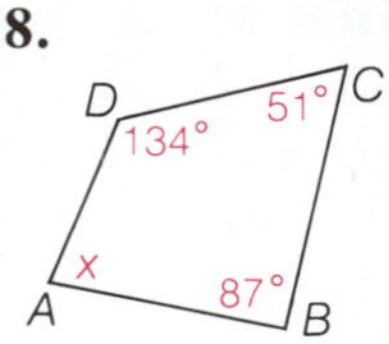

5. x + 44 + 109 + 89 = 360; 118° 6. x + 115 + 105 + 39 = 360; 101° 7. x + 150 + 86 + 75 = 360; 49° 8. x + 134 + 51 + 87 = 360; 88°

B Write and solve an equation to find each angle measure. See additional answers

9. 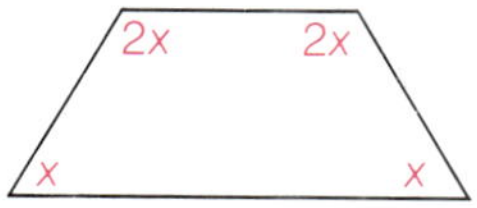

10. 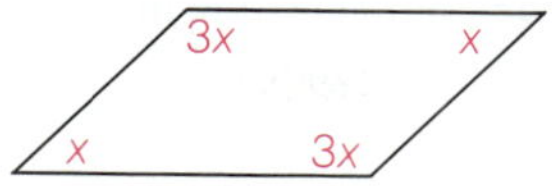

11. 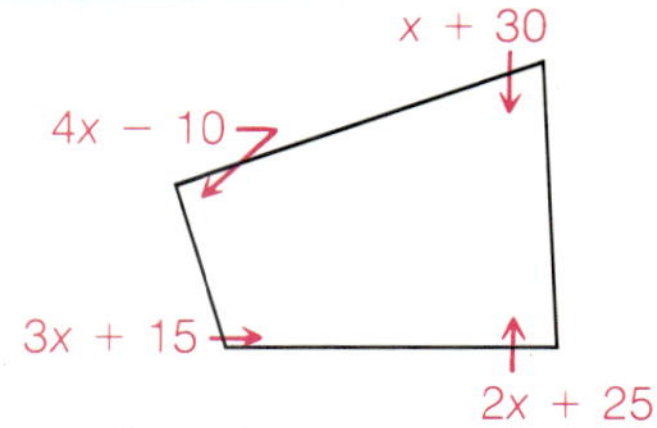

12. Name a quadrilateral that is not regular but has all sides the same length. rhombus

13. Name a quadrilateral that is not regular but has all angles the same measure. rectangle

C Extending Thinking Skills

14. Draw a quadrilateral, a pentagon, and a hexagon. Find the number of diagonals from a single vertex needed to divide each figure into triangles. What generalization can you make from your answer?

The number of diagonals needed is 3 less than the number of sides. The number of triangles formed is 2 less than the number of sides.

Thinking skill: 14. Generalizing

Mixed Review

Give the greatest common factor (GCF). **15.** 20, 32, 8 4 **16.** 15, 21, 36 3

17. What percent of 70 is 105? 150% **18.** What percent of 200 is 1? $\frac{1}{2}$%

NUMBERS TO ALGEBRA

You can find a general formula for the sum of the measures of the angles of a polygon with n sides by generalizing from the number patterns shown below.

Numbers			Algebra
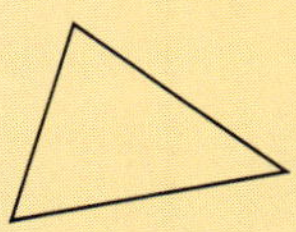	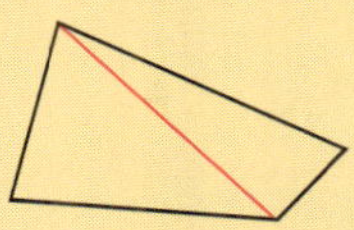	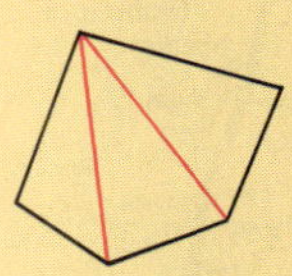	Polygon with n sides
3 sides 1 × 180	4 sides 2 × 180	5 sides 3 × 180	n sides $(n - 2) \times 180$

1. Find the sum of the angle measures of a polygon with 6 sides. 4(180) = 720°
2. Find the sum of the angle measures of a polygon with 12 sides. 10(180) = 1800°
3. Find the sum of the angle measures of a polygon with 150 sides. 148(180) = 26,640°

11-7 Using a Road Map

Objective: To solve applied problems, using road maps, mileage charts, and formulas.
ALL: 1–13

A road map gives information about distances between cities. On the map below, the interstate highway is shown in blue. Notice the red arrowheads at exits 18 and 33. About halfway between these two arrowheads is a red number 40. This means the distance between the arrowheads is 40 miles. Notice the small numbers in black between each pair of exits. Between exits 23 and 28 is a small black 8. This means the distance between these two exits is 8 miles.

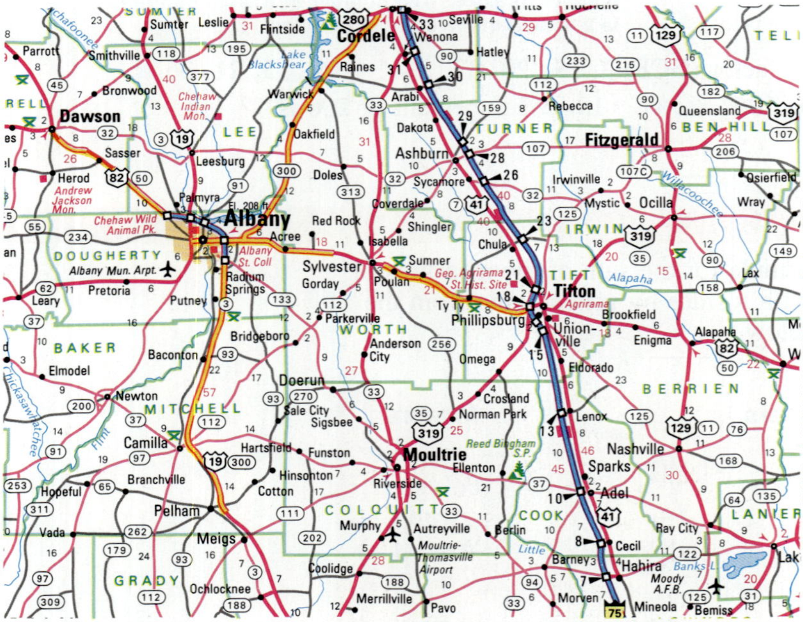

You might wish to have students choose from among the techniques pencil and paper, mental math, estimation, and calculator. Have them choose the most appropriate technique for each problem.

Problems

Solve. Use the map and the formula Distance = rate × time as needed.

1. How far is it from Moultrie to Sylvester along highway 33? (Note the red arrows at these towns and the red numbers between them.) **27 miles**

2. How far is it from Tifton to Ocilla on Highway 319? **20 miles**

3. How far is it from Albany to Tifton on Highway 82? **39 miles**

4. How far will a car traveling at 55 mi/h drive in $1\frac{1}{2}$ hours? 82.5 mi

5. Would a car traveling at 55 mi/h be able to drive from exit 18 to exit 33 in less than 45 minutes? yes

6. How long does it take to travel 125 miles at 55 mi/h? approx. 2 hr, 16 min

7. The distance from San Francisco to New York is about 2800 miles. About how long will it take for a jet to fly from San Francisco to New York at an average speed of 620 mi/h? approx. $4\frac{1}{2}$ hr

8. A truck driver averages 50 mi/h. If he drives $10\frac{1}{2}$ hours a day, about how many days will it take him to travel 1500 miles? 3 days

9. A family drove a distance of about 3000 miles from Boston, MA, to Los Angeles, CA. They averaged 500 miles a day. Traveling costs were $35 a day for food and $40 for a motel each night. If it cost $0.17 per mile to operate their car, what was their total expense for the trip? $960

Use the mileage chart to answer problems 10–12.

	Chicago	Dallas	St. Louis	Wash., D.C.
Chicago		921	289	709
Dallas	921		655	1307
St. Louis	289	655		862
Wash., D.C.	709	1307	862	

10. How many miles is it from Washington, D.C. to Dallas and back? 2614 mi

11. How many miles is it from Dallas to St. Louis and back? 1310 mi

12. How many more miles is it to travel through St. Louis on the way from Dallas to Chicago than to go directly? 23 mi

13. ***Data Search*** For a driver averaging 50 miles per hour, how long would it take to drive from Kansas City, Missouri, to Denver, Colorado?

What's Your Decision? This requires use of road maps or a road atlas.

Suppose you are planning to travel from New Orleans, LA, to Detroit, MI on interstate highways. To travel the minimum number of miles, what route should you take? If you average 50 miles per hour, how many hours will you drive? Answer will vary.

11-8 Circles and Circumference

Objective: To find the circumference, radius, or diameter of a circle.
MIN: 1–18; REG: 1–18 even, 19–23; MAX: 1–18 even, 19–25; All: Mixed Review

A **circle** is the set of all points in a plane that are a fixed distance from a point called the **center**. Any segment that joins the center to a point on the circle is called a **radius** (*r*) of the circle. The **diameter** (*d*) of the circle is a segment that passes through the center and has endpoints on the circle.

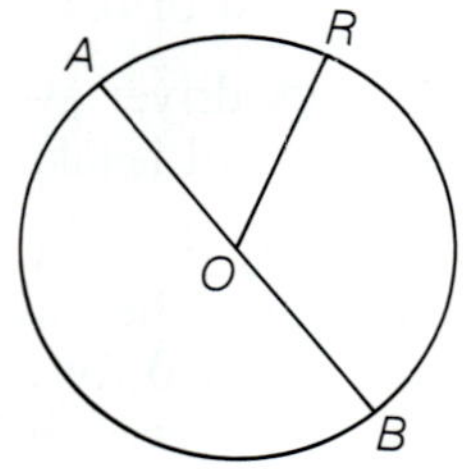

O is the center
$\overline{OR}$ is a radius
$\overline{AB}$ is a diameter

The distance around a circle is called the **circumference** (*C*) of the circle. The ratio $\frac{C}{d}$ is the same number for all circles, and is represented by the Greek letter π. The number $\pi \approx 3.14$.

Discuss the fact that π is an irrational number.

Formula: Circumference of a Circle

The circumference of a circle is π times the diameter, or 2 times π times the radius.

$C = \pi d$ or $C = 2\pi r$

Example 1

Find the circumference of a circle with diameter 4 cm. Use 3.14 for π.

Solution $C = \pi d$

$C \approx 3.14 \times 4$ Substitute 3.14 for π and 4 for d.

$C \approx 12.56$ cm

Practice Find the circumference of a circle with the given diameter or radius.

a. diameter = 3.2 cm **b.** radius = 8.2 cm

a. 10.048 cm **b. 51.496 cm**

Example 2

Find the radius of a circle with circumference 9 cm. Use 3.14 for π.

Solution $C = 2\pi r$

$9 \approx 2(3.14)r$ Substitute 9 for C and 3.14 for π.

$r \approx \frac{9}{6.28} \approx 1.43$

Practice **a.** Find the radius of a circle with circumference 12 m. **1.91 m**

b. Find the diameter of a circle with circumference 18 cm. **5.73 cm**

Oral Exercises

For the figure at right:

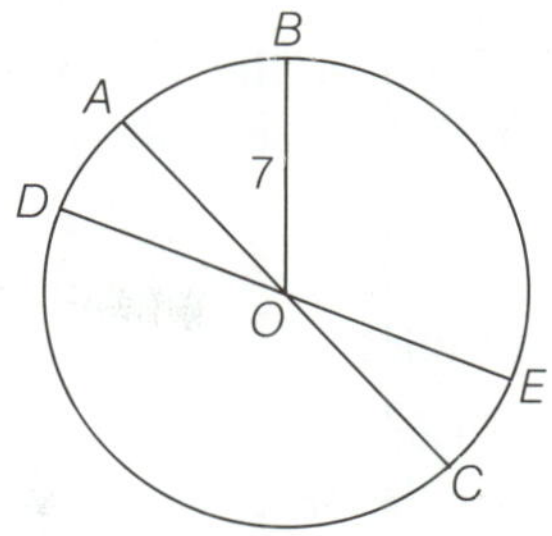

1. name the center. O
2. name three radii. OD, OA, OE, OB, OC
3. give the length of a radius. 7
4. name a diameter. AC
5. name another diameter. DE
6. give the length of a diameter. 14

Exercises

A Find the circumference of a circle with the given diameter or radius. Use 3.14 for π.

1. diameter = 5 cm 15.7 cm
2. diameter = 12 cm 37.68 cm
3. diameter = 20 cm 62.8 cm
4. diameter = 4.8 cm 15.072 cm
5. radius = 6.1 m 38.308 m
6. radius = 8.3 m 52.124 m

Find the diameter of a circle with the given circumference (C) or radius (r).

7. $C = 5$ cm 1.59 cm
8. $C = 8$ mm 2.548 mm
9. $C = 12$ m 3.822 m
10. $r = 35.7$ cm 71.4 cm
11. $C = 4.6$ m 1.465 m
12. $r = 99.8$ m 199.6 m

Find the radius of a circle with the given circumference (C) or diameter (d).

13. $C = 9$ cm 1.433 cm
14. $C = 4$ m 0.637 m
15. $C = 15$ cm 2.389 cm
16. $d = 452.8$ cm 226.4 cm
17. $C = 14.8$ m 2.357 m
18. $d = 32.84$ m 16.42 m

B Use 3.14 for π in Exercises 19–23.

19. Estimate the circumference of a 28-inch diameter bicycle wheel. Find the circumference with a calculator. 87.92 in.

20. How many feet does a bicycle travel with each revolution of a 28-inch diameter wheel? How many revolutions does the wheel make to travel one mile (5280 feet/mile)? approx $7\frac{1}{3}$ feet; 720 revolutions

Find the perimeter of each figure. All curves are parts of circles.

21.

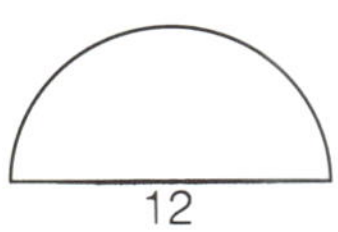

30.84

22.

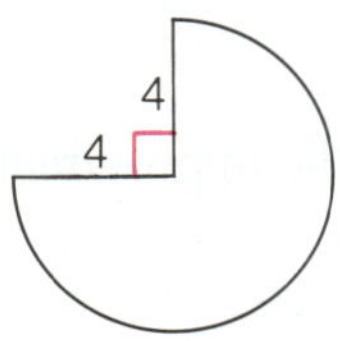

26.84

23.

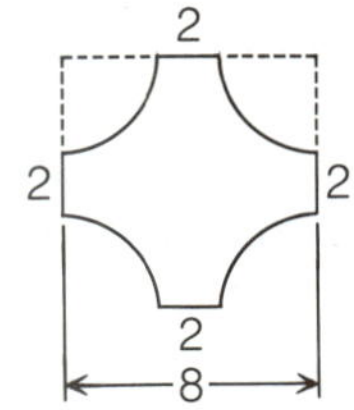

26.84

C Extending Thinking Skills

24. If the radius of a circle doubles, by how much does the circumference increase? It doubles, too.

Thinking skill: 24. Reasoning

25. The radius of the earth is approximately 6380 kilometers. Imagine that a steel ring is looped tightly around the equator. Suppose the steel ring is made 6 meters longer. Will this make the ring loose enough so that:

an ant could walk under the ring?

a finger would fit under the ring?

a cat could walk under the ring?

a truck could be driven under the ring?
A cat could walk under. (The radius will increase by 0.955 m.)

Mixed Review

Solve. There are 36 students in Mrs. Ward's class. 20 students are boys. 6 students are left-handed.

26. What is the ratio of boys to girls? $\frac{5}{4}$

27. How many more boys than girls are in the class? 4

28. What percentage of the students are left-handed? 16.67%

29. What is the ratio of left-handed to right-handed students? $\frac{1}{5}$

COMPUTER ACTIVITY

This program calculates an approximation for π by calculating the perimeter of a polygon that is inscribed in the circle.

```
10  PRINT "HOW MANY SIDES DO YOU WANT
    THE APPROXIMATING POLYGON TO HAVE?"
20  INPUT S
30  LET NM=INT(S/4)
40  LET AR=0:X0=0:Y=1:X=0:ST=1/NM
50  FOR N=1 TO NM
60  LET X1=X0+N*ST:LET Y1=SQR(1-X1*X1)
70  LET L=SQR((X1-X)*(X1-X)+(Y1-Y)*(Y1-Y))
80  LET AR=AR+L
90  LET X=X1:Y=Y1
100 NEXT N
110 PRINT "PI IS APPROXIMATELY ";2*AR
120 END
```

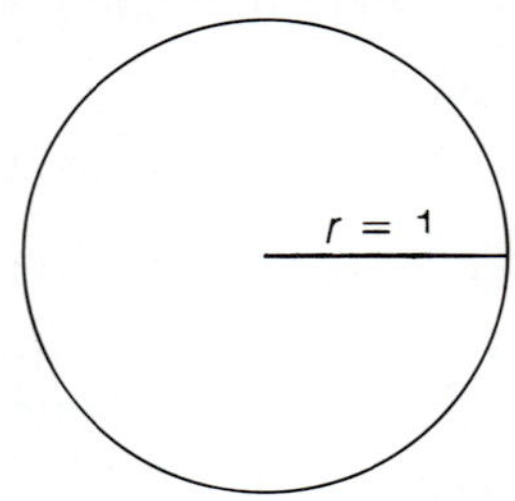

The circumference is π. The perimeter of this polygon is approximately π.

About how many sides are required for the approximation of π to round to:

1. 3.14 52

2. 3.141 500

3. 3.1415 1000

11-9 Congruent Figures

Objective: To write statements about congruent figures, vertices, and sides; to use SAS, SSS, and ASA properties to show congruence in triangles.
MIN: 1–18; REG: 1–18 even, 19, 20; MAX: 1–18 even, 19–22; ALL: Mixed Review

Two figures are **congruent** if they are identical in size and shape. Segments are congruent if they have the same length. Angles are congruent if they have the same measure. The symbol ≅ means "is congruent to."

Imagine that one congruent polygon is lifted and placed upon the other. The matching vertices are called **corresponding vertices,** and the matching sides are called **corresponding sides**. In the figures below, vertex A corresponds to vertex G and $\overline{CD}$ corresponds to $\overline{EH}$.

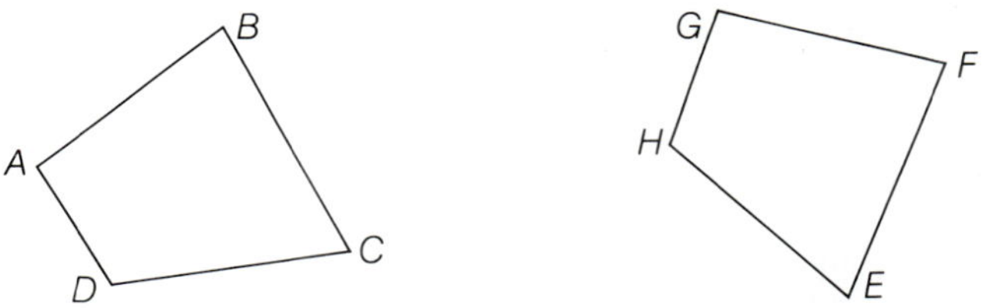

For any two congruent figures, corresponding vertices are congruent and corresponding sides are congruent. When naming two figures as congruent, list corresponding vertices in the same order. For example, if $\triangle ABC \cong \triangle DEF$, then A and D, B and E, C and F are corresponding vertices.

Example 1

$\triangle ABC \cong \triangle DEF$. Complete each statement.

$\angle A \cong$? $\quad \overline{AB} \cong$?

$\angle B \cong$? $\quad \overline{AC} \cong$?

$\angle C \cong$? $\quad \overline{BC} \cong$?

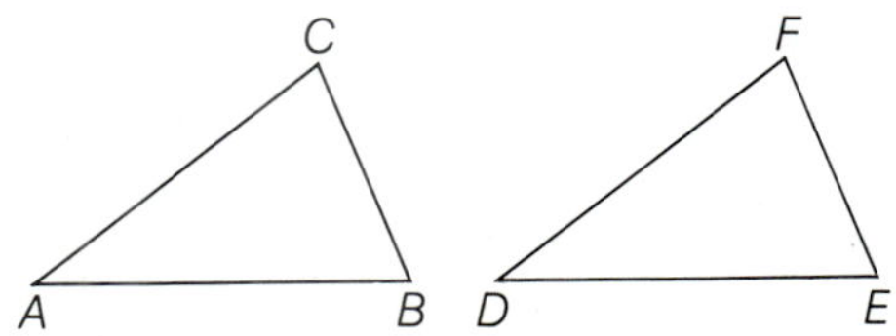

Solution $\angle A \cong \angle D$, $\angle B \cong \angle E$, $\angle C \cong \angle F$, $\overline{AB} \cong \overline{DE}$, $\overline{AC} \cong \overline{DF}$, $\overline{BC} \cong \overline{EF}$.

Practice Write all statements about congruent angles and sides.
$ABCDEF \cong GHIJKL$
see additional answers

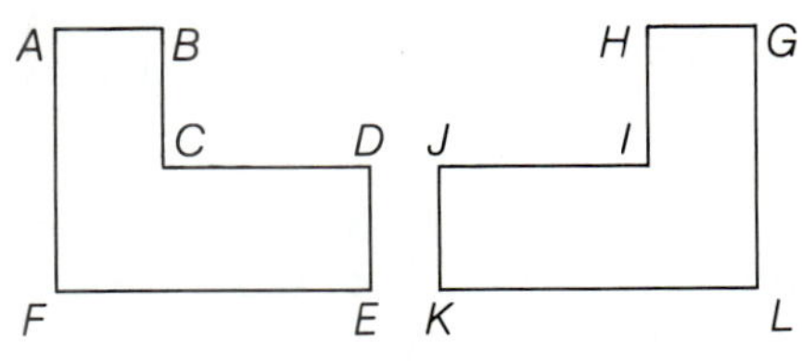

When two triangles are congruent, they have six pairs of congruent parts. To be sure that two triangles are congruent, you do not need to show that all six congruence statements are true. You can use any of the following three congruence properties to decide. The red marks in the figures below show the corresponding sides and angles.

Congruence Properties

1. **Side-Angle-Side (SAS)**
If two sides and the included angle (the angle between the two sides) of one triangle are congruent to two sides and the included angle of another triangle, then the two triangles are congruent.

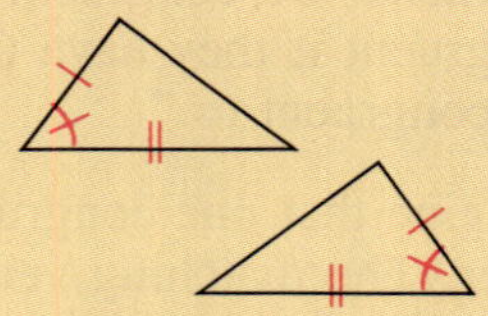

2. **Side-Side-Side (SSS)**
If the three sides of one triangle are congruent to the three sides of another triangle, then the two triangles are congruent.

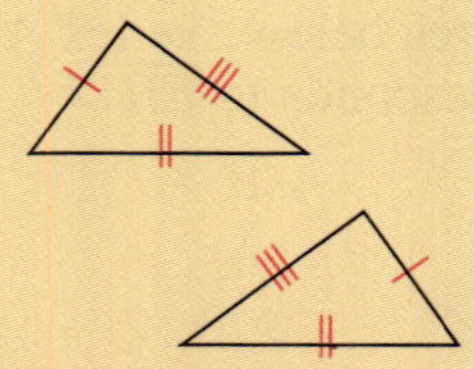

3. **Angle-Side-Angle (ASA)**
If two angles and the included side of one triangle are congruent to two angles and the included side of another triangle, then the two triangles are congruent.

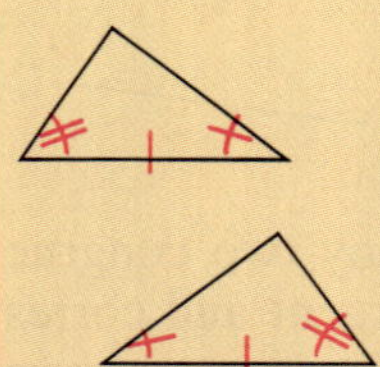

Example 2

Use SAS, SSS, or ASA property to show that the pair of triangles at right is congruent.

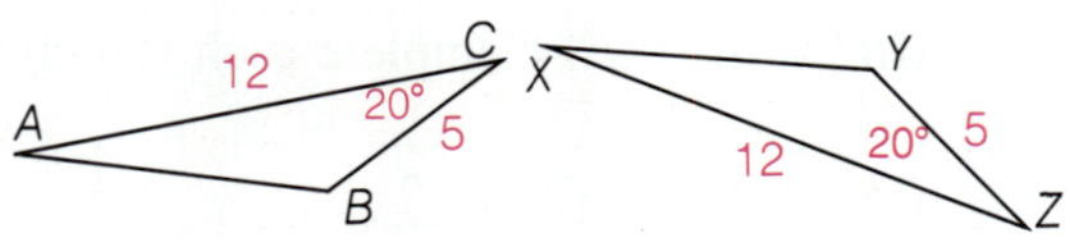

Solution $\overline{AC} \cong \overline{XZ}$

$\angle C \cong \angle Z$

$\overline{BC} \cong \overline{YZ}$ Since lengths of two sides and the included angle are given, use the SAS property.

$\triangle ABC \cong \triangle XYZ$ by the SAS property.

Practice Use the SAS, SSS, or ASA property to show that the pair of triangles is congruent.

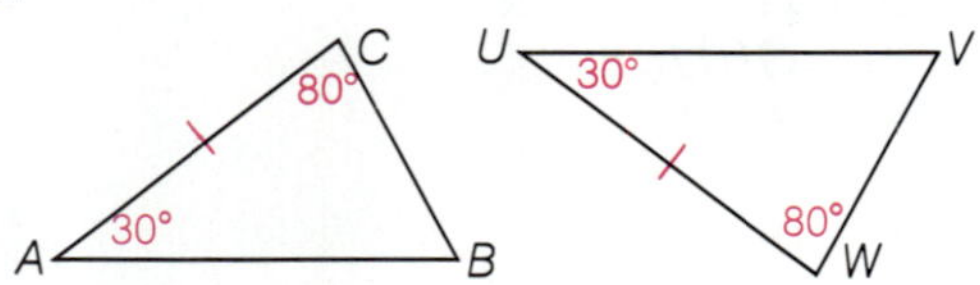

$\angle C \cong \angle W$ $\overline{AC} \cong \overline{UW}$ $\angle A \cong \angle U$
$\triangle ABC \cong \triangle UVW$ **by ASA property**

Oral Exercises

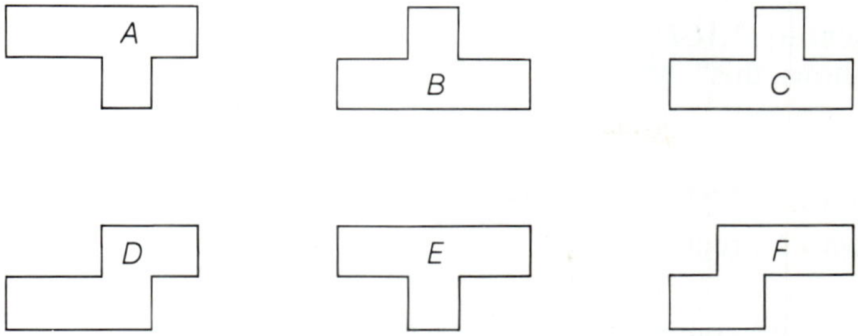

1. Which polygon seems to be congruent to polygon A? C
2. Which polygon seems to be congruent to polygon E? B
3. Which polygon seems to be congruent to polygon D? F

Exercises

A $\triangle RST \cong \triangle UVW$. Complete each statement.

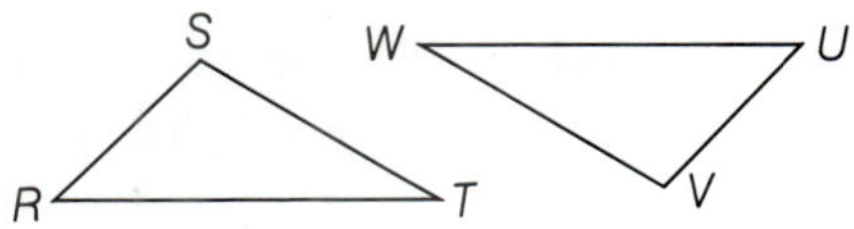

1. $\angle R \cong$? $\angle U$
2. $\overline{RS} \cong$? $\overline{UV}$
3. $\angle W \cong$? $\angle T$
4. $\overline{VW} \cong$? $\overline{ST}$

ABCDEF is congruent to *RSTUVW*.
Complete each statement.

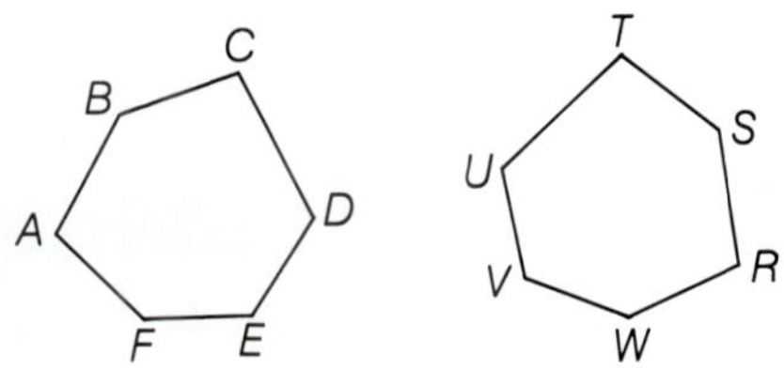

5. $\angle A \cong$? $\angle R$
6. $\angle U \cong$? $\angle D$
7. $\overline{EF} \cong$? $\overline{VW}$
8. $\overline{ST} \cong$? $\overline{BC}$
9. $\angle S \cong$? $\angle B$
10. $\angle E \cong$? $\angle V$
11. $\overline{CD} \cong$? $\overline{TU}$
12. $\angle D \cong$? $\angle U$
13. $\overline{RS} \cong$? $\overline{AB}$
14. $\overline{AB} \cong$? $\overline{RS}$

Use the SAS, SSS, or ASA property to show that each pair of triangles is congruent.

15. 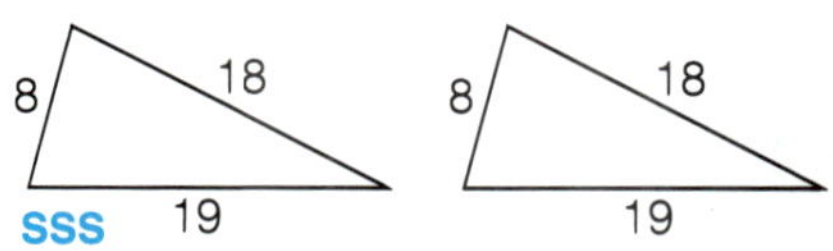

SSS

16. 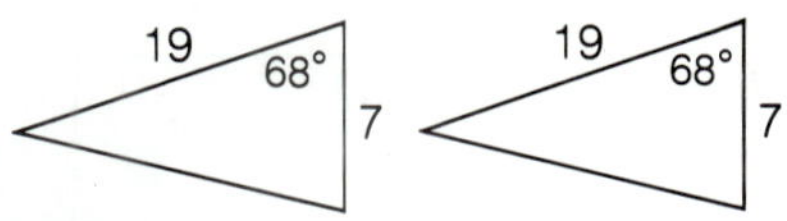

SAS

17. 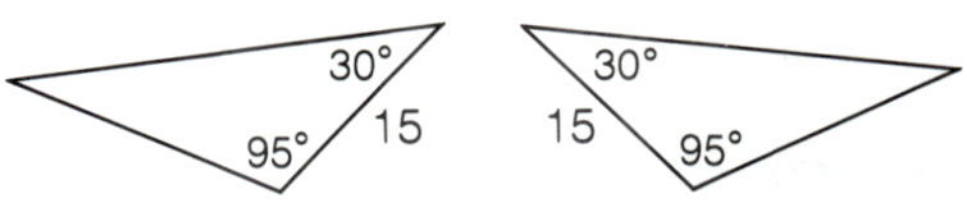

ASA

18. 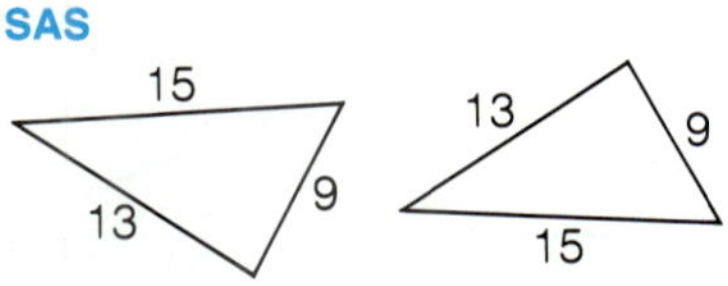

SSS

B The figure at right is a regular hexagon.

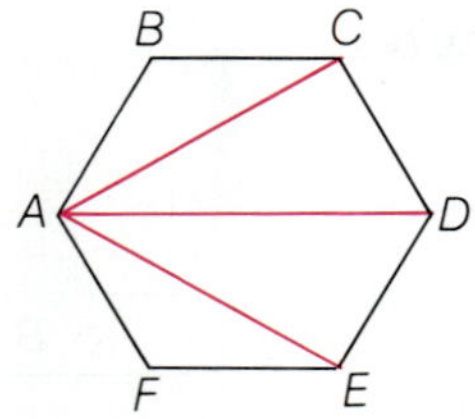

19. Name a triangle congruent to $\triangle ABC$. $\triangle AFE$
What congruence property shows this? SAS

20. Name a triangle congruent to $\triangle ACD$. $\triangle AED$
What congruence property shows this? SSS

C Extending Thinking Skills

Thinking skills: 21, 22. Using space visualization

The five-pointed star is drawn in a regular pentagon.

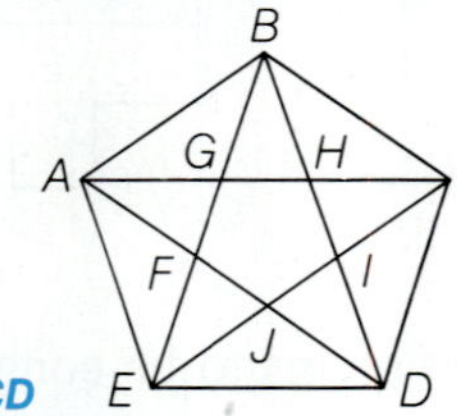

21. Name all the triangles that are congruent to $\triangle ABD$. $\triangle BCE$, $\triangle CDA$, $\triangle DEB$, $\triangle EAC$

22. List five triangles in this figure that are not congruent to each other. $\triangle AGF$, $\triangle ABG$, $\triangle ABF$, $\triangle ABE$, $\triangle ACD$
One could also list the five triangles having in common point *B* or *C* or *D* or *E*.

Mixed Review

Write as a decimal. Use a bar for a repeating decimal.

23. $\frac{3}{16}$ 0.1875 **24.** $\frac{2}{3}$ $0.\overline{6}$ **25.** $\frac{5}{12}$ $0.41\overline{6}$ **26.** $\frac{2}{5}$ 0.4 **27.** $\frac{2}{11}$ $0.\overline{18}$

Solve and check. **28.** $45c - 17 = 140.5$ 3.5 **29.** $5m \div 6 = 10$ 12

30. $m + 1.7m = 0.54$ 0.2 **31.** $r - 16 = -373$ −357 **32.** $26t - 14 = 18.5$ 1.25

Write as a fraction in lowest terms. **33.** $2\frac{1}{2}\%$ $\frac{1}{40}$ **34.** 84% $\frac{21}{25}$

Write as a percent. **35.** $\frac{45}{10}$ 450% **36.** $\frac{7}{12}$ $58.\overline{3}\%$ **37.** $1\frac{2}{5}$ 140% **38.** $\frac{1}{8}$ 12.5%

ESTIMATION

Use estimation to decide whether each pair of triangles is congruent.

1.
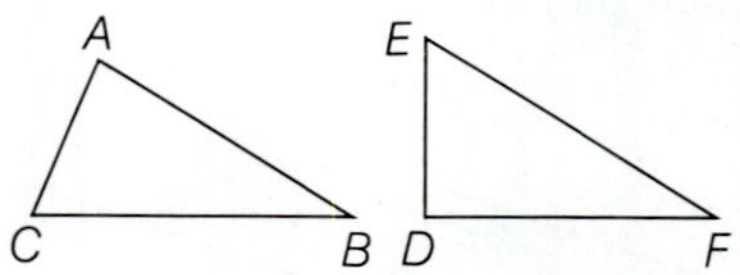

no

2.
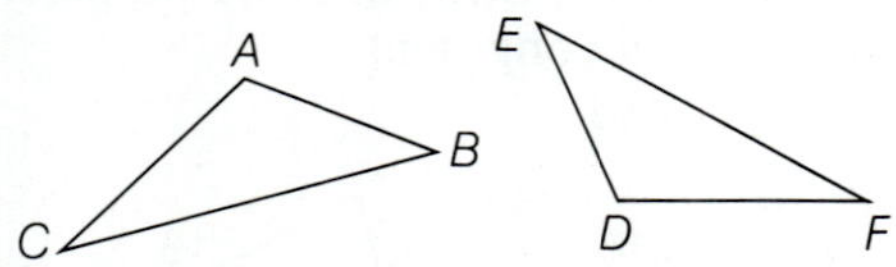

yes

3.
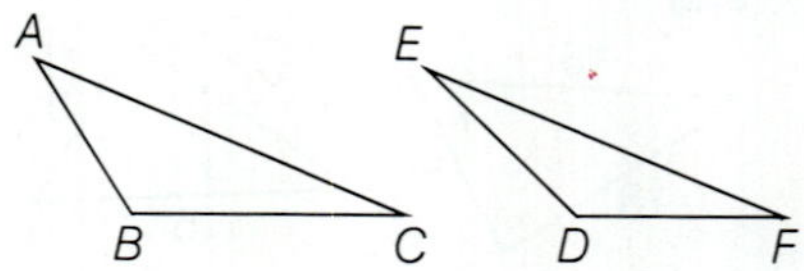

no

4.
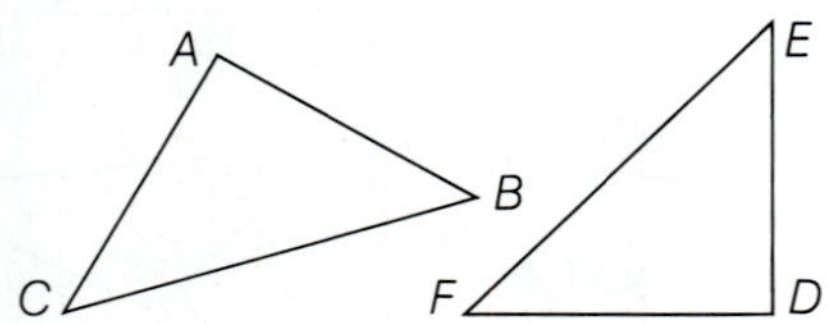

yes

PROBLEM SOLVING: WRITING EQUATIONS

11-10 Practice Solving Problems

Objective: To practice solving word problems by writing and solving equations.
MIN: 1–4; REG: 1–4 even, 5–8; MAX: 1–4 even, 5–10; ALL: Mixed Review

Problems

A Solve by writing an equation.

1. This week Lisa worked only 25% of the hours she usually works in a week. She worked 9 hours this week. How many hours does she usually work in a week? $0.25x = 9$ 36 hours

2. Tammy bought 2 tapes for the same price. With a "rebate" of \$1.50, the total cost for both came to \$7.50. What was the original cost for each tape?
$2x - 1.50 = 7.50$ \$4.50

3. The length of one side of a rectangular lot is $2\frac{1}{2}$ times as long as the width. How long is each side if the perimeter is 56 m? $2x + 2(2.5x) = 56$ 8, 8, 20, and 20

4. One angle of a triangular-shaped sign is 45°. The second angle is 2 times the measure of the third angle. What are the measures of the three angles?
$45 + x + 2x = 180$ 45°, 90°, 45°

B

5. Sally worked at a restaurant for 5 hours on Saturday and earned \$7.50 in tips. The total of her wages and tips was \$31.25. What is her hourly wage?
$5w + 7.50 = 31.25$ \$4.75

6. Arnie and Alicia each collected the same number of aluminum cans in a clean-up project. When they turned their cans in, they were told they had collected 25% of the total number of cans brought in that day. Arnie and Alicia each collected 158 cans. What was the total number of cans brought in that day? $2(158) = 0.25x$ 1264 cans

7. A child's theater ticket costs 25% less than an adult's ticket. A family paid \$22 for 2 adult's tickets and 1 child's ticket. How much is a child's ticket?
$2x + 0.75x = 22$ child's \$6

8. The cost to develop film at a store decreases for more than one roll. The second roll costs 10% less than the first roll and the third roll costs 20% less than the first roll. The cost for 3 rolls is \$23.76. What is the cost for 1 roll?
$x + 0.9x + 0.8x = 23.76$ \$8.80

C Extending Thinking Skills Answers will vary.

9. Write a word problem that would be solved using the equation $\frac{2}{3}x = 4$.

10. Write a word problem that would be solved using the equation $x - 15 = 120$.

Thinking skills: 9, 10. Formulating a problem

Mixed Review

Solve and check. **11.** $4.5t + 1.9t = 3.2$ 0.5 **12.** $-16 = 21m + 5$ −1

13. $14x = 4x - 30$ −3 **14.** $r + 63 = 8r$ 9 **15.** $17c + 4c = 84$ 4

11-11 Geometric Constructions

Objective: To construct segments, angles, triangles, perpendicular lines and angle bisectors.
MIN: 1–9; REG: 1–8 even, 10–18; MAX: 1–8 even, 10–20; ALL: Mixed Review

To do a geometric construction, you use only a compass and straightedge. Below are some basic constructions. **You might wish to provide cardboard to use as backing sheets for anchoring compasses.**

Example 1

Construct a copy of $\overline{AB}$.

Solution

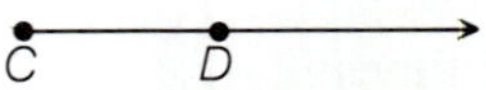

Use a straightedge to draw $\overrightarrow{CD}$.

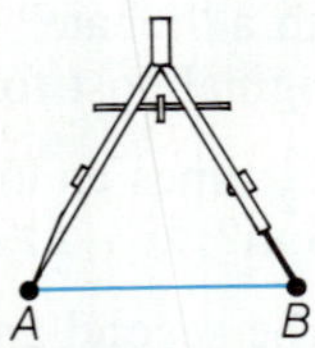

Place the sharp tip of the compass at A and the writing tip at B.

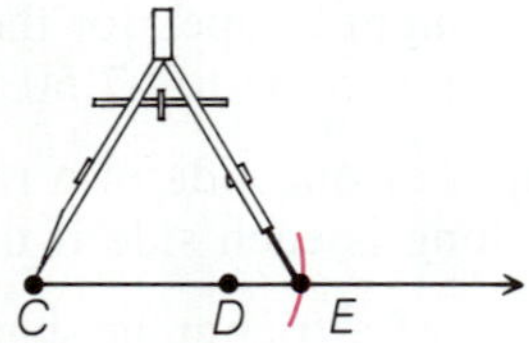

With the same setting, place the sharp tip of the compass at C and draw an arc. Label point E. $\overline{CE} \cong \overline{AB}$.

Practice Draw a segment and construct a copy of the segment. **Check students' work.**

Example 2

Construct a copy of $\angle ABC$.

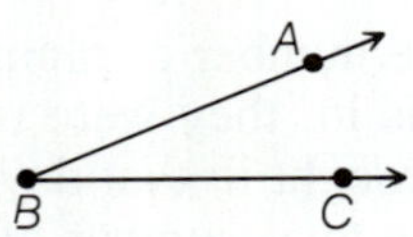

Remind students to use a straightedge when drawing lines, segments and rays.

Solution

Use a straightedge to draw $\overrightarrow{DE}$.

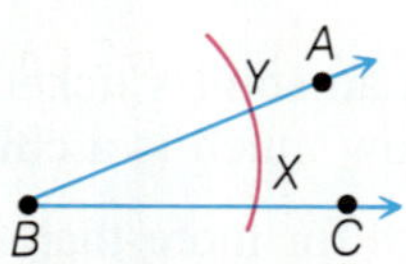

With the sharp tip of the compass at B, draw an arc on $\angle ABC$. Label points X and Y.

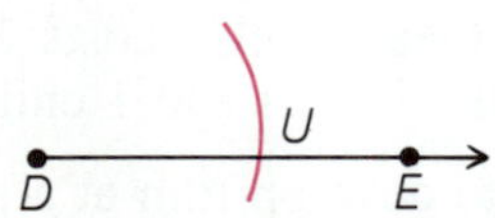

With the same setting, place the sharp tip of the compass at D and draw an arc. Label point U.

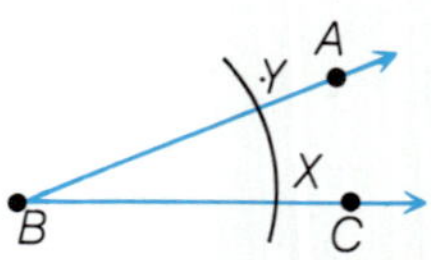

Place the sharp point on X and the writing point on Y.

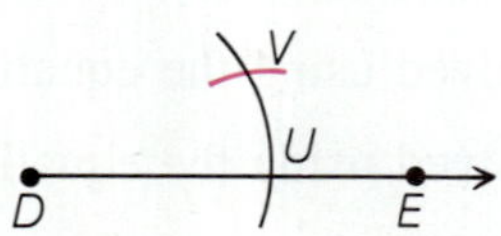

With the same setting, place the sharp point at U and draw an arc. Label point V.

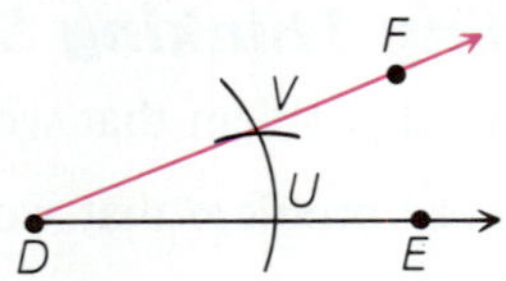

Draw ray DF through point V. $\angle EDF \cong \angle CBA$

Practice Draw an angle and construct a copy of it. **Check students' work.**

Example 3

Construct a line perpendicular to the given line through point P.

Solution

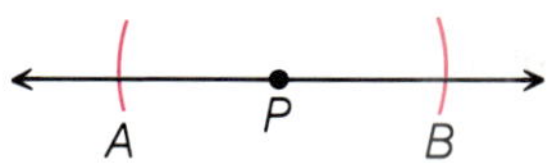

Place the sharp point of the compass at P and draw 2 arcs with the same setting, intersecting the line at A and B.

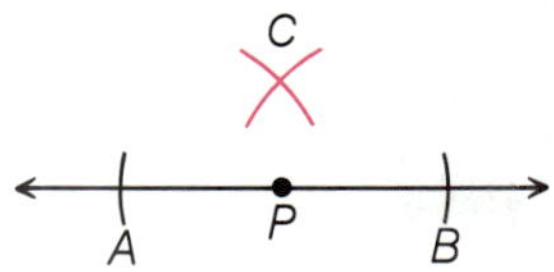

Place the sharp point at A, open it to beyond P, and draw an arc. With the same setting place the sharp point at B and draw an arc intersecting the first arc. Label point C.

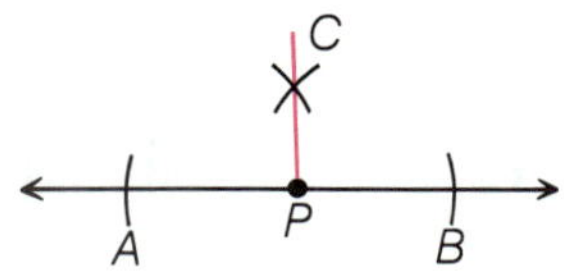

Draw line PC. It is perpendicular to line AB.

Practice Draw a line and locate a point on the line. Construct a line perpendicular to the line through the given point on the line.

Check students' work.

Example 4

Construct a triangle with side lengths a, b, and c.

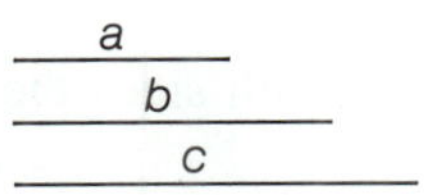

Solution

Draw a ray and copy the segment of length b.

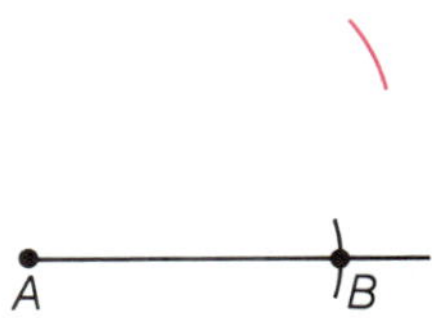

Set the compass opening by length c. Then place the sharp point at A and draw an arc.

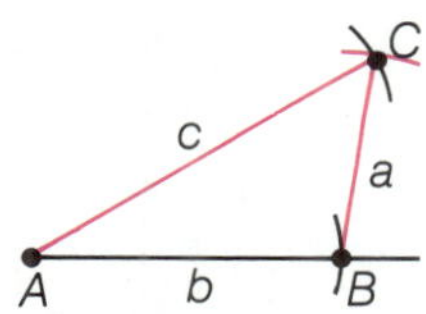

Set the compass opening by length a. Then place the sharp point at B and draw an arc to form point C. Draw segments $\overline{AC}$ and $\overline{BC}$.

Practice Construct an equilateral triangle. **Check students' work.**

An **angle bisector** is a ray that divides an angle into two congruent angles.

Example 5

Draw an angle ABC and bisect it.

Solution

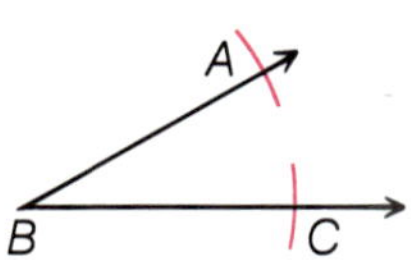

Place the sharp point of the compass at B and draw an arc intersecting the sides of the angle at A and C.

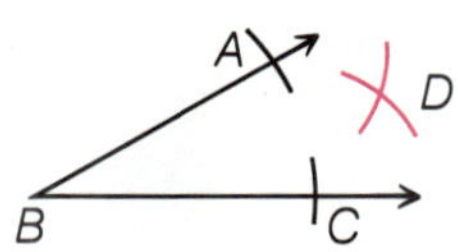

Open the compass more than half the distance from C to A and draw arcs from A and C. Label the point D.

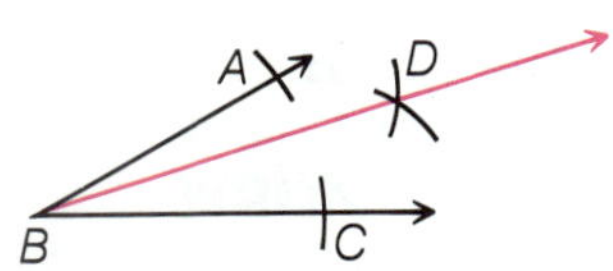

Draw ray $\overrightarrow{BD}$. It is the bisector of the angle.

Practice Construct a right angle and bisect it to form a 45° angle. **Check students' work.**

Exercises You may want to have students trace the figures before copying them.

A Copy the segments and construct a perpendicular at point M. Check students' work.

1. 

2. 

3. 

Copy the angle and construct a bisector.

4. 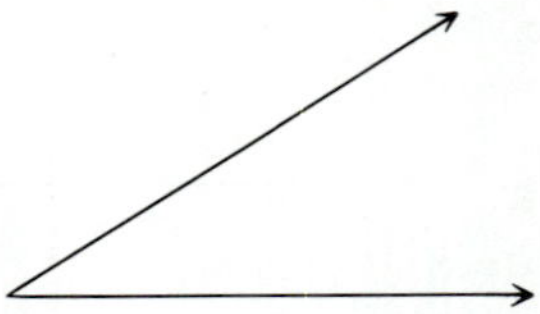

5. 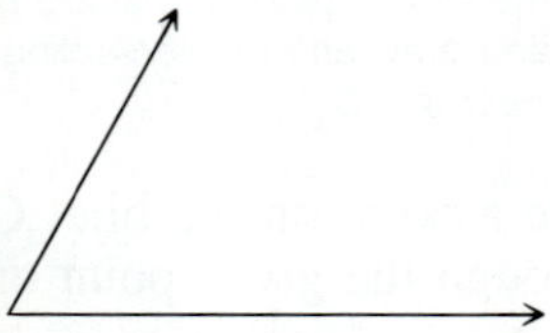

6. 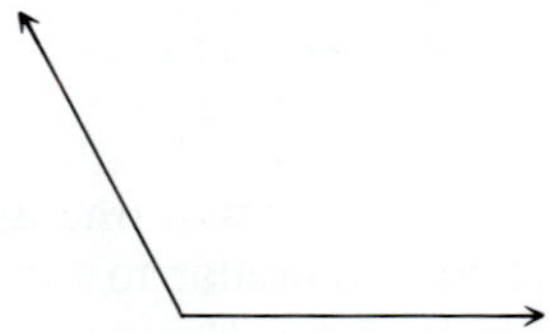

Construct a triangle with sides the lengths of the segments given.

7. 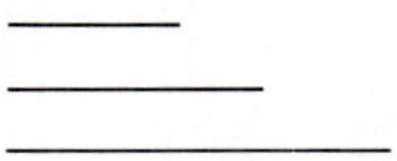

8. 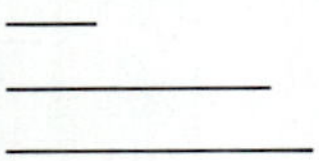

9. 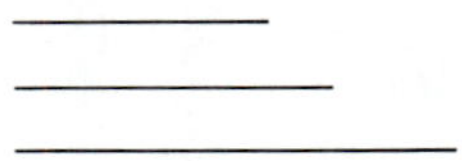

B Construct. Check students' work. Note that they must start with an equilateral triangle for Exercises 11, 12, 14, and 15.

10. a 45° angle

11. a 30° angle

12. a 60° angle

13. a $22\frac{1}{2}$° angle

14. a 15° angle

15. a $7\frac{1}{2}$° angle

16. a square

17. an isosceles right triangle

18. a pair of perpendicular lines

C Extending Thinking Skills Check students' work.

19. Draw a circle. Construct points A, B, C, D, E, F on it to make a regular hexagon.

20. Draw a circle. Construct points A, B, C, D, E, F, G, H on it to make a regular octagon.

Thinking skills: 19, 20. Reasoning

Mixed Review

Solve and check. **21.** $2x + 10 = 4x$ 5

22. $14 - 9x = -4$ 2

23. $36 = 15c - 24$ 4

24. What percent of 90 is 65? 72.$\overline{2}$%

25. 38 is 200% of what number? 19

PROBLEM SOLVING: STRATEGIES

11-12 Problems with More than One Answer

Objective: To solve nonroutine problems; to recognize and provide all answers for problems with more than one answer.
MIN: 1–2; REG: 1–4; MAX: 1–6

Some problems have more than one answer. You need to check solutions carefully to see whether a problem could have other answers. Consider the following.

Problem Some members of the school band were thirsty after a parade. They stood in a circle and passed around 31 cups of lemonade. Each person in order took 1 cup until no cups were left. The tuba player took the first and the last cup. He may have had more. How many band members were in the group?

You could use the **Guess, Check, Revise** strategy to find a solution. You might guess that there were 20 members in the group. When you checked this guess, you would see that if the tuba player took the first cup, the cups would be gone before he could take the last cup. Since about half the number of members as cups would work, you might revise the guess to 15. You could **Draw a Picture** and verify that 15 is an answer.

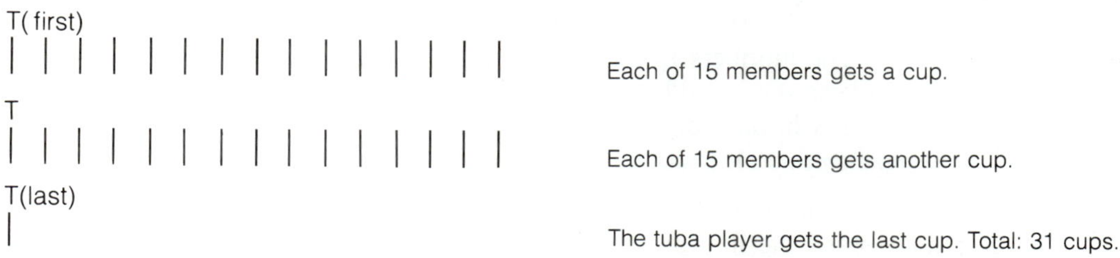

To be sure your solution is correct, you must check to see whether any other answers are possible. The problem above has several answers. You can **Make a Table** of answers and **Look for a Pattern** to help you describe a solution.

Number of Band Members	2	3	4	5	6	7	8	9	10	. . .	15	. . .	30
Is this a solution?	yes	yes	no	yes	yes	no	no	no	yes	no	yes	no	yes

Any number that is a factor of 30 will work. There could have been 2, 3, 5, 6, 10, 15, or 30 band members in the group.

Problem-Solving Strategies	
Choose the Operations	Write an Equation
Guess, Check, Revise	Simplify the Problem
Draw a Picture	Make an Organized List
Make a Table	Use Logical Reasoning
Look for a Pattern	Work Backwards

This chart shows the strategies presented so far.

Problems

Solve. Check to see whether the problem has more than one answer.

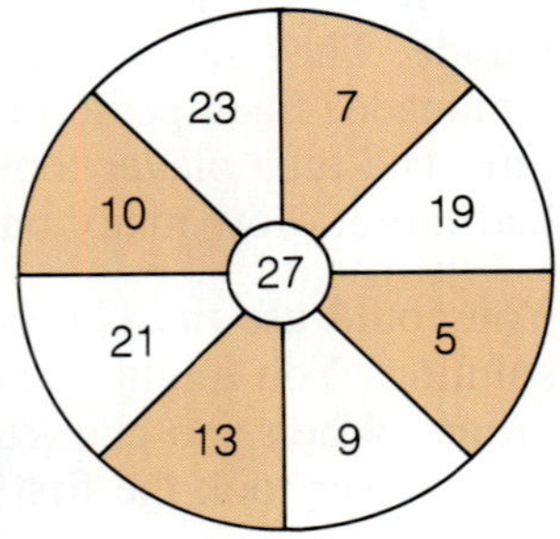

1. In a dart contest, Jennifer hit the dart board with 4 darts. Each dart hit a different number. Her total was 58. Which numbers did she hit? **23, 7, 19, 9 *or* 23, 5, 9, 21 or 27, 5, 7, 19, or 21, 19, 13, 5**

2. Several pairs of guinea pigs each gave birth to two new pairs (a second generation). Then each of those new pairs gave birth to two new pairs (a third generation). There were 48 pairs of guinea pigs in the fourth generation. How many pairs were in the first generation? **6**

3. A boy asked a girl for her telephone number. She described it as follows: "It contains all the digits 1–7. Each digit is used only once. The 3 numbers in the first part (prefix) are in order from smallest to largest and add up to 14. The numbers in the second part (suffix) are in order from smallest to largest and add up to 14. The number does not begin with 1." Using this information, could he figure out her phone number? **No (Three numbers are possible: 356–1247, 347–1256, and 257–1346).**

4. Five different-sized singing groups were needed for a musical program. The director formed the first group. Then she doubled the number of people in the first group and added one to make the second group. She doubled the number of people in the second group and added one more to make the third group. She used the same procedure for the fourth and fifth groups. The fifth group had 95 people in it. How many were in the first group? **5**

5. A cashier often gave change for $1 so people could use the soda machine. Just for fun, he tried to find all the ways to do this, using no more than 4 of any coin and no coin smaller than a nickel or bigger than a quarter. How many ways are there? **5 ways**

6. Ned had a higher bowling score than Jane, and Jane had a lower score than Marty. Jeff had a higher score than both Ned and Marty. Dee had a lower score than Marty but a higher score than Ned. Who had the highest score? **Jeff**

Enrichment

Symmetry

A figure has **reflectional symmetry** if, when it is traced and folded in half, one half falls exactly on the other half. The line on which the figure is folded is called **line of symmetry**. A figure has **rotational symmetry** if a tracing of it can be turned, or rotated, around a point less than a full revolution and fall exactly upon itself.

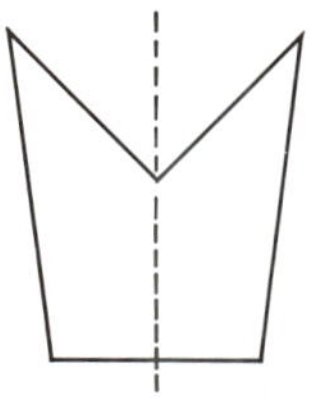

one line of symmetry

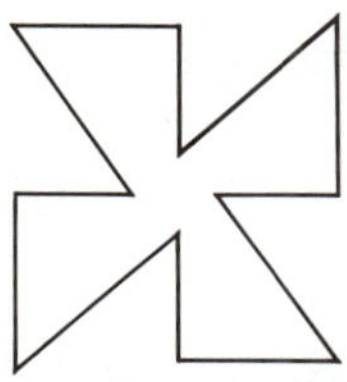

rotational symmetry

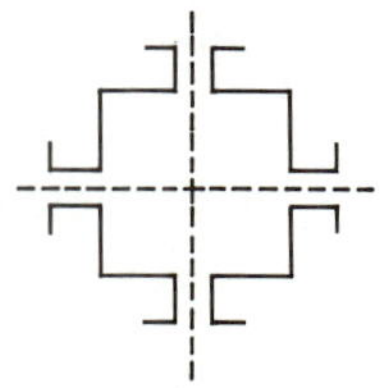

two lines of symmetry and rotational symmetry

Trace each figure and draw all lines of symmetry.

1. 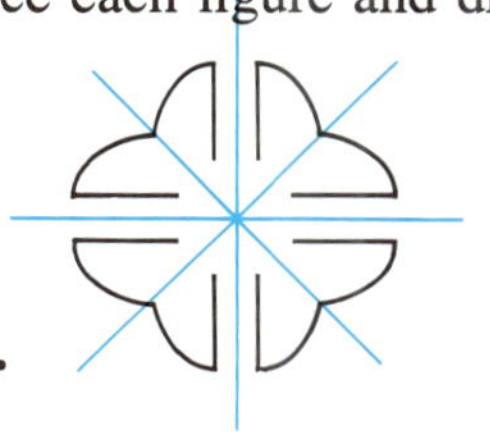

2. 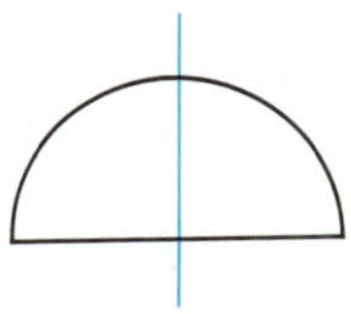

3. 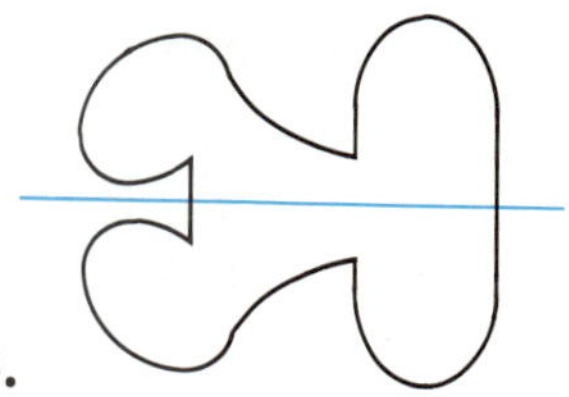

Which figures have rotational symmetry? **4 and 6**

4. 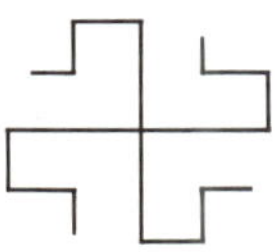

5. 

6. 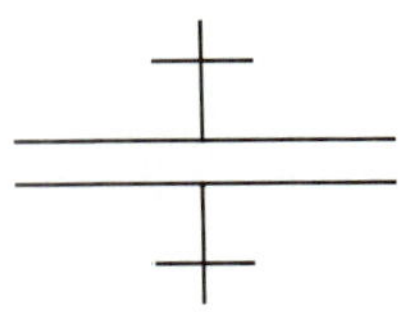

Trace each figure and draw all lines of symmetry. If the figure has rotational symmetry, how many degrees is the smallest rotation that turns the figure onto itself? **7. 90° rotation 8. 90° rotation 9. 60° rotation**

7. 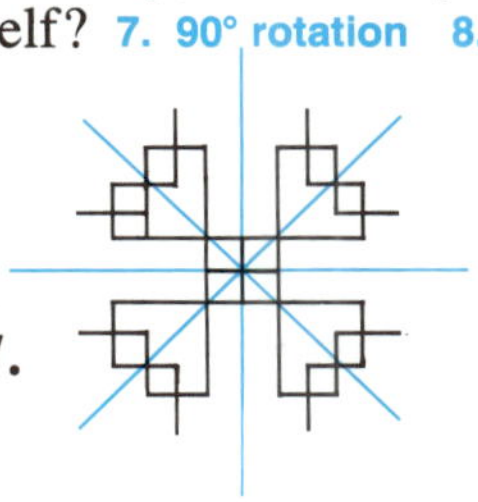

8. 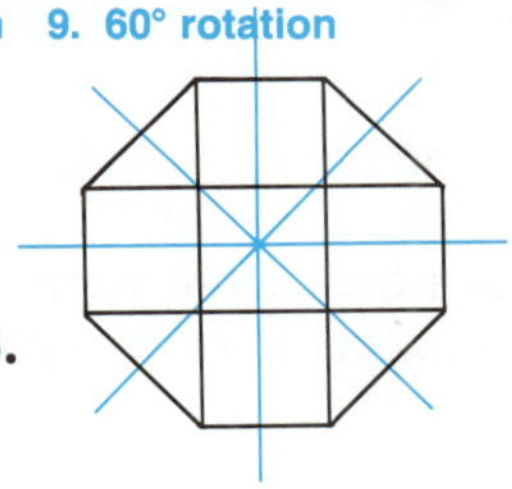

9. 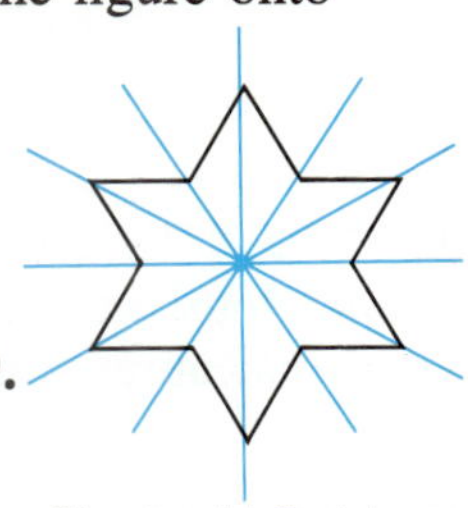

Chapter 11 Review

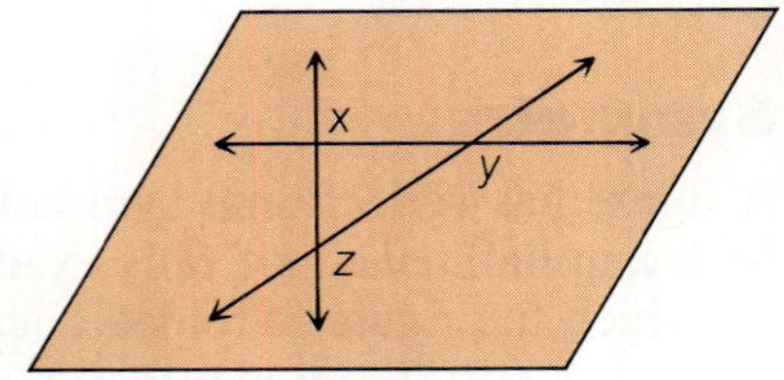

11-1 Refer to the figure for Exercises 1–3.

1. Name a point. **X, Y, or Z**

2. Name a line. **$\overleftrightarrow{XZ}$, $\overleftrightarrow{XY}$, or $\overleftrightarrow{YZ}$**

3. Name a segment that includes point Z. **$\overline{XZ}$ or $\overline{ZY}$**

11-2 Estimate the length of each segment to the nearest centimeter.

4. $\overline{AB}$ **2 cm** **5.** $\overline{AC}$ **8 cm**

11-3–11-4 Refer to the figure for Exercises 8–11.

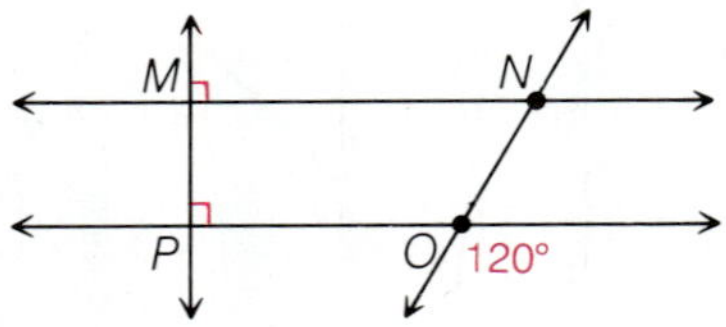

6. Name an acute angle. **∠MNO**

7. Name an obtuse angle. **∠NOP**

8. Find the measure of angle PON. **120°**

11-5 Classify each triangle by the measures of its angles and the lengths of its sides.

9. $\triangle RST$ has all 60 degree angles and all sides are equal in length. **acute, equilateral**

10. $\triangle LMN$ has one right angle and two sides equal in length. **right, isosceles**

11-6 Name the polygon.

11. 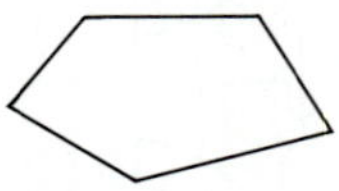 **pentagon**

12. 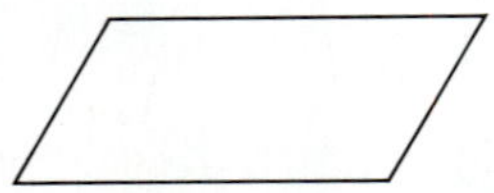 **parallelogram**

11-7 Solve. Use the formula *Distance* = *rate* × *time*.

13. If a driver averages 50 miles per hour, how long would it take to drive 215 miles? **4.3 hours**

11-8 Find the circumference of a circle with the given diameter or radius. Use 3.14 for π.

14. diameter = 3.5 m **10.99 m** **15.** radius = 6 cm **37.68 cm**

11-9 Use the SAS, SSS, or ASA property to show that the pair of triangles is congruent. **ASA**

16. 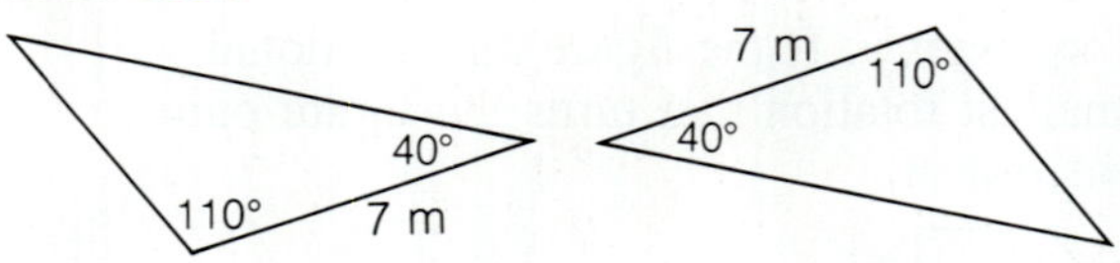

11-10 Solve by writing an equation.

17. Bob and Betty each sold 17 ads for the school paper. Together they sold 40% of the total number of ads in the paper. How many ads were in the paper? **85**

Chapter 11 Test

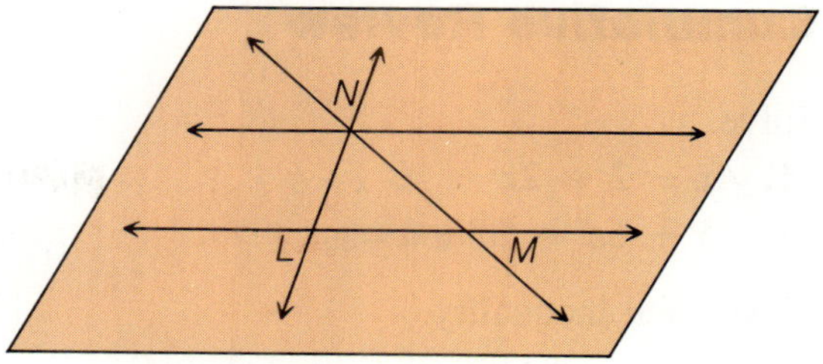

Refer to the figure for Exercises 1–3.

1. Name a point. L, M, or N

2. Name a line. $\overleftrightarrow{LM}$, $\overleftrightarrow{LN}$, or $\overleftrightarrow{NM}$

3. Name a segment that includes point L. $\overline{LM}$ or $\overline{LN}$

Estimate the length of each segment to the nearest centimeter.

4. $\overline{TU}$ 5 cm **5.** $\overline{SU}$ 9 cm

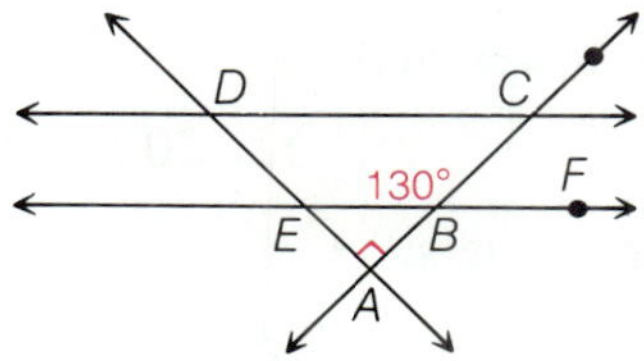

Refer to the figure for Exercises 8–11.

6. Name an acute angle. $\angle EDC$, $\angle CBF$, or $\angle DCB$

7. Name an obtuse angle. $\angle DEB$ or $\angle CBE$

8. Find the measure of $\angle CBF$. 50°

Classify each triangle by the measures of its angles and the lengths of its sides.

9. $\triangle EFG$ has a 90 degree angle and no sides are equal in length. right, scalene

10. $\triangle VWX$ has an obtuse angle and two sides equal in length. obtuse, isosceles

Name the polygon.

11. 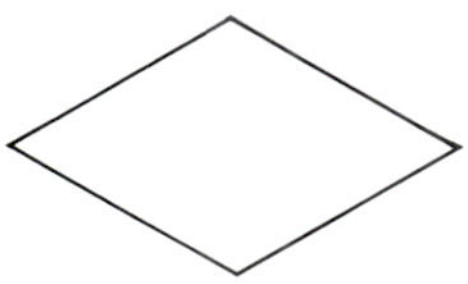 rhombus

12. 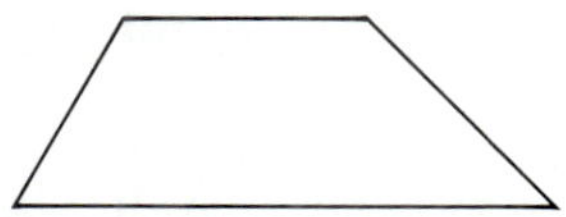 trapezoid

Solve. Use the formula *Distance* = *rate* × *time*.

13. Albert averaged 11 miles per hour on a $3\frac{1}{2}$ hour bicycle ride. How many miles did he ride? $38\frac{1}{2}$ miles

Find the circumference of a circle with the given diameter or radius. Use 3.14 for π.

14. diameter = 7 cm 21.98 cm

15. radius = 3.2 m 20.096 m

Use the SAS, SSS, or ASA property to show that the pair of triangles is congruent. SAS

16. 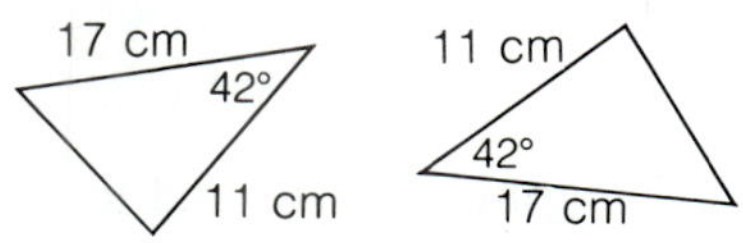

Solve by writing an equation.

17. Sandy attended 65% of the high school basketball games. If she attended 13 games, what was the total number of games played by the basketball team? 20 games.

Cumulative Review

Solve.

1. $4x - 2 = 2x + 16$ $x = 9$
2. $12f = 16f - 16$ $f = 4$
3. $25 + 6d = 5d$ $d = -25$
4. $30k + 8 = 22k + 24$ $k = 2$

Solve each inequality.

5. $7 > p - 6$ $p < 13$
6. $6m < -48$ $m < -8$
7. $9n - 5n > 20$ $n > 5$
8. $\frac{w}{3} - 6 > 12$ $w > 54$
9. $1 - 3x < 10$ $x > -3$
10. $2x - 3x > -8$ $x < 8$

Write each ratio in lowest terms.

11. 7:17 7:17
12. 30 to 40 3 to 4
13. $\frac{72}{18}$ 4

Solve. Check your answers.

14. $\frac{a}{35} = \frac{2}{7}$ $a = 10$
15. $\frac{7}{8} = \frac{y}{56}$ $y = 49$
16. $\frac{2}{3} = \frac{16}{p}$ $p = 24$

Solve by using a proportion.

17. The ratio of Amy's height to Rita's height is 11 to 9. Amy is 66 inches tall. How tall is Rita? 54 inches
18. The ratio of trumpets to trombones in a band is 4 to 3. The band has 16 trumpets. How many trombones does it have? $\frac{4}{3} = \frac{16}{n}$ 12 trombones

Simplify each rate using a proportion.

19. 159 km/3 h 53 km/h
20. $600/5 days $120/day
21. 192 beats/3 min 64 beats/min

Solve.

22. Paul charges $90 for 5 hours of labor. What is his hourly rate? $18/h
23. Shari drove 204 miles on 6 gallons of gas. How many miles per gallon did she get? 34 mi/gal

Solve by using a proportion or writing an equation.

24. 25% of 84 21
25. 120% of 50 60
26. 12.5% of 72 9

Find the percent.

27. What percent of 20 is 9? 45%
28. 25 is what percent of 125? 20%
29. What percent of 96 is 24? 25%
30. 18 out of 120 is what percent? 15%

Find the missing number. Round to the nearest tenth if necessary.

31. 7 is 5% of what number? 140
32. 80% of what number is 64? 80
33. 25% of what number is 19? 76
34. 36 is 75% of what number? 48

Chapter 12

Area and Volume Formulas

12-1 Area of Rectangles and Parallelograms

Objective: To use formulas to find the areas of rectangles and parallelograms.
MIN: 1–11; REG: 1–10 even, 12–24; MAX: 1–10 even, 12–26; ALL: Mixed Review

The **area** of a region is the number of unit squares needed to cover the region. The unit square used to measure the area of the rectangle below is a square centimeter. Since 15 square centimeters (cm^2) are used to cover the rectangle, its area is 15 cm^2.

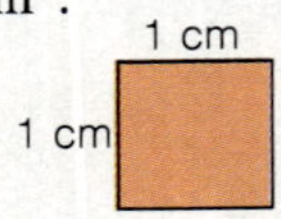

1 cm^2

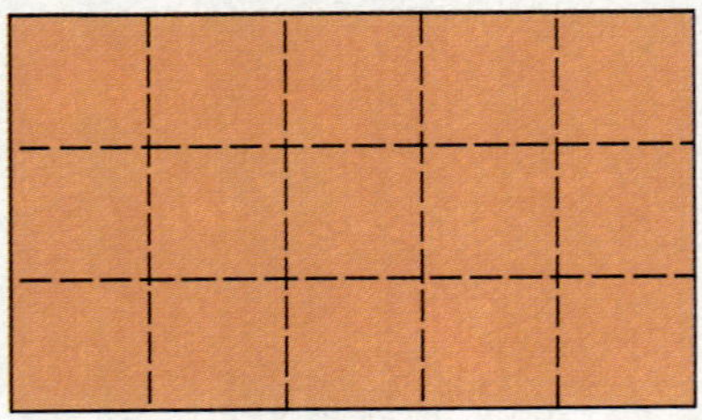

Area = 15 cm^2

In this rectangle, there are 3 rows of 5-unit squares. Note that the product of the length and the width of the rectangle gives its area, $5 \times 3 = 15$, and the area is 15 cm^2. This suggests the following formula for the area of a rectangle.

Formula: Area of a Rectangle

The area of a rectangle is equal to the length times the width.

$A = lw$.

Example 1

Find the area of the rectangle.

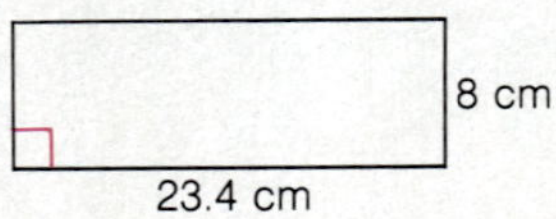

Solution

$A = lw$

$A = 23.4 \times 8$ Replace l with 23.4 and w with 8.

$A = 187.2$ The area is 187.2 cm^2.

Practice Find the area of the rectangle.

132.86 m^2

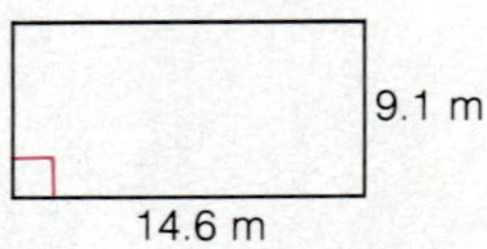

The formula for the area of a parallelogram is similar to that for a rectangle. The length of one side of a parallelogram is called the *base* (b). The perpendicular distance between a pair of parallel bases is called the *height* (h). Any parallelogram can be cut into two pieces and rearranged to form a rectangle. The area of the parallelogram is equal to the area of the resulting rectangle.

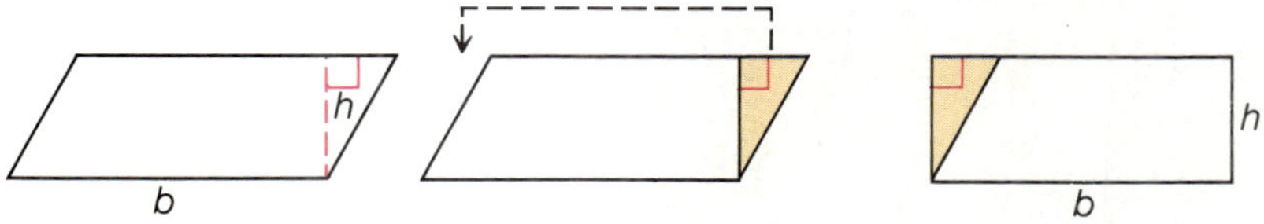

This suggests the following formula.

Formula: Area of a Parallelogram

The area of a parallelogram is equal to the base times the height.

$A = bh$

Example 2

Find the area of the parallelogram.

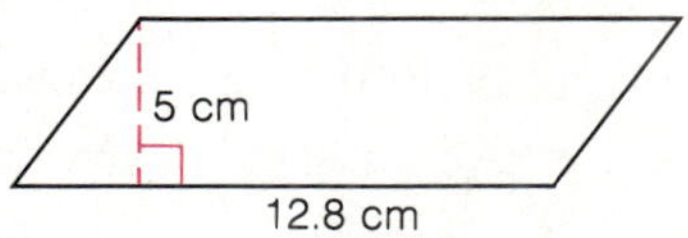

Solution $A = bh$

$A = 12.8 \times 5$ Replace b with 12.8 and h with 5.

$A = 64$ The area is 64 cm^2.

Practice Find the area of a parallelogram with the dimensions given.

a. 24 m long and 9 m high **216 m^2** **b.** 48.3 cm long and 51.5 cm high **2487.45 cm^2**

Oral Exercises

Give the dimensions you would use to find the area.

1.

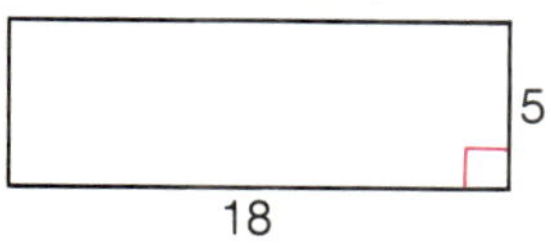

2.

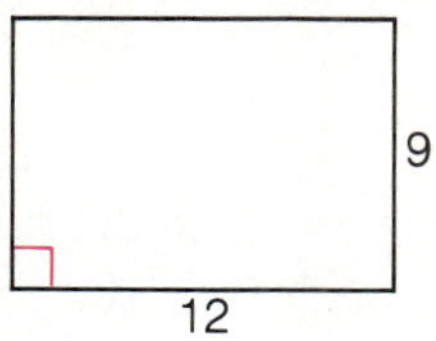

3.

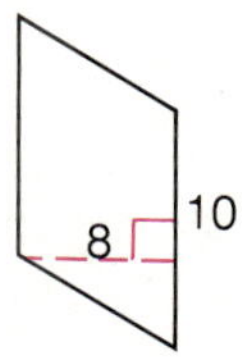

4.

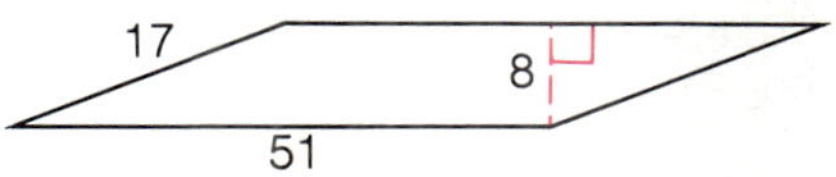

5.

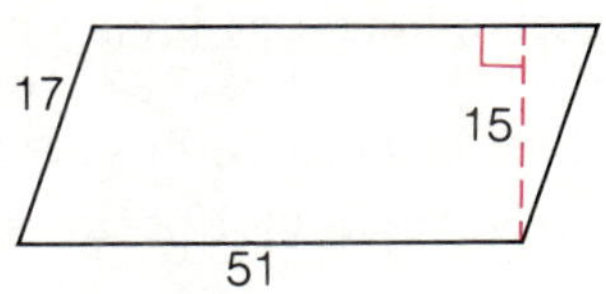

1. 18, 5
2. 12, 9
3. 8, 10
4. 8, 51
5. 51, 15

Exercises

A Find the area of each figure. Assume all units are centimeters.

1.

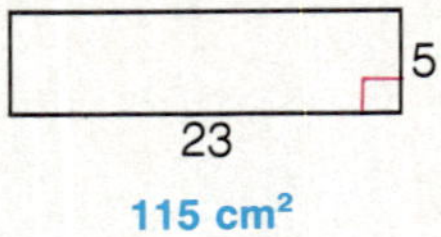

2.

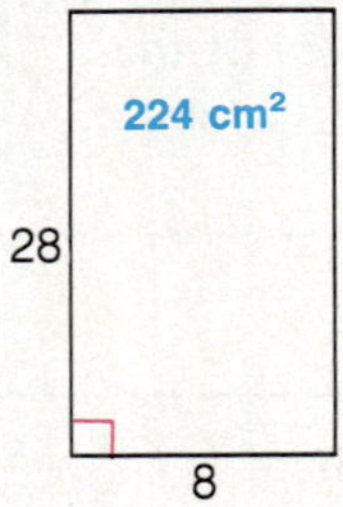

3.

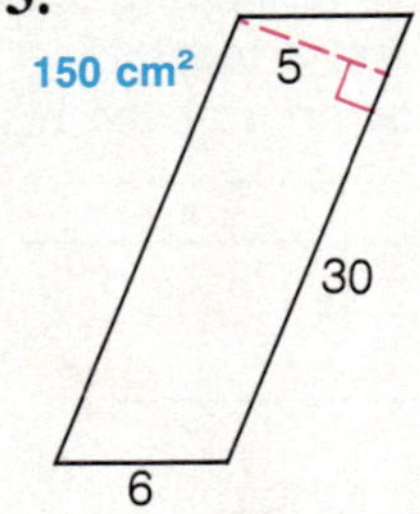

4.

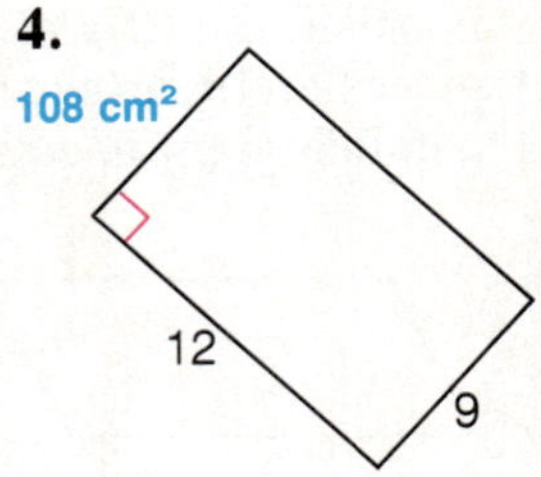

5.

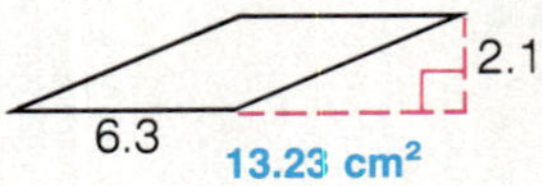

6.

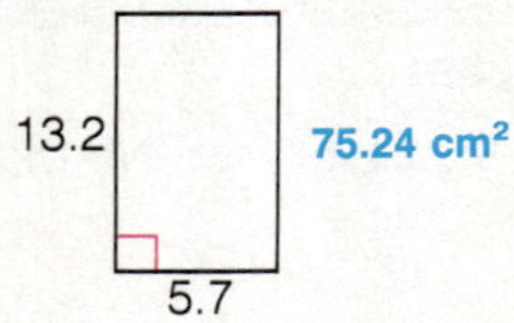

7.

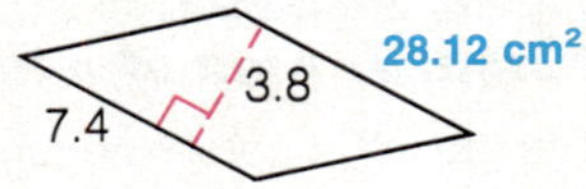

Find the area of a parallelogram with the dimensions given.

8. base = 14.3 cm, height = 5.8 cm **9.** base = 123 mm, height = 18 mm

10. base = 1.56 km, height = 0.23 km **11.** base = 2.51 m, height = 6.89 m

12. $b = 21.3$ m, $h = 12$ m 255.6 m² **13.** $b = 73$ cm, $h = 14.1$ cm 1029.3 cm²

8. 82.94 cm² 9. 2214 mm² 10. 0.3588 km² 11. 17.2939 m²

B Find the area of each parallelogram. All units are centimeters.

14. 100 cm²
15. 66 cm²
16. 54 cm²

14.

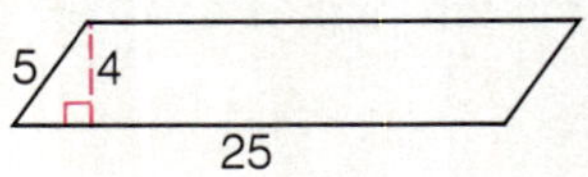

15.

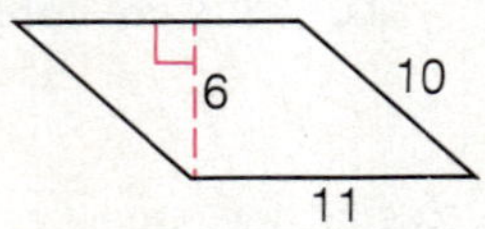

16.

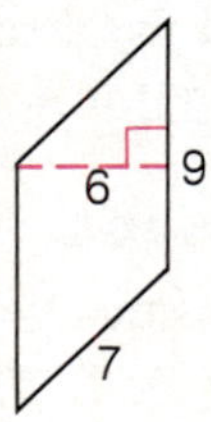

Find the base or height of a parallelogram with the dimensions given.

17. area = 4851 cm^2, height = 21 cm 231 cm **18.** area = 172.2 m^2, base = 12.3 m 14 m

19. area = 43.8 m^2, base = 36.5 m 1.2 m **20.** area = 24.91 mm^2, height = 5.3 mm 4.7 mm

Any unit of length can be converted to a unit of area. The length units inch, foot, and yard become the area units square inch (in^2), square foot (ft^2), and square yard (yd^2). To calculate area, you must use the same unit for length and width. Use the smaller unit in your answer.

21. A rectangle has length of 12.5 in. and width of 0.5 ft. What is its area? 75 in²

22. A rectangle has area 45 in^2. If its length is 9 inches, what is its width? 5 in

23. A gallon of paint covers about 450 ft^2. How many gallons are needed to paint the walls of a room that is 32 ft long, 16 ft wide, and 12 ft high? (The windows and doors take up 33 ft^2.) About $2\frac{1}{2}$ gallons

24. Carpet costs $18/$yd^2$. What is the cost to carpet a room 12 ft by 16 ft? $384

C Extending Thinking Skills

25. A rectangle has a length of 25 cm and a perimeter of 70 cm. Find its area. 250 cm^2

26. Complete the table to find which rectangle with perimeter 100 has the largest area. What generalization can you make from your answer? $l = 25, w = 25$
A square has the greatest area of all rectangles.

perimeter	100	100	100	100	100	100	100
width	5	10	15	20	? 25	? 30	35
length	45	40	? 35	? 30	? 25	? 20	? 15
area	225	400	? 525	? 600	? 625	? 600	? 525

Thinking skills: 25. Reasoning 26. Generalizing

Mixed Review

Evaluate for $n = -\frac{1}{3}$. **27.** $2n + \frac{1}{2}$ $-\frac{1}{6}$ **28.** $\frac{4}{5}n$ $-\frac{4}{15}$ **29.** $-6n$ 2 **30.** $\frac{3}{4} - n$ $1\frac{1}{12}$

31. 16 is what percent of 24? $66\frac{2}{3}\%$

32. 28 is what percent of 24? $116\frac{2}{3}\%$

33. 15 is what percent of 24? 62.5%

34. 21 is what percent of 24? 87.5%

CALCULATOR ACTIVITY

You can estimate the area of a square by rounding the length of one side before squaring the length. This gives a different result than you would get if you squared the exact length of one side first and then rounded the answer.

1. Use a calculator to complete this table for finding the area of a square.

Length of side	square, then round		round the side, then square	
23.4	547.56	548	23	529
117.3	13,759.29	13,759	117	13,689
259.2	67,184.64	67,185	259	67,081
98.7	9,741.69	9,742	99	9,801

2. Which method gives the more accurate area? Square, then round

12-2 Area of Triangles and Trapezoids

Objective: To use formulas to find the areas of triangles and trapezoids.
MIN: 1–22; REG: 1–22 even, 23–26; MAX: 1–22 even, 23–27; ALL: Mixed Review

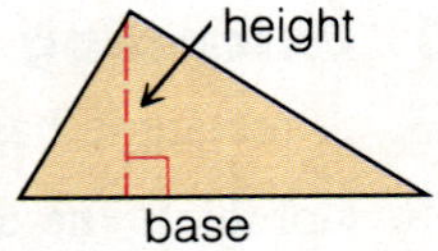

Any side of a triangle can be called a **base.** The **height** is the perpendicular distance from a vertex to the base.

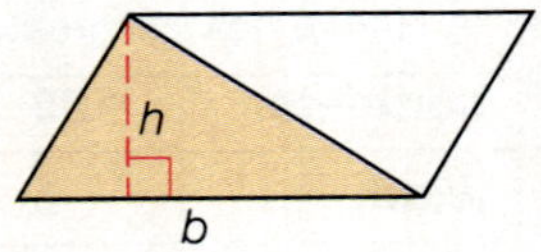

Study the figures at right. Since two congruent triangles can form a parallelogram, the area of the triangle must be one half the area of the parallelogram that has the same base and height.

Formula: Area of Triangle

The area of a triangle is equal to $\frac{1}{2}$ the base times the height.

$$A = \frac{1}{2}bh$$

Since any side of a triangle can be a base, each triangle has three bases and three heights. When calculating the area of a triangle, you must use the correct height for the chosen base.

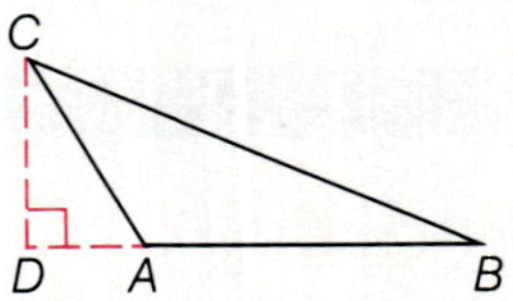

$\overline{AB}$ is a base
$\overline{CD}$ is a height

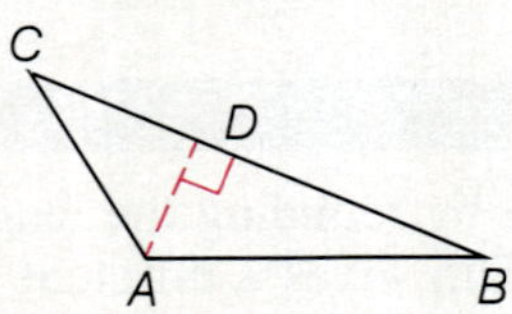

$\overline{BC}$ is a base
$\overline{AD}$ is a height

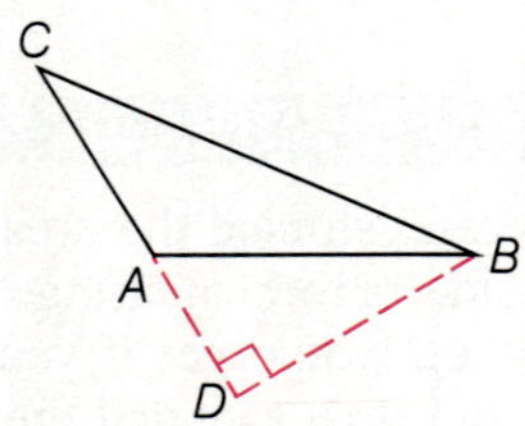

$\overline{AC}$ is a base
$\overline{BD}$ is a height

Example 1

Find the area of the triangle.

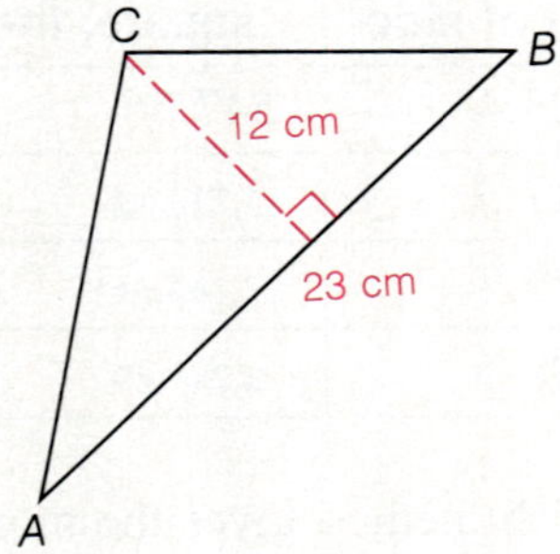

Solution $A = \frac{1}{2}bh$

$A = \frac{1}{2}(23 \times 12)$ Replace *b* with 23 and *h* with 12.

$= 138$

The area is 138 cm^2.

Practice Find the area of each triangle.

a. 20 cm^2

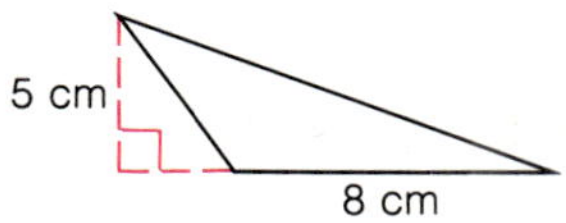

b. 33 cm^2

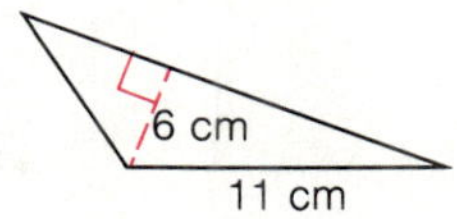

You learned earlier that a trapezoid is a quadrilateral with one pair of parallel sides, called bases. Since two congruent trapezoids can form a parallelogram, the area of one of the trapezoids is one half the area of the resulting parallelogram.

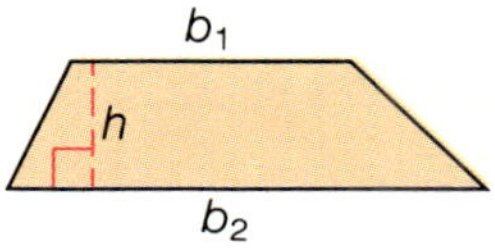

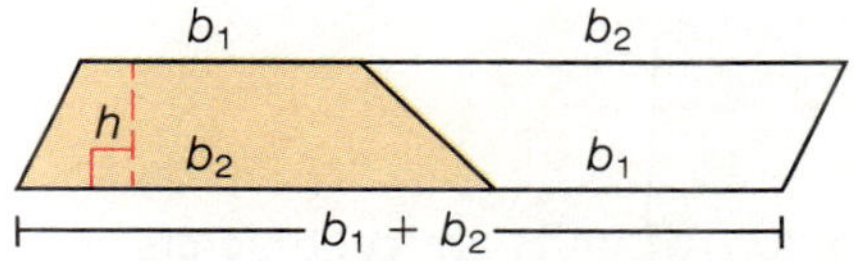

If the bases of the trapezoid have lengths b_1 and b_2, the parallelogram formed has a pair of sides with length $b_1 + b_2$.

Formula: Area of a Trapezoid

The area of a trapezoid is equal to $\frac{1}{2}$ the height times the sum of the bases.

$$A = \frac{1}{2}h(b_1 + b_2)$$

Example 2

Find the area of the trapezoid.

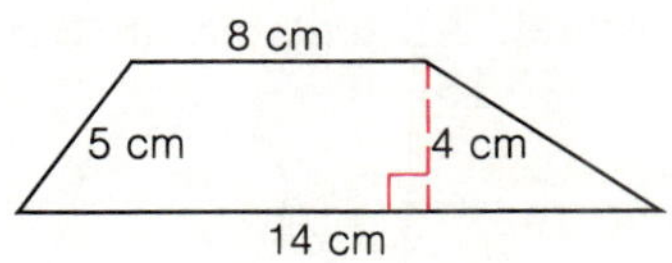

Solution $A = \frac{1}{2}h(b_1 + b_2)$

$A = \frac{1}{2}(4)(8 + 14)$ Notice that 5 is not the height, since the side is not perpendicular to the two bases.

$= 44$

The area is 44 cm^2.

Practice Draw a trapezoid and label it with the lengths given. All units are meters. Find the area.

a. $b_1 = 9$, $b_2 = 23$, $h = 29$ 464 m^2 **b.** $b_1 = 5$, $b_2 = 28$, $h = 14$ 231 m^2

Oral Exercises

Select the base or height that you could use with the given base or height.

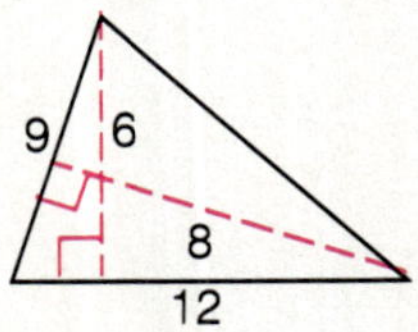

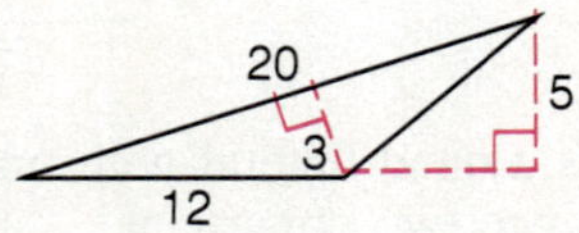

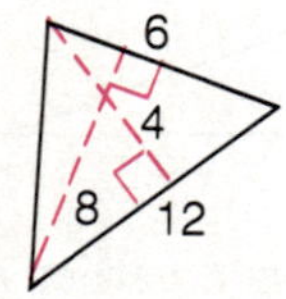

1. base = 12, height = ? 6
   height = 8, base = ? 9
2. height = 3, base = ? 20
   base = 12, height = ? 5
3. base = 12, height = ? 4
   base = 6, height = ? 8

Exercises

Find the area of each triangle. All units are centimeters.

1. 31.5 cm^2

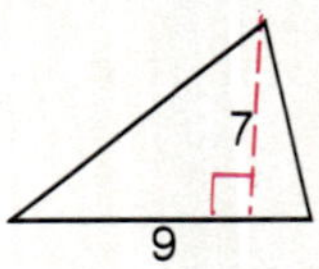

2. 48 cm^2

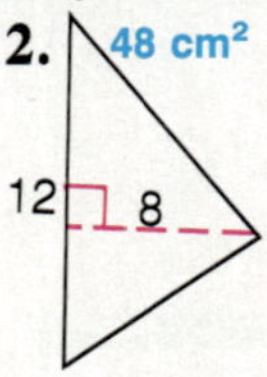

3. 63 cm^2

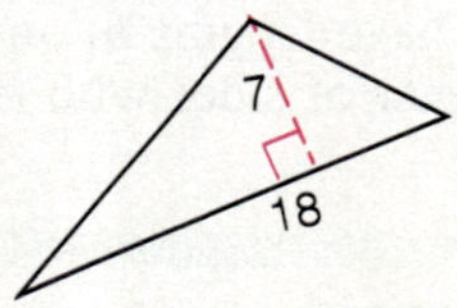

4. 20 cm^2

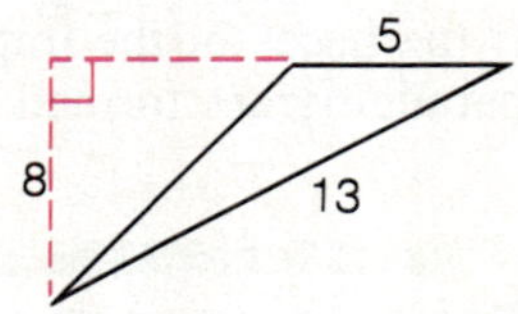

Draw a triangle and label it with the base and height given. Find the area.

5. b = 12 cm; h = 5 cm **6.** b = 15 m; h = 8 m **7.** b = 5 m; h = 18 m 45 m^2

8. b = 23 m; h = 9 m **9.** b = 18 m; h = 2 m **10.** b = 34 cm; h = 2 cm

11. b = 15 cm; h = 17 cm **12.** b = 40 m; h = 23 m **13.** b = 48 cm; h = 73 cm

5. 30 cm^2 6. 60 m^2 8. 103.5 m^2 9. 18 m^2 10. 34 cm^2 11. 127.5 cm^2 12. 460 m^2 13. 1,752 cm^2

Find the area of each trapezoid. All units are meters.

14. 44 m^2

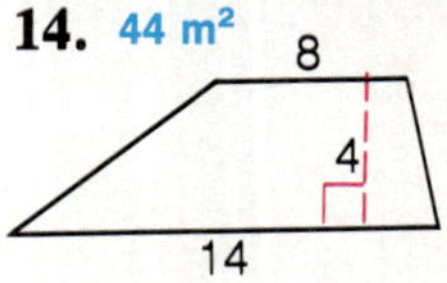

15. 81 m^2

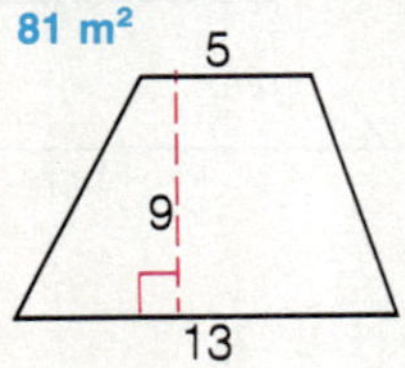

16. 72 m^2

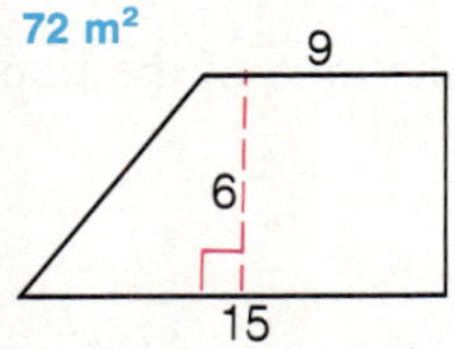

Draw a trapezoid and label it with the lengths given. Find the area.

17. b_1 = 8 cm, b_2 = 15 cm; h = 5 cm **18.** b_1 = 9 cm, b_2 = 17 cm; h = 5 cm

19. b_1 = 5 m, b_2 = 8 m; h = 18 m **20.** b_1 = 9 cm, b_2 = 16 cm; h = 12 cm

21. b_1 = 4 m, b_2 = 1.2 m; h = 0.23 m **22.** b_1 = 24 m, b_2 = 4 m; h = 2 m

B You might wish to have students do the Numbers to Algebra activity before these problems.

23. A triangle has area 30 cm^2 and a height of 5 cm. Find the length of the base. 12 cm

24. A triangle has area 50 m^2 and a base of 20 meters. Find the height. 5 m

25. A trapezoid has bases of 15 cm and 48 mm. The area is 79.2 cm^2. Find the height. 8 cm

26. A trapezoid has nonparallel sides of length 7 cm and 9 cm and a perimeter of 40 cm. One base is twice as long as the other base. Find the lengths of the bases. $b_1 = 8$ cm, $b_2 = 16$ cm

C *Extending Thinking Skills*

27. Use an equilateral triangle as the unit of area to find the pattern. Complete the next three columns of this table. What is the area of an equilateral triangle with side length n? n^2

Side length	1	2	3	4	5	6
		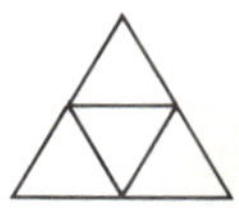	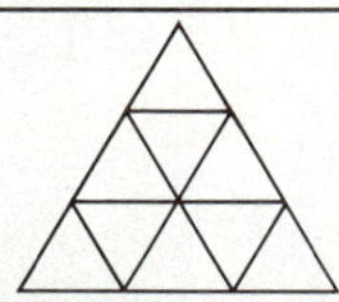			
Area	1	4	9	16	25	36

Thinking skill: 27. Finding a pattern

Mixed Review

Solve and check. **28.** $3.4t + 5.2 = -3.3$ -2.5 **29.** $3(y + 2) = 3$ -1

30. Evaluate the formula $2a + 3(b + 4)$ for $a = 17$, $b = 9$. 73

NUMBERS TO ALGEBRA

Replacing variables in a formula with specific numbers can help you see how to solve a formula for one variable in terms of the other variables. The example below shows how to solve the formula $A = \frac{1}{2}bh$ for h by using numbers for a triangle with an area of 45 square units and a base of 9 units.

Numbers		Algebra	
$A = \frac{1}{2}bh$		$A = \frac{1}{2}bh$	
$45 = \frac{1}{2}(9)h$	Replace A with 45 and b with 9.		
$2(45) = 9h$	Multiply both sides by 2.	$2A = bh$	Multiply both sides by 2.
$\frac{2(45)}{9} = h$	Divide both sides by 9 to solve for h.	$\frac{2A}{b} = h$	Divide both sides by b to solve for h.

1. Solve the formula $A = bh$ for h. $h = \frac{A}{b}$

2. Solve the formula $P = 2l + 2w$ for w. $w = \frac{P - 2l}{2}$ or $w = \frac{P}{2} - l$

12-3 Area of Circles

Objective: To use formulas to find the areas of circles.
MIN: 1–12; REG: 1–12 even, 13–23; MAX: 1–12 even, 13–24; ALL: Mixed Review

An exploratory lesson opener can be found on page 553.

A city engineer uses the formula $A = \pi r^2$ to calculate the amount of water that a pipe carries.

The circumference, C, of a circle with radius r is found by using the formula $C = 2\pi r$. This formula is used to develop a formula for the area of a circle.

In the figure below, a circle of radius r has been divided into pie-shaped pieces. The pieces have been rearranged into a parallelogram-like shape. This "parallelogram" has a height r and a base that is about half the circumference of the circle. Therefore, the approximate area is $(\pi r)r$, or πr^2. Since the area of the "parallelogram" would be the same as the area of a circle, the area of the circle is also πr^2.

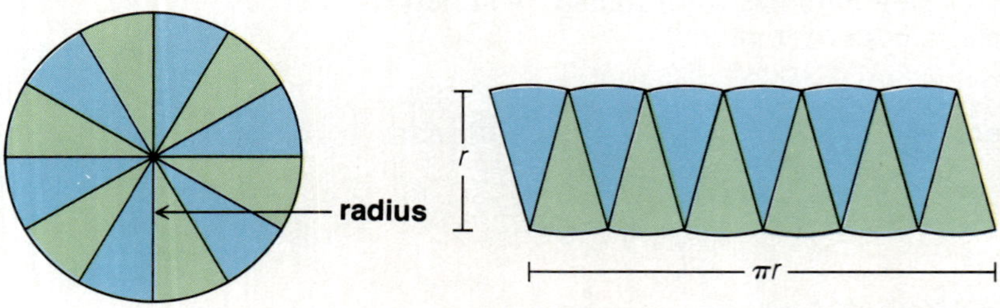

Formula: Area of a Circle

The area of a circle is equal to π times the radius squared.

$$A = \pi r^2$$

Example

Find the area of a circle with radius 5 cm.

Solution $A = \pi r^2$

$\approx (3.14)5^2$ Replace π with 3.14 and r with 5.

$A \approx 78.5$ The area is approximately 78.5 cm^2.

Practice Find the area of the circle. **a.** radius = 3 ft 28.26 ft^2 **b.** diameter = 8 m 50.24 m^2

Oral Exercises

Give the radius and diameter of each figure.

1.

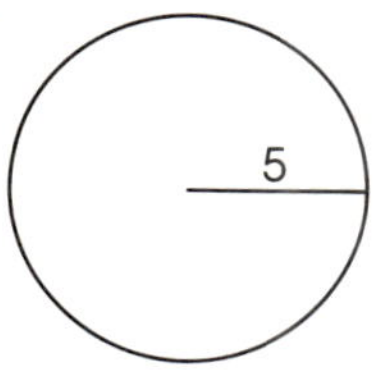

$r = 5, d = 10$

2.

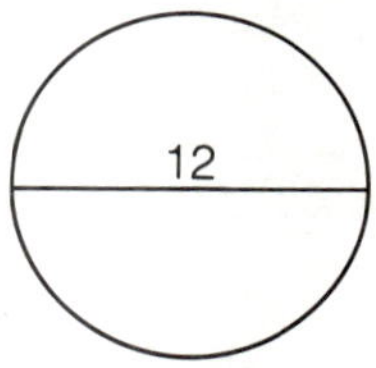

$r = 6, d = 12$

3.

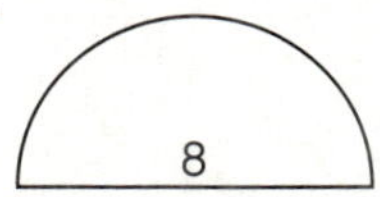

$r = 4, d = 8$

4.

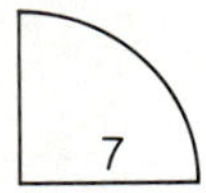

$r = 7, d = 14$

Exercises

1. 78.5 cm^2 2. 200.96 cm^2 3. 78.5 cm^2 4. 1,962.5 m^2

A Find the area of each circle. Use 3.14 as an approximation for π.

1. $r = 5$ cm **2.** $r = 8$ cm **3.** $d = 10$ cm **4.** $r = 25$ m

5. $d = 4.8$ cm **6.** $r = 3.9$ m **7.** $d = 500$ m **8.** $r = 250$ cm

9. $r = 50$ cm **10.** $r = 4.8$ m **11.** $d = 3$ m **12.** $d = 12$ mm

5. 18.0864 cm^2 6. 47.7594 m^2 7. 196,250 m^2 8. 196,250 cm^2

9. 7850 cm^2 10. 72.3456 m^2 11. 7.065 m^2 12. 113.04 mm^2

B

13. Estimate, then give the area of a microcomputer disk with diameter of:

a. 8 inches 50.24 in^2 **b.** 5 inches 19.625 in^2 **c.** $3\frac{1}{2}$ inches 9.61625 in^2

14. Estimate whether the area of an 8-inch disk is about 2 times, 3 times, 4 times, or 5 times the area of a $3\frac{1}{2}$-inch disk. Check your estimate with a calculation. About 5 times.

Find the area of each shaded region. Express the answer in terms of π. Assume that semicircular regions are half-circles.

15. 12π

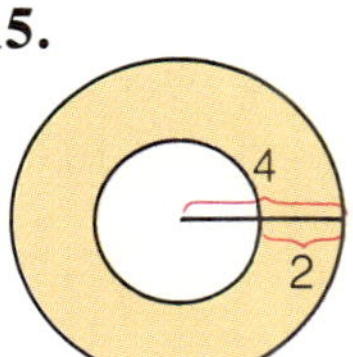

16. $72 + \frac{9}{2}\pi$

17. $144 + \frac{25\pi}{2}$

Use a calculator to find the area of the circle in Exercises 18–23. 18. 860,852.85 cm^2

18. $r = 523.6$ cm **19.** $r = 83.5$ m 21,892.865 m^2 **20.** $r = 14.2$ cm 633.1496 cm^2

21. $d = 25.2$ cm 498.5064 cm^2 **22.** $d = 4.5$ m 15.89625 m^2 **23.** $d = 18.8$ cm 277.4504 cm^2

C *Extending Thinking Skills*

24. Mr. Green divides a square-mile field into quarters with a circular irrigation system on each quarter. Ms. Peabody has one large circular irrigation system. Guess which farmer irrigates the greater percentage of land. Check your guess with a calculation.

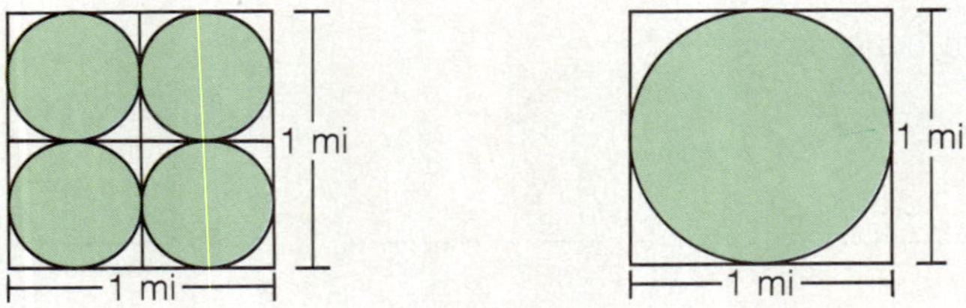

They irrigate the same percentage.

Area = $\frac{\pi}{4}$

Thinking skill: 24. Reasoning

Mixed Review

Solve by writing an equation. Leonetti's Fruit Market is selling apples for 25¢/lb and oranges for 35¢/lb.

25. What is the cost of 3 lbs of apples and 3 lbs of oranges? $3(0.25) + 3(0.35) = x$ \$1.80

26. How many lbs of apples can you buy for \$3.50? $3.50 = 0.25x$ 14 lb

Write as a fraction in lowest terms. **27.** 24% $\frac{6}{25}$ **28.** 15% $\frac{3}{20}$ **29.** 11% $\frac{11}{100}$

Give the least common multiple (LCM). **30.** 2, 3, 4, 8 24 **31.** 2, 4, 6, 12 12

Evaluate for $x = 2$, $y = 24$. **32.** x^2y 96 **33.** $3(x - y)$ −66 **34.** $x(y - 1)$ 46

Solve and check. **35.** $1.4 - n = 0.28$ 1.12 **36.** $4m + 9 = m$ −3 **37.** $6(y + 1) = 6$ 0

ESTIMATION

Estimate the ratio of the area of the smaller circle to the area of the larger circle. Then check your estimate by calculating.

1.

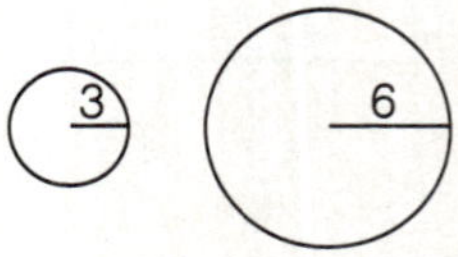

$\frac{9}{36} = \frac{1}{4}$

2.

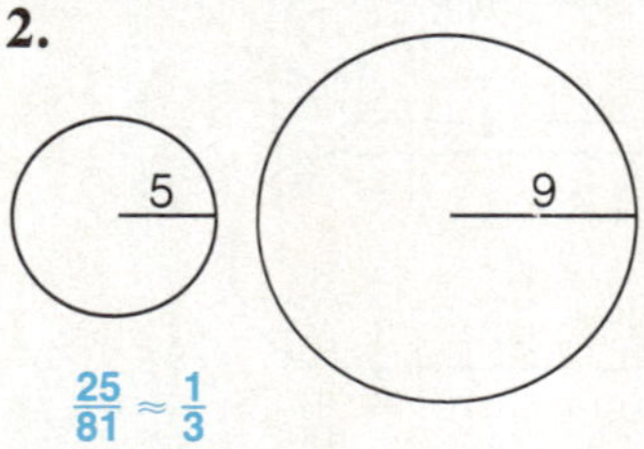

$\frac{25}{81} \approx \frac{1}{3}$

3.

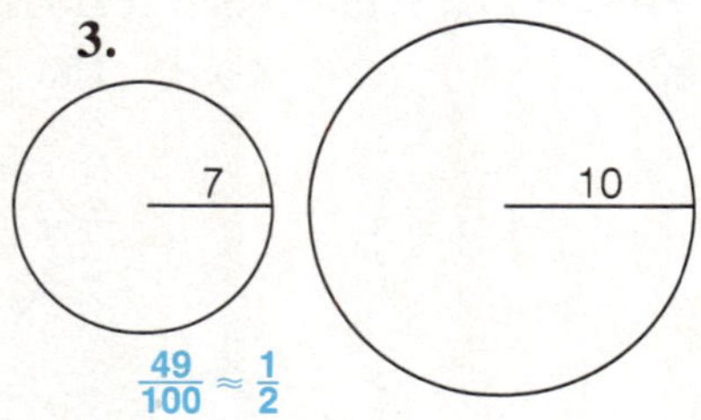

$\frac{49}{100} \approx \frac{1}{2}$

12-4 Area of Figures of Irregular Shape

Objective: To find areas of figures of irregular shapes by dividing them into familiar shapes.
MIN: 1–6; REG: 1–6 even, 7–9; MAX: 1–6 even, 7–11; ALL: Mixed Review

An exploratory lesson opener can be found on page 553.

An architect will often design a floor plan for a house that is irregular in shape. To find the area of such a figure, divide it into smaller regions that have familiar shapes. Find the area of each region and then add the areas together.

Example

Find the area of the figure.

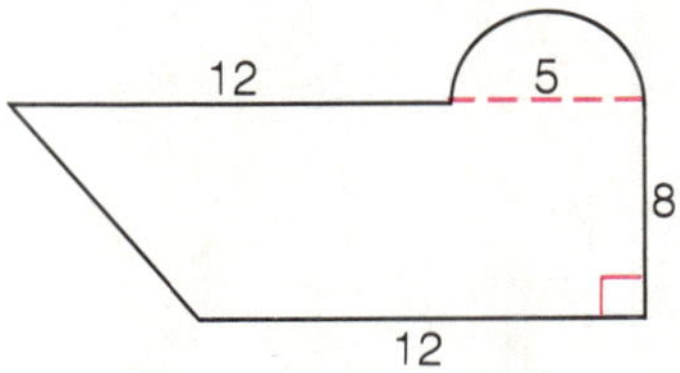

Solution

$A_T = \frac{1}{2}(8)(12 + 17)$ First find the area A_T of the trapezoid, using the formula $\frac{1}{2}h(b_1 + b_2)$.

$= 116$ square units

$A_{SC} = \frac{1}{2}\pi\left(\frac{5}{2}\right)^2$ Next find the area A_{SC} of the semicircle. The area of a semicircle is $\frac{1}{2}$ the area of the whole circle. The radius is $\frac{5}{2}$.

$\approx \frac{25}{8}(3.14) = 9.8125$ square units

$$A = A_T + A_{SC} \approx 116 + 9.8125 = 125.8125 \text{ square units.}$$

Practice Find the area of each figure. All units are meters.

a. 239 m²

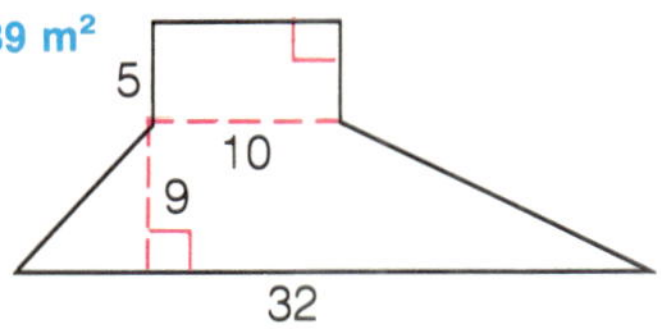

b. 222.13 m²

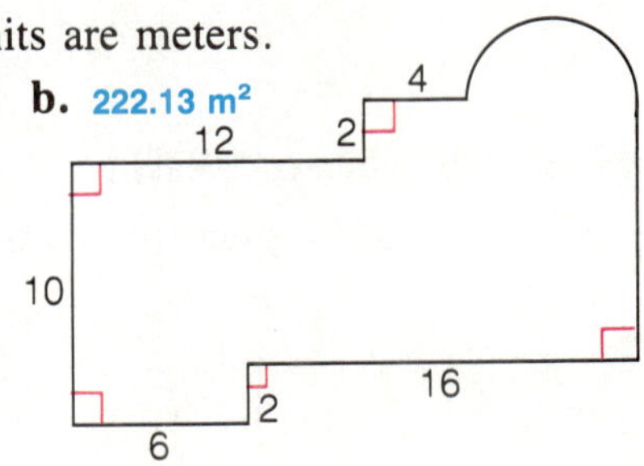

Oral Exercises

First name the shapes into which each figure could be divided. Then find lengths x and y in each figure.

1.

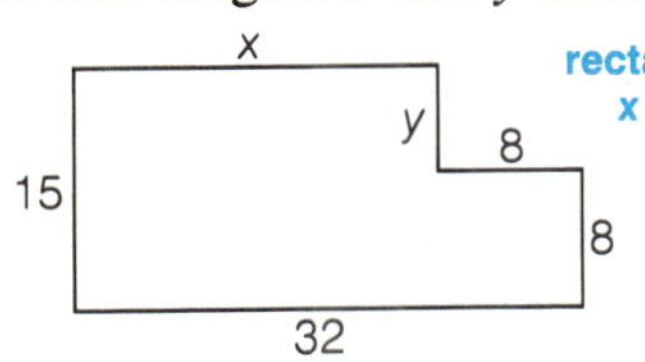

rectangle + square
$x = 24$, $y = 7$

2.

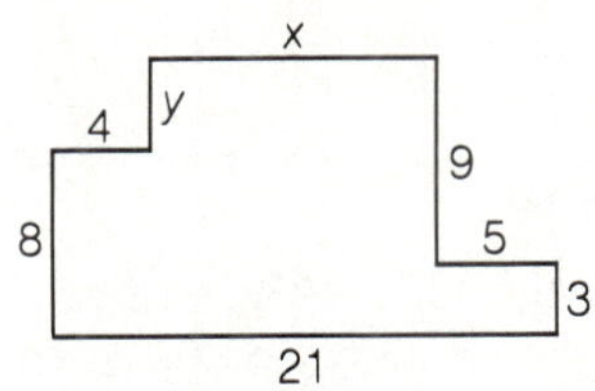

3 rectangles
$x = 12$, $y = 4$

Exercises

A Find the area of each figure. All units are meters.

1. 108 m²

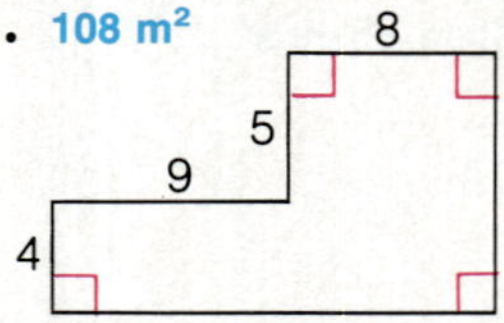

2. 43 m²

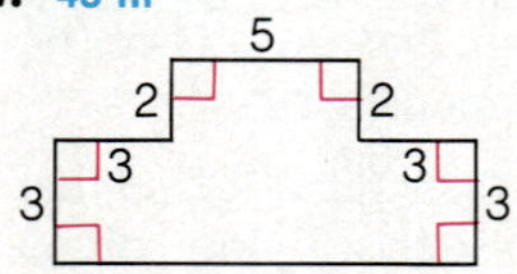

3. 48 m²

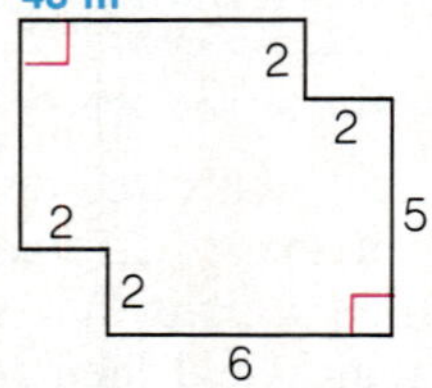

4. 236 m²

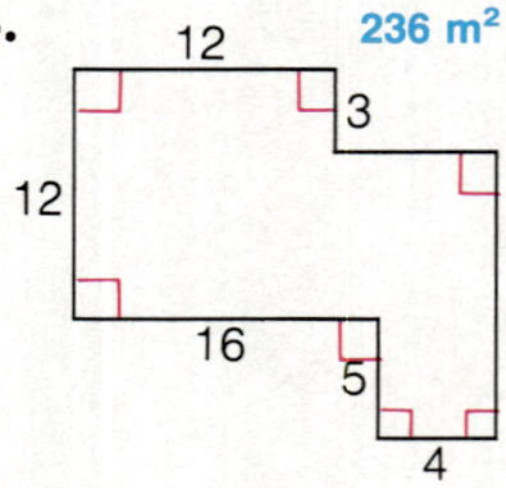

5. 119.28 m²

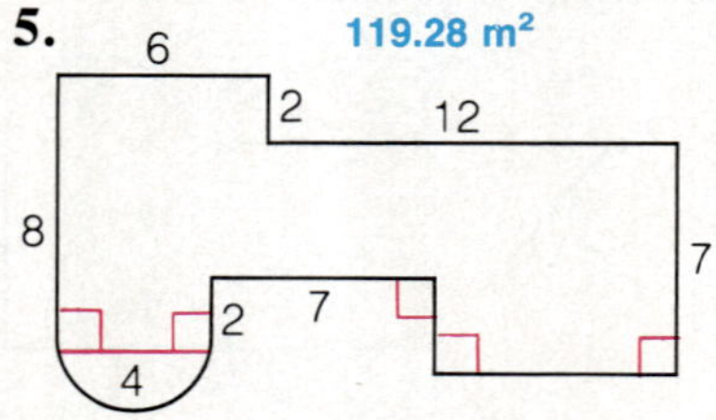

6. 76.13 m²

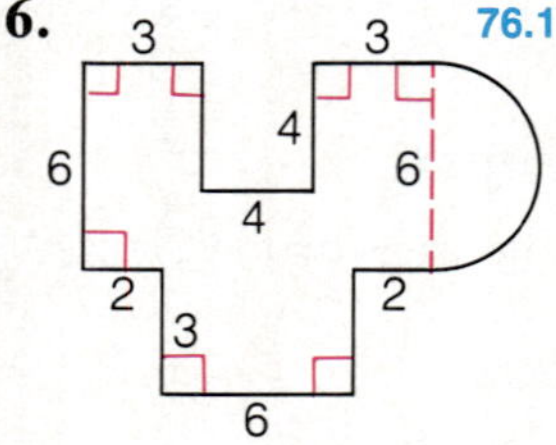

B Find the area of each figure. All units are feet.

7. 72 ft²

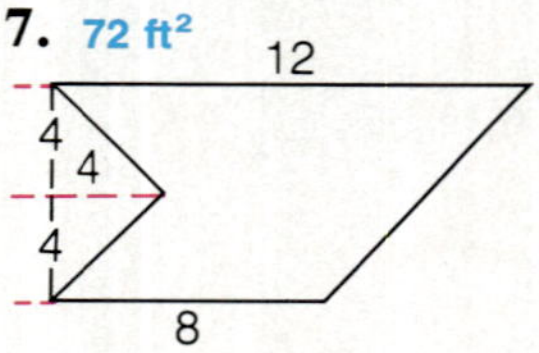

8. 58.26 ft²

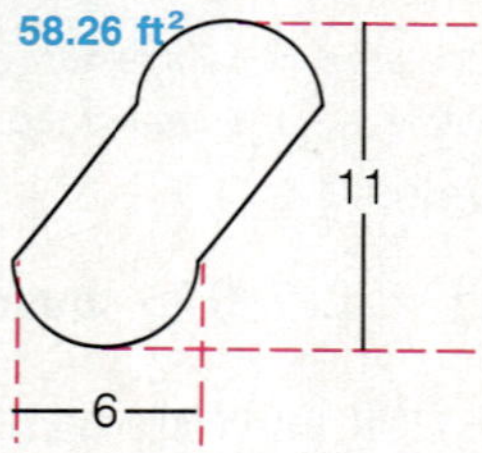

9. 188 ft²

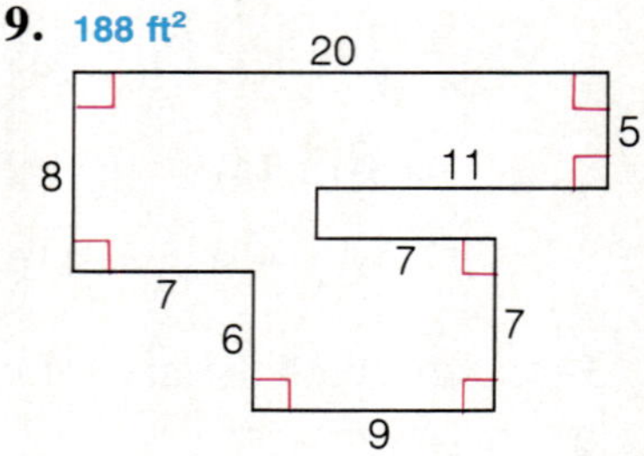

C Extending Thinking Skills

Find the area of each shaded region. All units are meters.

10. 136 m²

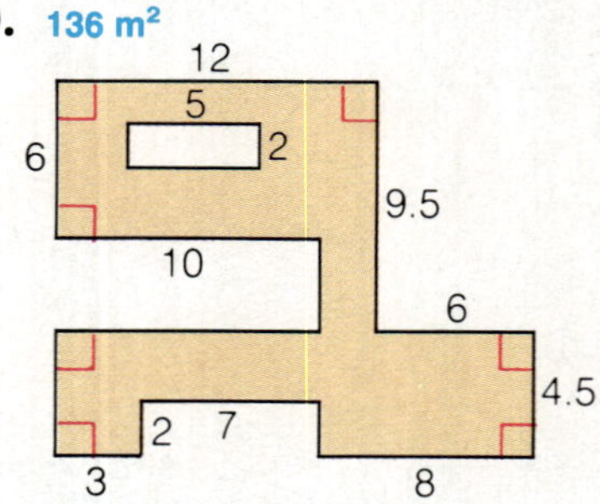

11. 320.11 m²

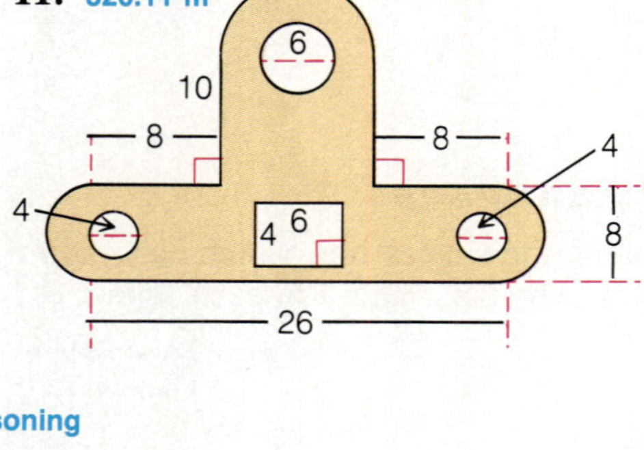

Thinking skills: 10, 11. Using space visualization/reasoning

Mixed Review

Write as a decimal. **12.** 16% 0.16 **13.** 12.5% 0.125 **14.** 120% 1.2 **15.** 7% 0.07

Solve and check. **16.** $4(2n + 1) = 10$ $\frac{3}{4}$ **17.** $0.3n + 16.5 - 3n = 3$ 5

12-5 Practice Solving Problems

Objective: To solve word problems by writing and solving equations.
MIN: 1–5; REG: 1–7; MAX: 1–5 even, 6–9; ALL: Mixed Review

Problems

Solve by writing an equation.

A

1. Joanna broke the old high-jump record, set in 1980, by $1\frac{3}{4}$ in. The new record is 6 ft $5\frac{1}{2}$ in. What was the old record? $72 + 5\frac{1}{2} - 1\frac{3}{4} = x$ 6 ft $3\frac{3}{4}$ in

2. Mr. Bertolini saved \$48 on a tape deck by buying it at 30% off the regular price. What was the regular price of the tape deck? $48 = 0.3x$ \$160

3. Popcorn costs \$0.25 more than a cup of juice. What is the price of each if they cost \$1.65 together? $s + (s + 0.25) = 1.65$ popcorn \$0.95, juice \$0.70

4. Teresa bought 3 pairs of hiking shorts. Each pair cost the same. The sales tax on the 3 pairs of shorts came to \$2.25. The total cost, including tax, was \$39.75. What was the cost of each pair of shorts? $3h + 2.25 = 39.75$ \$12.50

5. Lynn got a \$4 rebate for buying two hair dryers. Each cost the same. The cost for both dryers after the rebate was \$21. What was the cost for one? $2d = 21 + 4$ \$12.50

B

6. Patrick worked 6 hours on Monday and 8 hours on Wednesday. He did not work the rest of the week. A total of \$13 was taken out of his paycheck that week for taxes and insurance. His "take home" pay was \$39.50. What was his hourly wage? $(6 + 8)w = 39.50 + 13$ \$3.75

7. An airline has a holiday special. When 2 people travel together, the cost of the second ticket is $\frac{1}{2}$ the price of a full-fare ticket. Jayne paid a total of \$210 for 2 tickets. What was the price of a full-fare ticket? $x + 0.5x = 210$ \$140

C Extending Thinking Skills

8. Write a word problem that would be solved using the equation $0.5x + 1.25 = 7.5$.

9. Write a word problem that would be solved using the equation $x + 2x + 3x = 120$.

Thinking skills: 8, 9. formulating a problem

Mixed Review

Use a proportion to find: **10.** 12% of 20 **11.** 15% of 120 **12.** 110% of 30

Solve and check. **13.** $a + 2.5a = -7$ -2 **14.** $3n < 10 - 2n$ $n < 2$

10. $\frac{12}{100} = \frac{x}{20}$; 2.4 11. $\frac{15}{100} = \frac{x}{120}$; 18 12. $\frac{110}{100} = \frac{x}{30}$; 33

12-6 Volume of Prisms and Cylinders

Objective: To use formulas to find the volumes of prisms and cylinders.
MIN: 1–7; REG: 1–6 even, 8–22; MAX: 1–6 even, 8–25; ALL: Mixed Review

The **volume** of a solid is the number of cubic units needed to occupy the amount of space the solid occupies. The rectangular solid at right has 8 cubic centimeters in the base layer and a height of 2 layers, making a volume of $8 \times 2 = 16$ cubic centimeters. This is written 16 cm^3.

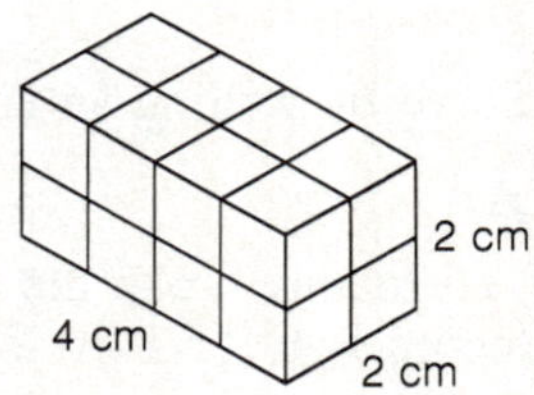

A **prism** is a solid that has a pair of bases that are congruent and parallel; its sides are parallelograms. A prism with triangles as bases is called a triangular prism. A prism with hexagons as bases is called a hexagonal prism.

Formula: Volume of a Prism

The volume of a prism is equal to the area of a base times the height.

$$V = Bh$$

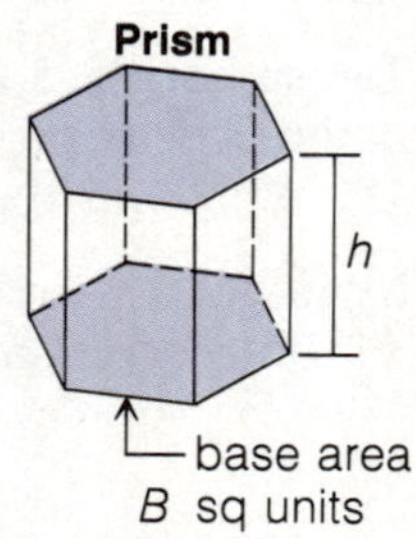

Example 1

Find the volume of the triangular prism.

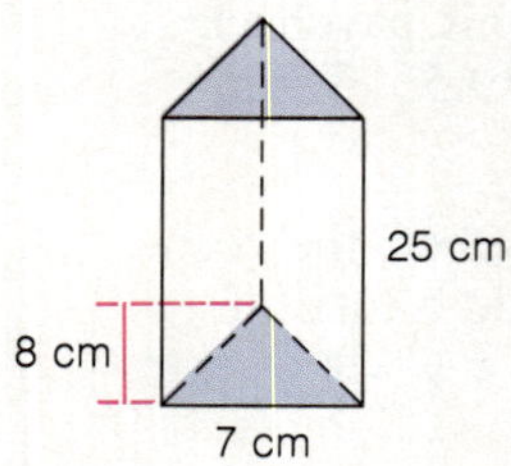

Solution

$V = Bh$

$= \frac{1}{2} \times 7 \times 8 \times 25$ *The bases are triangles.* $B = \frac{1}{2} \times 7 \times 8.$

$V = 700$ cm^3

The volume of the prism is 700 cm^3.

Practice Find the volume of the prism at right.
1528.8 cm^3

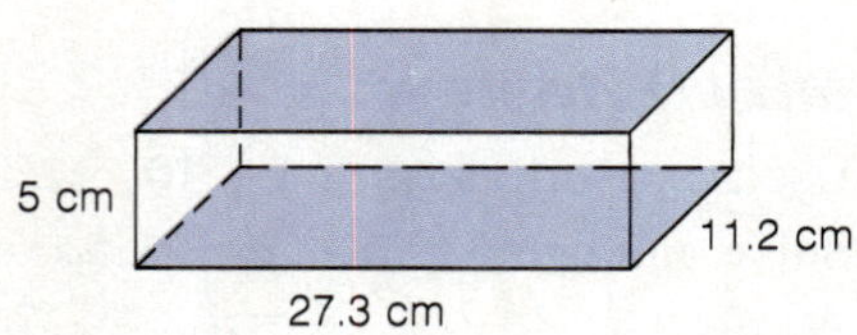

A **cylinder** has a pair of parallel and circular bases with the same radius.

Formula: Volume of a Cylinder

The volume of a cylinder is equal to the area of the base (πr^2) times the height.

$$V = \pi r^2 h$$

Cylinder

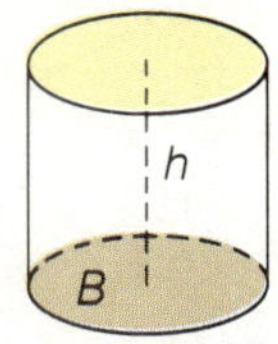

area of base is πr^2 sq units

Example 2

Find the volume of the cylinder at right. Use 3.14 for π.

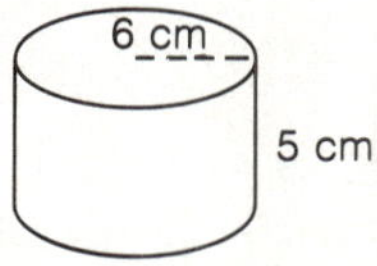

Solution $V = \pi r^2 h$

$\approx (3.14) \times 6^2 \times 5$

$V \approx 565.2$

The volume is approximately 565.2 cm^3.

Practice Find the volume of a cylinder with the dimensions given.

a. $r = 10$ cm, $h = 8$ cm **b.** $r = 5$ cm, $h = 1.8$ cm **141.3 cm^3**

Note that the units *r* and *h* must be consistent.

a. 2512 cm^3

Oral Exercises

Give the height of each prism.

1. **10 cm**

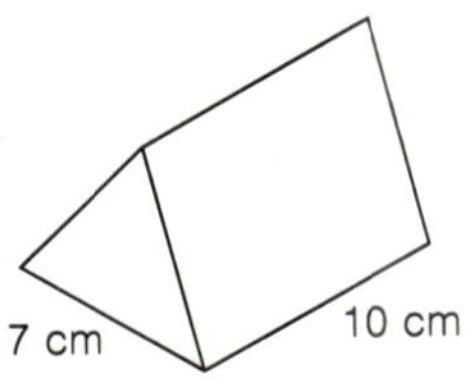

2. **9 cm**

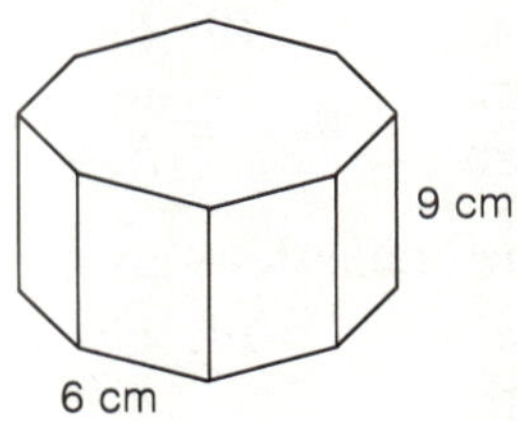

3. **6 ft**

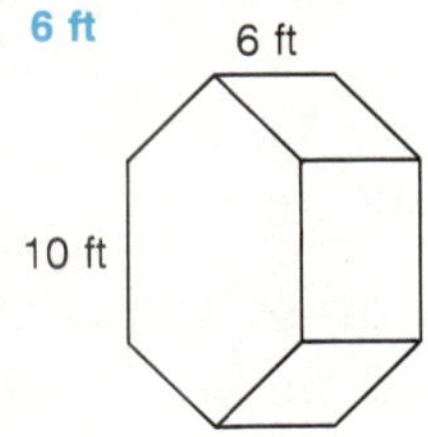

Give the radius and height of each cylinder.

4.

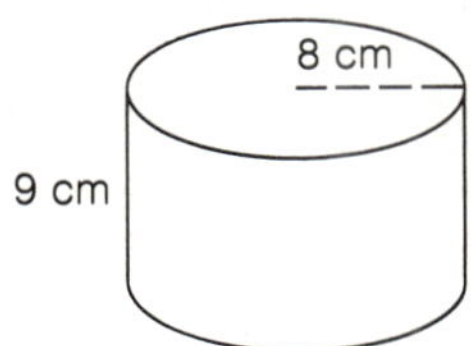

***h* = 9 cm**
***r* = 8 cm**

5.

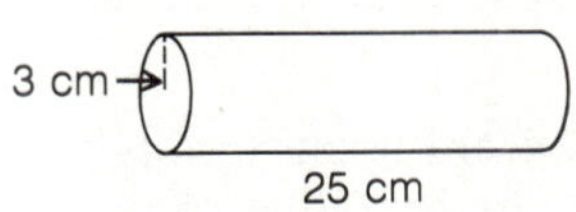

***h* = 25 cm**
***r* = 3 cm**

6.

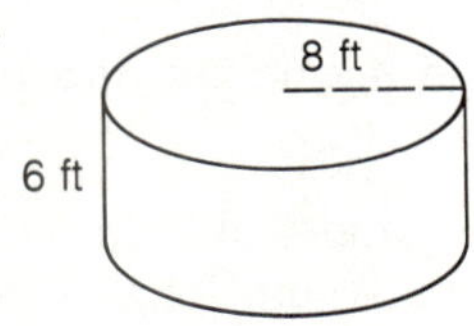

***r* = 8 ft**
***h* = 6 ft**

Exercises

A Find the volume of each prism or cylinder. Use 3.14 for π.

1. $352\ cm^3$

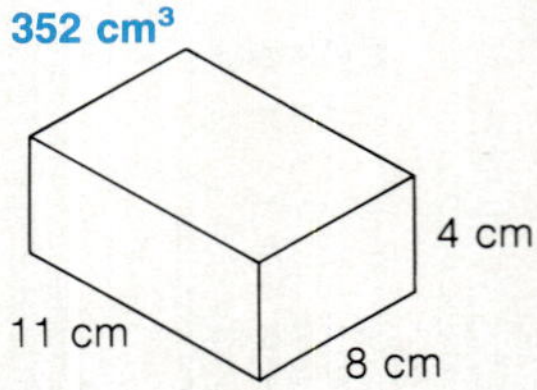

2. $724.5\ cm^3$

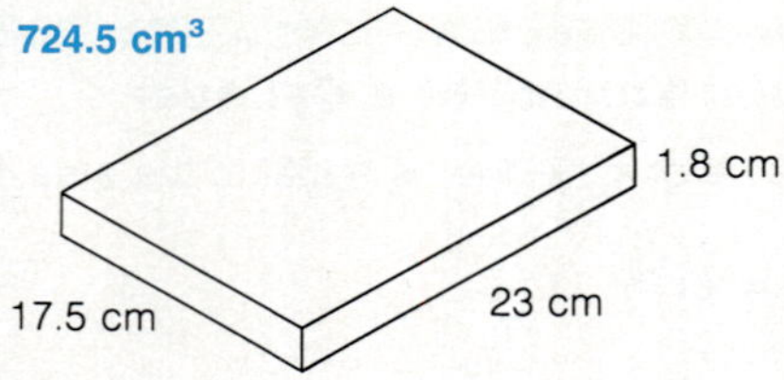

3. $2009.6\ ft^3$

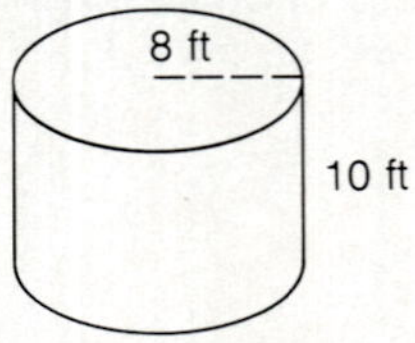

4. $763.02\ m^3$

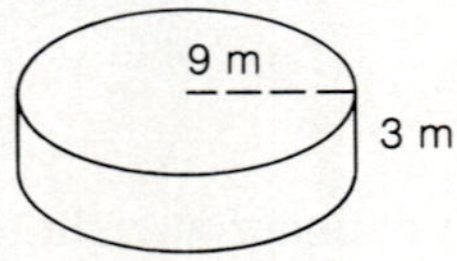

5. $B = 25\ cm^2$
$h = 12$ cm $300\ cm^3$

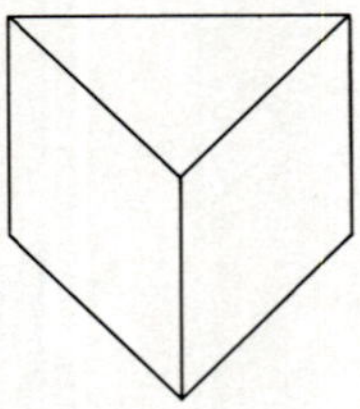

6. $B = 192\ in^2$
$h = 24$ in $4608\ in^3$

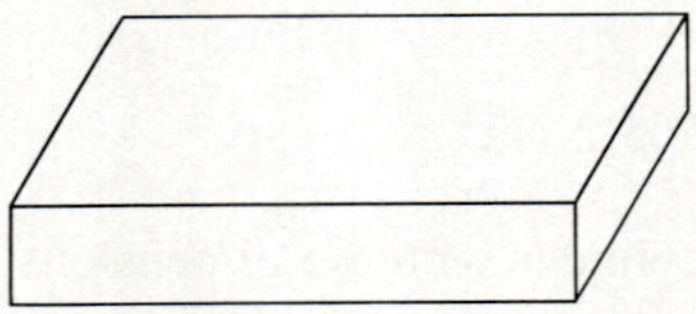

7. $B = 126\ ft^2$
$h = 4$ ft $504\ ft^3$

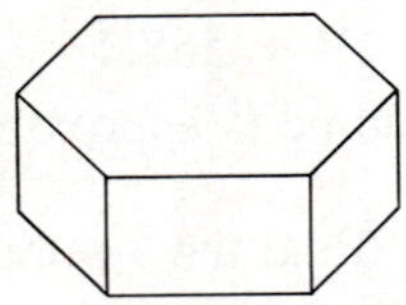

B For Exercises 8–19, be sure to use consistent units when computing. Express the answer in the smaller unit. Use the formula $V = lwh$ to find the volume of a rectangular prism with the dimensions given.

8. $108{,}000\ cm^3$
9. $72{,}000\ cm^3$
10. $3{,}626.532\ cm^3$

8. $l = 5$ m
$w = 12$ cm
$h = 18$ cm

9. $l = 25$ cm
$w = 8$ dm
$h = 36$ cm

10. $l = 23.4$ cm
$w = 12.3$ cm
$h = 12.6$ cm

11. $l = 7$ m $28.98\ m^3$
$w = 2.3$ m
$h = 1.8$ m

Find the volume of a cylinder with the dimensions given. Use 3.14 for π. See additional answers.

12. $r = 12$ cm
$h = 8$ cm

13. $r = 11$ m
$h = 19$ m

14. $r = 2$ m
$h = 3$ m

15. $r = 4$ dm
$h = 12$ dm

16. $r = 2.4$ cm
$h = 3.1$ cm

17. $r = 56$ cm
$h = 2$ m

18. $r = 8$ m
$h = 40$ cm

19. $r = 3$ cm
$h = 2$ dm

20. Find the base area of a cylinder with volume $500\ cm^3$ and height 5 cm. $100\ cm^2$

21. Find the height of a prism with volume $125\ m^3$ and a base area of $5\ m^2$. 25 m

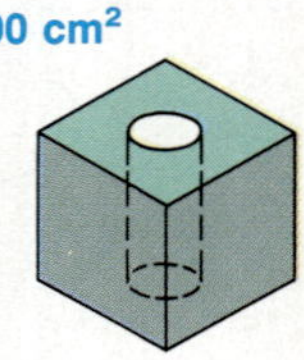

22. A cylindrical hole with radius 4 cm is cut through a solid cube whose edges are 10 cm long. What is the total volume of this solid? $497.6\ cm^3$

C *Extending Thinking Skills*

Mentally count unit cubes to find the volume of each figure. The hidden back view of each looks like the corner of a box.

23.

20 cubic units

24.

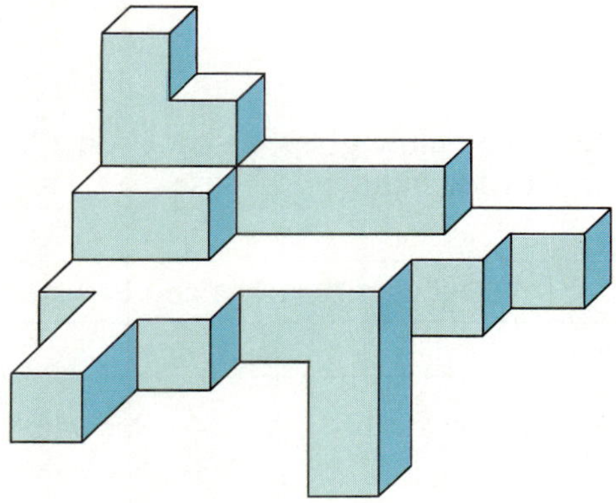

34 cubic units

Thinking skill: 23, 24. Using space visualization

Mixed Review

Give the greatest common factor (GCF). **25.** 18, 42, 90 6 **26.** 15, 72, 111 3

Write as a decimal. **27.** $3\frac{4}{5}$ 3.8 **28.** $-1\frac{3}{4}$ −1.75 **29.** $\frac{15}{12}$ 1.25 **30.** $\frac{1}{3}$ $0.\overline{3}$

Write as a percent. **31.** $\frac{16}{5}$ 320% **32.** $\frac{24}{8}$ 300% **33.** $-\frac{3}{4}$ −75% **34.** $\frac{7}{10}$ 70%

Solve and check. **35.** $30 = 5(x - 2)$ 8 **36.** $9 = 2.5 + y$ 6.5 **37.** $3t + 0.25 = 1$ 0.25

Reduce. **38.** $\frac{9}{6}$ $1\frac{1}{2}$ or $\frac{3}{2}$ **39.** $\frac{35}{7}$ 5 **40.** $\frac{16}{64}$ $\frac{1}{4}$

ESTIMATION

Estimate the volume of each cylinder by rounding π down to 3 and rounding r^2 or h up. For example, for a cylinder with $r = 7$ and $h = 10$, volume $= \pi r^2 h \approx 3 \times 50 \times 10 = 1500$. Estimated volume: 1500 cubic units.

1.

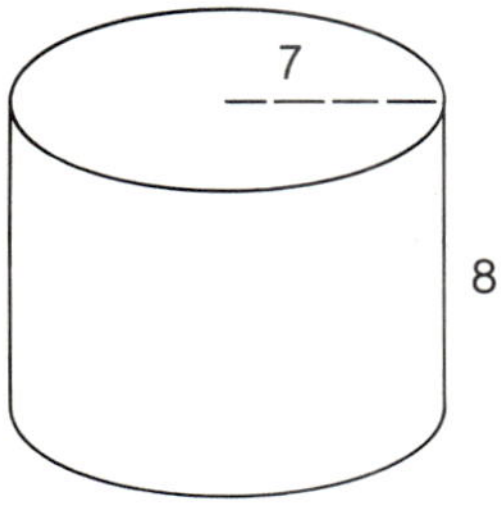

1,200 cubic units

2.

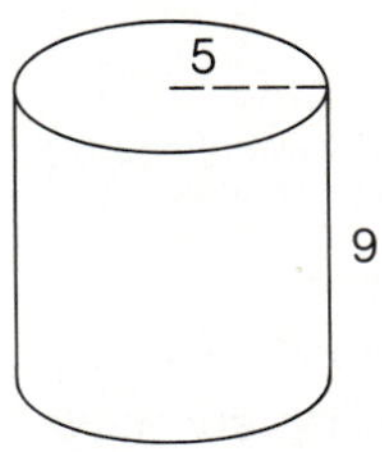

750 cubic units

3.

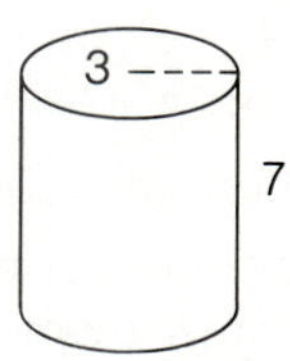

210 cubic units

PROBLEM SOLVING: APPLICATIONS

12-7 Building a House

Objective: To solve applied problems involving area and volume.
ALL: 1–8

The table below gives costs of some building materials and services for building a house.

Building Materials and Services	
Item	**Cost**
Excavation	\$ 1.30/cubic yard
Carpet	12.00/square yard
Concrete	45.00/cubic yard
Wallpaper	18.00/roll
Roofing plywood	8.85 per 4 × 8 foot sheet

To find costs for materials and services, you need to find the area or volume of the space being built. You may need to change the units of a dimension to make them consistent.

Example

How many cubic yards of concrete are needed to pour a floor 81 feet long, 42 feet wide, and 4 inches thick?

Solution $V = lwh$

$= 81 \times 42 \times \frac{1}{3}$ ft^3 — Express 4 inches as $\frac{1}{3}$ foot. Replace the l with 81, w with 42, and h with $\frac{1}{3}$.

$= 1{,}134$ ft^3

$= \frac{1{,}134}{27}$ yd^3 — **Discuss why 27 ft^3 = 1 yd^3.** There are 27 ft^3 in one yd^3. To change cubic feet into cubic yards, divide by 27.

$= 42$ yd^3

The volume is 42 yd^3.

Problems

You might wish to have students choose from among the techniques pencil and paper, mental math, estimation, and calculator. Have them choose the most appropriate technique for each problem.

Use the data in the table above to solve.

1. An excavator digs a hole 34 feet by 120 feet by 4 feet for the foundation of a building. How many cubic yards of earth are moved? What is the cost?
604.4 yd^3 are moved. Cost: \$785.78

2. What is the cost for a concrete floor 56 ft long, 40 ft wide, and 4 inches thick? **\$1244.44 (or \$1244.45, since money is rounded *up*.)**

3. How many cubic yards of concrete are needed for a wall 48 inches high, 125 feet long and 10 inches thick? How much will the concrete cost?
15.4321 yd^3; \$694.44 (or \$694.45, since money is rounded *up*.)

Use this figure for Exercises 4 and 5.

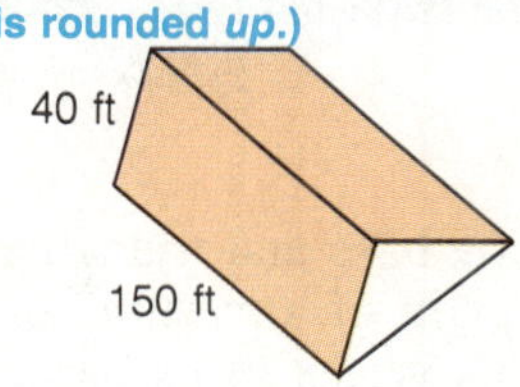

4. A roof is 40 feet from edge to peak and 150 feet long. What is the area of the roof (the shaded region on both sides)? **12,000 ft^2**

5. How much will the plywood for the roof in Problem 4 cost? **\$3318.75**

6. Wallpaper comes in rolls 20 inches wide and 24 feet long. How much will the wallpaper cost for a wall 8 feet high by 12 feet wide? The wall has no windows or doors. **\$54**

7. Estimate the cost of carpeting a two-story house with a rectangular foundation that measures 30 feet by 42 feet. **\$3360.**

8. ***Data Search*** Find the cost of a gallon of interior paint at a store near your school. Estimate the cost of the paint needed for two coats of paint on the walls of your classroom. **Answers will vary.**

What's Your Decision?

The Malone family wants to carpet a room that is shaped as shown. The carpet salesman says they need to order 30 square yards. At \$12/yd this would cost them \$360. They find a pre-cut carpet piece that is 12 ft by 19 ft for \$250. Should the Malones buy the pre-cut piece and save \$110, or order the 30 square yards? Why? **Answers may vary. The area of the pre-cut piece is 228 ft^2, and their room is 230 ft^2, so the pre-cut piece will not fit.**

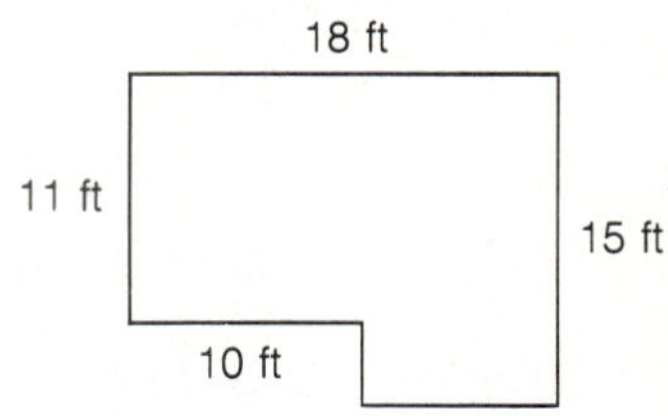

12-8 Volume of Pyramids and Cones

Objective: To use formulas to find the volumes of pyramids and cones.
MIN: 1–22; REG: 1–22 even, 23–35; MAX: 1–22 even, 23–37; ALL: Mixed Review

A **pyramid** has a polygon as a base and triangular sides. A **cone** has one circular base. The height of a pyramid or a cone is the perpendicular distance from its vertex to its base.

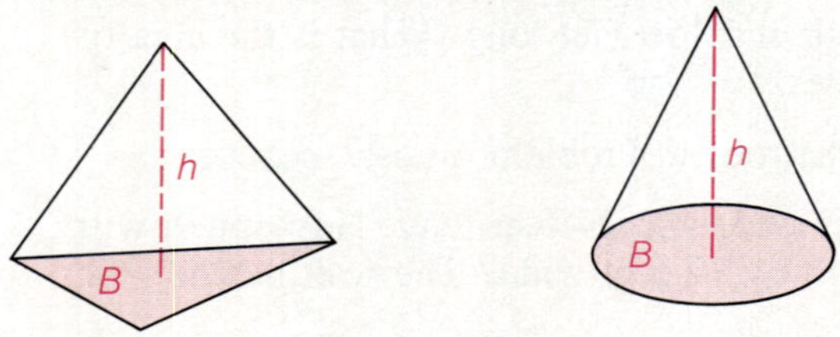

Pyramid **Cone**

Note that a cone's volume is $\frac{1}{3}$ the volume of a cylinder with the same base and height. There is a similar relationship between a pyramid and a prism.

The volume of a cone or a pyramid can be found by the same formula.

Formula: Volume of a Pyramid or Cone

The volume of a pyramid or a cone is equal to one-third the area of the base times the height.

$$V = \frac{1}{3}Bh$$

Example

Find the volume of a cone with radius 4 cm and height 8 cm.

Solution $V = \frac{1}{3}Bh$ — The volume of a cone is $\frac{1}{3}$ the area of the base times the height.

$\approx \frac{1}{3} \times 3.14 \times 4^2 \times 8$ — The area of the base is πr^2. Substitute 3.14 for π and 4 for r. Substitute 8 for h.

$= \frac{401.92}{3}$

$= 133.97$

The volume is 133.97 cm^3.

Practice Find the volume of a pyramid or cone with the base area or radius and the height given.

a. $B = 27\ cm^2$, $h = 22$ cm **198 cm^3**

b. $r = 6$ m, $h = 5$ m **188.4 m^3**

Oral Exercises

Is the statement true or false?

1. A cube is a pyramid. F
2. The volume of a cone is $\frac{1}{3}$ the area of the base times the height. T
3. Some pyramids have rectangular sides. F
4. All pyramids have round bases. F
5. A pyramid always has a square base. F

Exercises

A Find the volume of each figure.

1.

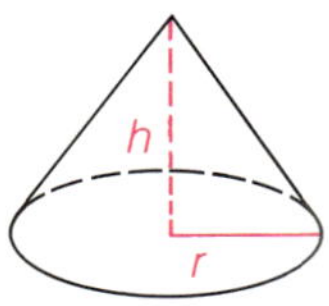

$r = 5$ cm
$h = 8$ cm 209.$\overline{3}$ cm³

2.

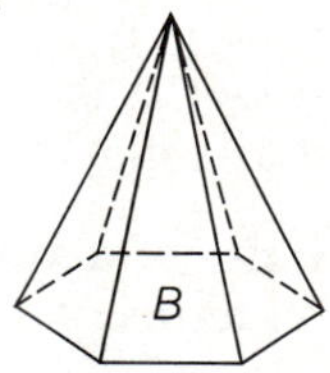

$B = 14\ \text{cm}^2$
$h = 10$ cm 46.$\overline{6}$ cm³

3.

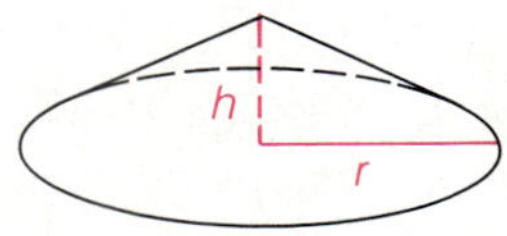

$r = 9$ cm
$h = 4$ cm 339.12 cm³

4.

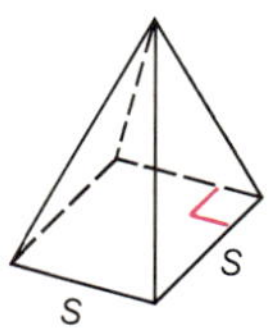

$s = 9$ cm
$h = 12$ cm 324 cm³

5.

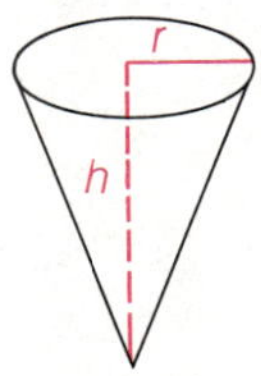

$r = 4$ in
$h = 9$ in 150.72 in³

6.

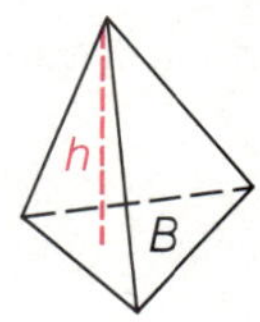

$B = 9\ \text{cm}^2$
$h = 5$ cm 15 cm³

Find the volume of a cone with the radius and height given. Round your answer to the nearest tenth.

7. $r = 11$ cm, $h = 8$ cm 1013.2 cm³
8. $r = 12$ m, $h = 4$ m 602.9 m³
9. $r = 5$ dm, $h = 9$ dm 235.5 dm³
10. $r = 4$ cm, $h = 8$ cm 134 cm³
11. $r = 5$ cm, $h = 12$ cm 314 cm³
12. $r = 4$ cm, $h = 124$ cm 2076.6 cm³
13. $r = 4$ mm, $h = 24$ mm 401.9 mm³
14. $r = 5$ m, $h = 10$ m 261.7 m³

Find the volume of a pyramid with the base area and height given. Round your answer to the nearest tenth. 15. 158.3 cm³

15. $B = 25\ \text{cm}^2$, $h = 19$ cm
16. $B = 96\ \text{m}^2$, $h = 23$ m 736 m³
17. $B = 47\ \text{m}^2$, $h = 12$ m 188 m³
18. $B = 12\ \text{cm}^2$, $h = 5$ cm 20 cm³
19. $B = 46\ \text{cm}^2$, $h = 45$ cm 690 cm³
20. $B = 31\ \text{cm}^2$, $h = 6$ cm 62 cm³
21. $B = 270\ \text{m}^2$, $h = 47$ m 4230 m³
22. $B = 75\ \text{m}^2$, $h = 12$ m 300 m³

B In the table below, B represents the area of the base, h the height, and V the volume of a pyramid. Find the unknown variable in each column.

	23.	24.	25.	26.	27.	28.	29.
Base area B	12π	35	?	?	36	5.3	?
height h	6	?	12	32	12	?	4.5
volume V	?	105	288	160π	?	63.6	10.35
	24π	9	72	15π	144	36	6.9

30. What is the height of a prism if the volume is 162 m^3 and the area of the base is 9 m^2? 54 m

31. A cone and a cylinder have the same base. If the cone's height is 6 times the height of the cylinder, how much greater is the volume of the cone? 2 times greater

Use the formula $V = \frac{4}{3}\pi r^3$ to find the **volume of a sphere** with the dimensions given. Give your answers in terms of π.

32. $r = 8$ cm $682.\overline{6}\ \pi\ cm^3$

33. $r = 6$ cm $288\ \pi\ cm^3$

34. $r = 7$ m $457.\overline{3}\ \pi\ m^3$

35. $r = 5.1$ m $176.868\ \pi\ m^3$

C *Extending Thinking Skills*

These spheres are in pyramid-shaped stacks. Find the pattern and give the next 3 numbers.

36.

1

5

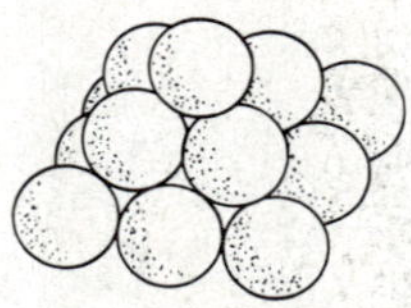

14

30 55 91

37.

1

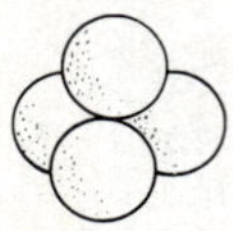

4

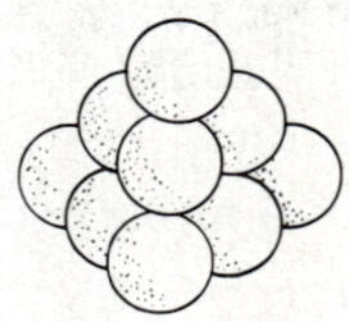

10

20 35 56

Thinking skills: 36, 37. Using space visualization/Finding a pattern

Mixed Review

Write as a decimal. **38.** 8% 0.08 **39.** 135% 1.35 **40.** 420% 4.2 **41.** 19.4% 0.194

Write in scientific notation. **42.** 16,500 **43.** 0.00016 **44.** 42,300,000

Solve and check. **45.** $y \div \left(-\frac{3}{4}\right) = \frac{3}{2}$ $-\frac{9}{8}$ **46.** $\frac{4}{5}c = \frac{1}{20}$ $\frac{1}{16}$ **47.** $2x + 5 = -12$ -8.5

42. $1.65(10^4)$ 43. $1.6(10^{-4})$ 44. $4.23(10^7)$

12-9 Surface Area

Objective: To find the surface areas of pyramids and cylinders.
MIN: 1–15; REG: 1–14 even, 16–20; MAX: 1–14 even, 16–22; ALL: Mixed Review

The cylinder and prism shown below have been cut apart and laid flat. The surface area of each is equal to the area of the corresponding flat regions.

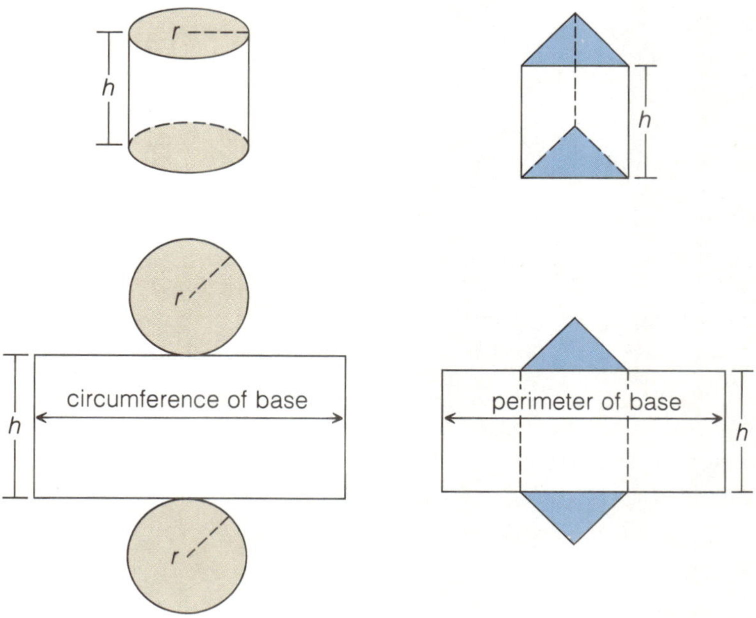

For a prism or a cylinder, surface area is equal to the area of the bases plus the area of the sides.

Example 1

Find the surface area of the cylinder at right. Use 3.14 for π.

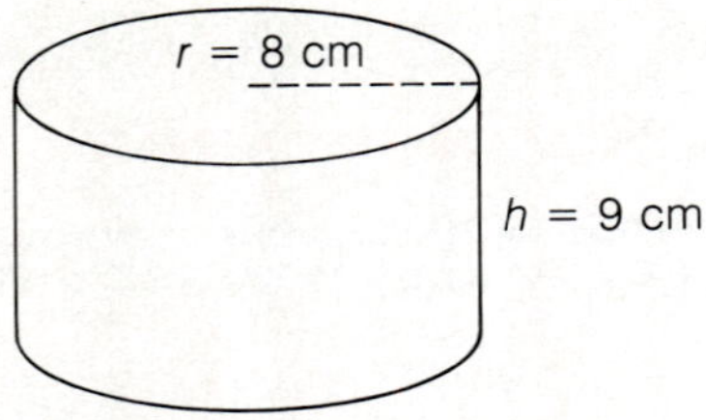

Solution Area of one base $= \pi 8^2$

$\approx 3.14 \times 64$

≈ 200.96

Area of sides = circumference of base × height

$= 2\pi \times 8 \times 9 \approx 2 \times 3.14 \times 8 \times 9 \approx 452.16$

Surface area = area of bases + area of sides

$\approx 2 \times 200.96 \text{ cm}^2 + 425.16 \text{ cm}^2 \approx 854.08 \text{ cm}^2$

Practice Find the surface area of a cylinder with the radius and height given.

a. $r = 5$ cm, $h = 12$ cm **533.8 cm²**

b. $r = 8$ m, $h = 15$ m **1155.52 m²**

Example 2

Find the surface area of the rectangular prism at right.

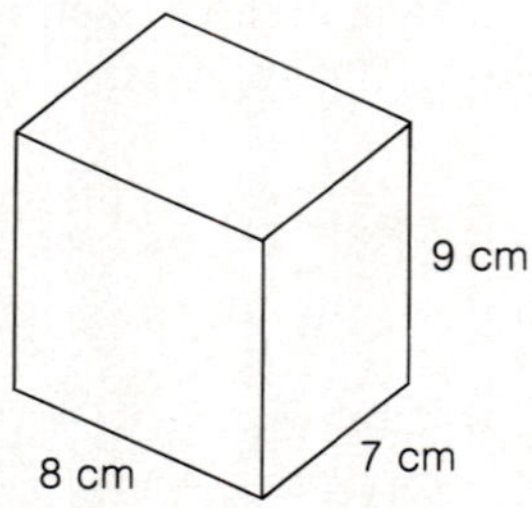

Solution

Area of one base $= 8 \times 7 = 56$

Area of sides = (perimeter of base) × 9

$= 30 \times 9$

$= 270$

Surface area = area of the bases + area of the sides

$= 2 \times 56 \text{ cm}^2 + 270 \text{ cm}^2 = 382 \text{ cm}^2$

Practice Find the surface area of the prism at right.

788 cm²

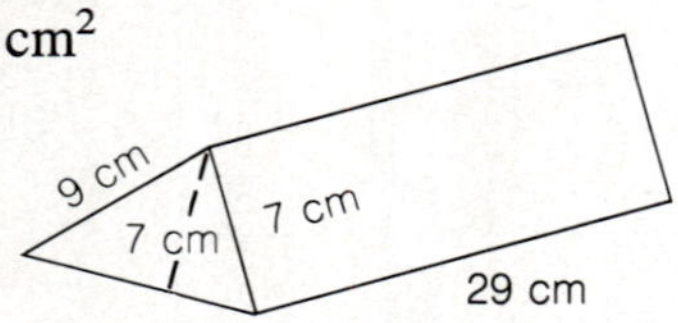

Oral Exercises

Give the height and the perimeter or circumference of the base for each figure.

1. $c = 10\pi$, $h = 7$

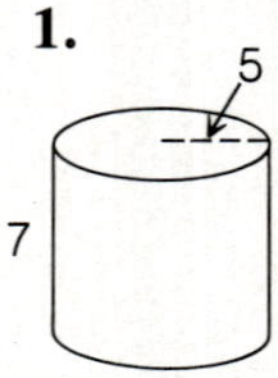

2. $p = 22$, $h = 15$

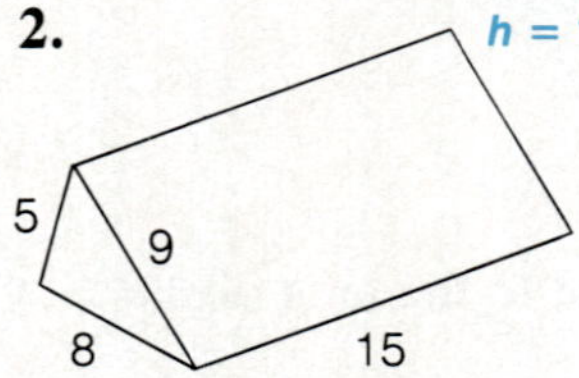

3. $c = 16\pi$, $h = 21$

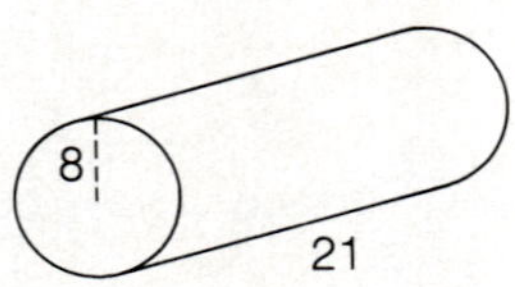

4. $p = 36$, $h = 9$

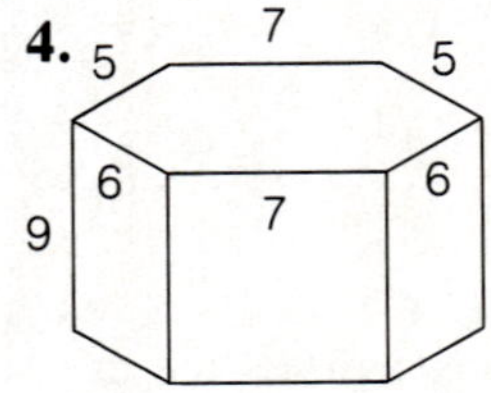

Exercises

A Find the surface area of each cylinder.

1. 527.52 cm²

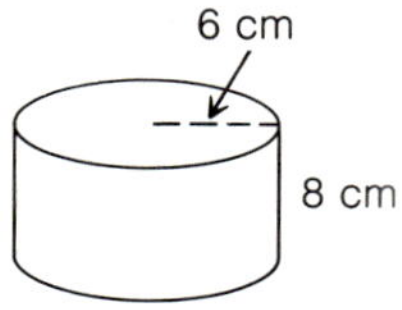

2. 502.4 cm²

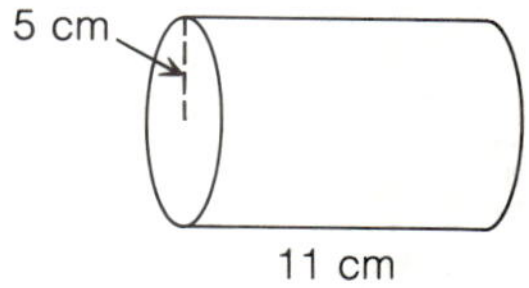

3. 427.04 cm²

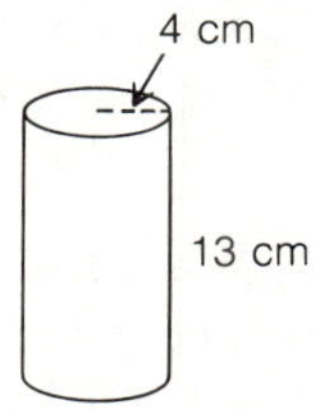

4. 226.08 cm²

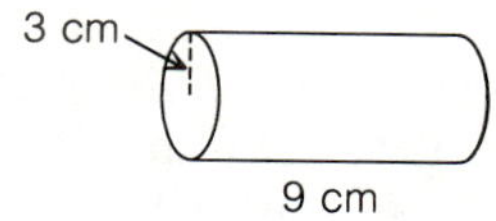

Find the surface area of a cylinder with the radius and height given.
See additional answers.

5. $r = 8$ cm, $h = 12$ cm
6. $r = 3$ cm, $h = 5$ cm
7. $r = 9$ m, $h = 3$ m
8. $r = 2.1$ m, $h = 8.3$ m

9. $r = 5$ dm, $h = 9$ dm
10. $r = 11$ cm, $h = 25$ cm
11. $r = 12$ m, $h = 30$ m
12. $r = 8$ cm, $h = 4$ cm

Find the surface area of each prism.

13.

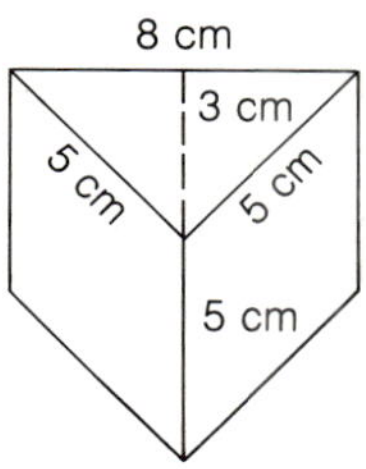

138 cm²

14.

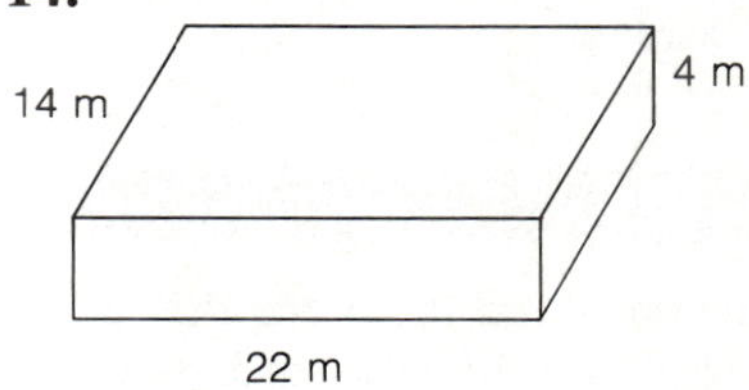

904 m²

15.

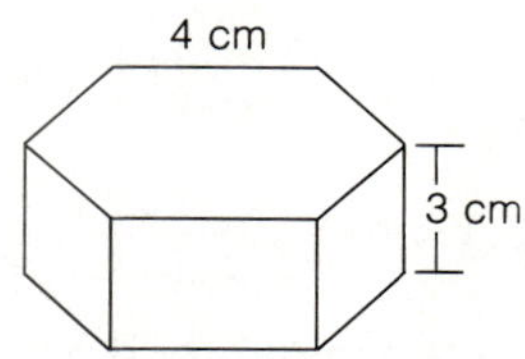

base is a regular hexagon with area 32 cm²

136 cm²

B

16. A cylinder with height 8 cm has a volume of 288π cm³. What is its surface area? 527.52 cm²

17. A cylinder with radius 16 cm has a volume of 2560π cm³. What is its surface area? 2612.48 cm²

18. A prism with a square base has a height of 9 m and a volume of 576 m³. What is its surface area? 416 m²

The surface area of a sphere is equal to 4π times the radius squared. The formula is $A = 4\pi r^2$. Use this formula for Exercises 19–20.

19. A spherical satellite that is 2 m in diameter is covered with a thin layer of a reflective material. What is the surface area of this satellite? 12.56 m²

20. A spherical storage tank with a 30-foot radius is to be painted. If the paint covers 350 square feet per gallon, how many gallons of paint are needed? 32.3

C *Extending Thinking Skills*

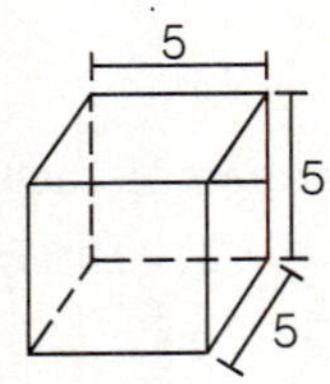

21. The hollow cube at right has a removable top and is to be painted on all surfaces, inside and out. If each edge is 5 feet long, how many square feet of surface will be painted? 300 ft^2

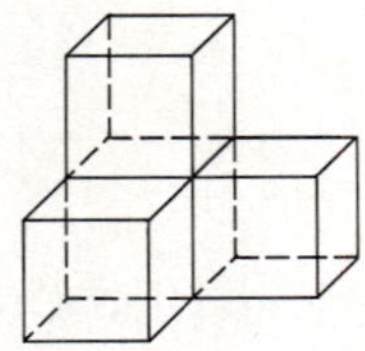

22. The solid at right is made up of 4 cubes glued together. If an edge of each cube is 2 feet, how many square feet of surface can be painted? 72 ft^2
Thinking skills: 21, 22. Using space visualization

Mixed Review

Write as a fraction in lowest terms. **23.** 36% $\frac{9}{25}$ **24.** 65% $\frac{13}{20}$ **25.** 26% $\frac{13}{50}$

Solve by writing an equation. **26.** In 1970, a haircut cost \$2.50. Now it costs \$7.00. What is the percentage of increase in the price?

27. 11 is 25% of what number? 44 **28.** 12 is 150% of what number? 8

Solve and check. **29.** $24c + 17 = 65$ 2 **30.** $14 = -r + 2$ −12 **31.** $6x = 16 - 2x$ 2

26. $h = \frac{7.00 - 2.50}{2.50} \cdot 100$ 180% increase

COMPUTER ACTIVITY

This program will ask you to input a number for the radius of a sphere. It will then calculate the volume and surface area of the sphere.

```
10 REM: CALCULATING VOLUME AND SURFACE AREA OF SPHERES
20 PRINT "TYPE A DECIMAL NUMBER THAT IS THE RADIUS OF A
   SPHERE"
30 INPUT R
40 LET V=4/3*3.14*R*R*R
50 LET S=4*3.14*R*R
60 PRINT "THE VOLUME OF A SPHERE OF RADIUS ";R;" IS ";V
70 PRINT
80 PRINT "THE SURFACE AREA OF A SPHERE OF RADIUS ";R;
   " IS ";S
90 END
```

1. Use the program to find the volume and surface area of these planets:
a. Mercury, radius 1516 miles **b.** Mars, radius 2108 miles
c. Earth, radius 3963 miles See additional answers.

2. Find the lines in the program that use the formulas for volume and surface area of a sphere. Write the formulas. $V = \frac{4\pi r^3}{3}$ $S = 4\pi r^2$

PROBLEM SOLVING: STRATEGIES

12-10 Using Special Insight

Objective: To solve nonroutine problems that require special insight.
MIN: 1–2; REG: 1–4; MAX: 1–6

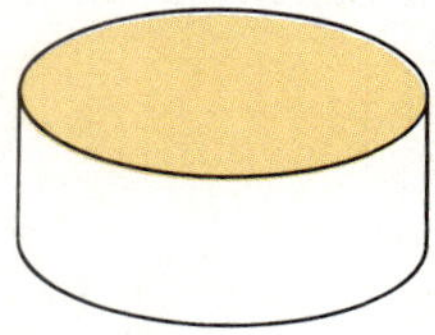

Solving some problems requires special insight. If you use the strategies and your first attempts are not successful, try thinking about the problem in a different way. A new point of view may help a solution idea to pop into your head. Try the following problem.

Problem A waitress had always used 4 straight cuts to cut a cylinder of cheese into 8 identical pieces. One day, she suddenly realized that she could do it with only 3 straight cuts! How did she do it?

When looking for a solution to the problem, you might first think about cutting the cheese with 4 cuts, as in figure A below. Then you might think about ways to make 3 straight cuts across the top, as in figure B. After several unsuccessful tries, you might try to think about the problem in a different way. Perhaps, suddenly, you will decide to try horizontal cuts as well as vertical cuts, and find the solution shown in figure C.

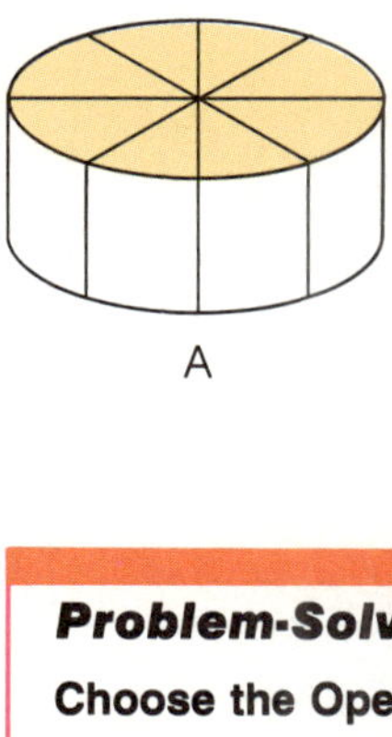

A

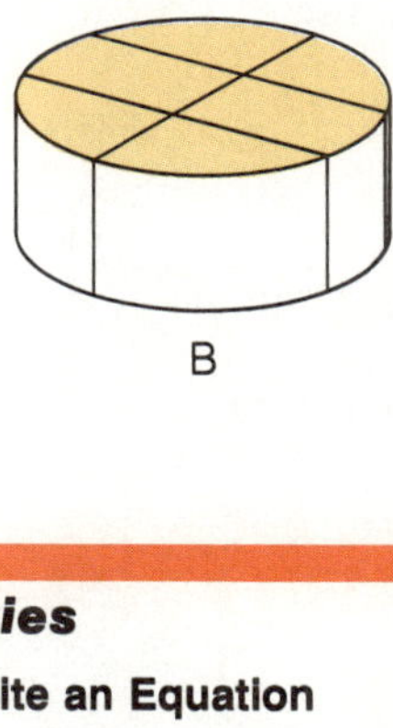

B

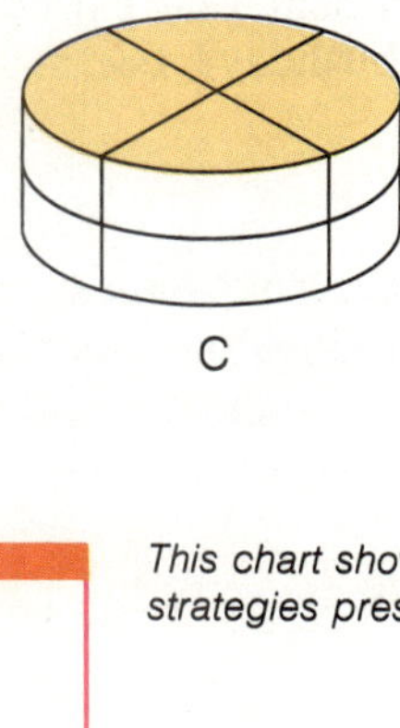

C

Problem-Solving Strategies

Choose the Operations	Write an Equation
Guess, Check, Revise	Simplify the Problem
Draw a Picture	Make an Organized List
Make a Table	Use Logical Reasoning
Look for a Pattern	Work Backwards

This chart shows the strategies presented.

Problems

Solve. Allow yourself time to think about problems in different ways.

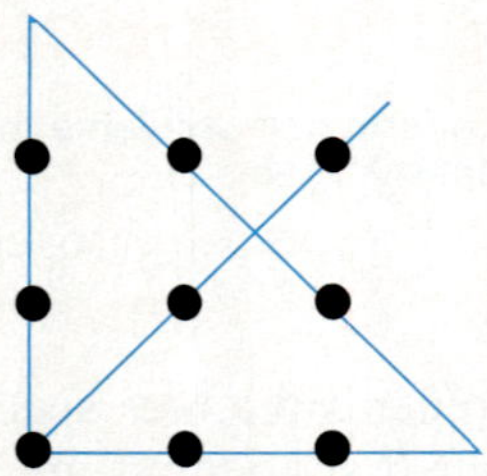

1. Copy these nine dots. Draw four connected straight lines through all nine dots without lifting your pencil or retracing any part of a line.

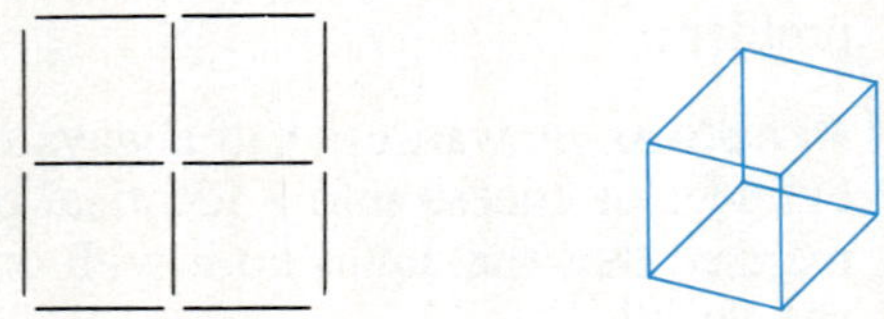

2. Place these 12 toothpicks together to form six squares. Make a drawing of your answer.

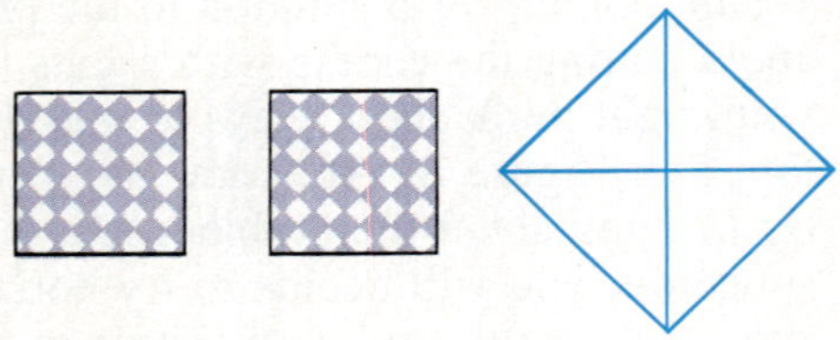

3. How can these two square quilts be cut and sewn back together to form one square quilt with twice the area of one quilt?

4. A very honest teenager said "Two days ago I was 13, but next year I'll be 16." When is the teenager's birthday? Dec. 31

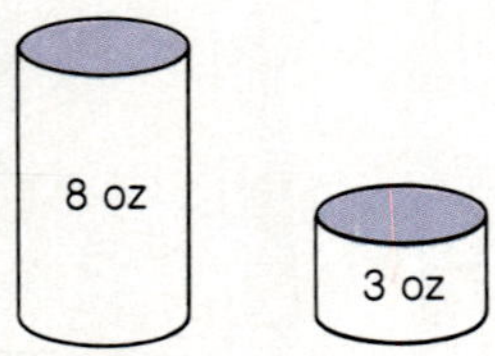

5. You have an 8-ounce can and a 3-ounce can, with no markings on either can. How can you measure out 4 ounces of juice using only these two cans? See additional answers.

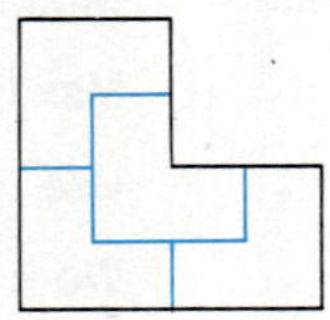

6. Where should you put the fences to divide this field into 4 congruent fields that are the same shape as the original field?

Enrichment

Pick's Formula for Area

The area of polygon *ABCDE* can be found by dividing it as shown and adding together the areas of the smaller regions.

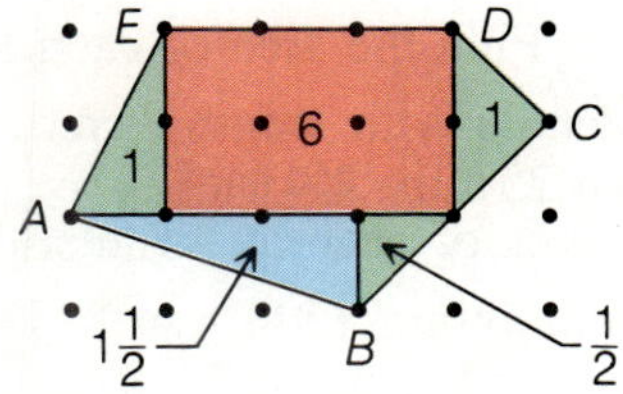

Area = 10

Another way to find the area of this polygon is to use Pick's Formula. The formula uses the number of dots on the boundary (***b***) and the number of dots in the interior (***i***) of the polygon.

The shaded part of the table below has been filled in for polygon *ABCDE*. **Pick's Formula is $A = \frac{b}{2} + i - 1$.**

Polygon	b	i	$\frac{b}{2} + i$	Area
ABCDE	8	7	11	10
Exercise 1	10	5?	10?	9?
Exercise 2	10?	4?	9	8?
Exercise 3	8?	2?	6?	5

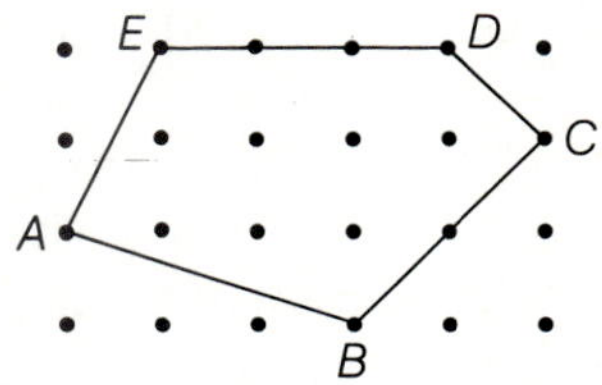

Complete the table above for the polygons in Exercises 1–3 and discover Pick's Formula.

1. 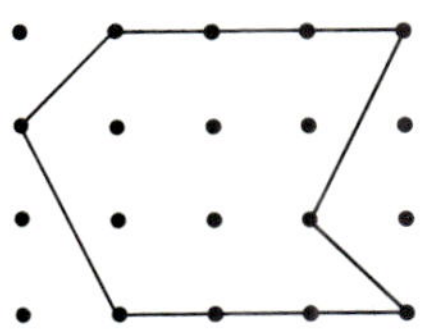

2. 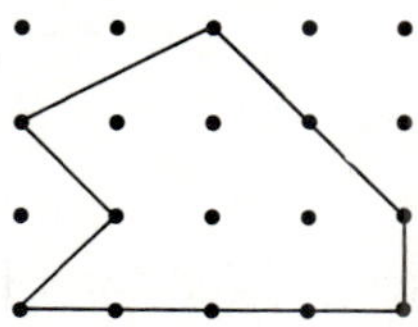

3. 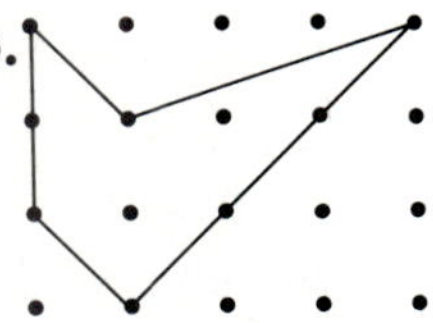

Use Pick's Formula to find the areas of each.

4. 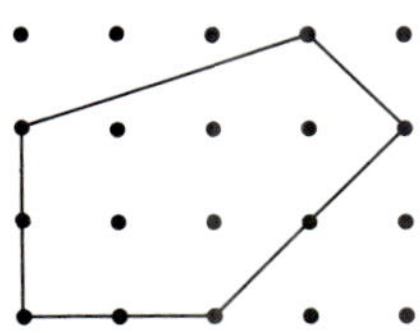

5. 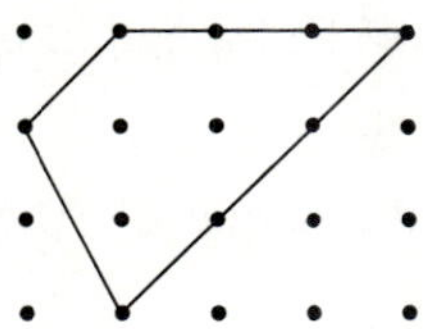

6. 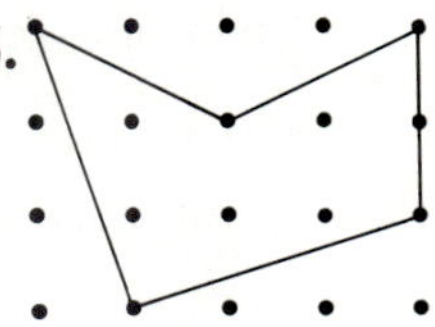

8 units²

6 units²

7 units²

Chapter 12 Review

12-1–12-2 Find the area of each figure.

1. Rectangle with length 8 mm, width 25 mm 200 mm^2

2. Triangle with base 16 cm, height 16 cm 128 cm^2

Find the area of each quadrilateral.

3. Parallelogram with base = 14.4 m, height = 6.8 m 97.92 m^2

4. Trapezoid with $b_1 = 16$ cm, $b_2 = 12$ cm, $h = 7$ cm 98 cm^2

12-3 Find the area of each circle. Use 3.14 for π.

5. $r = 9$ m 254.34 m^2

6. $d = 3.4$ km 9.0746 km^2

12-4 Find the area of the figure.

7. 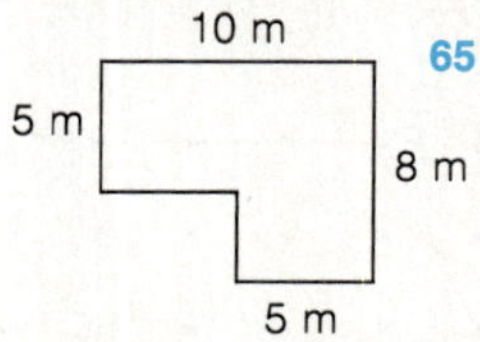

65 m^2

12-5 Solve by writing an equation.

8. Irene bought 3 adult tickets and one child's ticket for \$16.50. If the child's ticket cost \$3.00, what was the cost of each adult ticket? $3a + 3 = 16.50$; \$4.50

12-6 Find the volume of the prism or cylinder. Use 3.14 for π.

9. 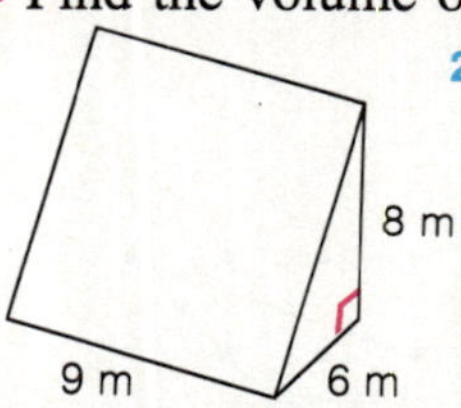

216 m^3

10. 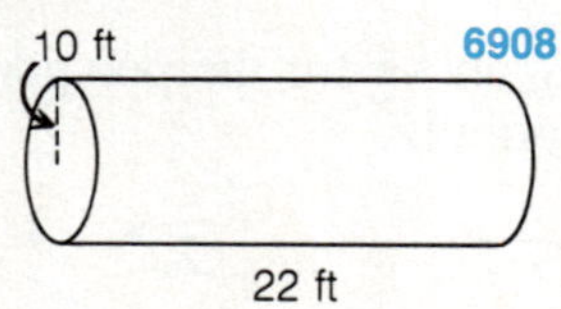

6908 ft^3

12-7 Solve each problem.

11. How many cubic yards of concrete are needed for a patio 42 feet long, 27 feet wide, and 6 inches thick? 21 yd^3

12-8 Find the volume. Use 3.14 for π.

12. A cone with a radius of 11 feet and a height of 4 feet 506.5867 ft^3

13. A pyramid with a base area of 32 dm^2 and a height of 16 dm $170.\overline{6}$ dm^3

12-9 Find the surface area. Use 3.14 for π.

14. A cylinder with a radius of 8 cm and a height of 30 cm 1909.12 cm^2

15. 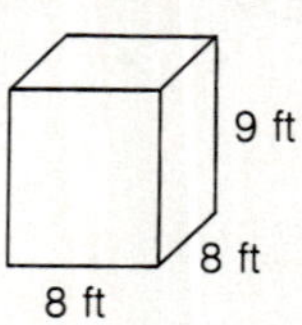

416 ft^2

Chapter 12 Test

Find the area of each figure.

1. Rectangle with length 7 cm, width 36 cm **252 cm²**

2. Triangle with base 34 m, height 8 m **136 m²**

Find the area of each quadrilateral.

3. Parallelogram with base = 146 mm, height = 36 mm **5256 mm²**

4. Trapezoid with $b_1 = 18$ m, $b_2 = 7$ m, $h = 12$ m **150 m²**

Find the area of each circle. Use 3.14 for π.

5. $r = 4$ m **50.24 m²**

6. $d = 30$ km **706.5 km²**

Find the area of the figure.

7. 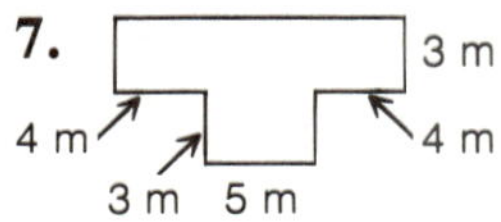

54 m²

Solve by writing an equation.

8. Brian saved \$40 on a suit at a 25%-off sale. What was the regular price of the suit? **0.25 t = \$40 \$160.00**

Find the volume of the prism or cylinder. Use 3.14 for π.

9. 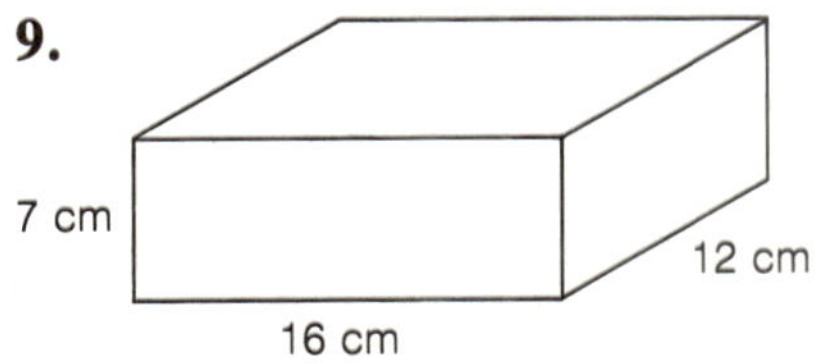

1344 cm³

10. 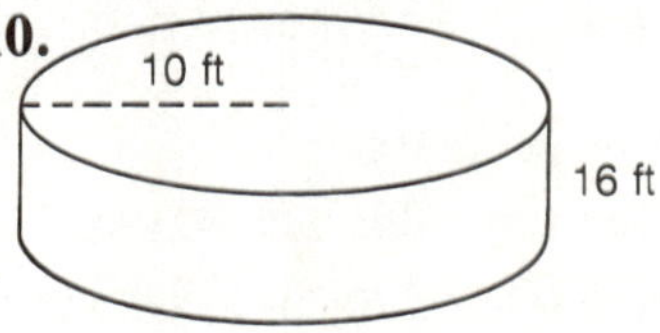

5024 ft³

Solve each problem.

11. If concrete costs \$45 per cubic yard, how much will it cost for concrete for a foundation 48 feet long, 30 feet wide, and 8 inches thick? **\$1600.00**

Find the volume. Use 3.14 for π.

12. A cone with a radius of 7 inches and a height of 15 inches **769.3 in³**

13. A pyramid with a base area of 80 square m and a height of 22 m **$586.\overline{6}$ m³**

Find the surface area. Use 3.14 for π.

14. A cylinder with a radius of 6 feet and a height of 24 feet **1130.4 ft²**

15. 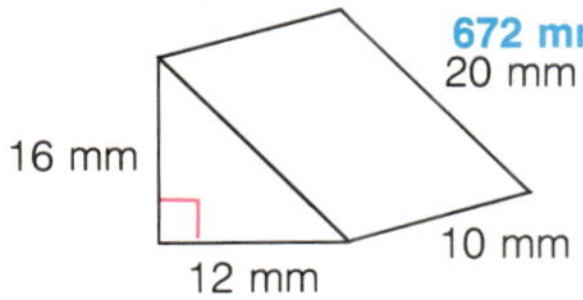

672 mm²

Cumulative Review

Write each as a fraction in lowest terms.

1. 34% $\frac{17}{50}$
2. 5% $\frac{1}{20}$
3. $8\frac{1}{2}$% $\frac{17}{200}$

Use a proportion to write each fraction as a percent.

4. $\frac{2}{5}$ 40%
5. $\frac{13}{20}$ 65%
6. $\frac{5}{8}$ 62.5%
7. $\frac{625}{100}$ 625%
8. $\frac{14}{5}$ 280%
9. $\frac{25}{250}$ 10%

Write each percent as a fraction or mixed number in lowest terms.

10. 275% $2\frac{3}{4}$
11. $\frac{3}{4}$% $\frac{3}{400}$
12. $12\frac{1}{2}$% $\frac{1}{8}$

Write each as a decimal.

13. 7% 0.07
14. $22\frac{1}{5}$% 0.222
15. 425% 4.25

Write each as a percent.

16. 0.60 60%
17. 0.085 8.5%
18. 8.98 898%

Find the percent of increase.

19. Original amount = \$120; new amount = \$180 50%
20. Original amount = \$90; new amount = \$108 20%

Find the percent of decrease.

21. Original amount = \$85; new amount = \$68 20%
22. Original amount = \$276; new amount = \$69 75%

Solve.

23. Juan bought a computer program for \$240.00. If the sales tax was 5.5%, what was the total amount that Juan paid? \$253.20

24. Bill bought a camera priced at \$125. The sales tax was 6%. What was the total cost of the camera? \$132.50

Use the figure at right for Exercises 25–29.

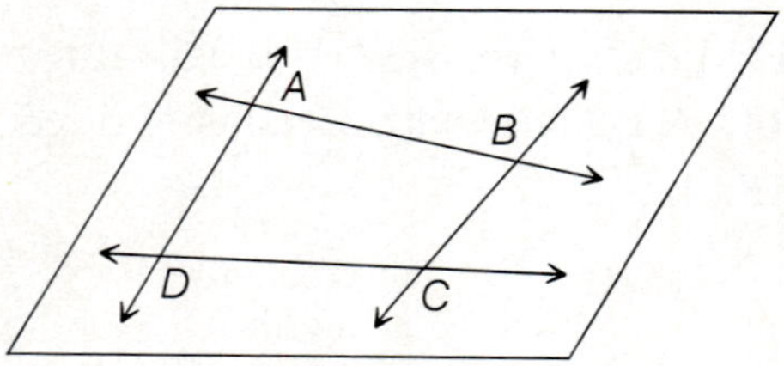

25. Name a point. *A, B, C, or D*
26. Name a line. $\overleftrightarrow{AB}$, $\overleftrightarrow{AD}$, $\overleftrightarrow{DC}$, or $\overleftrightarrow{BC}$
27. Name a plane. ABC, ABD, ACD, BCD
28. Name a segment that includes point *B*. $\overline{AB}$ or $\overline{BC}$
29. Name a ray with vertex *A*. $\overrightarrow{AB}$ or $\overrightarrow{AD}$

Chapter 13
Probability, Statistics, and Graphs

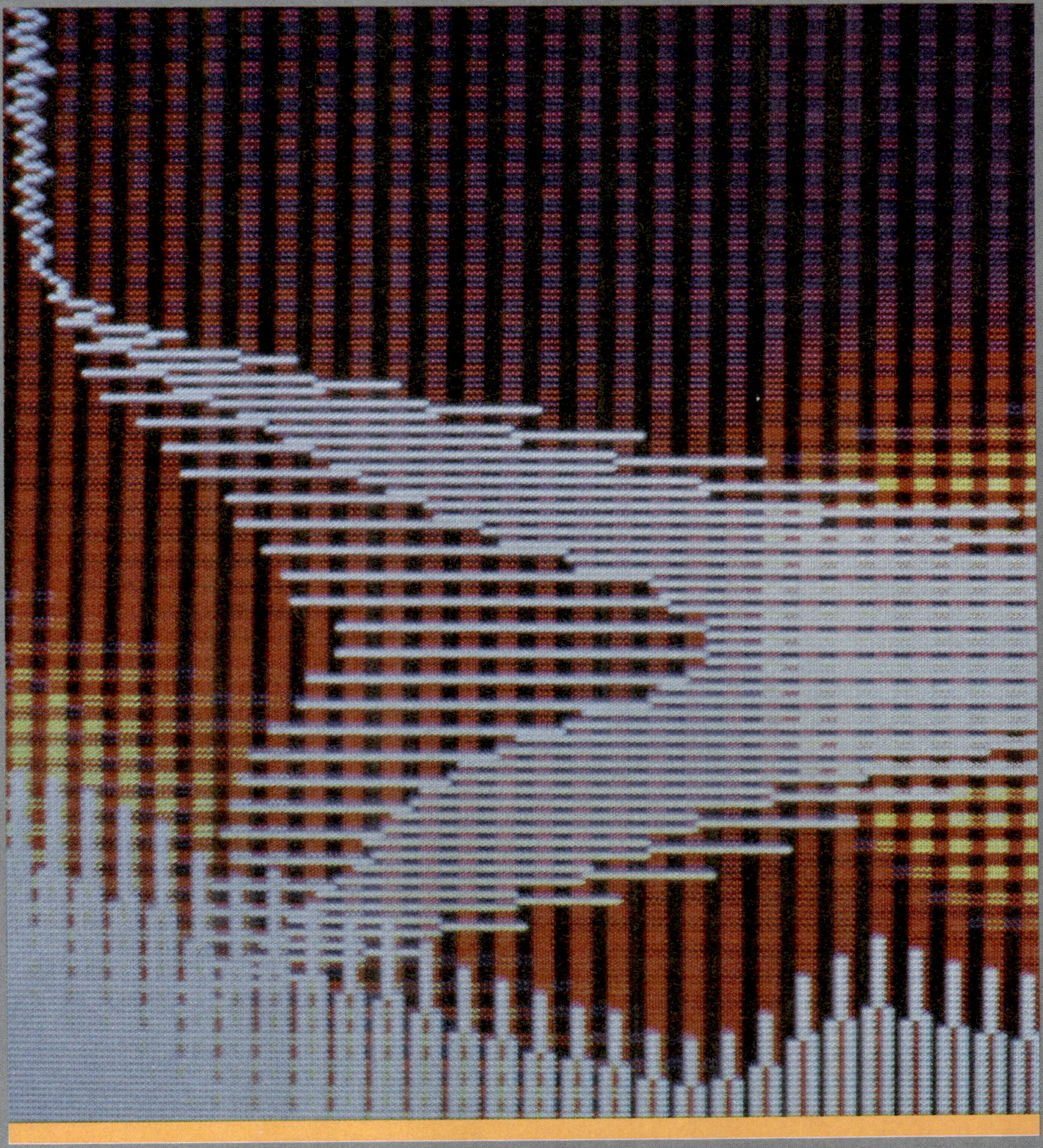

13-1 Counting Principle

Objective: To use diagrams and the counting principle to find the total number of outcomes of an event.
MIN: 1–10; REG: 1–10 even, 11–14; MAX: 1–10 even, 11–15; ALL: Mixed Review

A school is having an election for student council president, vice-president, and secretary. The ballot below shows one possible **outcome** of the election. One way to show all the possible outcomes is to make a tree diagram as shown below. Each choice for president can be matched with two choices for vice-president. And each choice for president and vice-president can be matched with 2 choices for secretary. Each "branch" shows a possible outcome.

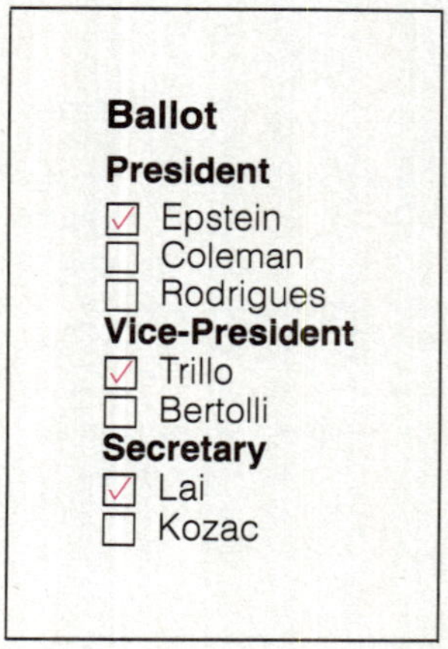

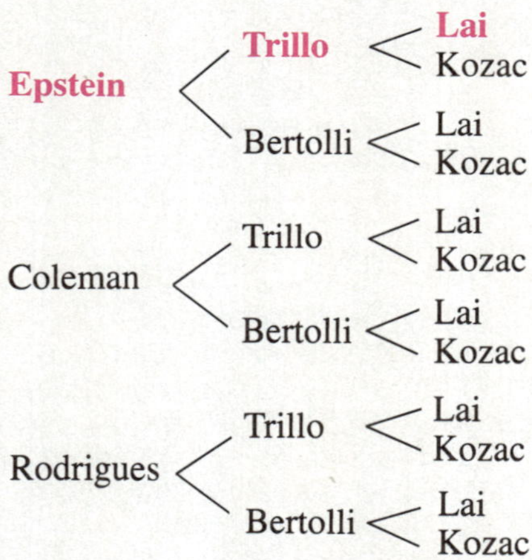

This election has 12 possible outcomes.

Since there are 3 choices for president, 2 choices for vice-president, and 2 choices for secretary, there are $3 \times 2 \times 2 = 12$ possible outcomes for the election. When you multiply to find the total number of outcomes, you are using the **counting principle.**

Counting Principle

To find the total number of choices for an event, multiply the number of choices for each part.

Example

Use the counting principle to solve.

A singer has 4 blouses and 5 pairs of jeans to use in concerts. How many different ways can she dress for a concert?

Solution $4 \times 5 = 20$ The total number of choices equals the product of the number of blouses and the number of pairs of jeans.

She has 20 choices for her outfit.

Practice Use the counting principle to solve.

Reggie needs to get glasses. He can choose from 3 different colors for the frame, and 4 tints for the glass. How many different kinds of glasses could Reggie select? 12

Oral Exercises

Use the tree diagram showing total possible outcomes for a Frozen Yogurt Special to answer questions 1–2.

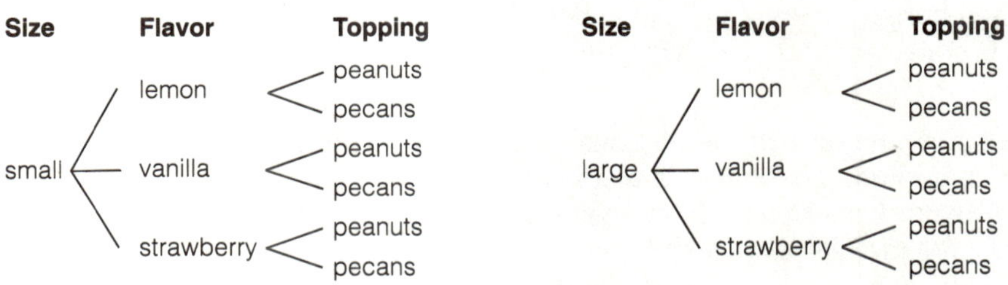

1. How many choices are there for: size? yogurt flavor? topping? a small yogurt special? a large yogurt special? a yogurt special? 2; 3; 2; 6; 6; 12

2. What equation can you use to find the total number of choices? $2 \cdot 3 \cdot 2 = 12$

Exercises

A Use the counting principle to find the total number of outcomes.

1. Select 1 sandwich and 1 drink. Sandwiches: hamburger, chicken, cheese. Drinks: juice, milk, tea. 9

2. Select 1 shirt and 1 tie. Shirts: white, blue, yellow, cream. Ties: striped, light blue, navy. 12

3. Select 1 exterior car color and 1 interior car color. Exterior colors: white, blue, tan, red. Interior colors: cream, black, red, blue, light green. 20

4. Select 1 club and 1 sport. Clubs: debate, drama, art, math. Sports: tennis, golf, volleyball. 12

5. Select 1 section of the airplane and 1 seat location. Sections: first class, business, coach. Locations: window, middle, aisle. 9

6. Select 1 watch style and 1 watchband. Watch style: digital, standard, Roman numerals. Watchband: metal, leather, canvas. 9

7. Select 1 ticket for a level and section in a theater. Level: main floor, balcony. Section: front, middle, back. 6

8. Select a 2-digit football jersey number. First digit: 5, 6, 7, 8, 9. Second digit: any number from 0 through 9. 50

9. Select a 2-digit basketball jersey number. Choose any of the numbers 0 through 9 for each digit. 100

10. Select colors for a room. Wall colors: white, yellow, pink, blue. Rug colors: white, yellow, red, blue, brown. 20

B Find the total number of outcomes.

11. Select 1 president, 1 vice-president, and 1 secretary. President: Sunji, Carlos. Vice-president: Denise, Kim, Kerry, Celia. Secretary: Mark, Inez. 16

12. Select 1 telephone style, 1 color, and 1 format. Style: wall, desk, antique. Color: green, white, tan, black, brown. Format: touch-tone, rotary. 30

13. Select 1 size of hamburger, 1 topping, and 1 bun. Sizes: single, double, triple. Toppings: mushrooms, pickles, onions, lettuce. Bun: regular, whole wheat. 24

14. The winners of each conference in the National Basketball Association play to determine the champion. There are 11 teams in the Eastern Conference and 12 teams in the Western Conference. How many outcomes are possible for the two teams that play for the championship? 132

C Extending Thinking Skills Thinking skill: 15. Organizing

15. Ms. Ortega wants a one-way airline ticket from Chicago to San Francisco with a stop in Denver. She wants a minimum of 1 hour and a maximum of 3 hours between flights. How many choices for flights does she have? 3

Flight Number	CHICAGO TO DENVER Depart	Arrive
#63	7:05 a.m.	8:20 a.m.
#1122	8:15	9:30
#34	10:10	11:25
#125	12:01 p.m.	1:16 p.m.
#456	1:40	2:55

Flight Number	DENVER TO SAN FRANCISCO Depart	Arrive
#1156	7:50 a.m.	8:51 a.m.
#92	8:50	9:51
#8	9:50	10:51
#54	12:50 p.m.	1:51 p.m.
#32	1:50	3:30

Mixed Review

Solve and check. **16.** $6(2m + 3) = 78$ 5 **17.** $4c + 2c = 5c + 2$ 2

What percent of 50 is: **18.** 40? 80% **19.** 75? 150% **20.** 15? 30%

13-2 Permutations

Objective: To find the number of permutations of n objects and the number of permutations of n objects taken r at a time.
MIN: 1–18; REG: 1–18 even, 19–25; MAX: 1–18 even, 19–27; ALL: Mixed Review
An exploratory lesson opener can be found on page 553.

A **permutation** is an arrangement of objects in a particular order. The tree diagram at right shows permutations of 3 books. Any of the three books can be chosen for the first position. Either of the two books left can be chosen for the second position. One book is left for the third position. Using the counting principle, there are $3 \times 2 \times 1 = 6$ permutations of the books. The product $3 \times 2 \times 1$ can be written as **3!** This is read as "**3 factorial**."

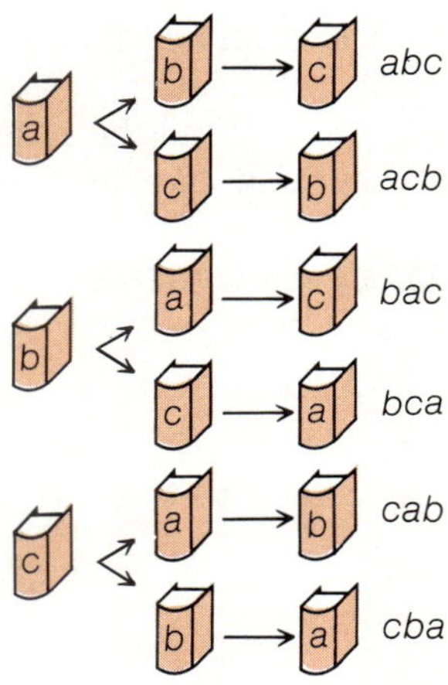

Permutations

To find the number of permutations of n objects, find the product of the numbers **1** through n.

$n! = n \times (n-1) \times \ldots \times 3 \times 2 \times 1$

Example 1

Find the number of permutations of the letters A, B, C, D.

Solution $4! = 4 \times 3 \times 2 \times 1 = 24$ Since there are 4 letters, the number of permutations is equal to 4 factorial.

Practice Find the number of permutations. **a.** Tom, Dick, Harry 6 **b.** 1, 3, 5, 7, 9 120

The list at right shows the arrangement of two numbers that can be made from the numbers 2, 4, 6, and 8. The list shows all the permutations of two numbers with 2 as the first number, then the permutations with 4 as the first number, and so on. There are 4 choices for the first number. Then, after the first number is chosen, there are 3 choices for the second number. Using the counting principle, $4 \times 3 = 12$ two-digit numbers can be made. You can use the steps below to find the total number of ways to order part of a group.

24 26 28
42 46 48
62 64 68
82 84 86

Permutations of n Objects Taken r at a Time

To find the number of permutations of n objects taken r at a time, carry out the product $n(n-1)(n-2)\ldots$ to r factors.

You may want to introduce the formula $\frac{n!}{(n-r)!}$

Example 2

Find the number of permutations. Five students ran in a 100-meter dash. In how many ways could first, second, and third place be won?

Solution

$5 \times 4 \times 3 = 60$ You want to find the number of permutations of 5 objects taken 3 at a time. There are 5 choices for first place, 4 choices for second, and 3 choices for third.

The prizes could be awarded in 60 ways.

Practice Find the number of permutations. Four people applied for 2 jobs, painter and carpenter. In how many ways could the jobs be filled? 12

Oral Exercises

Use $n! = n \times (n - 1) \times \ldots \times 3 \times 2 \times 1$ to restate each expression.

1. $5!$ $5 \cdot 4 \cdot 3 \cdot 2 \cdot 1$

2. $6!$ $6 \cdot 5 \cdot 4 \cdot 3 \cdot 2 \cdot 1$

3. $11!$ $11 \cdot 10 \cdot 9 \cdot 8 \cdot 7 \cdot 6 \cdot 5 \cdot 4 \cdot 3 \cdot 2 \cdot 1$

4. $1 \times 2 \times 3 \times 4 \times 5 \times 6 \times 7$ $7!$

5. $1 \times 2 \times 3 \times \ldots \times 9 \times 10$ $10!$

6. $1 \times 2 \times 3 \times \ldots \times 19 \times 20$ $20!$

Exercises

A Find the number of permutations.

1. 1, 2 2

2. Sue, Kim, Heather 6

3. 12, 14, 16 6

4. math, history, English, art 24

5. quick, Jane, run 6

6. m, u, s, i, c 120

7. Ace, King, Queen 6

8. North, South, East, West 24

9. 10, 20, 30, 40 24

10. m, i, l, k 24

11. Lincoln, Washington, Jefferson 6

12. winter, spring, summer, fall 24

13. 0, 1, 2, 3, 4, 5, 6, 7, 8, 9 3,628,800

14. Huey, Dewey, Louie, Donald 24

15. Harpo, Chico, Groucho, Zeppo 24

16. soup, salad, potatoes, turkey, pie 120

Solve.

17. How many 3-letter arrangements can be made from the letters A, B, C, D, E? 60

18. How many 3-song arrangements can be made from a list of 7 songs? 210

B Solve.

19. How many 2-letter arrangements are possible from the 26 letters? 676
You can use the same letter twice in an arrangement.

20. How many 2-digit track jersey numbers can be made from 3, 5, 7, and 9? 16
You can use the same number twice in an arrangement.

21. How many ways can first, second, and third place be given to 10 players? 720

22. How many 6-digit zip codes would be possible if no digit could be repeated? 151,200

23. In how many ways can 6 track teams be assigned to 6 lanes on a track? 720

24. A band, a float, and an antique car are to be in a parade. In how many different ways can they be arranged in the parade? 6

25. How many different license plates are possible with 2 letters followed by 4 numbers if no letter or number is used twice? 3,276,000

C Extending Thinking Skills Thinking skill: 26, 27. Generalizing

26. How many different 3-digit telephone area codes are possible if the first digit cannot be a 0 or a 1? 800

27. How many different 7-digit phone numbers are possible in each area code if the following restrictions apply: the first digit cannot be a 0 or a 1, the middle digit in the prefix cannot be a 0 or a 1, and the 3-digit prefix cannot be 555 or 911? 6,390,000

Mixed Review

Simplify. **28.** $(6 + 3t) + 15$ $3t + 21$ **29.** $m + 6m - 4m$ $3m$ **30.** $12 + 3c - 9$ $3c + 3$
31. $26n - 19n + 2n$ $9n$ **32.** $3 + 6r - 12$ $6r - 9$ **33.** $4x + 3y + 2x - 9y$ $6x - 6y$

Give the circumference. Use 3.14 for π. **34.** radius = 4 25.12 **35.** diameter = 5 15.7

Solve and check. **36.** $3(x + 4) - 3 = 15$ 2 **37.** $0.4c + 0.7c = 22$ 20
38. $14 = 2t - 8$ 11 **39.** $6m = m - 10$ -2 **40.** $27 + 6(y - 2) = 3$ -2

NUMBERS TO ALGEBRA

You can use variables to write expressions for consecutive numbers.

Numbers	Algebra
The next 3 even numbers after 4.	The next 3 even numbers after an even number n.
4, 6, 8, 10	$n, (n + 2), (n + 4), (n + 6)$

The numbers 27, 28, and 29 are consecutive whole numbers. The numbers 8, 10, and 12 are consecutive even numbers.

Write expressions for each of the following.

1. The next 4 consecutive whole numbers after whole number n. $(n + 1), (n + 2), (n + 3), (n + 4)$
2. The 2 whole numbers just before a whole number x. $(x - 1), (x - 2)$
3. The 2 even numbers just before an even number y. $(y - 2), (y - 4)$
4. Numbers that are 5, 10, and 15 times greater than a whole number m. $5m, 10m, 15m$

13-3 Combinations

Objective: To find the number of combinations of n objects taken r at a time.
MIN: 1–20; REG: 1–20 even, 21–24; MAX: 1–20 even, 21–27; ALL: Mixed Review

Candidates (choose 2)
(A) Alvarez
(B) Barnes
(C) Carlsen
(D) Dickinson
(E) Ebel

Tanya's class must select 2 students out of 5 candidates as class representatives. How many different pairs can be selected from the 5 candidates?

You learned in the last lesson that there are $5 \cdot 4 = 20$ ways to select 2 out of 5 people. Here are the 20 ways.

AB	**AC**	**AD**	**AE**
BA	**BC**	**BD**	**BE**
CA	**CB**	**CD**	**CE**
DA	**DB**	**DC**	**DE**
EA	**EB**	**EC**	**ED**

Alvarez–Barnes (AB) and Barnes–Alvarez (BA) are both included in the list above. Since the *order* of the students does not matter, you need to divide the total number of permutations by the number of permutations of 2 people ($2! = 2 \cdot 1 = 2$). The list below shows all 10 ways in which 2 out of the 5 people can be selected. An arrangement of objects in which the *order of the objects does not matter* is called a **combination**.

AB	**AC**	**AD**	**AE**
	BC	**BD**	**BE**
		CD	**CE**
			DE

You can use the following steps to find the number of combinations.

Combinations

To find the number of combinations of n objects taken r at a time:

- Find the number of permutations of n objects taken r at a time.
- Divide this number by the number of permutations of r objects.

The formula is
$$\frac{n!}{(n-r)!r!}$$
You must let $0! = 1$ when using the formula

Example

Find the number of combinations. 3 bracelets from 5.

Solution

$\frac{5 \cdot 4 \cdot 3}{3 \cdot 2 \cdot 1} = \frac{5 \cdot 4 \cdot \cancel{3}}{\cancel{3} \cdot 2 \cdot 1}$ Find the number of permutations of 5 objects taken 3 at a time. Divide that by the number of permutations of 3 objects.

$= 10$

There are 10 combinations of bracelets.

Practice Find the number of combinations.

a. 5 ties from a box of 7 ties. 21

b. 2 cards from a pile of 8 different cards. 28

Oral Exercises

Tell whether each is a permutation or a combination. Do not solve.

1. Ways 5 pictures can be arranged in a row on a wall. P
2. Ways 3 people can be selected from a group of 5. C
3. Ways first and second violin players can be chosen from 6 players. P
4. Ways 4 sweaters can be chosen out of 7. C
5. Ways 5 chairs can be arranged in a row. P
6. Possible finishing orders of 7 people in a race if there are no ties. P
7. Ways 3 representatives can be selected from 21 students. C

Exercises

A Find the number of combinations.

1. 2 letters from A, B, C 3
2. 3 people from Blake, Carol, Annie, Phil 4
3. 3 numbers from 10, 20, 30 1
4. 2 symbols from #, $, ?, %, * 10
5. 3 letters from M, A, T, H 4
6. 2 colors from red, orange, green, blue 6
7. 2 digits from 0 through 5 15
8. 3 numbers from 1, 2, 3, 4, and 5 10
9. 1 digit from 0 through 9 10
10. 4 people from Carl, Jim, Sue, Tim 1
11. 1 shirt from a box of 8 shirts 8
12. 5 colors from a list of 5 choices 1
13. 7 people from 9 people 36
14. 4 records from a list of 7 35
15. 2 blouses from 8 blouses 28
16. 5 movies from 8 in town 56
17. 6 books from 10 books 210
18. 4 cheerleaders from a group of 10 210
19. 3 shirts from a box of 12 220
20. 5 stamps from a book of 12 792

B

21. A planning committee for a talent show is to have 4 members. Twelve apply for the committee. How many committee selections are possible? 495

22. There are 10 movies in town but you have money to see only 3. How many choices of 3 movies do you have? 120

23. A shoe store has 8 kinds of running shoes. The window holds only 6 shoes. How many different displays can be put in the window? 28

24. The winner of a contest can select any 5 record albums from a collection of 10. How many choices for 5 albums does the winner have? 252

C Extending Thinking Skills

25. How many teams of 4 boys and 3 girls can be made from 5 boys and 7 girls? 175

26. A table tennis club has 10 members. How many games would be played if each member played each other member 1 time? 45

27. If seventy different committees of the same size could be formed from a group of 8 people, how many people would be on a committee? 4

Thinking skills: 25. Generalizing 26. Using logic 27. Testing a conjecture

Mixed Review

Solve and check. **28.** $\frac{6}{7} = \frac{n}{14}$ 12 **29.** $\frac{5}{8} = \frac{20}{x}$ 32 **30.** $\frac{20}{48} = \frac{y}{12}$ 5

31. $-18 = 0.45t$ −40 **32.** $1.3 - x = 2.1$ −0.8 **33.** $6.3 = 2.1m$ 3

What percent of 40 is: **34.** 15? 37.5% **35.** 10? 25% **36.** 56? 140%

Evaluate for $x = 2$, $y = 4$. **37.** $2xy$ 16 **38.** $4x - 2y$ 0 **39.** $x(x - y)$ −4

40. $3x - 2y + 9$ 7 **41.** $6(x - 9) + 3y$ −30 **42.** $4.5x - 2.1y$ 0.6

43. Give the perimeter of a square with a side with length of 4.5 cm. 18 cm

13-4 Probability

Objective: To find the probability of simple events and mutually exclusive events.
MIN: 1–25; REG: 1–25 even, 26–30; MAX: 1–25 even, 26–31; ALL: Mixed Review
An exploratory lesson opener can be found on page 554.

A die has sides numbered 1 through 6. If you toss a die once there are 6 possible outcomes: 1, 2, 3, 4, 5, or 6. Each outcome is **equally likely**. Since there are 6 equally likely outcomes, the **probability** of each outcome is $\frac{1}{6}$. If there are n equally likely outcomes from an activity, the probability of each outcome is $\frac{1}{n}$. An outcome or a combination of outcomes is called an **event**.

Probability

To find the probability of an event **A**, written ***P*(A)**, divide the number of ways for the event to occur by the total number of possible outcomes.

Example 1

Find the probability of each for one toss of a die.

a. $P(\text{even number})$ **b.** $P(\text{a number} < 6)$ **c.** $P(\text{a number} > 6)$

Solution

a. $P(\text{even number}) = \frac{3}{6} = \frac{1}{2}$ There are 3 even numbers. There are 6 possible outcomes. Reduce $\frac{3}{6}$ to $\frac{1}{2}$.

b. $P(\text{a number} < 6) = \frac{5}{6}$

c. $P(\text{a number} > 6) = \frac{0}{6} = 0$ It is impossible to throw a number greater than 6. The probability of an impossible event is 0.

Practice Find the probability of each for one toss of a die.

a. $P(\text{a multiple of } 2)$ $\frac{3}{6} = \frac{1}{2}$ **b.** $P(0 < \text{a number} < 7)$ 1 **c.** $P(\text{not an even number})$ $\frac{3}{6} = \frac{1}{2}$

If two events cannot occur at the same time, they are **mutually exclusive**. In one toss of a die, the events "an even number" and "an odd number" are mutually exclusive, since no number is both even and odd. The events "a number > 3" and "an even number" are not mutually exclusive, since 4 and 6 are both greater than 3 and even. This relationship is stated in the following formula.

Formula: Mutually Exclusive Events

If two events are **mutually exclusive**, the probability of one or the other occurring is equal to the sum of the probabilities of each one occurring. If A and B are mutually exclusive events,

$$P(\text{A or B}) = P(\text{A}) + P(\text{B})$$

Stress that "or" signals the *addition* of separate probabilities.

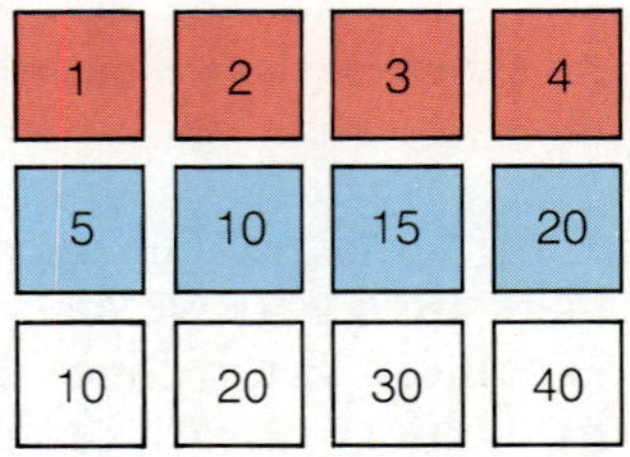

Example 2

Find the probability of each.

a. $P(\text{red or white})$ **b.** $P(\text{red or a multiple of 10})$

Solution

a. $P(\text{red or white}) = \frac{1}{3} + \frac{1}{3} = \frac{2}{3}$

The events are mutually exclusive. $P(\text{red}) = \frac{1}{3}$ and $P(\text{white}) = \frac{1}{3}$

b. $P(\text{red or a multiple of 10}) = \frac{1}{3} + \frac{6}{12}$

$= \frac{4}{12} + \frac{6}{12}$

$= \frac{10}{12} = \frac{5}{6}$

The events are mutually exclusive. $P(\text{red}) = \frac{1}{3}$ and $P(\text{multiple of 10}) = \frac{6}{12}$

Practice Find the probability of each, using the picture above.

a. $P(\text{number} < 4 \text{ or number} > 20)$ $\frac{5}{12}$

b. $P(\text{prime number or a multiple of 10})$ $\frac{9}{12} = \frac{3}{4}$

Oral Exercises

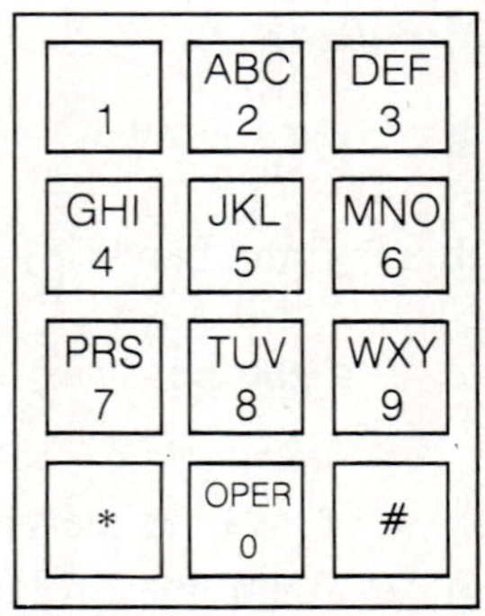

Suppose you closed your eyes and pushed one button on the telephone at random. What is the probability of pushing each?

1. 0 (zero) $\frac{1}{12}$

2. an even number $\frac{5}{12}$

3. an odd number $\frac{5}{12}$

4. a number less than 5 $\frac{1}{3}$

5. a prime number $\frac{1}{3}$

6. a factor of 4 $\frac{1}{4}$

7. a vowel $\frac{5}{12}$

8. a consonant $\frac{2}{3}$

9. the letter Z 0

10. a button with no number $\frac{1}{6}$

To prepare students for 1–12, show them a deck of cards and explain suits, "face" cards, etc.

Exercises

A Suppose you draw 1 card from a deck of 52 playing cards. Find each probability.

1. $P(\text{red card})$ $\frac{1}{2}$

2. $P(\text{black card})$ $\frac{1}{2}$

3. $P(\text{ace})$ $\frac{1}{13}$

4. $P(\text{heart})$ $\frac{1}{4}$

5. $P(5)$ $\frac{1}{13}$

6. $P(\text{queen})$ $\frac{1}{13}$

7. P(king of diamonds) $\frac{1}{52}$ 8. P(ace of hearts) $\frac{1}{52}$ 9. P(not a black card) $\frac{1}{2}$

10. P(red 7) $\frac{1}{26}$ 11. P(not a heart) $\frac{3}{4}$ 12. P(black face card) $\frac{3}{26}$

Suppose there are 15 balls in a bag. All are the same size. Four are red, 6 are white, and 5 are blue. You select one ball from the bag. Find each probability.

13. P(red) $\frac{4}{15}$ 14. P(white) $\frac{2}{5}$ 15. P(blue) $\frac{1}{3}$

16. P(not blue) $\frac{2}{3}$ 17. P(not white) $\frac{3}{5}$ 18. P(not red) $\frac{11}{15}$

Use the spinner at right. Find each probability.

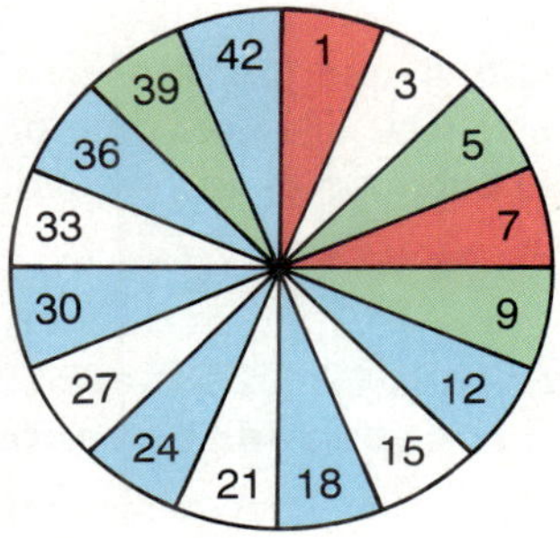

19. P(blue or white) $\frac{11}{16}$

20. P(blue or prime) $\frac{9}{16}$

21. P(multiple of 5 or multiple of 4) $\frac{3}{8}$

22. P(odd or blue) 1

23. P(even or green) $\frac{9}{16}$

24. P(green or blue or white) $\frac{7}{8}$

25. P(multiple of 10 or green or red) $\frac{3}{8}$

B Solve.

26. One postcard will be drawn out of 500 received for a ticket to a concert. You mailed 25 postcards. What are your chances of getting the ticket? $\frac{1}{20}$

27. Janis and 4 of her friends are among 40 candidates for a visit to Washington, D.C. One student will be selected at random. What is the probability that the student selected will be Janis or one of her friends? $\frac{1}{8}$

28. A radio station is giving away 20 prizes: 8 movie tickets, 5 albums, 4 cassettes, and 3 concert tickets. Each prize has a number from 1 to 20. You have to pick a number from 1 to 20 to determine which prize is yours. What is the probability you will get an album or a cassette? $\frac{9}{20}$

29. Lisa and her brother are among 12 girls and 18 boys nominated for band positions. One girl and one boy will be selected at random. What is the probability Lisa or her brother will be selected? $\frac{5}{36}$

C Extending Thinking Skills

Thinking skills: 30., 31. Organizing

30. One die has a blank face instead of a 1-dot face. Another die has a blank face instead of a 4-dot face. What is the probability of rolling a sum of 7 with the dice? $\frac{1}{9}$

31. A lost-and found office had 3 identical jackets. One was yours. You and two others arrived to claim your jackets. The clerk handed out the three jackets at random. What is the probability each of you got your own jacket? $\frac{1}{6}$

Mixed Review

Give the least common multiple (LCM). **32.** 3, 5, 6 30 **33.** 4, 6, 9 36

Solve and check. **34.** $2m - 6 < -2$ $m < 2$ **35.** $4(c + 1) = 28$ 6 **36.** $6 = 0.8t$ 7.5

37. $19 - 3y = 3y + 1$ 3 **38.** $x + 3 = 2x$ 3 **39.** $3m - 5m = 16$ -8

13-5 Independent and Dependent Events

Objective: To distinguish between and calculate the probability of independent and dependent events.
MIN: 1–24; REG: 1–24 even, 25–32; MAX: 1–24 even, 25–36; ALL: Mixed Review

Suppose you first spin the spinner with numbers and then the spinner with letters. The outcome of the first spin does not influence the outcome of the second. The events are **independent**.

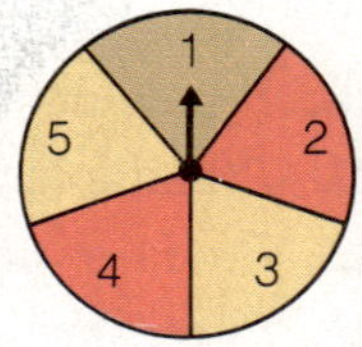

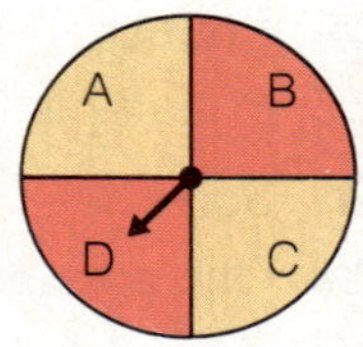

Formula: Independent Events

When two events are independent, the probability of both occurring is equal to the product of the probabilities of each event occurring.

$$P(\text{A and B}) = P(\text{A}) \times P(\text{B})$$

Stress that the word "and" signals the multiplication of the probabilities.

Example 1

For the two spinners above, find P(even number and vowel).

Solution

$P(\text{even number}) = \frac{2}{5}$ $P(\text{vowel}) = \frac{1}{4}$

$P(\text{even number and vowel}) = \frac{2}{5} \times \frac{1}{4} = \frac{1}{10}$ Since the two events are independent, multiply their probabilities.

Practice Find each probability for the spinners above.

a. P(odd number and consonant) $\frac{9}{20}$ **b.** $P(\text{number} < 5 \text{ and a B})$ $\frac{1}{5}$

Suppose you select 1 ball and then another from a bag filled with 8 yellow balls and 4 green balls. The probability that the second ball you select is yellow depends on whether you replaced the first ball before selecting the second. If you did replace it, the two events are independent. If you did *not* replace it, the probability would depend on the outcome of the first draw. When an event's outcome is influenced by the outcome of a previous event, it is a **dependent event**.

To find the **probability of dependent events**, find the probability of the first event. Then, taking the first event into consideration, find the probability of the second event. Repeat this for each dependent event. Then multiply the probabilities.

Example 2

For the bag of marbles at right, find P(red then yellow).

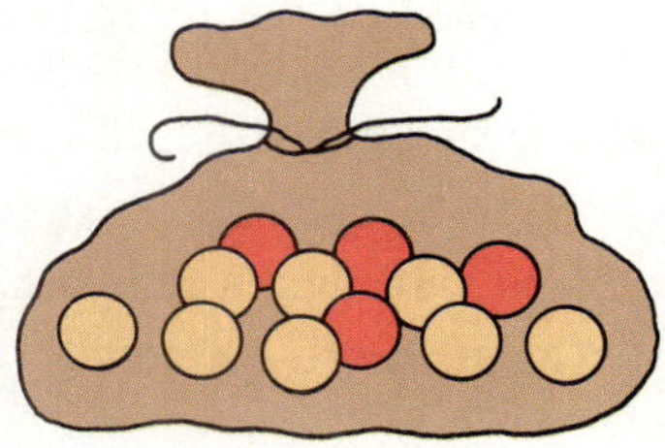

Solution

$P(\text{red}) = \frac{4}{12}$ 4 out of 12 marbles are red.

$P(\text{then yellow}) = \frac{8}{11}$ If a red marble is selected first, only 11 marbles remain in the bag. 8 of these are yellow.

$P(\text{red then yellow}) = \frac{4}{12} \times \frac{8}{11}$ Multiply the probabilities of the two events.

$= \frac{32}{132} = \frac{8}{33}$

Practice For the bag shown above, find each probability.

a. P(yellow then red) $\frac{8}{33}$ **b.** P(yellow then yellow) $\frac{14}{33}$

Oral Exercises

State whether the pairs of activities are dependent or independent events.

1. Tossing heads on 1 coin and tails on another coin. I

2. Taking 1 white sock from a drawer and then, without replacing it, taking another. D

3. Seeing a blue motorcycle yesterday and a blue motorcycle today. I

4. Picking a king of clubs from a deck and then, without replacing it, picking a king of hearts from the same deck. D

5. Throwing a 5 with a die and then throwing a 4 on the second throw of the same die. I

Exercises

A Find the probability for the spinner and for the cube numbered 1–6. Assume that the first outcome is for the spinner and the second is for the cube.

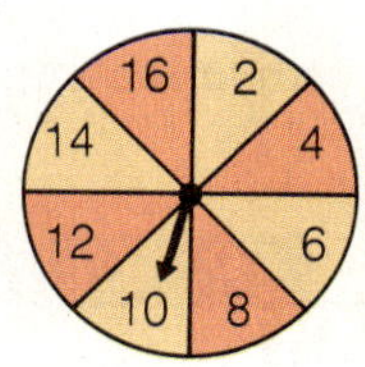

1. P(8 and 3) $\frac{1}{48}$

2. P(14 and odd) $\frac{1}{16}$

3. P(multiple of 4 and even) $\frac{1}{4}$

4. P(12 and a number <6) $\frac{5}{48}$

5. P(even and odd) $\frac{1}{2}$

6. P(a number >12 and a number >4) $\frac{1}{12}$

7. P(multiple of 5 and 5) $\frac{1}{48}$

8. P(multiple of 6 and a number <2) $\frac{1}{24}$

9. P(even and even) $\frac{1}{2}$

10. P(a number >10 and a number <6) $\frac{5}{16}$

11. P(odd and odd) 0

12. P(divisible by 7 and prime) $\frac{1}{16}$

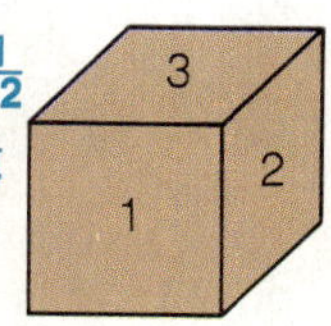

Suppose you have a standard deck of 52 playing cards. You select two cards at random without replacing the first. Find the probability of each.

13. P(king of spades then ace of hearts) $\frac{1}{2652}$
14. P(red card then black card) $\frac{13}{51}$
15. P(red card then a red card) $\frac{25}{102}$
16. P(ace then an ace) $\frac{1}{221}$
17. P(heart then a diamond) $\frac{13}{204}$
18. P(spade then a spade) $\frac{1}{17}$
19. P(a 4 then a queen) $\frac{4}{663}$
20. P(4 of hearts then 3 of hearts) $\frac{1}{2652}$
21. P(a king then a jack) $\frac{4}{663}$
22. P(queen then queen) $\frac{1}{221}$
23. P(heart then another heart) $\frac{1}{17}$
24. P(club then a diamond) $\frac{13}{204}$

B Solve.

You have a drawer with 3 white shirts, 4 gray shirts, 5 green shirts, and 4 blue shirts. You reach in and select three shirts at random without replacing them. Find each probability.

25. P(white then gray then green) $\frac{1}{56}$
26. P(blue then green then gray) $\frac{1}{42}$
27. P(green then green then blue) $\frac{1}{42}$
28. P(blue then blue then blue) $\frac{1}{140}$

Suppose you replace each shirt each time in the situation above. Find each probability.

29. P(white then blue then green) $\frac{15}{1024}$
30. P(gray then blue then green) $\frac{5}{256}$
31. P(green then blue then gray) $\frac{5}{256}$
32. P(gray then gray then gray) $\frac{1}{64}$

C Extending Thinking Skills

Thinking skills: 33–36. Generalizing

The probability that person A will pass gym class is $\frac{5}{6}$, that B will pass is $\frac{3}{4}$, and that C will pass is $\frac{2}{3}$. If the probability that A will pass gym is $\frac{5}{6}$, then the probability A will not pass gym is $\frac{1}{6}$. Find each probability.

33. all 3 will *not* pass the course $\frac{1}{72}$
34. A and C will pass but B will not $\frac{5}{36}$
35. A and B will pass but C will not $\frac{5}{24}$
36. C will pass but A and B will not pass $\frac{1}{36}$

Mixed Review

Write as a percent. **37.** $\frac{3}{5}$ 60% **38.** $1\frac{3}{8}$ 137.5% **39.** $\frac{9}{15}$ 60% **40.** $\frac{11}{4}$ 275%

Give the greatest common factor (GCF). **41.** 6, 24, 51 3

CALCULATOR ACTIVITY

Use a calculator to help you give each probability as a percent. Round to the nearest whole percent.

1. Of 60 previous days with atmospheric conditions like today's, it has rained on 35. What is the probability of rain today? 58%

2. Of 150 previous days with atmospheric conditions like today's, it has rained on 20. What is the probability of rain today? 13%

PROBLEM SOLVING: WRITING EQUATIONS

13-6 Practice Solving Problems

Objective: To write and solve equations to solve word problems.
MIN: 1–5; REG: 1–5 even, 6–8; MAX: 1–5 even, 6–10; ALL: Mixed Review

Exercises

A Solve by writing an equation.

1. A bricklayer makes \$15.75 an hour. She makes 3 times as much per hour as she did when she started. What was her starting hourly wage?
$3s = 15.75$ \$5.25

2. Fred's pay was 4 times his assistant's. Together, they earned \$675 for building a greenhouse. How much did each earn?

2. $a + 4a = 675$
assistant \$135
Fred \$540

3. A shovel was priced at \$3.74 less than a saw. A carpenter bought one shovel and one saw and paid \$24.50. What was the price of the shovel? $s + (s + 3.74) = 24.50$ \$10.38

4. The length of a garage is 3 times the width. The perimeter is 144 ft. What is the length and width?
$2w + 2(3w) = 144$ $w = 18$ ft, $l = 54$ ft

5. Students raised money to visit a museum. This year, they earned \$50 more than twice the amount they raised last year. This year they collected \$1000. How much more did they collect this year than last year? $50 + 2y = 1000$ \$475

B

6. Three boys saved a total of \$75 for carpentry tools. Tim saved \$10 more than Kyle, and Jeff saved \$5 more than Kyle. How much did each boy save?
$k + (k + 10) + (k + 5) = 75$ Kyle \$20, Tim \$30, Jeff \$25

7. The area of the floor in a closet is 36 ft^2. The length of the closet is 4 times the width. What is the length and width of the closet? $w(4w) = 36$ $w = 3$ ft, $l = 12$ ft

8. A worker repaired a fireplace. Since he did the job on a Saturday, his fee was \$20 plus twice his normal hourly wage. He did the job in 5 hours and charged \$87.50. What was his regular hourly wage? $5(2r) + 20 = 87.50$ \$6.75

C Extending Thinking Skills Answers will vary

9. Write a word problem that could be solved by the equation $2x - 5 = 43$.

10. Write a word problem that could be solved by the equation $x + 3x = 58$.

Thinking skills: 9, 10. Formulating a problem

Mixed Review

Write <, >, or = for each □. **11.** -0.633 [<] -0.632 **12.** -0.4 [<] 0.5

Solve and check. **13.** $0.45c = 1.8$ 4 **14.** $t - 0.6t = 3.6$ 9

13-7 Frequency Tables, Range, and Mode

Objective: To make a frequency table for a set of data; to find range and mode.
MIN: 1–12; REG: 1–12 even, 13; MAX: 1–12 even, 13–16; ALL: Mixed Review

Some students wanted to know the most common starting salary paid by restaurant chains in their city. They surveyed 33 restaurants and recorded their data in this **frequency table**.

Hourly Wage	Tally	Frequency
\$3.35	𝍸	5
\$3.50	𝍸 𝍸 I	11
\$3.75	𝍸 I	6
\$4.00	𝍸 II	7
\$4.10	IIII	4

The **range** of a set of data is the difference between the greatest and the least numbers in the set.

$$\begin{array}{r l} \$4.10 & \leftarrow \text{greatest hourly wage} \\ -\ 3.35 & \leftarrow \text{least hourly wage} \\ \hline \$0.75 & \leftarrow \text{range} \end{array}$$

The **mode** is the number or item that appears most frequently. Eleven restaurants pay \$3.50 an hour, so the mode is \$3.50.

Example

Make a frequency table for the data. Find the range and mode. Student admission prices to 20 movie theaters: \$3, \$2.75, \$3, \$3.50, \$2.25, \$2.75, \$3, \$4, \$3.50, \$3, \$3, \$3.50, \$2.25, \$3.50, \$3, \$3, \$2.25, \$3, \$4, \$3.

Solution

Ticket Price	Tally	Frequency
\$2.25	III	3
\$2.75	II	2
\$3.00	𝍸 IIII	9
\$3.50	IIII	4
\$4.00	II	2

Range = \$4.00 − \$2.25 = \$1.75
Greatest price is \$4; lowest price is \$2.25.

Mode = \$3.00
\$3 occurs most frequently in the table.

Practice Make a frequency table for the data. Find the range and mode.
Ages of girls on a soccer team: $16\frac{1}{2}$, 17, 15, 16, $15\frac{1}{2}$, $16\frac{1}{2}$, $17\frac{1}{2}$, 17, $15\frac{1}{2}$, $16\frac{1}{2}$, $16\frac{1}{2}$, 18, $16\frac{1}{2}$, $17\frac{1}{2}$, $16\frac{1}{2}$, $16\frac{1}{2}$, 16, 16, 17, 16, $16\frac{1}{2}$. **range = 3, mode = $16\frac{1}{2}$**

Oral Exercises

In a survey, English students asked people to try to say five tongue twisters. They made a frequency table showing which ones people were able to say.

Tongue Twister	Tally	Frequency
1. The sinking steamer sunk.	𝍸 𝍸 𝍸 𝍸 𝍸	
2. Rubber baby buggy bumpers	𝍸 𝍸 𝍸 𝍸 𝍸 𝍸 𝍸 I	
3. Toy boat (repeat 6 times)	𝍸 𝍸 𝍸 𝍸 𝍸 𝍸 𝍸 𝍸 I	
4. The sixth sick sheik's sixth sheep's sick.	𝍸 𝍸 III	
5. She sells seashells by the seashore.	𝍸 𝍸 𝍸 𝍸 𝍸 𝍸 𝍸 𝍸 𝍸 II	

1. How many people were able to say each tongue twister? **1:25, 2:36, 3:41, 4:13, 5:47**

2. Which tongue twister could the most people say? **#5**

3. Which tongue twister could the fewest people say? **#4**

Exercises **See Solutions Manual for frequency tables.**

A Make a frequency table for each set of data. Find the range and mode.

1. Runs scored in baseball games: 1, 1, 1, 1, 1, 2, 2, 2, 2, 2, 2, 2, 3, 3, 3, 4, 4, 4, 4, 5, 5, 5, 6, 6, 7, 7, 7, 7, 8, 8, 8, 9, 9, 10. **Range 9, mode 2**

2. High temperature, in degrees Celsius: 25, 25, 26, 26, 26, 27, 29, 29, 30, 31, 31, 31, 31, 31, 31, 32, 32, 32, 33, 33, 34, 35, 35. **Range 10, mode 31**

3. Grades on a math test: 72, 73, 73, 73, 73, 73, 73, 76, 76, 80, 80, 84, 84, 84, 84, 84, 86, 88, 88, 88, 90, 90, 98, 100. **Range 28, mode 73**

4. Games won: 45, 46, 46, 46, 46, 46, 54, 62, 62, 62, 62, 63, 63, 65, 65, 65, 65, 70, 70, 70, 70, 70, 70, 76, 79. **Range 34, mode 70**

5. Miles run: 14, 14, 14, 16, 17, 17, 17, 17, 20, 20, 20, 20, 20, 34, 34, 34. **Range 20, mode 20**

6. Ages of airplane pilots: 37, 37, 38, 38, 38, 39, 39, 40, 40, 40, 42, 43, 43, 43, 45, 45, 50, 50, 50, 52, 52, 52, 52, 52, 52, 52, 53, 55. **Range 18, mode 52**

7. Chalk length in mm: 39, 76, 27, 98, 18, 52, 21, 76, 45, 98, 19, 76, 93, 76. **Range 80, mode 76**

8. High school seniors in several schools: 175, 210, 124, 150, 275, 110, 210, 325, 265, 173, 423, 212, 314, 226, 239, 219. **Range 313, mode 210**

9. Percent correct on a test: 65%, 45%, 58%, 82%, 90%, 76%, 82%, 97%, 69%. **Range 52, mode 82%**

10. Students on the honor roll: 65, 43, 67, 56, 80, 54, 43, 72, 54, 47, 43, 64, 80. **Range 37, mode 43**

11. People on a waterslide each hour: 35, 45, 24, 34, 50, 32, 19, 28, 34, 17, 28, 27, 36, 41, 16, 29, 24, 35, 32, 41, 36, 35, 29, 34, 27, 38, 42, 37, 28, 28, 34, 44. **Range 34, mode 28 and 34**

12. Skiers on the slopes: 42, 27, 16, 15, 22, 23, 34, 17, 28, 33, 30, 34, 41, 26, 22, 30, 18, 35, 31, 21, 23, 22, 40, 36, 37, 34, 28, 33, 2, 22, 18, 36, 23, 16, 22, 34, 27, 42, 17, 22, 40, 12, 22, 32, 19, 25. **Range 40, mode 22**

B

13. Make a frequency table for the data. Use intervals of \$2, with \$1.00–2.99 as the first interval. Find the range and mode. **Range: \$5 Mode: \$1–\$2.99**

Videocassette rental fees of different stores: \$4.50, \$1.99, \$5.75, \$1, \$1.50, \$5, \$1, \$2, \$3, \$3.80, \$2, \$2.75, \$5, \$4.25, \$5, \$2.50, \$3.75, \$2, \$3, \$2, \$2.75, \$2.60, \$2, \$3.25, \$4, \$3.10, \$1.69, \$3, \$2, \$6, \$3.25, \$2, \$3.50, \$4.75, \$4, \$2.50.

C *Extending Thinking Skills* **Thinking skills: 14–16. Using logic**

14. Add two numbers so that the mode changes. 18, 18, 18, 17, 16, 16, 15, 14 **16, 16**

15. Add one of the numbers 20 through 24 so that the mode does not change. 24, 24, 24, 23, 23, 22, 22, 21, 21, 20, 20 **24**

16. Add two numbers so the range becomes 8. 9, 8, 8, 7, 6, 6, 6, 5, 3, 3, 3, 3, 2, 2
10 and any number 2–10. or 1 and any number 1–9.

Mixed Review

Solve and check. **17.** $9 < 2c + 1$ **$c > 4$** **18.** $4.5y = -18$ **−4** **19.** $-6t = 3$ **−0.5**
20. $a \div 4 = 9$ **36** **21.** $11y + 16 = -17$ **−3** **22.** $12m = 10m + 12$ **6**

ESTIMATION

Find the range and mode for the data shown in the graph. Estimate the frequency of each response. **Range ≈ 14, mode = 4-year college**

20 13 16 9 6

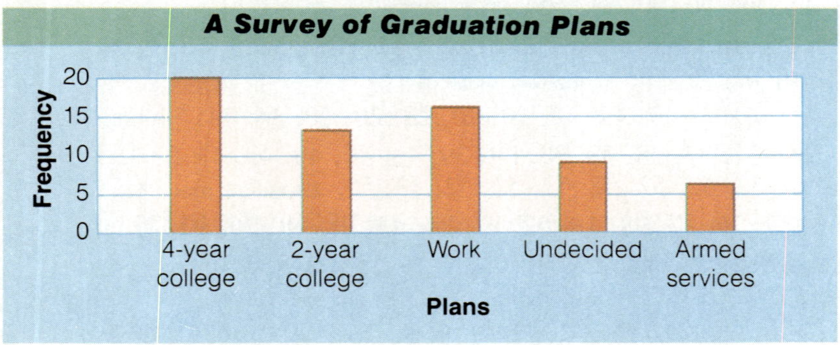

13-8 Mean and Median

Objective: To find the mean and median of a set of data.
MIN: 1–22; 25–26; REG: 1–22 even, 23–26; MAX: 1–22 even, 23–26; ALL: Mixed Review

In section 13-7, you learned how you find the mode. Two other measures used to determine which number in a list of numbers is "most typical" are **mean** and **median**. To find the **mean** or **average** of a list of numbers, find the total and divide by the number of items. The **median** is the middle number in a list of numbers arranged in order. If there are two middle numbers, the median is the mean of the two middle numbers.

Advertising by Restaurant Chains	
Restaurant	**Amount Spent in 3 Months**
A	\$45 million
B	\$20 million
C	\$15 million
D	\$11 million
E	\$11 million
F	\$ 6 million

Example

Find the mean and median for the numbers in the chart above.

Solution

$$\text{Mean} = \frac{45 + 20 + 15 + 11 + 11 + 6}{6}$$ Find the sum of the 6 numbers and divide by 6.

$$= \frac{108}{6}$$

$$\text{Mean} = 18$$ The mean is 18 million dollars.

45 20 15 11 11 6 — List the numbers in order.

$$\text{Median} = \frac{15 + 11}{2}$$ The median is the mean of the two middle numbers.

$$\text{Median} = \frac{26}{2} = 13$$ The median is 13 million dollars.

Practice Find the mean and median. Round to the nearest tenth if necessary. 11, 14, 12, 12, 16, 18, 10, 20, 22 **mean = 15 median = 14**

Oral Exercises

Tell whether each is true or false.

1. The mean and median are always the same. F
2. An equal number of numbers in a list are above and below the median. T
3. The mean is always one of the numbers in a list of numbers. F
4. The median is always one of the numbers in a list of numbers. F
5. Mean and average are the same. T
6. Items should be in order to find the mean. F
7. Items should be in order to find the median. T
8. If there is 1 middle number in a list, it is the median. T
9. To find the mean, find the total and divide by the number of items. T

Exercises

A Find the mean and median for each set of numbers. Round to the nearest tenth if necessary.

1. 4, 2, 3, 4, 6, 8, 2 mean 4.1, median 4
2. 10, 14, 9, 12, 14, 13, 8 mean 11.4, median 12
3. 23, 25, 20, 18, 26, 28, 22, 23 mean 23.1, median 23
4. 24, 35, 20, 16, 45, 50 mean 31.7, median 29.5
5. 42, 60, 42, 42, 60, 40, 70, 45, 45 mean 49.6, median 45
6. 12, 56, 43, 34, 58, 62, 71 mean 48, median 56
7. 12, 8, 8, 24, 15, 3, 4, 9, 10, 16 mean 10.9, median 9.5
8. 100, 125, 145, 132, 150, 143 mean 132.5, median 137.5
9. 6, 8, 2, 6, 8, 10, 8, 2, 4, 8, 8, 2, 10, 2, 8, 4, 6, 6, 8, 10 mean 6.3, median 7
10. 20, 40, 50, 10, 20, 30, 30, 10, 10, 30, 40, 50, 70, 40, 30, 20 mean 31.25, median 30
11. 125, 250, 225, 75, 75, 200, 275, 175, 100, 200, 125, 250, 150 mean 171.2, median 175
12. 500, 250, 250, 750, 200, 350, 400, 150, 600, 350, 100, 200 mean 341.7, median 300

Find the mean and median. Round answers to the nearest hundredth if necessary.

13. 2.6, 3.8, 7, 4.5, 4.6, 7.8, 5.1, 8.1, 4.6, 5.6, 6.2 mean 5.45, median 5.1
14. 7.5, 1.25, 6, 8.25, 11, 12.2, 14, 6.5, 14 mean 8.97, median 8.25
15. 0.8, 0.25, 0.006, 0.26, 0.08, 0.045, 0.007 mean 0.21, median 0.08
16. 5, 3, 6, 4, 7, 8, 9, 10, 3, 18, 24, 26, 30, 14, 14, 14, 25, 36 mean 14.22, median 12
17. 2.8, 3.5, 4.07, 5.009, 4, 3.2, 2.25, 3.86, 1.4 mean 3.34, median 3.5
18. 273, 485, 233, 225, 486, 387, 250, 239, 428 mean 334, median 273

Solve. Round answers to the nearest hundredth if necessary.

19. Jared made \$2, \$5, \$4, \$5, \$1, and \$3 by doing odd jobs. What was the average amount he received? **\$3.33**

20. The Falcons scored the following numbers of runs in their games last week: 6, 4, 1, 8, 2, 5, 4. Find the mean and median number of runs scored. **mean 4.3, median 4**

21. Carrie bowled 5 games. Her scores were 95, 110, 88, 102, and 130. What was her average score? **105**

22. The math teachers at Jefferson School are 24, 23, 35, 40, 42, 50, 32, and 36 years old. What is the median age for these teachers? **35.5**

B

23. Restaurants in Middletown paid different starting hourly wages. A new restaurant opened and wanted to offer a starting wage greater than the median wage paid by others. What would be the lowest wage the new restaurant could offer? Wages: \$3.65, \$4.50, \$4.25, \$5.00, \$3.65, \$4.75 **\$4.38**

24. In 29 at-bats in a World Series, Roberto Clemente had 7 singles, 2 doubles, 1 triple, and 2 home runs. What was his batting average for the World Series? Batting average is the number of hits divided by the times at bat. Round the decimal to the thousandths place. **0.414**

C *Extending Thinking Skills* Thinking skills: 25. Estimating 26. Testing a conjecture

25. Estimate the mean for the following numbers mentally. 87, 195, 213, 230, 298, 305, 479, 550 **≈300**

26. Derek's score on each math exam increased by 5 points from the previous exam. The mean for the 5 exams he took was 70. What were his exam scores? **60, 65, 70, 75, 80**

Mixed Review

What percent of 24 is: **27.** 16? **$66\frac{2}{3}$%** **28.** 9? **37.5%** **29.** 12? **50%** **30.** 21? **87.5%**

Evaluate for $x = 3$, $y = 4$. **31.** $\frac{x}{y}$ **$\frac{3}{4}$** **32.** $\frac{2x}{3y}$ **$\frac{1}{2}$** **33.** $\frac{x}{3} + \frac{y}{2}$ **3** **34.** $\frac{1 - x}{1 - y}$ **$\frac{2}{3}$**

Solve and check. **35.** $12 = 3m + 15$ **−1** **36.** $m - 0.3m = 2.1$ **3**

Write <, >, or = for each □. **37.** $\frac{2}{3} \square \frac{18}{25}$ **<** **38.** $-\frac{5}{16} \square -\frac{3}{16}$ **<** **39.** $\frac{9}{4} \square \frac{53}{24}$ **>**

40. $1.106 \square 1.016$ **>** **41.** $1 - 0.73 \square 0.27$ **=** **42.** $-0.62 \square -0.63$ **>**

PROBLEM SOLVING: APPLICATIONS

13-9 Managing a Computer Store

Objective: To solve applied problems using information from graphs.
MIN: 1–15; REG: 1–15; MAX: 1–15

A manager of a computer store keeps data on sales of computers and of computer materials. The data are often shown in graphs.

Schools in City with at Least One Computer

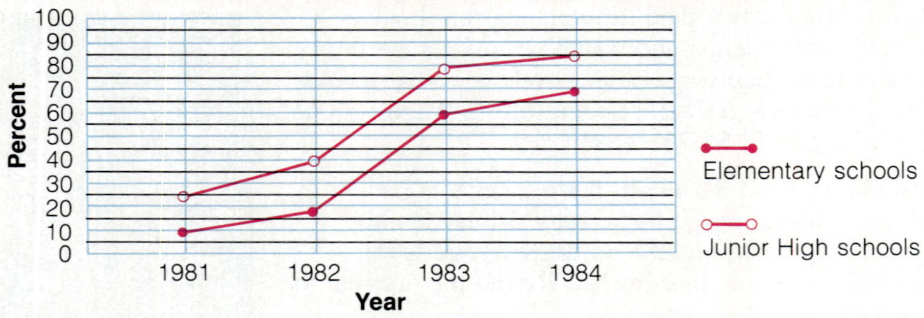

Problems

You might wish to have students choose from among the techniques pencil and paper, mental math, estimation, and calculator. Have them choose the most appropriate technique for each problem.

Use the graphs to solve.

1. What percentage more junior high schools than elementary schools had at least 1 computer in 1983? **20%**

2. By how many percentage points did the elementary schools having at least 1 computer increase from 1981 to 1984? **60%**

3. By how many percentage points did the junior high schools having at least 1 computer increase from 1981 to 1984? **60%**

4. What was the mean percentage of elementary and junior high schools having at least 1 computer in 1981? **17.5%**

5. What was the mean percentage of elementary and junior high schools having at least 1 computer in 1984? **77.5%**

6. How many fewer computers costing less than $500 were sold in 1984 than were sold in 1983? **8**

7. How many more computers costing $500 to $1000 than computers costing more than $1000 were sold in 1984? **20**

8. What was the total number of computers sold in 1983? In 1984? **1983, 98; 1984, 101**

9. What was the mean number of computers sold in 1983? What was the mean number sold in 1984? **1983, 32.67 1984, 33.67**

10. What was the approximate percentage of increase in sales of computers costing more than $1000 from 1983 to 1984? **40%**

11. What was the approximate percentage of decrease in sales of computers costing less than $500 from 1983 to 1984? **25%**

12. How much more money was made on game software than on business software? **$150**

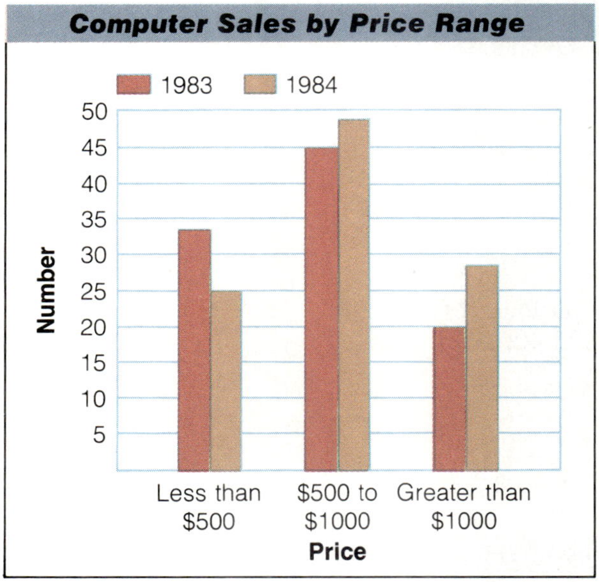

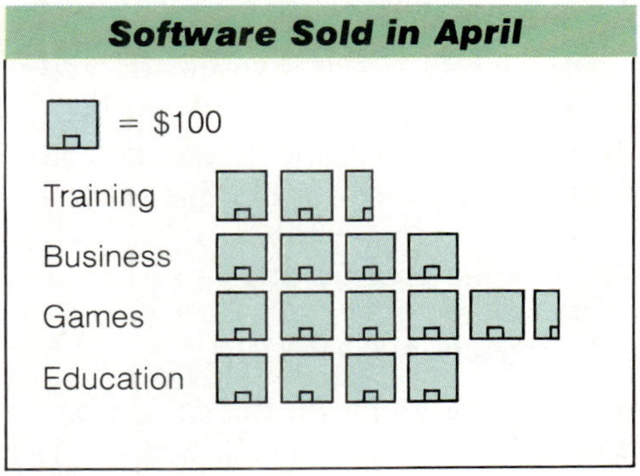

13. How much less money was made on training software than on education software? **$150**

14. What were the average sales for computer software in April? **$400**

15. ***Data Search*** Find the cost of 5 different computers in a computer store in your city. Find the average cost of the computers. **Answers will vary.**

What's Your Decision?

Your school wants to buy 20 new computers. The regular price at two stores is the same, $1500 per computer. Store A will give the school 5 computers at no charge if it buys 15 at the regular price. Store B will give the school 15% off the cost of each computer. A 5% sales tax must be paid at either store. Where should your school buy the computers?

Store A ($22,500 + tax).
Store B would be $25,500 + tax.

13-10 Making Bar Graphs and Line Graphs

Objective: To make bar graphs and line graphs.
MIN: 1–4, 6; REG: 1–6; MAX: 1–6; ALL: Mixed Review

Bar graphs and line graphs are useful for comparing data. You can use these steps to make bar graphs and line graphs.

1. Determine the scale.
   - Find the greatest and least values for the data.
   - Select a scale to fit your data.
2. Draw and label the horizontal and vertical sides of the graph.
3. Plot the points for the line graph or draw the bars for the bar graph.
4. Give the graph a title.

Example 1

Make a line graph for the data. Room rates at Pines Motel: 1940, \$4.75; 1950, \$8.75; 1960, \$13.40; 1970, \$25.40; 1980, \$39.75.

Solution Greatest amount = \$39.75
Least amount = \$4.75
Let the scale = \$5 per mark

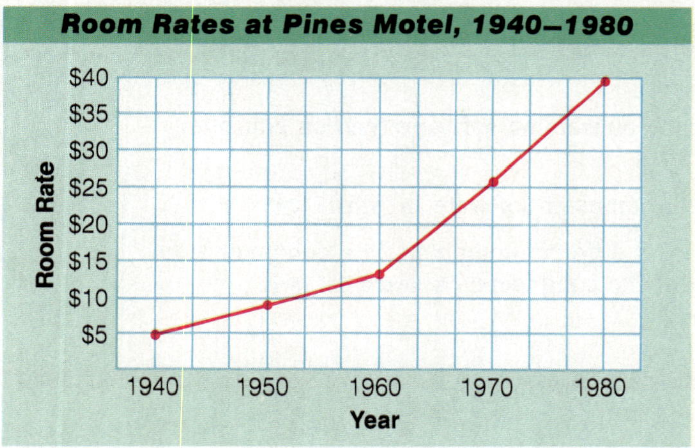

Practice Make a line graph for the data. Median heights of boys: age 11, 142 cm; age 12, 150 cm; age 13, 158 cm; age 14, 163 cm; age 15, 169 cm; age 16, 176 cm; age 17, 177 cm. **See Solutions Manual.**

Example 2

Make a bar graph for the data. Airline tickets sold on one day: Northeast, 9; Southeast, 3; Midwest, 6; Southwest, 4; Rockies, 3; West Coast, 7.

Solution Greatest number = 9

Least number = 3

Let the scale = 1 ticket per mark

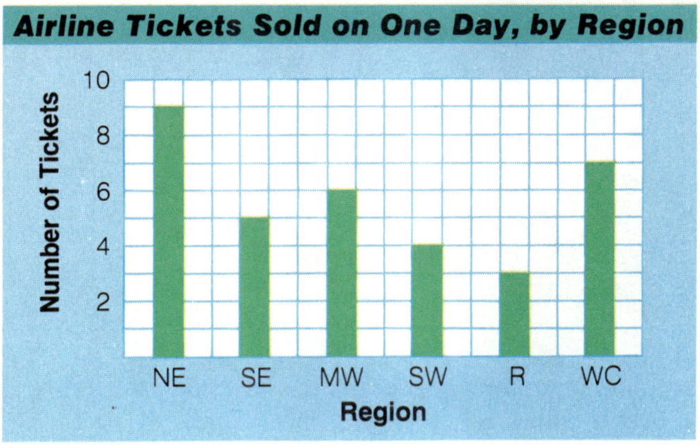

Practice Make a bar graph for the data. Take-offs and landings at busy airports each year: Chicago, 600,000; Atlanta, 580,000; Los Angeles, 480,000; Denver, 475,000; Dallas, 440,000. **See Solutions Manual.**

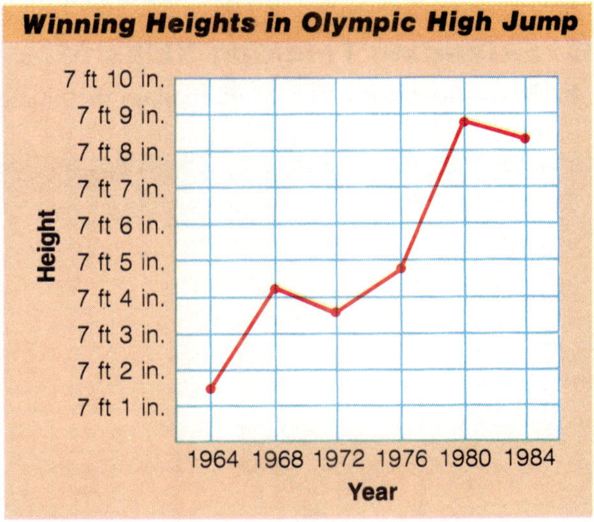

Oral Exercises

Use the graph to answer the questions.

1. How many inches does each mark on the vertical scale represent? **1**

2. How many years does each mark on the horizontal scale represent? **4**

3. In what year was the highest jump? The lowest jump? **1980, 1964**

4. Between which years did the winning height drop? **1968–'72 and 1980–'84**

5. Between which years was the change in the winning height greatest? **1976–1980**

Exercises See Solutions Manual.

A Make a bar graph for the data in Exercises 1 and 2.

1. Number of students using a computer lab at lunchtime. Monday, 7; Tuesday, 6; Wednesday, 4; Thursday, 5; Friday, 5. Let each mark on the scale represent 1 student.

2. Average wind velocities, in miles per hour: Boston, 12.5; Chicago, 10.3; Honolulu, 11.8; San Francisco, 10.5; Washington, D.C., 9.4. Let each mark on the scale represent 0.5 mi/h.

Make a line graph for the data in Exercises 3 and 4.

3. Average monthly rainfall in a town, in centimeters: January, 1.8; February, 2.2; March, 4.0; April, 7.6; May, 8.2; June, 5.4; July, 3.6; August, 2.8; September, 4.2; October, 4.2; November, 2.6; December, 2.0. Let each mark on the scale represent 0.4 cm.

4. Price of gold, per ounce: 1972, \$59; 1973, \$98; 1974, \$160; 1975, \$161; 1976, \$125; 1977, \$148; 1978, \$194; 1979, \$308; 1980, \$613; 1981, \$460; 1982, \$376. Let each mark on the scale represent \$50.

B See Solutions Manual.

5. Make a line graph for the data. Then make another line graph, using the same data but changing the scale so the graph is almost a straight line.

Leaded Regular Gasoline Prices, 1973–1982									
1973	1974	1975	1976	1977	1978	1979	1980	1981	1982
\$0.40	\$0.53	\$0.57	\$0.59	\$0.62	\$0.63	\$0.86	\$1.19	\$1.31	\$1.26

C Extending Thinking Skills Thinking skill: 6. Formulating problem

6. Make up a story that could explain the changes in the graph below. Answers will vary.

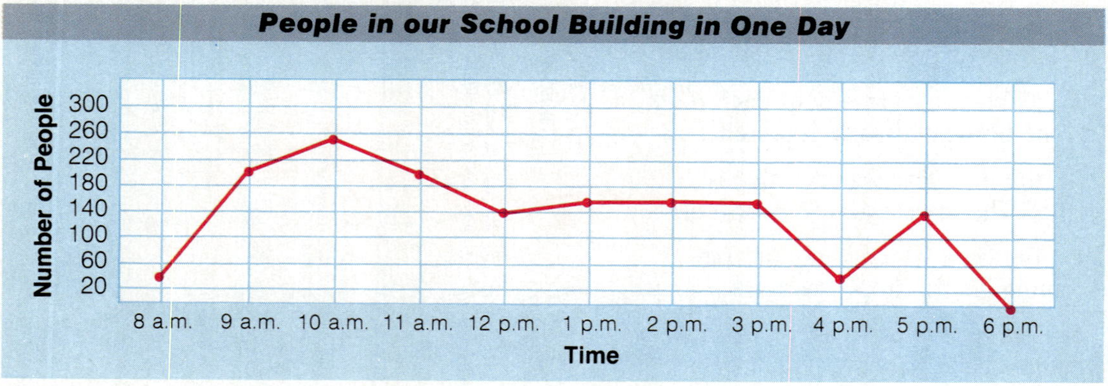

Mixed Review

Simplify. **7.** $16n + 3n - 9 - 10n$ **9n − 9** **8.** $2(a + 6) - 8$ **2a + 4**

Solve and check. **9.** $2t = 16 - 2t$ **4** **10.** $3c = c + 5$ **2.5** **11.** $-42 = 6y$ **−7**

12. $2m < 3m + 1$ **m > − 1** **13.** $t + 2t = 21$ **7** **14.** $-18 = 6 + 4n$ **−6**

15. $\left(\frac{3}{16}\right)y = -\frac{3}{4}$ **−4** **16.** $r - \frac{1}{2} = \frac{5}{16}$ $\frac{13}{16}$ **17.** $\frac{m}{6} = \frac{25}{15}$ **10** **18.** $\frac{2}{3} = z + \frac{4}{5}$ $-\frac{2}{15}$

A square has one side = 6 m. **19.** Give the perimeter. **24 m** **20.** Give the area. **36 m²**

21. Give the perimeter of a regular pentagon with one side = 7.5 cm. **37.5 cm**

22. Give the perimeter of an equilateral triangle with one side = 11 m. **33 m**

ESTIMATION

Estimate the winning times for this race for the next three years. **Answers will vary.**

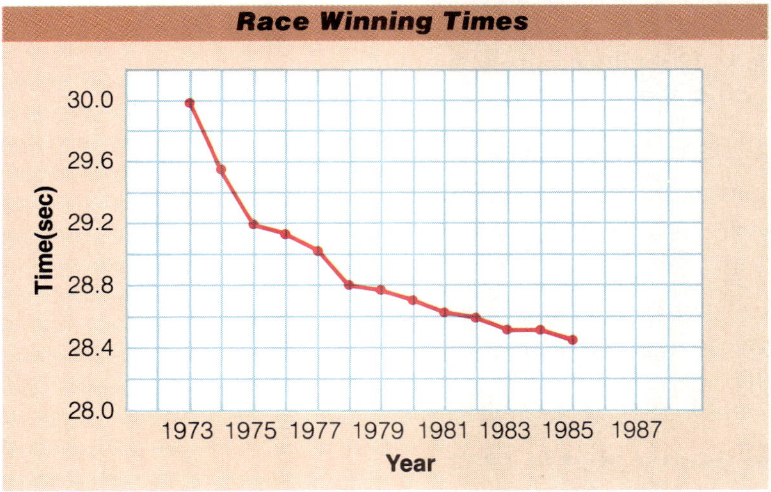

13-11 Making Pictographs

Objective: To make pictographs.
MIN: 1–5; REG: 1–5 even, 6–7; MAX: 1–5 even, 6–8; ALL: Mixed Review

You can follows these steps to make a pictograph.

1. Determine the scale:
   - Find the greatest and least values for the data.
   - Select a scale to fit the data.
2. Round the data; draw and label the pictures for the graph.
3. Show the scale for the graph.
4. Give the graph a title.

Example

Make a pictograph for the data. Coins made at the Denver Mint in 1982: half-dollars, 6 million; quarters, 120 million; dimes, 54 million; nickels, 19 million; pennies, 60 million.

Solution Greatest number = 120. Least number = 6.

Let ($) = 10 million coins and (half $) = 5 million coins.

Rounded data: Half-dollars, 5 — 6 is closer to 5 than to 10.
Quarters, 120
Dimes, 55 — 54 is closer to 55 than to 60.
Nickels, 20 — 19 is closer to 20 than to 15.
Pennies, 60

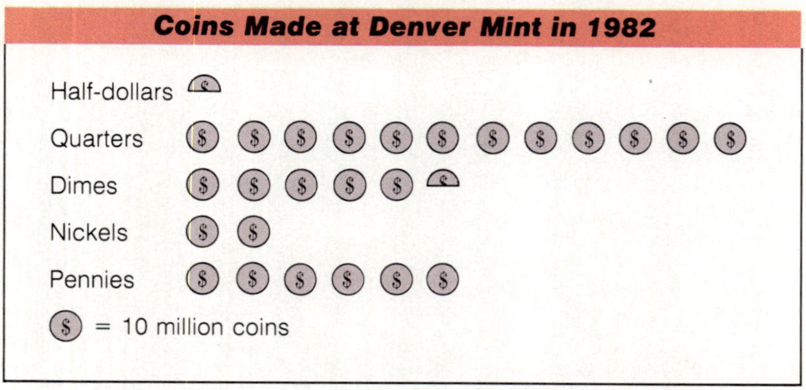

Practice Make a pictograph for the data. Favorite breakfast meals: cereal, 88; eggs, 39; pancakes, 129; toast, 138; fruit, 70. **See Solutions Manual.**

Oral Exercises

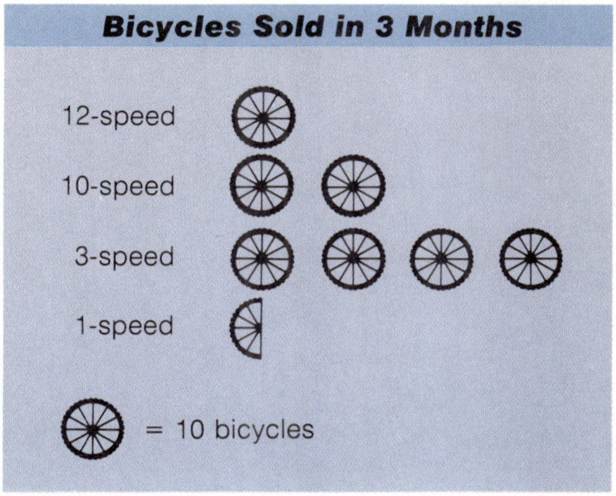

1. What does each wheel represent?
2. What does each half-wheel represent?
3. Which type of bicycle was sold most?
4. How many 3-speed bikes were sold?
5. How many 1-speed bikes were sold?
6. How many 10-speed bikes were sold?

1. 10 bicycles 2. 5 bicycles 3. 3-speed 4. 40 5. 5 6. 20

Exercises

A Make a pictograph for the data. **See Solutions Manual.**

1. Favorite sports at the Simmonsville Racquet Club: racquetball, 18 members; tennis, 32 members; handball, 12 members; squash, 10 members. Let each picture represent 4 people.

2. Campers on Labor Day weekend: Camp Pine, 125; Camp Timber, 75; Tent City, 150; Mountain Trees, 105. Let each picture represent 25 campers.

3. Favorite forms of exercise: running, 125 people; walking, 58 people; swimming, 29 people; bicycling, 46 people; exercise class, 15 people.

4. Populations of cities: New York, 7,072,000; Tokyo, 8,336,000; Mexico City, 9,191,000; Sydney, 3,281,000; London, 6,696,000; Bombay, 8,227,000; Nairobi, 1,048,000; Moscow, 7,831,000. Let each picture represent 1 million people.

5. Size (in acres) of National Parks in Alaska: Wrangell-St. Elias, 8,945,000; Gates of the Arctic, 7,500,000; Denali, 4,700,000; Katmai, 3,716,000; Glacier Bay, 3,225,000; Lake Clark, 2,875,000; Kobuk Valley, 1,750,000. Let each picture represent 1 million acres.

B

6. Use the graph on campers during Labor Day weekend (Exercise 2) to find the mean number of people at a camp. **113.75**

7. Use the graph on populations of cities (Exercise 4) to find the mean and median number of people living in the cities.
(If rounded to 500,000) mean ≈ 6,500,000
median ≈ 7,500,000

C Extending Thinking Skills Thinking skill: 8. Formulating a problem

8. Make up two different sets of data that, when rounded, would be shown by the graph below. Answers will vary.

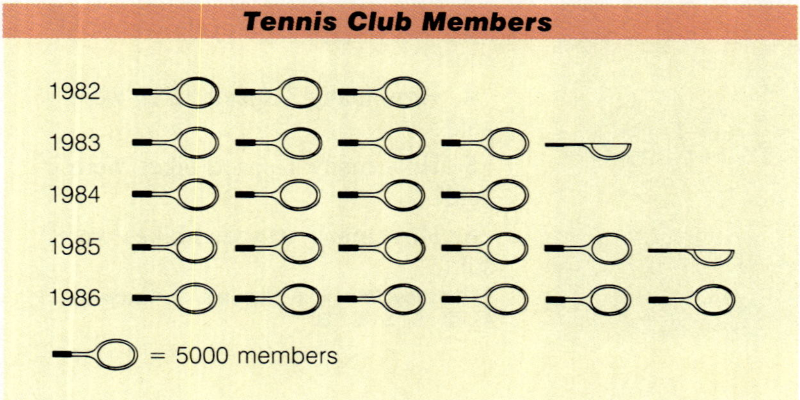

Mixed Review

Write <, >, or = for each □. **9.** −0.33 □ −0.32 (<) **10.** 9.01 □ 9.1 (<)

11. $3(n - 4)$ □ $3n - 3(4)$ (=) **12.** 6(2) □ −6(−2) (=) **13.** *ab* □ *ba* (=)

Solve and check. **14.** $9.6n + 1.4n = 27.5$ 2.5 **15.** $3.5c = 14.7$ 4.2

What percent of 64 is: **16.** 36? 56.25% **17.** 20? 31.25% **18.** 72? 112.5% **19.** 4? 6.25%

COMPUTER ACTIVITY

The program below will compute the mean temperature for a week (7 days). Line 30: Remind students to separate their inputs with a comma.

```
10 PRINT "MEAN TEMPERATURE FOR THE LAST WEEK"
20 PRINT "ENTER SEVEN TEMPERATURES."
30 INPUT T1,T2,T3,T4,T5,T6,T7
40 AV=(T1+T2+T3+T4+T5+T6+T7)/7
45 PRINT "THE MEAN TEMPERATURE WAS "; AV
50 IF AV>65 THEN GOTO 80
60 PRINT "PLEASE GET WARMER!"
70 GOTO 90
80 PRINT "GREAT WEATHER!"
90 END
```

1. Use the program to find the mean temperature for the last week. Answers will vary.
2. Change the program to find the mean temperature for 3 days.
3. Change the program to find the mean score on 5 math tests.

PROBLEM SOLVING: STRATEGIES

13-12 Practice Solving Problems

Objective: To solve nonroutine problems, using combinations of strategies.
MIN: 1–2; REG: 1–4; MAX: 1–8;

Most problems are solved by some combination of strategies. Notice how strategies are used in the example below.

Problem Amanda wants to lay out a patio in a design like the one shown below. She has 50 bricks to use. How many bricks should she place in the middle row to use the greatest number of bricks?

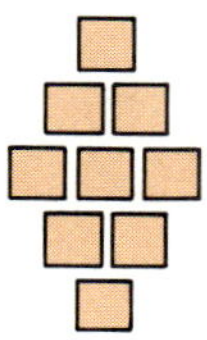

To solve, you can first **simplify the problem** to find the total bricks used if 1 brick is in the middle row, then the total with 2 bricks in the middle row, 3 bricks in the middle row, and so on. You can **draw a picture** to determine the number of bricks used each time. You can **make a table** to record this information, then **look for a pattern**.

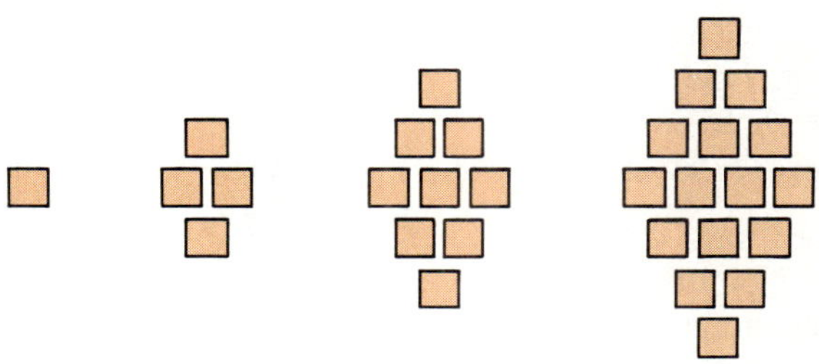

Number in Middle Row	1	2	3	4	5	6	7
Total Number of Bricks Used	1	4	9	16	25	36	49

The total number of bricks used is equal to the square of the number in the middle row. If there were 7 in the middle row, 49 would be used in all.

To use the greatest number of bricks, Amanda should place 7 bricks in the middle row.

This chart shows the strategies presented.

Problem-Solving Strategies	
Choose the Operations	Write an Equation
Guess, Check, Revise	Simplify the Problem
Draw a Picture	Make an Organized List
Make a Table	Use Logical Reasoning
Look for a Pattern	Work Backwards

Problems

Solve.

1. What is the greatest number of pieces into which a pizza can be cut with 5 straight cuts if the pieces cannot be stacked for cutting? **16**

2. A box holding 40 pencils weighs 135 g. The same box holding 20 of the same pencils weighs 75 g. What is the weight of the box? **15 g**

3. In how many different ways can a football team score 18 points? **14**

touchdown = 6 points　　field goal = 3 points
point after a touchdown = 1 point　　safety = 2 points

4. Three boys were playing a game. They decided to play 3 more rounds. The player with the most points at the end of 3 rounds would be the winner. For each round there would be one loser and two winners, and the loser would have to double the points each winner had by giving his own points away. Each boy won twice and lost once. The game ended in a tie, with each boy having 40 points. How many points did each boy have before the last 3 rounds?
Boy 1 : 20 points; Boy 2 : 65 points; Boy 3 : 35 points

5. Jane, Al, and Sally were selling school sweatshirts to raise money for a class trip. Al sold 1 less shirt than Jane, and Sally sold half as many shirts as Jane. Together the three sold 89 shirts. How many did each sell? **Jane: 36　Sally: 18　Al: 35**

6. Ms. Davies gave a math quiz with 20 questions. She gave 2 points for each correct answer and subtracted 3 points for each incorrect answer. Yolanda answered all 20 questions and got a score of 0. How many did she get right and how many did she get wrong? **12 right, 8 wrong**

7. A drink in a vending machine costs 50¢. The machine takes nickels, dimes, and quarters. How many different ways could you pay for a drink? **10**

8. A restaurant has 2 types of tables. One type seats 4 people and one type seats 6. One night, 114 people were eating at 24 tables in the restaurant, and no table with people had empty seats. How many tables of each type were used?
9 tables of 6, 15 tables of 4.

Enrichment

A Probability Experiment

George Buffon (1707–1788), a French biologist, created the *Buffon Needle Problem*. In Buffon's house, the joints between the floorboards made parallel lines. Buffon took a needle shorter than the width of a board and dropped it on the floor over and over again. Each time, he recorded whether the needle crossed a line or did not cross a line. His results are, oddly enough, related to the number π.

You can conduct a similar experiment. On a large sheet of paper, draw 5 parallel lines 3 cm apart. Drop a piece of toothpick 2 cm long on the paper from a height of about 50 cm, giving it a slight twist as you drop it. Record whether the toothpick lands across a line or between the lines. Repeat this 50 times.

Students may combine results with a classmate's to have a greater number of trials.

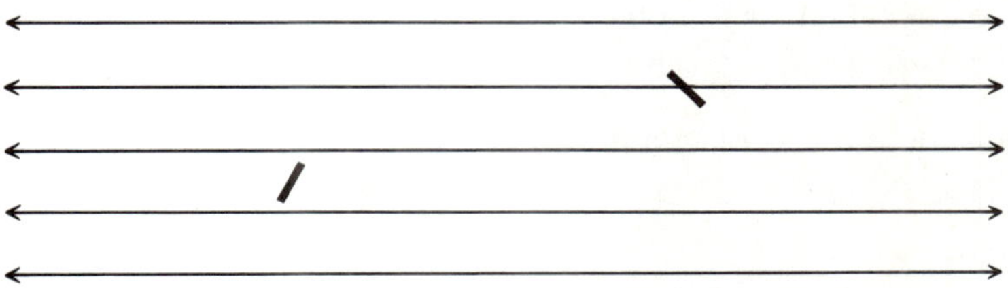

Let l represent the length of the toothpick, b the distance between lines, and P the probability the toothpick will fall across a line (number of times the toothpick fell across a line divided by the number of tosses). Buffon concluded that the following formula gives a formula for π.

$$\pi = \frac{2l}{bP}$$

Using the length of the toothpick and the distance between lines given above,

$$\pi = \frac{2l}{bP} = \frac{2 \cdot 2}{3 \cdot P} = \frac{4}{3P}$$

1. Conduct this experiment and calculate a value for π. How close did you come to 3.14, an approximate value for π? **Answers will vary.**

2. Suppose the toothpick fell across a line on 21 out of 50 tosses. How close is the value to the true value of π? **3.17; close**

3. Change the distance between lines to 9 cm and the length of the toothpick to 4 cm and repeat this experiment. How do the results compare to the first experiment? **Answers will vary.**

Chapter 13 Review

13-1 Use the counting principle to find the total number of outcomes.

1. Selection: 1 size and 1 flavor 6
Sizes: 6 ounces, 12 ounces
Flavors: strawberry, peach, apple

13-2 Find the number of permutations.

2. f, l, a, v, o, r 720

3. 2, 4, 6, 8, 10 120

13-3 Find the number of combinations.

4. 7 people try out for a part in a play. How many ways can the actor and an understudy be chosen? 42

5. Select 2 records from a rack of 6 records. 15

13-4 There are 14 names in a hat. 8 are boys' names and 6 are girls' names. You select one name without looking. Find each probability.

6. $P(\text{boy})$ $\frac{8}{14} = \frac{4}{7}$

7. $P(\text{not boy})$ $\frac{6}{14} = \frac{3}{7}$

8. $P(\text{boy or girl})$ $\frac{14}{14} = 1$

13-5 Find the probability of each for one toss of two dice.

9. $P(1 \text{ and } 6)$ $\frac{1}{36}$

10. $P(3 \text{ and odd})$ $\frac{1}{12}$

13-6 Solve by writing an equation.

11. Carla paid a total of $34.88 for a scarf and a pair of gloves. The scarf cost $2.90 less than the gloves. What was the cost of the scarf? $g + (g - 2.90) = 34.88$; $15.99

13-7–13-8 Use the data to solve problems 12–15. Plant height in cm: 16, 18, 22, 22, 23, 25.

12. Find the range. 9

13. Find the mode. 22

14. Find the median. 22

15. Find the mean. 21

13-9 **16.** Make a line graph for the data. Library books loaned: Monday, 12; Tuesday, 8; Wednesday, 25; Thursday, 18; Friday, 21. See Solutions Manual.

13-10 Use the bar graph to solve.

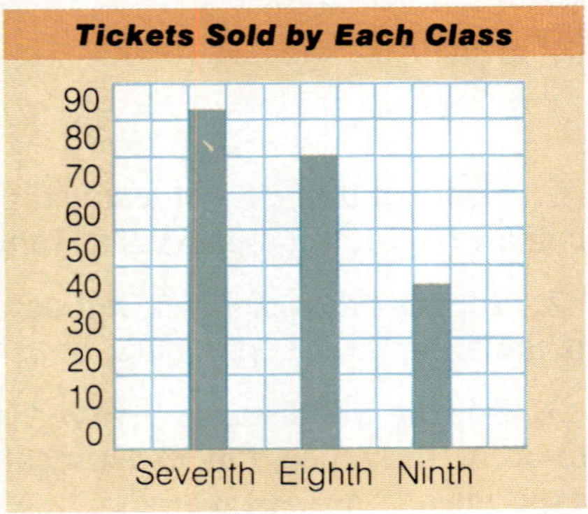

17. How many tickets did the seventh grade class sell? about 92

18. Which class sold the most tickets? 7th

19. Which class sold about 48 tickets? 9th

13-11 Make a pictograph for the data.

20. Trees on schoolgrounds: maple, 9; birch, 15; pine, 6; oak, 12. See Solutions Manual.

Chapter 13 Test

Use the counting principle to find the total number of outcomes.

1. Selection: 1 date and 1 show time 18

Dates: June 10, June 11, June 17, June 18, June 24, June 25

Times: 6:00 p.m., 8:00 p.m., 10:00 p.m.

Use a formula to find the number of permutations.

2. c, o, m, p, u, t, e, r 40,320

3. left, right, forward, back 24

Use a formula to find the number of combinations.

4. 6 dogs are in a contest. How many ways can the winner and runner-up be chosen? 30

5. Select 3 kites from a display of 8 kites. 56

There are 10 tennis balls in a bag. 2 are green, 3 are white, and 5 are yellow. You select one ball from the bag without looking. Find each probability.

6. P(yellow) $\frac{5}{10} = \frac{1}{2}$

7. P(not green) $\frac{8}{10} = \frac{4}{5}$

8. P(green or white) $\frac{5}{10} = \frac{1}{2}$

Find the probability of each for one toss of two dice.

9. P(2 and 5) $\frac{1}{36}$

10. P(a number $>$ 5 and an even number) $\frac{1}{12}$

Solve by writing an equation. **11.** The length of a room is 4 times its width. The perimeter is 155 ft. Find the dimensions of the room. $2(4w) + 2w = 155$; $l = 62$ ft $w = 15.5$ ft

Use the data to solve problems 12–15.

Team members' ages: 20, 20, 20, 22, 23, 25, 25, 26, 26.

12. Find the range. 6

13. What is the mode? 20

14. What is the median? 23

15. Find the mean. 23

16. Make a bar graph for the data. Mountain heights in feet: McKinley, 20,320; Whitney, 14,494; Mauna Kea, 13,796; Granite Peak, 12,799; Rainier, 14,410. See Solutions Manual.

Use the line graph to answer each question.

17. During which month were Jean's scores highest? May

18. Between which months did Jean's scores drop? May–June

19. Between which months was the change in Jean's score the greatest? April–May

20. Make a pictograph for the data. Records sold: Monday, 12; Tuesday, 20; Wednesday, 18; Thursday, 26; Friday, 22. See Solutions Manual.

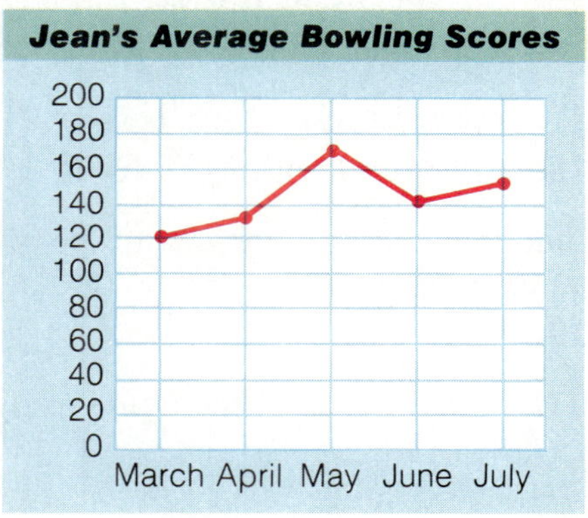

Jean's Average Bowling Scores

Cumulative Review

Find the discount and the sale price. Round to the nearest cent if necessary.

1. Regular price = \$60
Discount = 30%
\$18, \$42

2. Regular price = \$1200
Discount = 9%
\$108, \$1092

Find the commission. Round to the nearest cent if necessary.

3. Sale price = \$66
Commission = 4%
\$2.64

4. Sale price = \$90
Commission = 8%
\$7.20

Find the interest and the total amount. Round to the nearest cent if necessary.

5. \$750 at 15% per year for 3 years \$337.50, \$1087.50

6. \$600 at 2% per year for 4 months \$4, \$604

Refer to the figure for Exercises 7–10.

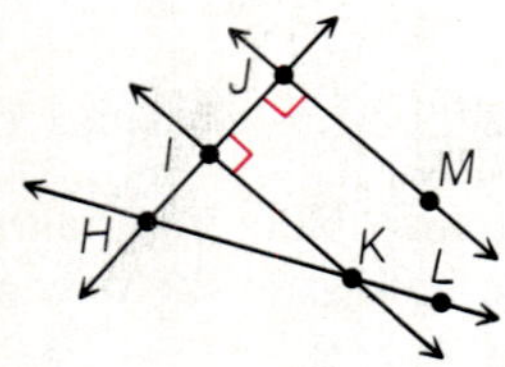

7. Name an acute angle. $\angle IHK$, $\angle IKH$

8. Name an obtuse angle. $\angle IKL$

9. Name a right angle. $\angle MJI$, $\angle JIK$, $\angle HIK$

10. Name a pair of parallel lines. $\overleftrightarrow{JM}$ and $\overleftrightarrow{IK}$

Classify each triangle according to the measures of its angles and the lengths of its sides.

11. $\angle CDE$ has a 90-degree angle and two sides equal in length. Right, Isosceles

12. $\angle PQR$ has a 112-degree angle and no two sides equal in length. Obtuse, Scalene

13. Name the polygon.

square

14. Write and solve an equation to find the measure of $\angle M$.

65°

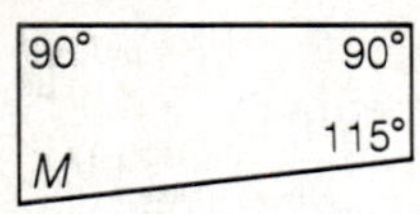

Find the area of each figure.

15. A rectangle with width 9 ft and length 22 ft. 15. 198 ft^2

16. A triangle with base 14 m and height 5 m. 16. 35 m^2

17. a parallelogram with base = 212 mm, height = 40 mm 8480 mm^2

18. a trapezoid with b_1 = 16 m, b_2 = 8 m, h = 12 m 144 m^2

Find the area of a circle with the given radius or circumference. Use 3.14 as an approximation for π.

19. r = 6 m 113.04 m^2

20. d = 1.5 km 1.76625 km^2

21. Find the area of the figure at right.

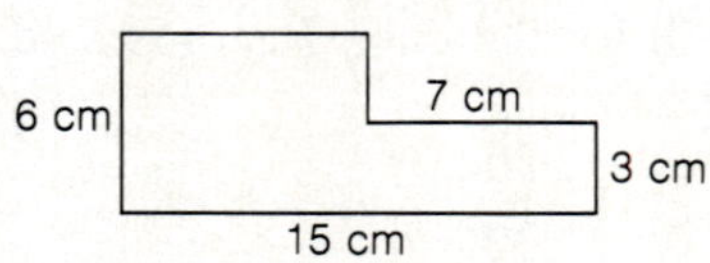

21. 69 cm^2

Chapter 14

Square Roots and Special Triangles

14-1 Square Roots

Objective: To find square roots and evaluate expressions including square roots.
MIN: 1–37; REG: 1–37 even, 38–45; MAX: 1–37 even, 38–47; ALL: Mixed Review

An exploratory lesson opener can be found on page 554.

The **square root** of 36 is 6 since $6 \cdot 6 = 36$. In general, the square root of a number x is the number y if $y^2 = x$. The symbol $\sqrt{\ }$, called a **radical sign**, is used to indicate the square root. We write $\sqrt{36} = 6$.

Each positive number has two square roots. For example, 64 has square roots of 8 and -8 since $8^2 = 64$ and $(-8)^2 = 64$.

8 is the positive square root of 64, written as $\sqrt{64} = 8$.

-8 is the negative square root of 64, written as $-\sqrt{64} = -8$.

Emphasize that the radical sign refers to the positive square root. The minus sign preceding the radical denotes the negative square root.

Example 1

Find $\sqrt{49}$.

Solution $\sqrt{49} = 7$

Practice Find the square root.

a. $-\sqrt{25}$ -5 **b.** $\sqrt{64}$ 8

Whenever the square root of a number is not an integer it is an *irrational* number. This means the square root cannot be written as a fraction, but can be approximated as a decimal. For example, $\sqrt{7} \approx 2.646$.

Example 2

Between what two consecutive integers does $\sqrt{18}$ lie?

Solution $4 < \sqrt{18} < 5$ Since $4^2 = 16$ and $5^2 = 25$, $\sqrt{18}$ is between 4 and 5.

Practice Between what two consecutive integers does each square root lie?

a. $\sqrt{28}$ $5 < \sqrt{28} < 6$ **b.** $\sqrt{53}$ $7 < \sqrt{53} < 8$

When numbers are added or subtracted under a single radical sign, carry out the operation before finding the square root.

Example 3

Evaluate each expression. **a.** $\sqrt{81} - \sqrt{36}$ **b.** $\sqrt{29 + 7}$

Solution

a. $\sqrt{81} - \sqrt{36} = 9 - 6$
$= 3$

Find the square root of each number before subtracting.

b. $\sqrt{29 + 7} = \sqrt{36}$
$= 6$

Since both numbers are under the same radical sign, add before finding the square root.

Practice Evaluate each expression. **a.** $\sqrt{49} + \sqrt{121}$ 18 **b.** $\sqrt{121 - 40}$ 9

Oral Exercises

State whether the following are true or false.

1. $\sqrt{25} = 5$ T **2.** $\sqrt{9} = 3$ T **3.** $\sqrt{36} = -6$ F **4.** $-\sqrt{16} = 4$ F

5. $\sqrt{36} = 6$ T **6.** $\sqrt{16} = -4$ F **7.** $-\sqrt{25} = -5$ T **8.** $\sqrt{49} = 7$ T

Exercises

A Find the square root.

1. $\sqrt{25}$ 5 **2.** $-\sqrt{81}$ −9 **3.** $\sqrt{4}$ 2 **4.** $\sqrt{9}$ 3 **5.** $\sqrt{144}$ 12

6. $\sqrt{64}$ 8 **7.** $-\sqrt{16}$ −4 **8.** $-\sqrt{4}$ −2 **9.** $\sqrt{16}$ 4 **10.** $\sqrt{100}$ 10

Between what two consecutive integers does each square root lie?

11. $\sqrt{7}$ 2,3 **12.** $\sqrt{12}$ 3,4 **13.** $\sqrt{17}$ 4,5 **14.** $-\sqrt{5}$ −2,−3 **15.** $\sqrt{43}$ 6,7

16. $\sqrt{39}$ 6,7 **17.** $-\sqrt{42}$ −6,−7 **18.** $\sqrt{38}$ 6,7 **19.** $\sqrt{29}$ 5,6 **20.** $-\sqrt{41}$

21. $\sqrt{24}$ 4,5 **22.** $\sqrt{37}$ 6,7 **23.** $\sqrt{85}$ 9,10 **24.** $\sqrt{57}$ 7,8 **25.** $\sqrt{112}$ 10,11

20. −6,−7

Evaluate each expression.

26. $\sqrt{25} + \sqrt{81}$ 14 **27.** $\sqrt{41 - 5}$ 6 **28.** $\sqrt{55 - 6}$ 7

29. $\sqrt{74 + 26}$ 10 **30.** $\sqrt{81} - \sqrt{100}$ −1 **31.** $\sqrt{221 - 100}$ 11

32. $\sqrt{23 + 58} - \sqrt{36}$ 3 **33.** $\sqrt{81 - 17} - \sqrt{81}$ −1 **34.** $\sqrt{16 + 9} - \sqrt{9}$ 2

35. $\sqrt{12 + 13} + \sqrt{49}$ 12 **36.** $\sqrt{81} - \sqrt{45 - 20}$ 4 **37.** $\sqrt{64} - \sqrt{36 + 28}$ 0

B Write $<$, $>$, or $=$ in place of □ to make a true statement.

38. $\sqrt{25} + \sqrt{36}$ □ $\sqrt{25 + 36}$ $>$ **39.** $\sqrt{81} - \sqrt{4}$ □ $\sqrt{81 - 4}$ $<$

40. $\sqrt{16 - 9}$ □ $\sqrt{16} - \sqrt{9}$ $>$ **41.** $\sqrt{16 + 9}$ □ $\sqrt{16} + \sqrt{9}$ $<$

A square root can be multiplied by another number.
For example, $6\sqrt{25} = 6 \times 5 = 30$. In Exercises 42–43, write $<$, $>$, or $=$ in place of $\square$ to make a true statement.

42. $2\sqrt{16} \square \sqrt{32}$ $>$

43. $3\sqrt{9} \square \sqrt{27}$ $>$

44. Evaluate the expression $3\sqrt{a} + \sqrt{b}$ for $a = 25$ and $b = 81$. 24

45. Evaluate the expression $\sqrt{\frac{a}{b}}$ for $a = 243$ and $b = 3$. 9

C Extending Thinking Skills

Thinking skill: 46, 47. Finding a pattern

Find the pattern and give the next two terms in the sequence.
Then complete the nth term.

$\sqrt{5^4} = 25, \sqrt{6^4} = 36, \sqrt{n^4} = n^2$

46. $\sqrt{2^4} = 4, \sqrt{3^4} = 9, \sqrt{4^4} = 16,$ _?_, _?_, $\sqrt{n^4} =$ _?_

47. $\sqrt{2^6} = 8, \sqrt{3^6} = 27, \sqrt{4^6} = 64,$ _?_, _?_, $\sqrt{n^6} =$ _?_

$\sqrt{5^6} = 125, \sqrt{6^6} = 216, \sqrt{n^6} = n^3$

Mixed Review

Use the figure at right for Exercises 48–52.

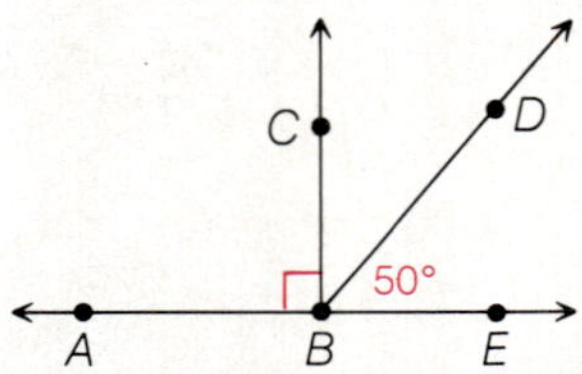

48. Name a pair of supplementary angles.

49. Name a pair of complementary angles.

50. Name a right angle.

Write an equation to find the measure of: **51.** $\angle CBD$ 40° **52.** $\angle ABD$ 130°

Give the area. **53.** A rectangle with height = 6 cm, width = 11 cm 66 cm²

Give the mean, median, and mode. **54.** 11, 13, 12, 11, 9, 14, 11

48. $\angle ABC$, $\angle CBE$ or $\angle ABD$, $\angle DBE$ 49. $\angle CBD$, $\angle DBE$ 50. $\angle ABC$ or $\angle CBE$ 51. $90 = \angle CBD + 50$; 40
52. $180 = \angle ABD + 50$; 130 54. mean 11.57; median 11; mode 11

NUMBERS TO ALGEBRA

Check to see that the number examples are true. Then replace a and b with other perfect squares to test the generalization given.

Numbers	Algebra
$\sqrt{4} \cdot \sqrt{9} = \sqrt{4 \cdot 9}$ $\sqrt{4} \cdot \sqrt{16} = \sqrt{4 \cdot 16}$ $\sqrt{9} \cdot \sqrt{16} = \sqrt{9 \cdot 16}$	For non-negative numbers a and b, $\sqrt{a}\sqrt{b} = \sqrt{ab}$

Use the generalization above to express each product as an integer.

1. $\sqrt{2} \cdot \sqrt{32}$ 8
2. $\sqrt{18} \cdot \sqrt{2}$ 6
3. $\sqrt{3} \cdot \sqrt{48}$ 12
4. $\sqrt{5} \cdot \sqrt{20}$ 10
5. $\sqrt{2} \cdot \sqrt{50}$ 10
6. $\sqrt{8} \cdot \sqrt{18}$ 12
7. $\sqrt{3} \cdot \sqrt{27}$ 9
8. $\sqrt{6} \cdot \sqrt{24}$ 12

14-2 Finding Square Roots

Objective: To find approximations of square roots that are irrational using square root tables or a calculator with a $\sqrt{\ }$ key.
MIN: 1–27; REG: 1–27 even, 28–40; MAX: 1–27 even, 28–42; ALL: Mixed Review

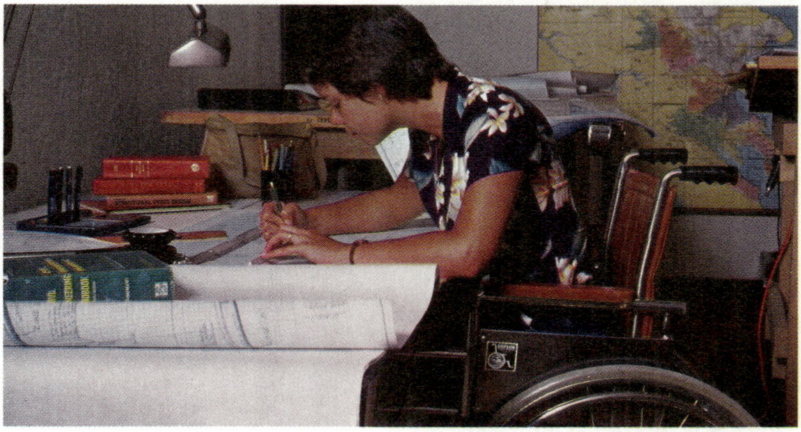

Engineers evaluate formulas such as $t = \sqrt{\frac{d}{16}}$ to calculate the time for an object to travel d feet.

To find approximations of square roots, you can use square root tables or a calculator with a $\sqrt{\ }$ key.

A more complete square root table is given on p. 556.

Number n	Square Root $\sqrt{n}$	Square n^2
1	1	1
2	1.414	4
3	1.732	9
4	2	16
5	2.236	25
6	2.449	36
7	2.646	49
8	2.828	64
9	3	81
10	3.162	100
11	3.317	121
12	3.464	144
13	3.606	169
14	3.742	196
15	3.873	225
16	4	256

Example 1

Use the table to find an approximation for $\sqrt{13}$.

Solution $\sqrt{13} \approx 3.606$

Look down the left column of the table to find 13. Then look across to the square root column to read 3.606.

Practice Use this table to find an approximate value for each.

a. $\sqrt{15}$ 3.873 **b.** $\sqrt{11}$ 3.317

Example 2

Use a calculator with a $\sqrt{\ }$ key to find an approximation for $\sqrt{34}$.

Solution 5.83095

Input	Display
34 [√] [=]	5.83095

Practice Use a calculator with a $\sqrt{\ }$ key to find an approximation for each.

a. $\sqrt{53}$ 7.2801098 **b.** $\sqrt{43.7}$ 6.6105975

Oral Exercises

Give the closest integer estimate for each square root.

1. $\sqrt{12}$ 3 **2.** $\sqrt{17}$ 4 **3.** $\sqrt{24}$ 5 **4.** $\sqrt{39}$ 6 **5.** $\sqrt{15}$ 4

6. $\sqrt{27}$ 5 **7.** $\sqrt{80}$ 9 **8.** $\sqrt{50}$ 7 **9.** $\sqrt{10}$ 3 **10.** $\sqrt{99}$ 10

Exercises

A Use the table on page 556 to find each square root.

1. $\sqrt{2}$ 1.414 **2.** $\sqrt{5}$ 2.236 **3.** $\sqrt{14}$ 3.742 **4.** $\sqrt{17}$ 4.123 **5.** $\sqrt{19}$ 4.359

6. $\sqrt{27}$ 5.196 **7.** $\sqrt{10}$ 3.162 **8.** $\sqrt{47}$ 6.856 **9.** $\sqrt{93}$ 9.644 **10.** $\sqrt{39}$ 6.245

11. $\sqrt{32}$ 5.657 **12.** $\sqrt{18}$ 4.243 **13.** $\sqrt{28}$ 5.292 **14.** $\sqrt{76}$ 8.718 **15.** $\sqrt{87}$ 9.327

Use a calculator with a $\sqrt{\ }$ key to find an approximation for each square root. Round your answer to the nearest thousandth.

16. $\sqrt{46}$ 6.782 **17.** $\sqrt{93}$ 9.644 **18.** $\sqrt{421}$ 20.518 **19.** $\sqrt{57}$ 7.550

20. $\sqrt{948}$ 30.790 **21.** $\sqrt{95.8}$ 9.788 **22.** $\sqrt{84.25}$ 9.179 **23.** $\sqrt{8392}$ 91.608

24. $\sqrt{38.9}$ 6.237 **25.** $\sqrt{42.93}$ 6.552 **26.** $\sqrt{38.64}$ 6.216 **27.** $\sqrt{57.23}$ 7.565

B Square the number in parentheses. Write $<$, $>$, or $=$ for $\square$.

28. $(3.606)^2 \square 13$ $>$ **29.** $(1.414)^2 \square 2$ $<$ **30.** $(1.732)^2 \square 3$ $<$

31. $(2.828)^2 \square 8$ $<$ **32.** $(3.317)^2 \square 11$ $>$ **33.** $(3.873)^2 \square 15$ $>$

34. $(5.764)^2 \square 33$ $>$ **35.** $(4.444)^2 \square 20$ $<$ **36.** $(6.602)^2 \square 44$ $<$

37. Find the time (t) in seconds it takes for a parachutist to free-fall the distance (d) 3600 feet. Use the formula $t = \sqrt{\frac{d}{16}}$. 15 seconds

Evaluate each formula.

38. $r = \sqrt{\frac{a}{\pi}}$ for $a = 121\pi$. $r = 11$

39. $D = \sqrt{a^2 + b^2 + c^2}$ where $a = 3$, $b = 7$, and $c = \sqrt{63}$. Hint: $c^2 = 63$. $D = 11$

C Extending Thinking Skills

Thinking skill: 40, 41. Testing a conjecture

40. Choose nonzero whole numbers for a, b, and c so that $\sqrt{a} \cdot \sqrt{b} = \sqrt{c}$. Answers will vary. Examples: $\sqrt{4}\sqrt{9} = \sqrt{36}$; $\sqrt{4}\sqrt{4} = \sqrt{16}$; $\sqrt{9}\sqrt{36} = \sqrt{324}$

41. Find two solutions to $\sqrt{x} = x$. $\sqrt{0} = 0$; $\sqrt{1} = 1$

Mixed Review

Use the figure at right for Exercises 42–45.
Write and solve an equation to find each.

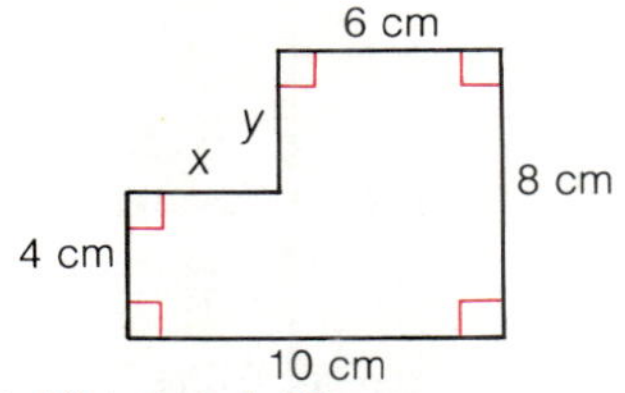

42. The length of x. $x + 6 = 10$; 4 cm

43. The length of y. $4 + y = 8$; 4 cm

44. The perimeter of the figure. $P = 4 + 6 + 8 + 10 + 4 + 4$; 36 cm

45. The area of the figure. $A = (10 \cdot 8) - (4 \cdot 4)$; 64 cm^2

Give the probability of drawing from a standard deck of 52 playing cards:

46. a diamond $\frac{1}{4}$ **47.** the king of spades $\frac{1}{52}$ **48.** a red ten $\frac{1}{26}$

Solve and check. **49.** $5m - 2m = 27$ 9 **50.** $4.5c + 6.9c = 45.6$ 4

51. $11 = 9x + 6.5$ 0.5 **52.** $3t = 12 - 5t$ 1.5 **53.** $21m + 16 = 79$ 3

CALCULATOR ACTIVITY

On a calculator without a $\sqrt{\ }$ key, you can find an approximate square root of a number by using a method of averaging. The method is explained below for $\sqrt{38}$.

- Locate the square root between a pair of consecutive integers. $6 < \sqrt{38} < 7$
- Select an estimated square root. $\sqrt{38} \approx 6.2$
- Divide by the estimate. Find the quotient to one more place than is desired in the final answer. $6.2\overline{)38.000}$ = 6.129
- Find the average of the estimated square root and the quotient from the division. $\frac{6.2 + 6.129}{2} = 6.164$
- Use this average as a new divisor. Continue this process until the quotient and the divisor agree in the hundredths place. $6.164\overline{)38.00000}$ = 6.164

Use the method of averaging to find an approximate square root.

1. $\sqrt{6}$ 2.45 **2.** $\sqrt{11}$ 3.32 **3.** $\sqrt{31}$ 5.57 **4.** $\sqrt{18}$ 4.24 **5.** $\sqrt{42}$ 6.48

14-3 Solving Equations: Using Square Roots

Objective: To solve equations of the form $x^2 = k$.
MIN: 1–16; REG: 1–16 even, 17–24; MAX: 1–16 even, 17–28; ALL: Mixed Review

To solve equations of the form $x^2 = 25$, you need to use the property shown by the examples below.

$$\sqrt{5^2} = \sqrt{25} = 5 \qquad \sqrt{7^2} = \sqrt{49} = 7$$

The general property is stated as follows.

The Square Root Property

For any number x, $\sqrt{x^2} = |x|$.

When both sides of an equation are positive, you can find the square root of both sides and the equality is maintained.

Example 1

Solve and check. $x^2 = 25$

Solution $x^2 = 25$

$\sqrt{x^2} = \sqrt{25}$ Undo squaring by finding the square root of both sides.

$|x| = \sqrt{25}$ Use the property $\sqrt{x^2} = |x|$ on the left side.

$x = 5$ or $x = -5$ **Emphasize that these equations have two solutions since there is both a positive and a negative square root.**

Check $5^2 = 25$ and $(-5)^2 = 25$

Practice Solve and check. **a.** $x^2 = 49$ **±7** **b.** $x^2 = 100$ **±10**

Example 2

Solve and check. $w^2 = 17$

Solution $w^2 = 17$

$\sqrt{w^2} = \sqrt{17}$

$|w| = \sqrt{17}$

$w = \sqrt{17}$ or $w = -\sqrt{17}$ When $\sqrt{w^2}$ is not an integer, write the answer using the $\sqrt{\ }$ symbol.

Check $(\sqrt{17})^2 \stackrel{?}{=} 17$ Replace x with $\sqrt{17}$ and $-\sqrt{17}$ in $x^2 = 17$.

$(-\sqrt{17})^2 \stackrel{?}{=} 17$

$17 = 17$ ✓

Practice Solve and check. **a.** $x^2 = 121$ ± 11 **b.** $x^2 = 39$ $\pm\sqrt{39}$ **c.** $x^2 = 73$ $\pm\sqrt{73}$

Oral Exercises

State whether or not the equation has integer solutions.

1. $x^2 = 16$ yes **2.** $x^2 = 64$ yes **3.** $x^2 = 85$ no **4.** $x^2 = 48$ no **5.** $x^2 = 25$ yes

6. $x^2 = 121$ yes **7.** $x^2 = 144$ yes **8.** $x^2 = 81$ yes **9.** $x^2 = 65$ no **10.** $x^2 = 36$ yes

Exercises

A Solve and check.

1. $x^2 = 16$ ± 4 **2.** $x^2 = 36$ ± 6 **3.** $x^2 = 81$ ± 9 **4.** $x^2 = 47$ $\pm\sqrt{47}$

5. $x^2 = 121$ ± 11 **6.** $x^2 = 38$ $\pm\sqrt{38}$ **7.** $x^2 = 56$ $\pm\sqrt{56}$ **8.** $x^2 = 121$ ± 11

9. $x^2 = 132$ $\pm\sqrt{132}$ **10.** $x^2 = 144$ ± 12 **11.** $x^2 = 53$ $\pm\sqrt{53}$ **12.** $x^2 = 99$ $\pm\sqrt{99}$

13. $x^2 = 113$ $\pm\sqrt{113}$ **14.** $x^2 = 75$ $\pm\sqrt{75}$ **15.** $x^2 = 55$ $\pm\sqrt{55}$ **16.** $x^2 = 33$ $\pm\sqrt{33}$

B Solve and check.

17. $x^2 - 25 = 0$ ± 5 **18.** $x^2 + 7 = 43$ ± 6 **19.** $x^2 + 11 = 111$ ± 10 **20.** $x^2 + 15 = 40$ ± 5

21. $x^2 - 9 = 40$ ± 7 **22.** $x^2 + 34 = 70$ ± 6 **23.** $x^2 + 3 = 52$ ± 7 **24.** $x^2 - 8 = 41$ ± 7

C Extending Thinking Skills

Find two solutions to each equation. Thinking skill: 25–28. Testing a conjecture

25. $x^2 = x$ 0,1 **26.** $x^2 = x + 2$ −1,2 **27.** $x^2 = -x$ −1,0 **28.** $x^4 = 16$ ± 2

Encourage students to guess a solution and verify the guess.

Mixed Review

What percent of 60 is: **29.** 33? 55% **30.** 45? 75% **31.** 12? 20% **32.** 20? $33\frac{1}{3}$%

Solve and check. **33.** $16t + 13t = 15t + 28$ 2 **34.** $4.2c - 11 = 1.6$ 3

ESTIMATION

To estimate a square root of a number that is not an integer, you can choose compatible numbers. For example, $\sqrt{26} \approx \sqrt{25} = 5$, so $\sqrt{26} \approx 5$.

Use compatible numbers to estimate these square roots.

1. $\sqrt{37}$ ≈6 **2.** $\sqrt{80}$ ≈9 **3.** $\sqrt{99}$ ≈10 **4.** $\sqrt{10}$ ≈3 **5.** $\sqrt{15}$ ≈4

14-4 Pythagorean Theorem

Objective: To use the Pythagorean theorem to find unknown lengths of sides of right triangles.
MIN: 1–24; REG: 1–24 even, 25–36; MAX: 1–24 even, 25–40; ALL: Mixed Review

The longest side of a right triangle, the side opposite the right angle, is called the **hypotenuse**. The two shorter sides are called the **legs** of the triangle.

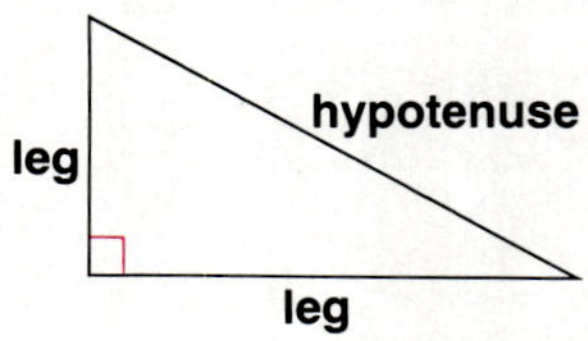

The triangle in the figure below has sides that are 3, 4, and 5 units long. The areas of the squares that have been constructed on each side show that the square of the hypotenuse is equal to the sum of the squares of the two legs.

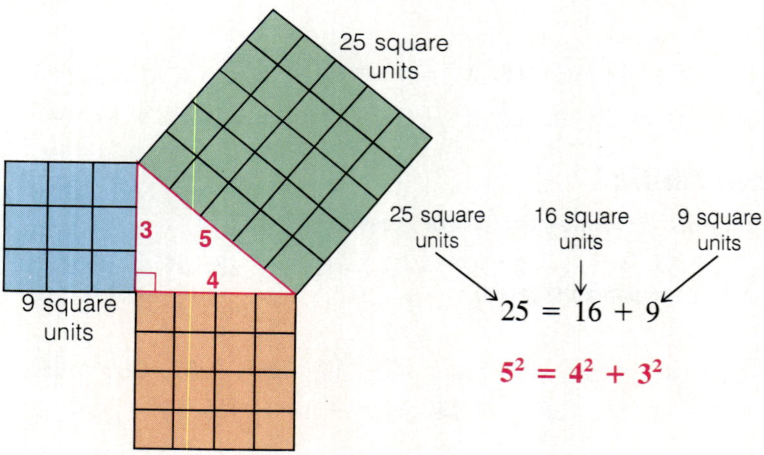

$$25 = 16 + 9$$

$$5^2 = 4^2 + 3^2$$

This suggests the following theorem.

The Pythagorean Theorem

In a right triangle, the square of the length of the hypotenuse is equal to the sum of the squares of the lengths of the two legs. If c is the length of the hypotenuse and a and b are the lengths of the two legs,

$$c^2 = a^2 + b^2$$

Example 1

Find the length of the hypotenuse. Express your answer in radical form.

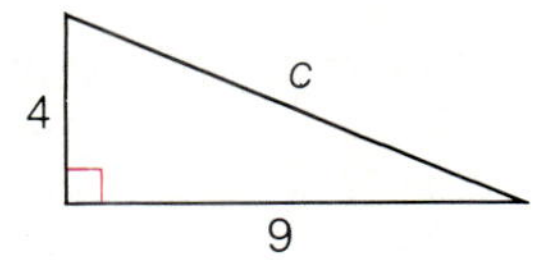

Solution

$c^2 = a^2 + b^2$ Use the Pythagorean theorem.

$c^2 = 4^2 + 9^2$ Substitute 4 for *a* and 9 for *b*.

$c^2 = 16 + 81$ Square terms and add.

$c^2 = 97$

$c = \sqrt{97}$ Since *c* represents a length, its value is the positive square root. The length of the hypotenuse is $\sqrt{97}$.

Practice Draw a right triangle with hypotenuse *c* and legs *a* and *b*.

a. Find length *c* if $a = 5$ and $b = 8$. $\sqrt{89}$

b. Find length *c* if $a = 7$ and $b = 15$. $\sqrt{274}$

Example 2

Find the length of leg *a*. Express your answer in radical form.

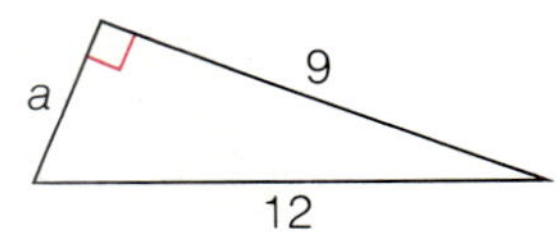

Solution

$a^2 + b^2 = c^2$ Use the Pythagorean theorem.

$a^2 + 9^2 = 12^2$ Substitute 9 for *b* and 12 for *c*.

$a^2 + 81 - 81 = 144 - 81$

$a^2 = 63$

$a = \sqrt{63}$ The length of leg *a* is $\sqrt{63}$.

Practice Draw a right triangle with hypotenuse *c* and legs *a* and *b*.

a. Find the length of leg *a* if $c = 15$ and $b = 8$. $\sqrt{161}$

b. Find the length of leg *a* if $b = 6$, $c = 14$. $\sqrt{160}$

Oral Exercises

State the equation given by the Pythagorean Theorem.

1.
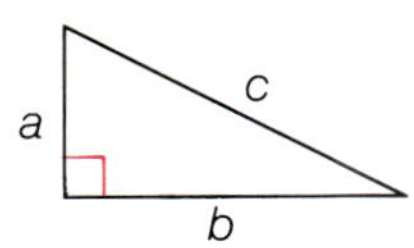

$c^2 = a^2 + b^2$

2.
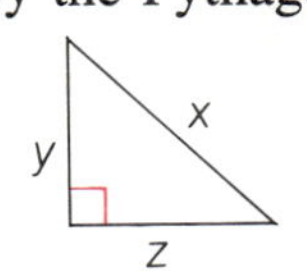

$x^2 = y^2 + z^2$

3.
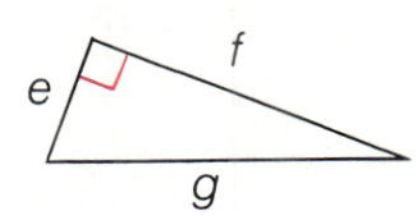

$g^2 = e^2 + f^2$

4.
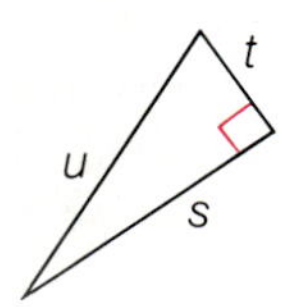

$u^2 = t^2 + s^2$

Which statements about legs *a* and *b* and hypotenuse *c* of a right triangle are true and which are false?

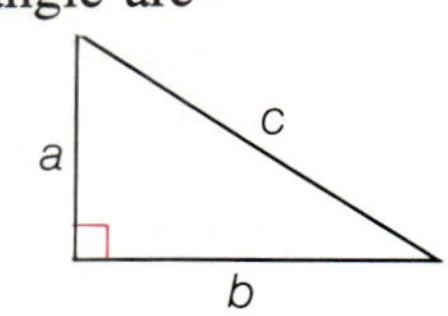

5. $c^2 = a^2 + b^2$ T **6.** $a^2 = b^2 + c^2$ F **7.** $b^2 = c^2 - a^2$ T

8. $a^2 = b^2 - c^2$ F **9.** $b^2 = a^2 - c^2$ F **10.** $a^2 = c^2 - b^2$ T

Exercises

A Find the length of the hypotenuse. **1.** 15 **2.** $\sqrt{193}$ **3.** $\sqrt{130}$ **4.** $\sqrt{202}$

1.

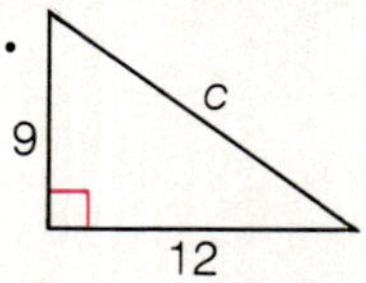

2.

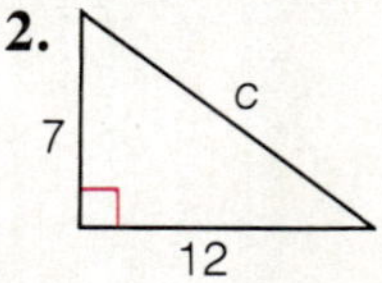

3.

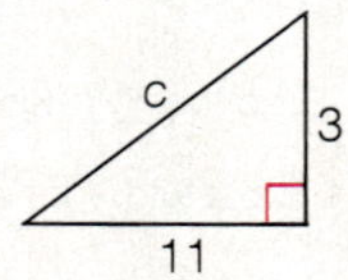

4.

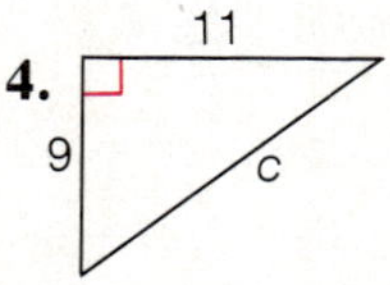

Draw a right triangle with legs a and b and hypotenuse c to use for Exercises 5–12. Find the length of c. If c is not an integer, express in radical form.

5. $a = 8$ and $b = 12$. Find c. $\sqrt{208}$

6. $a = 8$ and $b = 15$. Find c. 17

7. $a = 12$ and $b = 5$. Find c. 13

8. $a = 9$ and $b = 40$. Find c. 41

9. $a = 10$ and $b = 16$. Find c. $\sqrt{356}$

10. $a = 15$ and $b = 6$. Find c. $\sqrt{261}$

11. $a = 12$ and $b = 20$. Find c. $\sqrt{544}$

12. $a = 7$ and $b = 9$. Find c. $\sqrt{130}$

Find the missing length of leg a. **13.** $\sqrt{119}$ **14.** $\sqrt{19}$ **15.** 8 **16.** $\sqrt{72}$

13.

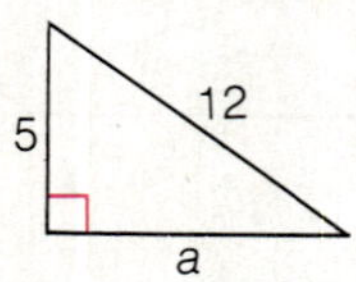

14.

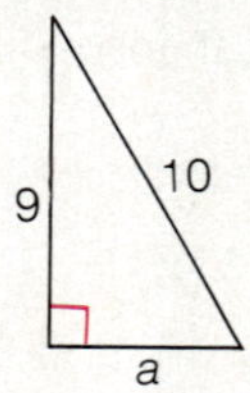

15.

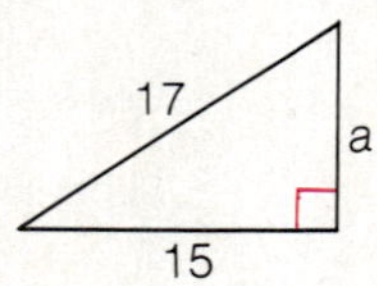

16.

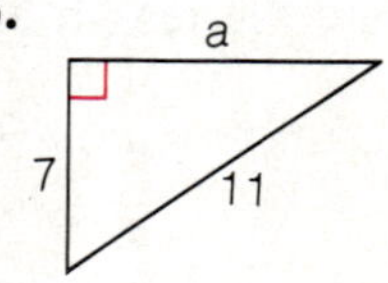

Draw a right triangle with legs a and b and hypotenuse c to use for Exercises 17–24. Find the length.

17. $a = 6$ and $c = 13$. Find b. $\sqrt{133}$

18. $a = 10$ and $c = 20$. Find b. $\sqrt{300}$

19. $b = 13$ and $c = 15$. Find a. $\sqrt{56}$

20. $b = 6$ and $c = 12$. Find a. $\sqrt{108}$

21. $a = 18$ and $c = 23$. Find b. $\sqrt{205}$

22. $a = 12$ and $c = 15$. Find b. 9

23. $b = 5$ and $c = 14$. Find a. $\sqrt{171}$

24. $b = 8$ and $c = 20$. Find b. $\sqrt{336}$

B Solve.

25. The base of a 32-foot ladder is 10 feet from the building. How high above the ground is the top of the ladder? $\sqrt{924}$ or 30.4 ft

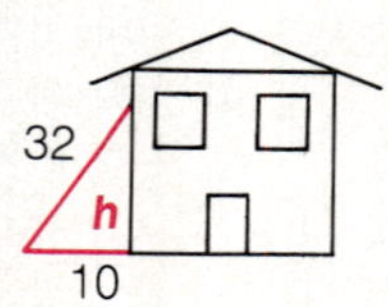

26. Movers are trying to take a large table through a door with these dimensions. They want to know the length of diagonal $\overline{AB}$. What is the length? $\sqrt{7,696}$ or 87.73″

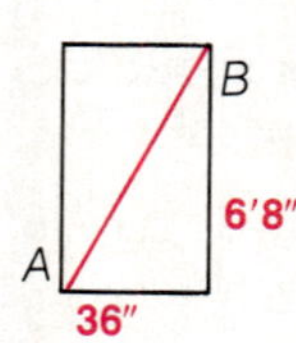

27. A road climbs 800 feet as you travel a horizontal distance of 3 miles. How much longer is the road surface than the horizontal distance? (5280 ft = 1 mile) 20.19 ft

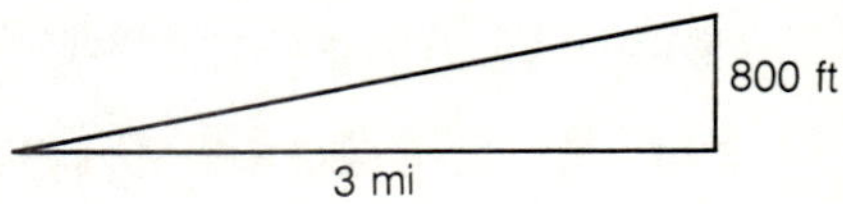

28. A building 36 feet wide has a roof that rises 4 feet vertically for each 12-foot horizontal change. The roof has a 2-foot overhang. How long is length AB? $18^2 + 6^2 = x^2$, $x \approx 18.97$ $AB \approx 20.97$ ft

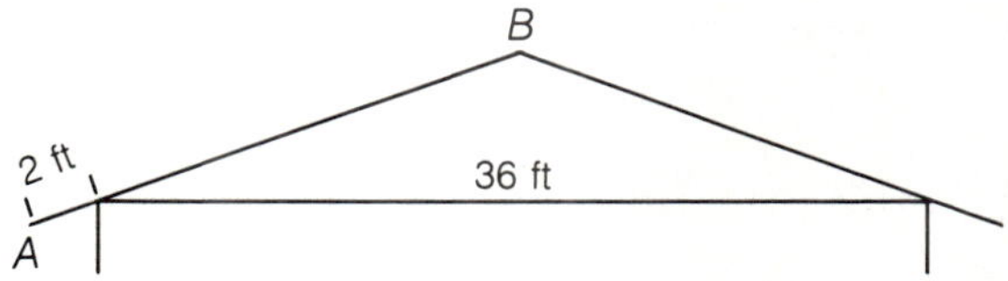

Three whole numbers a, b, and c are called **Pythagorean triples** if $c^2 = a^2 + b^2$. For example, 3, 4, and 5 are Pythagorean triples since $5^2 = 3^2 + 4^2$. Which of these are Pythagorean triples?

29. 40, 50, 60 no **30.** 5, 12, 13 yes **31.** 7, 15, 21 no **32.** 8, 15, 17 yes
33. 9, 11, 17 no **34.** 7, 24, 25 yes **35.** 20, 21, 29 yes **36.** 9, 29, 36 no

C Extending Thinking Skills

Thinking skill: 37–40 Reasoning

Suppose that u and v are any two whole numbers. Let x, y, and z be the following three numbers: $x = 2uv$, $y = u^2 - v^2$, $z = u^2 + v^2$.

For which of these values of u and v are x, y, and z Pythagorean triples?

37. $u = 4, v = 1$ yes **38.** $u = 4, v = 3$ yes **39.** $u = 5, v = 2$ yes **40.** $u = 6, v = 5$ yes

Have students pick other values for u and v and decide whether the formula will always give Pythagorean triples.

Mixed Review

Simplify. **41.** $6t + 9 - 4t - 1$ **42.** $9c - 6c - 12$ **43.** $2n + 4n + 12$

Solve and check. **44.** $3.6y = 5y - 0.7$ 0.5 **45.** $9x + 6 = 8.7$ 0.3

41. $2t + 8$
42. $3c - 12$
43. $6n + 12$

MENTAL MATH

These equations show a way to square mentally a 2-digit number ending in 5. Study the pattern.

$25 \cdot 25 = (20 \cdot 30) + 25 = 625$

$35 \cdot 35 = (30 \cdot 40) + 25 = 1225$

$45 \cdot 45 = (40 \cdot 50) + 25 = 2025$

Use mental math to find each product.

1. $55 \cdot 55$ 3025 **2.** $65 \cdot 65$ 4225 **3.** $75 \cdot 75$ 5625 **4.** $85 \cdot 85$ 7225 **5.** $95 \cdot 95$ 9025

14-5 Sports Playing Areas

Objective: To solve word problems.
MIN: 1–9; REG: 1–9; MAX: 1–9

Many sports are played on rectangular areas. The table below lists the dimensions commonly used for several sports.

You might wish to have students choose from among the techniques pencil and paper, mental math, estimation, and calculator. Have them choose the most appropriate technique for each problem.

Playing Area Dimensions	
Sport	**Dimensions**
Boxing	20 ft × 20 ft
Karate	26 ft × 26 ft
Wrestling	39 ft 3 in × 39 ft 3 in
Judo	52 ft 6 in × 52 ft 6 in
Basketball	28 yd × 15 yd 9 in
Ice Hockey	200 ft × 100 ft
U.S. Football	120 yd × 53 yd
Soccer	110 yd × 80 yd

Problems

Use the data from the table to solve.

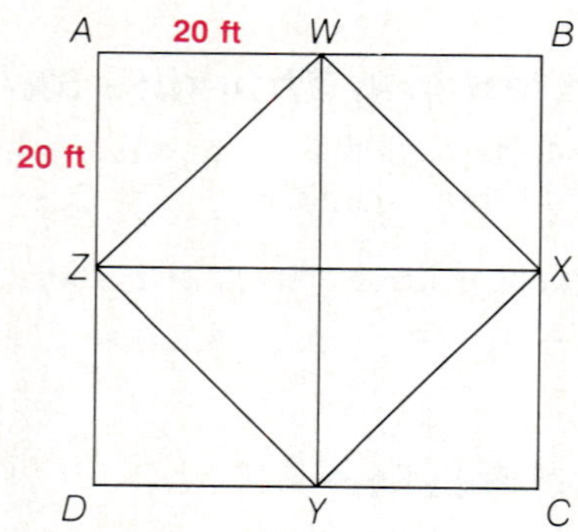

1. This gymnasium floor has overlapping areas drawn to allow for several different sports. The square *ABCD* is divided into quarters for use as four boxing areas. Is the square *WXYZ* large enough for karate? **each side ≈ 28.28′; square is large enough**

2. Suppose that in the figure above, $AW = AZ = 39$ ft 3 in and the square is used for wrestling. Is square *WXYZ* large enough to be used for judo? **$ZW \approx 55.51'$; yes**

3. Suppose that in the figure above, $AW = AZ = 26$ ft and the square is used for karate. Is square *WXYZ* large enough to be used for wrestling? **$ZW \approx 36.77'$; too small**

4. When a basketball court *QRST* is laid out, *QRST* must be a rectangle. A way of checking this is to measure the diagonals *QS* and *RT* to make sure they are the same length. How long are these diagonals? **95.65′, or 31.88 yd**

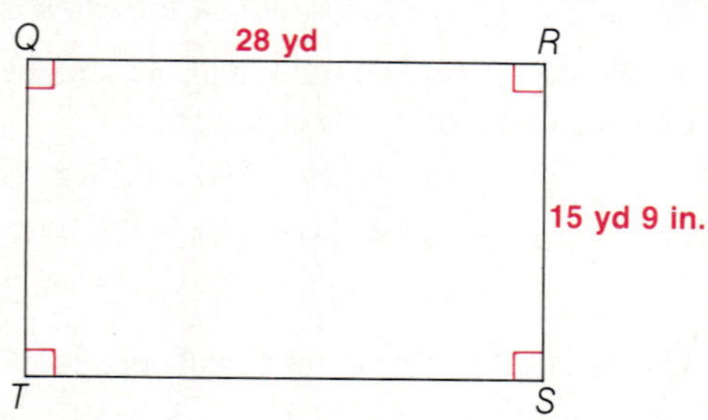

5. What is the length of the diagonal of a soccer field? (This is the longest possible kick that would keep the ball in play.) **136.01 yd**

6. How many basketball courts could fit on a football field? **14.89**

7. A football field *ABCD* and a soccer field *WXYZ* are to be laid out in overlapping fashion, as shown (O is the center of both fields). The people painting the lines want to know the length *AW*. What is it? **≈14.39 yd ($\sqrt{207.25}$)**

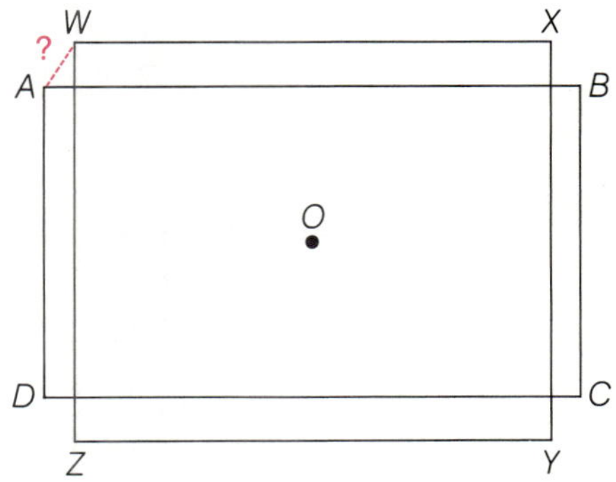

8. The ratio of length to width for the ice hockey field is 2 to 1. Which playing field listed in the table has the greatest length to width ratio?
U.S. Football, 2.26 : 1

9. *Data Search* Find the cost of grass seed in your area and estimate the cost of grass seed for a soccer field and a football field.
Answers will vary.

What's Your Decision?

A school board bought a piece of land with the dimensions shown. On the land, they want to lay out football and soccer practice fields and 6 basketball courts. How would you arrange these playing fields on the land?
Answers will vary.

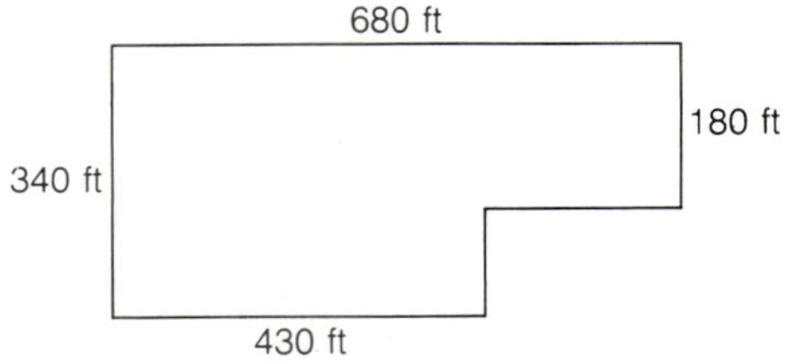

14-6 Similar Triangles

Objective: To determine lengths of corresponding sides of similar triangles.
MIN: 1–4, 5–11 even; REG: 1–11; MAX: 1–13; ALL: Mixed Review

An exploratory lesson opener can be found on page 554.

Triangles that are the same shape but not necessarily the same size are called **similar triangles**. Two triangles that are similar have matching congruent angles, called corresponding angles. Corresponding angles are shown by matching arc symbols.

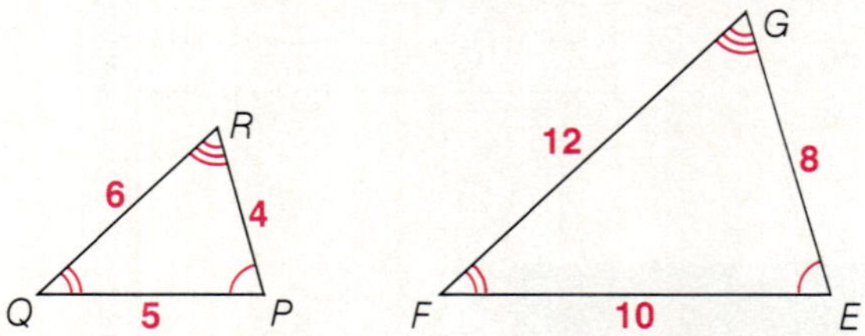

Corresponding angles
$\angle P$ and $\angle E$
$\angle Q$ and $\angle F$
$\angle R$ and $\angle G$

The symbol ~ means *is similar to*. We write $\triangle \boldsymbol{PQR} \sim \triangle \boldsymbol{EFG}$. Lengths of corresponding sides of similar triangles are proportional. In the figures above, the three ratios of corresponding sides all equal $\frac{1}{2}$.

Emphasize that when writing a statement such as $\triangle PQP \sim \triangle EFG$ you always give vertices for corresponding angles in the same order.

$$\frac{QR}{FG} = \frac{6}{12} = \frac{1}{2} \qquad \frac{PR}{EG} = \frac{4}{8} = \frac{1}{2} \qquad \frac{PQ}{EF} = \frac{5}{10} = \frac{1}{2}$$

You can use this idea to find lengths of sides of similar triangles.

Example

$\triangle QPR \sim \triangle YZX$. Use a proportion to find r.

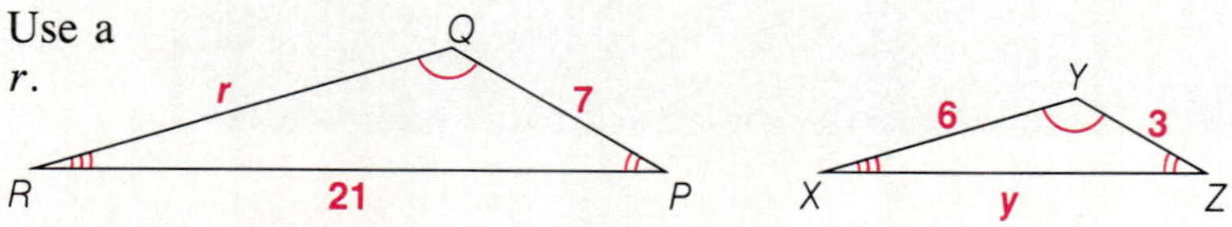

Solution

$\frac{PQ}{ZY} = \frac{QR}{YX}$ — PQ corresponds to ZY. QR corresponds to YX. Lengths of corresponding sides are proportional.

$\frac{7}{3} = \frac{r}{6}$ — Substitute lengths of sides in the proportion.

$7 \cdot 6 = 3 \cdot r$

$42 = 3r$

$14 = r$

Practice $\triangle ABC \sim \triangle XYZ$

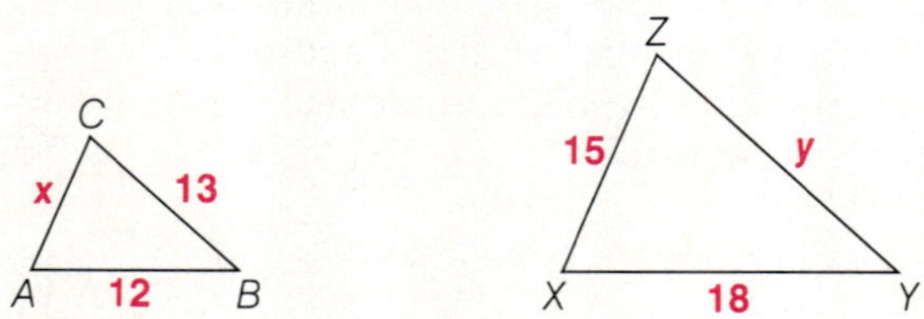

Use a proportion to find:

a. x $\frac{x}{15} = \frac{12}{18}$, 10 **b.** y $\frac{y}{13} = \frac{18}{12}$, 19.5

Oral Exercises 2–3. See additional answers.

Name the corresponding angles and sides for these similar triangles.

1. 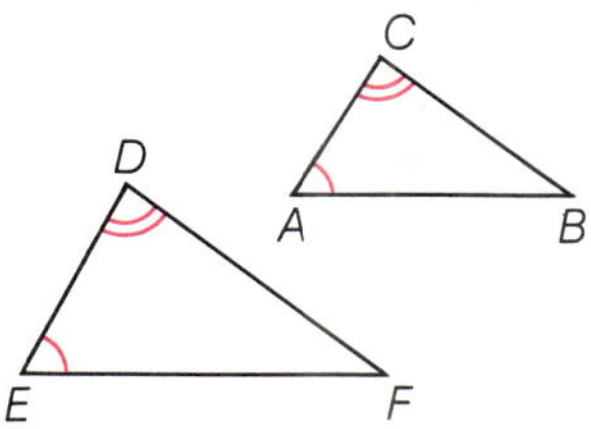

$\angle A$ and $\angle E$ $\overline{AB}$ and $\overline{EF}$
$\angle C$ and $\angle D$ $\overline{BC}$ and $\overline{FD}$
$\angle B$ and $\angle F$ $\overline{CA}$ and $\overline{DE}$

2. 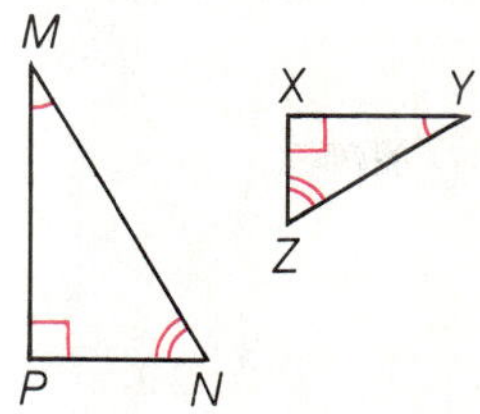

3. 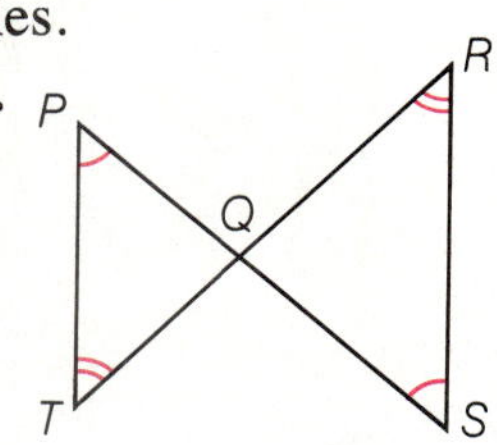

Exercises

A Use proportions to find the unknown lengths for these similar triangles.

1. 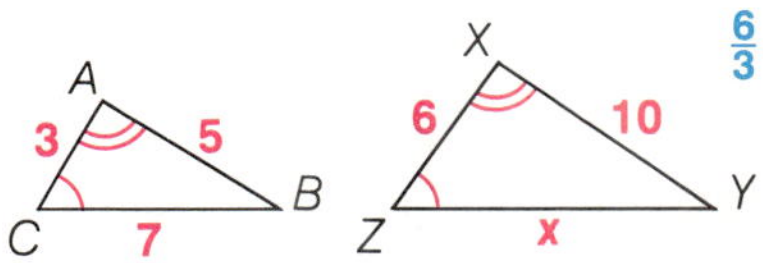

$\frac{6}{3} = \frac{x}{7}$, 14

2. 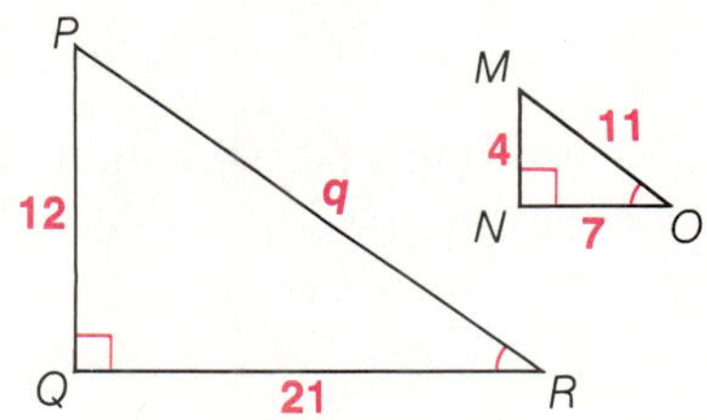

$\frac{12}{4} = \frac{q}{11}$, 33

3. 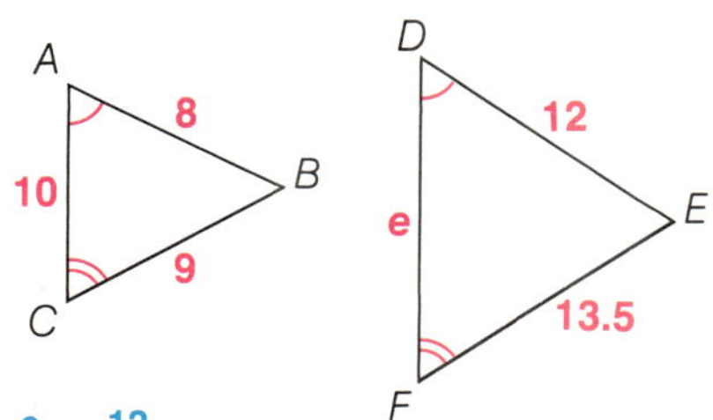

$\frac{e}{10} = \frac{12}{8}$, 15

4. 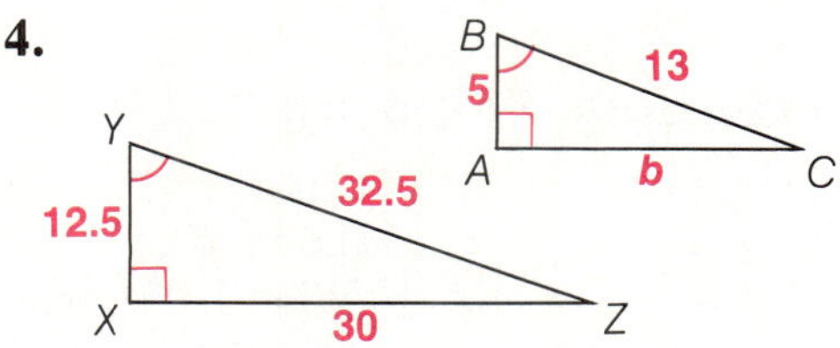

$\frac{b}{30} = \frac{5}{12.5}$, 12

B Use a proportion to find the unknown length for these similar triangles.

5. 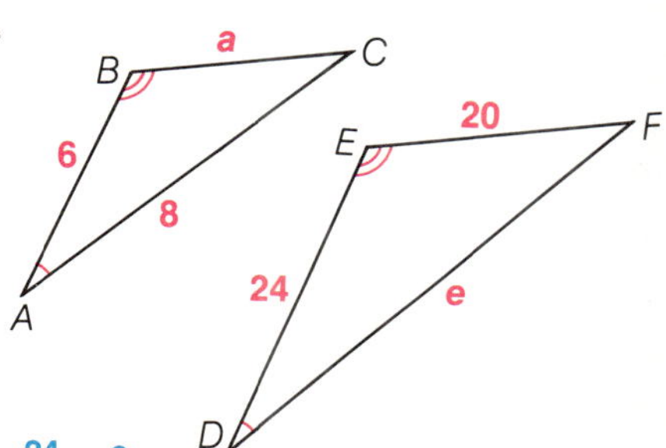

$\frac{24}{6} = \frac{e}{8}$, 32

$\frac{a}{20} = \frac{6}{24}$, 5

6. 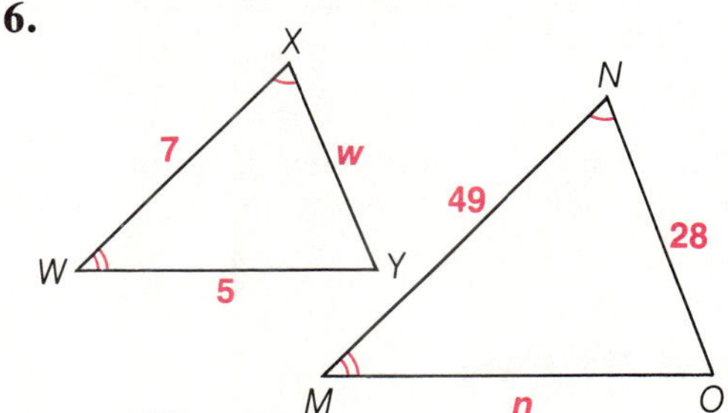

$\frac{n}{5} = \frac{49}{7}$, 35

$\frac{w}{28} = \frac{7}{49}$, 4

7.

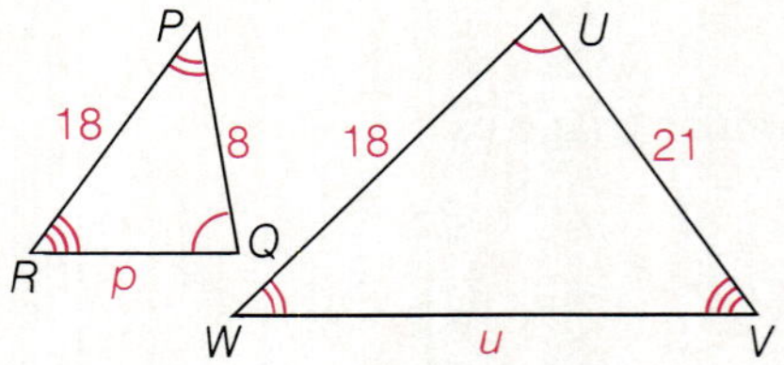

$\frac{p}{21} = \frac{8}{18}, 9\frac{1}{3}$ $\frac{u}{18} = \frac{18}{8}, 40.5$

8.

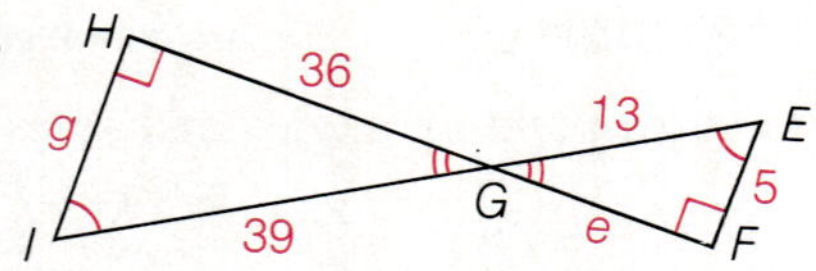

$\frac{e}{36} = \frac{13}{39}, 12$

$\frac{g}{5} = \frac{39}{13}, 15$

9. Find the distance (d) across the pond.

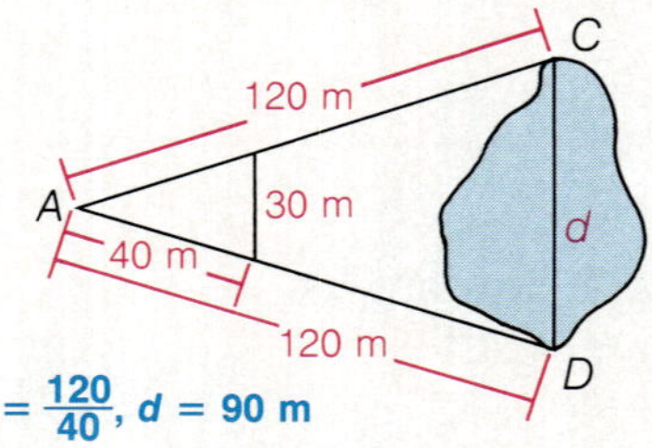

$\frac{d}{30} = \frac{120}{40}, d = 90$ m

10. Find the height (h) of the wall.

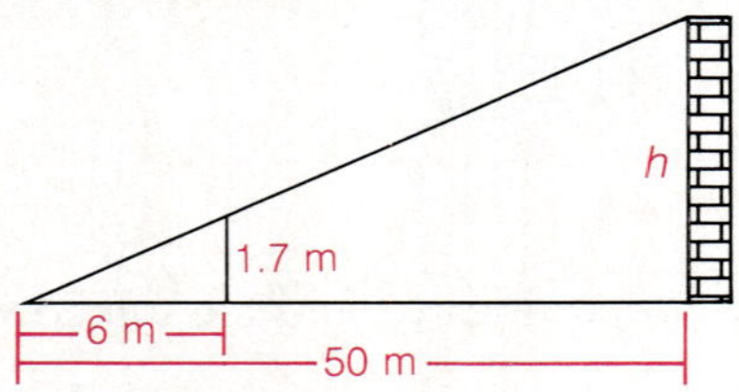

$\frac{h}{1.7} = \frac{50}{6}, h = 14.1\overline{6}$

11. A surveyor needs to find the length AB, but the garage is in the way. $\triangle ABC \sim \triangle DEC$. What is the length of AB? $\frac{AB}{22} = \frac{36}{12}, \overline{AB} = 66$ m

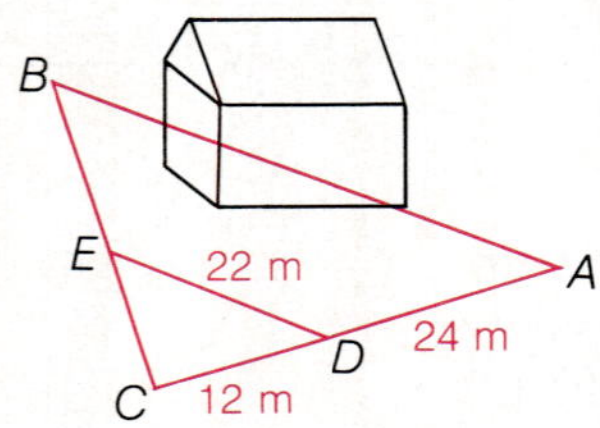

C Extending Thinking Skills

Thinking skills: 12. Reasoning 13. Generalizing

12. Triangle ABC at right is similar to a larger triangle PQR (not shown). The ratio of corresponding sides is 2 : 1. What is the area of each triangle? Area $\triangle ABC = 150$ Area $\triangle PQR = 600$

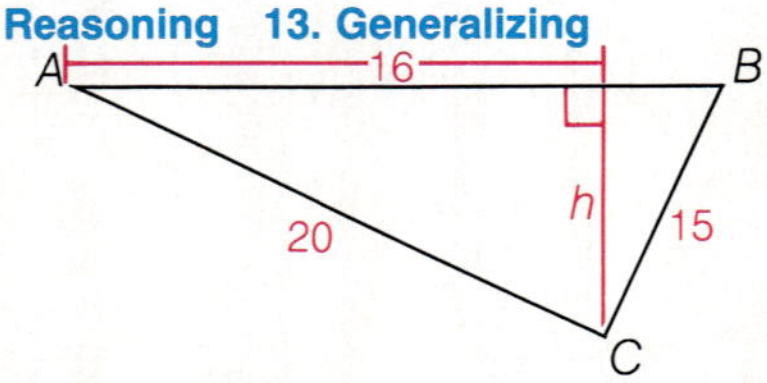

13. This table shows the areas of rectangles of given lengths and widths.

length	2	4	8	. . .	x	$2x$
width	3	6	12	. . .	y	$2y$
area	6	24	96	. . .	xy	$4xy$

Complete the following generalization. If the length and width of a rectangle are both doubled, the area of the rectangle is increased by a factor of __?__ 4

Mixed Review

Reduce to simplest terms. **14.** $\frac{18}{84}$ $\frac{3}{14}$ **15.** $\frac{52}{65}$ $\frac{4}{5}$ **16.** $\frac{45}{63}$ $\frac{5}{7}$ **17.** $\frac{54}{6}$ 9

Solve and check. **18.** $4n + 9n = 10 + 12n$ 10 **19.** $2.4y = y + 7$ 5

14-7 Special Triangles

Objective: To find lengths of sides of special triangles
MIN: 1–8; REG: 1–8 even, 9–13; MAX: 1–8 even, 9–15; ALL: Mixed Review

The two acute angles of an isosceles right triangle are both 45° angles. An isosceles right triangle is called a **45°-45° right triangle**.

You can use the Pythagorean theorem to show that the length of the hypotenuse of the 45°-45° right triangle ABC is $\sqrt{2}$ when the legs are length 1.

A line drawn diagonally across a baseball infield divides it into two congruent, isosceles right triangles.

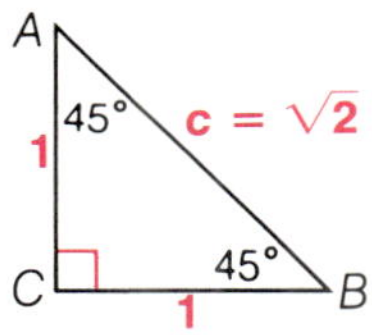

$$c^2 = 1^2 + 1^2$$
$$c^2 = 2$$
$$c = \sqrt{2}$$

$\triangle DEF$ below is a 45°-45° right triangle with legs a units long. We use the fact that $\triangle ABC \sim \triangle DEF$ to write the proportion below.

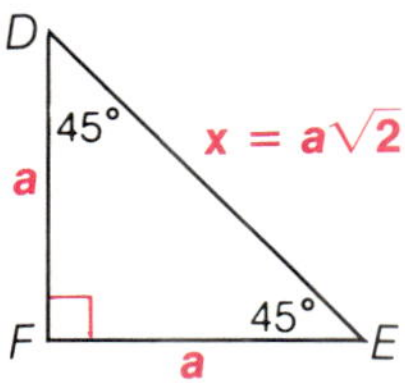

$$\frac{a}{1} = \frac{x}{\sqrt{2}} \quad \text{or} \quad x = a\sqrt{2}.$$

45°-45° Right Triangle

If a 45°-45° right triangle has legs $\boldsymbol{a}$ units long, then the hypotenuse is $\boldsymbol{a\sqrt{2}}$ units long.

Example 1

Find length x.

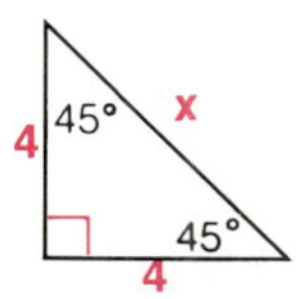

Solution $x = 4\sqrt{2}$ x is equal to the length of the leg times $\sqrt{2}$.

Practice Draw a 45°-45° triangle. Find the length of the hypotenuse if the length of each leg is 5. $x = 5\sqrt{2}$

An altitude of an equilateral triangle bisects an angle and a base to form two right triangles. Each of these triangles has a 30° angle and a 60° angle and is called a **30°-60° right triangle**.

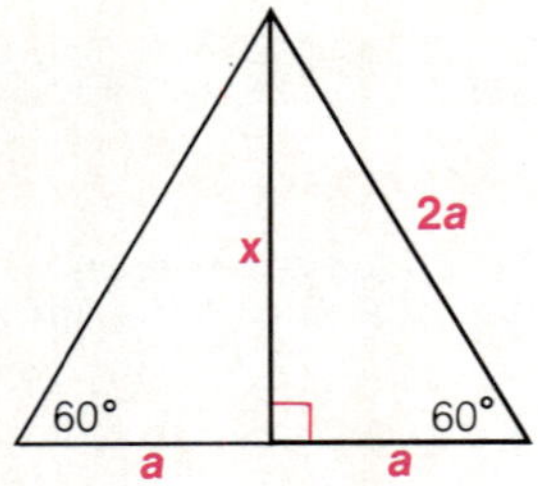

The short leg of the 30°-60° right triangle is half as long as the hypotenuse. If the hypotenuse is length **2*a***, the short leg has length ***a***.

You can use the Pythagorean theorem to find the length of the long leg of the 30°-60° right triangle.

$$(2a)^2 = x^2 + a^2$$

$$4a^2 = x^2 + a^2$$

$$3a^2 = x^2$$

$$\sqrt{3}a = x$$

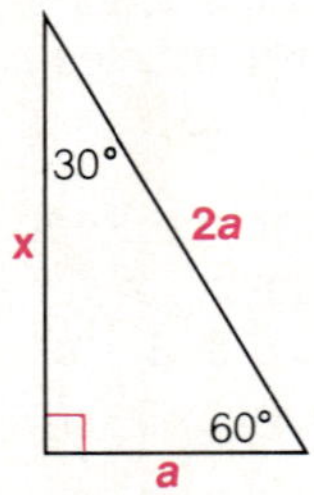

30°-60° Right Triangles

If a 30°-60° right triangle has a shorter leg ***a*** units long, then the longer leg is ***a*√3** units long and the hypotenuse is **2*a*** units long.

Example 2

Find lengths x and y.

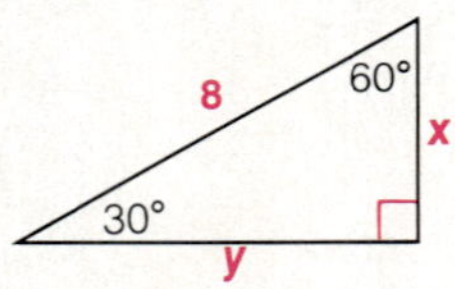

Solution $x = \frac{8}{2} = 4$ The short leg of a 30°–60° right triangle is half the length of the hypotenuse.

$y = 4\sqrt{3}$ The long leg of a 30°–60° right triangle is $\sqrt{3}$ times the length of the short leg.

Practice Draw a 30°-60° right triangle. Find the lengths of the legs if the hypotenuse is 5. $x = 2.5$; $y = 2.5\sqrt{3}$

Oral Exercises

Is it possible for a right triangle to have side lengths as shown?

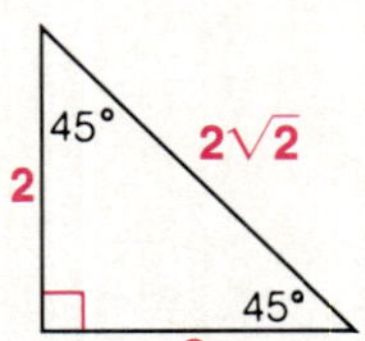

1. yes

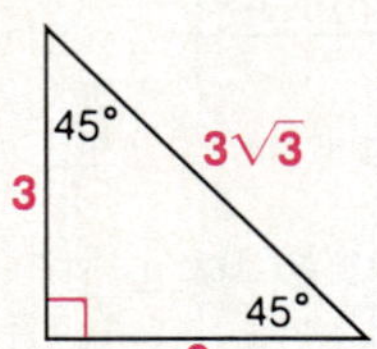

2. no

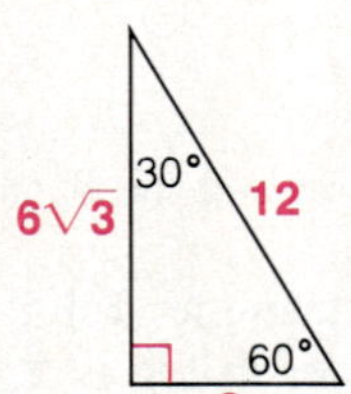

3. yes

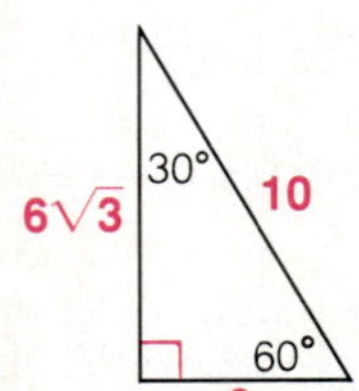

4. no

Exercises

A Find the unknown lengths.

1. 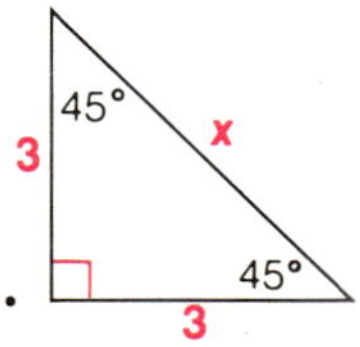

$x = 3\sqrt{2}$

2. 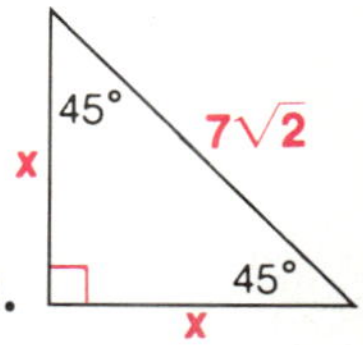

$x = 7$

3. 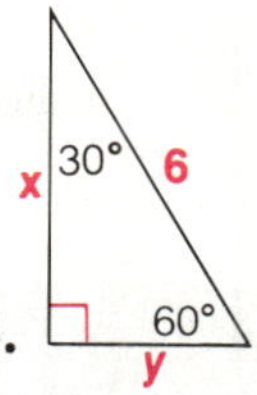

$x = 3\sqrt{3}$ $y = 3$

4. 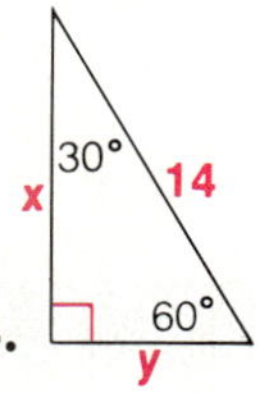

$x = 7\sqrt{3}$ $y = 7$

Draw a 45°–45° right triangle for Exercises 5–6.

5. Find the length of the hypotenuse if the length of each leg is 5. $5\sqrt{2}$

6. Find the length of the hypotenuse if the length of each leg is 4.3. $4.3\sqrt{2}$

Draw a 30°–60° right triangle for Exercises 7–8.

7. Find the lengths of the legs if the length of the hypotenuse is 25. $12.5\sqrt{3}$, 12.5

8. Find the lengths of the legs if the length of the hypotenuse is $2\sqrt{3}$. $\sqrt{3}$, 3

B Solve. Use the approximations $\sqrt{2} \approx 1.414$ and $\sqrt{3} \approx 1.732$.

Refer to the planter box at right for Exercises 9 and 10. The end of a planter box is shaped like a trapezoid, with dimensions as shown.

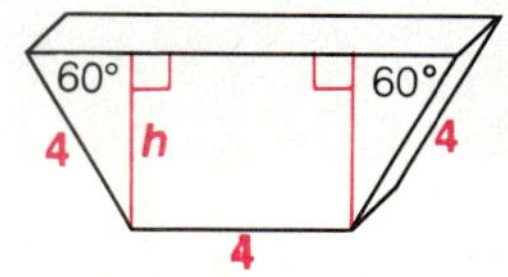

9. Find the height of the planter. 3.464

10. Find the length of the upper edge of the end of the planter. 8

11. On a baseball diamond, the distance between bases is 90 ft. Find the approximate distance from home plate to second base. 127.26 ft

12. The end of a tent is shaped like an equilateral triangle. If the base of the tent is 5 feet across, how high is the tent? 4.33 ft

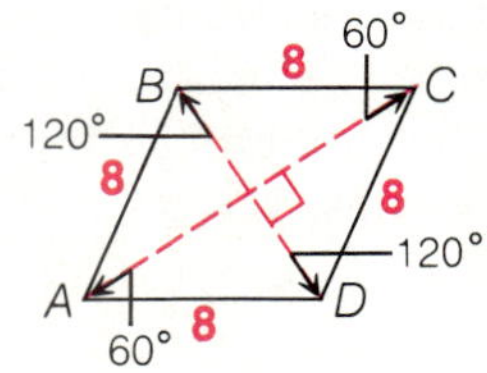

13. *ABCD* is a rhombus with all sides of length 8. Diagonals *AC* and *BD* bisect the angles and are perpendicular to each other. Find the length of each diagonal.
BD = 8 *AC* = $8\sqrt{3} = 13.856$

C Extending Thinking Skills

14. Triangle *ABC* is drawn on the surface of a box whose length, width, and height are all 1. What is the length of *AB*? $\sqrt{2}$

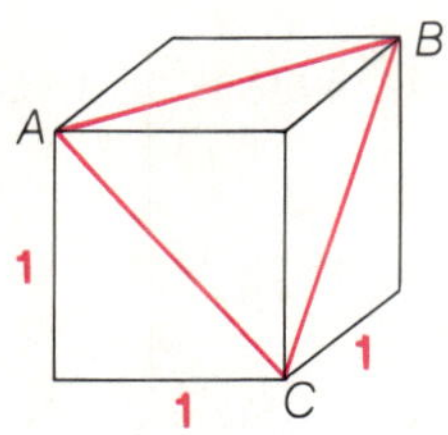

15. What is the area of $\triangle ABC$? $\frac{\sqrt{3}}{2}$

Thinking skill: 14, 15. Using space visualization

Mixed Review

Solve and check. **16.** $35 = 5c + 2c$ 5 **17.** $m + 16 = 17$ 1

18. $10 + 2m = 13$ 1.5 **19.** $t + 6t = -42$ −6 **20.** $14 - 2m = 18$ −2

14-8 Trigonometric Ratios

Objective: To find and use sine, cosine, and tangent.
MIN: 1–6; REG: 1–9 even, 10–25; MAX: 1–9 even, 10–31; ALL: Mixed Review

You can use trigonometric ratios to find a length that cannot be measured directly.

A table of trigonometric ratios appears on page 557.

When you want to find a distance that cannot be measured directly, such as the height of a building, you can use an indirect method of measurement. One such method is to use trigonometric ratios.

Trigonometric ratios are ratios of lengths of sides of right triangles. Three of these ratios are called the **sine,** the **cosine,** and the **tangent.** They are defined for an acute angle A in a right triangle as follows.

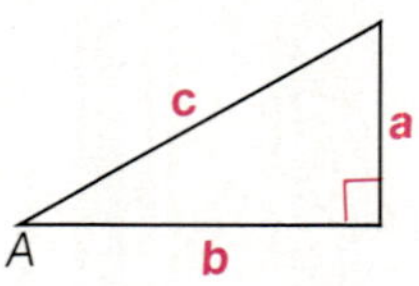

$$\text{sine of } \angle A \text{ or } \sin A = \frac{\text{length of side opposite } \angle A}{\text{length of hypotenuse}} = \frac{a}{c}$$

$$\text{cosine of } \angle A \text{ or } \cos A = \frac{\text{length of side adjacent to } \angle A}{\text{length of hypotenuse}} = \frac{b}{c}$$

$$\text{tangent of } \angle A \text{ or } \tan A = \frac{\text{length of side opposite } \angle A}{\text{length of side adjacent to } \angle A} = \frac{a}{b}$$

Since ratios of corresponding sides of similar triangles are equal, these three ratios have the same value for any acute angle congruent to angle A.

Example 1

Find sin A, cos A, and tan A for $\angle A$.

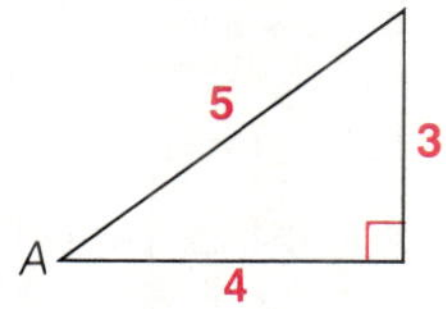

Solution

$\sin A = \frac{3}{5}$ Sine is the length of the opposite side divided by the length of the hypotenuse.

$\cos A = \frac{4}{5}$ Cosine is the length of the adjacent side divided by the length of the hypotenuse.

$\tan A = \frac{3}{4}$ Tangent is the length of the opposite side divided by the length of the adjacent side.

Practice Find sin A, cos A, and tan A for $\angle A$.

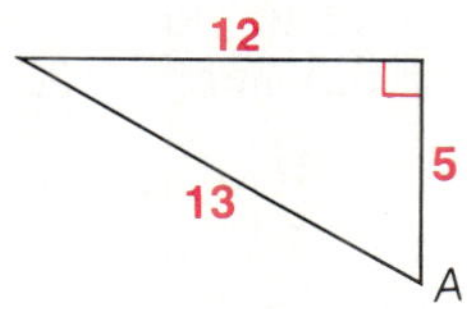

$\sin A = \frac{12}{13}$
$\cos A = \frac{5}{13}$
$\tan A = \frac{12}{5}$

Example 2

The measure of angle A is 37° at a point 155 feet from the base of the flagpole. Tan 37° ≈ 0.75. Find the height of the flagpole.

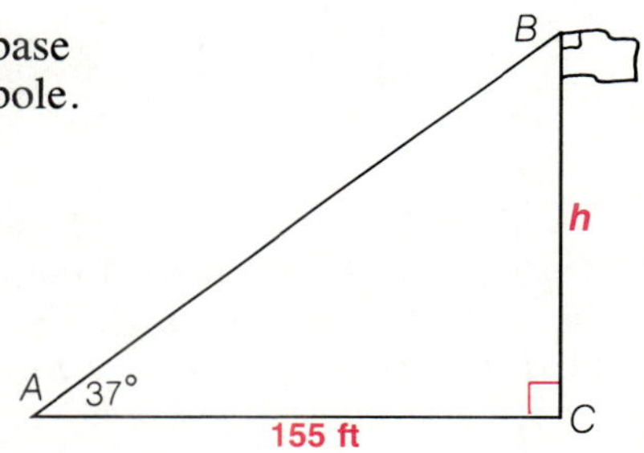

Solution $\tan 37° = \frac{h}{155}$

$$h = 155(\tan 37°)$$
$$h \approx 155(0.75)$$
$$\approx 116.25 \text{ ft.}$$ Substitute the value of tan 37°.

Practice Find the length of AB in the figure above.
$AB^2 = 116.25^2 + 155^2$, $AB \approx 193.75$

Oral Exercises

State whether each equation is true or false.

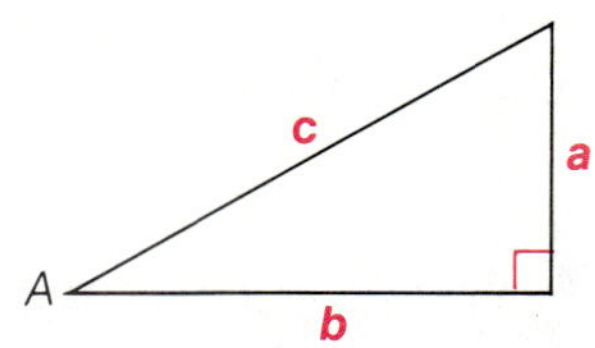

1. $\sin A = \frac{a}{b}$ F **2.** $\cos A = \frac{b}{c}$ T **3.** $\tan A = \frac{b}{a}$ F

4. $\sin A = \frac{a}{c}$ T **5.** $\cos A = \frac{b}{a}$ F **6.** $\tan A = \frac{a}{b}$ T

Exercises **1–6. See additional answers.**

A Find sin A, cos A, and tan A for $\angle A$.

1. 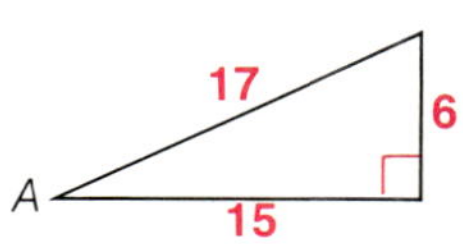

2. 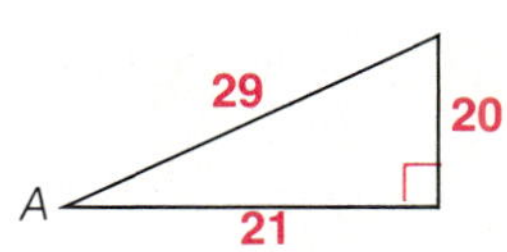

3. 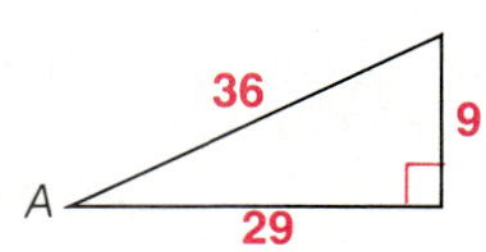

4. 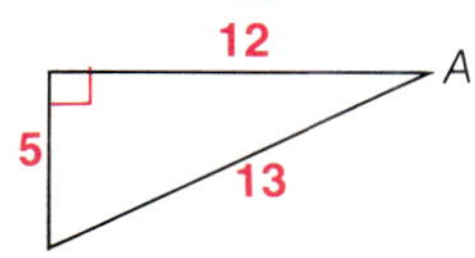

5. 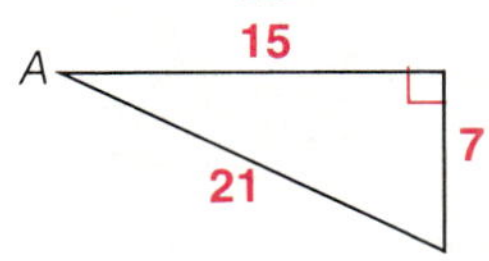

6. 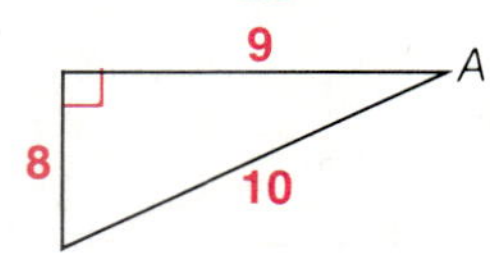

7. The angle from the ground to the top of the World Trade Center in New York City is 48° at a distance of 1220 ft from the building. Tan 48° ≈ 1.11. Find the height of the building.

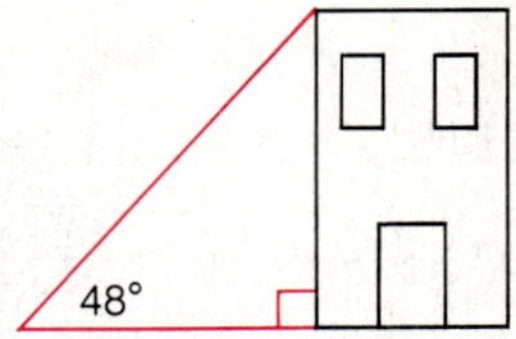

tan 48° = $\frac{h}{1,220}$ h = 1,354.2 ft

8. Sears Tower in Chicago is 1454 ft tall. Suppose point A is 1000 ft from the base of the tower. What is the tangent of the angle at A formed by the ground and the line of vision to the top of the tower? tan A = 1.454

9. A surveyor measured BC to be 125 ft. Sin 29° ≈ 0.48. Find the distance AB across the lake. sin 29° = $\frac{125}{AB}$ AB = 260.42 ft

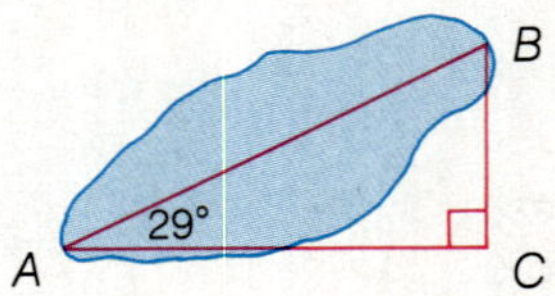

10. A surveyor wants to find the distance between mountain peaks A and B. He finds a point C, 228 ft from point A, so that $\angle ACB$ is a right angle. The measure of $\angle BAC$ is 89°. Find the distance AB if cos 89° ≈ 0.017. cos 89° = $\frac{228}{AB}$ AB ≈ 13,412 ft

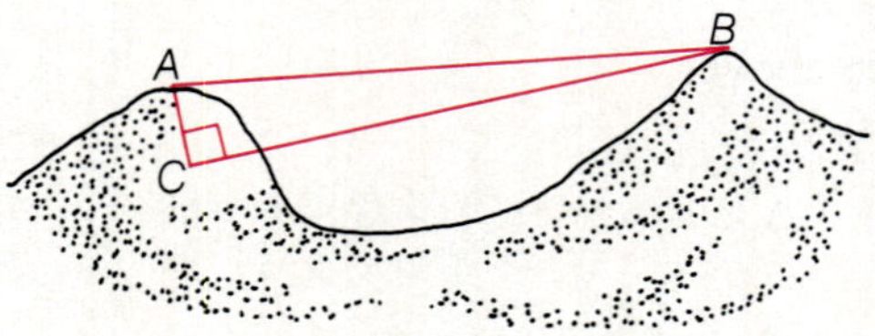

B Draw a 45°–45° right triangle with legs of length 2. Draw a 30°–60° right triangle with a short leg of length 1. Find the lengths of all other sides. Use these figures to find the trigonometric ratios in Exercises 11–18. Leave your answer in radical form.

11. sin 30° $\frac{1}{2}$ **12.** tan 45° 1 **13.** cos 60° $\frac{1}{2}$ **14.** tan 30° $\frac{1}{\sqrt{3}}$ or $\frac{\sqrt{3}}{3}$

15. cos 45° $\frac{1}{\sqrt{2}}$ or $\frac{\sqrt{2}}{2}$ **16.** sin 60° $\frac{\sqrt{3}}{2}$ **17.** sin 45° $\frac{1}{\sqrt{2}}$ or $\frac{\sqrt{2}}{2}$ **18.** tan 60° $\sqrt{3}$

Find sin A, cos A, and tan A for $\angle A$. Use the Pythagorean Theorem to find x.
See additional answers.

19. 25
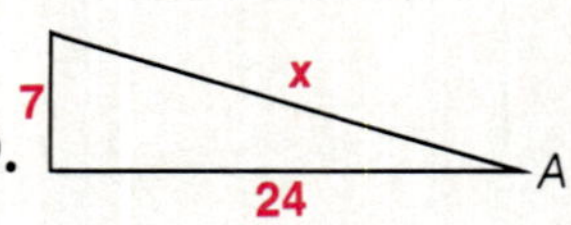

20. 14.42
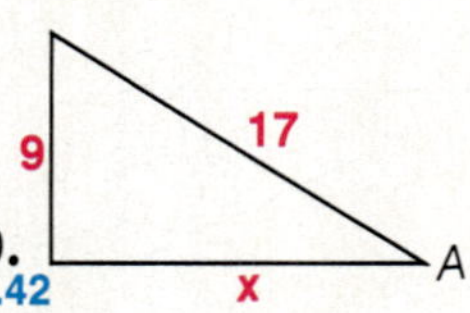

21. 5.74
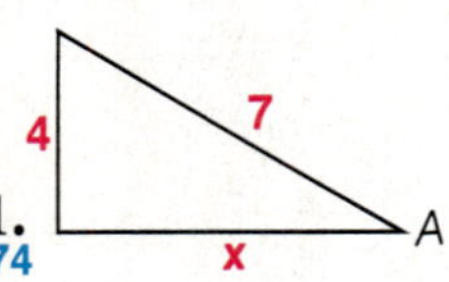

Use your answers to Exercises 11–18 to evaluate each expression.

22. $5 + x \sin 30° \text{ for } x = 4$ 7

23. $y \cos 45° \text{ for } y = 4\sqrt{2}$ 4

24. $x \tan 45° + 2x \sin 30° \text{ for } x = 5$ 10

25. $5x \sin 30° \text{ for } x = 12$ 30

C Extending Thinking Skills Thinking skill: Using reverse reasoning

Use the results from Exercises 11–18 to solve the equations.

26. $\sin x = \frac{1}{2}$ $x = 30°$

27. $\cos x = \frac{\sqrt{3}}{2}$ $x = 30°$

28. $\tan x = \sqrt{3}$ $x = 60°$

29. $\sin 2x = \frac{\sqrt{3}}{2}$ $x = 30°$

30. $\tan 2x = 1$ $x = 22.5°$

31. $\cos 3x = \frac{1}{2}$ $x = 20°$

Mixed Review

Use the counting principle to find the total number of choices.

32. Select one soup and one salad from 4 soups and 5 salads. 20

Give the area. **33.** A square with one side = 4 cm. 16 cm^2

34. A circle with diameter = 6 cm. Use 3.14 for π. 28.26 cm^2

35. A triangle with base = 7 in, height = 8 in. 28 in^2

Give the mean. **36.** 12, 15, 17, 12, 13 13.8 **37.** 6, 2, 5, 4, 6, 0, 6 4.14

Give the least common multiple (LCM). **38.** 4, 5, 6, 12 60 **39.** 2, 3, 4, 8 24

Solve and check. **40.** $3m - 12 = -6$ 2 **41.** $9y - 6y = 12$ 4

42. $10 = 4c - 2c$ 5 **43.** $9 - 3c = 0$ 3 **44.** $12 + 3n = 9$ −1

COMPUTER ACTIVITY

This program will list the values of the trigonometric ratios for all angles from 1 degree to 45 degrees.

```
10 PRINT TAB(1)"ANGLE";"SINE(A)"; TAB(19)"COSINE(A)";
   TAB(31)"TANGENT(1)"
20 PRINT
30 FOR N=1 TO 45
40 PRINT N;" DEG ";SIN(N*3.14159/180);" ";
   COS(N*3.14159/180);" ";TAN(N*3.14159/180)
50 NEXT N
60 END
```

1. Load and run this program.

2. Change the program to find the ratios for all angles from 1 degree to 90 degrees. Load and run the new program. Change: 30 for N = 1 to 90

PROBLEM SOLVING: WRITING EQUATIONS

14-9 Practice Solving Problems

Objective: To write and solve equations to solve word problems.
MIN: 1–6; REG: 1–6 even, 7–8; MAX: 1–6 even, 7–10; ALL: Mixed Review

A Solve by writing an equation.

1. A field is in the shape of a right triangle. The length of the longest side of the field is twice the length of the shortest side. The other side is 149 m long. The perimeter of the field is 407 m. How long is each side? $407 = 149 + x + 2x$; 86 m, 149 m, 172 m

2. The total cost of a bicycle is \$140. Glenda put \$20 down on the bicycle and will pay the rest in 3 equal payments. How much will each payment be? $140 = 20 + 3x$; \$40

3. The Redbirds were only 2 points short of getting double the Gray Sox score in a basketball game. The Redbirds scored 86 points. How many points did the Gray Sox score? $86 + 2 = 2s$; $s = 44$

4. Bob rode his bike from his house to a house 5 miles north. He rode at the rate of $\frac{1}{4}$ mile per minute. How far from home was he after 20 minutes? $d = \frac{1}{4}(20)$; 5 mi

5. A board 6 m long is to be cut into 2 pieces. The length of one piece must be 0.5 m more than twice the length of the other. How long should each piece be? $6 = x + (2x + 0.5)$; 1.83 m and 4.17 m

6. Mary made \$4.75 an hour and \$15 in overtime pay the first week of her job. Her total salary for the week was \$181.25. How many regular hours did she work? $4.75h + 15 = 181.25$; 35 h

B Solve by writing an equation.

7. A carpenter measured the sides of a triangular wooden frame for a house. Each side measured one foot more than the previous side. The perimeter of the frame was 51 feet. How long was each side of the frame? $51 = x + (x + 1) + (x + 1 + 1)$, 16 ft, 17 ft, 18 ft

8. 50% of the cost of a television set was \$60 more than 25% of its cost. How much did the television cost? $0.5c - 60 = 0.25c$, $c = \$240$

C Extending Thinking Skills

Thinking skill: 9, 10 Formulating a problem

9. Write a word problem that could be solved by the equation $25x - 15 = 65$.
Answers will vary.

10. Write a word problem that could be solved by the equation $x + 12 = 4x$.
Answers will vary.

Mixed Review

Solve and check. **11.** $9x + 3x = 48$ 4 **12.** $r = 16 + 3r$ -8 **13.** $x + 3x = 8$ 2

Write the prime factorization, using exponents. **14.** 504 $2^3 \cdot 3^2 \cdot 7$ **15.** 495 $3^2 \cdot 5 \cdot 11$

PROBLEM SOLVING: STRATEGIES

14-10 Practice Solving Problems

Objective: To solve nonroutine problems.
MIN: 1–4; REG: 1–7; MAX: 1–10

Problem-Solving Strategies

Choose the Operations	Write an Equation
Guess, Check, Revise	Simplify the Problem
Draw a Picture	Make an Organized List
Make a Table	Use Logical Reasoning
Look for a Pattern	Work Backwards

This chart shows the strategies presented in previous chapters.

Problems

Solve. **1. Assume all tanks are empty. Assume all tanks are $\frac{1}{4}$ full. Answer falls between.**

1. The owner of a car rental business buys gasoline at $1.05 per gallon. Each of the 25 rental cars holds 15 gallons. If the tank in each car is between empty and one-quarter full, about how much would it cost to fill all 25 cars with gasoline? **$\$295.31 \le x \le \$393.75$**

2. A shipment of 3 boxes of machine parts weighed a total of 65 kg. The heaviest box weighed 3 times as much as the second box, which was 3 times heavier than the lightest box. How much did each box weigh? **5, 15, and 45 kg**

3. The owner of a furniture factory set up a plan for training new employees. The first training session was to be for 1 employee. Each of the following sessions was to hold 2 more people than the previous one. The factory had 30 new employees. How many training sessions were needed? **1 + (1 + 2) + (3 + 2) + (5 + 2) . . . 6 sessions**

4. Three baseball teams played each other one time at home and one time away. The Blue Birds never defeated the Lions. The Spartans never lost a home game. The Spartans lost 2 games. Find how many of these games each team won and lost.

	won	lost
B	1	3
S	2	2
L	3	1

5. A printer who made envelopes and stationery for Companies A, B, and C mixed up the shipments. Each company received another company's envelopes and yet another company's stationery. Company A received Company C's stationery. Whose envelopes and whose stationery did each company receive? **See additional answers.**

6. By mistake, three groups are all scheduled to meet at the same time on the same day. They are the Debate Club, the Student Council, and the Math Club. Every member of every club cannot attend, because some students belong to more than one of the clubs. Half of the Math Club and all of the Debate Club belong to the Student Council. Five students are members of all three groups. There are 28 members of the Math Club, 11 members of the Debate Club, and 35 members of the Student Council. The Student Council's meeting is official only if 60% of its members attend. If none of the Math Club and Debate Club members who also belong to the Student Council attend the Student Council meeting, will the Student Council meeting be official? **no**

7. A secret-service agent has been told to find all possible routes from the airport to the state capitol building. Each route has to be studied and the safest one selected. The figure below shows the streets from the airport to the capitol. All streets are one-way, as shown by the arrows. How many different routes are possible? **21 routes**

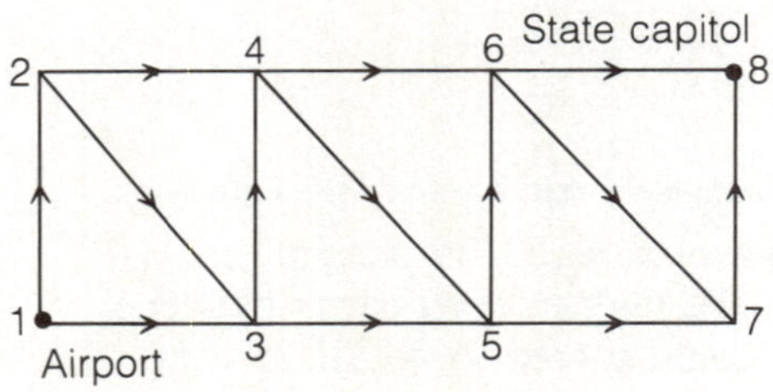

8. Ms. Wells had a stack of math papers to grade. She graded $\frac{1}{4}$ of the papers during lunch break. She graded $\frac{1}{2}$ of what was left after school. She took the rest of the papers home with her and graded only $\frac{1}{6}$ of those. She had 20 papers left to grade the next morning. How many papers did she have in the beginning? **64**

9. At Peter's Pizza Pan, customers can order thin or deep-dish pizza with whole-wheat or regular crust. They can choose either one or two of the following toppings: pepperoni, sausage, mushrooms. How many different pizzas could a customer order? **24**

10. A paint salesperson visits a store every 8th working day. Another salesperson visits the same store every 6th working day. The store is open Monday through Saturday. Today is Tuesday, May 2nd. Both visited the store today. On what date will they both visit the store again? Can one of them change his schedule and never meet the other in the store again? Explain. **May 30. They could each visit the store every six days, but start on different days.**

Enrichment

Sequences and Square Roots

Because the number $\sqrt{2}$ is an irrational number, it cannot be written exactly as a decimal or a fraction. The terms of the sequence below are approximations of $\sqrt{2}$. Each term of the sequence is a closer approximation than the term before it.

$$\frac{3}{2}, \frac{7}{5}, \frac{17}{12}, \frac{41}{29}, \frac{99}{70}, \ldots$$

$$\frac{99}{70} \longrightarrow \frac{99 + 2(70)}{99 + 70} = \frac{239}{169}$$

These calculations show that the fractions get closer and closer to $\sqrt{2}$.

You can use a calculator to confirm that the square of each term of this sequence is as shown.

$$\left(\frac{3}{2}\right)^2 = 2.25 \quad \left(\frac{7}{5}\right)^2 = 1.96 \quad \left(\frac{17}{12}\right)^2 = 2.0069 \quad \left(\frac{41}{29}\right)^2 = 1.9988$$

$$\left(\frac{99}{70}\right)^2 = 2.00020$$

You can see that $\frac{99}{70}$ is approximately $\sqrt{2}$.

Find the next term in each sequence. Then use a calculator to find what square root each sequence approximates.

1. $\frac{4}{3}, \frac{10}{7}, \frac{24}{17}, \frac{58}{41}, \frac{140}{99}, \ldots$ $\frac{338}{239}$ $\sqrt{2}$

2. $\frac{3}{2}, \frac{9}{5}, \frac{24}{14}, \frac{66}{38}, \frac{180}{104}, \ldots$ $\frac{492}{284}$ $\sqrt{3}$

3. $\frac{4}{3}, \frac{13}{7}, \frac{34}{20}, \frac{94}{54}, \frac{256}{148}, \ldots$ $\frac{700}{404}$ $\sqrt{3}$

4. $\frac{2}{1}, \frac{5}{3}, \frac{14}{8}, \frac{38}{22}, \frac{104}{60}, \ldots$ $\frac{284}{164}$ $\sqrt{3}$

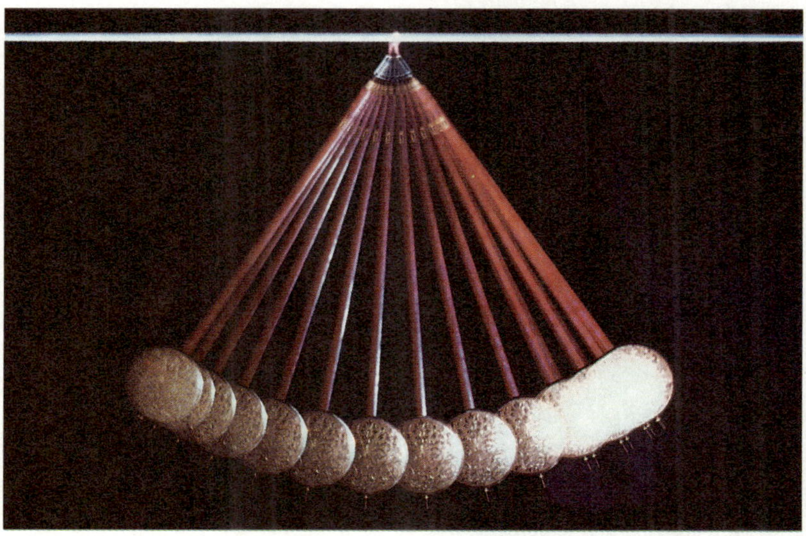

For a pendulum with length L, the time (T) of one period is given by the formula $T = 2 \cdot \sqrt{L}$.

Chapter 14 Review

14-1 Between what two consecutive integers does each square root lie?

1. $\sqrt{33}$ 5 and 6

2. $\sqrt{75}$ 8 and 9

Evaluate each expression.

3. $\sqrt{81} - \sqrt{49}$ 2

4. $\sqrt{140 - 19}$ 11

14-2 Use a calculator with a $\sqrt{\ }$ key to find an approximation for each square root.

5. $\sqrt{7}$ 2.65

6. $\sqrt{117}$ 10.82

14-3 Solve and check. **7.** $x^2 = 169$ 13, −13 **8.** $x^2 = 12$ 3.46, −3.46

14-4 Find the length of a to the nearest hundredth.

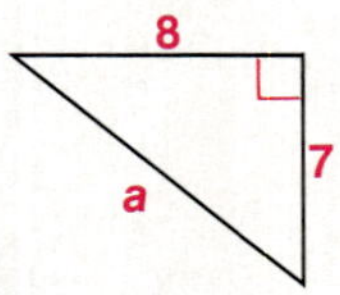

9.
10.63

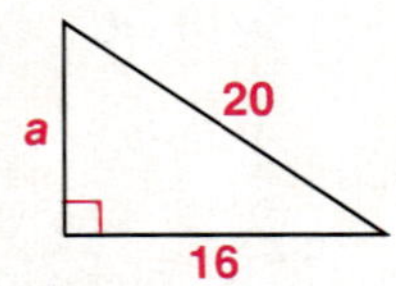

10.
12

14-5 Solve. **11.** A standard-size base in baseball is 15 inches long and 15 inches wide. What is the length of the diagonal of a base? $\sqrt{450}$ in. or 21.21 in.

14-6 Use proportions to find the unknown length for these similar triangles.

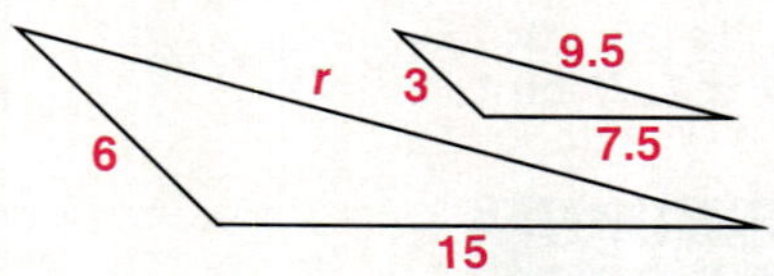

12.
19

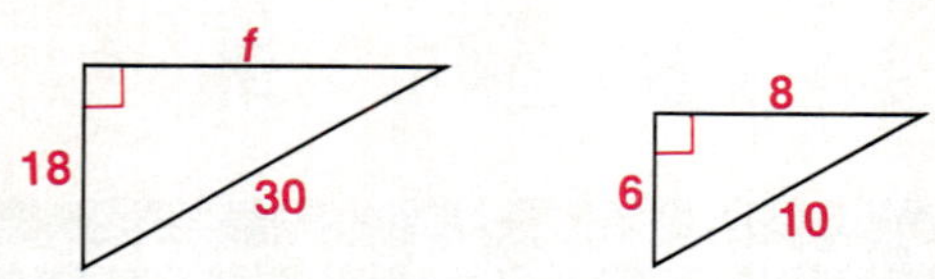

13.
24

14-7 Find the unknown length.

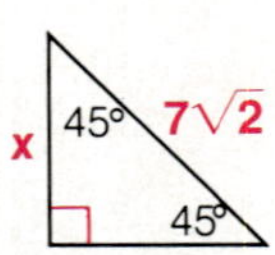

14.
7

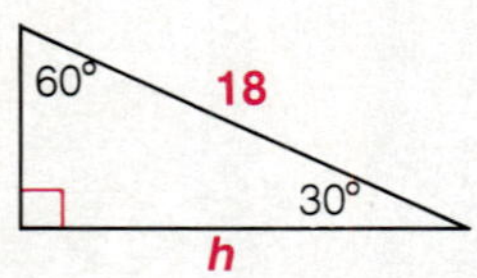

15.
$9\sqrt{3}$

14-8 Find each ratio for $\angle A$.

16. $\sin A$ $\frac{30}{42} = \frac{5}{7}$

17. $\cos A$ $\frac{14}{42} = \frac{1}{3}$

18. $\tan A$ $\frac{30}{14} = \frac{15}{7} = 2\frac{1}{7}$

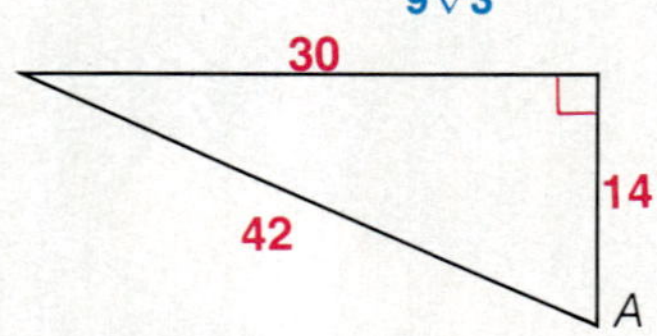

14-9 Solve by writing an equation. **19.** A garden with an area of 870 square feet is to be divided into 2 sections. One section will be 50 square feet larger than the other. What will be the area of each section?
$x + (x + 50) = 870$ 410 ft^2, 460 ft^2

Chapter 14 Test

Between what two consecutive integers does each square root lie?

1. $\sqrt{61}$ 7 and 8 **2.** $\sqrt{18}$ 4 and 5

Evaluate each expression.

3. $\sqrt{25} - \sqrt{100}$ −5 **4.** $\sqrt{33 + 16}$ 7

Use a calculator with a $\sqrt{}$ key to find an approximation for each square root.

5. $\sqrt{11}$ 3.32 **6.** $\sqrt{96}$ 9.8

Solve and check. **7.** $x^2 = 1$ 1, −1 **8.** $x^2 = 43$ 6.56, −6.56

Find the length of a to the nearest hundredth.

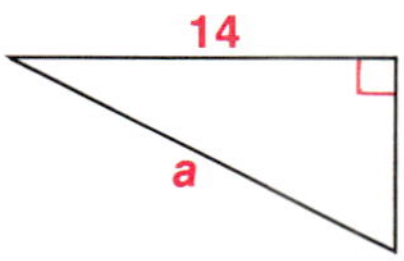

9.
15.65

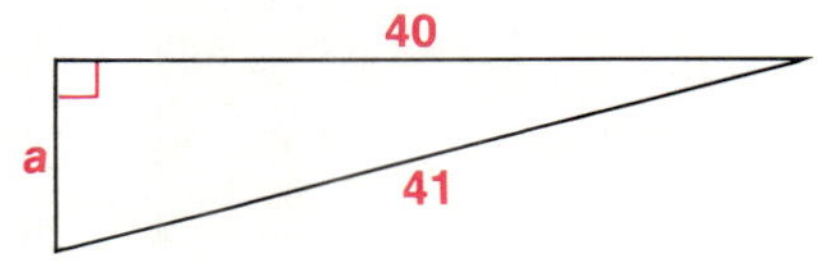

10.
9

Solve. **11.** A window is 6 feet wide and 4 feet high. What is the length of the diagonal of the window? 7.2 ft

Use proportions to find the unknown lengths for these similar triangles.

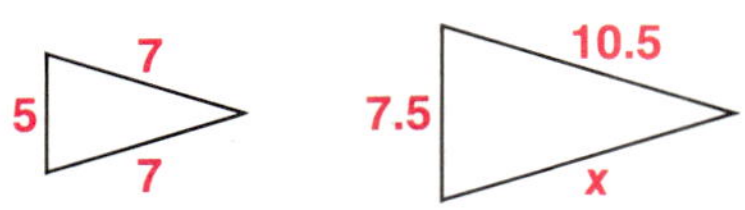

12.
10.5

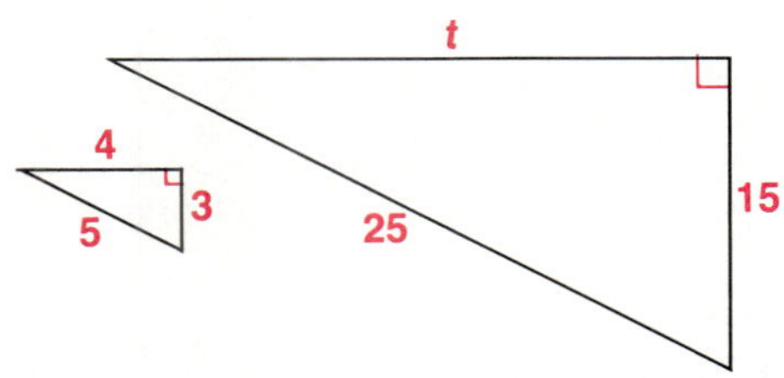

13.
20

Find the unknown length.

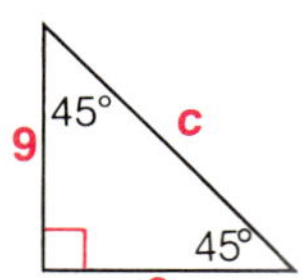

14.
$9\sqrt{2}$

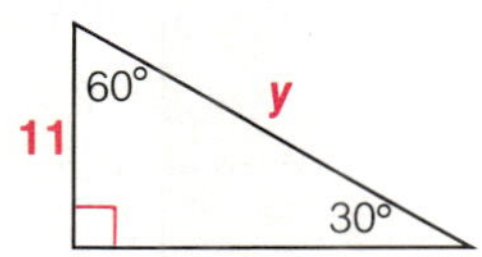

15.
22

Find each ratio for $\angle A$.

16. sin A $\frac{21}{29}$

17. cos A $\frac{20}{29}$

18. tan A $\frac{21}{20}$

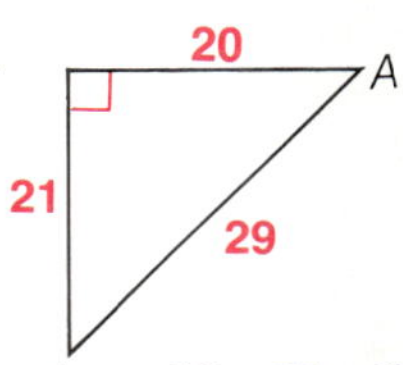

Solve by writing an equation. **19.** The Wilson family drove 623 miles to visit relatives. They drove 51 miles more than $\frac{1}{2}$ of the distance on the first day. How far did they drive the second day?
260.5 miles $\frac{623}{2} + 51 + d = 623$

Cumulative Review

Solve the problem. Use the formula Distance = rate × time.

1. An airplane averaged 470 miles per hour on a $4\frac{1}{2}$ hour flight. How far did it fly? 2115 miles

Find the circumference of a circle with the given diameter or radius. Use 3.14 for π.

2. diameter = 8 cm 25.12 cm

3. radius = 1.2 m 7.536 m

Use the SAS, SSS, or ASA property to show that each pair of triangles is congruent.

4.

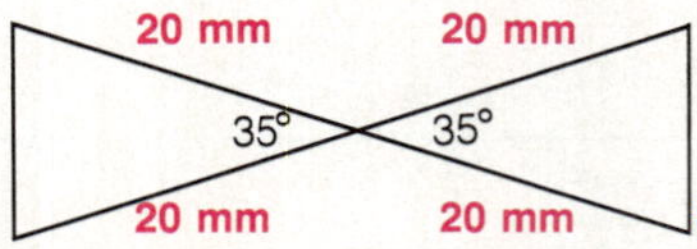

SAS

5.

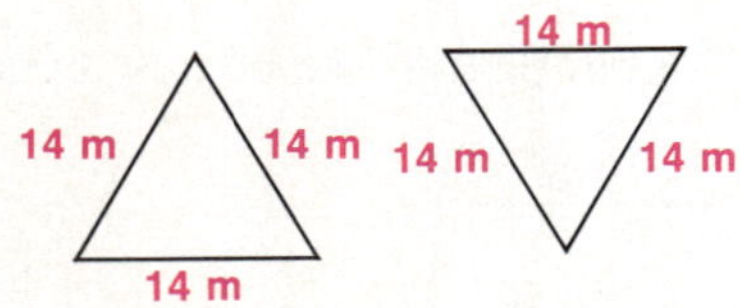

SSS

Solve by writing an equation. **6.** Mr. Dunlop bought 15% of the total number of tickets sold to a concert. If Mr. Dunlop bought 180 tickets, how many tickets were sold in all? $0.15n = 180, n = 1200$

Find the volume of the prism or cylinder. Use 3.14 for π.

7.

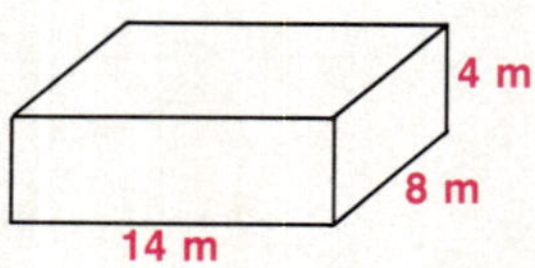

448 m³

8.

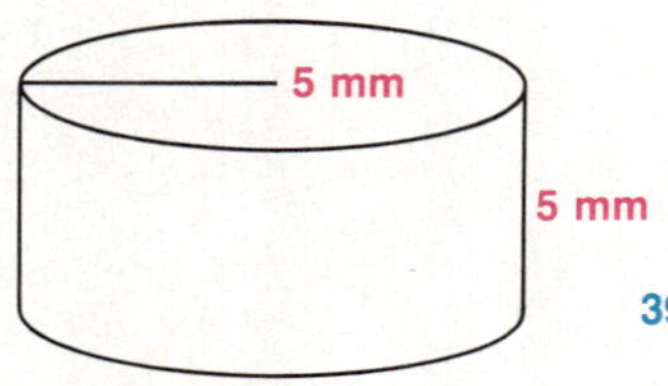

392.5 mm³

Solve.

9. How many cubic yards of concrete are needed for a driveway 54 feet long, 12 feet wide and 6 inches thick? 12 yd³

There are 12 tennis balls in a bag: 4 are green, 3 are white, 5 are yellow. You select one ball from the bag without looking. Find each probability.

10. P(yellow) $\frac{5}{12}$

11. P(not green) $\frac{8}{12} = \frac{2}{3}$

12. P(green or white) $\frac{7}{12}$

You throw a die twice. Give the probability of each.

13. P(3 and 4) $\frac{1}{36}$

14. P (first throw > 4, second throw an even number) $\frac{1}{6}$

There are 9 balls in a bag: 5 red and 4 blue. Find each probability if you pull one ball from the bag and then pull another ball from the bag without replacing the first.

15. P(red then blue) $\frac{5}{18}$

16. P(blue then blue) $\frac{1}{6}$

Chapter 15
Graphs of Equations and Inequalities

15-1 Graphing on the Number Line

Objective: To graph solutions of inequalities on a number line.
MIN: 1–20; REG: 1–20 even, 21–28; MAX: 1–20 even, 21–30; ALL: Mixed Review
You may want to review the development of inequalities in Chapter 8.

The **real numbers** are the numbers represented by the points on the number line. Solutions to inequalities are graphed on the number line. **Emphasize that the real numbers consist of the rational and irrational numbers together.**

Example 1

Solve $2x - 5 < 1$. Graph the solutions on a number line.

Solution $2x - 5 < 1$ To solve the inequality, you need to get x by itself on one side.

$2x - 5 + 5 < 1 + 5$ Add 5 to both sides of the inequality.

$\frac{2x}{2} < \frac{6}{2}$ Divide both sides by 2.

$x < 3$

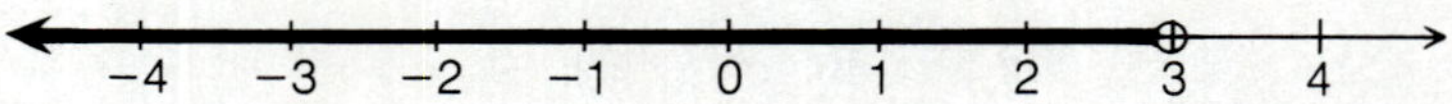

This solution is "all real numbers less than 3." Shade the number line to the left of 3. The open circle means the number 3 is not included.

Practice Solve. Graph the solutions on a number line.

a. $3x + 2 > -6$ **b.** $3x + 4 < 7$

$x > -\frac{8}{3}$ $x < 1$

a.

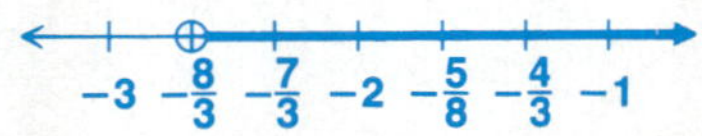

b.

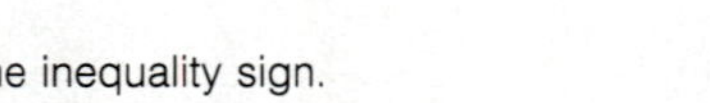

Example 2

You may want to review the method of checking inequalities, presented in Chapter 8.

Solve $-2x + 1 \leq 5$. Graph the solutions on a number line.

Solution $-2x + 1 \leq 5$

$-2x + 1 - 1 \leq 5 - 1$ Subtract 1 from both sides of the inequality.

$-2x \leq 4$

$\frac{-2x}{-2} \geq \frac{4}{-2}$ Divide both sides by −2. Reverse the inequality sign.

$x \geq -2$

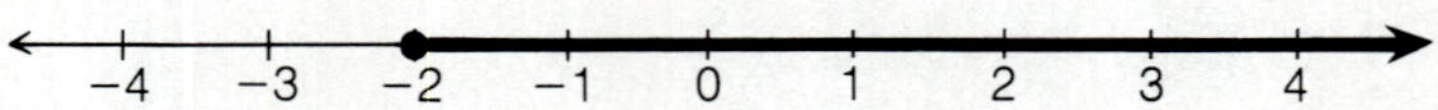

Use a closed circle on the graph to show that −2 is included in the solution.

Practice Solve. Graph the solutions on a number line.

a. $-3x + 7 \leq 13$ **b.** $-2x + 5 \geq 3$

$x \geq -2$ $x \leq 1$

a.

b.

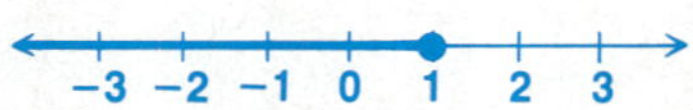

Oral Exercises

Give an inequality for each solution shown.

1.  $x \geq -3$

2.  $x < 1$

3.  $x < 3$

Exercises 15. $x \leq \frac{-9}{5}$

A Solve. Graph the solutions on a number line. For graphs, see additional answers.

1. $x < 3$ **2.** $x > -1$ **3.** $x \leq 2$ **4.** $x \geq -2$

5. $4x > 2$ $x > \frac{1}{2}$ **6.** $-3x \geq -1$ $x \leq \frac{1}{3}$ **7.** $-x > -5$ $x < 5$ **8.** $\frac{1}{5}x < 1$ $x < 5$

9. $x + 2 \leq -3$ $x \leq -5$ **10.** $x - 5 \leq -4$ $x \leq 1$ **11.** $-x + 5 < 2$ $x > 3$ **12.** $3x + 5 \geq -4$ $x \geq -3$

13. $2x + 3 \leq 5$ $x \leq 1$ **14.** $x - 2 > -4$ $x > -2$ **15.** $5x - 3 \leq -12$ **16.** $-2x + 1 \geq 5$ $x \leq -2$

17. $\frac{1}{4}x - 1 \geq -17$ $x \geq -64$ **18.** $-2x - 5 \leq 3$ $x \geq -4$ **19.** $\frac{1}{3}x + 1 < 4$ $x < 9$ **20.** $-3x + 5 < -4$ $x > 3$

B Solve. Graph the solutions on a number line. See additional answers.

Example: $-2 < x \leq 4$. Solution:

The solution is all numbers both greater than -2 and less than or equal to 4.

21. $-1 < x \leq 3$ **22.** $1 \leq x \leq 3$ **23.** $-2 < x < 3$ **24.** $-3 \leq x < -1$

25. $0 \leq x < 2$ **26.** $-3 < x \leq 4$ **27.** $-1 \leq x \leq 2$ **28.** $-3 < x \leq 0$

29, 30. See additional answers.

C Extending Thinking Skills Thinking skills: 29. Using logic 30. Testing a conjecture

29. Graph on a number line all real numbers that satisfy both $-2 < x \leq 3$ *and* $1 \leq x < 4$. $1 \leq x \leq 3$

30. Determine the graph of the solutions to $|x - 5| < 3$. $2 < x < 8$ Encourage students to guess a solution and verify the guess.

Mixed Review

Evaluate for $a = 3$, $b = -1$. **31.** $2(a - b)$ 8 **32.** $a - 3b$ 6 **33.** $b(ba)$ 3

Solve and check. **34.** $21t - 15t = 4t + 8$ 4 **35.** $7y = 6 + 2y$ 1.2

15-2 Graphing on the Coordinate Plane

Objective: To graph an ordered pair on the coordinate plane.
MIN: 1–22; REG: 1–22 even, 23–37; MAX: 1–22 even, 23–39; ALL: Mixed Review

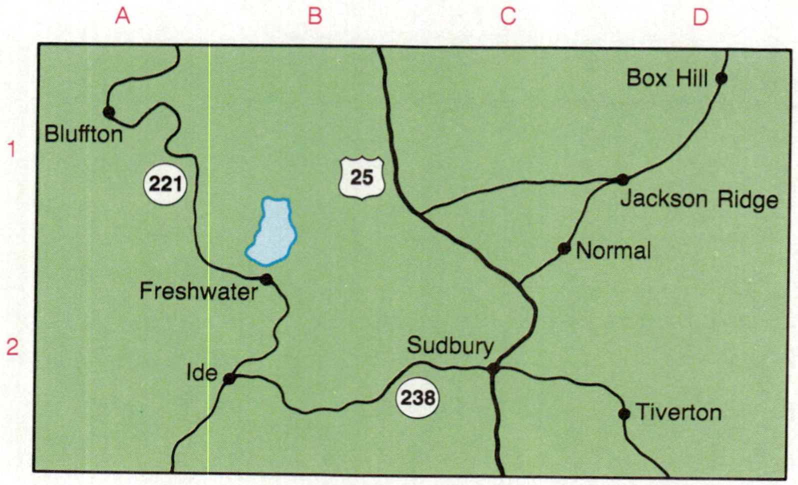

A letter and number pair is used to identify a position on this map. The town of Sudbury is located near the point with coordinates C-2.

In a **rectangular coordinate system**, two perpendicular number lines, called **axes**, intersect at a point called the **origin**. The horizontal axis is called the ***x*-axis** and the vertical axis is called the ***y*-axis**. These two axes allow each point to be named by an **ordered pair** of numbers called the **coordinates** of the point.

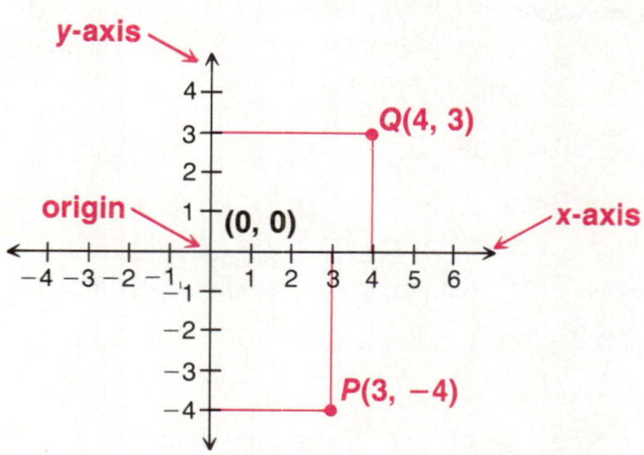

The coordinates of the origin are (0, 0). The coordinates of point P are $(3, -4)$ and the coordinates of point Q are (4, 3). The ***x*-coordinate** of point P is 3 and the ***y*-coordinate** of point P is -4.

Example 1

Give the coordinates of points A, B, and C.

Solution

$A(2, 4)$

$B(-3, 3)$

$C(3, -4)$

Practice Give the coordinates of points D, E, F, and G.
$D(-3, -5)$ $E(5, 2)$ $F(5, -2)$ $G(-4, 2)$

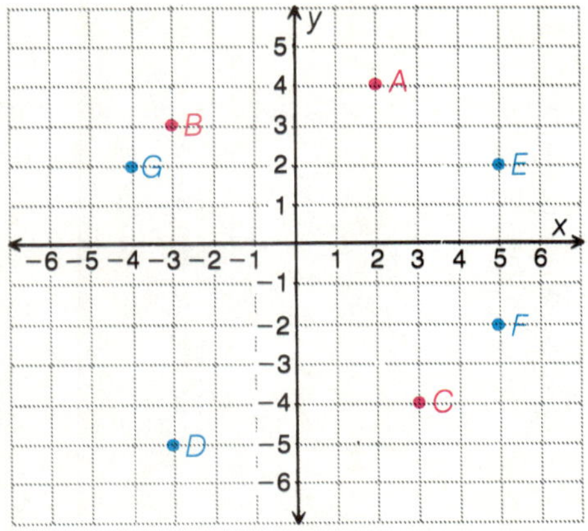

Example 2

Graph points $A(2, 3)$, $B(-3, 5)$, and $C(4, -1)$.

Solution

To graph point A at (2, 3), begin at the origin and move 2 units to the right and 3 units up.

To graph point B at (−3, 5), begin at the origin and move 3 units to the left and 5 units up.

To graph point C at (4, −1), begin at the origin and move 4 units to the right and 1 unit down.

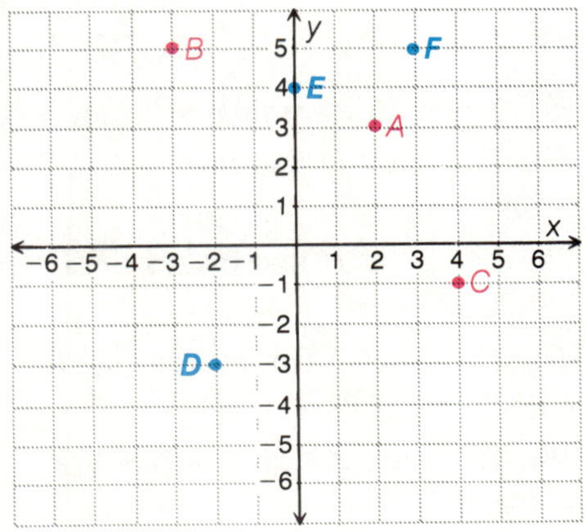

Practice Graph points $D(-2, -3)$, $E(0, 4)$, and $F(3, 5)$.

Oral Exercises

State the x-coordinate and the y-coordinate for each point.

1. (3, −4) **x = 3, y = −4**
2. (−5, 4) **x = −5, y = 4**
3. (−1, 4) **x = −1, y = 4**
4. (2, −3) **x = 2, y = −3**
5. (−12, −7) **x = −12, y = −7**
6. (−3, 0) **x = −3, y = 0**

Describe the moves from the origin you would make to graph these points.

7. (−2, 5) **left 2, up 5**
8. (3, 5) **right 6, up 5**
9. (4, −2) **right 4, down 2**
10. (1, −1) **right 1, down 1**
11. (7, −6) **right 7, down 6**
12. (0, 4) **up 4**

Exercises

A Give the coordinates of each point.

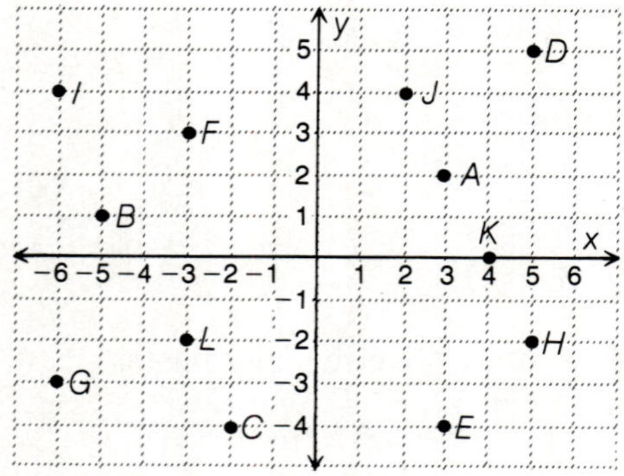

1. *A* (3, 2)
2. *B* (−5, 1)
3. *C* (−2, −4)
4. *D* (5, 5)
5. *E* (3, −4)
6. *F* (−3, 3)
7. *G* (−6, −3)
8. *H* (5, −2)
9. *I* (−6, 4)
10. *J* (2, 4)
11. *K* (4, 0)
12. *L* (−3, −2)

Graph each point. See additional answers.

13. $A(3, 5)$
14. $B(-2, 7)$
15. $C(4, -3)$
16. $D(4, -4)$
17. $E(-2, -3)$
18. $F(-5, 1)$
19. $G(6, 3)$
20. $H(1, -5)$
21. $I(-4, -4)$
22. $J(2, 3)$
23. $K(0, 4)$
24. $L(-3, 0)$

B Name the coordinates for each point.

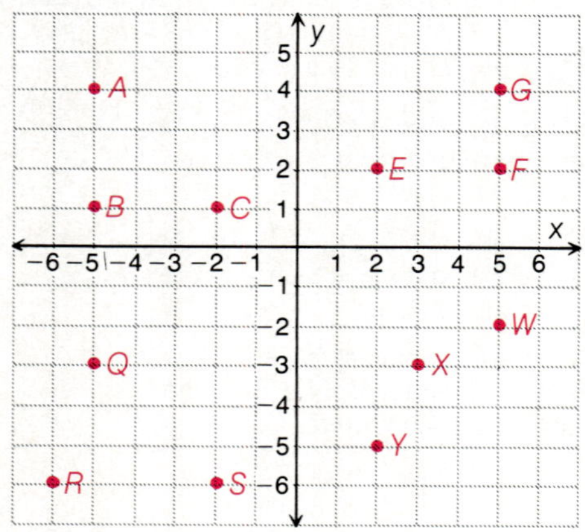

25. The point *D* that makes *ABCD* a square. (−2, 4)

26. The point *H* that makes *EFGH* a rectangle. (2, 4)

27. The point *T* that makes *QRST* a parallelogram. (−1, −3)

28. The point *Z* that makes *WXYZ* a rhombus. (4, −4)

On a single pair of axes draw a continuous line from point *A* to points *B*, *C*, *D*, *E*, and *F* for each of Exercises 29–37. What do you discover? Forms letters: MATH IS FUN.

29. $A(-6, 9)$; $B(-5.5, 11)$; $C(-5, 10)$; $D(-4.5, 11)$; $E(-4, 9)$
30. $A(-3, 9)$; $B(-2, 11)$; $C(-1, 9)$; $D(-1.5, 10)$; $E(-2.5, 10)$
31. $A(1, 11)$; $B(3, 11)$; $C(2, 11)$; $D(2, 9)$
32. $A(4, 11)$; $B(4, 9)$; $C(4, 10)$; $D(6, 10)$; $E(6, 11)$; $F(6, 9)$
33. $A(-1.5, 7)$; $B(-0.5, 7)$; $C(-1, 7)$; $D(-1, 5)$; $E(-1.5, 5)$; $F(-0.5, 5)$
34. $A(3, 7)$; $B(1, 7)$; $C(1, 6)$; $D(3, 6)$; $E(3, 5)$; $F(1, 5)$
35. $A(-2, 3)$; $B(-4, 3)$; $C(-4, 1)$; $D(-4, 2)$; $E(-3, 2)$

36. $A(-1, 3)$; $B(-1, 1)$; $C(1, 1)$; $D(1, 3)$

37. $A(2, 1)$; $B(2, 3)$; $C(4, 1)$; $D(4, 3)$

Thinking skills: 38. Generalizing 39. Testing a conjecture

C Extending Thinking Skills

38. Graph the pairs of points (2, 1) and (1, 2); (−2, 3) and (3, −2); (3, −4) and (−4, 3); (2, 4) and (4, 2); (3, 5) and (5, 3); (−4, −2) and (−2, −4). From the pattern formed, make a generalization about the way the graphs of (a, b) and (b, a) are related. (a, b) and (b, a) are symmetrical about the line $y = x$. Graph, see additional answers.

39. Find an ordered pair (x, y) that is a solution to $|x - 2| + |y - 3| = 0$. (2, 3)

Mixed Review

Evaluate each expression. 40. $\sqrt{169 - 144} + \sqrt{9}$ 8 41. $\sqrt{(104 - 79)} + \sqrt{16}$ 9

Give the mean and mode. 42. 4, 9, 4, 8, 6 mean 6.2, mode 4 43. 17, 12, 15, 11, 12, 14, 16, 12 mean 13.625; mode 12

NUMBERS TO ALGEBRA

You can find the distance between a point and the origin O (0, 0) by using the Pythagorean theorem. In the example below, we demonstrate the method for the specific point $A(4, 3)$ and for the general point $A(x, y)$.

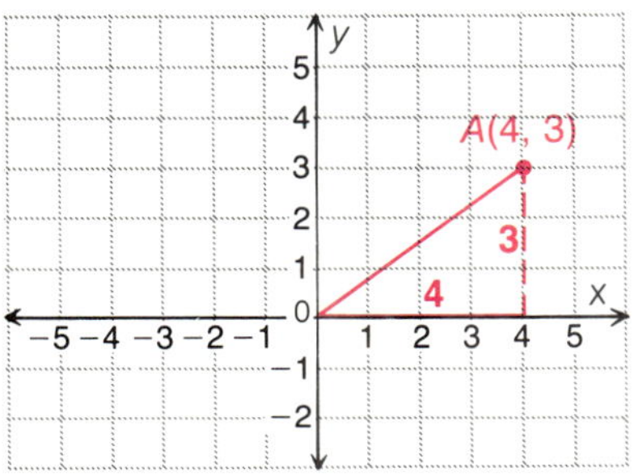

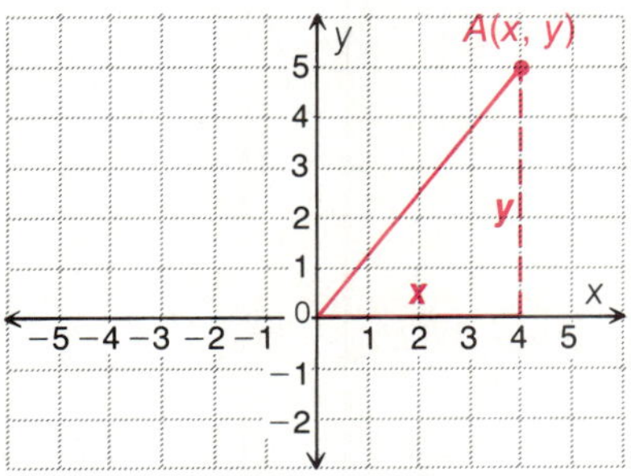

$$OA = \sqrt{4^2 + 3^2} = \sqrt{25} = 5$$

$$OA = \sqrt{x^2 + y^2}$$

Find the distance between the origin O (0, 0) and each point A.

1. $A(5, 12)$ 13 2. $A(8, 15)$ 17 3. $A(8, 6)$ 10 4. $A(2, 3)$ $\sqrt{13}$

15-3 Graphing Linear Equations

Objective: To find and graph the solutions of a linear equation.
MIN: 1–20; REG: 1–20 even, 21–28; MAX: 1–20 even, 21–30; ALL: Mixed Review

The graph of each of the equations $y = 2x + 1$, $2x + y = 5$, and $3x - 4y = 17$ is a line. Any equation that can be written in the form $y = ax + b$, where a and b are numbers and x and y are variables, is called a **linear equation**. An ordered pair is a solution for a linear equation if the equation is true when the coordinates of the ordered pair are substituted for x and y.

Example 1

Which of the ordered pairs (2, 7) and (−1, 2) is a solution for the linear equation $y = 3x + 1$?

Solution

$$y = 3x + 1$$
$$7 \stackrel{?}{=} 3(2) + 1$$ To determine whether (2, 7) is a solution, substitute 2 for x and 7 for y.
$$7 \stackrel{?}{=} 6 + 1$$
$$7 = 7$$

(2, 7) is a solution.

$$y = 3x + 1$$
$$2 \stackrel{?}{=} 3(-1) + 1$$ To determine whether (−1, 2) is a solution, substitute −1 for x and 2 for y.
$$2 \stackrel{?}{=} -3 + 1$$
$$2 \neq -2$$

(−1, 2) is not a solution.

Practice Which of the ordered pairs (2, −3) and (−2, −2) is a solution to the linear equation $y = 4x - 11$? **(2, −3)**

A linear equation has an infinite number of solutions. To find solutions for an equation such as $y = 2x + 1$, substitute a number for x and solve for y.

Substitute 1 for x	Substitute 2 for x	Substitute 3 for x
↓	↓	↓
$y = 2x + 1$	$y = 2x + 1$	$y = 2x + 1$
$y = 2(1) + 1$	$y = 2(2) + 1$	$y = 2(3) + 1$
$y = 3$	$y = 5$	$y = 7$
(1, 3) is a solution.	(2, 5) is a solution.	(3, 7) is a solution.

You can use a table to show a set of solutions for a linear equation.

Example 2

Make a table of solutions for $y = 2x + 3$ when $x = -2, -1, 0, 1, 2$, and 3.

Solution $y = 2x + 3$

x	y
-2	-1
-1	1
0	3
1	5
2	7
3	9

Substitute each value of x into the equation and solve for the corresponding value of y. Enter this value of y into the table beside the value of x.

$y = 2x - 3$

x	y
-3	-9
-2	-7
-1	-5
0	-3
1	-1

Practice Make a table of solutions for $y = 2x - 3$ for $x = -3, -2, -1, 0$, and 1.

When the ordered pairs found in a solution table are graphed, the graphed points fall on a straight line. By drawing the line through this set of graphed points, you *graph the linear equation*. The line represents *all* solutions to the equation.

Example 3

Make a table of solutions for $y = x + 1$. Graph the equation.

Solution $y = x + 1$

x	y
-2	-1
-1	0
0	1
1	2
2	3
3	4

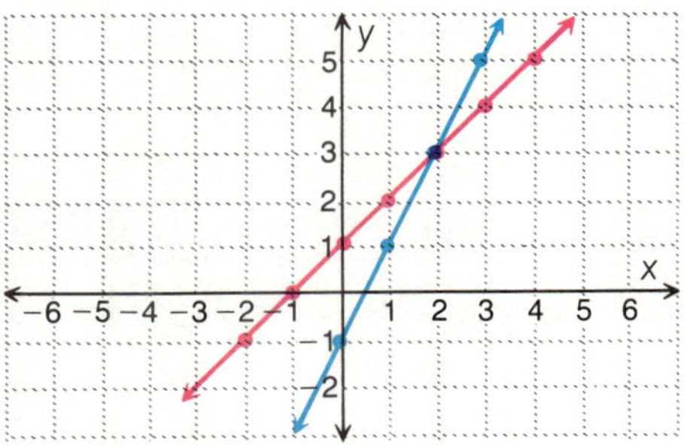

$y = 2x - 1$

x	y
-2	-5
-1	-3
0	-1
1	1
2	3

Practice Make a table of solutions for $y = 2x - 1$. Graph the equation.

For graph, see above.

Oral Exercises

These equations are written in the form $y = ax + b$. Give the values of a and b for each equation. 1. $a = 3, b = 5$ 2. $a = 2, b = -4$ 3. $a = -1, b = 17$ 4. $a = 3, b = 5$

1. $y = 3x + 5$ **2.** $y = 2x - 4$ **3.** $y = -x + 17$ **4.** $y = 3x + 5$

5. $y = -x + 6$ **6.** $y = -2x + 3$ **7.** $y = 5x + 7$ **8.** $y = -x$

5. $a = -1, b = 6$ 6. $a = -2, b = 3$ 7. $a = 5, b = 7$ 8. $a = -1, b = 0$

Exercises

A Which of the ordered pairs is a solution for the equation?

1. (2, 3) or (3, 9) for $2x + y = 7$ (2, 3)

2. (1, 5) or (2, 6) for $3x - y = 0$ (2, 6)

3. (−3, 4) or (3, −4) for $3x + 2y = -1$ (−3, 4)

4. (5, 7) or (7, 8) for $x - 2y = -9$ *(5, 7) and (7, 8)*

For each equation, make a table of solutions for $x = -2, -1, 0, 1, 2, 3$, and 4.

5. $y = 2x - 1$ **6.** $y = 3x + 2$ **7.** $-2x + 5 = y$ **8.** $y = -x + 3$

9. $y = 4x - 7$ **10.** $y = -2x + 3$ **11.** $y = x - 12$ **12.** $y = 2x - 5$

See additional answers.

Make a table of solutions and graph each linear equation. See additional answers.

13. $y = x + 1$ **14.** $y = 2x - 3$ **15.** $y = -x + 2$ **16.** $y = 3x - 4$

17. $y = x + 3$ **18.** $y = -x - 2$ **19.** $y = -2x + 1$ **20.** $y = -x + 3$

B Rewrite the equation in the form $y = ax + b$. Make a table of solutions and graph the equation. See additional answers.

21. $4x + y = 4$ **22.** $-2x + y = 6$ **23.** $2x + y = 8$ **24.** $y - 2x = 1$

25. $3x = 2y + 4$ **26.** $3y + 6x = 6$ **27.** $2x = 4y$ **28.** $3x - 4y = 8$

C Extending Thinking Skills

Thinking skills: 29, 30. Testing a conjecture
Encourage students to guess solutions and confirm their guesses.

29. Find an ordered pair that is a solution to both $x + y = 2$ and $x - y = 0$. (1, 1)

30. At what points does the graph of $\frac{x}{2} + \frac{y}{3} = 1$ cross the x- and y-axes? (2, 0), (0, 3)

Mixed Review

Write each as a fraction. **31.** 60% **32.** 11% **33.** 75% **34.** 240%

Solve and check. **35.** $0.75m + 4 = 13$ 12 **36.** $3t + 11t - 16 = 12$ 2

31. $\frac{3}{5}$ 32. $\frac{11}{100}$ 33. $\frac{3}{4}$ 34. $2\frac{2}{5}$

COMPUTER ACTIVITY

This program prints 20 solutions to a linear equation.

```
10  PRINT "THIS PROGRAM GIVES 20 SOLUTIONS TO A"
20  PRINT "LINEAR EQUATION OF THE FORM Y = AX + B"
30  PRINT
40  INPUT "CHOOSE A VALUE FOR A AND A VALUE FOR B:";A, B
50  PRINT "X";TAB(12);"Y=";A;"X + ";B
60  PRINT
70  FOR N=1 TO 20
80  PRINT N;TAB(16);A*N+B
90  NEXT N
100 END
```

1. Type and run the program for the equation $y = 3x + 5$.

2. Type and run the program for $y = 17x - 236$.

15-4 Practice Solving Problems

Objective: To write and solve equations to solve word problems.
MIN: 1–5; REG: 1–5 even, 6–8; MAX: 1–5 even, 6–10; ALL: Mixed Review

Problems

A Solve by writing an equation. 1. $\frac{1}{2}x + 175 = 2465$ $4580.00

1. Ms. Ericson bought a used car for $175 more than half its original price. She paid $2465 for the car. What was its original price?

2. Sally has 8 more than twice as many coins in her collection as she did when her grandmother gave her the first coins in her collection. She now has 120 coins. How many did her grandmother give her? $2x + 8 = 120$ 56 coins

3. A gasoline station charges $1.16 per gallon and $2.50 for a car wash. Celia's total charge for gas and a wash was $12.94. How many gallons of gasoline did she buy? $1.16x + 2.5 = 12.94$ 9 gal.

4. An amusement park had $\frac{1}{3}$ as many customers on Sunday as on Saturday. The total number of customers for the two days was 1648. How many customers did the park have on each of the two days? $x + \frac{1}{3}x = 1648$ Sat 1236, Sun 412

5. Mr. Walton weighed 186 pounds. He planned a diet that would help him lose 2 pounds per week. He wanted to get his weight down to 160 pounds. How many weeks should he plan to diet? $186 - 2x = 160$ 13 weeks

B 6. $1.25 + 1.75x + 2 = 16$ ≈ 7.3 miles

6. A taxi driver charges $1.25 plus $1.75 for each mile traveled. To pay his cab fare and a $2 tip, Mr. Bush gave the driver $16. How far did he travel?

7. Tim and Ginny work in different restaurants. They both worked the same number of hours and earned the same amount last week. Tim worked for $6 per hour and could keep no tips. Ginny worked for $4 an hour, but was allowed to keep tips totaling $76. How many hours did each work? $4x + 76 = 6x$ 38 hours

8. A three-member relay team ran the relay in 186 seconds. Lisa ran her lap in 9 seconds less than Sara. Kay took 3 seconds more than Sara to run her lap. What was the lap time for each runner? $x + (x - 9) + (x + 3) = 186$ $3x - 6 = 186$
Sara 64 sec., Lisa 55 sec., Kay 67 sec.

C Extending Thinking Skills

Thinking skill: 9–10. Formulating a Problem

9. Write a word problem that could be solved using the equation $\frac{1}{2}x + 3 = 75$.

10. Write a word problem that could be solved using the equation $x + 7 = 12$.

Answers will vary.

Mixed Review

Give the greatest common factor (GCF). **11.** 78, 54, 24 6 **12.** 9, 27, 15 3

Solve and check. **13.** $19 - 4m = 11$ 2 **14.** $t + 0.9t = 9.5$ 5

15-5 Direct and Inverse Variation

Objective: To recognize direct and inverse variation.
MIN: 1–10; REG: 1–10 even, 11–13; MAX: 1–10 even, 11–16; ALL: Mixed Review
An exploratory lesson opener can be found on page 555.

A car driven 40 km/h will travel 40 km in 1 hour, 80 km in 2 hours, 120 km in 3 hours, and so on. You can show this with the equation $d = 40t$, where d is distance, 40 is rate, and t is time. In this situation, as the value of one variable *increases at a constant rate,* the value of the other variable *also increases at a constant rate*. The second variable **varies directly** with the first.

Suppose a car is to travel 40 km. Traveling at the rate of 40 km/h, it would take 1 hour. At 80 km/h, it would take $\frac{1}{2}$ hour, at 120 km/h it would take only $\frac{1}{3}$ hour, and so on. You can show this relationship as $t = \frac{40}{r}$ or $rt = 40$, where t is time, 40 is distance, and r is rate. In this situation, as the value of one variable *increases* at a constant rate, the value of the other variable *decreases* and the product of the variables is a constant. The second variable **varies inversely** with the first variable. Direct and inverse variation are defined as follows.

Direct Variation

A variable **y varies directly with x** if there is a positive number k such that $y = kx$.

Inverse Variation

A variable **y varies inversely with x** if there is a positive number k such that $y = \frac{k}{x}$ or $xy = k$.

Example 1

Does y vary directly with x? If it does, find the value of k.

Point out that for direct variation: 1) x increases as y increases, 2) y increases at a constant rate

x	y
1	3
2	6
3	9
4	12

Solution

$3 = k(1)$ The pair (1, 3) satisfies $y = kx$ if $k = 3$.

$6 = 3(2)$ Check to see whether other pairs satisfy $y = 3x$.

$9 = 3(3)$

$12 = 3(4)$ The pairs (2, 6), (3, 9), and (4, 12) do satisfy $y = 3x$.

y varies directly with x, and $k = 3$.

Practice Does y vary directly with x? If it does, find the value of k.

a. yes, $k = 5$.
b. no, $k = -2$ (not positive).

a.

x	y
1	5
2	10
3	15
4	20
5	25

b.

x	y
1	−2
2	−4
3	−6
4	−8
5	−10

Example 2

Does y vary directly with x? If it does, find the value of k.

x	y
1	2
2	3
3	6
4	9
5	12

Solution

$2 = k(1)$ The pair (1, 2) satisfies $y = kx$ if $k = 2$.

$3 \neq 2(2)$ Check to see whether other pairs satisfy $y = 2x$.

$6 = 2(3)$

$9 \neq 2(4)$

$12 \neq 2(5)$ The pairs (2, 3), (4, 9), and (5, 12) do not satisfy $y = 2x$.

y does not vary directly with x.

Practice Does y vary directly with x? **no** If it does, find the value of k.

x	y
1	5
2	7
3	9
4	3
5	6

Example 3

Point out that for inverse variation,
1. as x increases, y decreases
2. xy is a constant.

Does y vary inversely with x? If it does, find the value of k.

x	y
1	24
2	12
3	8
4	6
5	$\frac{24}{5}$

Solution

$24 = \frac{k}{1}$ The pair (1, 24) satisfies $y = \frac{k}{x}$ if $k = 24$.

$12 = \frac{24}{2}$ Check to see whether other pairs satisfy $y = \frac{24}{x}$.

$8 = \frac{24}{3}$

$6 = \frac{24}{4}$

$\frac{24}{5} = \frac{24}{5}$ All pairs do satisfy $y = \frac{24}{x}$.

y varies inversely with x, and $k = 24$.

Practice Does y vary inversely with x? If it does, find the value of k.

a.

x	y
1	5
2	4
3	3
4	2
5	1

a. no

b.

x	y
1	1
2	$\frac{1}{2}$
3	$\frac{1}{3}$
4	$\frac{1}{4}$

b. yes $k = 1$

Oral Exercises

For each question, state whether y varies directly with x, y varies inversely with x, or neither case is true.

1. $y = 2x$ dir. **2.** $xy = 5$ inv. **3.** $xy = 14$ inv. **4.** $y = 4x$ dir.

5. $x = \frac{1}{y}$ inv. **6.** $x = \frac{y}{5}$ dir. **7.** $y = \frac{3}{x}$ inv. **8.** $y = \pi x$ dir.

Exercises

A Does y vary directly with x? If it does, find the value of k.

1.

x	y
1	1
2	2
3	3
4	4
5	5

yes. $k = 1$

2.

x	y
1	2
2	4
3	6
4	8
5	10

yes. $k = 2$

3.

x	y
1	2
2	4
3	5
4	6
5	7

no.

4.

x	y
1	4
2	8
3	12
4	16
5	20

yes. $k = 4$

5.

x	y
1	−2
2	4
3	−6
4	8
5	−10

no.

Does y vary inversely with x? If it does, find the value of k.

6.

x	y
1	12
2	10
3	8
4	6
5	4

no

7.

x	y
1	120
2	60
3	40
4	30
5	24

yes. $k = 120$

8.

x	y
1	12
2	6
3	4
4	3
5	$\frac{12}{5}$

yes. $k = 12$

9.

x	y
1	2
2	1
3	$\frac{2}{3}$
4	$\frac{1}{2}$
5	$\frac{2}{5}$

yes. $k = 2$

10.

x	y
1	20
2	10
3	5
4	3
5	1

no.

B

11. A car averages 25 miles per gallon of gas. Does the total number of miles traveled vary directly or inversely with the number of gallons of gas used? **directly**

12. A truck driver drives 150 miles between rest stops. Does his average speed vary directly or inversely with the time it takes to travel the 150 miles? **inversely**

13. The tax on Mr. Jones's income is 20%. Does the amount of tax he pays vary directly or inversely with the amount he earns? **directly**

C Extending Thinking Skills

See additional answers.

14. Copy and complete the table at right so that y varies directly with x.

15. Copy and complete the table at right so that y varies inversely with x.

16. A variable y varies directly with the square of a variable x if $y = kx^2$ when a constant k is a positive number. Complete the table at right so that y varies directly with the square of x.

x	y
1	?
2	?
3	?
4	80

Thinking skill: 14–16. Reasoning

Mixed Review

Compute. **17.** 1.2×25 30 **18.** $36 + (-25)$ 11 **19.** 3.6×25 90

20. $3^2 + 3^3$ 36 **21.** $3^2 \cdot 3^3$ 243 **22.** 4.16×10^3 4160 **23.** 2^{-2} $\frac{1}{4}$ *or* 0.25

Solve and check. **24.** $12.4 = 36m - 5.6$ 0.5 **25.** $3c - 1.7c = 3.12$ 2.4

CALCULATOR ACTIVITY

The constant multiplier 29 can be stored in the memory to make completing this table of direct variations easier. The key sequence below puts the constant 29 in the memory, then calculates the first entry of the table.

Use the following key sequence:

29 [M+] 30 [×] [MR] [=] Display: 870

$y = 29x$

x	y	
30	870	
60	?	1740
90	?	2610
120	?	3480
150	?	4350
180	?	5220
210	?	6090

1. Use the key sequence above to continue completing the table.

2. Use a calculator to complete a table of direct variation for $y = 123x$. Use values of x with multiples of 35 up to 280. See additional answers.

3. Find a key sequence that includes the [MR] key to complete a table for the inverse variation described by $y = \frac{123}{x}$. See additional answers.

15-6 Slope and Intercepts

Objective: To determine the slope, given the equation for a line.
MIN: 1–28; REG: 1–28 even, 29–31; MAX: 1–28 even, 29–34; ALL: Mixed Review
An exploratory lesson opener can be found on page 555.

A roof that increases 4 feet vertically for every 12 feet of horizontal distance has a slope of $\frac{4}{12}$.

Note that slope = $\frac{\text{rise}}{\text{run}} = \frac{\text{change in } y}{\text{change in x}}$

The **slope** of a line gives the idea of the steepness of a line. The slope is **the change in *y* divided by the change in *x***. The change in *y* is called the **rise**. The change in *x* is called the **run**. To find the slope of a line, choose any two points on the line and calculate the change in the *y*-coordinate and *x*-coordinate for these points.

From *A* to *B*:

$$\text{slope} = \frac{\text{change in } y}{\text{change in } x} = \frac{2}{4} = \frac{1}{2}$$

From *B* to *C*:

$$\text{slope} = \frac{\text{change in } y}{\text{change in } x} = \frac{3}{6} = \frac{1}{2}$$

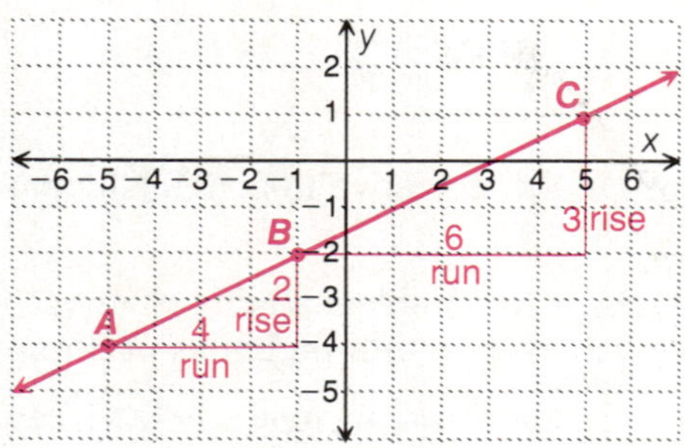

Example 1

Graph the line that contains the points $A(-2, 3)$ and $B(6, -3)$. Find the slope.

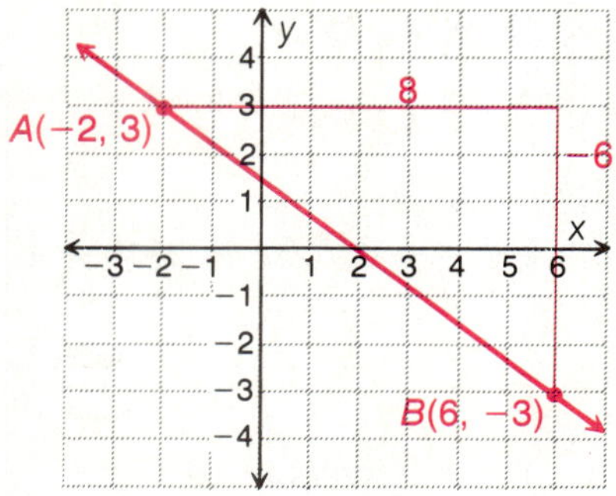

Solution

$$\text{slope} = \frac{\text{change in } y}{\text{change in } x}$$ Subtract in the same order.

$$\text{slope of } \overleftrightarrow{AB} = \frac{-3 - 3}{6 - (-2)}$$

$$= \frac{-6}{8} = -\frac{3}{4}$$

Practice Graph the line that contains the points given. Find the slope.

a. $A(1, 2)$ and $B(4, 8)$ **b.** $C(-3, 1)$ and $D(5, 3)$ $\frac{2}{8} = \frac{1}{4}$

a. $\frac{6}{3} = 2$ **See additional answers.**

You can find the slope of the line for a linear equation by using any two points that make the equation true.

Example 2

Find the slope of the line $y = 3x - 1$.

Solution

When $x = 1$, $y = 3(1) - 1 = 2$ Find two points by using any two values of x and calculating the corresponding values of y. (1, 2) is on the graph.

When $x = 2$, $y = 3(2) - 1 = 5$ (2, 5) is on the graph.

$$\text{slope} = \frac{\text{change in } y}{\text{change in } x}$$

$$= \frac{5 - 2}{2 - 1} = \frac{3}{1} = 3$$ Make sure you begin with the same point in calculating both the change in y and the change in x.

Practice Find the slope of the line.

a. $y = -2x + 1$ -2 **b.** $2x - 3y = 6$ $\frac{2}{3}$

The ***x*-intercept** is the x-coordinate of the point where the line crosses the x-axis. The ***y*-intercept** is the y-coordinate where the line crosses the y-axis. Since all points on the y-axis have an x-coordinate of 0, you find the y-intercept by letting $x = 0$. Since all points on the x-axis have y-coordinate of 0, you find the x-intercept by letting $y = 0$.

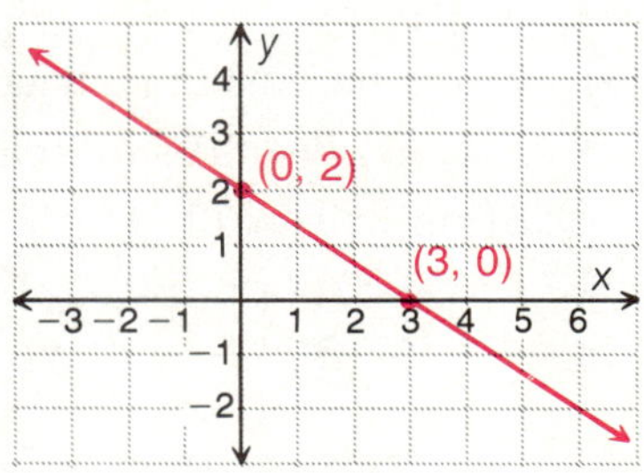

The line is the graph of $2x + 3y = 6$.

The x-intercept is 3.

The y-intercept is 2.

Example 3

Find the x- and y-intercepts for the line $3x + 2y = 12$.

Solution

Let $y = 0$. $3x + 2(0) = 12$ — The y-coordinate of the x-intercept is 0. So let $y = 0$ and solve for x.

$$3x = 12$$
$$x = 4$$

The x-intercept is 4. — The pair (4, 0) satisfies the equation $3x + 2y = 12$.

Let $x = 0$. $3(0) + 2y = 12$ — The x-coordinate of the y-intercept is 0. So let $x = 0$ and solve for y.

$$2y = 12$$
$$y = 6$$

The y-intercept is 6. — The pair (0, 6) satisfies the equation $3x + 2y = 12$.

Practice Find the x- and y-intercepts. **a.** $-2x + y = 6$ **b.** $3x - 2y = 10$.

a. x-intercept is -3; y-intercept is 6 b. x-intercept is $\frac{10}{3}$; y-intercept is -5

Oral Exercises

For each line, give the slope, the x-intercept, and the y-intercept.

1. 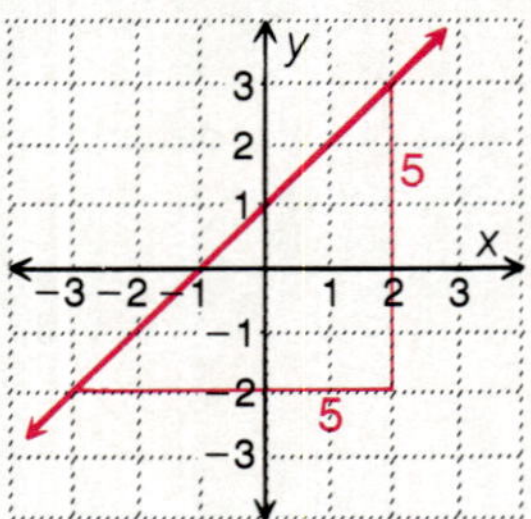

slope $= \frac{5}{5} = 1$ x-int. $= -1$ y-int. $= 1$

2. 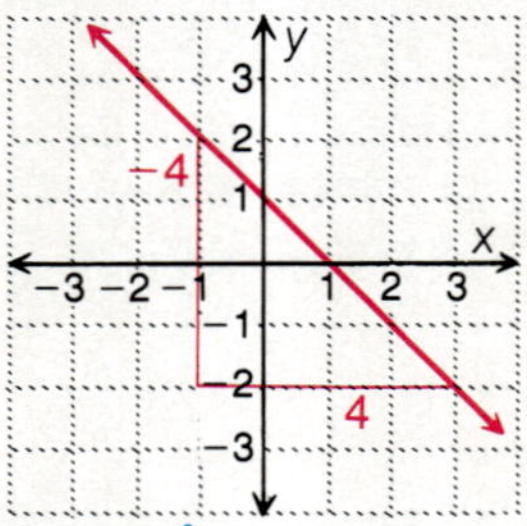

slope $= \frac{-4}{4} = -1$ x-int $= 1$ y-int $= 1$

3. 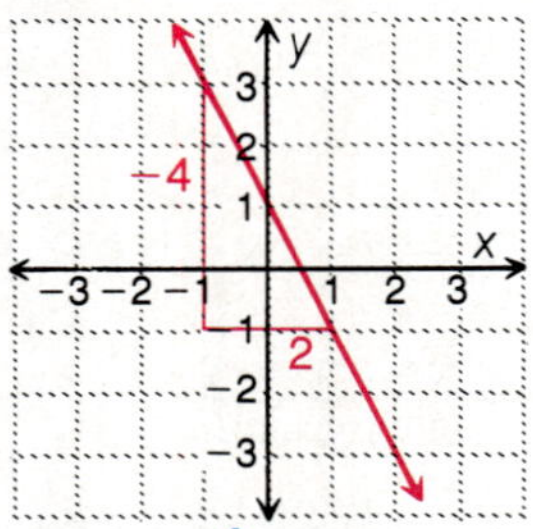

slope $= \frac{-4}{2} = -2$ x-int. $= \frac{1}{2}$ y-int. $= 1$

Exercises

1–12 See additional answers.

A Graph the line that contains the given points and find the slope.

1. $A(1, 1), B(4, 3)$ **2.** $A(1, 3), B(5, 8)$ **3.** $A(-2, -3), B(4, 6)$

4. $A(-3, -1), B(3, 1)$ **5.** $A(-2, 2), B(2, -2)$ **6.** $A(-1, 3), B(1, -3)$

7. $A(-2, 8), B(3, -2)$ **8.** $A(-2, 3), B(4, -6)$ **9.** $A(8, 4), B(1, 3)$

10. $A(3, 5), B(-2, -3)$ **11.** $A(-4, -6), B(-1, 3)$ **12.** $A(3, -5), B(-3, 8)$

Find the slope of each line.

13. $y = 2x - 1$ **2** **14.** $y = -3x + 7$ **−3** **15.** $2x - y = 5$ **2** **16.** $3x + 2y = 24$ $-\frac{3}{2}$

17. $x + 3y = 6$ $-\frac{1}{3}$ **18.** $2x - 5y = 10$ $\frac{2}{5}$ **19.** $2x - 7 = y$ **2** **20.** $3y = 2x - 5$ $\frac{2}{3}$

Find the x- and y-intercepts. 21–31. See additional answers.

21. $x + y = 5$ **22.** $2x + y = 4$ **23.** $x - 3y = 6$ **24.** $2x - 3y = 24$

25. $4x + 5y = 20$ **26.** $3x - 4y = 12$ **27.** $-y + 2x = 5$ **28.** $5x + 3y = 15$

B Graph the equation. How do the numbers in the equation relate to the slope?

29. $y = 2x - 3$ **30.** $y = -3x + 5$ **31.** $y = \frac{1}{2}x + 1$

32. A long grade on a highway through the mountains has a vertical rise of 7 ft for each 100 ft of horizontal distance. How many feet of rise are there in one mile (5280 ft.) of horizontal change? 369.6 ft

33. A ladder is to reach a point 30 feet above the ground. How far from the building should its base be if its slope is 2.5? 12 ft.

C Extending Thinking Skills

Thinking skill: 34. Using reverse reasoning

34. Suppose a line through (1, 2) has slope 3. Find values for b, c, and d so that (2, b), (3, c), and (4, d) are points on the line. $b = 5$ $c = 8$ $d = 11$

Mixed Review

Give the least common multiple (LCM). **35.** 2, 4, 7 28 **36.** 4, 6, 8 24

Solve and check. **37.** $3t = 15 - 2t$ 3 **38.** $406 - 5y = 371$ 7

ESTIMATION

The graph of the equation that gives the relationship between temperature in degrees Fahrenheit and degrees Celsius is a line with slope $\frac{5}{9}$. That means a change in temperature of 9°F is a change of 5°C. The equation is $C = \frac{5}{9}(F - 32)$.

Emphasize that in Celsius you count by 5s beginning with zero and that in Fahrenheit you count by 9s beginning at 32.

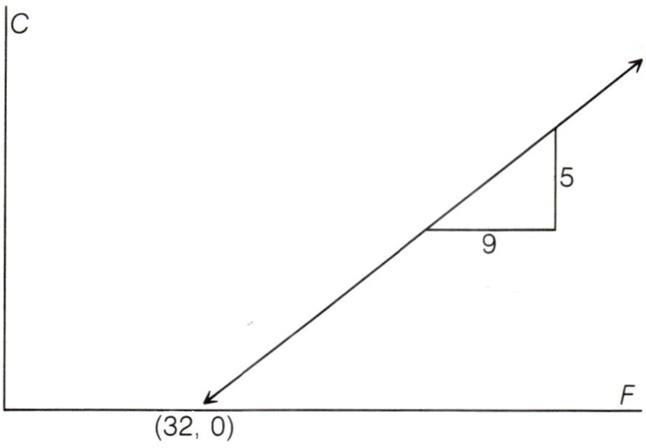

F	C	
change by 9	change by 5	
32	0	← water freezes
41	5	
50	10	
59	15	
68	20	
77	25	

Use the table above to estimate each temperature in degrees Celsius.

1. 55°F ≈ _?_°C 13 **2.** 70°F ≈ _?_°C 21 **3.** 45°F ≈ _?_°C 7 **4.** 23°F ≈ _?_°C −5

PROBLEM SOLVING: APPLICATIONS

15-7 School Newspaper Advertising

Objective: To practice solving word problems.
MIN: 1–12; REG: 1–12; MAX: 1–12

The staff of the school newspaper sells advertising to help pay printing costs. The pages are set up in 3 columns. Each column is 4 inches wide and 14 inches long. A "column inch" is an area 1 column wide by 1 inch long.

Classified Ads	6 lines per column inch	\$ 0.10 per line
Merchant Ads	less than half page half page full page	\$ 0.60 per column inch \$12.00 \$22.00
Ads for School Activities		no charge

Problems

You might wish to have students choose from among the techniques pencil and paper, mental math, estimation, and calculator. Have them choose the most appropriate technique for each problem.

1. How much would 1 column of classified ads earn for the paper? \$8.40

2. How much would a 2-inch classified ad earn for the paper? \$1.20

3. How much is earned by 8 inches of classified ads? \$4.80

4. How much more is earned by a full page of classified ads than by two half-page merchant ads? \$1.20

5. How much is earned by a merchant ad 3 inches long by 2 columns wide? \$3.60

6. How much is earned by a merchant ad 4 inches long by 3 columns wide? \$7.20

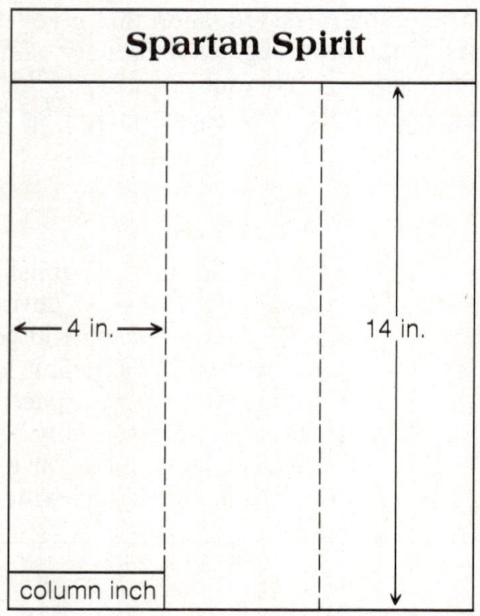

7. How much would the page of ads shown at right earn for the paper? **$22.80**

8. How much more would the page at right earn if the merchant ad 4 inches long by 1 column wide were replaced with a 4-inch classified ad? **they would earn the same**

9. The typesetter charges the *Spartan Spirit* $3.00 more to prepare a page of classified ads than it charges to prepare one full-page merchant ad. After paying the extra typesetting cost, how much more has the paper earned from a full page of classified ads than from a full-page merchant ad? **$0.20**

10. The *Spartan Spirit* provides a free ad 4 inches long by 2 columns wide for the school dance on the day of each of 5 home football games. It also provides a half-page ad for the Junior Prom and two half-page ads for the Senior Prom. How much would these ads cost at the merchant advertising rates? **$60**

11. The circulation of the *Spartan Spirit* (the number of copies printed and distributed) is 1550. How much per copy does a merchant pay for a full-page ad? **About 1.4¢**

12. *Data Search* Find the cost of a full-page merchant ad in a local newspaper. Find the circulation of the paper. How much per copy does a merchant pay for a full-page ad? **Answers will vary.**

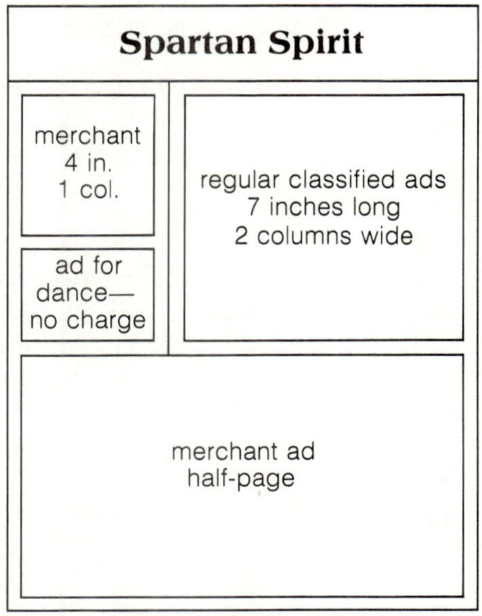

What's Your Decision?

The six-page *Spartan Spirit* is published 17 times per school year. About 50% of the space is devoted to ads that earn money. Printing costs are increasing by $50 per issue. The following ideas for reducing the budget deficit have been proposed:
a. increase the paid advertising space to 60% per issue; **b.** raise advertising rates 15%; **c.** print only 13 issues per year; **d.** charge each of the 1500 students $1.50 for a year's subscription. As the *Spartan Spirit* Managing Editor, which idea would you prefer? Why? **Answers will vary.**

15-8 Graphing Systems of Equations

Objective: To find the solution of a system of equations by graphing.
MIN: 1–9; REG: 1–9 even, 10–14; MAX: 1–9 even, 10–16; ALL: Mixed Review

Two equations with the same two variables are called a **system of equations**. Any ordered pair that is a solution to both equations is called a solution of the system of equations. To find a solution of a system, find the point of intersection of the graphs of the two equations.

Example

Graph the equations to solve the system. Check the solution.

$$x + y = 2$$
$$-x + 2y = 1$$

Solution

Graph each equation on the same coordinate system.

Decide what point lies on both lines.

The solution appears to be (1, 1).

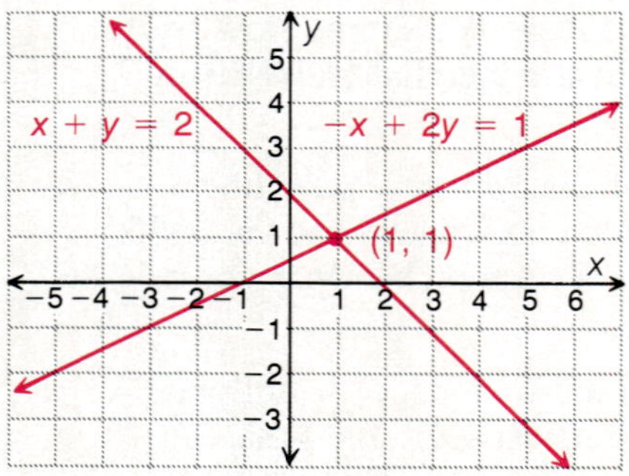

Check

$x + y = 2$ $\quad$ $-x + 2y = 1$

$1 + 1 = 2\checkmark$ $\quad$ $-1 + 2(1) = 1\checkmark$

Check to see that the point you have selected satisfies both equations. The solution is (1, 1).

Practice Graph the equations to solve the system. $x - y = 1$, $x + y = 3$

(2, 1). For graph, see additional answers.

Oral Exercises

Is the ordered pair given a solution for the system of equations?

1. (1, 2) $\quad 2x + y = 4$; $\quad 3x - 2y = -1$ **yes**

2. (−2, 3) $\quad 2x + y = 1$; $\quad -x + 2y = 8$ **no**

3. (3, 1) $\quad x + 2y = 5$; $\quad 2x - y = 2$ **no**

4. (2, −1) $\quad -3x + y = -7$; $\quad 3x + 5y = 1$ **yes**

Exercises Graphs: see additional answers.

A Graph the equations to find a solution for the system. Check the solution.

1. $x - y = 1$
$x + 2y = 1$ (1, 0)

2. $x + y = 2$
$2x - y = 1$ (1, 1)

3. $x - y = 0$
$2x + y = 6$ (2, 2)

4. $x + y = 0$
$x - y = -2$ (−1, 1)

5. $2x - y = -1$
$x + y = -2$ (−1, −1)

6. $2x + y = 5$
$-2x + y = 1$ (1, 3)

7. $x + y = 3$
$x - y = 1$ (2, 1)

8. $-2x + y = -4$
$2x + y = 0$ (1, −2)

9. $x - 2y = 1$
$2x - y = 5$ (3, 1)

B

A system of equations has *no solutions* if the lines of the system have the same slope, making the lines parallel. A system of equations has an *infinite number of solutions* if the graphs of the two linear equations are the same line. Decide whether each system has no solutions, an infinite number of solutions, or a one-point solution. 10. infinite 11. none 12. infinite

10. $3x - 2y = 2$
$4y - 6x = -4$

11. $2x + y = 3$
$2x + y = 5$

12. $3x + 2y = 6$
$4y - 12 = -6x$

13. What value of k will make $2x + 3y = k$ and $4x + 6y = k$ have the same set of solutions? $k = 0$

14. Translate to a system of equations and solve. The sum of two numbers is 6 and their difference is 2. Find the two numbers.
$x + y = 6$
$x - y = 2$ $x = 4, y = 2$
or $x + y = 6$
$y - x = 2$ $x = 2, y = 4$

C Extending Thinking Skills

15. Cheap-at-the-Price Rent-a-car charges \$20 a day and \$0.15 per mile. More-for-Your-Money Rent-a-car charges \$15 a day and \$0.20 per mile. Which company charges the least for 75 miles in one day? For what mileage is the cost the same? More-for-Your-Money. 100 miles.

16. Guess and check to find the value of k that makes (1, 2) a solution of the system.
$3x - y = 1$ $k = 2$
$-x + ky = 3$

Mixed Review

Find the volume. Use 3.14 for π. **17.** A cylinder with radius 4 cm, height 5 cm 251.2 cm³
18. A cube with one side = 9 in. 729 in³ **19.** A cone with radius 5 m, height 9 m. 235.5 m³
What percent of 36 is: **20.** 12? $33\frac{1}{3}\%$ **21.** 27? 75% **22.** 4? $11\frac{1}{9}\%$ **23.** 54? 150%
Solve and check. **24.** $12t - 10t = 26$ 13 **25.** $495 = 27m + 90$ 15
26. $1640 + 32x = 360$ −40 **27.** $9y + 4.75 = 31.3$ 2.95 **28.** $3c - 2.4c = 3$ 5

15-9 Graphing Inequalities

Objective: To graph the solutions of a linear inequality.
MIN: 1–9; REG: 1–9 even, 10–17; MAX: 1–9 even, 10–20; ALL: Mixed Review

The graph of a linear equation such as $y = x + 2$ divides the plane into two regions, one above the line and the other below the line. The line is the **boundary** of the two regions.

Point (1, 4) is a solution to $y > x + 2$, and is a point in the region **above** the line.

Point (1, 3) is a solution to $y = x + 2$, and is a point **on** the line.

Point (1, 2) is a solution to $y < x + 2$, and is a point in the region **below** the line.

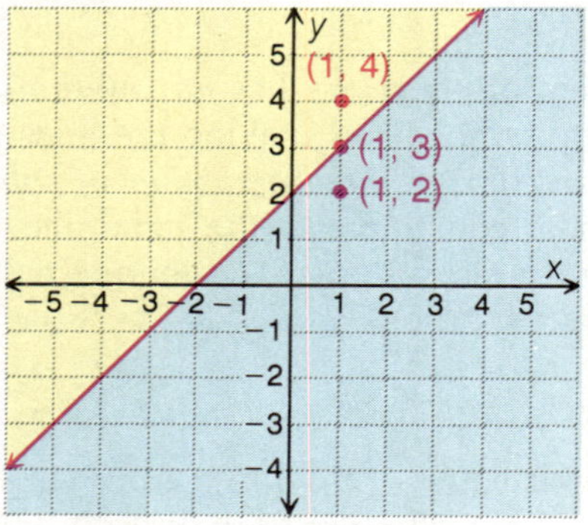

The region above the line is the set of all points that are solutions to the inequality $y > x + 2$. The region below the line is the set of all points that are solutions to the inequality $y < x + 2$. To graph a linear inequality, shade the graph to indicate the region that represents the solutions. If the inequality is $\leq$ or $\geq$, the boundary line is part of the region and is drawn as a solid line. If the inequality is $<$ or $>$, the boundary line is *not* a part of the region, and is drawn as a dashed line.

Example

Graph the inequality $y < x - 1$.

Solution

First, graph the boundary line $y = x - 1$.

The graph is the region below the line $y = x - 1$.

Since the inequality is $<$, draw the boundary line as a dashed line rather than a solid line.

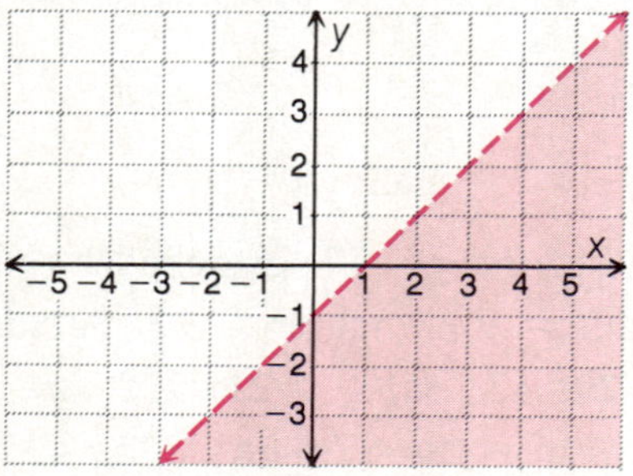

Practice Graph the inequality.

a. $y > 2x - 1$ **b.** $y \leq \frac{1}{2}x + 1$ **See additional answers.**

Oral Exercises

When the inequality is graphed, will the boundary line be drawn as a solid line or a dashed line?

1. $y < 4x - 5$ dashed
2. $3x + 2 > y$ dashed
3. $-x \geq 4 + y$ solid
4. $-2 - y \geq x$ solid
5. $3x - 2y > 1$ dashed
6. $y - 7 < \frac{1}{2}x$ dashed

Exercises

A Graph the inequality. See additional answers

1. $y < 3x - 2$
2. $y \geq -x + 2$
3. $y > -2x + 1$
4. $x \geq 1$
5. $y < 5$
6. $y < \frac{1}{2}x + 3$
7. $y < 2x - 5$
8. $x + 4 \geq y$
9. $y < 3 - x$

B Graph the inequality. See additional answers. Encourage students to write the boundary equation in the form $y = ax + b$ when possible.

10. $2x + 3y \geq 6$
11. $3(x + 2y) \geq 6$
12. $x < 2y + 1$
13. $3 \leq \frac{6}{x}$
14. $2x + 7 > y - 3x$
15. $4y - 3x < 12$

16. Jim and Sally are a two-person relay team in a bicycle marathon. Jim averages 20 mi/h, and Sally averages 18 mi/h. The team with the current record rode 872 miles. Write an inequality that must be satisfied for Jim and Sally to set a new record if Jim rides x hours and Sally rides y hours. $20x + 18y > 872$

17. Suppose corn seeds cost \$5/lb and wheat seeds cost \$7/lb. A farmer has a budget of \$500 for seeds. She plants x pounds of corn and y pounds of wheat. Write an inequality that must be satisfied for her to stay within the budget. $5x + 7y \leq 500$

C Extending Thinking Skills

To find the solution for a system of linear inequalities, you graph each inequality and identify the region common to both graphs. Graph each system of linear inequalities.

Example: Graph the system $x \geq 1$
$y \leq 2$

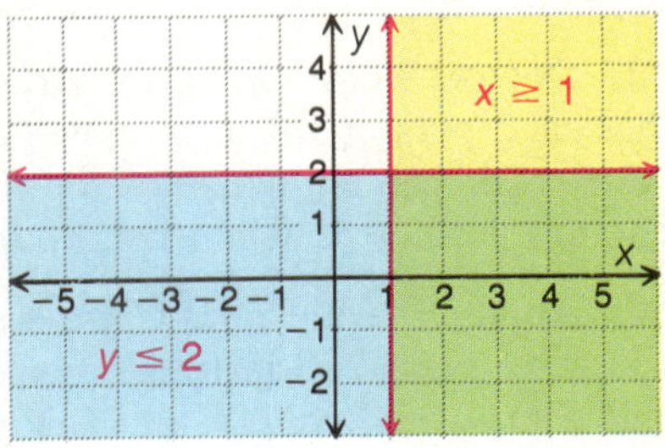

Graph each system of linear inequalities.

18. $x \leq 3$, $y \geq -1$
19. $y \leq x + 2$, $y \leq -x + 1$
20. $y \geq \frac{1}{2}x - 2$, $y \leq -2x + 6$

Thinking skill: 18–20. Reasoning

Mixed Review

21. Find the diameter and area of a circle with radius = 6 in. Use 3.14 for π.

Solve and check. 22. $13t + 16 = 3$ −1 23. $4m + 17 - 7m + 5m = 21$ 2

21. $d = 12$ in area = 113.04 in^2

PROBLEM SOLVING: STRATEGIES

15-10 Practice Solving Problems

Objective: To solve nonroutine problems.
MIN: 1–2; REG: 1–4; MAX: 1–6

Problem-Solving Strategies	
Choose the Operations	**Write an Equation**
Guess, Check, Revise	**Simplify the Problem**
Draw a Picture	**Make an Organized List**
Make a Table	**Use Logical Reasoning**
Look for a Pattern	**Work Backwards**

This chart shows the strategies presented in previous chapters.

Problems

Solve.

1. A meeting room has a floor area of 625 ft^2 and an 8-foot high ceiling. Office guidelines recommend at least 200 ft^3 per person. What should be the maximum number of people allowed to meet in this room? **25**

2. A car rental agency charges $21.50 per day plus $0.13 a mile. Mrs. Higby needs to rent a car for 1 day. Her company will pay up to $50 per day for a rental car. What is the maximum number of miles she can drive in one day and stay below the $50 limit? **219**

3. Doctors knew that a flu virus could spread very rapidly. They estimated that each person who had it would give it to 4 other people every day. On April 12th, the first case of the flu was reported. If the flu spread as the doctors predicted, when would about 20,000 people have had the flu? **April 19**

4. Angie, Brenda, and Clara weighed themselves 2 at a time. Brenda and Clara weighed 208 lb together. Angie and Brenda weighed 222 lb, and Angie and Clara weighed 216 lb. How much did each one weigh?
Angie 115 lb, Brenda 107 lb, Clara 101 lb

5. A rich man left his money to two nieces and a nephew. To the older niece he gave half of his money. He gave the younger niece $\frac{1}{3}$ what he gave the older niece. He gave the nephew twice as much as he gave the younger niece. He gave the nephew $3.5 million. How much did he give to all three? **$10.5 million**

6. A grocer is arranging grapefruit in his produce department so that it will form a pyramid. He forms the base of the pyramid using 64 grapefruit, with 8 grapefruit on each side. How many will he need to build the whole pyramid? **204**

Enrichment

Telephone Rate Function

A telephone directory lists various rates for long-distance calls. The table below lists daytime rates for direct-dialed calls at \$0.58 for the first minute and \$0.39 for each additional minute. It shows a relationship between the length of a call in minutes and the cost of the call. The pairs of points in the table are graphed on the right.

time/min	total cost
0.5	\$0.58
1	0.58
1.5	0.97
2	0.97
2.5	1.36
3	1.36
3.5	1.75
4	1.75
4.5	2.14
5	2.14

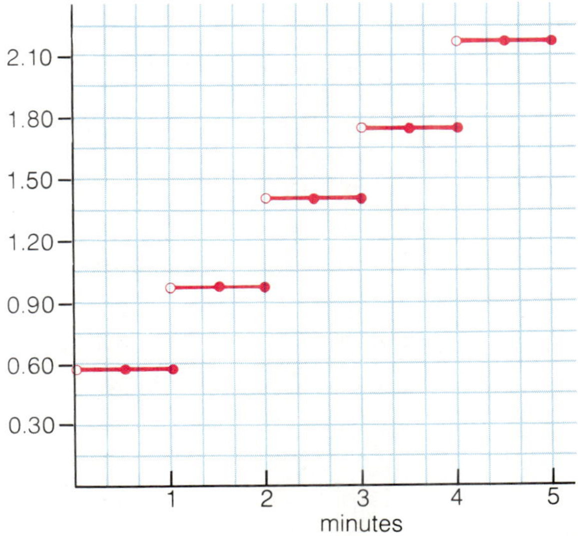

The cost of a call stays the same for any part of a minute. For example, from the beginning of the call to the end of the first minute, the cost is \$0.58. On the graph, this is a horizontal line from 0 to 1. As soon as the call goes over 1 minute in length, the cost jumps to \$0.97, and the line on the graph also jumps. The graph shows a jump at the end of each minute.

See additional answers.

1. Evenings from 5 p.m. to 11 p.m., rates are reduced to \$0.34 for the first minute and \$0.24 for each additional minute. Complete a table of values and make a graph showing the relationship between time and cost.

2. All day Saturday and Sunday, rates for direct-dialed calls are reduced to \$0.23 for the first minute and \$0.16 for each additional minute. Complete a table of values and make a graph.

3. For calls handled by the operator, a \$1.55 charge is added to the regular charges. Complete a table of values and make a graph for Saturday and Sunday calls handled by the operator. Refer to Exercise 2 for the regular charges.

Chapter 15 Review

15.1 Graph the solution on a number line. See additional answers.

1. $x < 0$

2. $x + 1 \geq 0$

15.2 Give the coordinates of each point.

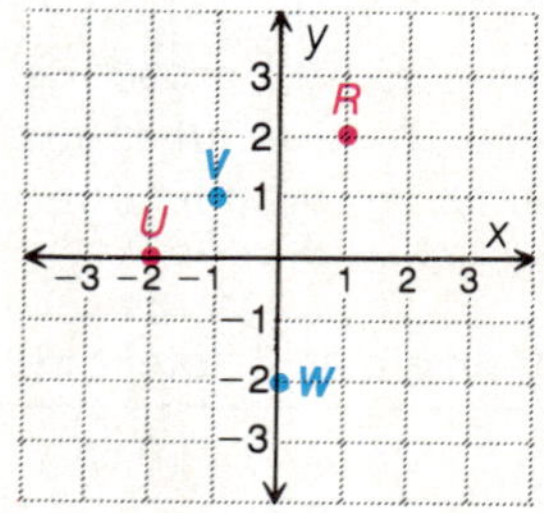

3. R (1, 2)

4. U (−2, 0)

Graph the points on a pair of axes.

5. $V(-1, 1)$

6. $W(0, -2)$

15.3 **7.** Which of the ordered pairs (4, 9) or (3, 5) is a solution for the linear equation $y = 2x + 1$? (4, 9)

Make a table of solutions for the equation.

8. $y = -x - 1$ Use $x = -2, -1, 0, 1, 2, 3$. See additional answers.

15.4 Solve. **9.** Ms. Pierce bought a coat for \$30 more than half its original price. If she paid \$110 for the coat, what was the original price? $30 + \frac{c}{2} = 110$ \$160

15.5 Does y vary directly with x? If it does, find the value of k.

Does y vary inversely with x? If it does, find the value of k.

10. yes $k = 4$

x	y
1	4
2	8
3	12
4	16
5	20

11. no

x	y
1	2
2	4
3	−6
4	−8
5	−10

12. yes $k = 20$

x	y
1	20
2	10
3	$6.\overline{6}$
4	5
5	4

13. yes $k = 15$

x	y
1	15
2	7.5
3	5
4	3.75
5	3

15.6 Find the slope of the line.

14. A line that contains the points $C(-2, -2)$, $D(2, 2)$ slope = 1

15. $y = 5x + 2$ slope = 5

Find the intercepts for the line $2x + y = 4$.

16. x-intercept 2

17. y-intercept 4

15.7 Solve. **18.** Cabins at Duck Lake rent for \$250 for the first week and \$25 for each additional day. How much would it cost to rent a cabin for 10 days? \$325.

15.8 **19.** Graph the equations to solve the system. $-x + 2y = 0$, $x + y = 3$ (2, 1)

15.9 **20.** Graph the inequality $y \geq 2x - 1$. Is point $(0, -3)$ a solution? no

Chapter 15 Test

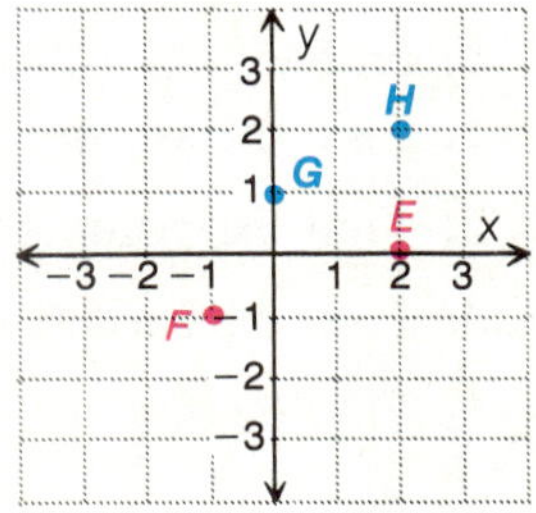

Graph the solution on a number line. See additional answers.

1. $x > -2$ **2.** $2x - 1 > -1$

Give the coordinates of each point.

3. E (2, 0) **4.** F (−1, −1)

Graph the points on a pair of axes.

5. G (0, 1) **6.** H (2, 2)

7. Which of the ordered pairs (−3, −6) or (−3, 6) is a solution for the linear equation $2x + y = 0$? (−3, 6)

Make a table of solutions for the equation.

8. $y = -x + 3$ Use $x = -2, -1, 0, 1, 2, 3$. See additional answers.

Solve. **9.** Bob sold a radio at a garage sale for \$17. This was \$3 more than $\frac{1}{7}$ of its original price. What was the original price of the radio? $\frac{p}{7} + 3 = 17$ \$98

Does y vary directly with x? If it does, find the value of k.

10. no

x	y
1	3
2	4
3	5
4	6
5	7

11. yes $k = 5$

x	y
1	5
2	10
3	15
4	20
5	25

Does y vary inversely with x? If it does, find the value of k.

12. yes $k = 24$

x	y
1	24
2	12
3	8
4	6
5	4.8

13. no

x	y
1	3
2	2
3	1
4	0
5	−1

Find the slope of the line.

14. A line contains the points $A(-1, 2)$, $B(2, -4)$ slope = −2

15. $3x + 4 = y$ slope = 3

Find the intercepts for the line $9 + 3y = x$.

16. x-intercept 9 **17.** y-intercept −3

Solve. **18.** A house-painting company charges \$180 to paint two rooms, and \$80 for each additional room. How much would it cost to have 5 rooms painted? \$420.

19. Graph the equation to solve the system $x + 2y = 1$, $x - y = -2$ (−1, 1)

20. Graph the inequality $y < (2 - x)$. Is point (4, −1) a solution? no

Cumulative Review

For problems 1–4, use 3.14 for π.

1. Find the volume of a cone with a radius of 6 inches and a height of 19 inches. 715.92 in.^{3}

2. Find the volume of a pyramid with a base area of 64 square m and a height of 20 m. $426\frac{2}{3}$ m^3

3. Find the surface area of a cylinder with a $r = 4$ ft, $h = 11$ ft. 376.80 ft^2

4. Find the surface area of the figure at right. 208 m^2

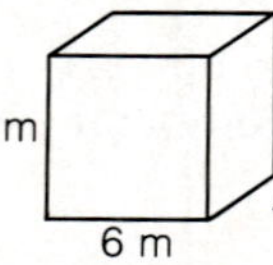

Use the counting principle to find the total number of outcomes.

5. Selection: 1 color and 1 size. Colors: Blue, Red, Orange, Green, Black; Sizes: Small, Medium, Large, Extra Large 20

Find the number of permutations.

6. t, u, e, s, d, a, y 5040

7. stop, start, slow, fast 24

Solve. **8.** 5 drawings are entered in a drawing contest. How many ways can a winner and a runner-up be chosen? 20

Find the number of combinations. **9.** 3 books from a shelf of 9 books. 84

Use the data to solve problems 10–13.

Points scored by Pamela in basketball games: 14, 14, 10, 18, 15, 14, 13

10. Find the range. 8

11. Find the mode. 14

12. Find the median. 14

13. Find the mean. 14

Between what two consecutive integers does each square root lie?

14. $-\sqrt{93}$ −9 and −10

15. $\sqrt{34}$ 5 and 6

Use a calculator with a $\sqrt{}$ key to find an approximation for each square root.

16. $\sqrt{28}$ 5.29

17. $\sqrt{80}$ 8.94

Find the length of a.

18.

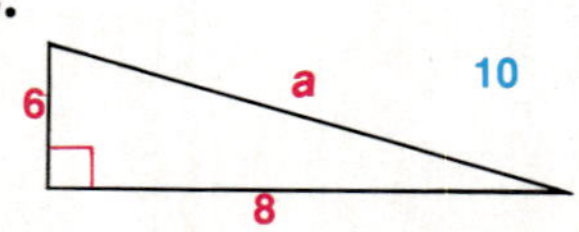

19.

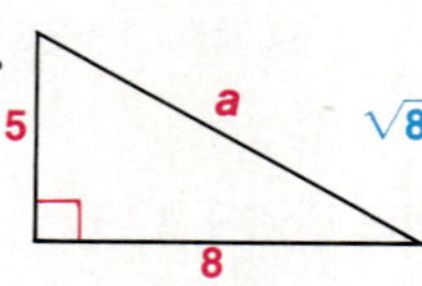

$\sqrt{89}$ or ≈ 9.43

Find each ratio for angle A.

20. sin A $\frac{5}{13}$

21. cos A $\frac{12}{13}$

22. tan A $\frac{5}{12}$

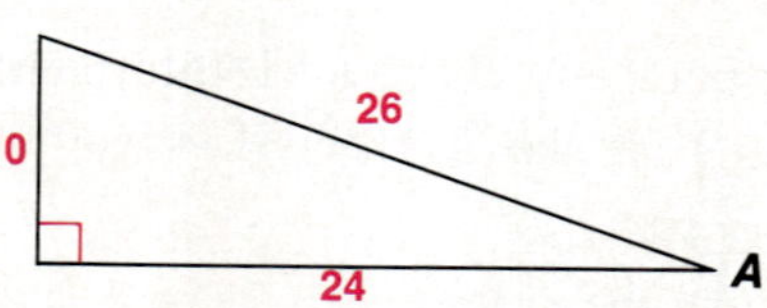

Introducing Polynomials

Objectives: To write monomials and polynomials; to identify binomials and trinomials.
MIN: 1–24; REG: 1–28; MAX: 1–33

Explore

Look for patterns to give the area of each tile.

(1)

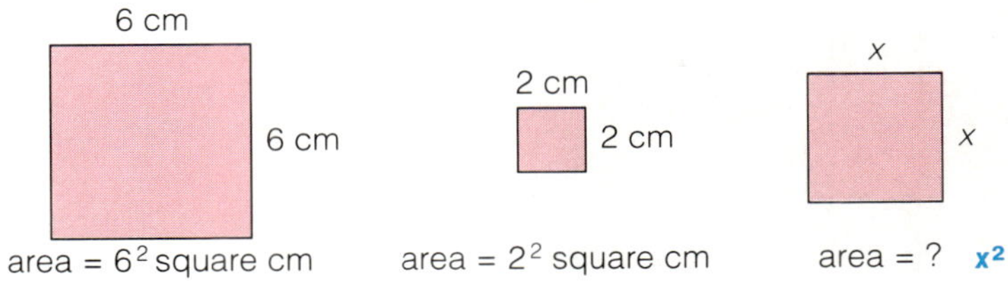

(2)

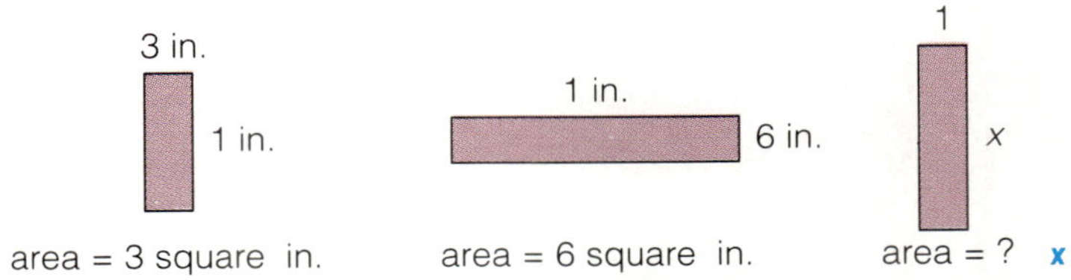

- If the side of a square is n units long, what is the area of the square in terms of n? **n^2 square units**
- If one side of a rectangle is 1 unit long and the other side is n units wide, what is the area of the triangle in terms of n? **n square units**

Monomials

A **monomial** is an expression that is either a numeral, a variable, or a product of numerals and variables with whole number exponents. If the monomial is a numeral, we call it a **constant**. You can use algebra tiles like those in the Explore activity to model monomials.

Example

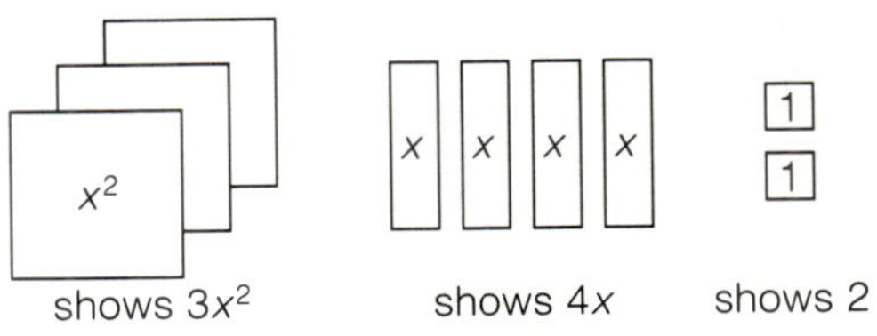

Try This Use algebra tiles to model each monomial.

1. $6x$ **2.** $4x^2$ **3.** 5

4. x^2 **5.** $3x$ **6.** 8

Try This Write a monomial for each model.

7. 

$2x$

8. 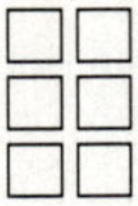

6

9. 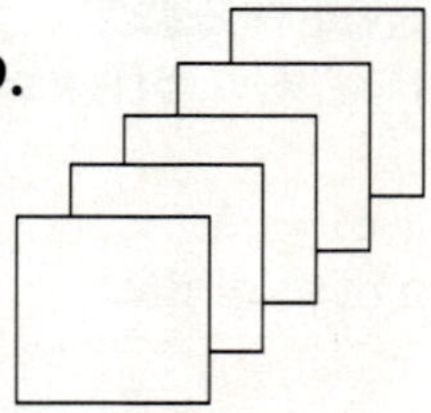

$5x^2$

10. 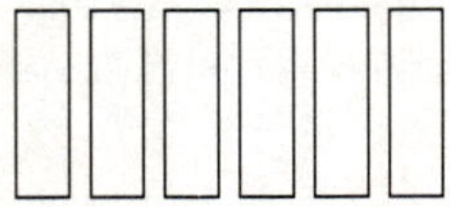

$6x$

Polynomials

A **polynomial** is a monomial or a sum of monomials. You can use algebra tiles to model polynomials.

Example

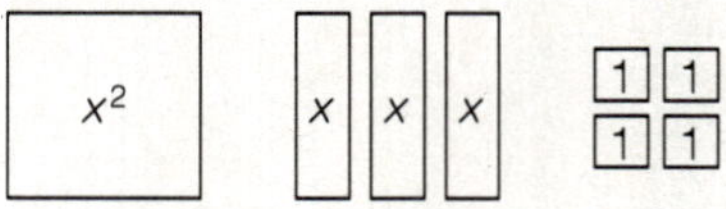

$x^2 + 3x + 4$

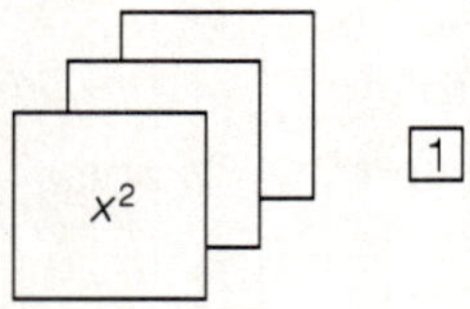

$3x^2 + 1$

In a polynomial, each monomial is called a **term**. $x^2 + 3x + 4$ has three terms and $3x^2 + 1$ has two terms.

Try This Use algebra tiles to model each polynomial. Tell how many terms each polynomial has.

11. $4x + 3$ 2 **12.** $x^2 + x$ 2 **13.** $2x^2 + x + 5$ 3 **14.** $5x^2 + 4$ 2

Try This Write the polynomial for each model.

15. 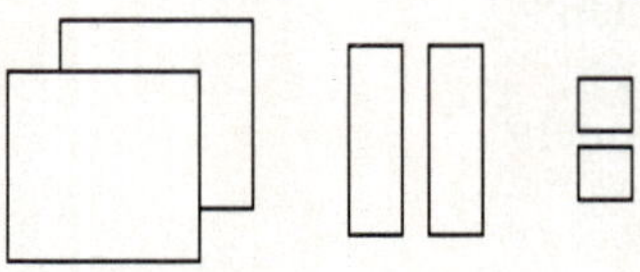

$2x^2 + 2x + 2$

16. 

$5x + 1$

17. 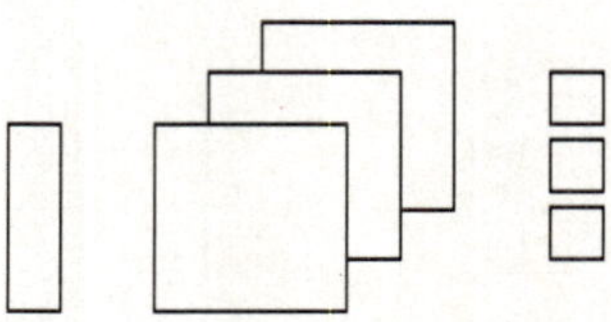

$3x^2 + x + 3$

18. 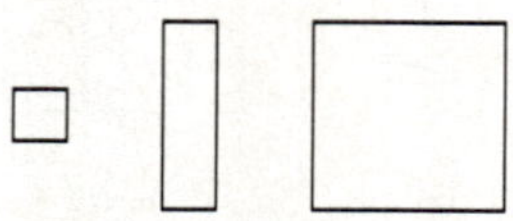

$x^2 + x + 1$

Binomials and Trinomials

A polynomial with exactly two terms is called a **binomial**. A polynomial with exactly three terms is called a **trinomial**.

Example

binomials	$4y + 5$	$x^2 + 2$	$3b^3 + b^2$
	$2 + 6y$	$7 + 8x^2$	$5x^4 + 3x^2$
trinomials	$x^2 + 2x + 4$	$2x^3 + y + 4$	$2a + 2b + 2c$
	$5 + 2a + a^2$	$3x^2 + x + 2$	$x^2 + 3y + z^3$

Try This Tell whether each expression is a binomial or a trinomial.

19. $h^2 + 4h + 3$ trinomial **20.** $6 + t^3$ binomial **21.** $2x + 12$ binomial **22.** $f + 5 + 4f^2$ trinomial

23. Make up a trinomial using the variable y. Answers will vary. Ex: $y^2 + 3y + 4$

24. Make up a binomial using the variable p. Answers will vary. Ex: $p^3 + 2p$

Extending Thinking Skills

Rearrange algebra tiles, then write the polynomial for the model shown. Tell whether each is a monomial, a binomial, or a trinomial.

25.

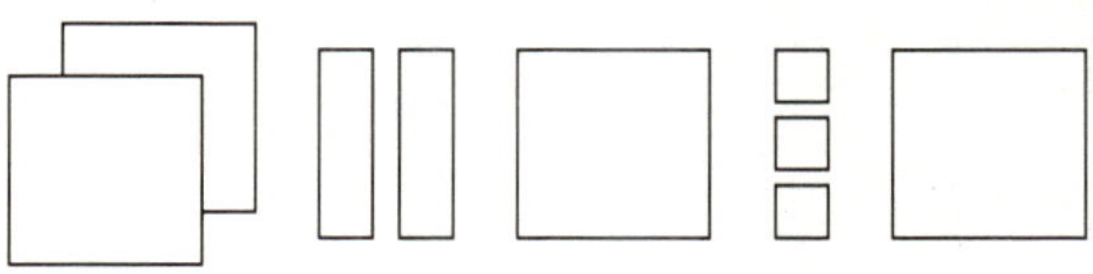

$4x^2 + 2x + 3$; trinomial

26.

$x^2 + 2x + 5$; trinomial

27. Each side of the cube has length x. What is the volume of the cube expressed using x? x^3

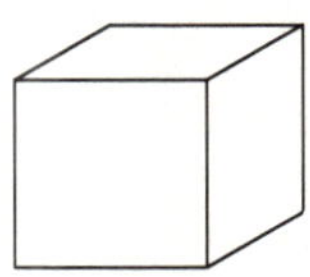

28. Write the polynomial for this picture. $2x^3 + x^2 + 2$

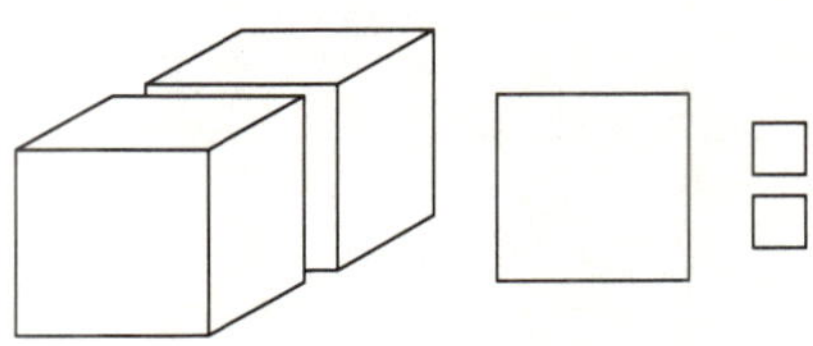

Which of the following are monomials? For those that are not, tell why.

29. $\frac{x}{3}$ yes **30.** $3x^{-1}$ no **31.** $y^2 + xy$ no **32.** $8ab^2c^5$ yes **33.** $2x^2 + 4$ no

Adding Polynomials

Objective: To add polynomials.
MIN: 1–31; REG: 1–16, 17–31 even, 39–44; MAX: 1–31 even, 32–47

Explore

You can use two-color algebra tiles to model polynomials. Same-size tiles represent **like terms**. Same-size light and dark tiles are opposites and cancel each other.

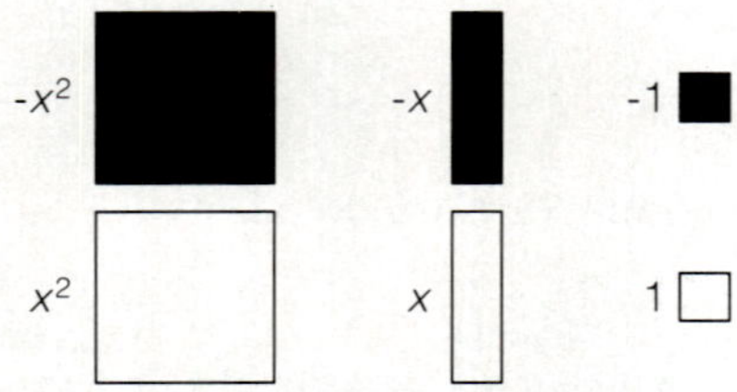

Work with a partner. One person places several algebra tiles on a table, some dark, some light, and some that represent like terms. The other person then combines the algebra tiles, canceling tiles when possible, and writes the polynomial that results. Repeat this until each person has written three polynomials.

Adding Polynomials Using Manipulatives

You can use algebra tiles to add polynomials.

Example

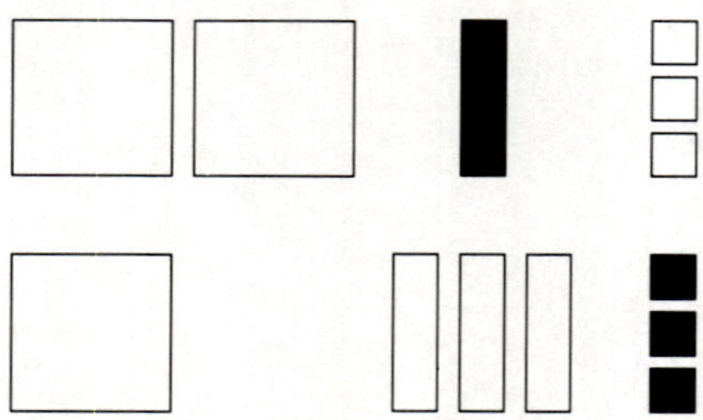

$2x^2 - x + 3$

$x^2 + 3x - 3$

To add these polynomials, combine the like terms.

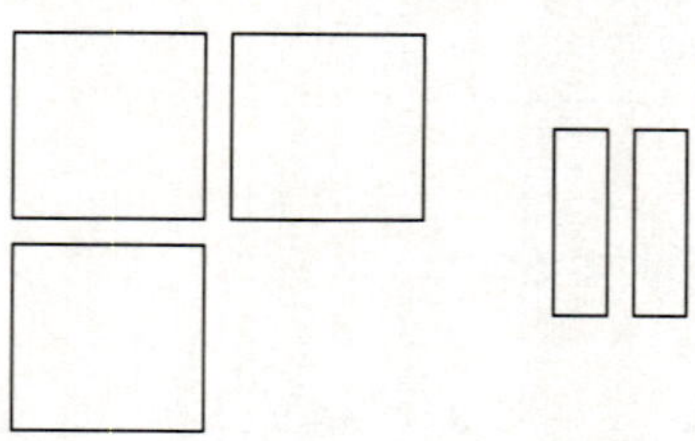

$$\begin{array}{r} 2x^2 - x + 3 \\ +\quad x^2 + 3x - 3 \\ \hline 3x^2 + 2x \phantom{{}-3} \end{array}$$

Try This Write the polynomial addition problem shown by each set of tiles. Use algebra tiles to find the sum. Write the sum as a polynomial.

1. 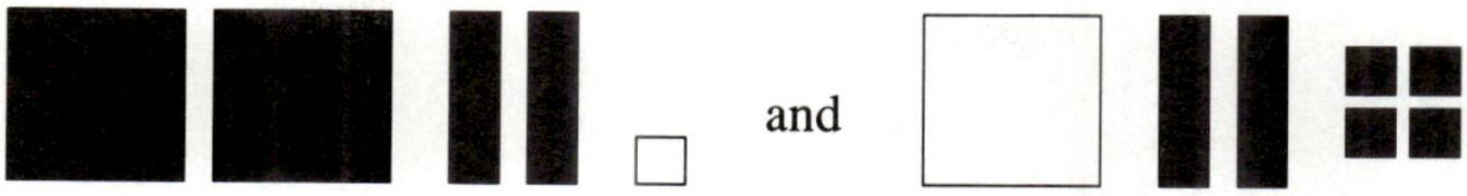

$-2x^2 - 2x + 1$ and $x^2 - 2x - 4$ **Sum: $-x^2 - 4x - 3$**

2. 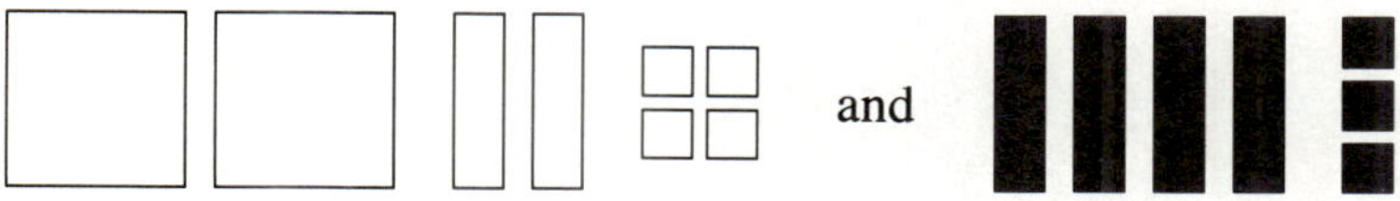

$2x^2 + 2x + 4$ and $-4x - 3$ **Sum: $2x^2 - 2x + 1$**

3. 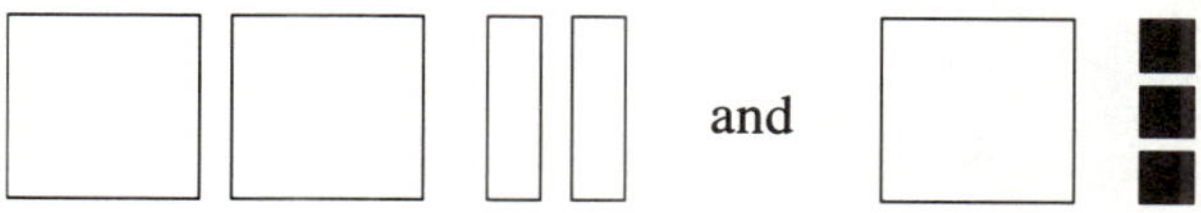

$2x^2 + 2x$ and $x^2 - 3$ **Sum: $3x^2 + 2x - 3$**

4. 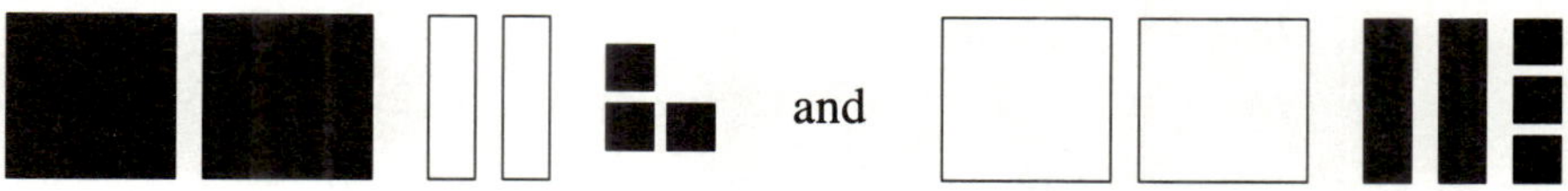

$-2x^2 + 2x - 3$ and $2x^2 - 2x - 3$ **Sum: -6**

Like Terms and Coefficients

In a polynomial, each monomial is called a term. In the term $3x^2$, 3 and x^2 are factors. The numerical factor is called the **coefficient** of the term. 3 is the coefficient of the term $3x^2$ and -5 is the coefficient of $-5x$. You saw in the Explore that same-size tiles represent like terms. **Like terms** in a polynomial are terms with exactly the same variable factors.

like terms ($3x$ and $-5x$)

$$3x^2 + 3x - x^2 + 2 - 5x$$

like terms ($3x^2$ and $-x^2$)

Try This Tell which are like terms.

5. $4x - 2 + x^2 - 2x$ **6.** $n^2 - 3n^2 + 5n^2 - 5n$ **7.** $x^2 - 2x + 4 + 4x^2 - 2x - 3$

8. $6 + 3y - 4 + 3y$ **9.** $3t^2 - 2t^2 + 6 + 12$ **10.** $(4m^2 + 5) + (m^2 - 5m)$

5. $4x$ and $-2x$ **6.** n^2, $-3n^2$, and $5n^2$ **7.** x^2 and $4x^2$; $-2x$ and $-2x$; 4 and -3 **8.** 6 and -4; $3y$ and $3y$ **9.** $3t^2$ and $-2t^2$; 6 and 12 **10.** $4m^2$ and m^2

Tell the coefficient of each term.

11. $5x$ **5** **12.** $3y^2$ **3** **13.** $-4p^2$ **-4** **14.** $-x$ **-1** **15.** $24n$ **24** **16.** $-x^2$ **-1**

Adding Polynomials

You can also add polynomials by adding like terms. To add like terms, you add the coefficients of the like terms.

Examples

$$\begin{array}{r} 4x^2 - 3x + 5 \\ + \quad x^2 + 2x + 3 \\ \hline 5x^2 - x + 8 \end{array}$$

$$(-3p^2 - 2p - 5) + (6p^2 - 2p + 7) = 3p^2 - 4p + 2$$

Try This Add. You may want to have students check their sums using algebra tiles.

17. $\begin{array}{r} 3b + 6 \\ + \; 4b - 5 \\ \hline \end{array}$ 7b + 1

18. $\begin{array}{r} 4x^2 + 2x + 3 \\ + \; 2x^2 + 6x + 2 \\ \hline \end{array}$ 6x² + 8x + 5

19. $\begin{array}{r} -c^2 + c \\ + \; c^2 - 4c + 3 \\ \hline \end{array}$ −3c + 3

20. $(5y + 3) + (2y - 4)$ 7y − 1

21. $(-y + 4) + (3y + 6)$ 2y + 10

22. $(2h^2 + 2) + (h^2 + 8)$ 3h² + 10

23. $(t^2 + 7t - 3) + (-3t^2 - 7t + 3)$ 2t²

24. $(-z^2 + z) + (-z^2 - z - 4)$ −2z² − 4

25. $(x^2 + 6x + 3) + (3x^2 - x - 5)$ 4x² + 5x − 2

26. $(2x^2 + x - 4) + (x^2 - 3x + 2)$ 3x² − 2x − 2

27. $(3n^2 - 3n) + (4n + 2)$ 3n² + n + 2

28. $(2 + 6m) + (4 - 4m)$ 6 + 2m

29. $(5y^2 - 3y - 3) + (2y^2 + y - 3)$ 7y² − 2y − 6

30. $(-x^2 - 6x - 9) + (x^2 - 4x - 1)$ −10x − 10

31. $(p^2 - 6) + (p^2 + 6)$ 2p²

Extending Thinking Skills

Tell whether each is always true, sometimes true, or never true.

32. A polynomial is a monomial. sometimes

33. A monomial is a polynomial. always

34. The coefficient of a term is the same as the exponent of the term. sometimes

35. A binomial has exactly two terms. always

36. Like terms have the same coefficients. sometimes

37. $-x$ is a negative number. sometimes

38. A trinomial is also a binomial. never

You can **simplify** an algebraic expression by adding like terms. Simplify each expression.

39. $4x + 3 - 2x - 6$ 2x − 3

40. $8 - 4y + 3y - 6$ 2 − y

41. $3w + 4w - 5 + 8w$ 15w − 5

42. $r - r - r + 3r$ 2r

43. $6m + 3 - 2m + 7$ 4m + 10

44. $4f - 5 + 3f + 7$ 7f + 2

Solve for x. First simplify the expression on the left side of the equal sign.

45. $3x + 4 - 2x + 7 = 24$ x = 13

46. $12 - 5x + 8x - 6 = 36$ x = 10

47. $12x + 2 - 8x + 5 = 47$ x = 10

Subtracting Polynomials

Objective: To subtract polynomials.
MIN: 1–28; REG: 1–10, 11–28 even, 34–35; MAX: 1–28 even, 29–37

Explore

Work with a partner. Follow these steps. *Partner 1:* Show a polynomial using algebra tiles. *Partner 2:* Write an algebraic expression that represents the polynomial. Then take away some of the algebra tiles. *Partner 1:* Write an algebraic expression for the new polynomial. Repeat this three times changing roles.

Subtracting Polynomials Using Manipulatives

You can use algebra tiles to subtract polynomials.

Example

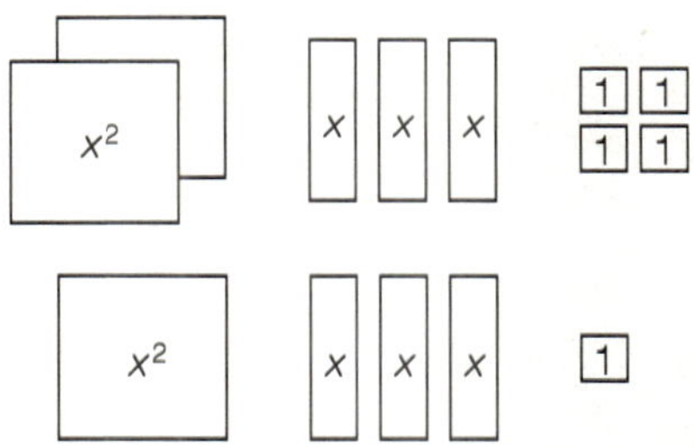

$(2x^2 + 3x + 4) - (x^2 + 3)$

Take away one x^2 and 3 units.

$(2x^2 + 3x + 4) - (x^2 + 3) = x^2 + 3x + 1$

Try This Show the first polynomial using algebra tiles. Take away the second polynomial. Write the polynomial for the tiles that are left.

1. $(4x^2 + 3x + 6) - (x^2 + x + 1)$ **$3x^2 + 2x + 5$**
2. $(8x + 2) - (2x + 2)$ **$6x$**
3. $(2x^2 + 4) - (2x^2 + 3)$ **1**
4. $(3x^2 + 4x + 5) - (2x^2 + 2x + 2)$ **$x^2 + 2x + 3$**
5. $(4x^2 + 6) - (2x^2 + 1)$ **$2x^2 + 5$**
6. $(2x^2 + 3x + 7) - (2x^2 + 3x + 6)$ **1**

Finding the Opposite of a Polynomial

You have learned that the expression $-x$ means **the opposite of x**. Similarly, the expression $-(2x + 4)$ means **the opposite of $(2x + 4)$**. To find the opposite of an expression, write the opposite of each term. You can use algebra tiles to show that $-(2x - 4)$ is the same as $-2x + 4$.

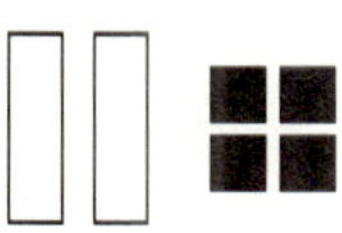 $2x - 4$

 $-(2x - 4) = -2x + 4$

Example Write an equivalent expression.

$$-(3x^2 - 3x + 6) = -3x^2 + 3x - 6$$

$$-(-x^2 + x - 10) = x^2 - x + 10$$

Try This Use algebra tiles to show the opposite of each expression. Write the algebraic expression you have modeled.

7. 

$x^2 - 2x + 3$

8. 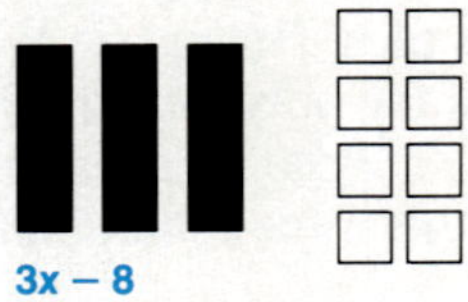

$3x - 8$

9. 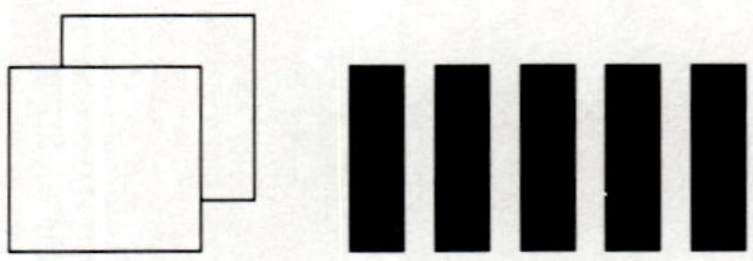

$-x^2 + 5x$

10. 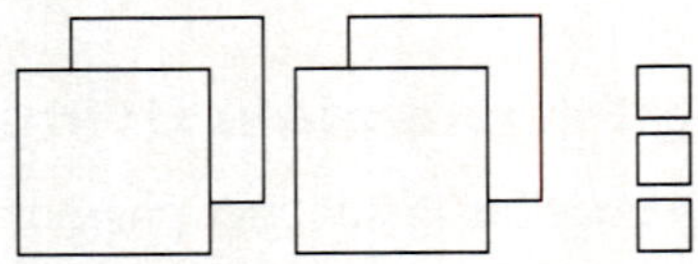

$-4x^2 - 3$

Write an equivalent expression.

11. $-(4x - 3)$ $-4x + 3$

12. $-(x^2 + 2x - 4)$ $-x^2 - 2x + 4$

13. $-(-3x^2 + 6)$ $3x^2 - 6$

14. $-(-x^2 - 4x + 12)$ $x^2 + 4x - 12$

Subtracting Polynomials by Adding the Opposite

You have learned that subtracting is the same as adding the opposite. You can use this idea to subtract polynomials.

Examples Write each as an addition problem.

$$(4x - 4) - (3x - 8) = (4x - 4) + (-3x + 8)$$

$$(2x^2 + 4x - 3) - (-3x^2 + 2x - 6) = (2x^2 + 4x - 3) + (3x^2 - 2x + 6)$$

After you have written the subtraction problem as an addition problem, you can add the polynomials by combining like terms.

Examples Find the difference.

$$(x + 6) - (3x - 4) = x + 6 + (-3x + 4) \quad \text{Write as an addition problem.}$$
$$= -2x + 10$$

$$(-x^2 - 4x + 7) - (4x^2 - 5) = -x^2 - 4x + 7 - 4x^2 + 5$$
$$= -5x^2 - 4x + 12$$

Try This Write each as an addition problem.

15. $(7x + 6) - (x + 2)$ $(7x + 6) + (-x - 2)$

16. $(-2y - 4) - (4y + 3)$ $(-2y - 4) + (-4y - 3)$

17. $(3m^2 + 3) - (2m^2 + 5)$ $(3m^2 + 3) + (-2m^2 - 5)$

18. $(5u^2 + 2u + 1) - (-u^2 - 4u)$ $(5u^2 + 2u + 1) + (u^2 + 4u)$

19. $(-4x^2 - 5) - (6x - 2)$ $(-4x^2 - 5) + (-6x + 2)$

20. $(t^2 + 2t - 3) - (t^2 - 2t - 3)$ $(t^2 + 2t - 3) + (-t^2 + 2t + 3)$

Find the difference.

21. $(7b - 4) - (b + 6)$ $6b - 10$

22. $(-7x - 5) - (-x - 4)$ $-6x - 1$

23. $(d^2 - 2d) - (6d + 9)$ $d^2 - 8d - 9$

24. $(2y^2 + 3y - 8) - (4y^2 - 2y - 1)$ $-2y^2 + 5y - 7$

25. $(-x^2 - x - 12) - (4x - 3)$ $-x^2 - 5x - 9$

26. $(4 + 8y) - (6 - 3y)$ $-2 + 11y$

27. $(3x + 5 - 2x^2) - (4x^2 + 4 - 5x)$ $-6x^2 + 8x + 1$

28. $(-m + m^2) - (-4 - 6m + m^2)$ $5m + 4$

Extending Thinking Skills

Which is correct for each expression when x can be replaced with any rational number?

29.	$-x$	A. Negative	B. Positive	C. Can't tell	C
30.	$\lvert x \rvert$	A. Negative	B. Positive	C. Can't tell	B
31.	$-\lvert x \rvert$	A. Negative	B. Positive	C. Can't tell	A
32.	$\lvert -x \rvert$	A. Negative	B. Positive	C. Can't tell	B
33.	$-\lvert(-x)\rvert$	A. Negative	B. Positive	C. Can't tell	A

34. What can be subtracted from $(-3x^2 - 4x + 5)$ to get $x^2 - 8$? $-4x^2 - 4x + 13$

35. Make up a polynomial subtraction problem with a difference of $-x^2 + 5x - 14$. Answers will vary. Ex: $(2x^2 - 5x - 12) - (3x^2 + 2)$

Solve.

36. $(3x - 6) - (2x + 8) = 19$ $x = 33$

37. $(-4x + 3) - (2x - 7) = 22$ $x = -2$

38. $(-7y - 9) - (3y + 8) = -7$ $y = -1$

39. $(3x^2 - 4x - 1) - (3x^2 + 5x - 12) = -7$ $x = 2$

40. $(6m + 3) - (-3m + 7) = 32$ $m = 4$

41. $(8 - 6g) - (-12 - 4g) = 30$ $g = -5$

Multiplying Polynomials

Objective: To multiply polynomials.
MIN: 1–31; REG: 1–16, 17–31 even, 32–34; MAX: 1–31 even, 32–37

Explore

- Lay out algebra tiles (x tiles) to show the length and width of a rectangle, as shown in the diagram. How many x^2 pieces does it take to complete this rectangle?
- Work with a partner and repeat the above process several times using different values for the length and width of the rectangle. Record the length, width, and number of x^2 pieces for each trial.

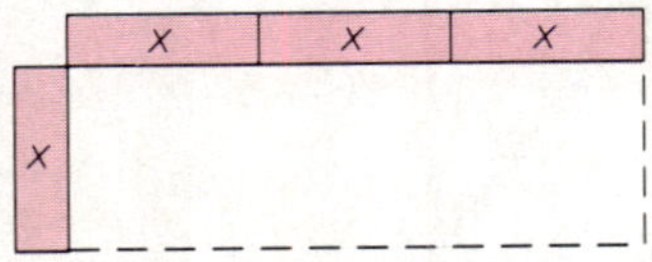

Multiplying Monomials Using Manipulatives

You can use algebra tiles to multiply two monomials.

Multiply $2x$ and $3x$.

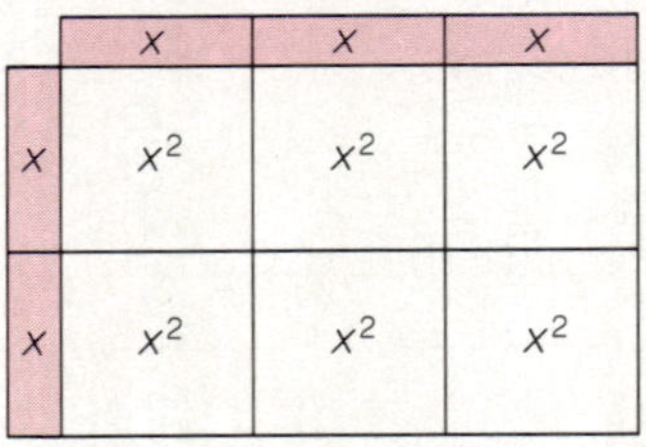

$(2x)(3x) = (2 \cdot 3)x^2 = 6x^2$

Multiply 3 and $2x$.

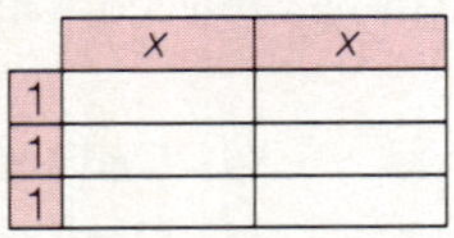

$(3)(2x) = (3 \cdot 2)x = 6x$

Write the multiplication problem shown by each model.

1.

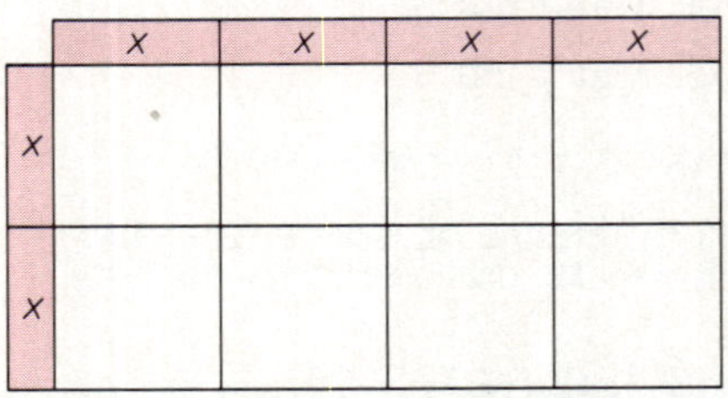

(2x)(4x) = 8x²

2.

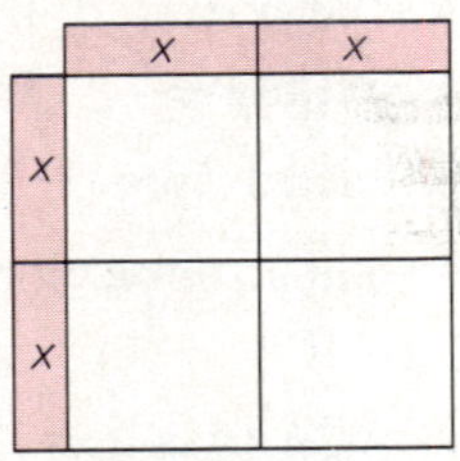

(2x)(2x) = 4x²

Multiply.

3. $3(4x)$ **12x**

4. $(3x)(5x)$ **15x²**

5. $(4x)(7x)$ **28x²**

6. $(5x)(7)$ **35x**

7. $(2x)(4x)$ **8x²**

8. $(5x)(6x)$ **30x²**

9. $(2)(5x)$ **10x**

10. $(3x)(2x)$ **6x²**

Multiplying Monomials and Polynomials Using Manipulatives

You can use algebra tiles to model multiplication of a monomial and a polynomial.

Example

Multiply $2x$ and $x + 3$.

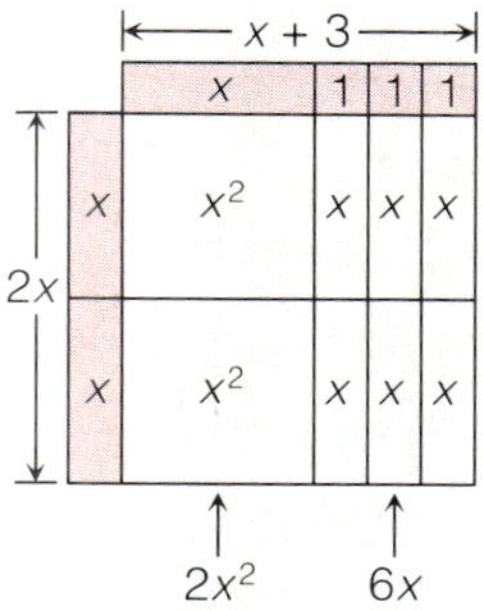

$$2x(x + 3) = 2 \cdot x^2 + (2 \cdot 3)x$$
$$= 2x^2 + 6x$$

Multiply $2x + 2$ and $2x$.

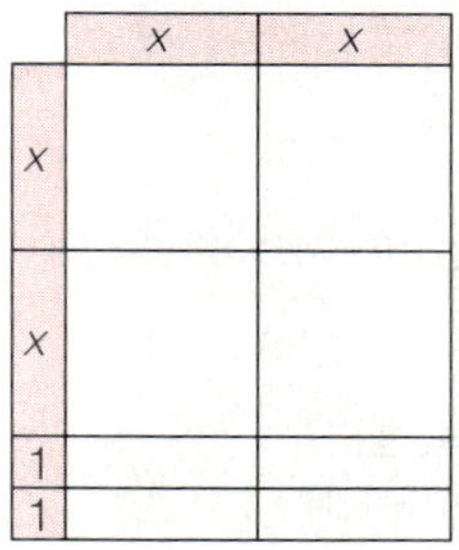

$$(2x + 2)2x = (2 \cdot 2)x^2 + (2 \cdot 2)x$$
$$= 4x^2 + 4x$$

Use algebra tiles to model each of these products.

11. $2x(2x + 1)$ $4x^2 + 2x$ **12.** $x(3x + 2)$ $3x^2 + 2x$ **13.** $3x(x + 5)$ $3x^2 + 15x$

14. $(x + 3)(3x)$ $3x^2 + 9x$ **15.** $(x + 2)2x$ $2x^2 + 4x$ **16.** $(2x + 3)x$ $2x^2 + 3x$

Multiplying Monomials and Polynomials

You can multiply a monomial and a polynomial by using the distributive property.

$$a(b + c) = ab + ac$$

Example Multiply.

$(4x)(x + 4) = (4x)(x) + (4x)(4)$ Use the distributive property.

$= 4x^2 + 16x$ Multiply the monomials.

Examples Multiply.

$$(5x)(3x-7) = (5x)(3x) - (5x)(7)$$
$$= 15x^2 - 35x$$

$$(3x)(4x^2+2x+7) = (3x)(4x^2) + (3x)(2x) + (3x)(7)$$
$$= 12x^3 + 6x^2 + 21x$$

Try This Multiply.

17. $(2y)(y+6)$ $2y^2 + 12y$
18. $(4x)(3x+2)$ $12x^2 + 8x$
19. $(p+2)(2p)$ $2p^2 + 4p$
20. $(2x+1)(3x)$ $6x^2 + 3x$
21. $(7m)(2m-3)$ $14m^2 - 21m$
22. $(4a)(4-a)$ $16a - 4a^2$
23. $(5x)(3-2x)$ $15x - 10x^2$
24. $(3y)(2y-7)$ $6y^2 - 21y$
25. $(12x)(3x-8)$ $36x^2 - 96x$
26. $(3a^2)(a-4)$ $3a^3 - 12a^2$
27. $(4x^2)(3x+5)$ $12x^3 + 20x^2$
28. $(4y)(y^2+3y+1)$
29. $(3x^2)(x^2+5x-3)$
30. $(4p^3)(2p-7)$
31. $(a^2)(3a^2-4a-3)$

28. $4y^3 + 12y^2 + 4y$ 29. $3x^4 + 15x^3 - 9x^2$ 30. $8p^4 - 28p^3$ 31. $3a^4 - 4a^3 - 3a^2$

Extending Thinking Skills

When you multiply two monomials with the same variable, you add exponents.

$$x^n \cdot x^m = x^{n+m}$$

Multiply.

32. $x^{19} \cdot x^{12}$ x^{31}
33. $x^2 \cdot x^p$ x^{2+p}
34. $x^r \cdot x^7$ x^{r+7}

You can multiply two binomials by using the distributive property more than once. Study the examples below.

$$(x+2)(x+3) = \underset{\boxed{1}}{(x+2)(x)} + \underset{\boxed{2}}{(x+2)(3)}$$ Use the distributive property to multiply $(x + 2)$ by each term in $x + 3$.

$$= (x \cdot x) + (2 \cdot x) + (x \cdot 3) + (2 \cdot 3)$$ Use the distributive property on both parts $\boxed{1}$ and $\boxed{2}$.

$$= x^2 + 2x + 3x + 6$$ Simplify.

$$= x^2 + 5x + 6$$ Combine like terms.

Multiply.

35. $(x+3)(x+4)$ $x^2 + 7x + 12$
36. $(x+2)(2x-7)$ $2x^2 - 3x - 14$
37. $(x+5)(2x+7)$ $2x^2 + 17x + 35$

Dividing Polynomials by Monomials

Objective: To divide monomials and polynomials by a monomial.
MIN: 1–22; REG: 1–26; MAX: 1–22 even, 23–33

You have studied how to simplify a fraction by canceling like factors from the numerator and the denominator.

Example

$$\frac{105}{21} = \frac{\cancel{3} \cdot 5 \cdot \cancel{7}}{\cancel{3} \cdot \cancel{7}} = 5$$

Since the bar of a fraction indicates division, you have completed a division problem when you have simplified this fraction.

Dividing a Monomial by a Monomial

You can divide one monomial by another by factoring and canceling as shown in the numerical example above.

Examples

$$\frac{6x^3}{3x} = \frac{2 \cdot \cancel{3} \cdot \cancel{x} \cdot x \cdot x}{\cancel{3} \cdot \cancel{x}}$$ Factor the numerator and denominator and cancel like factors.

$$= 2x^2$$ Simplify.

$$\frac{42x^3}{14x^2} = \frac{\cancel{2} \cdot 3 \cdot \cancel{7} \cdot \cancel{x} \cdot \cancel{x} \cdot x}{\cancel{2} \cdot \cancel{7} \cdot \cancel{x} \cdot \cancel{x}} = 3x$$

$$\frac{3x^3}{9x} = \frac{\cancel{3} \cdot \cancel{x} \cdot x \cdot x}{\cancel{3} \cdot 3 \cdot \cancel{x}} = \frac{1}{3}x^2$$

You can also divide two monomials by dividing the coefficients and using the properties of exponents as explained on page 221.

Examples

$$\frac{3a^6}{15a^2} = \frac{3}{15} \cdot \frac{a^6}{a^2}$$ Write as division of coefficients and division of exponents.

$$= \frac{1}{5} \cdot a^{6-2}$$ Subtract the exponents.

$$= \frac{1}{5}a^4$$ Simplify.

$$\frac{-36y^2}{4y} = \frac{-36}{4} \cdot \frac{y^2}{y} = -9 \cdot y^{2-1} = -9y$$

Try This Divide.

1. $\frac{9y^2}{3y}$ **3y**
2. $\frac{6x^3}{2x^2}$ **3x**
3. $\frac{16x^4}{4x^2}$ **4x²**
4. $\frac{35n^6}{5n^4}$ **7n²**
5. $\frac{24x^5}{6x}$ **4x⁴**
6. $\frac{18y^7}{3y^6}$ **6y**
7. $\frac{21x^4}{3x}$ **7x³**
8. $\frac{34p}{17}$ **2p**
9. $\frac{5a^4}{25a^3}$ $\frac{1}{5}a$
10. $\frac{x^3}{2x^3}$ $\frac{1}{2}$
11. $\frac{10c^4}{2c}$ **5c³**
12. $\frac{18y^5}{6y^5}$ **3**

Dividing a Polynomial by a Monomial

To divide a polynomial by a monomial divide each term of the polynomial by the monomial and simplify. Since a fraction bar indicates division, division may be written as a fraction.

Example Divide $2x^2 - 8x + 6$ by 2.

$$(2x^2 - 8x + 6) \div 2 = \frac{2x^2 - 8x + 6}{2}$$ Write as a fraction.

$$= \frac{2x^2}{2} - \frac{8x}{2} + \frac{6}{2}$$ Divide each term of the polynomial by the denominator.

$$= x^2 - 4x + 3$$ Simplify.

Example Divide $9y^3 + 6y^2 + 15y$ by $3y$.

$$(9y^3 + 6y^2 + 15y) \div 3y = \frac{9y^3 + 6y^2 + 15y}{3y}$$

$$= \frac{9y^3}{3y} + \frac{6y^2}{3y} + \frac{15y}{3y}$$

$$= 3y^2 + 2y + 5$$

Divide.

13. $(6x^2 + 3x + 21) \div 3$ $2x^2 + x + 7$
14. $(5b^3 + 25b - 10) \div 5$ $b^3 + 5b - 2$
15. $(8r^5 - 4r^2 + 12) \div 4$ $2r^5 - r^2 + 3$
16. $(x^3 + 3x^2) \div x$ $x^2 + 3x$
17. $(x^4 - 4x^3 + x^2) \div x^2$ $x^2 - 4x + 1$
18. $(9y^3 + 21y) \div y$ $9y^2 + 21$
19. $(6s^3 + 2s^2) \div 2s$ $3s^2 + s$
20. $(14d^3 - 21d^2) \div 7d$ $2d^2 - 3d$
21. $(3t^5 - 15t^3 + 6t^2) \div 3t^2$ $t^3 - 5t + 2$
22. $(21x^6 + 35x^5 + 105x^4) \div 7x^3$ $3x^3 + 5x^2 + 15x$

Extending Thinking Skills

When dividing a polynomial by a monomial, if the exponent of the divisor is greater than the exponent of any term, you can write the answer as a polynomial plus a fraction. Study the example below.

$$\begin{aligned}(x^3 + 2x^2 + 3) \div x &= \frac{x^3 + 2x^2 + 3}{x} \\ &= \frac{x^3}{x} + \frac{2x^2}{x} + \frac{3}{x} \\ &= (x^2 + 2x) + \frac{3}{x}\end{aligned}$$

Divide.

23. $(k^2 + 3k + 5) \div k$ $(k + 3) + \frac{5}{k}$

24. $(3x^2 - 5x + 7) \div x$ $(3x - 5) + \frac{7}{x}$

25. $(5x^3 + 15x - 25) \div 5x$ $(x^2 + 3) - \frac{5}{x}$

26. $(4y^3 + 3y^2 + 7y) \div y^2$ $(4y + 3) + \frac{7}{y}$

Tell whether each is always true, sometimes true, or never true.

27. When you divide one monomial by another monomial the quotient is a monomial. **sometimes**

28. When you divide a polynomial by a monomial the quotient is a monomial. **sometimes**

29. When you divide a polynomial by a monomial the quotient is a polynomial. **sometimes**

30. When you divide one monomial by another monomial the quotient is a number. **sometimes**

Division is sometimes described in terms of multiplication as follows:

$$\boldsymbol{a \div b = c} \text{ if and only if } \boldsymbol{a = bc \ (b \neq 0)}.$$

Verify this relationship for the following quotients. **Check students work.**

31. $(x^3 + 3x^5 - x^2) \div x$

32. $(8z^2 - 4z + 1) \div z^2$

33. By substituting a numerical value for x, decide whether this division is correct.

$$\frac{2x}{4x^2 + 6x} = \frac{2x}{4x^2} + \frac{2x}{6x} = \frac{1}{2x} + \frac{1}{3}$$

Not correct; numerical values will vary. Ex: For $x = 3$, $\frac{1}{9} \neq \frac{1}{2}$

Conducting a Survey

A **survey** is a technique for collecting data involving fact or opinion. Completed surveys are often used to present information which can help make decisions. Conducting a survey involves the following steps:

- **Choose** the topic or issue to be surveyed.
- **Plan** the questions to be asked.
- **Ask** the questions and collect the data.
- **Organize** and **interpret** the data.
- **Present** the findings.

Work with a group to conduct the following survey. Follow the procedures suggested.

Survey Topic

Colors of high-top sneakers that are most popular among students of your age group.

Planning the Survey

- Decide what colors of high-top sneakers you will consider.
- Make a *questionnaire* you can give to students. List all of the colors of sneakers you will consider. Decide how students indicate their favorite color or colors on the questionnaire.
- Suppose you cannot give the questionnaire to all students at your grade level. Decide which students will be your *sample* — that is, the ones to whom you will give the questionnaire.

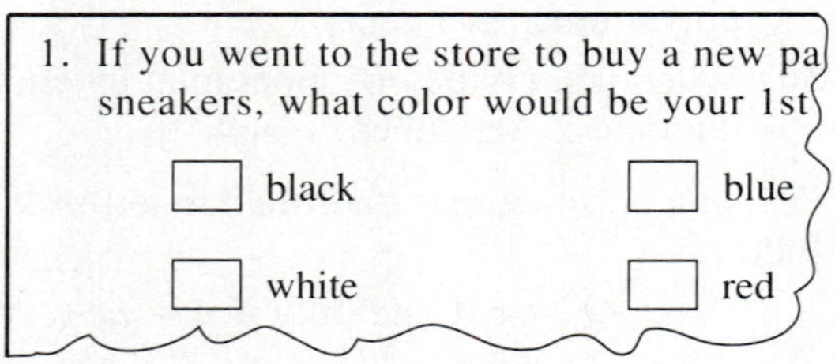

Collecting Data

- Give the questionnaire to the sample you selected.

Organizing the Data

- Make and complete a *frequency table*, like the one below, to record the *data* from your survey.

Sneaker Color	1st choice	2nd choice	3rd
white	𝍸 ||	𝍸	
black			

- Make and complete a bar graph, like the one below, to show the data pictorially.

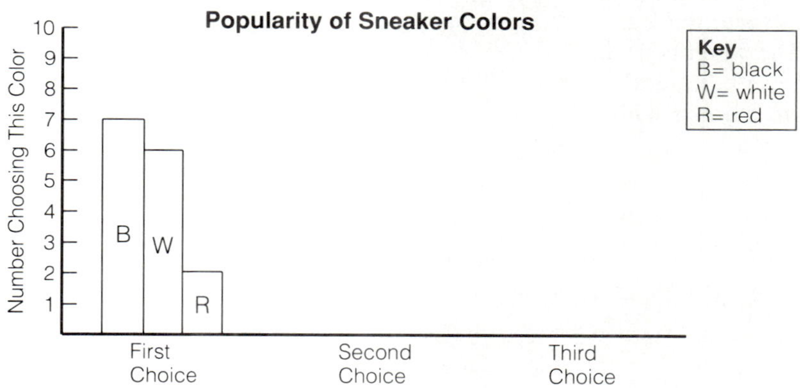

Interpreting the Data

- What color of high top sneakers was chosen most? least? **Answers will vary.**
- Do you think there is any doubt about which 2 colors of sneaker are most popular? Give reasons to support your answer.

Presenting Your Findings

- Write a letter to a sneaker company that summarizes the findings from your survey. Use the frequency table and the graph in your letter. Include recommendations for the manufacturing of their sneakers with regard to color.

Try This

Work in a group to choose a survey topic of your own and conduct a survey. Generally follow the procedure above, but make your own decisions about how to best organize the data and to present the findings.

Stem-and-Leaf Diagrams

A **stem-and-leaf diagram** is a method of displaying numerical data that shows all of the original numbers and has the visual impact of a tally. A stem-and-leaf diagram is easy to construct, easy to read, and is a useful means of comparing data.

Making a Stem-and-Leaf Diagram

Here is how to construct a stem-and-leaf diagram for the data at the right. To make the diagram, we split each numeral into two parts, a stem and a leaf. In this diagram, the stem numbers are tens digits and the leaf numbers are ones digits. The weights in the bottom row, 37, 35, 39, 38, 39, 38, and 37 are read by putting the stem number with each of the leaf numbers.

Weights in Kilograms of Boys and Girls in a Class.	
Boys Weights	**Girls Weights**
41, 37, 35, 46, 50, 44, 40, 41, 45, 37, 39, 43, 45, 38, 51, 45	50, 39, 51, 50, 49, 48, 47, 38, 46, 52, 45, 49, 53, 47, 52, 52

Stem	Leaf
5	0, 1, 0, 1, 0, 2, 3, 2, 2
4	1, 6, 4, 0, 1, 5, 3, 5, 5, 9, 8, 7, 6, 5, 9, 7
3	7, 5, 7, 9, 8, 9, 8

Try This

1. The data below shows student heights in centimeters. Make a stem-and-leaf diagram to show the data. You should use 2-digit stem numbers.

Boys Heights
156, 139, 164, 149, 147, 176, 134, 153, 172, 163, 168, 138, 170, 157, 155, 145

Girls Heights
164, 128, 136, 147, 129, 136, 145, 161, 149, 171, 164, 173, 165, 173, 156, 159

2. A **ranked** stem-and-leaf diagram has numbers in order from least to greatest. Rank the stem-and-leaf diagram you made in problem 1.

1.

Stem	Leaf
17	6, 2, 0, 1, 3, 3
16	4, 3, 8, 4, 1, 4, 5
15	6, 3, 7, 5, 6, 9
14	9, 7, 5, 7, 5, 9
13	9, 4, 8, 6, 6
12	8, 9

2.

Stem	Leaf
12	8, 9
13	4, 6, 6, 8, 9
14	5, 5, 7, 7, 9, 9
15	3, 5, 6, 6, 7, 9
16	1, 3, 4, 4, 4, 5, 8
17	0, 1, 2, 3, 3, 6

Making Stem-and-Leaf Diagrams with Two Sets of Data

A stem-and-leaf diagram can also be used to display and compare two sets of data. This diagram shows the weight data from the previous page separated into weights for boys and girls. The stem numbers are in the center column. The leaf numbers are in the outside columns.

Try This

3. Make a stem-and-leaf diagram to display and compare the heights of boys and girls in a class. Use the height data given on the previous page.

Boys		Girls
6, 2, 0	17	1, 3, 3
4, 3, 8	16	4, 1, 4, 5
6, 3, 7, 5	15	6, 9
9, 7, 5	14	7, 5, 9
9, 4, 8	13	6, 6
	12	8, 9

Interpreting the Diagrams

- Here is one interpretation someone might make from the diagram on the previous page: The largest group of weights are in the 40s.
- Here is one intepretation someone might make from the diagram at the top of this page: The average girl's weight is in the upper 40s.

Work in groups to generate as many observations as possible about the two diagrams. Discuss your observations and compare them with the observations of other groups.

Try This

4. Explain how the diagrams suggest each of the interpretations above. **Answers will vary.**
5. Look at your diagrams for heights and make some statements that interpret the data. Explain your interpretations. **Answers will vary.**
6. Make stem-and-leaf diagrams for the heights and weights of the students in your class. Use the diagrams to make statements that interpret the data. **Answers will vary.**

Glyphs

A **glyph** is a picture or symbol that shows selected data at a glance. It allows us to combine and visually display several bits of information.

A **circle-ray glyph** is a circle with rays connected on the outside. The length of the rays give information about the data. A **facial glyph** is a glyph that uses a picture of a face to display information. Glyphs using other types of recognizable objects are also useful.

Making a Circle-Ray Glyph

Teri received the following grades for the past 4 quarters.

Math C, C, B, A
Social Studies C, B, B, B
Science F, D, C, B
English C, B, A, A

To make glyphs to show the grades, she let A = 4, B = 3, C = 2, D = 1 and F = 0. Then she let $\frac{1}{8}$ inch equal 1 grade point, and drew the rays on the glyphs.

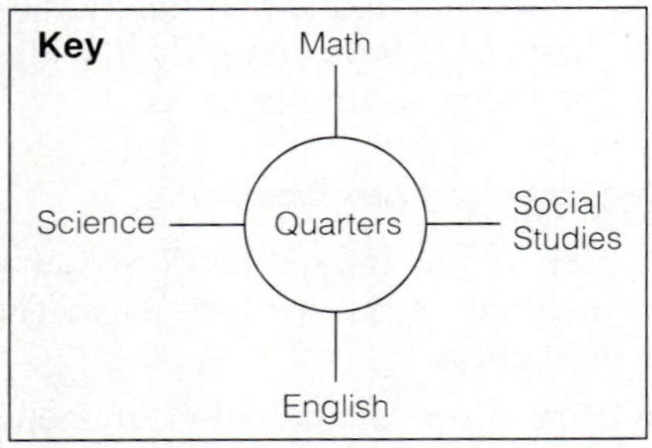

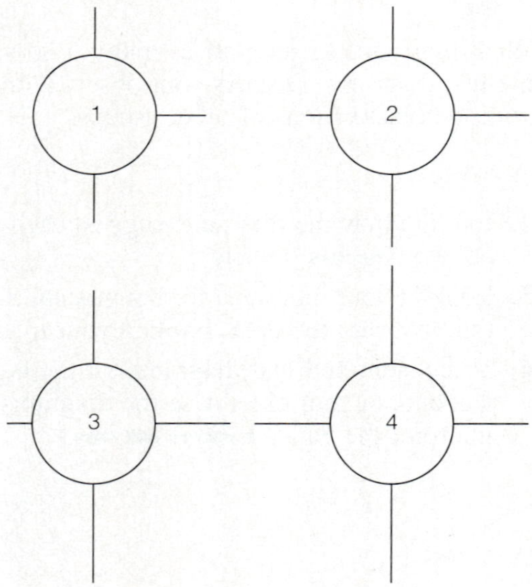

Interpreting the Circle-Ray Glyphs

Use only the glyphs and the key to answer these questions.

- During which quarter did Teri make the worst grades? the best grades? **First; fourth**
- Do you think her grades improved from quarter to quarter? Explain. **Yes**
- Did Teri get an F? If so, during which quarter? **Yes, first quarter**

Try This

1. Make a circle-ray glyph for each student to show the four test scores shown below. **Note: remind students that they need to make a key.**
   Bob: 59, 85, 95, 80
   Jane: 65, 90, 79, 81
   Carlos: 76, 94, 99, 86
2. Look at the glyphs and make statements to interpret the data. The top student is the one with the longest combined length of rays. Do the glyphs show the top student? **Yes, Carlos**
3. Choose some data of your own, such as sports data, and make glyphs to show it. **Answers will vary.**

Making a Facial Glyph

Matt and his friends made these facial glyphs to show some things about themselves. They used three levels, *below average, average,* and *above average* to determine the size of the facial characteristics that show the data. The key shows the average size.

Key

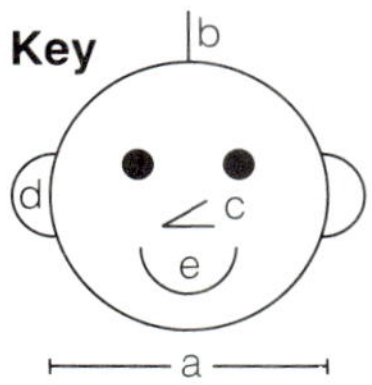

a. weight (circle size)
b. height (segment length)
c. heart beats per minute at rest (angle size)
d. cholesterol level (semi-circle size)
e. breaths per minute at rest (arc length)

Matt   Jay   Todd

Interpreting the Facial Glyphs

- Who is short with above average weight? **Jay**
- Who is tall and thin? **Todd**
- Who has the lowest cholesterol level and the slowest heartbeat? **Matt**
- Who do you think is probably in the best physical shape? Explain. **Matt; Jay is overweight with high cholesterol; Todd is too thin and has too fast a heartbeat.**

Try This

4. Use the key above and make a facial glyph for a person with average weight, above average height, very low cholesterol level, above average heartbeat count per minute when at rest, and below average breaths per minute at rest.

5. Use the picture in the key above, but make the following changes in interpretation:

a. height b. arm span c. shoe size d. belt size e. neck size

Instead of using below average, average, and above average, choose three categories for each measure, such as:

height: up to 139 cm; 140 cm–159 cm; 160 cm and above

Make a glyph to show your data. Compare it with the glyphs of others in your class. **Answers will vary.**

6. Select some data, create a key, and make a facial glyph to show the data. Make some statements that interpret the glyph. **Answers will vary.**

Chapter 1 Extra Practice

Evaluate each numerical expression.

1. $(23 + 8) - 14$ 17 **2.** $12 + (16 - 7)$ 21 **3.** $9 - (21 + 13)$ −25

4. $44 + (17 - 9)$ 18 **5.** $(38 + 7) - 12$ 33 **6.** $(46 - 9) + 27$ 64

Evaluate each algebraic expression.

7. $b - 2$ for $b = 11$ 9 **8.** $8 + v$ for $v = 5$ 13

9. $3 + (x - 4)$ for $x = 7$ 6 **10.** $(21 - t) + 2$ for $t = 4$ 19

11. $r + z$ for $r = 13$, $z = 4$ 17 **12.** $(a + 7) - 8$ for $a = 6$ 5

Write as an algebraic expression.

13. 7 decreased by b $7 - b$ **14.** 3 added to p $p + 3$ **15.** 6 more than r $r + 6$

Write an expression for each question.

Chip rides his horse 11 miles every day. How far will he ride if

16. he rides 3 miles less? $11 - 3$ **17.** he rides m miles farther? $11 + m$

Use the basic properties. Write an equivalent expression.

18. $z + 0$ $0 + z$ **19.** $(2 + r) + 6$ $2 + (r + 6)$ **20.** $4 + b$ $b + 4$

Solve the equation for the replacement set given.

21. $g - 8 = 5$ {12, 13, 14, 15} 13 **22.** $6 + b = 17$ {9, 10, 11, 12} 11

Solve and check.

23. $n - 18 = 7$ 25 **24.** $x + 8 = 13$ 5 **25.** $t - 9 = 4$ 13

26. $43 + s = 72$ 29 **27.** $t - 35 = 63$ 98 **28.** $212 = x - 27$ 239

29. $b + 345 = 567$ 222 **30.** $111 = r - 343$ 454 **31.** $654 = 321 + a$ 333

Solve.

32. Robin played tennis 2 hours less today than she did yesterday. She played for 3 hours today. How many hours did she play yesterday? 5 hours

33. Bruce has 7 magazine subscriptions. All but 4 of them were gifts from his family. How many subscriptions did he get as gifts? 3 subscriptions

Round to the nearest ten and estimate the value of the variable.

34. $a = 26 + 47$ 80 **35.** $67 + 234 = b$ 300 **36.** $z = 257 - 79$ 180

Round to the nearest hundred and estimate the value of the variable.

37. $r = 968 + 497$ 1500 **38.** $p = 423 + 1254$ 1700 **39.** $2873 - 952 = t$ 1900

Use front-end estimation to estimate the value of the variable. Possible answers are given.

40. $768 + 269 + 548$ 1600 **41.** $4584 - (1955 + 2462)$ 170

Chapter 2 Extra Practice

Simplify.

1. $72 \div (5 + 3)$ 9
2. $9 \div [(4 + 1) - 2]$ 3
3. $(5 \times 6) \div (7 + 3)$ 3

Evaluate.

4. $3d$ for $d = 9$ 27
5. $\frac{6w}{9}$ for $w = 3$ 2
6. $a(4 + b)$ for $a = 7$ $7(4 + b)$ or $28 + 7b$
7. $\frac{8x}{z}$ for $z = 2$ $\frac{8x}{2}$ or $4x$
8. $(f - g)$ for $f = 9$, $g = 4$ 5
9. $t(10 - v)$ for $v = 9$ $t(1)$ or t

Write as an algebraic expression.

10. 9 divided by s $\frac{9}{s}$
11. 4 less than twice r $2r - 4$
12. 7 multiplied by x $7x$

Use the commutative property to write an equivalent expression.

13. rd dr
14. $m(3)$ $3m$
15. $6s$ $s(6)$

Use the associative property to write an equivalent expression.

16. $13(3x)$ $(13 \cdot 3)x$
17. $(11b)t$ $11(bt)$
18. $(r6)7$ $r(6 \cdot 7)$

Use the distributive property to write an equivalent expression.

19. $d(4 + 7)$ $4d + 7d$
20. $2(n + 9)$ $2n + 18$
21. $(x + 5)6$ $6x + 30$

Use the basic properties to simplify.

22. $12(3d)$ $36d$
23. $5x(4)$ $20x$
24. $17(a + 9)$ $17a + 17 \cdot 9$ or $17a + 153$
25. $4b + 3b$ $7b$
26. $15z + z$ $16z$
27. $24r - 7r$ $17r$

Solve.

28. $9n = 63$ 7
29. $\frac{56}{x} = 8$ 7
30. $\frac{t}{4} = 6$ 24

Solve and check.

31. $6x = 72$ 12
32. $121 = 11d$ 11
33. $\frac{d}{13} = 4$ 52
34. $207 = 23f$ 9

Solve.

35. Nancy used 280 calories while exercising for 35 minutes. How many calories did she use each minute? 8 calories/minute

36. Formula: $V = lwh$. V is the volume of a box in cubic centimeters, l is the length of the box in centimeters, w is the width of the box in centimeters, and h is the height of the box in centimeters. What is the volume of a box with length 5 cm., width 6 cm., and height 8 cm.? 240 cm^3

Use rounding and choose compatible numbers to estimate the value of each variable.
Possible answers are given.

37. $\frac{420}{38} = x$ 10
38. $\frac{561}{82} = z$ 7
39. $\frac{2493}{52} = t$ 50

Chapter 3 Extra Practice

Find the absolute value.

1. $|-23|$ 23 **2.** $|-14|$ 14 **3.** $|1|$ 1

Use inequality symbols. Order from least to greatest.

4. 7, −6, 3 $-6 < 3 < 7$ **5.** 0, 4, −12 $-12 < 0 < 4$ **6.** 3, −3, −33 $-33 < -3 < 3$

Find the sum.

7. $-5 + (-5)$ −10 **8.** $-4 + 8$ 4 **9.** $-9 + 3$ −6

10. $12 + (-6)$ 6 **11.** $14 + -5$ 9 **12.** $-17 + 6$ −11

Evaluate.

13. $-15 + x$ for $x = -1$ −16 **14.** $s + (-8)$ for $s = 19$ 11

Subtract.

15. $9 - 14$ −5 **16.** $-5 - 7$ −12 **17.** $2 - 21$ −19

18. $8 - (-5)$ 13 **19.** $-26 - 5$ −31 **20.** $14 - (-3)$ 17

Evaluate.

21. $x - 9$ for $x = 4$ −5 **22.** $-17 - n$ for $n = -8$ −9

Solve and check.

23. $b + 37 = 15$ −22 **24.** $25 + x = -2$ −27 **25.** $r - 47 = 35$ 82

26. $43 = 78 + c$ −35 **27.** $d - (-56) = 77$ 21 **28.** $f + (-24) = 68$ 92

Solve.

29. Tom has \$235 in his checking account and \$450 in bills to pay. How much does he need to deposit to keep his account from being overdrawn? \$215

Find the product.

30. $9(6)$ 54 **31.** $7(-3)$ −21 **32.** $-4(-8)$ 32

33. $-7(2)$ −14 **34.** $8(-6)$ −48 **35.** $-9(-4)$ 36

Find the quotient.

36. $-9 \div 3$ −3 **37.** $-56 \div (-7)$ 8 **38.** $121 \div (-11)$ −11

39. $92 \div (-4)$ −23 **40.** $-57 \div 3$ −19 **41.** $-18 \div (-6)$ 3

Solve and check.

42. $75 = -3n$ −25 **43.** $-9y = 72$ −8 **44.** $12 = -\frac{x}{3}$ −36 **45.** $55 = -b$ −55 **46.** $-k = -1$ 1

Write an equation.

47. 17 multiplied by a number q is 85. $17q = 85$

48. 34 less than a number p is 71. $p - 34 = 71$

Chapter 4 Extra Practice

Write as a decimal.

1. $8(10) + 6(1) + 3\left(\frac{1}{100}\right)$ **86.03**
2. $7\left(\frac{1}{100}\right) + 3\left(\frac{1}{1000}\right) + 2(1)$ **2.073**

Write $<$, $>$, or $=$ for each $\square$.

3. $3.23 \square 3.32$ **<**
4. $7.41 \square 7.414$ **<**
5. $5.05 \square 5.050$ **=**

Estimate the value of x by rounding.

6. $x = 49.82 + 27.89$ **80**
7. $x = 64.79 + 575.13$ **640**
8. $x = 99.64 - 64.82$ **40**
9. $x = 749.86 - 244.91$ **500**

Add or subtract.

10. $4.87 + 3.9$ **8.77**
11. $13.56 + 24.94 - 17.31$ **21.19**
12. $-5.3 + 8.44$ **3.14**
13. $72.4 + (-6.9) + 21.78$ **87.28**
14. $-21.65 + (-4.2)$ **−25.85**
15. $74.68 - (-29.38)$ **104.06**

Solve and check.

16. $c - 2.96 = 3.84$ **6.8**
17. $13.62 + d = -4.85$ **−18.47**
18. $m - 8.34 = -2.79$ **5.55**
19. $-41 + b = 17.32$ **58.32**

Estimate the product by rounding.

20. 30.21×4.7 **150**
21. 4.12×24.89 **100**
22. 0.79×19.8 **20**

Find the product.

23. 4.2×11.83 **49.686**
24. 6.7×-5.92 **−39.664**
25. 14.3×-0.03 **−0.429**
26. -3752×0.001 **−3.752**

Solve.

27. A tennis court is 12.3 meters long. How many centimeters long is the court? **1230 cm**
28. Kim's dog weighs 5500 grams. How many kilograms does it weigh? **5.5 kg**

Solve and check.

29. $3.6r = 194.4$ **54**
30. $-0.04x = 16$ **−400**
31. $1.7s = 16.66$ **9.8**
32. $\frac{b}{0.23} = 5.9$ **1.357**
33. $\frac{f}{-1.4} = 13.8$ **−19.32**
34. $\frac{d}{4.2} = -3.7$ **−15.54**

Solve by writing an equation.

35. Ivan got a $13.79 discount on a baseball glove with an original price of $54.98. How much did he pay for the glove? **$41.19**
36. Alan has 54.3 meters of insulation. He needs a total of 100 meters to insulate his attic. How much more insulation does he need to get? **45.7 meters**

Chapter 5 Extra Practice

Give the first five nonzero multiples of each number.

1. 7 7, 14, 21, 28, 35 **2.** 20 20, 40, 60, 80, 100 **3.** 60 60, 120, 180, 240, 300

Find all the factors of the given number.

4. 18 1, 2, 3, 6, 9, 18 **5.** 42 1, 2, 3, 6, 7, 14, 21, 42 **6.** 81 1, 3, 9, 27, 81

Tell whether the number is divisible by 2, by 3, and by 5.

7. 98 2 **8.** 12345 3, 5 **9.** 86 2

10. 156 2, 3 **11.** 272 2 **12.** 985 5

Find the solution by writing and solving an equation.

13. Gloria is making 96 sandwiches for a party. Sandwich rolls come in packages of 8. How many packages does she need to buy? $8p = 96$ 12

14. Alfred has \$250 more in his savings account than in his checking account. If he has \$735 in his savings account, how much does he have in his checking account?
$735 - c = 250$ \$485

State whether the number is prime or composite.

15. 57 composite **16.** 79 prime **17.** 96 composite

Write in expanded form. Simplify if possible.

18. 2^5 $2 \cdot 2 \cdot 2 \cdot 2 \cdot 2 = 32$ **19.** $(-4)^3$ $(-4)(-4)(-4) = -64$ **20.** r^7 $r \cdot r \cdot r \cdot r \cdot r \cdot r \cdot r$

Write using exponents.

21. $6 \cdot 6 \cdot 6 \cdot 6$ 6^4 **22.** $b \cdot b \cdot b \cdot b \cdot b$ b^5 **23.** $(-8)(-8)(-8)$ $(-8)^3$

Multiply. Give the answer in exponent form.

24. $x^2 \cdot x^3$ x^5 **25.** $7^4 \cdot 7^1$ 7^5 **26.** $(-4)^4(-4)^5$ $(-4)^9$

Simplify.

27. $2^4 + 2^6$ 80 **28.** $(-6)^2 + (-6)^3$ -180

Use exponents to show the prime factorization of each number.

29. 72 $2^3 \cdot 3^2$ **30.** 210 $2 \cdot 3 \cdot 5 \cdot 7$ **31.** 54 $2 \cdot 3^3$ **32.** 225 $3^2 \cdot 5^2$

Find the Greatest Common Factor (GCF) of each pair of numbers.

33. 42, 91 7 **34.** 21, 27 3 **35.** 70, 182 14

36. Formula: $D = r \cdot t$. D is the distance in miles, r is rate in mi/hour, and t is time of ride in hours. Find how far Hank drove his truck in $4\frac{1}{2}$ hours at 53 mi/hour. 238.5 miles

Find the Least Common Multiple (LCM) of each pair of numbers.

37. 5, 12 60 **38.** 18, 24 72 **39.** 7, 11 77

Chapter 6 Extra Practice

Write an equivalent fraction by replacing the variable with a whole number.

1. $\frac{4}{9} = \frac{x}{54}$ 24　**2.** $\frac{m}{7} = \frac{35}{49}$ 5　**3.** $-\frac{3}{11} = \frac{p}{66}$ −18

Reduce each fraction to lowest terms.

4. $\frac{24}{36}$ $\frac{2}{3}$　**5.** $-\frac{11}{99}$ $-\frac{1}{9}$　**6.** $\frac{105}{133}$ $\frac{15}{19}$

Write each as a mixed numeral.

7. $\frac{21}{8}$ $2\frac{5}{8}$　**8.** $-\frac{41}{13}$ $-3\frac{2}{13}$　**9.** $\frac{246}{25}$ $9\frac{21}{25}$

Write each as an improper fraction in lowest terms.

10. $6\frac{7}{9}$ $\frac{61}{9}$　**11.** $-4\frac{3}{8}$ $-\frac{35}{8}$　**12.** $-5\frac{8}{10}$ $-\frac{29}{5}$

13. Graph $-\frac{6}{4}, \frac{2}{3}, \frac{1}{4}$, and $\frac{3}{2}$ on a number line.

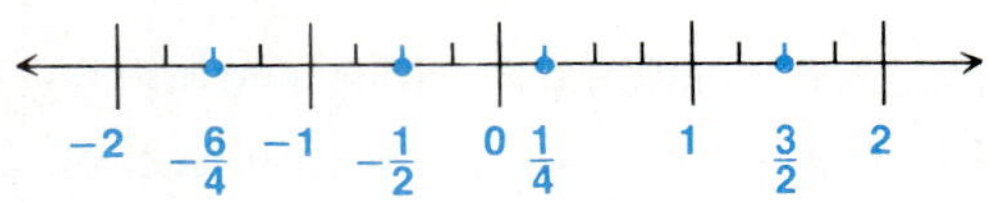

Write >, <, or = for each □.

14. $\frac{4}{8}$ □ $\frac{3}{7}$ >　**15.** $-\frac{2}{3}$ □ $-\frac{5}{9}$ <　**16.** $\frac{26}{39}$ □ $\frac{2}{3}$ =　**17.** $\frac{7}{11}$ □ $\frac{3}{4}$ <

Add or subtract. Reduce to lowest terms.

18. $\frac{6}{13} + \frac{3}{13}$ $\frac{9}{13}$　**19.** $-\frac{8}{18} + \frac{14}{18}$ $\frac{1}{3}$　**20.** $-\frac{8}{20} - \frac{7}{20}$ $-\frac{3}{4}$

21. $\frac{1}{4} + \frac{1}{5}$ $\frac{9}{20}$　**22.** $\frac{4}{7} - \frac{1}{2}$ $\frac{1}{14}$　**23.** $-\frac{3}{8} - \left(-\frac{2}{5}\right)$ $\frac{1}{40}$

24. $3\frac{5}{6} - 2\frac{3}{8}$ $1\frac{11}{24}$　**25.** $-5\frac{3}{4} + 1\frac{5}{12}$ $-4\frac{1}{3}$　**26.** $4\frac{2}{5} - 7\frac{3}{7}$ $-3\frac{1}{35}$

27. Mr. Drucker walked $2\frac{7}{8}$ miles, stopped for a drink, then walked $3\frac{1}{6}$ miles more. How far did he walk? $6\frac{1}{24}$ miles

28. Alita had $6\frac{1}{2}$ buckets of paint. After she painted the living room, she had $2\frac{2}{3}$ buckets of paint left. How much did she use? $3\frac{5}{6}$ buckets

Solve and check.

29. $x - \frac{3}{8} = \frac{5}{12}$ $\frac{19}{24}$　**30.** $\frac{3}{7} + y = -\frac{3}{14}$ $-\frac{9}{14}$　**31.** $\frac{9}{10} = b - \frac{2}{5}$ $\frac{13}{10}$

Find the solution by writing and solving an equation.

32. Ms. Gomez bought 5 yards of material to sew a dress. When she was finished, she had $1\frac{5}{12}$ yards of material left. How much material did she use to make the dress? $3\frac{7}{12}$ yards

Chapter 7 Extra Practice

Multiply. Reduce to lowest terms.

1. $-\frac{3}{8}\left(\frac{4}{5}\right)$ $-\frac{3}{10}$

2. $-\frac{5}{9}\left(-\frac{3}{5}\right)$ $\frac{1}{3}$

3. $-2\frac{3}{4}\left(-3\frac{4}{7}\right)$ $9\frac{23}{28}$

4. $6\frac{4}{9}\left(-\frac{3}{4}\right)$ $-4\frac{5}{6}$

Divide. Reduce to lowest terms.

5. $\frac{5}{12} \div \left(-\frac{3}{4}\right)$ $-\frac{5}{9}$

6. $-\frac{7}{10} \div \left(-\frac{5}{7}\right)$ $\frac{49}{50}$

7. $4\frac{1}{2} \div \left(-2\frac{3}{4}\right)$ $-1\frac{7}{11}$

8. $-6\frac{4}{7} \div 7\frac{2}{3}$ $-\frac{6}{7}$

Solve and check.

9. $-\frac{x}{5} = -\frac{18}{30}$ 3

10. $\frac{1}{4}y = \frac{7}{8}$ $3\frac{1}{2}$

11. $-4\frac{1}{5} = 1\frac{2}{5}m$ -3

12. $\frac{2}{3}z = \frac{-5}{7}$ $-1\frac{1}{14}$

Find the solution by writing and solving an equation.

13. Maria lives $4\frac{8}{10}$ miles from school. She rode her moped $\frac{2}{3}$ of the way there before she ran out of gas. How far did she have to walk to get to school? $1\frac{3}{5}$ miles

Solve.

14. If $\frac{2}{3}$ cup of soy sauce is used to make 2 cups of teriyaki sauce, how much soy sauce is used to make 10 cups of teriyaki sauce? $3\frac{1}{3}$ cups

15. Gil placed 28 dominoes end-to-end to form a line 63 inches long. How long was each domino? $2\frac{1}{4}$ inches

Write each as a decimal. Use a bar for repeating decimals.

16. $\frac{3}{8}$ 0.375

17. $\frac{4}{9}$ $0.\overline{4}$

18. $\frac{7}{25}$ 0.28

Simplify. Write the expression with exponents.

19. $\frac{9^7}{9^3}$ 9^4

20. $\frac{(-6)^6}{(-6)^4}$ $(-6)^2$

21. $\frac{x^8}{x^5}$ x^3

Write each expression without exponents.

22. 2^{-6} $\frac{1}{64}$

23. $\frac{5^3}{5^7}$ $\frac{1}{625}$

24. $\frac{(-3)^5}{(-3)^3}$ 9

Write each in decimal form.

25. 3.9×10^4 39,000

26. 2.4×10^{-2} 0.024

27. 5.38×10^{-5} 0.0000538

Write each in scientific notation.

28. 4270 4.27×10^3

29. 67,000 6.7×10^4

30. 200,000,000 2.0×10^8

31. 0.0035 3.5×10^{-3}

32. 0.0001001 1.001×10^{-4}

33. 0.246 2.46×10^{-1}

Chapter 8 Extra Practice

Solve.

1. $4d - 3 = 45$ 12

2. $33 = -2x + 3$ −15

3. $5(7 + w) = -10$ −9

4. $\frac{a - 11}{9} = -3$ −16

Write an equation.

5. Paul has 24 marbles, which is 4 more than twice as many marbles as Eric has. How many marbles does Eric have? $2e + 4 = 24$

6. Erica's age, 7, is 1 less than twice Brian's age. How old is Brian? $2b - 1 = 7$

Solve by writing an equation.

7. Mr. Fritz has been a teacher for 17 years. That is 5 fewer than twice the number of years Mrs. Weiner has been a teacher. How many years has Mrs. Weiner been a teacher? $2w - 5 = 17$ 11 years

8. Steve had a box of spaghetti that was $\frac{2}{3}$ full. He used 4 oz, leaving 12 oz in the box. How many ounces did the box hold? $\frac{2}{3}x - 4 = 12$ 24 oz

Solve and check.

9. $2x + 7x = 54$ 6

10. $-121 = 8p + 3p$ −11

11. $108 = 17w - 5w$ 9

12. $-11b - 12b = -92$ 4

Solve and check.

13. $3(x + 2) + 6x = 60$ 6

14. $12d + (4 - 5d) = -52$ −8

15. $8m + 6 - 2(3m) = 16$ 5

16. $c + 5(3c - 5) = 23$ 3

17. $3p + 2(p - 4) + 16 = 38$ 6

18. $6r - 6 - 2(-2r) = 74$ 8

Solve and check.

19. $156 - 9t = 4t$ 12

20. $11z + 7 = 31 - z$ 2

21. $6y + 11 = 5y$ −11

22. $-6 + r = -20 - r$ −7

Solve the problem.

23. Margaret invited 3 more than $\frac{4}{5}$ of her class to a party. She invited 63 people to her party. How many people are in her class? 75

Solve and check.

24. $5d > 40$ $d > 8$

25. $-3 + r < 12$ $r < 15$

Solve.

26. $17 < 4a - 3$ $a > 5$

27. $-7b + 10b > -18$ $b > -6$

Chapter 9 Extra Practice

Write each ratio as a fraction in lowest terms.

1. 72 to 88 $\frac{9}{11}$ **2.** 12:21 $\frac{4}{7}$ **3.** $\frac{68}{16}$ $\frac{17}{4}$

Solve and check.

4. $\frac{9}{x} = \frac{27}{42}$ 14 **5.** $\frac{y}{11} = \frac{30}{66}$ 5 **6.** $\frac{3}{8} = \frac{27}{c}$ 72

Simplify the rate.

7. \$75/5 days \$15/day **8.** 217 km/3.5 h 62 km/h **9.** 396 steps/18 floors 22 steps/floor

Solve.

10. Florida and New York are 2100 km apart. If it takes 3 hours to fly this distance, how fast does the plane travel? 700 km/h

Write an equation.

11. Connie wants to tile her bedroom floor and her kitchen floor. Her bedroom is twice as large as her kitchen. If Connie needs 144 tiles for both rooms, how many are used in each room? $2x + x = 144$

Solve by writing an equation.

12. Juan had half the number of Bs on his report card as he had As. He was taking 6 courses all together, and got no lower than a B. How many of each grade did he get?
$\frac{1}{2}x + x = 6$ or $x + 2x = 6$ 4 As, 2Bs

Find the scale dimensions of each if the scale is 1 cm = 0.5 m.

13. desk: 1 m × 2 m 2 cm × 4 cm **14.** bookcase: 1.5 m × 2.5 m 3 cm × 5 cm

Solve.

15. Grape juice costs \$1.99 for the 32 ounce bottle and \$2.69 for the 48 ounce bottle. Which size is the better buy? 48 ounce

Write each as a fraction in lowest terms.

16. 22% $\frac{11}{50}$ **17.** 4% $\frac{1}{25}$ **18.** 8.5% $\frac{17}{200}$

Use a proportion to write each as a percent.

19. $\frac{5}{8}$ 62.5% **20.** $\frac{7}{25}$ 28% **21.** $\frac{9}{5}$ 180%

Write each percent as a fraction or mixed number in lowest terms.

22. 175% $1\frac{3}{4}$ **23.** $\frac{4}{5}\%$ $\frac{1}{125}$ **24.** $3\frac{1}{2}\%$ $\frac{7}{200}$

Write as a decimal. **25.** $14\frac{1}{4}\%$ 0.1425 **26.** 350% 3.5

Write each as a percent. **27.** 0.072 $7\frac{1}{5}\%$ **28.** 9.68 968%

Chapter 10 Extra Practice

Use a proportion or an equation to solve.

1. 60% of 185 111 **2.** 5% of 80 4 **3.** 4% of 55 2.2

Use a proportion or an equation to solve.

4. What percent of 90 is 27? 30%
5. 63 is what percent of 105? 60%
6. What percent of 160 is 120? 75%
7. 9 out of 36 is what percent? 25%

Use a proportion or an equation to solve. Round to the nearest tenth if necessary.

8. 14 is 7% of what number? 200
9. 40% of what number is 35? 87.5
10. 15% of what number is 45? 300
11. 22 is 25% of what number? 88

Solve by writing an equation.

12. A pair of jeans costs 75% as much as a pair of corduroy pants. The total cost for the jeans and cords is $42. What is the cost of each? $0.75x + x = 42$ $18, $24

Find the percent of increase or decrease.

13. Original amount = $75; new amount = $60. 20% decrease

14. Original amount = $60; new amount = $75. 25% increase

Solve.

15. When Juanita was on vacation, it rained 35% of the time. She was on vacation for 20 days. How many days did it rain? 7 days

Find the discount and sales price for each. Round to the nearest cent if necessary.

16. Regular price = $45
Discount = 40% $18, 27

17. Regular price = $2,450
Discount = 6% $147, $2303

Find the commission.

18. Sale price = $11,250
Commission percent = 4% $450

19. Sale price = $225
Commission percent= 6% $13.50

Find the interest and total amount. Round to the nearest percent if necessary.

20. $600 at 15% per year for 3 years. $270, $870

21. $800 at 12% per year for 4 months. $32, $832

22. $300 at 3% per month for 6 months. $54, $354

23. Make a circle graph to show the data in the table.
Rackets Sold:

Badminton	Racquetball	Squash	Tennis	Other
5%	31%	19%	43%	2%

Badminton	18°
Racquetball	111.6°
Squash	68.4°
Tennis	154.8°
Other	7.2°

Chapter 11 Extra Practice

Refer to the figure for Exercises 1–5.

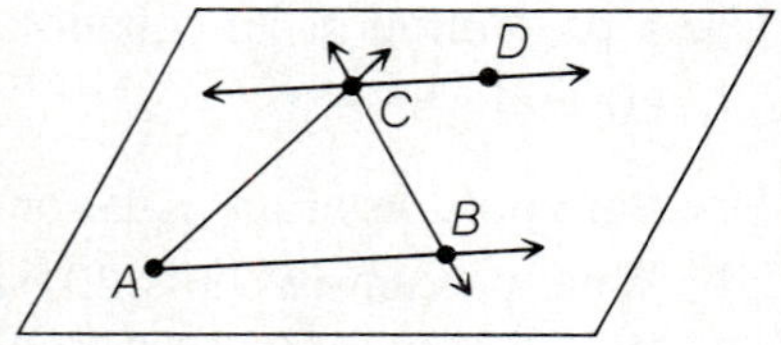

1. Name a point. *A, B, C, or D*
2. Name a line. $\overleftrightarrow{BC}$ or $\overleftrightarrow{CD}$
3. Name a plane. *ABC*
4. Name a segment that includes point *C*. $\overline{AC}$, $\overline{BC}$, $\overline{CD}$
5. Name a ray. $\overrightarrow{AB}$, $\overrightarrow{AC}$, $\overrightarrow{CB}$, $\overrightarrow{CD}$, $\overrightarrow{BC}$, $\overrightarrow{DC}$

Estimate the length of each segment to the nearest centimeter.

6. $\overline{XY}$ 4 cm   7. $\overline{XZ}$ 6 cm

Refer to the figure for Exercises 8–11.

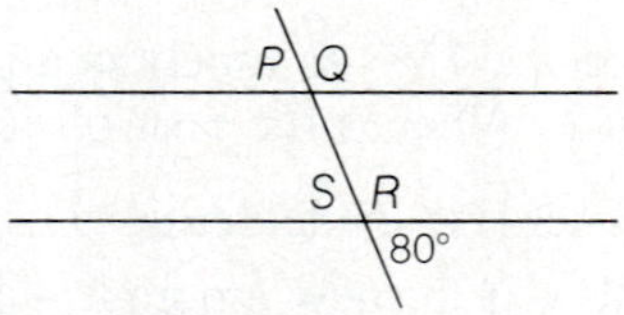

8. Name an acute angle. $\angle P$, $\angle S$
9. Name an obtuse angle. $\angle Q$, $\angle R$
10. Find the measure of angle *R*. 100°

Classify each triangle by the measures of its angles and the lengths of its sides.

11. $\triangle ABC$ has a 120 degree angle and two sides equal in length. obtuse, isosceles
12. $\triangle RST$ has three sides equal in length and three acute angles. acute, equilateral

Name the polygon.

13.

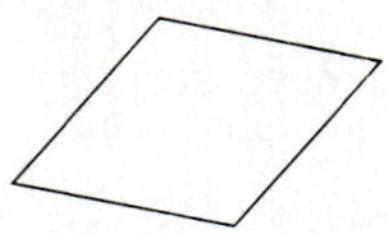

rhombus

14.

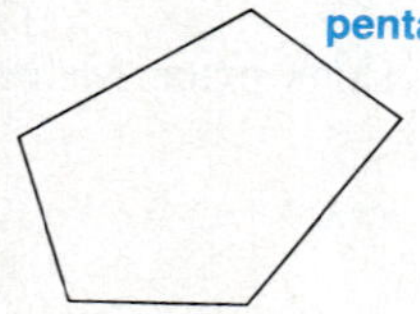

pentagon

Solve. Use the formula *Distance* = *rate* × *time*.

15. Hans drove $4\frac{1}{2}$ hours and traveled 243 miles. What was his average speed? 54 miles/hour

Find the circumference of a circle with the given diameter or radius. Use 3.14 for π.

16. radius = 4.3 cm 27.004 cm
17. diameter = 7 m 21.98 m

Use the SAS, SSS, or ASA property to show that each pair of triangles is congruent.

18.

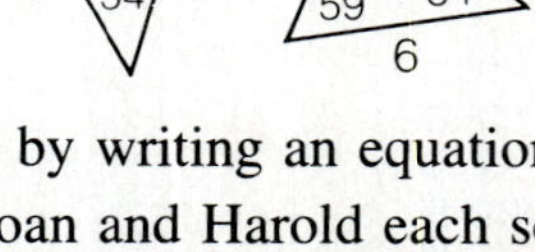

ASA

19.

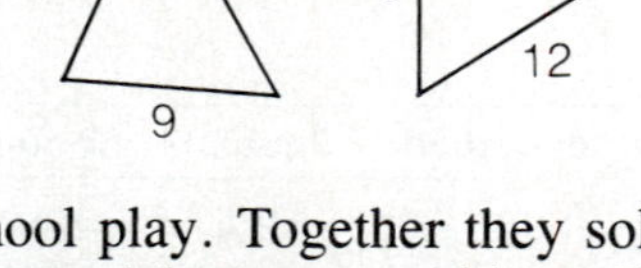

SSS

Solve by writing an equation.

20. Joan and Harold each sold 23 tickets to the school play. Together they sold 40% of the total number of tickets sold to the play. How many tickets were sold in all? 115

Chapter 12 Extra Practice

Find the area of each figure.

1. Rectangle with length 15 m, width 6 m **90 m²**

2. Triangle with base 38 ft, height 8 ft **152 ft²**

Find the area of each quadrilateral.

3. Parallelogram with base = 11.2 m, height = 9.4 m **105.28 m²**

4. Trapezoid with $b_1 = 9$ cm, $b_2 = 5.5$ cm, $h = 13$ cm **94.25 cm²**

Find the area of each circle. Use 3.14 for π.

5. $r = 6$ cm **113.04 cm²**

6. $d = 5.8$ m **26.4074 m²**

Find the area of the figure.

7. 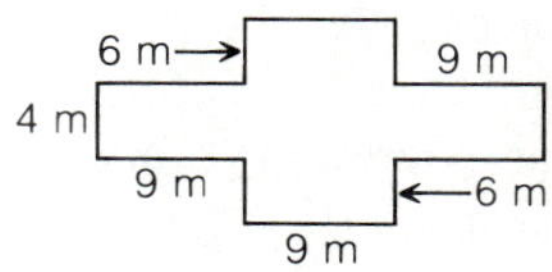

216 m²

Solve by writing an equation.

8. Linda bought 4 medium-sized drinks and 1 large drink for \$3.20. If the large drink cost \$0.80, how much did each medium-sized drink cost? **$4m + 0.8 = 3.20$ 60¢**

Find the volume of the prism or cylinder. Use 3.14 for π.

9. 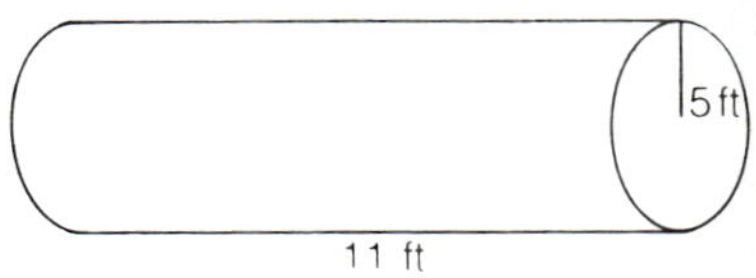

863.5 ft.³

10. 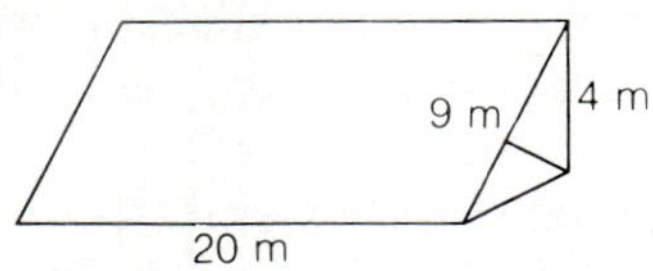

360 m³

Solve.

11. Concrete is on sale for \$30 per cubic yard. How much will it cost to buy concrete to cover a foundation 38 feet long, 30 feet wide, and 6 inches thick? **\$633.33**

Find the volume. Use 3.14 for π.

12. Cone with a radius of 6 cm and a height of 14 cm **527.52 cm³**

13. Pyramid with a base area of 45 square feet and a height of 12 feet **180 ft³**

Find the surface area. Use 3.14 for π.

14. Cylinder with a radius of 2 m and a height of 7 m **113.04 m²**

15. 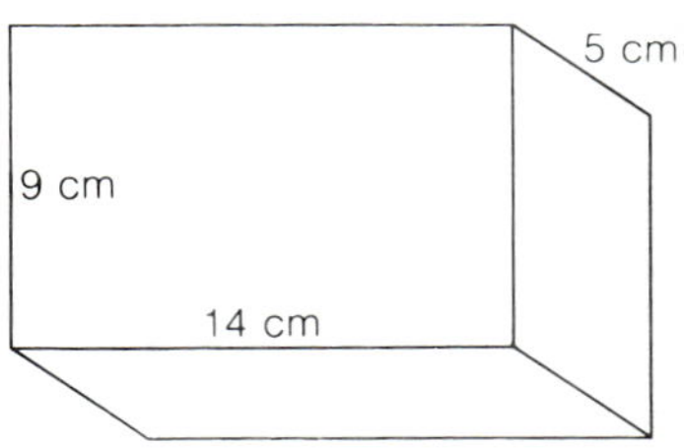

482 cm²

Chapter 13 Extra Practice

Use the counting principle to find the total number of outcomes.

1. Select 1 tennis racket. Composition: wood, metal, graphite, fiberglass. Head size: small, midsize, oversize. Grip size: $4\frac{3}{8}$, $4\frac{1}{2}$, $4\frac{5}{8}$. **36**

Find the number of permutations.

2. m, a, t, h **24**

3. 1, 2, 4, 5, 7, 8 **720**

Find the number of combinations.

4. You may be scheduled to work 4 days out of a 7-day work week. **35**

5. Select 2 blouses from 5 possible choices. **10**

There are 18 marbles in a bag. 6 are clear, 4 are black, and 8 are red.

6. P(black) $\frac{4}{18} = \frac{2}{9}$

7. P(not clear) $\frac{12}{18} = \frac{2}{3}$

8. P(clear or red) $\frac{14}{18} = \frac{7}{9}$

Find the probability of each for one toss of two dice.

9. P(total of 2) $\frac{1}{36}$

10. P(a number < 3, a number > 3) $\frac{1}{6}$

Solve by writing an equation.

11. The length of a yard is 3 times its width. It takes 104 ft. of fencing to surround the yard. Find the dimensions of the yard. $2(x) + 2(3x) = 104$ $l = 39$ ft $w = 13$ ft

Use this data to solve Exercises 12–15.
Test grades: 71, 75, 75, 78, 82, 83, 86, 89, 93

12. Find the range. **22**

13. What is the mode? **75**

14. What is the median? **82**

15. Find the mean. $81.\overline{3}$

16. Make a bar graph for the data. Distance from Orlando in miles: Miami, 220; New York, 1050; Raleigh, 600; Syracuse, 1230. **See Solutions Manual.**

Use the line graph to answer each question

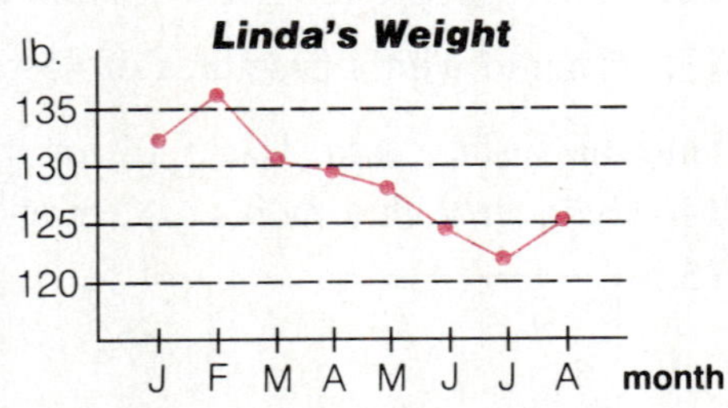

17. During which month was Linda's weight the lowest? **July**

18. Between which months did Linda's weight increase? **Jan and Feb, July and Aug.**

19. Between which months was the change in Linda's weight the greatest? **Feb and March**

20. Make a pictograph for the data. Cars sold: January, 75; April, 150; July, 120; October, 90. **See Solutions Manual.**

Chapter 14 Extra Practice

Between what two consecutive integers does each square root lie?

1. $\sqrt{45}$ 6, 7

2. $\sqrt{88}$ 9, 10

Evaluate each expression.

3. $\sqrt{36} - \sqrt{25}$ 1

4. $\sqrt{(155 - 11)}$ 12

Use a calculator with a $\sqrt{}$ key to find an approximation for each square root.

5. 10 3.16

6. 234 15.3

Solve and check. **7.** $x^2 = 196$ ±14 **8.** $x^2 = 15$ ±3.9

Find the unknown length.

9. 13

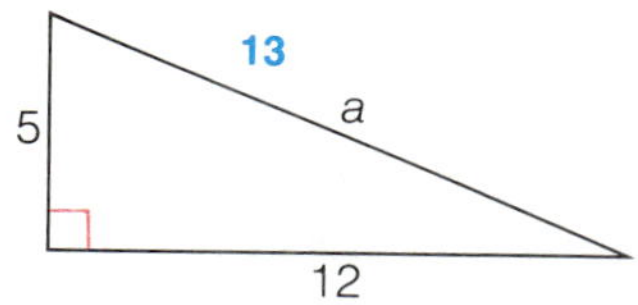

10. 10.2

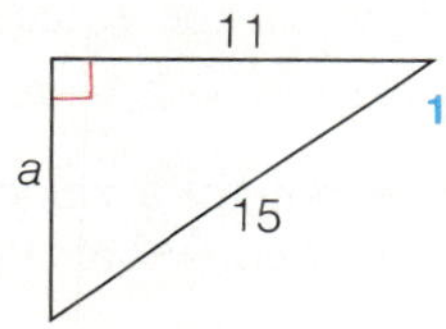

Solve.

11. A television screen is 7 inches tall and 9 inches wide. What is the length of the diagonal of the screen? 11.4 inches

Use proportions to find the unknown lengths for these similar triangles.

12. x = 12.5

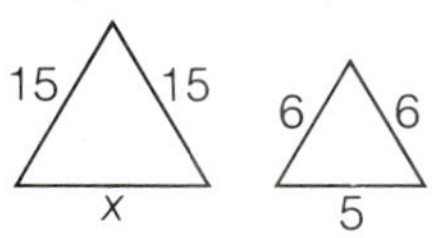

13. b = 36

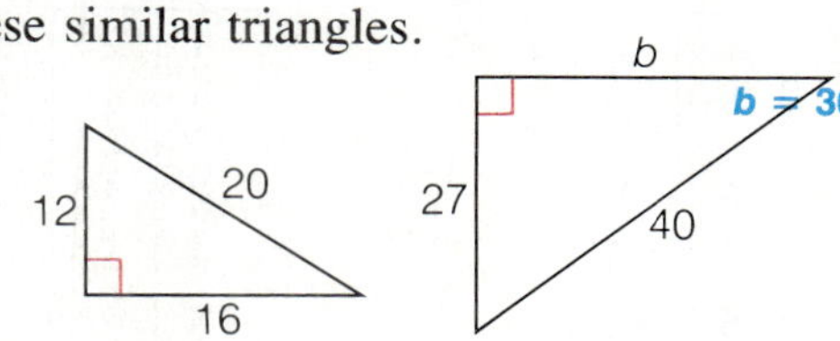

Find the unknown length.

14. 31

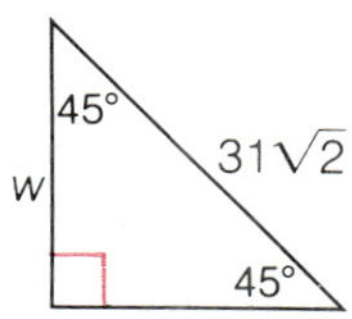

15. $7\sqrt{3}$

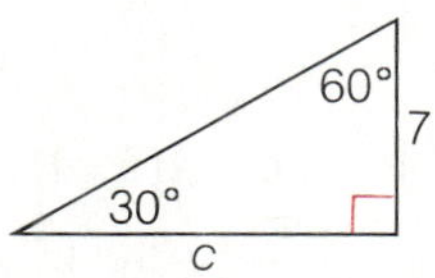

Find each ratio for $\angle X$.

16. $\cos X$ $\frac{32}{40} = \frac{4}{5}$

17. $\sin X$ $\frac{24}{40} = \frac{3}{5}$

18. $\tan X$ $\frac{24}{32} = \frac{3}{4}$

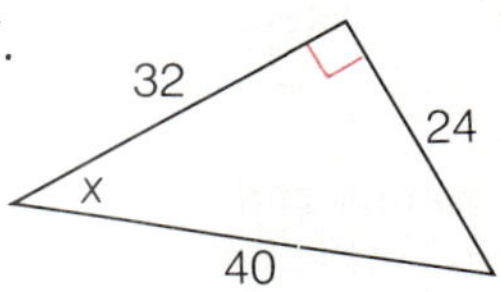

Solve by writing an equation.

19. Rick, Ray and Rose picked apples. Rose picked twice as many apples as Ray, Ray picked twice as many apples as Rick. In all, they picked 21 bushels of apples. How many bushels of apples did Rose pick? $r + 2r + 2(2r) = 21$ 12 bushels

Chapter 15 Extra Practice

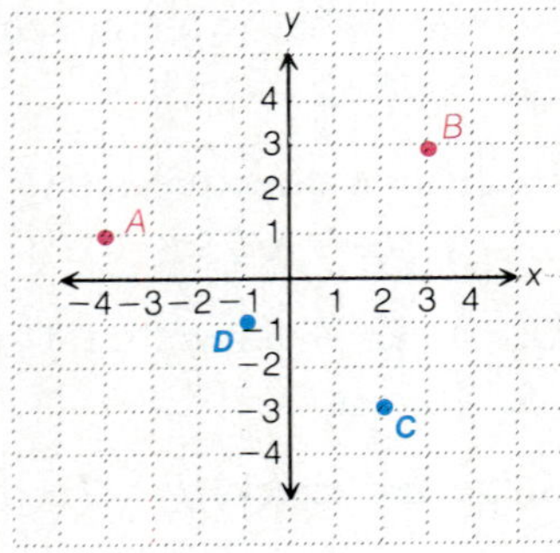

Graph each solution on a number line. See below.

1. $x \geq 0$

2. $2x - 2 < -6$

Give the coordinates of each point on the graph at right.

3. A (−4, 1)

4. B (3, 3)

Graph the point on a pair of axes.

5. $C(2, -3)$

6. $D(-1, -1)$

7. Which of the ordered pairs $(-2, 3)$ or $(-3, -2)$ is a solution for the linear equation $2x - 3y = 0$? (−3, −2)

Make a table of solutions for the equation.

8. $y = -2x + 1$ Use $x = -2, -1, 0, 1, 2, 3$ See additional answers.

Solve. **9.** Mr. Himes bought a camera on sale for \$115. This was \$25 more than 40% of its original price. What was the original price of the camera? \$225

Does y vary directly with x? If it does, find the value of k.

10.

x	y
1	2
2	4
3	6
4	8
5	10

yes $k = 2$

11.

x	y
1	2
2	3
3	4
4	5
5	6

no

Does y vary inversely with x? If it does, find the value of k.

12.

x	y
1	9
2	7
3	5
4	3
5	1

no

13.

x	y
1	12
2	6
3	4
4	3
5	2.4

yes $k = 12$

Find the slope of the line.

14. $3 - 7x = y$ −7

15. A line contains the points $A(3, 4)$, $B(-2, -6)$. 2

Find the intercepts for the graph of $8 + 4y = x$.

16. x-intercept 8

17. y-intercept −2

Solve.

18. Ace Carpet Cleaners charges \$75 to clean 3 rooms and \$20 for each additional room. How much would it cost to have 8 rooms cleaned? \$175

19. Graph the equation to solve the system. $x + y = 3$, $y - 3x = 3$. (0, 3)

20. Graph the inequality $y > x + 4$. Is point (2, 2) a solution? no

1. 

2. 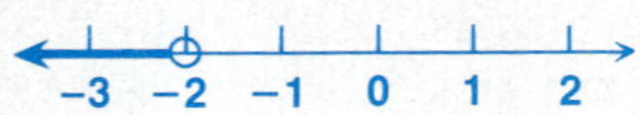

Explore

For use with Lesson 1-1

Use a calculator and try the following several times.

n	+	276	−	367	
enter any number	push ⊞	enter 276	push ⊟	enter 367	push ⊜

Now try to have your result equal these target numbers in as few tries as possible.

a) 100 **191** **b)** 0 **91** **c)** 9 **100** **d)** 58 **149**

Explore

Students should work in groups of two and alternate roles.

For use with Lesson 1-4

Place the same number of counters on each side of a scale so that the scale balances. Use a card to cover ***n*** counters on one side as shown. Have your partner find the value for ***n***, by removing visible counters if desired. Repeat this activity several times, changing ***n***, the total number of counters, or both.

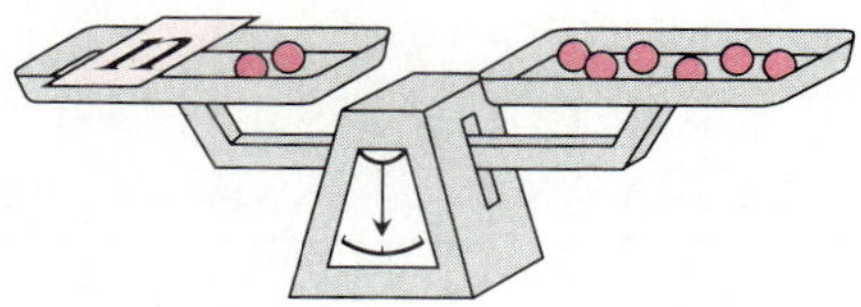

You may want to place a card under the counters on the right side of the scale to ensure that the scale remains balanced. Note that this activity mirrors a technique used for solving equations.

Explore

For use with Lesson 2-1

How many of the numbers 0 through 10 can you represent with numerical expressions that use the following symbols. Use each numeral in each expression. Use the other symbols as needed.

1 2 3 4 + − × ÷ ( )

Example $(4 \div 2) - (3 - 1) = 0$

$(4 \div 2) - (3 - 1) = 0$
$(4 \div 2) \div (3 - 1) = 1$
$(2 \times 3) - (1 \times 4) = 2$
$(4 - 1) \times (3 - 2) = 3$
$(4 \times 3) \div (2 + 1) = 4$
$(2 \times 4) - (3 \times 1) = 5$
$(4 \times 3) \div (2 \times 1) = 6$
$(3 + 4) \div (2 - 1) = 7$
$(2 + 3 + 4) - 1 = 8$
$(4 \times 3) - (2 + 1) = 9$
$(1 + 2) + (3 + 4) = 10$

Explore

Students should work in groups of two and alternate partners. *For use with Lesson 2-5*

Place an even number of counters on each side of a scale so that the scale balances. Use 2 cards, each covering ***n*** counters on one side as shown. Have your partner find the value for ***n***, grouping the visible counters as needed. Repeat this activity several times, changing the number of counters used, the number of groups, or both.

Remind students that each card should cover the same number of counters, and that there should be no counters left over.

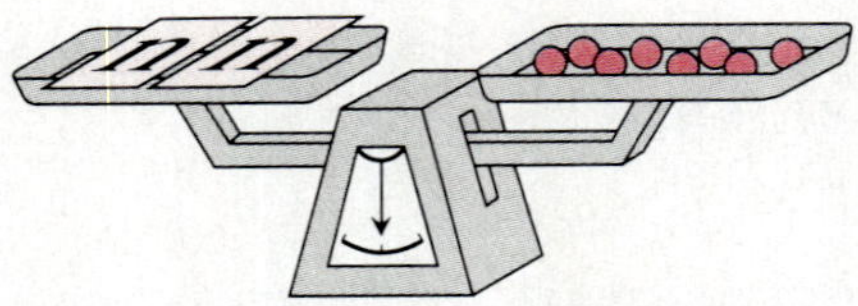

Students may find the answer using division, which should help them understand the technique for solving equations taught in the next section.

Explore

For use with Lesson 3-2

Black counters represent positive integers, and red counters represent negative integers. When a red counter comes together with a black counter, they cancel each other and both are removed!

Complete several rows of a table like the one below by putting the same kind of counters in each of two piles, combining the piles, and removing pairs of counters when possible. Talk about some rules you might invent for adding integers.

First Pile	Second pile	Results of combining the Piles	Equation
●●●● ●●●● +8	●●● −3	● ●●●● +5	+8 +(−3)= +5

Encourage students to talk about rules in their own words.

Explore

For use with Lesson 3-6

Start with 4 or 5 handfuls of counters in a bag. Use additional counters to follow the directions below. Talk about rules you might create for multiplying integers.

Action: Putting in 4 counters
Do this 3 times.
$+3(+4) = \square$ ← Does the bag have more (+) or less (−) counters How many more or less?

Action: Putting in 4 counters
Do the opposite of this 3 times.
$-3(+4) = \square$ ← Does the bag have more (+) or less (−) counters How many more or less?

Action: Taking out 4 counters
Do this 3 times.
$+3(-4) = \square$ ← Does the bag have more (+) or less (−) counters How many more or less?

Action: Taking out 4 counters
Do the opposite of this 3 times.
$-3(-4) = \square$ ← Does the bag have more (+) or less (−) counters How many more or less?

Explore

For use with Lesson 4-1

- These tiles represent a number.
  - a) How many hundreds are there? **3**
  - b) How many tens? **6**
  - c) How many ones? **7**
- Suppose the large square tile represents one unit. What number do these tiles represent? **3.67**
- Represent these numbers with tiles:
  a) 4.7 b) 2.61
  Check student's work.

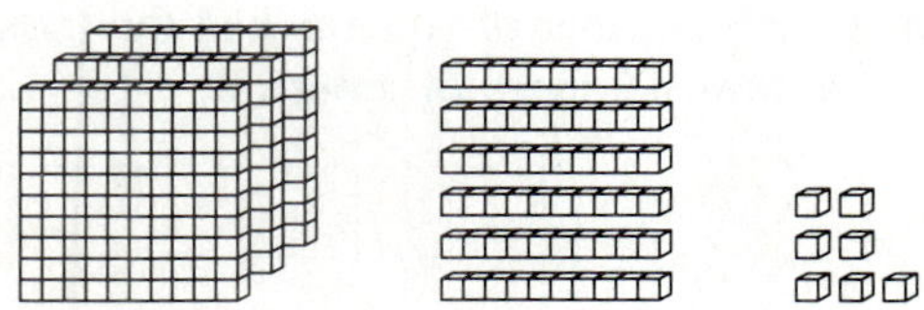

If blocks like those shown are not available, you can substitute other materials such as straws.

Explore

For use with Lesson 4-5

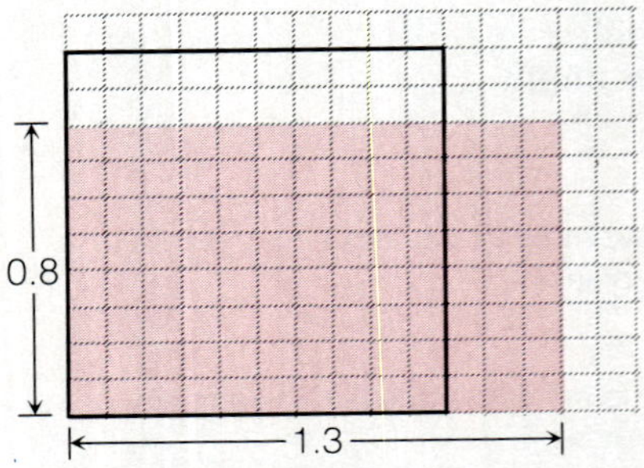

This represents 0.8 × 1.3

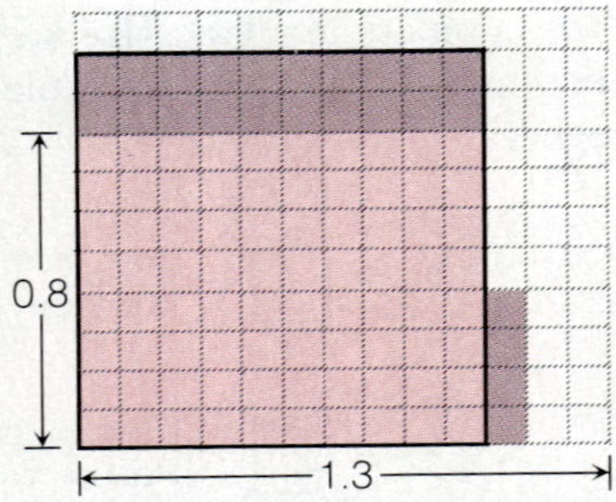

Rearrange shaded squares to show the product represents 1.04

Represent each product on graph paper using the procedure shown above.

a) 1.3 × 2.4 **3.12** **b)** 0.8 × 1.2 **0.96** **c)** 1.7 × 1.9 **3.23**

Explore

For use with Lesson 5-1

- How many different shaped rectangles with 16 squares can you draw on a sheet of graph paper? List the numbers that are lengths and widths of these rectangles. Do not repeat a number in your list. **16 and 1, 8 and 2, 4 and 4**
- Do this again for 24 squares and for 17 squares. **24: 24 and 1, 12 and 2, 8 and 3, 6 and 4** **17: 17 and 1**
- Which number of squares had the most numbers in the list? the fewest? **most: 24, fewest: 17**

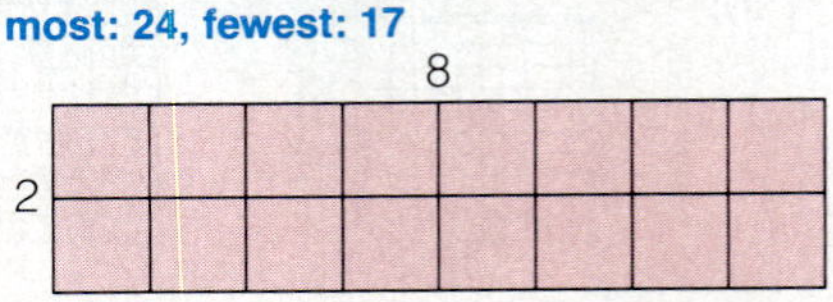

Explore

For use with Lesson 5-4

Here is the only shape rectangle with 7 squares that can be made from graph paper.

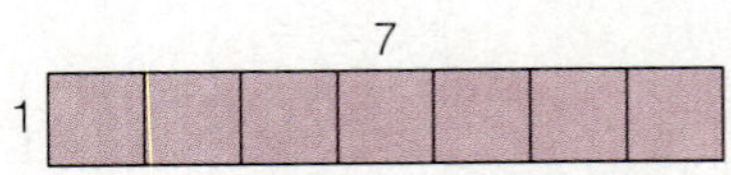

Use graph paper and show all the other numbers less than 30 that have only one rectangle. Talk about the properties of these numbers.
1, 2, 3, 5, 7, 11, 13, 17, 19, 23, 29; all are prime numbers

Explore

For use with Lesson 6-4

Draw pictures or use fraction pieces to show five fractions which are each less than $\frac{1}{2}$. Each should have a different denominator and none of the numerators should be 1.

Have students draw pictures to record their work if they use fraction pieces. Students' drawings could be taped on the chalkboard and then discussed as a class.

Explore

For use with Lesson 6-7

Suppose you have $3\frac{1}{3}$ pies left from a school sale, and a friend orders $1\frac{1}{2}$ pies. Use paper pies, like the ones at the right, to show how many pies will be left. Write a paragraph that describes what you did to find the answer.

Students' paragraphs should include the concept of finding a common denominator and may include the ideas of converting to improper fractions or borrowing.

The answer is $1\frac{5}{6}$ pies.

Explore

For use with Lesson 7-2

This picture shows that there are six $\frac{1}{3}$s in 2.

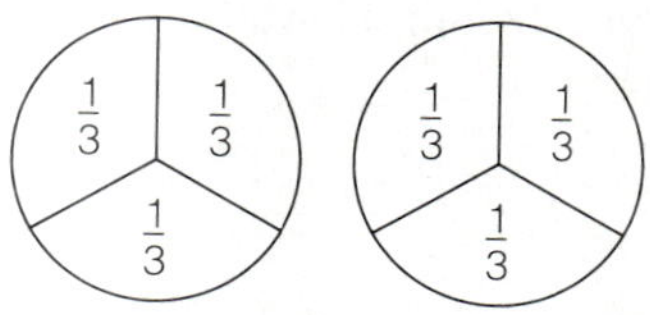

Draw pictures or use fraction pieces to answer each question.

- How many $\frac{1}{2}$s are in 2? **4**
- How many $\frac{1}{4}$s are in $\frac{1}{2}$? **2**
- How many $\frac{1}{6}$s are in $\frac{2}{3}$? **4**

Explore

For use with Lesson 7-8

Copy and complete to make a true equation. You can use whole numbers or fractions for replacements.

a) $32{,}000 = 3.2 \times$ ____ 10,000 **b)** $4.52 \times 100{,}000 =$ ____ 452,000

c) $0.15 = 1.5 \times$ ____ $\frac{1}{10}$ **d)** $8.05 \times \frac{1}{1000} =$ ____ $\frac{805}{10000}$

Explore

For use with Lesson 8-4

Consider the large rectangle with parts A and B. Write algebraic expressions for the areas of parts A and B. Write an equation that shows two ways of representing the total area of the rectangle.

Area of A = 3x
Area of B = 4x
Since the length of the large rectangle is 3 + 4, or 7, and its width is x, its area can be expressed as 7x. Its area can also be expressed as the sum of the areas of its parts, or 3x + 4x.

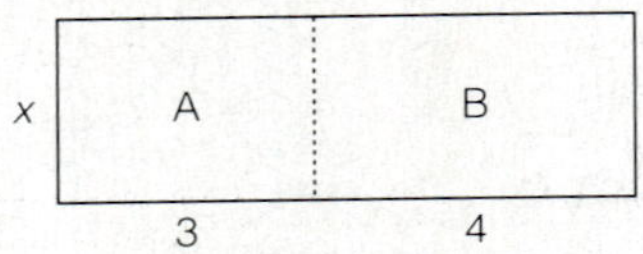

Work with a group to draw rectangles similar to the one above to show that each of the following are true.

a) $3x + 2x = 5x$ **b)** $5x + 4x = 9x$

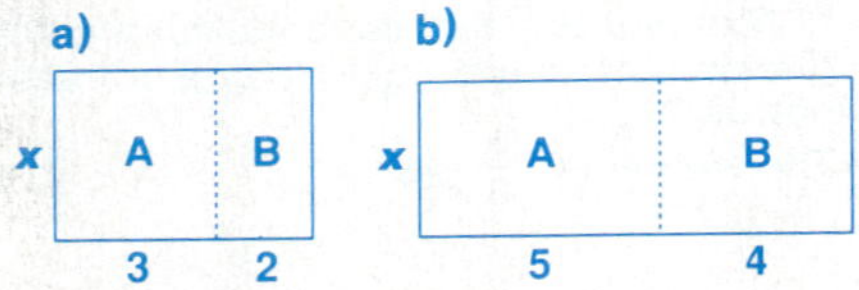

Explore

For use with Lesson 8-8

What conclusions can you draw about the number of counters under each card? Use counters and a balance scale, if available, to test your conclusions.

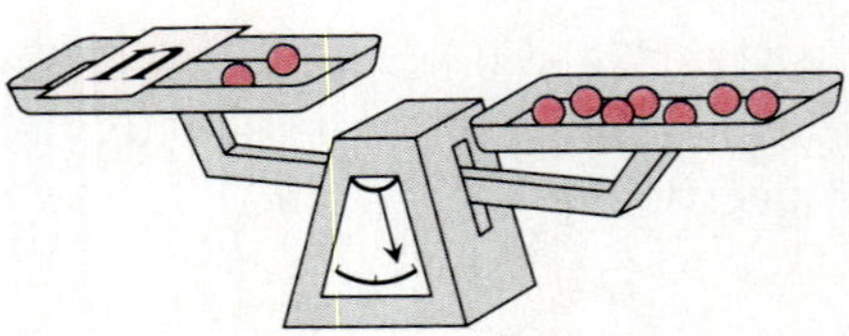

$n + 2 < 7$

n must be less than 5.

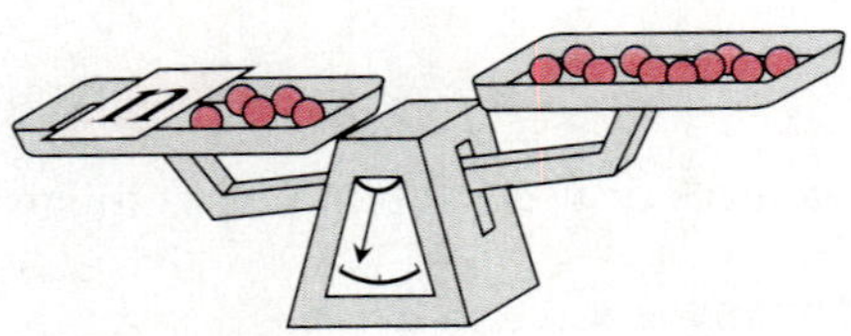

$n + 5 > 10$

n must be greater than 5.

If possible, have students demonstrate their answers using a scale.

Explore

For use with Lesson 9-8

Use a 10 by 10 grid to make a math figure such as a numeral or an operation sign. Your figure should be made up of whole squares only. Shade in your figure. What fraction of the squares is shaded? What fraction of the squares is not shaded?

Emphasize that there are a total of 100 squares. Answers will vary depending on the figure drawn. For the figure shown, $\frac{25}{100}$ or $\frac{1}{4}$ of the figure is shaded.

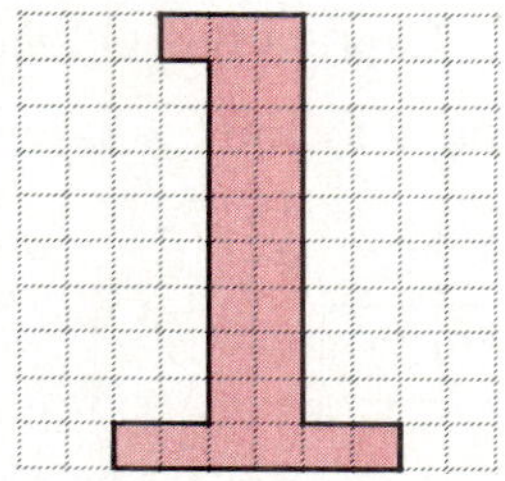

Explore

For use with Lesson 9-9

Some copying machines will make copies that are 120% the size of the original. What size do you think a 5″ by 5″ square would be when it's enlarged to 120% in a copying machine? Test your guess using a copying machine. **6″ by 6″**

Explore

For use with Lesson 10-1

The point that represents 50% of the length of each line segment is shown. Copy each segment. Estimate to mark the given point on the line segment. **Ask students to estimate what number the point represents.**

- Show 10% of 80.

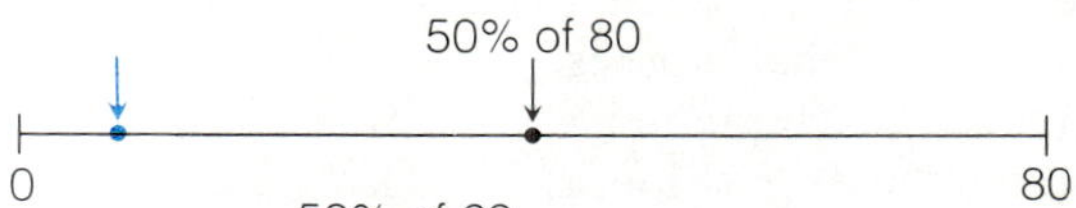

- Show 60% of 60.

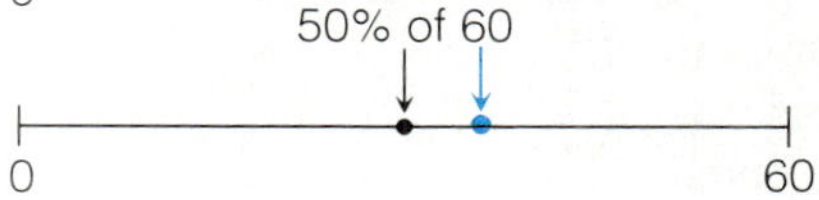

- Show 90% of 120.

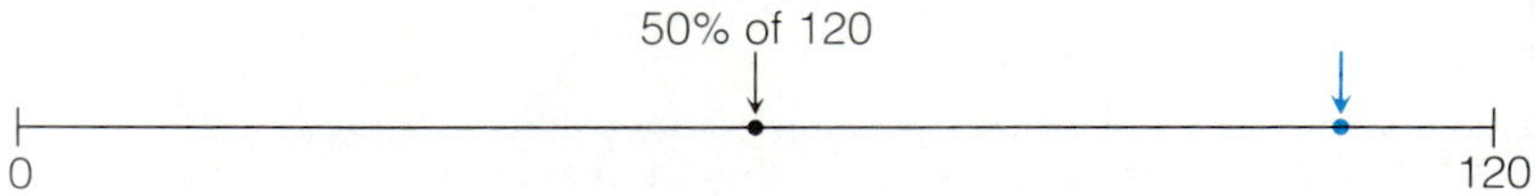

Explore

For use with Lesson 10-2

Copy and complete the ratio table to find an equal ratio with 100 as the second number. Express this ratio as a percent.

a) 6 is to 15 as what is to 100? **40%**

6	**12**	**4**	**20**	**40**
15	30	10	50	100

↑ ↑ equal ratios

b) 30 is to 40 as what is to 100? Make a ratio table like the one above to find out. Express the ratio as a percent. **75%**

Tables will vary. Ex:

30	**15**	**75**
40	**20**	**100**

Explore

For use with Lesson 11-3

Position a protractor on this figure so you can give the measures of 5 different angles without moving the protractor. Record your measures.

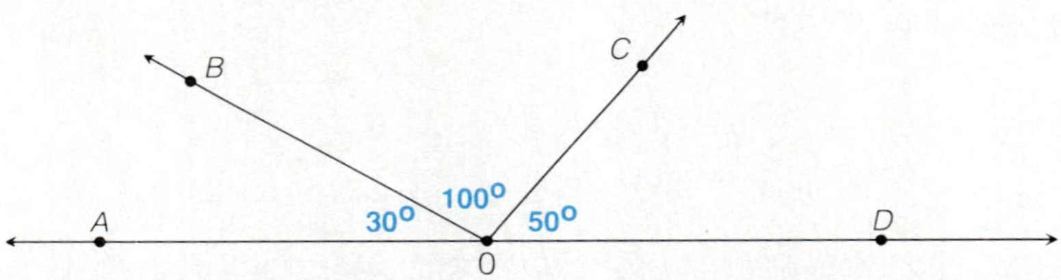

Review how to use a protractor. Encourage students to find these angle measures without moving the protractor between measurements. They should place the protractor in one position, then find the angle measures by finding differences. For example, $m\angle BOC = m\angle BOD - m\angle COD$.

Explore

For use with Lesson 11-4

Draw three pairs of parallel lines. Through each pair, draw a different line *r* that crosses both lines. Label the angles in each picture with numbers in the same relative position as the numbers on this picture. Measure all of the angles with a protractor and make a conjecture about which pairs have equal measures?

For each figure $m\angle 1 = m\angle 4 = m\angle 5 = m\angle 8$ and $m\angle 2 = m\angle 3 = m\angle 6 = m\angle 7$.

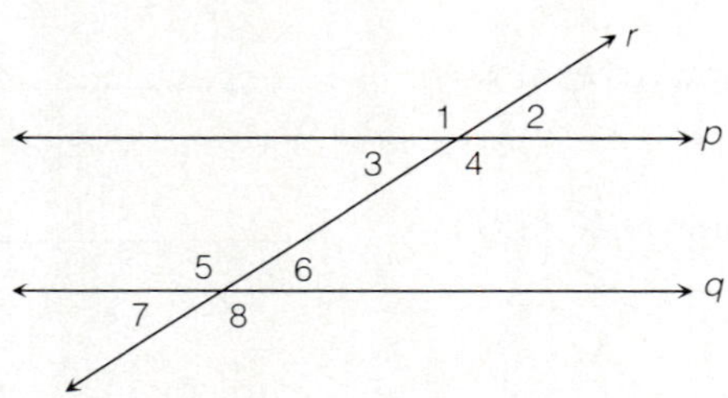

Encourage students to draw three figures in very different positions. Lead students to make generalizations they think are true.

Explore

For use with Lesson 12-3

On square grid paper draw a circle with a radius of 4 as shown. Estimate the area of the circle by counting the number of whole unit squares and fractions of unit squares in the circle.

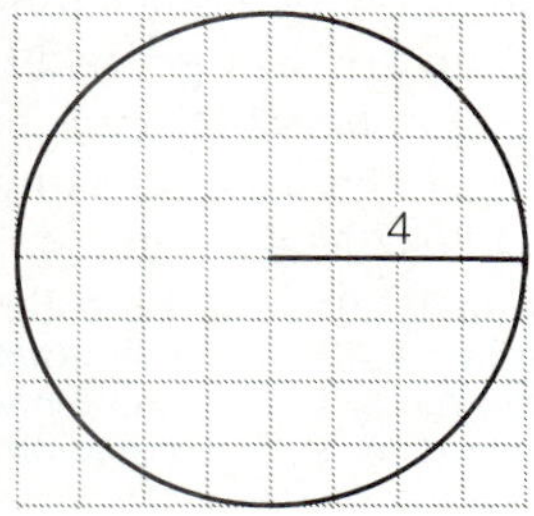

Is your estimate a little more than 3 times the radius squared, or a little more than 4 times the radius squared?

Students' estimates should be a little more than 3 times the radius squared, or a little more than 48.

Explore

For use with Lesson 12-4

Draw these polygons on graph paper. Divide each figure into rectangles and triangles, then find the area of each polygon by finding the area of the rectangles and triangles.

31 square units

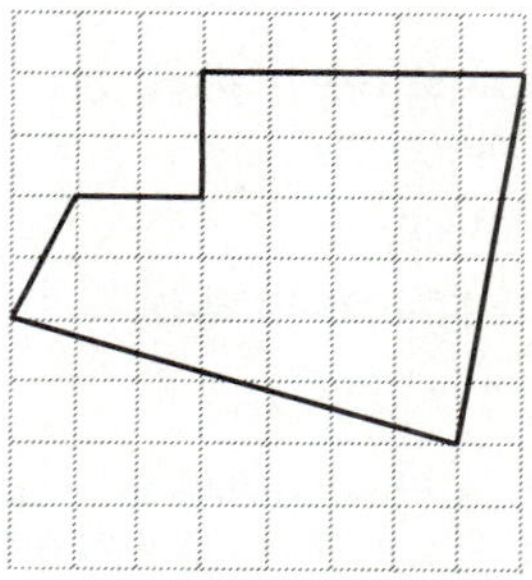

38 square units

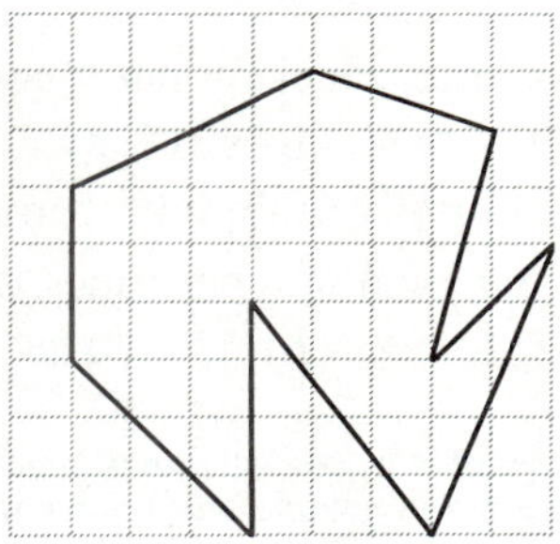

Encourage students to count squares rather than use formulas to find the area of each figure. This will help them focus on the concept of area. Have students compare their result and help them to realize that different ways of dividing each figure still produce the same result.

Explore

For use with Lesson 13-2

- Suppose you have 2 books — A and B. Use pieces of paper to represent each book, and find the number of ways these books can be arranged on a shelf side-by-side. Record your results.
- Suppose you have 3 books; use pieces of paper to represent each book, and find the number of ways they can be arranged. Record your results.
- Suppose you have 4 books; how many ways can they be arranged?
- Can you find a rule that tells the number of arrangements possible for any number of books? [Hint: Think about the counting principle.]

# of books	2	3	4	n
# of arrangements	2	6	24	$n!$

Explore

For use with Lesson 13-4

When you throw two fair number cubes (each numbered 1 to 6), the smallest sum you can get is 2 (1 + 1); the largest sum is 12 (6 + 6). All whole numbers between 2 and 12 are possible. Which sums have the best chance of coming up when you throw two fair number cubes? Make a guess. To check your guess, find all of the ways each sum can be rolled. **7 has the best chance**

1 combination: 2, 12 **2 combinations: 3, 11**
3 combinations: 4, 10 **4 combinations: 5, 9**
5 combinations: 6, 8 **6 combinations: 7**

Explore

For use with Lesson 14-1

- Guess, then check. What whole number times itself is:

  a) 576? **24** **b)** 961? **31** **c)** 2809? **53**

  Repeat until you find the correct answer.
- What decimal number times itself is approximately:

  a) 20? **≈4.47** **b)** 53? **≈7.20** **c)** 72? **≈8.48**

You may want to allow the use of a calculator for this activity. In order to discourage random guessing, have students work in groups and have each member explain how they got their estimate to the rest of the group.

Explore

For use with Lesson 14-6

- Draw these polygons on square grid paper as shown. Draw two other polygons that are the same shape but different sizes than the first two figures.
- Write a paragraph explaining how you drew the polygons to be certain they were the same shape.

Students should include the concepts of ratio and proportion in their answers.

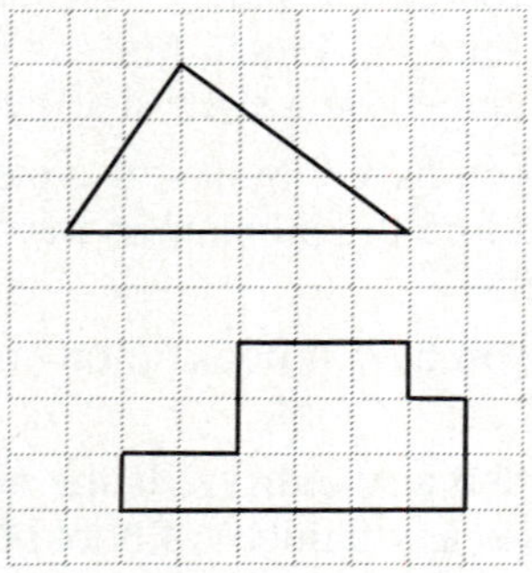

Explore

For use with Lesson 15-5

The equation distance = rate × time or $d = r \cdot t$ relates distance, rate, and time.

- Complete this table of values in which $d = 100$. As the time increases, what happens to the rate? **As time increases, rate decreases.**

$100 = r \cdot t$

time(t)	1	2	10	20	100
rate(r)	100	**50**	**10**	**5**	**1**

- Complete this table of values in which $r = 25$. As the time increases, what happens to the distance? **As time increases, distance increases.**

$d = 25 \cdot t$

time(t)	1	2	3	4	5
distance(d)	25	**50**	**75**	**100**	**125**

Graph both tables of values. Which graph is a line? Which graph is a curve?
The first table graphs a curve, the second table graphs a line.

Explore

For use with Lesson 15-6

Stretch a string from point A, the bottom left corner, to point B, the top right corner, of the chalkboard at the front of your classroom.

- Measure the horizontal distance from A to B, called the run.
- Measure the vertical distance from A to B, called the rise.
- Calculate the ratio, rise to run. This ratio is called the slope of the line.
- Find a rectangular shaped object in your classroom where the slope of a string from the bottom left corner to the top right corner is greater than the chalkboard slope. What is the slope of this string?
Answers will vary.

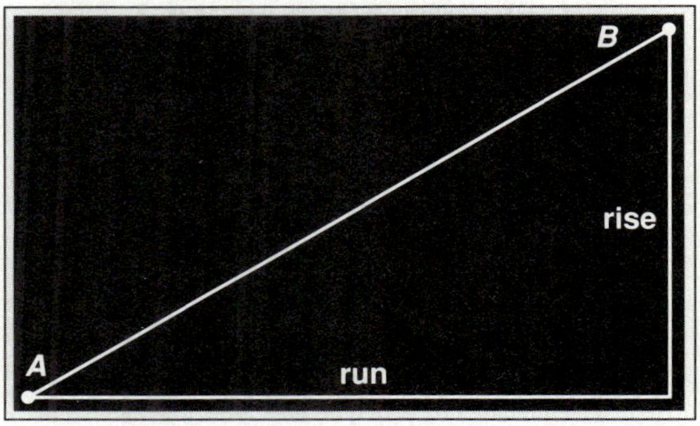

This activity allows students to discover the concept of slope through direct measurement. This experience with concrete materials will help students understand slope as it is described in more abstract terms.

Table 1: Squares and Square Roots

N	N^2	$\sqrt{N}$	N	N^2	$\sqrt{N}$
1	1	1	51	2,601	7.141
2	4	1.414	52	2,704	7.211
3	9	1.732	53	2,809	7.280
4	16	2	54	2,916	7.348
5	25	2.236	55	3,025	7.416
6	36	2.449	56	3,136	7.483
7	49	2.646	57	3,249	7.550
8	64	2.828	58	3,364	7.616
9	81	3	59	3,481	7.681
10	100	3.162	60	3,600	7.746
11	121	3.317	61	3,721	7.810
12	144	3.464	62	3,844	7.874
13	169	3.606	63	3,969	7.937
14	196	3.742	64	4,096	8
15	225	3.873	65	4,225	8.062
16	256	4	66	4,356	8.124
17	289	4.123	67	4,489	8.185
18	324	4.243	68	4,624	8.246
19	361	4.359	69	4,761	8.307
20	400	4.472	70	4,900	8.367
21	441	4.583	71	5,041	8.426
22	484	4.690	72	5,184	8.485
23	529	4.796	73	5,329	8.544
24	576	4.899	74	5,476	8.602
25	625	5	75	5,625	8.660
26	676	5.099	76	5,776	8.718
27	729	5.196	77	5,929	8.775
28	784	5.292	78	6,084	8.832
29	841	5.385	79	6,241	8.888
30	900	5.477	80	6,400	8.944
31	961	5.568	81	6,561	9
32	1,024	5.657	82	6,724	9.055
33	1,089	5.745	83	6,889	9.110
34	1,156	5.831	84	7,056	9.165
35	1,225	5.916	85	7,225	9.220
36	1,296	6	86	7,396	9.274
37	1,369	6.083	87	7,569	9.327
38	1,444	6.164	88	7,744	9.381
39	1,521	6.245	89	7,921	9.434
40	1,600	6.325	90	8,100	9.487
41	1,681	6.403	91	8,281	9.539
42	1,764	6.481	92	8,464	9.592
43	1,849	6.557	93	8,649	9.644
44	1,936	6.633	94	8,836	9.695
45	2,025	6.708	95	9,025	9.747
46	2,116	6.782	96	9,216	9.798
47	2,209	6.856	97	9,409	9.849
48	2,304	6.928	98	9,604	9.899
49	2,401	7	99	9,801	9.950
50	2,500	7.071	100	10,000	10

Table 2: Values of Trigonometric Functions

Degrees	Sin	Cos	Tan	Degrees	Sin	Cos	Tan
0°	0.0000	1.0000	0.0000				
1°	0.0175	0.9998	0.0175	46°	0.7193	0.6947	1.0355
2°	0.0349	0.9994	0.0349	47°	0.7314	0.6820	1.0724
3°	0.0523	0.9986	0.0524	48°	0.7431	0.6691	1.1106
4°	0.0698	0.9976	0.0699	49°	0.7547	0.6561	1.1504
5°	0.0872	0.9962	0.0875	50°	0.7660	0.6428	1.1918
6°	0.1045	0.9945	0.1051	51°	0.7771	0.6293	1.2349
7°	0.1219	0.9925	0.1228	52°	0.7880	0.6157	1.2799
8°	0.1392	0.9903	0.1405	53°	0.7986	0.6018	1.3270
9°	0.1564	0.9877	0.1584	54°	0.8090	0.5878	1.3764
10°	0.1736	0.9848	0.1763	55°	0.8192	0.5736	1.4281
11°	0.1908	0.9816	0.1944	56°	0.8290	0.5592	1.4826
12°	0.2079	0.9781	0.2126	57°	0.8387	0.5446	1.5399
13°	0.2250	0.9744	0.2309	58°	0.8480	0.5299	1.6003
14°	0.2419	0.9703	0.2493	59°	0.8572	0.5150	1.6643
15°	0.2588	0.9659	0.2679	60°	0.8660	0.5000	1.7321
16°	0.2756	0.9613	0.2867	61°	0.8746	0.4848	1.8040
17°	0.2924	0.9563	0.3057	62°	0.8829	0.4695	1.8807
18°	0.3090	0.9511	0.3249	63°	0.8910	0.4540	1.9626
19°	0.3256	0.9455	0.3443	64°	0.8988	0.4384	2.0503
20°	0.3420	0.9397	0.3640	65°	0.9063	0.4226	2.1445
21°	0.3584	0.9336	0.3839	66°	0.9135	0.4067	2.2460
22°	0.3746	0.9272	0.4040	67°	0.9205	0.3907	2.3559
23°	0.3907	0.9205	0.4245	68°	0.9272	0.3746	2.4751
24°	0.4067	0.9135	0.4452	69°	0.9336	0.3584	2.6051
25°	0.4226	0.9063	0.4663	70°	0.9397	0.3420	2.7475
26°	0.4384	0.8988	0.4877	71°	0.9455	0.3256	2.9042
27°	0.4540	0.8910	0.5095	72°	0.9511	0.3090	3.0777
28°	0.4695	0.8829	0.5317	73°	0.9563	0.2924	3.2709
29°	0.4848	0.8746	0.5543	74°	0.9613	0.2756	3.4874
30°	0.5000	0.8660	0.5774	75°	0.9659	0.2588	3.7321
31°	0.5150	0.8572	0.6009	76°	0.9703	0.2419	4.0108
32°	0.5299	0.8480	0.6249	77°	0.9744	0.2250	4.3315
33°	0.5446	0.8387	0.6494	78°	0.9781	0.2079	4.7046
34°	0.5592	0.8290	0.6745	79°	0.9816	0.1908	5.1446
35°	0.5736	0.8192	0.7002	80°	0.9848	0.1736	5.6713
36°	0.5878	0.8090	0.7265	81°	0.9877	0.1564	6.3138
37°	0.6018	0.7986	0.7536	82°	0.9903	0.1392	7.1154
38°	0.6157	0.7880	0.7813	83°	0.9925	0.1219	8.1443
39°	0.6293	0.7771	0.8098	84°	0.9945	0.1045	9.5144
40°	0.6428	0.7660	0.8391	85°	0.9962	0.0872	11.4301
41°	0.6561	0.7547	0.8693	86°	0.9976	0.0698	14.3007
42°	0.6691	0.7431	0.9004	87°	0.9986	0.0523	19.0811
43°	0.6820	0.7314	0.9325	88°	0.9994	0.0349	28.6363
44°	0.6947	0.7193	0.9657	89°	0.9998	0.0175	57.2900
45°	0.7071	0.7071	1.0000	90°	1.0000	0.0000	

Table 3: Geometric Formulas

Rectangle
Area: $A = \ell w$
Perimeter: $P = 2\ell + 2w$

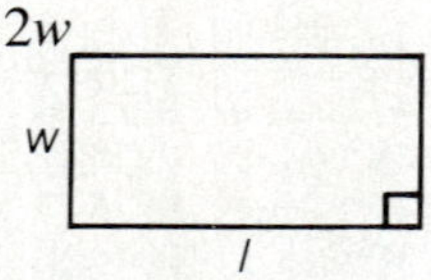

Parallelogram
Area: $A = bh$

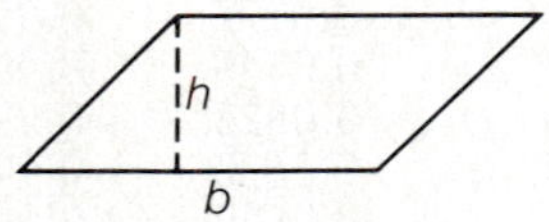

Square
Area: $A = s^2$
Perimeter: $P = 4s$

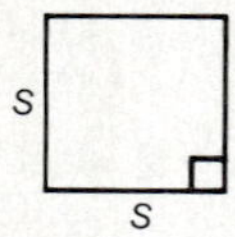

Trapezoid
Area: $A = \frac{1}{2}h(b_1 + b_2)$

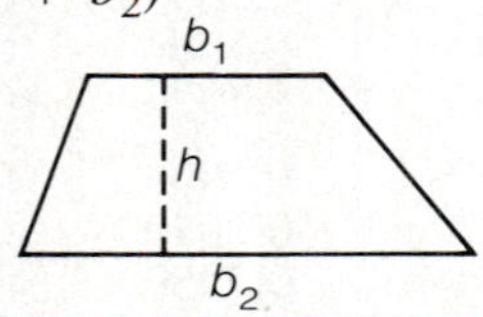

Triangle
Area: $A = \frac{1}{2}bh$
Sum of Angle Measures:
$A + B + C = 180°$

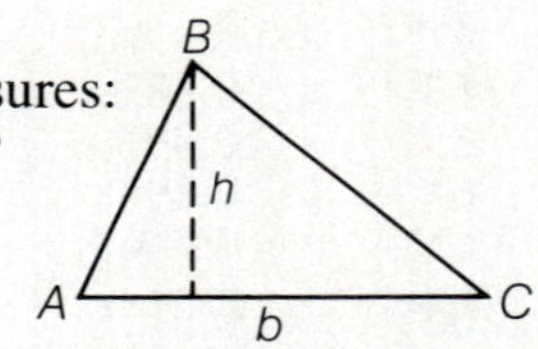

Circle
Area: $A = \pi r^2$
Circumference:
$C = \pi d = 2\pi r$

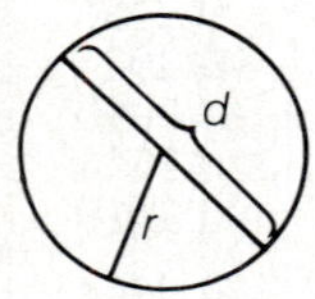

Right Triangle
Pythagorean Property:
$a^2 + b^2 = c^2$

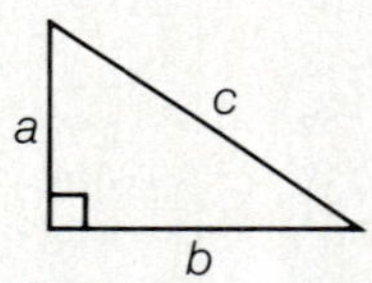

Rectangular Solid
Volume: $V = \ell wh$

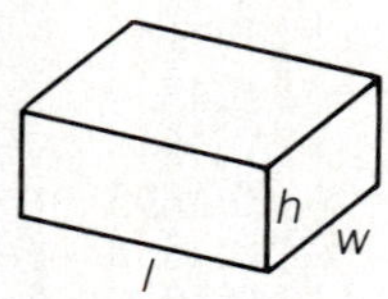

Pyramid
Volume $= Bh$
$B =$ base area

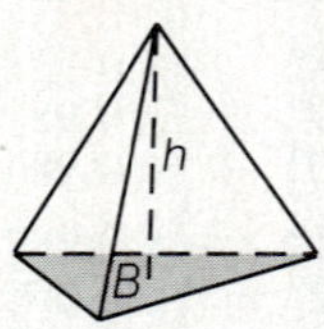

Cylinder
Volume: $V = \pi r^2 h$

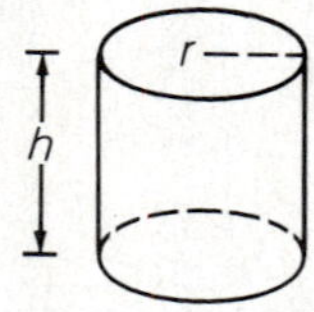

Prism
Volume: $V = Bh$
$B =$ base area

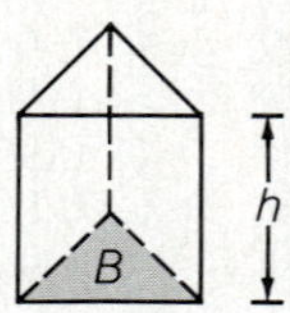

Cone
Volume: $V = \frac{1}{3}Bh$
$B =$ base area

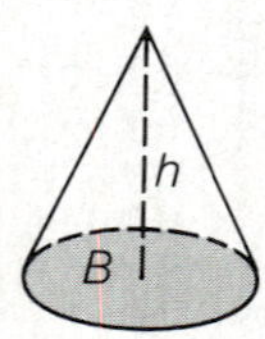

Glossary

Absolute value The absolute value of a number is its distance from 0 on a number line.

Acute angle An angle that has a measure less than 90°.

Additive identity Zero. When zero is added to a number, the result is that same number.

Additive inverse property *See* Inverse Property

Algebraic expression An expression that contains at least one variable; for example, $n + 7$.

Angle Two rays with a common endpoint called the vertex.

Area The measure of a plane region in terms of square units.

Associative property The sum or product of three or more numbers is the same regardless of grouping:
$(a + b) + c = a + (b + c)$ or
$(a \cdot b) \cdot c = a \cdot (b \cdot c)$

Bar graph A type of statistical graph used to show data.

Base (geometry) Any side of a polygon may be referred to as a base.

Base (numbers) In exponential notation a^n, a is the base.

Binomial A polynomial with exactly two terms.

Circle All the points in a plane that are a fixed distance from a point called the center.

Circle graph A type of statistical graph used to show the relationships of the parts to each other and to the total amount.

Circumference The distance around a circle.

Coefficient In any term, the coefficient is the numeric factor of the term of the number that is multiplied by the variable. For $-3x$, the coefficient is -3; for x^2, the coefficient is 1.

Combination A selection of a group of objects from a set without regard to order.

Common factor Any number that is a factor of each of the numbers in a set of numbers.

Commutative property The sum or product of any two numbers is the same regardless of the order in which they are added or multiplied:
$a + b = b + a$ or $a \cdot b = b \cdot a$

Complementary angles Two angles whose measures have a sum of 90°.

Composite number Any whole number greater than 1 that has more than two factors.

Cone A solid with exactly one circular base and exactly one vertex, not in the same plane as the base.

Congruent Two geometric figures are congruent if they have the same size and shape.

Constant A term with no variables.

Coordinate plane A plane containing two intersecting perpendicular number lines; used for graphing ordered number pairs.

Coordinates An ordered pair of numbers matched with a point in the coordinate plane.

Corresponding angles A pair of angles with equal measure, formed by the intersection of a transversal and a pair of parallel lines, with interiors on the same side as the transversal.

Corresponding parts The matching sides and angles in a pair of congruent figures.

Cosine ratio For an acute angle A in a right triangle, the ratio of the length of the adjacent side to the length of the hypotenuse.

Counting principle To find the total number of choices for an event, multiply the number of choices for each part.

Cross products For fractions $\frac{a}{b}$ and $\frac{c}{d}$, the products ad and bc.

Cylinder A solid having two congruent, circular bases in parallel planes.

Data Numerical information, usually about the real world.

Decimal system The place-value numeration system that uses the digits 0 through 9, and groups by tens.

Degree A unit used to measure angles. There are 90 degrees (90°) in a right angle.

Denominator For a fraction $\frac{a}{b}$, b is the denominator.

Dependent event An event whose outcome is influenced by the outcome of a previous event.

Diameter A segment that passes through the center of a circle and has endpoints on the circle.

Distributive property Connects multiplication and addition of whole numbers:

$a(b + c) = a \cdot b + a \cdot c$

Divisible A number is divisible by a second number if the quotient of the first divided by the second is a whole number and the remainder is 0.

Equation A statement of equality between two expressions.

Equivalent expressions Two algebraic expressions are equivalent if (and only if) they have the same value for any number that replaces the variable.

Evaluate an expression To evaluate an expression is to replace each variable in the expression with a given value and simplify the result.

Even numbers All whole numbers that are multiples of 2.

Expanded form (factored form) A number represented as the product of factors.

Exponent In exponential notation a^n, n is the exponent. The exponent tells how many times a, called the base, is used as a factor.

Factor A number that is multiplied by another number to yield a product.

Factor completely To express a number as the product of prime factors.

Formula An equation that shows a relationship between two or more variables.

Fraction A symbol in the form $\frac{a}{b}$, where a is the numerator and b is the denominator and $b \neq 0$.

Glyph A picture or symbol that shows selected data at a glance.

Greatest common factor (GCF) The Greatest Common Factor of two numbers is the greatest whole number that is a factor of both numbers.

Grouping symbols Parentheses (), brackets [], and the fraction bar (a division symbol) are grouping symbols, and indicate the order in which operations should be done.

Height In any geometric figure, the perpendicular distance from a vertex to a base.

Hypotenuse The longest side of a right triangle, opposite the right angle.

Identity property For addition, the sum of an addend and zero is the addend:

$a + 0 = 0 + a = a$

For multiplication, the product of a factor and one is the factor:

$a \cdot 1 = 1 \cdot a = a$

Improper fraction A fraction whose numerator is greater than or equal to its denominator.

Independent event An event whose outcome is not influenced by the outcome of a previous event.

Inequality A statement that uses the symbols $>$ (greater than), $<$ (less than), $\geq$ (greater than or equal to), or $\leq$ (less than or equal to), to compare two expressions.

Integers The whole numbers and their additive inverses.

Inverse property For addition, the sum of any number and its additive inverse is 0:

$a + -a = 0; \ -a + a = 0.$

For multiplication, the product of any rational number and its multiplicative inverse is 1:

$$\frac{a}{b} \cdot \frac{b}{a} = \frac{b}{a} \cdot \frac{a}{b} = 1$$

Irrational number A real number that cannot be expressed as the quotient of two integers. Irrational numbers have nonrepeating decimal representations.

Least common multiple (LCM) The smallest nonzero number that is a multiple of two or more given numbers.

Legs of a triangle The two sides of a right triangle that form the right angle.

Like terms Terms having exactly the same variables with exactly the same exponents.

Line A set of points determined by two points, and extending endlessly in both directions.

Line graph A type of statistical graph used to show changes over time.

Linear equation An equation that can be written in the form $y = ax + b$.

Lowest common denominator The least common multiple of the denominators of two or more fractions.

Lowest terms A fraction is in lowest terms when the only common factor of the numerator and the denominator is 1.

Mean The average; the sum of a set of numbers divided by the number of members in the set.

Median The middle number in a list of numbers given in order.

Metric system The system of measurement that uses the meter as the basic unit of length, the liter as the basic unit of capacity, and the gram as the basic unit of mass.

Mixed number A numeral that has a whole number part and a fraction part.

Mode In a list of numbers, the number that occurs most frequently.

Monomial An expression that is either a numeral, a variable, or the product of numerals and variables.

Multiple The product of two whole numbers is a multiple of each of the whole numbers.

Multiplicative identity One. When a number is multiplied by one, the result is that same number.

Multiplicative inverse property *See* Inverse Property

Negative number Any number that is less than zero.

Numerator For each fraction $\frac{a}{b}$, a is the numerator.

Obtuse angle An angle with measure greater than 90° and less than 180°.

Odd numbers All whole numbers that are not multiples of 2.

Opposites Numbers, such as 4 and -4, that are the same distance from 0 but are on opposite sides of 0 on the number line.

Ordered pair A pair of numbers in a particular order; the coordinates of a point in a plane.

Origin The point at which the axes of a graph cross; the point (0, 0) in the coordinate plane system.

Parallel lines Lines in the same plane that do not intersect.

Parallelogram A quadrilateral with both pairs of opposite sides parallel.

Percent Literally, "per one hundred"; represented by the symbol %.

Perimeter The sum of the lengths of the sides of a polygon.

Permutation An arrangement of objects in a particular order.

Perpendicular lines Two intersecting lines that form right angles.

Pi (π) The ratio of the circumference of a circle to its diameter. The decimal for π is unending and does not repeat. $\pi = 3.14159\ldots$

Pictograph A type of statistical graph using pictures to present data.

Place value The value given to the place a digit occupies in a numeral. In the decimal system, each place of a numeral has ten times the value of the place to its right.

Plane The geometric figure determined by three points that are not on the same line.

Point The simplest geometric figure; it shows a location.

Polygon A closed plane figure formed by three or more segments that intersect only at their endpoints so that exactly two segments meet at each point.

Polynomial A monomial or a sum of monomials.

Positive number Any number that is greater than zero.

Power A product in which each factor is identical; for example, 32 is the fifth power of 2.

Prime factorization The expression of a composite number as the product of prime factors.

Prime number A whole number greater than 1 that has exactly two factors, itself and 1.

Prism A solid that has a pair of parallel, congruent bases and rectangular sides that are parallelograms.

Probability The ratio of the number of times a certain outcome can occur to the number of total possible outcomes.

Proper fraction A fraction whose numerator is less than its denominator.

Property of equivalent fractions Multiplying or dividing both the numerator and denominator of a fraction by the same nonzero integer results in an equivalent fraction:
for all numbers a, b, and c, $(b \neq 0, c \neq 0)$,

$$\frac{a}{b} = \frac{a \cdot c}{b \cdot c} \text{ and } \frac{a}{b} = \frac{a \div c}{b \div c}$$

Proportion An equation stating that two ratios are equal.

Pyramid A solid that has three or more triangular sides and a polygon as a base.

Pythagorean theorem In a right triangle, the square of the length of the hypotenuse is equal to the sum of the squares of the lengths of the other two sides.

Quadrilateral A four-sided polygon.

Radical sign The symbol $\sqrt{\ }$ is called a radical sign and is used to indicate a square root.

Radius Any segment that joins the center to a point on a circle.

Range The difference between the greatest number and the least number in a set of data.

Rate A ratio that compares two different units.

Ratio A comparison of one number to another, expressed as a quotient.

Rational number Any number that can be written in the form $\frac{a}{b}$ where a and b are integers, $b \neq 0$.

Ray A part of a line that has one endpoint and extends endlessly in one direction.

Real numbers The rational numbers and the irrational numbers. There is a real number for each point of the number line.

Reciprocal Two numbers are reciprocals if their product is 1. A reciprocal is also called a multiplicative inverse.

Rectangle A quadrilateral with two pairs of parallel sides and four right angles.

Regular polygon A polygon in which all sides have the same measure and all angles have the same measure.

Repeating decimal A decimal with a sequence of digits that repeats endlessly.

Replacement set The set of numbers from which replacements for the variable in an equation are selected.

Right angle An angle that measures 90°.

Right triangle A triangle with a right angle.

Scientific notation A system of writing a number as the product of a power of 10, and a number greater than or equal to one but less than 10.

Segment Part of a line; two points and all the points between them.

Sequence A set of numbers in a particular order.

Simplify an expression To replace an algebraic (or numerical) expression with the simplest equivalent expression.

Sine ratio For an actue angle A in a right triangle, the ratio of the length of the opposite side to the length of the hypotenuse.

Slope The ratio of rise to run in the graph of any linear equation.

Solution set The collection of all the solutions to a given equation.

Sphere The set of all points in space at a fixed distance from a point called the center.

Square (geometry) A quadrilateral with sides of equal length and four right angles.

Square (numbers) The product of a number multiplied by itself is called the square of that number.

Square root If $a^2 = b$, then a is a square root of b.

Statistics The branch of mathematics that deals with the collection, organization, display, and interpretation of data.

Supplementary angles Two angles whose measures have a sum of 180°.

Surface area The area of the surfaces of a solid figure.

Survey A technique used for collecting data involving fact or opinion.

Tangent ratio For an acute angle A in a right triangle, the ratio of the length of the opposite side to the length of the adjacent side.

Term of a polynomial In a polynomial, each monomial is a term.

Terminating decimal A decimal with a finite number of digits; for example, 0.375.

Terms The parts of an algebraic expression that are separated by an addition or subtraction sign.

Transversal A line that intersects two or more lines.

Trapezoid A quadrilateral with one pair of parallel sides.

Triangle A three-sided polygon.

Trigonometric ratios Ratios of lengths of sides of right triangles. Three of these ratios are called the sine, the cosine, and the tangent.

Trinomial A polynomial with exactly three terms.

Variable A letter or other symbol used to reserve a place for a number in an expression or equation.

Vertex The point that the two rays of an angle have in common.

Vertical angles The two pairs of angles formed by the intersection of two lines.

Volume The measure of a solid region in terms of cubic units.

Whole number Any number in the set $\{0, 1, 2, 3, 4, 5, \ldots\}$.

***x*-axis** The horizontal axis in a rectangular coordinate plane system.

***x*-coordinate** The first number in an ordered pair; used to plot a point in a coordinate plane system.

***x*-intercept** The x-coordinate of the point where a line crosses the x-axis.

***y*-axis** The vertical axis in a rectangular coordinate plane system.

***y*-coordinate** The second number in an ordered pair; used to plot a point in a coordinate plane system.

***y*-intercept** The y-coordinate of the point where a line crosses the y-axis.

Zero property of multiplication The product of any number and zero is zero:
For all whole numbers a, $a(0) = 0$, $0(a) = 0$.

Selected Answers

Chapter 1
Expressions and Equations: Addition and Subtraction

1-1 pages 2–5
Practice **1a.** 9 **b.** 16 **c.** 14 **2a.** 24 **b.** 11 **3a.** 14 **b.** 4
Exercises **1.** 15 **3.** 0 **5.** 35 **7.** 30 **9.** 46 **11.** 15 **13.** 39 **15.** 5 **17.** 1 **19.** 10 **21.** 12 **23.** 33 **25.** 74 **27.** 50 **29.** 4, 6, 68 **31.** 4 **33.** 27 **35.** 90 **37.** 4 **39.** 31 **41.** 1 **43.** 2 **45.** 132 **47.** 38 **49.** 17 **51.** 8 **53.** 3 **55.** 22 **57.** yes **59.** $x = 9$; $y = 7$ **61.** 245 **63.** 101 **65.** 617 **67.** 859 **69.** 311 **71.** 348 **73.** 113 **75.** 2,242 **77.** 7,349 **79.** 1,099 **81.** 24,359 **83.** 10,836 **85.** 39,033 **87.** 8,835 **89.** 1,719 **91.** 2,556 **93.** 55,798 **95.** 4,164 **97.** 1,589
Calculator Activity **1.** 3,775 **3.** 2,355

1-2 pages 6–8
Practice **1a.** $10 - 6$ **b.** $9 - 4$ **c.** $8 + 7$ **d.** $8 + 7$ **2a.** $n + 5$ **b.** $n - 8$ **c.** $n + 7$ **d.** $n - 5$ **e.** $n - 3$ **3a.** $23 + 9$ **b.** $23 + y$ **c.** $23 - 14$ **d.** $23 - t$
Exercises **1.** $8 + 6$ **3.** $9 + 7$ **5.** $6 + 7$ **7.** $12 + 8$ **9.** $x + 6$ **11.** $n + 7$ **13.** $r + 8$ **15.** $y - 5$ **17.** $n - 9$ **19.** $n + 6$ **21.** $n + 16$ **23.** $n - 126$ **25.** $64 + 9$ **27.** $64 - 16$ **29.** $45 + 7$ **31.** $45 - 4$ **33.** $75 + 45$ **35.** $75 - 37$ **37.** $n + 1$ **39.** $n + 2$ **41.** $x + y = 31$, $x - y = 3$; $x = 17$, $y = 14$ **43.** 993 **45.** 436 **47.** 728, 524, 376
Mental Math **1.** 343 **3.** 325 **5.** 7,633

1-3 pages 9–12
Practice **1a.** $x + 45$ **b.** $5{,}138 + a$ **c.** $t + 479$ **2a.** $n + (145 + 68)$ **b.** $(125 + 75) + a$ **c.** $(m + 123) + 77$ **d.** $118 + (12 + c)$ **3a.** t **b.** a **c.** z **d.** n
Exercises **1.** $4 + y$ **3.** $n + 12$ **5.** $b + 137$ **7.** $y + 507$ **9.** $n + (2 + 8)$ **11.** $t + (9 + 7)$ **13.** $z + (16 + 34)$ **15.** $(17 + 8) + b$ **17.** $(38 + 12) + r$ **19.** y **21.** $7 + n$, comm. **23.** $n + (7 + 3)$, assoc. **25.** $23 + p$, comm. **27.** $s + 8$, comm. **29.** $x + y$, identity; or $(0 + x) + y$, assoc.; or $0 + (y + x)$, comm. **31.** $x = 0$ **33.** $z = 34$ **35.** yes, $b - a$ can be evaluated, $8 - 15 = -7$; no, commutative property is not true for subtraction because $7 \neq -7$ **37.** 198 **39.** 1,492 **41.** 299 **43.** 3 **45.** 1 **47.** 189 **49.** 1,519 **51.** 157 **53.** 26 **55.** 15 **57.** 1

1-4 pages 13–14
Practice **1a.** $n = 6$ **b.** $2 = b$ **c.** $y = 4$ **2a.** no solution **b.** no solution
Exercises **1.** $n = 9$ **3.** $n = 4$ **5.** $z = 9$ **7.** no solution **9.** $n = 20$ **11.** $p = 8$ **13.** no solution **15.** no solution **17.** all whole numbers **19.** answers will vary. Example: $m + 1 = 15$ **21.** $7 - 6$ **23.** 220, 230, 240

1-5 pages 15–17
Practice **1a.** $a = 3$ **b.** $n = 8$ **c.** $y = 8$ **2a.** $a = 36$ **b.** $y = 16$ **c.** $d = 14$
Exercises **1.** $n = 6$ **3.** $x = 2$ **5.** $c = 9$ **7.** $p = 5$ **9.** $t = 13$ **11.** $c = 11$ **13.** $n = 8$ **15.** $a = 6$ **17.** $n = 30$ **19.** $b = 70$ **21.** $x = 3$ **23.** $y = 19$ **25.** $a = 21$ **27.** $n = 13$ **29.** $n = 75$ **31.** $n = 756$ **33.** $c = 7{,}489$ **35.** $r = 5{,}365$ **37.** 7,500 buttons **39.** $431,400 **41.** 37,364 **43.** $y = 96 + 35$; $y = 131$ **45.** $x = 13$, $y = 11$ **47.** 100 **49.** $y + 28$
Mental Math **1.** 73 **3.** 102 **5.** 765

1-6 pages 18–20
Practice **1a.** $x = 317$ **b.** $x = 36$ **2a.** $m = 234$ **b.** $y = 235$
Exercises **1.** $n = 46$ **3.** $a = 38$ **5.** $c = 78$ **7.** $x = 115$ **9.** $x = 150$ **11.** $x = 197$ **13.** $c = 252$ **15.** $c = 864$ **17.** $n = 89$ **19.** $a = 187$ **21.** $x = 177$ **23.** $n = 966$ **25.** $a = 756$ **27.** $n = 6{,}659$ **29.** $a = 283$ **31.** $x = 156$ **33.** $y = 66$ **35.** $2{,}475 = 1{,}697 + n$; $n = 778$ **37.** $y = 6$; $n = 4$; $x = 11$ **39.** $b + a$ **41.** 1,313 **43.** 1,304 **45.** $y = 16$

1-7 pages 22–23
Problems **1.** $1,840 **3.** $1,576 **5.** $9,277 **7.** $10,404 **9.** AM/FM stereo radio and power steering

1-8 pages 24–26
Practice **1a.** 14,000 **b.** 72,000 **21.** 1,700 **b.** 6,500
Exercises **1.** $x \approx 150$ **3.** $y \approx 170$ **5.** $x \approx 960$ **7.** $n \approx 1{,}200$ **9.** $z \approx 1{,}600$ **11.** $c \approx 6{,}000$ **13.** $n \approx 1{,}500$ **15.** $x \approx 15{,}000$ **17.** $a \approx 80{,}000$ **19.** $x \approx 3{,}500$ **21.** $c \approx 40{,}000$ **23.** 94,000 mi **25.** 9,000 mi **25.** $F, G, H, I, J, K, L, M, N, O$ **29.** 325 **31.** $78 + (65 + 135)$ **33.** $x = 27$ **35.** $n = 92$ **37.** 1,022 **39.** 21,352

Chapter 1 Review page 30
1. 33 **2.** 12 **3.** 26 **4.** 35 **5.** 29 **6.** 10 **7.** 8 **8.** 4 **9.** 9 **10.** 11 **11.** 5 **12.** 8 **13.** $z + 1$ **14.** $p + 9$ **15.** $v - 2$ **16.** $9 + m$ **17.** $9 - 2$ **18.** $c + (1 + 3)$ **19.** $5 + k$ **2.** w or $w + 0$ **21.** 10 **22.** 8 **23.** 9 **24.** 13 **25.** 9 **26.** 11 **27.** 67 **28.** 86 **29.** 72 **30.** 0 **31.** 112 **32.** 22 km **33.** 17 **34.** 16,000 **35.** 2,000 **36.** 9,000 **37.** 1,500 **38.** 400 **39.** 2,800 **40.** 1,300 **41.** 7,500

Chapter 1 Cumulative Review page 32
1. 2 **3.** 34 **5.** $p \approx 19{,}000$ **7.** $k + 3$ **9.** $15 + 22$ **11.** $5 + r$ **13.** 6 or $0 + 6$ **15.** $a = 12$ **17.** $g = 10$ **19.** $a = 23$ **21.** $x = 19$ **23.** $n = 107$ **25.** $r = 215$ **27.** 22 hours **29.** $v \approx 80$ **31.** $d \approx 70$ **33.** $h \approx 200$ **35.** 15 **37.** 7

Chapter 2
Expressions and Equations: Multiplication and Division

2-1 pages 34–37
Practice **1a.** 3 **b.** 16 **2a.** 33 **b.** 26 **c.** 36 **3a.** 324 **b.** 9 **4a.** 5 **b.** 1 **c.** 2
Exercises **1.** 19 **3.** 6 **5.** 16 **7.** 2 **9.** 536 **11.** 49 **13.** 32 **15.** 50 **17.** 1,088 **19.** 4 **21.** 225 **23.** 60 **25.** 2 **27.** 4 **29.** 9 **31.** 66 **33.** 4 **35.** 50 **37.** 10 **39.** 54 **41.** 1,384 **45.** $25 = m$ **47.** $27 - x$ **49.** 6,500
Calculator Activity **1.** 569 **3.** 208 **5.** 40

2-2 pages 38–40
Practice **1a.** $24 \div 6$ **b.** $2 \cdot 8$ **c.** $12 \cdot 7$ **2a.** $6n$ **b.** $2n$ **c.** $n \div 5$ **3a.** $5t + 2n$ **b.** $7(n + 4)$ **4a.** $645 \cdot 8$ **b.** $645h$ **c.** $645 \div 5$ **d.** $645 \div b$
Exercises **1.** $8 \cdot 7$ **3.** $2 \cdot 46$ **5.** $48 \div 4$ **7.** $9n$ **9.** $15n$ **11.** $6x$ **13.** $n \div 5$ **15.** $35 + 2x$ **17.** $2n - 12$ **19.** $3r - 24$ **21.** $5n + 4$ **23.** $36(9)$ **25.** $36 \div 4$ **27.** $48(4)$ **29.** $48 \div 12$ **31.** $2n + 2$ **33.** $2n + 3$ **35.** Rams 38, Dolphins 13 **37.** $454 = n$ **39.** $r = 71$ **41.** $x + (7 + 3)$

2-3 pages 41–45
Practice **1a.** yx **b.** $6z$ **c.** $p(24)$ **2a.** $n(12 \cdot 7)$ **b.** $(25 \cdot 4)a$ **c.** $6(ab)$ **3a.** 0 **b.** n **c.** 0 **4a.** $6(r + 4) = 6r + 24$ **b.** $(5 + 3)n = 5n + 3n$
Exercises **1.** $4y$ **3.** dc **5.** $b(68)$ **7.** $r(15)$ **9.** $n(3 \cdot 8)$ **11.** $(8 \cdot 5)a$ **13.** y **15.** 1 **17.** $4x + 12$ **19.** $6y + 9y$ **21.** $30 + 5s$ **23.** $5z + 15$ **25.** yes **27.** 13 **29.** 36 **31.** no **37.** 29,165 **39.** 5,192 **41.** 19 **43.** 17 **45.** 23 **47.** 225 **49.** 841 **51.** 630 **53.** 505 **55.** 23,066
Numbers to Algebra **1.** $2(a + b)$ **3.** $6(n + 7)$ **5.** $p(3 + n)$

2-4 pages 46–48
Practice **1a.** $30x$ **b.** $63b$ **c.** $96y$ **2a.** $n + 32$ **b.** $z + 20$ **c.** $r + 32$ **3a.** $11n$ **b.** $40b$ **c.** $41z$ **d.** $6x$
Exercises **1.** $208c$ **3.** $48p$ **5.** $56x$ **7.** $81y$ **9.** $138b$ **11.** $1{,}564z$ **13.** $x + 16$ **15.** $b + 32$ **17.** $n + 23$ **19.** $b + 83$ **21.** $y + 203$ **23.** $12a$ **25.** $16z$ **27.** $23y$ **29.** $123c$ **31.** $30a$ **33.** $56c$ **35.** $35x$ **37.** 1 **39.** $\frac{2(x + 7 - 2) - 10}{2} = \frac{2x}{2} = x$ **41.** 25 **43.** 373
Mental Math **1.** 68 **3.** 92 **5.** 258

2-5 pages 49–51
Practice **1a.** 36 **b.** 8 **c.** 27 **2a.** 5 **b.** 54 **c.** 8
Exercises **1.** 9 **3.** 7 **5.** 36 **7.** 30 **9.** 5 **11.** 240 **13.** 30 **15.** 20 **17.** 12,500 **19.** 500 **21.** 4 **23.** 2 **25.** 6,615 **27.** 3,198 **29.** 121,156 **31.** $56,475 **33a.** 1,176 **b.** 1,272 **c.** 2,125 **35.** 4 **37.** 1 **39.** 19 **41.** 27
Estimation **1.** 900 **3.** 3,600 **5.** 800

2-6 pages 52–54
Practice **1a.** 13 **b.** 24 **2a.** 72 **b.** 32
Exercises **1.** 8 **3.** 11 **5.** 23 **7.** 12 **9.** 6 **11.** 3 **13.** 192 **15.** 205 **17.** 192 **19.** 512 **21.** 1,944 **23.** 2,622 **25.** 12 **27.** 17 **29.** 326 **31.** 9,747 **33.** 3 **35.** 4,860 **37.** 70 **39.** 204 **41.** 1,776 **43.** $(6 \cdot 9) + 13$ **45.** 22

2-7 page 56
Problems **1.** 270 **3.** 910 **5.** 385 **7.** lose

2-8 pages 57–58
Practice **1a.** 135 **b.** 150 **2.** 420
Exercises **1.** 140 **3.** 86 **5.** 272 **7.** 51 mi **9.** $277 **11.** 16 **13.** 4 **17.** $g = \frac{P - f}{2}$ **19.** 19

2-9 pages 59–60
Practice **1a.** 72,000 **b.** 30 **2a.** 80 **b.** 40
Exercises **1.** 2,400 **3.** 1,600 **5.** 32,000 **7.** 50,000 **9.** 3 **11.** 5 **13.** 40 **15.** 80 **17.** 480,000 **19.** 50 **21.** 80 **23.** 5 **25.** 7 **27.** 40 **29.** estimate: 3,000; exact: 3,234 **31.** $8,400 **33.** estimate: 9; exact: 8.68932 hours **35.** $\geq 3{,}350$ and $< 3{,}450$ **37.** 85 **39.** 66

Chapter 2 Review page 64
1. 8 **2.** 7 **3.** 22 **4.** $\frac{24}{y}$ **5.** $6c + 24$ **6.** 12 **7.** $2x$ **8.** $m \div 4$ **9.** $5b$ **10.** $4f$ **11.** ba **12.** $x(14)$ **13.** $(12 \cdot 3)p$ **14.** $7(tk)$ **15.** $v(3 \cdot 6)$ **16.** $4h + 24$ **17.** $7w + w$ **18.** $45 + 5n$ **19.** $20 + r$ **20.** $27m$ **21.** $24x$ **22.** $9y$ **23.** $11a$ **24.** $30p$ **25.** 6 **26.** 5 **27.** 35 **28.** 15 **29.** 51 **30.** 120 **31.** 184 **32.** 270 **33.** 300 mi **34.** 16,000 **35.** 200 **36.** 50

Cumulative Review page 66
1. 17 **3.** 37 **5.** 75 **7.** 9,000 **9.** 7,000 **11.** 200 **13.** 16,000 **15.** 300 **17.** 12,000 **19.** 9 **21.** $4 + f$ **23.** $981 + c$ **25.** $4z$ **27.** $a + 279$ **29.** $(m + 16b) + 580$ **31.** $44s$ **33.** $k + 7k$ **35.** $18 + 6p$ **37.** $54y$ **39.** $6v$ **41.** $45h$

Chapter 3
Integers

3-1 pages 68–71
Practice **1.** X: -9, Y: 4 **2a.** 9 **b.** 85 **3a.** $-16 < -8 < 6$ **b.** $-3 < 0 < |-6| < 9$
Exercises **1.** 2 **3.** 5 **5.** 8 **7.** 5 **9.** 123 **11.** 36 **13.** 199 **15.** 1 **17.** $-21 < 2$ **19.** $-6 < -1 < 4$ **21.** $-9 < -6 < -2 < -1$ **23.** $-24 < -23 < 8$ **25.** $-13 < -4 < 6 < 12 < |-18|$ **27.** -124 **29.** -12 **31.** 8 **33.** negative **35.** 0 **37.** negative **39.** 32, 47, 62 **41.** a square **43.** 189 **45.** 120 **47.** $99x$ **49.** 1 **51.** 6

3-2 pages 72–74
Practice **1a.** -30 **b.** 64 **c.** -60 **2a.** 2 **b.** 16 **c.** -25 **3a.** -20 **b.** 12
Exercises **1.** -12 **3.** -15 **5.** 4 **7.** 19 **9.** 8 **11.** 8 **13.** -6 **15.** 1 **17.** -9 **19.** -24 **21.** -12 **23.** 26 **25.** 52 **27.** -5 **29.** 201 **31.** 6 **33.** 20
Numbers to Algebra **1.** -8 **3.** 24

3-3 pages 75–77
Practice 1a. -5 **b.** -10 **c.** -11 **d.** 6 **2a.** 12 **b.** 14 **c.** -7 **3a.** -6 **b.** 6
Exercises 1. 6 **3.** -6 **5.** -6 **7.** -14 **9.** -22 **11.** 13 **13.** 25 **15.** 14 **17.** 8 **19.** -6 **21.** -8 **23.** 14 **25.** -49 **27.** 81 **29.** -15 **31.** -6 **33.** 16 **35.** -13 **37.** 2 **39.** -17, 15, 93 **41.** 1 **43.** -2 **45.** 23 **47.** $3 - 2 = 1$, but $2 - 3 = -1$ **49.** 14,776 ft **53.** 27 **55.** -2 **57.** 2
Calculator Activity 1. -431 **3.** 192 **5.** 13,511

3-4 pages 79–81
Practice 1a. -66 **b.** -118 **c.** -129 **2a.** 28 **b.** -16 **c.** -101 **3a.** -58 **b.** 69
Exercises 1. -111 **3.** -253 **5.** 78 **7.** -85 **9.** 309 **11.** -1712 **13.** -16 **15.** 145 **17.** -148 **19.** 442 **21.** 180 **23.** 200 **25.** -209 **27.** -311 **29.** -73 **31.** 581 **33.** -121 **35.** $y - 78 = -29$; $y = 49$ **37.** $A = 17$, $B = 7$ **39.** -7 **41.** $2(7 + n)$
Mental Math 1. 28 **3.** 210

3-5 pages 82–83
Problems 1. $46 **3.** $88; $663 **5.** $159 **7.** Overdrawn by $200

3-6 pages 84–86
Practice 1a. 54 **b.** 56 **c.** 45 **2a.** -20 **b.** -48 **c.** -36 **d.** -21
Exercises 1. 63 **3.** 32 **5.** 36 **7.** 14 **9.** 56 **11.** -45 **13.** 0 **15.** -81 **17.** 18 **19.** -48 **21.** 460 **23.** 1248 **25.** 7448 **27.** -4464 **29.** -120; odd; negative **31a.** negative **b.** positive **33.** -168 **35.** $-10,800$ **37.** -5040 **39.** -255 **41.** -41 **45.** -28°C **47.** $-1,255,616$ **49a.** 12, -24, 15 **b.** -10, -14, -18 **51.** -59 **53.** 43

Mental Math 1. 687,000 **3.** $-57,900$ **5.** 9,600

3-7 pages 87–88
Practice 1a. 5 **b.** 5 **c.** 9 **2a.** -6 **b.** -5 **c.** -8
Exercises 1. 7 **3.** 5 **5.** -9 **7.** -8 **9.** -9 **11.** 6 **13.** -9 **15.** 6 **17.** -18 **19.** -13 **21.** -25 **23.** -8 **25.** 3 **27.** 7 **29.** 8 **31.** -6 **33.** no **35.** -9 **37.** $-2,400$ **39.** $a = 8$, $b = -2$ **41.** 5 a.m. **43.** 57 **45.** $108c$ **47.** $13 + t$

3-8 pages 89–91
Practice 1a. -17 **b.** -26 **2a.** -432 **b.** -200 **3a.** 15 **b.** -36
Exercises 1. -12 **3.** 15 **5.** -23 **7.** 12 **9.** -153 **11.** 186 **13.** -270 **15.** -506 **17.** -6 **19.** -63 **21.** 46 **23.** -5082 **25.** -16 **27.** -36 **29.** 18 **31.** 6 **33.** 6 **35.** $7a = -224$, $8a = -512$, $9a = -1152$ **37.** 68, -68 **39.** 7 **41.** 216 **43.** 108 **45.** 11
Computer Activity 1. 6, -3

3-9 pages 93–94
Practice 1a. $x - 9 = 54$ **b.** $y + 47 = -112$ **2a.** $55h = 220$ **b.** $n \div 9 = 12$
Exercises 1. $n - 17 = 101$ **3.** $n + 35 = 372$ **5.** $56 + n = 104$ **7.** $n \div 25 = -12$ **9.** $9n = -171$ **11.** $n - 17 = 15$ **13.** $6d = 91$ **15.** $n + 9 = 72$ **17.** $34n = 272$ **19.** $c - 6000 = 59,820$ **21.** $n + 19 = 56$ **23.** A number divided by -4 is 9. **25.** 16 decreased by a number is 7. **27.** $n + (n + 1) + (n + 2) = 828$; $n = 275$ **29.** 31 **31.** -7 **33.** -13

Chapter 3 Review page 98
1. 6 **2.** 3 **3.** 17 **4.** $-8 < -2 < 2$ **5.** $-7 < -1 < 0$ **6.** $-8 < -5 < 0$ **7.** -6 **8.** -3 **9.** 3 **10.** -24 **11.** -8 **12.** 6 **13.** -16 **14.** -8 **15.** -13 **16.** -3 **17.** -17 **18.** 11 **19.** -7 **20.** -3 **21.** -12 **22.** -49 **23.** -17 **24.** 81 **25.** -93 **26.** -105 **27.** 71 **28.** $224 **29.** 9 **30.** 56 **31.** 25 **32.** -48 **33.** -8 **34.** -9 **35.** 9 **36.** -7 **37.** -4 **38.** 3 **39.** -24 **40.** -27 **41.** -3 **42.** -49 **43.** -24 **44.** -91 **45.** $10n = 110$ **46.** $n + (-15) = 60$

Cumulative Review page 100
1. $g + 9$ **3.** $v + 10$ **5.** $7y$ **7.** $16 - r$ **9.** 4 **11.** 47 **13.** 2 **15.** 64 **17.** 18 **19.** 15 **21.** 25 **23.** -10 **25.** qp **27.** $(20 \cdot 8)p$ **29.** $y + (2 + 9)$ **31.** $8a + 9a$ **33.** $10t$ **35.** $30d$ **37.** $6v$ **39.** -4 **41.** -2

Chapter 4 Decimals

4-1 pages 102–104
Practice 1a. 478.36 **b.** 0.093 **2a.** -2.74 **b.** 2.75 **3a.** > **b.** > **c.** =
Exercises 1. 540.76 **3.** 748.07 **5.** 578.3 **7.** 0.1 **9.** 1.33 **11.** -0.34 **13.** = **15.** > **17.** < **19.** 3.069, 3.07, 3.7 **21.** 6.010, 6.0101, 6.101 **23.** -5.404, -5.044, -5.040 **33.** -221 **35.** -25 **37.** -6

4-2 pages 105–107
Practice 1a. 7.09 **b.** 10 **c.** 0.9 **2a.** 73.5 **b.** 0.25 **c.** 4.8 **3a.** 33 **b.** 200
Exercises 1. 147 **3.** 127 **5.** 35 **7.** 20 **9.** 23.8 **11.** 42.7 **13.** 8.3 **15.** 17.0 **17.** 4.59 **19.** 0.28 **21.** 9.90 **23.** 8.89 **25.** 320 **27.** 40 **29.** 1300 **35.** 8 for $1.05 **37.** 2.113, 3.421, 5.534 **39.** 1.438, 2.323, 3.761 **41.** -18 **43.** -126 **45.** -72 m

4-3 pages 108–110
Practice 1a. 293.58 **b.** 43.595 **2a.** 57.9 **b.** 0.4 **3a.** -4.95 **b.** 14.04
Exercises 1. 48.12 **3.** 70.8 **5.** 6.03 **7.** 181.25 **9.** -12.63 **11.** 31.82 **13.** 924.82 **15.** 338.407 **17.** 121.72 **19.** 40.7274 **21.** 66.03 **25.** $8.76 million **27.** 15.89, 7.98, 13.3 **29.** 11.2, 8.38, 13 **33.** $615.92 - 231.15 = 384.77$ **35.** 21 **37.** -468 **39.** 2
Numbers to Algebra 1. $7x + 6y$ **3.** $11u + 5n$

4-4 pages 111–112
Practice 1. -17.24 **b.** 1.953 **2a.** 17.24 **b.** 3.107
Exercises 1. 7.9 **3.** 4.73 **5.** 113.2 **7.** 27.3 **9.** 601.5 **11.** 16.29 **13.** 369.01 **15.** 150.12 **17.** 50.17 **19.** 130.32

21. -130.32 **23.** 66.9 **25.** $17.36 = 8.24 + x$, $x = 9.12$ **29.** 51 **31.** 12
Mental Math **1.** 17 **3.** 40.45

4-5 pages 113–115
Practice **1a.** 120 **b.** 800 **c.** 1000 **d.** 20,000 **2a.** 8.652 **b.** 0.146 **3a.** -0.0111 **b.** -0.0462
Exercises **1.** 100 **3.** 4 **5.** 8,000 **7.** 0.189 **9.** 3.46 **11.** -0.3504 **13.** 258.03 **15.** 10.2453 **17.** -52.8 **19.** 25.056 **21.** 0.0219 **23.** 33.6 **25.** 8.86 **27.** 15.43 **29.** 62.6°F **31.** 134.06°F **33.** 24 **35.** $43c$ **37.** $-63k$ **39.** 12 **41.** $0.14 < 1.04 < 1.4$
Mental Math **1.** 27.75 **3.** 15.1 **5.** 14.4

4-6 pages 116–117
Problems **1.** 1.5 km **3.** 8 cm **5.** 1.8 cm **7.** 2 mm **9.** 1,000,000

4-7 pages 118–120
Practice **1a.** 3 **b.** 8 **2a.** 21.25 **b.** 9.3 **3a.** 114 **b.** 0.33 **c.** 7.1 **d.** 48.1 **4a.** 0.0025 **b.** 1.6 **c.** 0.045 **d.** 0.0013
Exercises **1.** 5 **3.** 3 **5.** 3 **7.** 0.46 **9.** 0.117 **11.** 4.19 **13.** 3.4 **15.** 16 **17.** 2190 **29.** 1.2 **21.** 11.3 **23.** 8.33 **25.** 1.6 **27.** 5 **31.** 1.05 **33.** -17.46 **35.** 4.83 **37.** $n + 0.95$

4-8 pages 121–123
Practice **1.** 19
Exercises **1.** $m = 2005 - 16$; 1989 **3.** $j - 17 = 53$; 70 cars **5.** $r + 4.52 = 29.32$; 24.80 seconds **7.** $0.15r = 6 \div 2$; 20 m **9.** $n + 17 = 15 \times 8$; 103 people **13.** 4.4 **15.** 3 **17.** 9
Calculator Activity **1.** \$76.49 **3.** \$133.33 **5.** \$478.79

4-9 pages 125–126
Practice **1a.** 150 **b.** -920 **2a.** 0.728 **b.** -18.56
Exercises **1.** 5.3 **3.** -7 **5.** -19.09 **7.** -15 **9.** -12 **11.** 20 **13.** -5.976 **15.** -17.28 **17.** -18 **19.** 17.5 **21.** 15.91 **23.** 113.82 **25.** $1.2x = 264$; $x = 220$ **27.** 3 **29.** -4.2 **31.** 6.6 **33.** -12

Chapter 4 Review page 130
1. 63.8 **2.** 5.072 **3.** $>$ **4.** $=$ **5.** $<$ **6.** 80 **7.** 1000 **8.** 62 **9.** 50 **10.** 8.43 **11.** 85.47 **12.** 5.88 **13.** 46.83 **14.** -28.75 **15.** 86.7 **16.** 1.46 **17.** -28.19 **18.** 29.41 **19.** 45.67 **20.** 200 **21.** 300 **22.** 300 **23.** 22.944 **24.** -1.386 **25.** 34.96 **26.** 1.462 **27.** 1750 cm **28.** 4000 g **29.** 21 **30.** -180 **31.** 8.8 **32.** 0.994 **33.** -3.63 **34.** -14.03 **35.** $37.90 - 10.75 = p$; $p = 27.15$ **36.** $88.6 + 97.4 = c$; $c =$ 186 kg

Cumulative Review page 132
1. $18 - 7$ **3.** $12 + 8$ **5.** $c - 9$ **7.** 8 **9.** 7 **11.** 7 **13.** -7 **15.** -7 **17.** 4 **19.** -11 **21.** 13 **23.** 2 **25.** 9 **27.** 48 **29.** -34 **31.** 265 **33.** 47 **35.** 6 **37.** \$843

Chapter 5 Number Theory

5-1 pages 134–136
Practice **1a.** 7, 14, 21 **b.** 9, 18, 27, 36, 45 **2a.** no **b.** yes **3a.** 1, 2, 3, 6, 9, 18 **b.** 1, 2, 4, 8, 16, 32
Exercises **1.** 3, 6, 9, 12 **3.** 7, 14, 21, 28 **5.** 6, 12, 18, 24 **7.** 9, 18, 27, 36 **9.** 4, 8, 12, 16 **11.** 20, 40, 60, 80 **13.** yes **15.** no **17.** yes **19.** no **21.** yes **23.** yes **25.** yes **27.** yes **29.** 1, 2, 4, 5, 10, 20 **31.** 1, 2, 3, 4, 6, 7, 12, 14, 21, 28, 42, 84 **33.** 1, 2, 3, 4, 5, 6, 10, 12, 15, 20, 30, 60 **35.** 1, 3, 5, 7, 15, 21, 35, 105 **37.** 91; 18, 91 **39.** 4, 9, 25, 49; squares **45.** 14.504 **47.** 20.2
Calculator Activity **1.** 1, 2, 3, 6, 13, 26, 39, 78 **3.** 1, 3, 5, 7, 15, 21, 35, 105

5-2 pages 137–139
Practice **1.** *a*, *c*, *d*, *f* **2.** *b*, *c* **3.** *a*, *b*
Exercises **1.** no **3.** yes **5.** yes **7.** no **9.** no **11.** yes **13.** no **15.** yes **17.** yes **19.** no **21.** yes **23.** yes **25.** no **27.** no **29.** yes **31.** yes **33.** yes **35.** yes **37.** yes **39.** yes **41.** yes **43.** no **45.** A number is divisible by 10 if its last digit is 0. **47.** 2520 **49.** 70; 2, 10, 14, 35, 70 **51.** $>$ **53.** $=$ **55.** -44.793 **57.** 26.163 **59.** 15.26 **61.** $2n + 11 = 27$
Mental Math **1.** yes **3.** no **5.** yes

5-3 pages 140–141
Practice **1.** $\frac{s}{6} = 84$; \$504
Exercises **b.** $\frac{c}{4} = 2.41$; \$9.64 **3.** $2r = 146$; 73 **5.** $3t = 453$; 151 **7.** $12c = 84$; \$7 **11.** -2 **13.** 20 **15.** 0 **17.** -3 **19.** -2.74 **21.** 19.264 **23.** -45.75

5-4 pages 142–143
Practice **1a.** com. **b.** prime **c.** com.
Exercises **1.** com. 2, 3 **3.** com. 3 **5.** com. 2, 3, etc. **7.** com. 2, 13 **9.** com. 2, 4, etc. **11.** com. 2, 17 **13.** prime **15.** prime **17.** prime **19.** com. 7, 11 **21.** 2, 3, 5, 7, 11, 13, 17, 19, 23 **23.** 6 **25.** 5 and 7, 11 and 13, 17 and 19, etc. **29.** 0.6125 **31.** 0.3 **33.** -19.36

5-5 pages 144–147
Practice **1a.** $3 \cdot 3 \cdot 3 \cdot 3 = 81$ **b.** $(-2)(-2)(-2)(-2) = 16$ **c.** $a \cdot a \cdot a$ **2a.** 10^3 **b.** $(-2)^2$ **c.** b^4 **3a.** $(-5)^6$ **b.** 10^7 **c.** x^5 **4a.** 1100 **b.** 650
Exercises **1.** $10 \cdot 10 \cdot 10 \cdot 10 \cdot 10 = 100{,}000$ **3.** $7 \cdot 7 \cdot 7 = 343$ **5.** $9 \cdot 9 \cdot 9 = 729$ **7.** $8 \cdot 8 = 64$ **9.** $n \cdot n \cdot n \cdot n$ **11.** 5^4 **13.** 10^4 **15.** 9^4 **17.** 3^6 **19.** 10^{10} **21.** 10^{14} **23.** 7^{27} **25.** 2^{11} **27.** 3^{14} **29.** p^4 **31.** c^{15} **33.** 48 **35.** 110 **37.** $8 + 4 = 12$ **39.** $5^3 = 125$ **41.** 10^2 **43.** 10^4 **45.** 65,536 **47.** 78,125 **49.** 123 **51.** -2, 4, -8, 16, -32, 64 **53.** positive **55a.** 100 **b.** A googol or 10^{100}. **57.** -0.75 **59.** 7.25 **61.** 2 **63.** 2.3 **65.** 4.7

5-6 pages 148–150
Practice **1a.** $2^3 \cdot 5$ **b.** $2 \cdot 3^3$ **2a.** $2^2 \cdot 7$ **b.** $2^3 \cdot 3^2$ **3a.** $2 \cdot 3 \cdot 7$ **b.** $3 \cdot 3 \cdot 5 \cdot 7$

Exercises **1.** $2^2 \cdot 3^2$ **3.** $2 \cdot 3 \cdot 5 \cdot 7$ **5.** $2^4 \cdot 3$ **7.** $2^2 \cdot 5 \cdot 7$ **9.** $2^2 \cdot 3$ **11.** $2^2 \cdot 3^2$ **13.** $3 \cdot 13^2$ **15.** $2 \cdot 3^2 \cdot 5$ **17.** $2^4 \cdot 3 \cdot 5$ **19.** $2 \cdot 3 \cdot 5 \cdot 11$ **21.** $3^2 \cdot 5^2$ **23.** $2 \cdot 3 \cdot 7 \cdot 11$ **25.** $3 \cdot 3 \cdot 3 \cdot 11$ **27.** $2 \cdot 2 \cdot 2 \cdot 3 \cdot 3 \cdot 3$ **29.** $2 \cdot 3 \cdot 7 \cdot 13 \cdot 17$ **31.** 5 **33.** 23 **35.** 2268 **37.** 137,200 **39.** 141,414; 232,323; 757,575; repeats the age three times. **41.** $2 \cdot 3 \cdot 5 \cdot 7 \cdot 11 \cdot 13 = 30{,}030$ **43.** $9(12 + n) = 135$ **45.** $36 + n = 5 \cdot 4$ **47.** 2.35

5-7 pages 151–153

Practice **1a.** 1, 2, 3, 6, 9, 18 GCF = 9 **b.** 1, 3, 4, 6, 12 GCF = 4 **2a.** $12 = 2 \cdot 2 \cdot 3$, $16 = 2 \cdot 2 \cdot 2 \cdot 2$ GCF = 4 **b.** $24 = 2 \cdot 2 \cdot 2 \cdot 3$, $30 = 2 \cdot 3 \cdot 5$ GCF = 6

Exercises **1.** 1, 2, 4, 8 GCF = 1 **3.** 1, 2, 4, 8, 16 GCF = 8 **5.** 1, 2, 3, 4, 6, 12 GCF = 12 **7.** 1, 2, 3, 4, 6, 8, 12, 16, 24, 48 GCF = 16 **9.** 1, 2, 3, 4, 6, 8, 12, 24 GCF = 8 **11.** 1, 2, 5, 10 GCF = 1 **13.** 1, 3, 9, 27 GCF = 9 **15.** 1, 2, 4, 8, 16, 32 GCF = 8 **17.** 15 **19.** 18 **21.** 15 **23.** 24 **25.** 21 **27.** 2 **29.** 72 **31.** 26 **33.** 6 **35.** 6 **37.** 8 **39a.** yes **b.** yes **c.** yes **41.** 63 **43.** $2t$ **45.** -1 **47.** $6x - 4$ **49.** 0 **51.** 30 **53.** 9

Numbers to Algebra **1.** xy **3.** $6xy^2$ **5.** $8x^2z^2$

5-8 pages 154–155

Problems **1.** 80 ft **3.** 65 ft **5.** 245 ft **7.** Braking Dist = 151.25 ft, Stopping Dist = 206.25 ft **9.** 146.25 ft **11.** 95 miles

5-9 pages 156–158

Practice **1a.** 9, 18, . . . LCM = 18 **b.** 12, 24, . . . LCM = 24 **2a.** $2 \cdot 2 \cdot 3 \cdot 5 = 60$ **b.** $2 \cdot 2 \cdot 2 \cdot 3 \cdot 5 = 120$

Exercises **1.** 10, 20, 30, 40, . . . LCM = 40 **3.** 10, 20, 30, . . . LCM = 30 **5.** 12, 24, . . . LCM = 24 **7.** 15, 30, 45, . . . LCM = 45 **9.** 21, 42, . . . LCM = 42 **11.** 15, 30, . . . LCM = 30 **13.** 14, 28, 42, . . . LCM = 42 **15.** 10, 20, . . . LCM = 20 **17.** 100 **19.** 210 **21.** 280 **23.** 210 **25.** 630 **27.** 240 **29.** 60 **31.** 180 **33.** 36 **35.** 1575 **37.** 280 **39.** 540 **41.** 120 cm **43.** 540 **45.** 1260 **47a.** $\frac{48}{2} = 24$, $\frac{48}{24} = 2$ **b.** $\frac{150}{5} = 30$, $\frac{150}{30} = 5$ **c.** $\frac{216}{6} = 36$, $\frac{216}{36} = 6$; yes **49.** $(-2)^3$ **51.** 18.82

Computer Activity **1.** 252

Chapter 5 Review page 162

1. 5, 10, 15, 20, 25 **2.** 10, 20, 30, 40, 50 **3.** 70, 140, 210, 280, 350 **4.** 1, 2, 4, 8, 16 **5.** 1, 2, 4, 8, 16, 32 **6.** 1, 5, 17, 85 **7.** 2 **8.** 3 **9.** 2, 3 **10.** 5 **11.** 2, 5 **12.** 2 **13.** $218 = 2x$; $x = 109$ **14.** $12c = 144$, $c = 12$ **15.** composite **16.** composite **17.** composite **18.** $8 \cdot 8 \cdot 8 = 512$ **19.** $(-6)(-6)(-6)(-6) = 1296$ **20.** $s \cdot s \cdot s \cdot s \cdot s \cdot s$ **21.** 6^5 **22.** a^3 **23.** $(-2)^4$ **24.** 8^3 **25.** $(-3)^6$ **26.** x^5 **27.** 36 **28.** 9990 **29.** $2^2 \cdot 3^2$ **30.** $3 \cdot 5^2$ **31.** $2 \cdot 3 \cdot 5^2$ **32.** $2 \cdot 3^3 \cdot 7$ **33.** 3 **34.** 13 **35.** 12 **36.** 40 mi/h **37.** 24 **38.** 48 **39.** 288

Cumulative Review page 164

1. 25 **3.** 6 **5.** $m - 18 = 20$ **7.** 220 **9.** 750 cm² **11.** 14 **13.** -3 **15.** -18 **17.** 3 **19.** -3 **21.** -9 **23.** 12 **25.** -11 **27.** -102 **29.** -11.55 **31.** 3.2 **33.** -4.5 **35.** 7.48

Chapter 6
Rational Numbers: Addition and Subtraction

6-1 pages 166–169

Practice **1a.** 36 **b.** 35 **2a.** yes **b.** yes **3a.** $\frac{1}{2}$ **b.** $\frac{2}{3}$

Exercises **1.** 14 **3.** 3 **5.** 1 **7.** 50 **9.** 65 **11.** 7 **13.** 35 **15.** 12 **17.** no **19.** yes **21.** yes **23.** yes **25.** $\frac{10}{13}$ **27.** $\frac{4}{7}$ **29.** $\frac{17}{19}$ **31.** $\frac{2}{3}$ **33.** $\frac{13}{14}$ **35.** $\frac{11}{63}$ **37.** $\frac{3}{7}$ **39.** $\frac{3}{5}$ **41.** $\frac{7}{15}$ **43.** $\frac{7}{9}$ **53.** yes **55.** no **57.** no **59.** yes **61.** false **65.** $5m$ **67.** $14r - 10$ **69.** $3t$ **71.** a^3 **73.** 5 **75.** -5.35 **77.** 6.56

Numbers to Algebra **1.** $\frac{2}{3}$ **3.** $\frac{7a}{8b}$ **5.** $\frac{a}{2}$ **7.** $\frac{4a}{bc}$ **9.** $\frac{3}{7}$ **11.** $\frac{12}{35a}$

6-2 pages 170–172

Practice **1a.** $3\frac{2}{3}$ **b.** 6 **c.** $5\frac{1}{7}$ **2a.** $\frac{19}{4}$ **b.** $\frac{16}{3}$ **c.** $\frac{73}{10}$

Exercises **1.** $4\frac{1}{6}$ **3.** 17 **5.** $8\frac{3}{11}$ **7.** 11 **9.** $3\frac{21}{50}$ **11.** 5 **13.** $11\frac{3}{5}$ **15.** $3\frac{1}{2}$ **17.** $\frac{31}{8}$ **19.** $\frac{41}{16}$ **21.** $\frac{52}{9}$ **23.** $\frac{26}{3}$ **25.** $\frac{137}{10}$ **27.** $\frac{223}{20}$ **29.** $\frac{86}{17}$ **31.** $\frac{791}{80}$ **33.** $4\frac{3}{5}$ **35.** $4\frac{5}{17}$ **37.** $13\frac{1}{5}$ **41.** $2 \cdot 3 \cdot 5$ **43.** 12.375 **45.** prime **47.** $n \div 5 = 40$; $n = 40$ **49.** -6.023

Computer Activity **a.** $\frac{1}{11}$ **b.** $\frac{4{,}123}{3{,}998{,}149}$ **c.** $\frac{1{,}950}{17}$

6-3 pages 173–175

Practice **1a.** $\frac{-5}{6}, -\frac{5}{6}, \frac{5}{-6}$ **b.** $\frac{-7}{3}, -\frac{7}{3}, \frac{7}{-3}$ **c.** $-\frac{11}{12}, \frac{-11}{12}, \frac{11}{-12}$ **d.** $\frac{-15}{9}, -\frac{15}{9}, \frac{15}{-9}$

Exercises **1.** $\frac{-3}{8}, -\frac{3}{8}, \frac{3}{-8}$ **3.** $\frac{-11}{16}, -\frac{11}{16}, \frac{11}{-16}$ **5.** $\frac{-5}{2}, -\frac{5}{2}, \frac{5}{-2}$ **7.** $\frac{-3}{2}, -\frac{3}{2}, \frac{3}{-2}$ **9.** $\frac{-3}{7}, -\frac{3}{7}, \frac{3}{-7}$ **11.** $\frac{-3}{16}, -\frac{3}{16}, \frac{3}{-16}$ **13.** $\frac{-9}{12}, -\frac{9}{12}, \frac{9}{-12}$ **15.** $\frac{-18}{24}, -\frac{18}{24}, \frac{18}{-24}$ **27.** $-\frac{3}{x}$ **29.** $\frac{h}{2}$ **31.** $\frac{6}{n}$ **33.** $\frac{12}{m}$ **35.** $\frac{3}{5}$ **37.** $\frac{15}{2}$ **39.** $-\frac{1}{13}, \frac{1}{15}, -\frac{1}{17}$ **41.** $4m + 8n$ **43.** -3.781

Estimation **1.** 5 **3.** 0 **5.** -4

6-4 pages 176–179

Practice **1a.** $>$ **b.** $<$ **2a.** $>$ **b.** $<$ **3a.** $>$ **b.** $>$

Exercises 1. $<$ 3. $>$ 5. $<$ 7. $>$ 9. $>$ 11. $>$ 13. $<$ 15. $>$ 17. $>$ 19. $>$ 21. $<$ 23. $<$ 25. $>$ 27. $>$ 29. $>$ 31. $<$ 33. $\frac{1}{5}, \frac{1}{4}, \frac{1}{3}$ 35. $\frac{4}{9}, \frac{6}{12}, \frac{2}{3}$ 37. $\frac{3}{8}, \frac{5}{6}, \frac{6}{7}$ 39. $>$ 41. $=$ 43. $<$ 45. $>$ 47. $\frac{5}{12}, \frac{5}{9}, \frac{5}{8}, \frac{5}{6}, \frac{5}{5}, \frac{5}{4}, \frac{5}{3}, \frac{5}{2}$ 49. 55 51. 3 53. 5.22

Calculator Activity 1. 0.8 3. 0.8125 5. −0.28

6-5 pages 180–182

Practice 1a. $\frac{3}{4}$ b. $\frac{2}{7}$ c. $-1\frac{1}{9}$ 2a. $\frac{1}{2}$ b. $-\frac{1}{4}$

Exercises 1. $\frac{1}{2}$ 3. $\frac{3}{11}$ 5. $\frac{1}{4}$ 7. $\frac{2}{11}$ 9. 2 11. $-\frac{4}{19}$ 13. $\frac{11}{15}$ 15. $-2\frac{1}{5}$ 17. $2\frac{8}{15}$ 19. $-1\frac{1}{4}$ 21. $\frac{15}{16}$ 23. $-\frac{2}{3}$ 25. $\frac{1}{3}$ 27. $\frac{5}{6}$ 29. $\frac{5}{9}$ 31. $\frac{2}{7}$ 33. $\frac{7}{12}$ 35. $\frac{7}{12}$ 37. 1 39. $\frac{11}{z}$ 41. $\frac{12}{x}$ 43. $\frac{42}{u}$ 45. $7\frac{1}{2}$ 47. $\frac{1}{4}$ mi or $\frac{2}{5}$ of the course 49. 1, 2, 3, 4, 6, 8, 12, 24 51. $\frac{1}{4}, \frac{1}{4}$ 53. 1.982

6-6 pages 183–185

Practice 1a. $1\frac{13}{60}$ b. $\frac{7}{10}$ 2a. $\frac{23}{24}$ b. $-\frac{9}{10}$

Exercises 1. $\frac{7}{8}$ 3. $-\frac{7}{12}$ 5. $-\frac{3}{10}$ 7. $-1\frac{1}{5}$ 9. $1\frac{7}{12}$ 11. $\frac{11}{20}$ 13. $1\frac{11}{24}$ 15. $\frac{3}{4}$ 17. $-\frac{41}{100}$ 19. $1\frac{17}{24}$ 21. $-\frac{5}{36}$ 23. $\frac{5}{28}$ 25. $-\frac{2}{9}$ 27. $\frac{11}{18}$ 29. 5 ft $1\frac{3}{4}$ in. 31. $-\frac{2}{5} + \frac{3}{4} = \frac{7}{20}$ 33. 1, 5, 25 35. 7, 7 37. est. ≈ 28, exact: 26 39. −37.31

Estimation 1. 2 3. $\frac{1}{4}$

6-7 pages 186–188

Practice 1a. $1\frac{3}{4}$ b. $9\frac{17}{40}$ c. $7\frac{1}{5}$ 2a. $2\frac{1}{4}$ b. $1\frac{1}{15}$ c. $4\frac{1}{4}$

Exercises 1. $1\frac{1}{15}$ 3. $2\frac{16}{21}$ 5. $4\frac{1}{60}$ 7. 6 9. $12\frac{1}{12}$ 11. $3\frac{9}{20}$ 13. $-6\frac{4}{7}$ 15. $-5\frac{35}{48}$ 17. $-1\frac{3}{4}$ 19. 3 21. −4 23. 7 25. 15 27. 36 29. 5 ft $2\frac{9}{10}$ in. 31. 120 33. 0

Mental Math 1. 0 3. $9\frac{5}{6}$ 5. $\frac{5}{12}$

6-8 pages 190–191

Problems 1. 31 3/4 3. 55 7/8 5. 23 1/2 7. $1,975 9. increase 2 1/4

6-9 pages 192–194

Practice 1a. $-\frac{1}{3}$ b. $\frac{1}{12}$ c. $\frac{5}{8}$ d. $1\frac{7}{15}$ 2a. $-\frac{19}{24}$ b. $-\frac{11}{45}$

Exercises 1. $1\frac{1}{9}$ 3. $-\frac{7}{30}$ 5. $-\frac{13}{30}$ 7. $\frac{1}{8}$ 9. $-\frac{1}{4}$ 11. $-\frac{47}{50}$ 13. $\frac{1}{2}$ 15. $-\frac{14}{15}$ 17. $-\frac{13}{40}$ 19. $\frac{1}{24}$ 21. $\frac{8}{15}$ 23. $-\frac{13}{35}$ 25. $-17\frac{11}{40}$ 27. $-1\frac{29}{60}$ 29. $1\frac{11}{36}$ 31. $-\frac{7}{12}$ 33. $-\frac{1}{2}$ 37. $\frac{1}{2}, \frac{1}{4}, \frac{1}{16}$ 39. −1.85 41. 1, 19 43. 1, 2, 3, 4, 6, 12

Mental Math 1. $3\frac{1}{4}$ 3. $3\frac{1}{5}$

6-10 pages 195–196

Exercises 1. $\frac{7}{8}$ in. 3. $\frac{17}{40}$ mi 5. 25 7. $\frac{1}{12}$ in. 11. $7 \cdot 7$ 13. $2 \cdot 3 \cdot 3 \cdot 5$ 15. 15 17. $\frac{64 + n}{12}$ 19. $7m$ 21. 84

Chapter 6 Review page 200

1. 60 2. 1 3. 20 4. $\frac{3}{4}$ 5. $-\frac{6}{13}$ 6. $\frac{12}{13}$ 7. $6\frac{1}{3}$ 8. $-11\frac{2}{11}$ 9. $6\frac{7}{30}$ 10. $\frac{36}{7}$ 11. $-\frac{42}{5}$ 12. $\frac{23}{5}$ 14. $=$ 15. $>$ 16. $>$ 17. $>$ 18. $\frac{5}{6}$ 19. $-\frac{2}{9}$ 20. $-\frac{9}{25}$ 21. $\frac{2}{3}$ 22. $-\frac{1}{3}$ 23. $-\frac{5}{8}$ 24. $4\frac{1}{12}$ 25. $3\frac{2}{3}$ 26. $12\frac{7}{10}$ 27. $1\frac{3}{4}$ lbs 28. $9\frac{5}{8}$ mi 29. $\frac{5}{8}$ 30. $-\frac{7}{10}$ 31. $1\frac{1}{4}$ 32. $1\frac{5}{8}$ yds

Cumulative Review page 202

1. 47.8 3. 5.17 5. $=$ 7. $>$ 9. $>$ 11. 80 13. 7,000 15. 60 17. 60 19. 200 21. 10 23. 50 25. 28 27. 7 29. 7, 14, 21, 28, 35 31. 1, 2, 4, 8, 16, 32 33. 1, 2, 3, 4, 5, 6, 10, 12, 15, 20, 30, 60 35. yes 37. no 39. yes 41. yes

Chapter 7 Rational Numbers: Multiplication and Division

7-1 pages 204–207

Practice 1a. $\frac{5}{16}$ b. $\frac{16}{27}$ 2a. $3\frac{1}{5}$ b. $-6\frac{1}{6}$

Exercises 1. $\frac{9}{20}$ 3. $\frac{5}{32}$ 5. $-\frac{5}{6}$ 7. $-\frac{7}{10}$ 9. $\frac{7}{9}$ 11. $1\frac{1}{4}$ 13. $\frac{5}{24}$ 15. $1\frac{1}{5}$ 17. $14\frac{5}{8}$ 19. $6\frac{3}{4}$ 21. $-12\frac{8}{9}$ 23. $23\frac{4}{7}$ 25. $-1\frac{2}{5}$ 27. 22 29. −16 31. $2\frac{7}{16}$ 33. $-\frac{11}{36}$ 35. $\frac{1}{2}$ 37. $-2\frac{2}{5}$ 39. $-\frac{3}{8}$ 41. 52 ft 43. 28 45. 32 47. 8 49. 12 51. 48 53. 8 55. $2c + 3t$ 57. 1 59. $-\frac{1}{5}$ 61. $-\frac{34}{5}$ 63. $\frac{87}{16}$ 65. −0.06 67. $\frac{7}{8}$ 69. $\frac{4}{15}$

Mental Math 1. 28 3. −46 5. 34 7. 101

7-2 pages 208–210

Practice 1a. −4 b. $\frac{1}{24}$ 2a. $-2\frac{2}{15}$ b. $1\frac{1}{2}$

Exercises 1. 3 3. $\frac{5}{7}$ 5. $-\frac{5}{12}$ 7. $-\frac{3}{16}$ 9. $1\frac{2}{3}$ 11. −2 13. $\frac{13}{14}$ 15. 6 17. 4 19. $3\frac{5}{21}$ 21. $-\frac{3}{10}$ 23. $\frac{1}{21}$ 25. 310 27. $6\frac{30}{43}$ 29. $-\frac{67}{128}$ 31. $4\frac{1}{12}$ 33. $\frac{1}{8}$ in. 35. $\frac{6}{5}, -\frac{1}{5}$ 37. −2.2

Estimation 1. >14 3. >3 5. <3

7-3 pages 211–213

Practice 1a. 24 **b.** −2 **c.** $\frac{1}{16}$ **d.** $\frac{1}{10}$ **2a.** $1\frac{2}{13}$ **b.** $-1\frac{29}{55}$

Exercises 1. $\frac{3}{16}$ **3.** $6\frac{1}{4}$ **5.** 5 **7.** −9 **9.** $-\frac{2}{9}$ **11.** 21 **13.** $\frac{6}{7}$ **15.** −8 **17.** $\frac{6}{11}$ **19.** $1\frac{17}{26}$ **21.** $-3\frac{3}{5}$ **23.** $-\frac{54}{335}$ **25.** $1\frac{5}{11}$ **27.** 15 **29.** 18 **31.** $2\frac{1}{4}$ **33.** $6\frac{3}{5}$ **35.** 4, −4 **37.** $8c$ **39.** −33.6

Computer Activity 1. 48.3333333 **3.** 177.015873

7-4 pages 215–216

1. 80 **3.** $16\frac{1}{2}$ lb **5.** 25 **7.** $106\frac{1}{4}$ lb **11.** $1\frac{1}{24}$ **13.** $\frac{7}{16}$ **15.** −23 **17.** $1\frac{1}{4}$ **19.** $\frac{15}{16}$

7-5 pages 217–218

Problems 1. 13″ **3.** 35″ **5.** 99 lb **7.** 4 yd³ **9.** $206\frac{5}{8}$″

7-6 pages 219–220

Practice 1a. $0.\overline{2}$ **b.** $0.1\overline{6}$

Exercises 1. $0.\overline{45}$ **3.** 0.35 **5.** $0.0\overline{6}$ **7.** 1.6875 **9.** $0.\overline{6}$ **11.** $0.\overline{4}$ **13.** 0.575 **15.** 0.375 **17.** 0.64 **19.** $2.\overline{6}$ **21.** 0.6 **23.** $3.6\overline{1}$ **25.** $0.41\overline{6}$ **27.** $0.\overline{27}$ **29.** 33 **31.** 219 **33.** 270 **37.** 60 **39.** 6.4 **41.** 2

7-7 pages 221–223

Practice 1a. 5^3 **b.** $(-3)^3$ **c.** 10^3 **2a.** y^5 **b.** m^4 **c.** z **3a.** $\frac{1}{125}$ **b.** $\frac{1}{4}$

Exercises 1. 3^3 **3.** $(-4)^2$ **5.** 10^4 **7.** 8^4 **9.** 6 **11.** x **13.** t^2 **15.** g **17.** x^5 **19.** n^2 **21.** $\frac{1}{16}$ **23.** $\frac{1}{16}$ **25.** $\frac{1}{10{,}000}$ **27.** $\frac{1}{25}$ **29.** 3^{-3} **31.** x **33.** 3 **35.** $(-3)^{-5}$ **37.** 5^{-1} **39.** $(-2)^2$ **41.** 3 **45.** 9 **47.** 4 **49.** 1.8 **51.** 7.5 **53.** −8.4

Numbers to Algebra 1. xy^2 **3.** $\frac{6n^2}{5m}$ **5.** $\frac{y^6}{x^2}$

7-8 pages 224–226

Practice 1a. 0.00018 **b.** 655.6 **c.** 0.04 **2a.** 4.567×10^3 **b.** 2.34×10^5 **c.** 5×10^7 **3a.** 2.06×10^{-2} **b.** 8×10^{-6} **c.** 2.004×10^{-1}

Exercises 1. 35,000 **3.** 405,000 **5.** 1.35×10^2 **7.** 3.45×10^5 **9.** 1.24×10^3 **11.** 4.55×10^6 **13.** 9.9×10^4 **15.** 7.8×10^{-2} **17.** 6.77×10^{-4} **19.** 1×10^{-6} **21.** 4.05×10^{-5} **23.** 1.01×10^{-1} **25.** 8×10^7 **27.** 3×10^1 **29.** 5.5×10^6 **31.** 2×10^5 **33.** 1.34×10^{-2} **35.** $\frac{1}{5}$ **37.** $\frac{3}{35}$ **39.** $\frac{7}{15}$ **41.** $10t$

Calculator Activity 1. 4.029×10^{18} **3.** 1.7192×10^{14} **5.** 1.53×10^{16}

Chapter 7 Review page 230

1. $\frac{1}{3}$ **2.** $-\frac{4}{7}$ **3.** $-5\frac{5}{7}$ **4.** 21 **5.** $-\frac{7}{16}$ **6.** $2\frac{1}{4}$ **7.** $2\frac{1}{2}$ **8.** $-\frac{2}{3}$ **9.** $\frac{1}{6}$ **10.** 4 **11.** $-\frac{8}{15}$ **12.** −3 **13.** 24 **14.** 15 **15.** $1\frac{1}{2}$ cups **16.** $0.6\overline{1}$ **17.** 0.45 **18.** $0.\overline{2}$ **19.** 8^3 **20.** $(-5)^7$ **21.** x^2 **22.** $\frac{1}{64}$ **23.** $\frac{1}{36}$ **24.** $\frac{1}{64}$ **25.** 4,700 **26.** 0.019 **27.** 0.000037 **28.** 4.36×10^5 **29.** 5.7×10^4 **30.** 8×10^7 **31.** 7.22×10^{-4} **32.** 1×10^{-6} **33.** 5×10^{-2}

Cumulative Review page 232

1. 67.54 **3.** 18.5 **5.** −51.4 **7.** $\frac{6}{13}$ **9.** $-\frac{3}{5}$ **11.** $-\frac{7}{15}$ **13.** $9\frac{7}{8}$ **15.** $8\frac{5}{6}$ **17.** 14.28 **19.** $-1\frac{1}{6}$ **21.** \$32.39 **23.** \$406 **25.** 3 7/8

Chapter 8
Equations and Inequalities

8-1 pages 234–236

Practice 1a. 16 **b.** −17 **c.** 18 **2a.** 34 **b.** −9

Exercises 1. 7 **3.** 10 **5.** 5 **7.** 80 **9.** 648 **11.** 203 **13.** 8 **15.** −17 **17.** 4.75 **19.** −62 **21.** 9 **23.** 14 **25.** $1\frac{1}{2}$ **27.** −5 **29.** 9 **31.** 16 **33a.** $\frac{c-b}{a}$ **b.** $\frac{c+b}{a}$ **35.** $\frac{8}{15}$ **37.** −3

8-2 pages 237–238

Practice 1a. $2j + 5 = 39$ **b.** $3s - 4 = 65$

Exercises 1. $2j + 25 = 200$ **3.** $67 = 3v + 4$ **5.** $94 = \frac{1}{2}m + 17$ **7.** $376 = \frac{3}{4}e + 56$ **13.** 4 **15.** $9\frac{3}{8}$ **17.** $10t + 800$

Calculator Activity 1. 56 **3.** 1,275

8-3 pages 239–240

Exercises 1. 100 **3.** 24 **5.** 42 **7.** 16°C **11.** 3.3 **13.** −1.25

8-4 pages 241–243

Practice 1a. 74 **b.** 30 **2a.** −6.5 **b.** 9

Exercises 1. 12 **3.** 4 **5.** 5 **7.** 4 **9.** −6 **11.** 3 **13.** 5 **15.** 5 **17.** 17 **19.** 8 **21.** 21 **23.** 16 **25.** −8 **27.** 3 **29.** 40 **31.** yes **33.** 75¢ **37.** 9 **39.** 9.6

8-5 pages 244–246

Practice 1a. 13 **b.** −4 **2a.** 12 **b.** −7

Exercises 1. −9 **3.** $20\frac{1}{2}$ **5.** 4 **7.** $6\frac{3}{4}$ **9.** 5 **11.** 2 **13.** $12\frac{1}{7}$ **15.** −48 **17.** 7 **19.** $8\frac{1}{2}$ **21.** −22 **23.** 14 **25.** 8 **27.** $-3\frac{1}{2}$ **29.** 45 **31.** Juana is 26; Mary is 78 **33.** 87, 88, 89 **35.** 0.75 **37.** 0.6 **39.** $3 \cdot 5$ **41.** $2 \cdot 2 \cdot 7$

Numbers to Algebra 1. $x - 5$ **3.** $5t - 7$ **5.** $-2b + 15$

8-6 pages 247–249

Practice 1a. 36 **b.** 3 **2a.** $\frac{2}{3}$ **b.** 1

Exercises 1. 2 **3.** 3 **5.** 6 **7.** 12 **9.** 10 **11.** 5 **13.** 11 **15.** 7 **17.** 2 **19.** −3 **21.** 1 **23.** $1\frac{3}{7}$ **25.** $-5\frac{1}{4}$ **27.** $-4\frac{1}{2}$ **29.** 1 **31.** 4 **33.** −72 **35.** 30 **37.** 12 **39.** < **41.** < **43.** 32 **45.** m^3

Mental Math 1. 3,599 **3.** 6,399 **5.** 4,896

8-7 pages 250–251
Problems **1.** \$2,800–\$4,200 **3.** \$6,800 **5.** yes **9.** 20,000 miles

8-8 pages 253–255
Practice **1a.** $n \le 1$ **b.** $x < 1\frac{1}{2}$ **2a.** $b > 16$ **b.** $p < -5$
Exercises **1.** $a < 6$ **3.** $c < 5$ **5.** $p > 21$ **7.** $t < -24$ **9.** $y < -5\frac{1}{2}$ **11.** $n < -4$ **13.** $s > 9$ **15.** $n > 7$ **17.** $m \le \frac{5}{12}$ **19.** $3 < x$ **21.** $x \le \frac{5}{2}$ **23.** 32 **25.** \$96,500 **27.** $\{-10, -9, -8, -7, -6, -5, -4, -3, 9, 10\}$ **29.** 0 **31.** 1 **33.** 1

8-9 pages 256–258
Practice **1a.** $n > 20$ **b.** $a \ge 4\frac{1}{2}$ **2a.** $r < -4$ **b.** $x > 3$ **3a.** $b < 1$ **b.** $c \le -\frac{1}{7}$
Exercises **1.** $a > 3$ **3.** $s < -1$ **5.** $r > -6$ **7.** $c > 12$ **9.** $p < 13$ **11.** $x < 8$ **13.** $n < 8$ **15.** $a \ge 3$ **17.** $n \ge 8$ **19.** $x < -\frac{3}{4}$ **21.** $a > 3$ **23.** $x \ge \frac{1}{7}$ **25.** $x > 12\frac{1}{2}$ **27.** $n < -10$ **29.** $x < 1$ **31.** $2n - 9 < 57$ **33.** $\frac{2}{3}n + 6 < 40$ **35.** $\frac{1}{2}n - 6 > 15$; 43 **37.** $2\frac{1}{2}q + 15{,}000 \ge 90{,}000$; \$30,000 **41.** 45 **43.** m^6n^4
Estimation **1.** less **3.** less **5.** more

Chapter 8 Review page 262
1. 10 **2.** −19 **3.** −23 **4.** 4 **5.** $5s + 3 = 33$ **6.** $2a + 2 = 100$ **7.** $3g - 2 = 22$; 8 **8.** 32 oz **9.** 6 **10.** −8 **11.** 12 **12.** 5 **13.** −8 **14.** 10 **15.** 4 **16.** 6 **17.** 8 **18.** 12 **19.** 7 **20.** 3 **21.** 75 **22.** $p < 7$ **23.** $p > -5$ **24.** $a > 4$ **25.** $f < -10$

Cumulative Review page 264
1. $9 \cdot 9 \cdot 9 \cdot 9 \cdot 9 = 59{,}049$ **3.** $t \cdot t \cdot t$ **5.** c^7 **7.** 4^5 **9.** v^4 **11.** 32 **13.** $5 \cdot 5$ **15.** $2 \cdot 2 \cdot 2 \cdot 5 \cdot 5$ **17.** 6 **19.** 30 **21.** 60 **23.** 0.875 **25.** 5^5 **27.** r^{-3} or $\frac{1}{r^3}$ **29.** $\frac{1}{9}$ **31.** 4.14×10^2 **33.** 1.2×10^7 **35.** 2.25×10^{-5} **37.** 27,800 **39.** 0.00708

Chapter 9
Ratio, Proportion, and Percent

9-1 pages 266–267
Practice **1a.** $\frac{1}{3}$ **b.** $\frac{5}{4}$ **c.** $\frac{3}{2}$
Exercises **1.** $\frac{3}{2}$ **3.** $\frac{4}{9}$ **5.** $\frac{3}{8}$ **7.** $\frac{12}{7}$ **9.** $\frac{3}{2}$ **11.** $\frac{8}{5}$ **13.** $\frac{4}{1}$ **15.** $\frac{16}{9}$ **17.** $\frac{5}{18}$ **19.** $\frac{1}{2}$ **21.** $\frac{1}{5}$ **23.** $\frac{17}{6}$ **25.** $\frac{x}{y}$, $x:y$ **27.** $3a:b$, $\frac{3a}{b}$ **29.** $\frac{m+n}{z}$, $m + n:z$ **31.** 2:1 **33.** 3:1,000 **35.** 24:425 **37.** 4:5:2:1 **39.** 2 **41.** $c > -6$ **43.** 0.375 **45.** 0.12 **47.** $1\frac{1}{14}$

9-2 pages 268–271
Practice **1a.** $\ne$ **b.** $=$ **2a.** 15 **b.** 63 **3.** 30
Exercises **1.** $=$ **3.** $\ne$ **5.** $\ne$ **7.** $=$ **9.** $=$ **11.** $=$ **13.** 18 **15.** 9 **17.** 18 **19.** 11 **21.** 35 **23.** 7 **25.** $15\frac{3}{4}$ **27.** 50 **29.** 20 **31.** 102 **33.** $\frac{10}{18} = \frac{5}{9}$ **35.** $\frac{7}{18} = \frac{77}{198}$ **37.** $\frac{4}{25} = \frac{32}{200}$ **39.** $\frac{3}{4} = \frac{18}{24}$ **41.** $\frac{10}{12} = \frac{15}{18}$ **43.** 15 **45.** 21 **49.** 20 min **51.** $\frac{3}{16}$ **53.** $\frac{5}{8}$ **55.** x^2 **57.** $\frac{3}{4}$ **59.** $-\frac{2}{3}$ **61.** 48 **63.** 9 **65.** 11
Numbers to Algebra **1.** 7 **3.** 25 **5.** 11

9-3 pages 272–273
Practice **1a.** 8.7 km/l **b.** 375 words/min **2.** \$12.75
Exercises **1.** \$8/h **3.** 250 words/min **5.** 937.5 km/h **7.** 3 people/car **9.** 66.5 rev/min **11.** 55 min/class **13.** 18.75 **15.** \$40 **17.** 100 beats/min **19.** 44 **23.** 1.5 **25.** −0.928 **27.** −0.08
Estimation **1.** ≈30 km/l **3.** ≈80¢/h **5.** ≈150 mi/h

9-4 pages 274–275
Practice **1a.** d **b.** $3d$ **2.** $n + 2n = 24$
Exercises **1.** k, $k + 7$ **3.** g, $\frac{1}{2}g$ **5.** $g + (g + 8) = 63$ **7.** $x + \frac{x}{2} = 110$ **9.** $p + (p + 25) = 297$ **13.** $a - 18$ **15.** −2 **17.** 6 **19.** −1

9-5 pages 276–277
Practice **1.** 12
Exercises **1.** $x + (x + 65) = 211$; 138 **3.** $f + (f - 64) = 354$; 145 male, 209 female **5.** $4r + r = 40$; large 32, regular 8 **7.** $f + 2f + (f + 12) = 48$; Fred 9, Jerry 18, Tom 21 **11.** 120 **13.** $\frac{19}{30}$ **15.** $\frac{7}{30}$

9-6 pages 279–281
Practice **1.** $L = 1.2$ m, $W = 0.8$ m **2.** 18 cm
Exercises **1.** 600 km **3.** 900 km **5.** 720 km **7.** 555 km **9.** 10 mm × 12.5 mm **11.** 2.5 mm × 7.5 mm **13.** 3.5 mm × 1.65 mm **15.** 5.375 m **17.** 3.75 m **19.** 11.6 cm × 6.4 cm **21.** c or d **23.** $21m$ **25.** $4t + 8$ **27.** $-\frac{2}{15}$ **29.** $\frac{3}{20}$ **31.** 25 **33.** 2,401 **35.** $-\frac{33}{5}$

9-7 pages 282–283
Problems **1.** 0.84, 0.81; Bob's **3.** 1.50, 1.20; A & B **5.** 2.50, 2.75; A & B **7.** Jones' **9.** Bob's **11.** about 1¢ more **13.** \$1.70

9-8 pages 284–286
Practice **1a.** $\frac{11}{20}$ **b.** $\frac{1}{25}$ **c.** $\frac{21}{100}$ **2a.** $\frac{11}{200}$ **b.** $\frac{11}{400}$ **c.** $\frac{1}{15}$ **3a.** 96% **b.** $83\frac{1}{3}$%
Exercises **1.** $\frac{9}{20}$ **3.** $\frac{1}{20}$ **5.** $\frac{1}{5}$ **7.** $\frac{1}{10}$ **9.** $\frac{37}{50}$ **11.** $\frac{3}{200}$ **13.** $\frac{2}{25}$

15. $\frac{23}{400}$ **17.** $\frac{63}{100}$ **19.** $\frac{1}{30}$ **21.** $\frac{39}{400}$ **23.** $\frac{149}{1,000}$ **25.** 75%

27. $16\frac{2}{3}$% **29.** $33\frac{1}{3}$% **31.** 80% **33.** 12% **35.** $62\frac{1}{2}$%

37. $83\frac{1}{3}$% **39.** $18\frac{3}{4}$% **41.** 34 **43.** $23\frac{9}{17}$% **45.** math

47. English **49.** 9 **51.** 1.6

Estimation **1.** ≈50% **3.** ≈$33\frac{1}{3}$% **5.** ≈$66\frac{2}{3}$%

9-9 pages 287–290

Practice **1a.** 125% **b.** 115% **2a.** $\frac{3}{4}$% **b.** $\frac{4}{5}$% **3a.** $\frac{9}{4}$ **b.** $\frac{41}{25}$

4a. $\frac{1}{200}$ **b.** $\frac{1}{800}$

Exercises **1.** 175% **3.** 225% **5.** 200% **7.** 400%

9. $133\frac{1}{3}$% **11.** $137\frac{1}{2}$% **13.** $\frac{1}{5}$% **15.** $\frac{3}{10}$% **17.** $\frac{4}{5}$% **19.** $\frac{3}{5}$%

21. 220% **23.** $\frac{3}{4}$% **25.** $1\frac{1}{10}$ **27.** 3 **29.** $1\frac{3}{4}$ **31.** 2 **33.** $\frac{1}{500}$

35. $\frac{3}{800}$ **37.** $2\frac{1}{4}$ **39.** $\frac{1}{40}$ **41.** 165% **43.** $33\frac{1}{3}$% **45.** no

47. $22\frac{2}{9}$%; $44\frac{4}{9}$% **49.** 3 **51.** $0.8\overline{3}$ **53.** 0.3 **55.** $8\frac{2}{3}$ **57.** $8\frac{3}{8}$

Computer Activity **1.** 83.3333333% **3.** 81.25%

5. 25%

9-10 pages 291–293

Practice **1a.** 0.05 **b.** 0.0825 **c.** 1.25 **2a.** 60% **b.** $8\frac{1}{2}$%

c. 214%

Exercises **1.** 0.4 **3.** 0.05 **5.** 0.02 **7.** 1.5 **9.** 0.9 **11.** 1

13. 0.045 **15.** 2 **17.** 0.12 **19.** 0.34 **21.** 0.0075

23. 0.05625 **25.** 62% **27.** 5% **29.** 80% **31.** $\frac{1}{10}$% **33.** $1\frac{1}{2}$%

35. $12\frac{1}{2}$% **37.** 150% **39.** $\frac{1}{2}$% **41.** $46\frac{1}{2}$% **43.** $7\frac{1}{2}$%

45. 438% **47.** $\frac{3}{5}$% **49.** 0.406 **51.** 0.353 **53.** $x(0.01)$

55. 12 **57.** 31 **59.** 1 **61.** −13 **63.** $c < 8$ **65.** 68 **67.** −3

69. 72

Calculator Activity **1.** 75% **3.** $16.\overline{6}$% **5.** $133.\overline{3}$%

7. 15%

Chapter 9 Review page 298

1. $\frac{3}{7}$ **2.** $\frac{8}{5}$ **3.** $\frac{14}{3}$ **4.** 25 **5.** 20 **6.** 30 **7.** 38 km/h **8.** $20/day

9. 13 kg/day **10.** 5 km/h **11.** $s + 3s = 44$

12. $\frac{1}{2}g + g = 24$; 8 boys, 16 girls **13.** 4 cm × 1 cm

14. 3 cm × 1 cm **15.** 5 ft² for $1.30 **16.** $\frac{7}{50}$ **17.** $\frac{7}{100}$

18. $\frac{3}{40}$ **19.** 75% **20.** 36% **21.** 12.5% **22.** 225% **23.** 160%

24. $\frac{3}{4}$% **25.** $2\frac{3}{4}$ **26.** $\frac{1}{250}$ **27.** $\frac{11}{400}$ **28.** 0.04 **29.** 0.085

30. 8100% **31.** $7\frac{1}{2}$%

Cumulative Review page 300

1. $\frac{84}{96}$ **3.** $-\frac{75}{110}$ **5.** $-\frac{2}{11}$ **7.** $6\frac{5}{6}$ **9.** $1\frac{49}{100}$ **11.** $-\frac{21}{4}$ **13.** >

15. < **17.** < **19.** $\frac{2}{5}$ **21.** −8 **23.** $\frac{5}{6}$ **25.** 3 **27.** 6

29. $72 = 6t + 6$

Chapter 10
Using Percent

10-1 pages 302–304

Practice **1a.** 6 **b.** 90 **c.** $\frac{18}{5}$ **2a.** 78 **b.** 3.375 **c.** 102

Exercises **1.** 12 **3.** 24 **5.** 45 **7.** 5.4 **9.** 23.52 **11.** 3

13. 8 **15.** 2.5 **17.** 0.378 **19.** 12 **21.** 15 **23.** 2.205

25. 0.37 **27.** 0.06 **29.** 8.25 **31.** $100 **33.** $240

35. $14,787.50 **37.** $6.30 **39.** $5.25 **41.** 50% **43.** 75%

45. −2 **47.** −5

Mental Math **1.** 45 **3.** 70 **5.** 15

10-2 pages 305–307

Practice **1a.** 20% **b.** $33\frac{1}{3}$% **2a.** $50\frac{5}{6}$% **b.** $13\frac{1}{3}$%

Exercises **1.** 15% **3.** 60% **5.** 80% **7.** $7\frac{1}{2}$% **9.** 40%

11. 20% **13.** $33\frac{1}{3}$% **15.** $33\frac{1}{3}$% **17.** $22\frac{2}{9}$% **19.** 60%

21. 80% **23.** $2\frac{1}{2}$% **25.** $12\frac{1}{2}$% **27.** 20% **29.** 25%

31. 44.4% **33.** 5% **35.** 52% **37.** no **39.** $x = 25$, $y = 50$

41. 0.25 **43.** 1.24 **45.** −3

Estimation **1.** 33% **3.** 67%

10-3 pages 308–310

Practice **1a.** 40 **b.** 200 **2a.** 62.9 **b.** 200

Exercises **1.** 32 **3.** 160 **5.** 500 **7.** 1,300 **9.** 300

11. 283.3 **13.** 29.4 **15.** 1,400 **17.** 266.7 **19.** 400 **21.** 18

23. 8 **25.** 62.5 **27.** 411.76 **29.** 764 **31.** 152.94 **33.** 800

35. 80 **37.** 180 yd³ **39.** 400 mL **41.** $\frac{3}{4}$, 75% **43.** $\frac{5}{1}$, 500%

45. $\frac{9}{2}$, 450% **47.** $\frac{25}{8}$

Calculator Activity **1.** 19.2 **3.** 200 **5.** 120

10-4 pages 311–312

Practice **1.** $35 = \frac{1}{2}x + 1$; 68 customers

Exercises **1.** $g + 3g = 76$; 19 **3.** $2x + 5 = 45$; 20

5. $20w + 64.50 = 146.50$; $4.10 **7.** $24.30 + 0.1s = s$;

$27 **11.** 36 **13.** 4 **15.** 34 **17.** $\frac{1}{64}$

10-5 pages 313–315

Practice **1.** $12\frac{1}{2}$% **2.** 48%

Exercises **1.** 50% **3.** 25% **5.** 20% **7.** 16% **9.** 75%

11. 43.8% **13.** 57.9% **15.** 33.3% **17.** 33.3% **19.** 117%

21. $6\frac{2}{3}$% **23.** $650 **25.** $74.25 **27.** 0.8 **29.** $2.\overline{6}$

31. 0.3125 **33.** $\frac{5}{16}$ **35.** $\frac{3}{38}$ **37.** −2 **39.** −1

10-6 pages 316–318

Practice **1a.** discount = $450, sale price = $1,800

b. discount = $105, sale price = $770 **2a.** $12

b. $31.25

Exercises **1.** $2.50, $7.50 **3.** $0.33, $11.67 **5.** $32.50, $227.50 **7.** $103.49, $586.46 **9.** $571.25, $3,998.75

11. \$1.35 **13.** \$0.71 **15.** \$165 **17.** \$23.39 **19.** 23.5% **21.** \$356, \$3,204, \$192.24 **23.** It doesn't matter. **25.** $\frac{2}{5}$ **27.** 0.25 **29.** 0.11 **31.** 6 **33.** 12 **35.** $r > -3$

10-7 pages 320–321
Problems **1.** \$49.50 **3.** \$375 **5.** \$7,987.50 **7.** \$100 **9.** \$8.97 **11.** \$257.82

10-8 pages 322–325
Practice **1a.** \$75, \$325 **b.** \$22.50, \$97.50 **2a.** \$15, \$140 **b.** \$180, \$2,180
Exercises **1.** \$3.60, \$123.60 **3.** \$24, \$424 **5.** \$100, \$600 **7.** \$101.25, \$476.25 **9.** \$114.75, \$539.75 **11.** \$121.63, \$816.63 **13.** \$144, \$544 **15.** \$76.80, \$396.80 **17.** \$336, \$1,036 **19.** \$1,500, \$11,500 **21.** \$33.75, \$333.75 **23.** \$12.60, \$102.60 **25.** \$7.50, \$2,007.50 **27.** \$936.30, \$14.05, \$950.35 **29.** \$570.04, \$370.04, \$5.56, \$375.60 **31.** \$178.24 **33.** $1\frac{1}{2}$% **35.** 4 months **37.** \$38.33 **39.** 12 **41.** c **43.** t
Numbers to Algebra **1.** $t = \frac{D}{r}$ **3.** $C = \frac{F - 32}{1.8}$

10-9 pages 326–328
Practice **1.** 4-year college 180°, work 99°, 2-year college 54°, armed services 18°, undecided 9°
Exercises **1.** adventure 180°, science fiction 72°, children's 54°, comedy 36°, other 18° **3.** walk 162°, bus 108°, bicycle 61°, car 29° **5.** popular 189°, soundtrack 76°, comedy 68°, classical 27° **7.** payments 228°, gas 66°, insurance 41°, repairs 19°, reg. 7° **9a.** 162° **b.** 126° **c.** 72° **11.** $\frac{7}{20}$ **13.** 3

Chapter 10 Review page 332
1. 54 **2.** 12 **3.** 2.25 **4.** 20% **5.** 40% **6.** 75% **7.** 25% **8.** 175 **9.** 102 **10.** 50 **11.** 700 **12.** sweater \$25, skirt \$20 **13.** 20% increase **14.** 40% increase **15.** 30% decrease **16.** \$1.80 **17.** \$5, \$20 **18.** \$70, \$1,680 **19.** \$2.85 **20.** \$10.58 **21.** \$48, \$348 **22.** \$16.80, \$176.80 **23.** mystery 36°, fiction 78°, nonfiction 126°, children's 72°, other 48°

Cumulative Review page 334
1. $-3\frac{3}{10}$ **3.** $\frac{1}{8}$ **5.** $-\frac{3}{16}$ **7.** 3.4 km **9.** 5 **11.** −13 **13.** 29 **15.** $3\frac{3}{4}$ **17.** 5 **19.** 4 **21.** 80 m **23.** 3 cm × 4 cm **25.** 16 oz

Chapter 11 Equations in Geometry

11-1 pages 336–337
Practice **1.** points: W, X, Y, Z; lines: $\overleftrightarrow{WY}$, $\overleftrightarrow{ZX}$, $\overleftrightarrow{WZ}$; segments: $\overline{YW}$, $\overline{ZW}$, $\overline{XY}$, $\overline{XZ}$, $\overline{YZ}$; rays: $\overrightarrow{YW}$, $\overrightarrow{WX}$, $\overrightarrow{ZW}$, etc.
Exercises **1.** any 3 of: A, B, C, D, E **3.** ACD, AED, etc. **5.** $\overrightarrow{AE}$, $\overrightarrow{AD}$, $\overrightarrow{AB}$, $\overrightarrow{AC}$ **7.** $\overrightarrow{DC}$, $\overrightarrow{DA}$ **9.** $\overline{AB}$, $\overline{AE}$, $\overline{AD}$ **11.** GCF, CGH **13.** AEF, FBC, CGH, HDA, FGH, ABC **15.** 4 **17.** $x < 2.5$ **19.** $x > 2$

11-2 pages 338–341
Practice **1a.** 3 cm **b.** 2 cm **2.** $31.5 + x = 47.3$ $BC = 15.8$ m **3.** 15 cm **4.** $x + 3.9 + 3.7 + 5.1 = 17$; $DC = 4.3$ cm
Exercises **1.** 3.2 cm **3.** 6.1 **5.** 7.2 cm **7.** 6.4 mm **9.** 4 m **11.** 35.8 cm **13.** 35 cm **15.** 16.3 m **17.** 4 cm, 4 cm, 6 cm **19.** 0.5 km **21.** $y = 11$, $x = 2$; 94 **23.** $y = 4$, $x = 10$; 104 **25.** 1.2 **27.** 2.625 **29.** $2\frac{4}{5}$ **31.** 1.3 **33.** $1\frac{2}{5}$ **35.** $1\frac{2}{25}$ **37.** $\frac{37}{100}$ **39.** 0.3 **41.** 150%

11-3 pages 342–345
Practice **1a.** $\angle CBA$, $\angle DAB$, $\angle DAC$, $\angle DCA$ **b.** $\angle ADC$ or $\angle ADB$ **2a.** 37° **b.** 16° **3a.** 61° **b.** 42°
Exercises **1.** $\angle TYZ$, $\angle TYS$, etc. **3.** $\angle XYS$, $\angle SYZ$, $\angle RYZ$, $\angle RYX$ **5.** $\angle XYS$, $\angle SYZ$, etc. **7.** 17° **9.** 13° **11.** 20.5° **13.** 37° **15.** 27° **17.** 73° **19.** $180 - x$ **21.** 22.5° **23.** 72°, 108° **25.** 18° west of due north **27.** $\frac{2}{5}$ **29.** 1.2
Estimation **1.** 15° **3.** 50°

11-4 pages 346–348
Practice **1.** 61° **2.** $m\angle 5 = 38°$, $m\angle 6 = 38°$, $m\angle 7 = 38°$, $m\angle 8 = 142°$
Exercises **1.** $m\angle 1 = 52°$, $m\angle 2 = 38°$ **3.** $m\angle 1 = 90°$, $m\angle 2 = 40°$ **5.** all = 100° **7.** $m\angle 1 = 95°$, $m\angle 2 = 95°$, $m\angle 3 = 85°$, $m\angle 4 = 95°$ **9.** 121° **11.** 20° **15.** 12.5% **17.** 60%

11-5 pages 349–351
Practice **1.** $\triangle XYZ$ is right and scalene, $\triangle BCD$ is obtuse and scalene **2.** 51°
Exercises **1.** isosceles and obtuse **3.** right and scalene **5.** $40 + 55 + a = 180$ $m\angle A = 85°$ **7.** $15 + 70 + a = 180$ $m\angle A = 95°$ **9.** 78° **11.** 130° **13.** 56° **15.** $x = \frac{180 - n}{2}$ **17.** $m < 3$ **19.** 85%
Calculator Activity **1.** 109 **3.** 69

11-6 pages 353–355
Practice **1a.** pentagon, not regular **b.** octagon, not regular
Exercises **1.** pentagon, not regular **3.** octagon, not regular **5.** 118° **7.** 49° **9.** 60°, 60°, 120°, 120° **11.** 110°, 60°, 85°, 105° **13.** rectangle **15.** 4 **17.** 150%
Numbers to Algebra **1.** 720 **3.** 26,640

11-7 pages 356–357
Problems **1.** 27 mi **3.** 39 mi **5.** yes **7.** $4\frac{1}{2}$ h **9.** \$960 **11.** 1,310 mi

11-8 pages 358–360
Practice **1a.** 10.048 cm **b.** 51.496 cm **2a.** 1.91 m **b.** 5.73 cm

Exercises **1.** 15.7 cm **3.** 62.8 cm **5.** 38.308 m **7.** 1.59 cm **9.** 3.822 m **11.** 1.465 m **13.** 1.433 cm **15.** 2.389 cm **17.** 2.357 m **19.** 87.92 in. **21.** 30.84 **23.** 26.84 **25.** A cat could walk under. **27.** 4 **29.** $\frac{1}{5}$
Computer Activity **1.** 52 **3.** 1,000

11-9 pages 361–364
Practice **1.** $A \cong G, B \cong H, C \cong I, D \cong J, E \cong K, F \cong L; \overline{AB} \cong \overline{GH}, \overline{BC} \cong \overline{HI}, \overline{CD} \cong \overline{IJ}, \overline{DE} \cong \overline{JK}, \overline{EF} \cong \overline{KL}, \overline{FA} \cong \overline{LG}$ **2.** $\overline{AC} \cong \overline{UW}, \angle C \cong \angle W, \angle A \cong \angle U, \triangle ABC \cong \triangle UVW$ by ASA
Exercises **1.** $\angle U$ **3.** $\angle T$ **5.** $\angle R$ **7.** $\overline{VW}$ **9.** $\angle B$ **11.** $\overline{TU}$ **13.** $\overline{AB}$ **15.** SSS **17.** ASA **19.** $\triangle AFE$, by SSS or SAS **21.** $\triangle BCE, \triangle CDA, \triangle DEB, \triangle EAC$ **23.** 0.1875 **25.** $0.41\overline{6}$ **27.** $0.\overline{18}$ **29.** 12 **31.** −357 **33.** $\frac{1}{40}$ **35.** 450% **37.** 140%
Estimation **1.** no **3.** no

11-10 page 365
Problems **1.** 36 **3.** 8 m, 8 m, 20 m, 20 m **5.** \$4.75 **7.** \$6 **11.** 0.5 **13.** −3 **15.** 4

11-11 pages 366–368
Exercises **21.** 5 **23.** 4 **25.** 19

Chapter 11 Review page 372
1. X, Y, or Z **2.** $\overleftrightarrow{XZ}, \overleftrightarrow{XY}$, or $\overleftrightarrow{YZ}$ **3.** $\overline{XZ}$ or $\overline{YZ}$ **4.** 2 cm **5.** 8 cm **6.** $\angle MNO$ **7.** $\angle NOP$ **8.** 120° **9.** acute and equilateral **10.** right and isosceles **11.** pentagon **12.** parallelogram **13.** 4.3 h or 4 h 18 min **14.** 10.99 m **15.** 37.68 cm **16.** ASA **17.** 85

Cumulative Review page 374
1. 9 **3.** −25 **5.** $p < 13$ **7.** $n > 5$ **9.** $x > -3$ **11.** 7:17 **13.** $\frac{4}{1}$ **15.** 49 **17.** 54 in. **19.** 53 km/h **21.** 64 beats/min **23.** 34 mi/gal **25.** 60 **27.** 45% **29.** 25% **31.** 140 **33.** 76

Chapter 12
Area and Volume Formulas

12-1 pages 376–379
Practice **1.** 132.86 m^2 **2a.** 216 m^2 **b.** 2,487.45 cm^2
Exercises **1.** 115 cm^2 **3.** 150 cm^2 **5.** 13.23 cm^2 **7.** 28.12 cm^2 **9.** 2,214 mm^2 **11.** 17.2939 m^2 **13.** 1,029.3 cm^2 **15.** 66 cm^2 **17.** 231 cm **19.** 1.2 m **21.** 75 in^2 **23.** 2.5 gal **25.** 250 cm^2 **27.** $-\frac{1}{6}$ **29.** 2 **31.** $66\frac{2}{3}\%$ **33.** $62\frac{1}{2}\%$

12-2 pages 380–383
Practice **1a.** 20 cm^2 **b.** 33 cm^2 **2a.** 464 m^2 **b.** 231 m^2
Exercises **1.** 31.5 cm^2 **3.** 63 cm^2 **5.** 30 cm^2 **7.** 45 m^2 **9.** 18 m^2 **11.** 127.5 cm^2 **13.** 1,752 cm^2 **15.** 81 m^2 **17.** 57.5 cm^2 **19.** 117 m^2 **21.** 0.598 m^2 **23.** 12 cm **25.** 8 cm **29.** −1
Numbers to Algebra **1.** $h = \frac{A}{b}$

12-3 pages 384–386
Practice **1a.** 28.26 ft^2 **b.** 50.24 m^2
Exercises **1.** 78.5 cm^2 **3.** 78.5 cm^2 **5.** 18.0864 cm^2 **7.** 196,250 m^2 **9.** 7,850 cm^2 **11.** 7.065 m^2 **13a.** 50.24 in^2 **b.** 19.625 in^2 **c.** 9.61625 in^2 **15.** 12π **17.** $144 + \frac{25\pi}{2}$ **19.** 21,892.865 m^2 **21.** 498.5064 cm^2 **23.** 277.4504 cm^2 **25.** \$1.80 **27.** $\frac{6}{25}$ **29.** $\frac{11}{100}$ **31.** 12 **33.** −66 **35.** 1.12 **37.** 0
Estimation **1.** $\approx\frac{1}{4}$ **3.** $\approx\frac{1}{2}$

12-4 pages 387–388
Practice **1a.** 239 m^2 **b.** 222.13 m^2
Exercises **1.** 108 m^2 **3.** 48 m^2 **5.** 119.28 m^2 **7.** 72 ft^2 **9.** 188 ft^2 **11.** 320.11 m^2 **13.** 0.125 **15** 0.07 **17.** 5

12-5 page 389
Problems **1.** 6 ft $3\frac{3}{4}$ in. **3.** juice \$0.70, popcorn \$0.95 **5.** \$12.50 **7.** \$140 **11.** 18 **13.** −2

12-6 pages 390–393
Practice **1.** 1,528.8 cm^3 **2a.** 2,512 cm^3 **b.** 141.3 cm^3
Exercises **1.** 352 cm^3 **3.** 2,009.6 ft^3 **5.** 300 cm^3 **7.** 504 ft^3 **9.** 72,000 cm^3 **11.** 29.98 m^3 **13.** 7,218.86 ft^3 **15.** 602.88 dm^3 **17.** 1,969,408 cm^3 **19.** 565.2 cm^3 **21.** 25 m **23.** 20 cubic units **25.** 6 **27.** 3.8 **29.** 1.25 **31.** 320% **33.** −75% **35.** 8 **37.** 0.25 **39.** 5
Estimation **1.** 1,200 cubic units **3.** 210 cubic units

12-7 pages 394–395
Problems **1.** 604.4 yd^3; \$785.78 **3.** 15.43 yd^3; \$694.44 **5.** \$3,318.75 **7.** \$3,360

12-8 pages 396–398
Practice **1a.** 198 cm^3 **b.** 188.4 m^3
Exercises **1.** $209\frac{1}{3}$ cm^3 **3.** 339.12 cm^3 **5.** 150.72 in^3 **7.** 1,013.2 cm^3 **9.** 235.5 dm^3 **11.** 314 cm^3 **13.** 401.9 mm^3 **15.** $158.\overline{3}$ cm^3 **17.** 188 m^3 **19.** 690 cm^3 **21.** 4,230 m^3 **23.** 24π **25.** 72 **27.** 144 **29.** 6.9 **31.** twice **33.** 288π cm^3 **35.** 176.868π m^3 **37.** 20, 35, 56 **39.** 1.35 **41.** 0.194 **43.** 1.6×10^{-4} **45.** −2 **47.** −8.5

12-9 pages 399–402
Practice **1a.** 533.8 cm^2 **b.** 1,155.52 m^2 **2.** 788 cm^2
Exercises **1.** 527.52 cm^2 **3.** 427.04 cm^2 **5.** 1,004.8 cm^2 **7.** 678.24 m^2 **9.** 439.6 dm^2 **11.** 3,165.12 m^2 **13.** 138 cm^2 **15.** 136 cm^2 **17.** 2,612.48 cm^2 **19.** 12.56 m^2 **21.** 300 ft^2 **23.** $\frac{9}{25}$ **25.** $\frac{13}{50}$ **27.** 44 **29.** 2 **31.** 2
Computer Activity **1a.** $V = 1.4587002\ E + 10$ $SA = 28866095.4$ **b.** $V = 3.9217527\ E + 10$ $SA = 55812419.8$ **c.** $V = 2.6057913\ E + 11$ $SA = 197259435$

Chapter 12 Review page 406
1. 200 mm^2 **2.** 128 cm^2 **3.** 97.92 m^2 **4.** 98 cm^2 **5.** 254.34 m^2 **6.** 9.0746 km^2 **7.** 65 m^2 **8.** \$4.50

9. 216 m^3 **10.** 6,908 ft^3 **11.** 21 yd^3 **12.** 506.5867 ft^3 **13.** $170.\overline{6}$ dm^3 **14.** 1,909.12 cm^2 **15.** 416 ft^2

Cumulative Review page 408

1. $\frac{17}{50}$ **3.** $\frac{17}{200}$ **5.** 65% **7.** 625% **9.** 10% **11.** $\frac{3}{400}$ **13.** 0.07 **15.** 4.25 **17.** 8.5% **19.** 50% **21.** 20% **23.** $253.20 **25.** *A*, *B*, *C*, or *D* **27.** *ABC*, *ABD*, *ACD*, or *BCD* **29.** $\overrightarrow{AB}$ or $\overrightarrow{AD}$

Chapter 13 Probability, Statistics, and Graphs

13-1 pages 410–412
Practice 1. 12
Exercises 1. 9 **3.** 20 **5.** 9 **7.** 6 **9.** 100 **11.** 16 **13.** 24 **15.** 3 **17.** 2 **19.** 150%

13-2 pages 413–415
Practice 1a. 6 **b.** 120 **2.** 12
Exercises 1. 2 **3.** 6 **5.** 6 **7.** 6 **9.** 24 **11.** 6 **13.** 3,628,800 **15.** 24 **17.** 60 **19.** 676 **21.** 720 **23.** 720 **25.** 3,276,000 **27.** 6,390,000 **29.** $3m$ **31.** $9n$ **33.** $6x - 6y$ **35.** 15.7 **37.** 20 **39.** −2
Numbers to Algebra 1. $(n + 1)$, $(n + 2)$, $(n + 3)$, $(n + 4)$ **3.** $(y - 2)$, $(y - 4)$

13-3 pages 416–418
Practice 1a. 21 **b.** 28
Exercises 1. 3 **3.** 1 **5.** 4 **7.** 15 **9.** 10 **11.** 8 **13.** 36 **15.** 28 **17.** 210 **19.** 220 **21.** 495 **23.** 28 **25.** 175 **27.** 4 **29.** 32 **31.** −40 **33.** 3 **35.** 25% **37.** 16 **39.** −4 **41.** −30 **43.** 18 cm

13-4 pages 419–421

Practice 1a. $\frac{3}{6} = \frac{1}{2}$ **b.** 1 **c.** $\frac{3}{6} = \frac{1}{2}$ **2a.** $\frac{5}{12}$ **b.** $\frac{9}{12} = \frac{3}{4}$

Exercises 1. $\frac{1}{2}$ **3.** $\frac{1}{13}$ **5.** $\frac{1}{13}$ **7.** $\frac{1}{52}$ **9.** $\frac{1}{2}$ **11.** $\frac{3}{4}$ **13.** $\frac{4}{15}$ **15.** $\frac{1}{3}$ **17.** $\frac{3}{5}$ **19.** $\frac{11}{16}$ **21.** $\frac{3}{8}$ **23.** $\frac{9}{16}$ **25.** $\frac{3}{8}$ **27.** $\frac{1}{8}$ **29.** $\frac{5}{36}$ **31.** $\frac{1}{6}$ **33.** 36 **35.** 6 **37.** 3 **39.** −8

13-5 pages 422–424

Practice 1a. $\frac{9}{20}$ **b.** $\frac{1}{5}$ **2a.** $\frac{8}{33}$ **b.** $\frac{14}{33}$

Exercises 1. $\frac{1}{48}$ **3.** $\frac{1}{4}$ **5.** $\frac{1}{2}$ **7.** $\frac{1}{48}$ **9.** $\frac{1}{2}$ **11.** 0 **13.** $\frac{1}{2,652}$ **15.** $\frac{25}{102}$ **17.** $\frac{13}{204}$ **19.** $\frac{4}{663}$ **21.** $\frac{4}{663}$ **23.** $\frac{1}{17}$ **25.** $\frac{1}{56}$ **27.** $\frac{1}{42}$ **29.** $\frac{15}{1,024}$ **31.** $\frac{5}{256}$ **33.** $\frac{1}{72}$ **35.** $\frac{5}{24}$ **37.** 60% **39.** 60% **41.** 3
Calculator Activity 1. 58%

13-6 page 425
Exercises 1. $5.25 **3.** $10.38 **5.** $525 **7.** $w = 3$ ft, $l = 12$ ft **11.** < **13.** 4

13-7 pages 426–428
Practice 1. range 3, mode $16\frac{1}{2}$

Exercises 1. range 9, mode 2 **3.** range 28, mode 73 **5.** range 20, mode 20 **7.** range 80, mode 76 **9.** range 52, mode 82% **11.** range 34, mode 28 and 34 **13.** range $5, mode $1–$2.99 **15.** 24 **17.** $c > 4$ **19.** −0.5 **21.** −3

13-8 pages 429–431
Practice 1. mean 15, median 14
Exercises 1. mean 4.1, median 4 **3.** mean 23.1, median 23 **5.** mean 49.6, median 45 **7.** mean 10.9, median 9.5 **9.** mean 6.3, median 7 **11.** mean 171.2, median 175 **13.** mean 5.45, median 5.1 **15.** mean 0.21, median 0.08 **17.** mean 3.34, median 3.5 **19.** $3.33

21. 105 **23.** $4.38 **25.** ≈300 **27.** $66\frac{2}{3}$% **29.** 50% **31.** $\frac{3}{4}$ **33.** 3 **35.** −1 **37.** < **39.** > **41.** =

13-9 pages 432–433
Problems 1. 20% **3.** 60% **5.** 77.5% **7.** 20 **9.** 1983: 32.67; 1984: 33.67 **11.** 25% **13.** $150

13-10 pages 434–437
Exercises 7. $9n - 9$ **9.** 4 **11.** −7 **13.** 7 **15.** −4 **17.** 10 **19.** 24 m **21.** 37.5 cm

13-11 pages 438–440
Exercises 9. < **11.** = **13.** = **15.** 4.2 **17.** 31.25% **19.** 6.25%

Chapter 13 Review page 444
1. 6 **2.** 720 **3.** 120 **4.** 42 **5.** 15 **6.** $\frac{4}{7}$ **7.** $\frac{3}{7}$ **8.** 1 **9.** $\frac{1}{36}$ **10.** $\frac{1}{12}$ **11.** $15.99 **12.** 9 **13.** 22 **14.** 22 **15.** 21 **17.** 92 **18.** 7th **19.** 9th

Cumulative Review page 446
1. $18, $42 **3.** $2.64 **5.** $337.50, $1,087.50 **7.** $\angle IHK$ or $\angle IKH$ **9.** $\angle MJI$ or $\angle JIK$ or $\angle HIK$ **11.** right and isosceles **13.** square **15.** 198 ft^2 **17.** 8,480 mm^2 **19.** 113.04 m^2 **21.** 69 cm^2

Chapter 14 Square Roots and Special Triangles

14-1 pages 448–450
Practice 1a. −5 **b.** 8 **2a.** $5 < \sqrt{28} < 6$ **b.** $7 < \sqrt{53} < 8$ **3a.** 18 **b.** 9
Exercises 1. 5 **3.** 2 **5.** 12 **7.** −4 **9.** 4 **11.** 2, 3 **13.** 4, 5 **15.** 6, 7 **17.** −6, −7 **19.** 5, 6 **21.** 4, 5 **23.** 9, 10 **25.** 10, 11 **27.** 6 **29.** 10 **31.** 11 **33.** −1 **35.** 12 **37.** 0 **39.** < **41.** < **43.** > **45.** 9 **47.** $\sqrt{5^6} = 125$; $\sqrt{6^6} = 216$; $\sqrt{n^6} = n^3$ **49.** $\angle CBD$ and $\angle DBE$ **51.** 40° **53.** 66 cm^2
Numbers to Algebra 1. 8 **3.** 12 **5.** 10 **7.** 9

14-2 pages 451–453
Practice 1a. 3.873 **b.** 3.317 **2a.** 7.2801098 **b.** 6.6105975
Exercises 1. 1.414 **3.** 3.742 **5.** 4.359 **7.** 3.162 **9.** 9.644 **11.** 5.657 **13.** 5.292 **15.** 9.327 **17.** 9.644 **19.** 7.550 **21.** 9.788 **23.** 91.608 **25.** 6.552 **27.** 7.565 **29.** $<$ **31.** $<$ **33.** $>$ **35.** $<$ **37.** 15 seconds **39.** 11 **41.** 0, 1 **43.** 4 cm **45.** 64 cm^2 **47.** $\frac{1}{52}$ **49.** 9 **51.** 0.5 **53.** 3
Calculator Activity 1. 2.45 **3.** 5.57 **5.** 6.48

14-3 pages 454–455
Practice 1a. ± 7 **b.** ± 10 **2a.** ± 11 **b.** $\pm\sqrt{39}$ **c.** $\pm\sqrt{73}$
Exercises 1. ± 4 **3.** ± 9 **5.** ± 11 **7.** $\pm\sqrt{56}$ **9.** $\pm\sqrt{132}$ **11.** $\pm\sqrt{53}$ **13.** $\pm\sqrt{113}$ **15.** $\pm\sqrt{55}$ **17.** ± 5 **19.** ± 10 **21.** ± 7 **23.** ± 7 **25.** 0, 1 **27.** -1, 0 **29.** 55% **31.** 20% **33.** 2
Estimation 1. ≈ 6 **3.** ≈ 10 **5.** ≈ 4

14-4 pages 456–459
Practice 1a. $\sqrt{89}$ **b.** $\sqrt{274}$ **2a.** $\sqrt{161}$ **b.** $\sqrt{160}$
Exercises 1. 15 **3.** $\sqrt{130}$ **5.** $\sqrt{208}$ **7.** 13 **9.** $\sqrt{356}$ **11.** $\sqrt{544}$ **13.** $\sqrt{119}$ **15.** 8 **17.** $\sqrt{133}$ **19.** $\sqrt{56}$ **21.** $\sqrt{205}$ **23.** $\sqrt{171}$ **25.** $\sqrt{924}$ or 30.4 ft **27.** 20.19 ft **29.** no **31.** no **33.** no **35.** yes **37.** yes **39.** yes **41.** $2t + 8$ **43.** $6n + 12$ **45.** 0.3
Mental Math 1. 3,025 **3.** 5,625 **5.** 9,025

14-5 pages 460–461
Problems 1. yes **3.** no **5.** 136.01 yd. **7.** ≈ 14.39 yd

14-6 pages 462–464
Practice 1a. 10 **b.** 19.5
Exercises 1. 14 **3.** 15 **5.** $a = 5$, $e = 32$ **7.** $p = 9\frac{1}{3}$, $u = 40\frac{1}{2}$ **9.** 90 m **11.** 66 m **13.** 4 **15.** $\frac{4}{5}$ **17.** 9 **19.** 5

14-7 pages 465–467
Practice 1. $5\sqrt{2}$ **2.** 2.5, $2.5\sqrt{3}$
Exercises 1. $3\sqrt{2}$ **3.** $3\sqrt{3}$, 3 **5.** $5\sqrt{2}$ **7.** $12.5\sqrt{3}$, 12.5 **9.** 3.464 ft **11.** 127.26 ft **13.** $BD = 8$, $AC = 8\sqrt{3}$ **15.** $\frac{\sqrt{3}}{2}$ **17.** 1 **19.** -6

14-8 pages 468–471
Practice 1. $\sin A = \frac{12}{13}$, $\cos A = \frac{5}{13}$, $\tan A = \frac{12}{5}$ **2.** ≈ 193.75
Exercises 1. $\cos A = \frac{15}{17}$, $\sin A = \frac{6}{17}$, $\tan A = \frac{6}{15} = \frac{2}{5}$ **3.** $\cos A = \frac{29}{36}$, $\sin A = \frac{9}{36} = \frac{1}{4}$, $\tan A = \frac{9}{29}$ **5.** $\sin A = \frac{7}{21} = \frac{1}{3}$, $\cos A = \frac{15}{21} = \frac{5}{7}$, $\tan A = \frac{7}{15}$ **7.** 1354.2 ft **9.** 260.42 ft **11.** $\frac{1}{2}$ **13.** $\frac{1}{2}$ **15.** $\frac{1}{\sqrt{2}}$ or $\frac{\sqrt{2}}{2}$ **17.** $\frac{1}{\sqrt{2}}$ or $\frac{\sqrt{2}}{2}$ **19.** $\sin A = \frac{7}{25}$, $\cos A = \frac{24}{25}$, $\tan A = \frac{7}{24}$, 25 **21.** $\sin A = \frac{4}{7}$, $\cos A = \frac{5.74}{7}$, $\tan A = \frac{A}{5.74}$, 5.74 **23.** 4 **25.** 30 **27.** 30° **29.** 30° **31.** 20° **33.** 16 cm^2 **35.** 28 in^2 **37.** 4.14 **39.** 24 **41.** 4 **43.** 3

14-9 page 472
Problems 1. 86, 149, 172 **3.** 44 **5.** 1.83 m and 4.17 m **7.** 16 ft, 17 ft, 18 ft **11.** 4 **13.** 2 **15.** $3^2 \cdot 5 \cdot 11$

Chapter 14 Review page 476
1. 5, 6 **2.** 8, 9 **3.** 2 **4.** 11 **5.** 2.65 **6.** 10.82 **7.** 13, -13 **8.** 3.46, -3.46 **9.** 10.63 **10.** 12 **11.** 21.21 in **12.** 19 **13.** 24 **14.** 7 **15.** $9\sqrt{3}$ **16.** $\sin A = \frac{30}{42} = \frac{5}{7}$ **17.** $\cos A = \frac{14}{42} = \frac{1}{3}$ **18.** $\tan A = \frac{30}{14} = \frac{15}{7} = 2\frac{1}{7}$ **19.** 410 ft^2, 460 ft^2

Cumulative Review page 478
1. 2,115 mi **3.** 7.536 m **5.** SSS **7.** 448 m^3 **9.** 12 yd^3 **11.** $\frac{8}{12} = \frac{2}{3}$ **13.** $\frac{1}{36}$ **15.** $\frac{5}{18}$

Chapter 15 Graphs of Equations and Inequalities

15-1 pages 480–481
Practice 1a. $x > -\frac{8}{3}$ **b.** $x < 1$ **2a.** $x \geq -2$ **b.** $x \leq 1$
Exercises 5. $x > \frac{1}{2}$ **7.** $x < 5$ **9.** $x \leq -5$ **11.** $x > 3$ **13.** $x \leq 1$ **15.** $x \leq -\frac{9}{5}$ **17.** $x \geq -64$ **19.** $x < 9$ **29.** $1 \leq x \leq 3$ **31.** 8 **33.** 3 **35.** 1.2

15-2 pages 482–485
Practice 1. $D(-3, -5)$; $E(5, 2)$; $F(5, -2)$; $G(-4, 2)$
Exercises 1. (3, 2) **3.** $(-2, -4)$ **5.** $(3, -4)$ **7.** $(-6, -3)$ **9.** $(-6, 4)$ **11.** (4, 0) **25.** $(-2, 4)$ **27.** $(-1, -3)$ **39.** (2, 3) **41.** 9 **43.** mean: 13.625; mode: 12
Numbers to Algebra 1. 13 **3.** 10

15-3 pages 486–488
Practice 1. $(2, -3)$
Exercises 1. (2, 3) **3.** $(-3, 4)$ **21.** $y = -4x + 4$ **23.** $y = -2x + 8$ **25.** $y = \frac{3}{2}x - 2$ **27.** $y = \frac{1}{2}x$ **29.** (1, 1) **31.** $\frac{3}{5}$ **33.** $\frac{3}{4}$ **35.** 12

15-4 page 489
Problems 1. \$4,580 **3.** 9 gal **5.** 13 weeks **7.** 38 hours **11.** 6 **13.** 2

15-5 pages 490–493
Practice 1a. yes, $k = 5$ **b.** no, $k = -2$ **2.** no **3a.** no **b.** yes, $k = 1$
Exercises 1. yes, $k = 1$ **3.** no **5.** no **7.** yes, $k = 120$ **9.** yes, $k = 2$ **11.** directly **13.** directly **17.** 30 **19.** 90 **21.** 243 **23.** 0.25 or $\frac{1}{4}$ **25.** 2.4

15-6 pages 494–497

Practice **1a.** 2 **b.** $\frac{2}{8} = \frac{1}{4}$ **2a.** -2 **b.** $\frac{2}{3}$ **3a.** x-int: -3; y-int: 6 **b.** x-int: $\frac{10}{3}$; y-int: -5

Exercises **1.** $\frac{2}{3}$ **3.** $\frac{3}{2}$ **5.** -1 **7.** -2 **9.** $\frac{1}{7}$ **11.** 3 **13.** 2 **15.** 2 **17.** $-\frac{1}{3}$ **19.** 2 **21.** (5, 0), (0, 5) **23.** (6, 0), (0, -2) **25.** (5, 0), (0, 4) **27.** (2.5, 0) (0, -5) **33.** 12 ft **35.** 28 **37.** 3 **39.** 24 **41.** 7

Estimation **1.** 13 **3.** 7

15-7 pages 498–499

Problems **1.** \$8.40 **3.** \$4.80 **5.** \$3.60 **7.** \$22.80 **9.** \$0.20 **11.** ≈1.4¢

15-8 pages 500–501

Practice **1.** (2, 1)

Exercises **1.** (1, 0) **3.** (2, 2) **5.** (-1, -1) **7.** (2, 1) **9.** (3, 1) **11.** none **13.** $k = 0$ **15.** More-For-Your-Money, 100 mi **17.** 251.2 cm^3 **19.** 235.5 m^3 **21.** 75% **23.** 150% **25.** 15 **27.** 2.95

15-9 pages 502–503

Exercises **11.** $y \geq -\frac{1}{2}x + 1$ **15.** $y < \frac{3}{4}x + 3$ **17.** $5x + 7y \leq 500$ **21.** $d = 12$ in, $a = 113.04$ in^2 **23.** 2

Chapter 15 Review page 506

3. (1, 2) **4.** (-2, 0) **7.** (4, 9) **9.** \$160 **10.** yes, $k = 4$ **11.** no **12.** yes, $k = 20$ **13.** yes, $k = 15$ **14.** 1 **15.** 5 **16.** 2 **17.** 4 **18.** \$325 **19.** (2, 1) **20.** no

Cumulative Review page 508

1. 715.92 in^3 **3.** 376.80 ft^2 **5.** 20 **7.** 24 **9.** 84 **11.** 14 **13.** 14 **15.** 5, 6 **17.** 8.94 **19.** 9.4 **21.** $\frac{12}{13}$

Index